Essential Genetics

Second Edition

Daniel L. Hartl
Harvard University

Elizabeth W. Jones
Carnegie Mellon University

JONES AND BARTLETT PUBLISHERS
Sudbury, Massachusetts
BOSTON TORONTO LONDON SINGAPORE

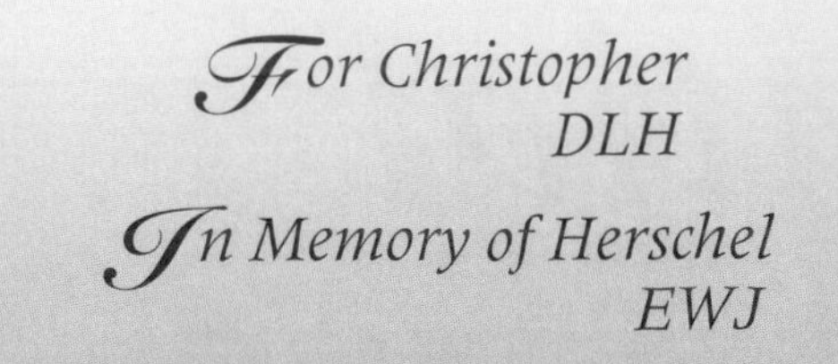

ABOUT THE AUTHORS

Daniel L. Hartl is a Professor of Biology and Chairman of Organismic and Evolutionary Biology at Harvard University. He received his B.S. degree and Ph.D. from the University of Wisconsin. His research interests include molecular genetics, molecular evolution, and population genetics. Elizabeth W. Jones is a Professor of Biological Sciences at Carnegie Mellon University. She received her B.S. degree and Ph.D. from the University of Washington in Seattle. Her research interests include gene regulation and the genetic control of cellular form. Currently she is studying the function and assembly of organelles in the yeast *Saccharomyces*.

ABOUT THE COVER

There are many examples of genetic diversity all around us. Even animals of the same species can appear quite different from each other. The two tigers on our cover, Prince Charles and Whiskers, are both Bengal tigers from the same litter. Prince Charles has a strikingly unusual white coat, while his sister Whiskers has the typical orange coat.
[Cover photograph © Ron Kimball Photography.]

ABOUT THE PUBLISHER

World Headquarters
Jones and Bartlett Publishers
40 Tall Pine Drive
Sudbury, MA 01776
978-443-5000
info@jbpub.com
www.jbpub.com

Jones and Bartlett Publishers Canada
P.O. Box 19020
Toronto, ON M5S 1X1 CANADA

Jones and Bartlett Publishers International
Barb House, Barb Mews
London W6 7PA UK

Chief Executive Officer: Clayton Jones
Chief Operating Officer: Don Jones, Jr.
Publisher: Tom Walker
V.P., Sales and Marketing: Tom Manning
V.P., Senior Managing Editor: Judith H. Hauck
Executive Editor: Brian L. McKean
Marketing Director: Rich Pirozzi
Director of Production: Anne Spencer
Manufacturing Director: Therese Bräuer
Interactive Media Director: Mike Campbell

Special Projects Editor: Mary Hill
Assistant Production Editor: Ivee Wong
Editorial/Production Assistant: Tim Gleeson
Web Designer: Mike DeFronzo
Interactive Technology Project Editor: W. Scott Smith
Book Design: Merce Wilczek
Cover Design: Anne Spencer
Art Development and Rendering: J/B Woolsey Associates
Composition and Book Layout: Thompson Steele, Inc.
Cover and Book Manufacture: Banta Company

Library of Congress Cataloging-in-Publication Data
Hartl, Daniel L.
Essential genetics / Daniel L. Hartl, Elizabeth W. Jones. —2nd ed.
p. cm.
Includes bibliographical references and index
ISBN 0-7637-0838-0
1. Genetics. I. Jones, Elizabeth W. II. Title.
QH430. H3732 1999
576.5—dc21 98-35331
CIP

Printed in the United States
03 02 01 00 99 9 8 7 6 5 4 3 2 1

Contents in Brief

Contents

CLASSICAL GENETIC ANALYSIS

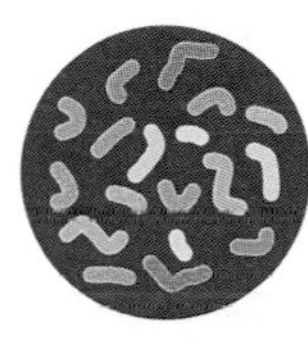

CHAPTER 5 VARIATION IN CHROMOSOME NUMBER AND STRUCTURE 156

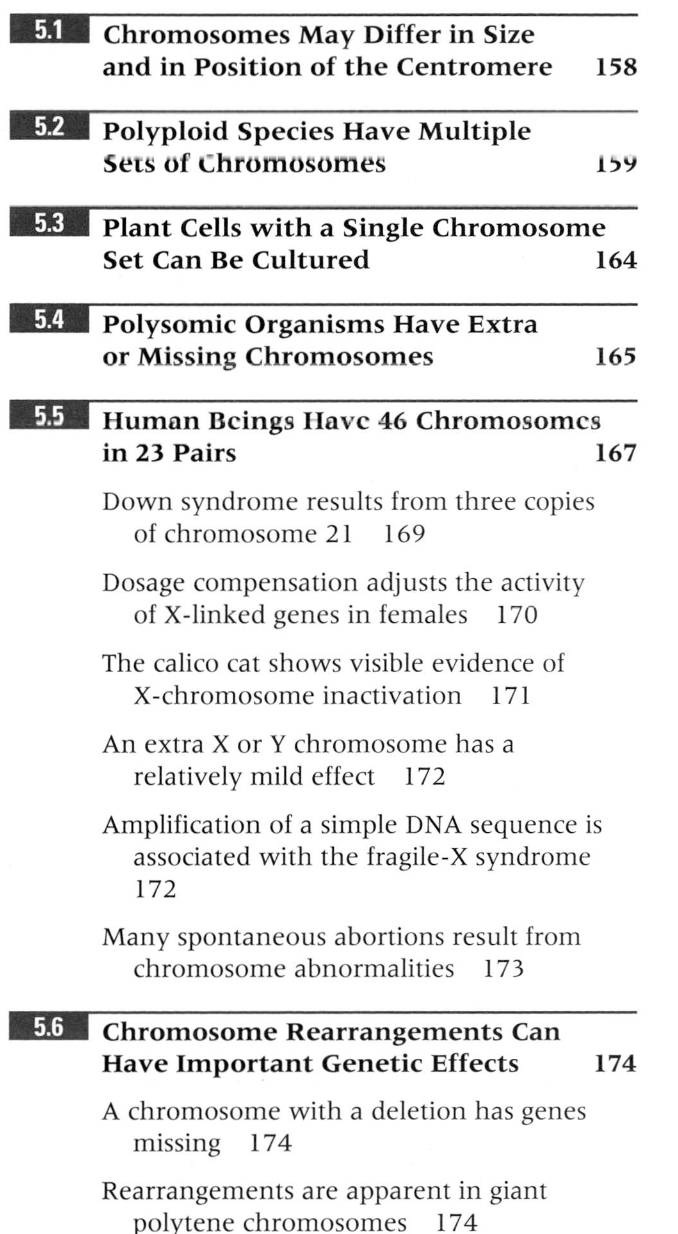

MOLECULAR GENETIC ANALYSIS

CHAPTER 6 THE CHEMICAL STRUCTURE AND REPLICATION OF DNA 194

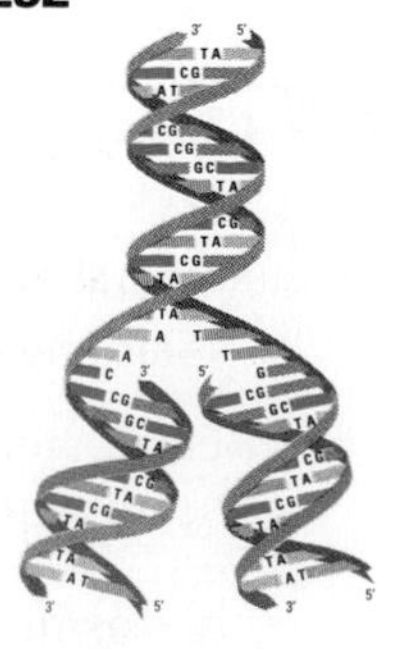

CHAPTER 7 MUTATION AND DNA REPAIR 234

CHAPTER 8 THE GENETICS OF BACTERIA AND VIRUSES 264

COMPLEX TRAITS AND EVOLUTIONARY GENETICS

CHAPTER 13 POPULATION GENETICS AND EVOLUTION 456

CHAPTER 14 THE GENETIC ARCHITECTURE OF COMPLEX TRAITS 492

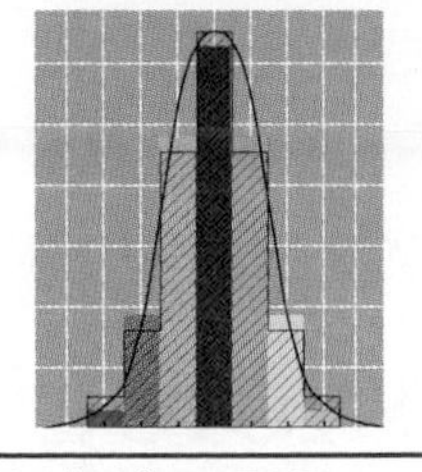

Papers Excerpted in Connections in Chronological Order

Karl Landsteiner, 1901
Anatomical Institute, Vienna, Austria
On Agglutination Phenomena in Normal Blood

Archibald E. Garrod, 1908
St. Bartholomew's Hospital, London, England
Inborn Errors of Metabolism

A. C. Allison, 1954
Radcliffe Infirmary, Oxford, England
Protection Afforded by Sickle-Cell Trait against Subtertian Malarial Infection

Joe Hin Tijo[1] and Albert Levan[2], 1956
[1]Estacion Experimental de Aula Dei, Zaragoza, Spain
[2]Institute of Genetics, Lund, Sweden
The Chromosome Number in Man

Vernon M. Ingram, 1957
Cavendish Laboratory, University of Cambridge, England
Gene Mutations in Human Hemoglobin: The Chemical Difference between Normal and Sickle-Cell Hemoglobin

Jerome Lejeune, Marthe Gautier, and Raymond Turpin, 1959
National Center for Scientific Research, Paris, France
Study of the Somatic Chromosomes of Nine Down Syndrome Children

Mary F. Lyon, 1961
Medical Research Council, Harwell, England
Gene Action in the X Chromosome of the Mouse (Mus musculus L.)

Francis Collins[1] and David Galas[2], 1993
[1]National Human Genome Research Institute, National Institutes of Health, Bethesda, Maryland
[2]Office of Health and Environmental Research, Department of Energy, Washington, DC
A New Five-Year Plan for the U.S. Human Genome Project

Frederick S. Leach and 34 other investigators, 1993
Johns Hopkins University, Baltimore, MD, and 10 other research institutions
Mutations of a mutS Homolog in Hereditary Nonpolyposis Colorectal Cancer

José M. Fernández-Cañon[1], Begoña Granadino[1], Daniel Beltrán-Valero de Bernabé[1], Monica Renedo[2], Elena Fernández-Ruiz[2], Miguel A. Peñlava[1], and Santiago Rodríguez de Córdoba[1], 1996
[1]Consejo Superior de Investigaciones Cientificas, Madrid, Spain
[2]Universidad Autónoma de Madrid, Madrid, Spain
The Molecular Basis of Alkaptonuria

Carl T. Montague and 14 other authors, 1997
University of Cambridge, Cambridge, UK, and 5 other research institutions
Congenital Leptin Deficiency Is Associated with Severe Early-Onset Obesity in Humans

David H. Skuse[1], Rowena S. James[2], Dorothy V. M. Bishop[3], Brian Coppin[4], Paola Dalton[2], Gina Aamodt-Leeper[1], Monique Barcarese-Hamilton[1], Catharine Creswell[1], Rhona McGurk[1], and Patricia A. Jacobs[2], 1997
[1]Institute of Child Health, London, England
[2]Salisbury District Hospital, Salisbury, Wiltshire, UK
[3]Medical Research Council Applied Psychology Unit, Cambridge, UK
[4]Princess Anne Hospital, Southampton, UK
Evidence from Turner's Syndrome of an Imprinted X-Linked Locus Affecting Cognitive Function

Didier Mazel, Broderick Dychinco, Vera A. Webb, and Julian Davies, 1998
University of British Columbia, Vancouver, Canada
A Distinctive Class of Integron in the Vibrio cholerae *genome*

Hugh E. Montgomery and 18 other investigators, 1998
University College, London, and 6 other research institutions
Human Gene for Physical Performance

Preface

Today's college students come to a course in genetics full of enthusiasm stimulated by social and ethical controversies related to genetics reported in the popular press. The challenges for the instructor are to sustain this enthusiasm, to kindle a desire to understand the principles of genetics, and to help students integrate genetic knowledge into a wider social and ethical context. We have written *Essential Genetics* to help instructors meet these challenges. It is designed for the shorter, less comprehensive introductory course in genetics. The brevity of the text fits the pace of what can be covered in a typical one-semester or one-quarter course. The topics have been carefully chosen to help students achieve the following learning objectives:

- Understand the basic processes of gene transmission, mutation, expression, and regulation
- Learn to formulate genetic hypotheses, work out their consequences, and test the results against observed data
- Develop basic skills in problem solving, including single-concept exercises, those requiring the application of several concepts in logical order, and numerical problems requiring some arithmetic for solution
- Gain some sense of the social and historical context in which genetics has developed as well as an appreciation of current trends
- Become aware of some of the genetics resources and information that are available through the World Wide Web

Many special features are included to make the book "user friendly." These are discussed individually below. Based on our own experiences in teaching genetics, they have been implemented and found to be effective in the classroom.

CHAPTER ORGANIZATION

The text is written in a clear, direct, lively, and concise manner. To help the student keep track of the main concepts without being distracted by details, each chapter begins with a list of **Key Concepts** written in simple declarative sentences, highlighting the most important concepts presented in the chapter. There is also an **Outline** showing the principal subjects to be discussed. The body of each chapter provides more detailed information and experimental evidence. An opening paragraph gives an overview of the chapter, illustrates the subject with some specific examples, and shows how the material is connected to genetics as a whole. The section and subsection **Headings** are in the form of complete sentences that encapsulate the main message. The text makes liberal use of **Numbered Lists** and **Bullets** to aid students in organizing their learning, as well as **Summary Statements** set apart from the main text in order to emphasize important principles. Each chapter also includes **The Human Connection.** This is a special feature highlighting a research paper in human genetics that reports a key experiment or raises important social, ethical, or legal issues. Each Human Connection has a brief introduction of its own, explaining the importance of the experiment and the context in which it was carried out. At the end of each chapter is a complete **Summary, Key Terms, geNETics on the web** exercises that guide students in the use of Internet resources in genetics, and several different types and levels of **Problems.** At the back of the book are **Answers** to some of the problems and a complete **Glossary.**

CONTENTS

The organization of the chapters is that favored by the majority of instructors who teach genetics. It is the organization we use in our own courses. An important feature is the presence of an introductory chapter providing a broad overview of the gene: what it is, what it does, how it changes, how it evolves. Today, most students learn about DNA in grade school or high school. In our teaching, we have found it rather artificial to pretend that DNA does not exist until the middle of the term. The introductory chapter therefore serves to connect the more advanced concepts that students are about to learn with what they already know. It also serves to provide each student with a solid framework for integrating the material that comes later.

Throughout each chapter, there is a balance between challenge and motivation, between observation and theory, and between principle and concrete example. Molecular, classical, and evolutionary genetics are integrated throughout. A number of points related to organization and coverage should be noted:

Chapter 1 is an overview of genetics designed to bring students with disparate backgrounds to a common level of understanding. This chapter enables classical, molecular, and evolutionary genetics to be integrated throughout the rest of the book. Included in Chapter 1 are the basic concepts of genetics: genes as DNA that function through transcription and translation, that change by mutation, and that affect organisms through inborn errors of metabolism. Chapter 1 also includes a discussion of the classical experiments demonstrating that DNA is the genetic material.

Chapters 2 through 5 are the core of Mendelian genetics, including segregation and independent assortment, the chromosome theory of heredity, mitosis and meiosis, linkage and chromosome mapping, tetrad analysis in fungi, and chromosome mechanics. An important principle of genetics, too often ignored or given inadequate treatment, is that of the complementation test and how complementation differs from segregation or other genetic principles. The complementation test is the experimental definition of a gene. Chapter 2 includes a clear and concise description of complementation, with examples, showing how complementation is used in genetic analysis to group mutations into categories (complementation groups), each corresponding to a different gene. Chapter 4 introduces the use of molecular markers in genetics, because these are the principal types of genetic markers in use today.

Chapters 6 and 7 deal with DNA, including the details of DNA structure and replication in Chapter 6 and mechanisms of mutation and DNA repair in Chapter 7, including chemical mutagens and new information on the genetic effects of the Chornobyl nuclear accident. A novel feature in Chapter 6 is a description of how basic research that revealed the molecular mechanisms of DNA replication ultimately led to such important practical applications as DNA hybridization analysis, DNA sequencing, and the polymerase chain reaction. These examples illustrate the value of basic research in leading, often quite unpredictably, to practical applications.

Chapter 8 deals with the principles of genetics in prokaryotes, with special emphasis on *E. coli* and temperate and virulent bacteriophages. There is an extensive discussion of mechanisms of genetic recombination in microbes, including transformation, conjugation, transduction, and the horizontal transfer of genes present in plasmids, such as F′ plasmids.

Chapters 9 through 12 focus on molecular genetics in the strict sense. Chapter 9 examines the details of gene expression, including transcription, RNA processing, and translation.

Chapter 10 deals with recombinant DNA and genome analysis. Included are the use of restriction enzymes and vectors in recombinant DNA, cloning strategies, transgenic animals and plants, and applications of genetic engineering. Also discussed are methods used in the analysis of complex genomes, such as the human genome, in which a gene that has been localized by genetic mapping to a region of tens of millions of base pairs must be isolated in cloned form and identified.

Chapter 11 is an integrative chapter dealing with genetic mechanisms of regulation. The first half of the chapter focuses on gene regulation in prokaryotes, the second half on that in eukaryotes.

Chapter 12 examines the genetic control of development. In addition to a discussion of developmental mechanisms in animals, there is a section on the genetic analysis of floral development in *Arabidopsis thaliana*. The material on pattern formation in *Drosophila* development is illustrated with spectacular color photographs by James Langeland, Stephen Paddock, and Sean Carroll, showing the discrete, overlapping patterns of gene expression as the embryo becomes segmented. The section on the development of flowers in plants is beautifully illustrated with photographs by Elliot Meyerowitz and John Bowman.

Chapters 13 and 14 deal with population and evolutionary genetics. The discussion of population genetics includes DNA typing in criminal investigations and paternity testing. The material on quantitative genetics includes a discussion of methods by which particular genes influencing quantitative traits (QTLs, or quantitative-trait loci) may be identified and mapped by linkage analysis. There is also a discussion of QTL identification through the study of "candidate" genes, as exemplified by the identification of the "natural Prozac" polymorphism in the human serotonin transporter gene.

Human genetics is integrated in every chapter of the book. *Essential Genetics* could qualify as a textbook in human genetics, were it not so much broader. The integration is necessary in a modern treatment of genetics, because human genetics plays such a central role in the field. Integrating human genetics is also a sensible way to teach genetics, since students typically have a great interest in the subject. Chapter 1 sets the stage with an examination of inborn errors of metabolism, with emphasis on phenylketonuria and alkaptonuria. These are superb cases with which to motivate a desire to understand gene structure, expression, and mutation. Chapter 2 includes human pedigree analysis for autosomal genes. Chapter 3 does human X-linked inheritance with special emphasis on the "Royal hemophilia." Chapter 4 emphasizes the study of molecular markers in human pedigrees. Human chromosomes and their disorders are a major part of Chapter 5. And on it goes throughout the entire book. Chapter 10 makes special reference to the human genome project and the research resources, such as Expressed Sequence Tags (ESTs) that have already emerged from it. Chapter 13 includes human DNA typing, the effects of inbreeding in humans, and the controversial mitochondrial "Eve." The discussion of QTLs and candidate genes in Chapter 14 brings into focus two of the most important current approaches for identifying the genetic basis of human disease. Throughout the book The Human Connections also serve to integrate human genetics with other aspects of genetics. In addition, human genetics is highlighted in the GeNETics on the Web exercises at the end of each chapter.

THE HUMAN CONNECTION

A unique and special feature of this book is found in boxes called The Human Connection, one in each chapter. They are our way of connecting genetics to the world of human genetics outside the classroom. All of the Connections include short excerpts from the original literature of genetics, usually papers, each introduced with a short explanatory passage. Many of the Connections are excerpts from classic materials, such as Garrod's book on inborn errors of metabolism, but by no means all of the "classic" papers are old papers. Half of The Human Connection papers are from the 1990's, which serves to emphasize the fast-moving pace of this field.

The pieces are called The Human Connection because each connects the material in the text to something that broadens or enriches its implications in regard to human beings. Some of the Connections raise issues of ethics in the application of genetic knowledge, social issues that need to be addressed, or issues related to the proper care of laboratory animals. They illustrate other things as well. Because each Connection names the place where the research was carried out, the student will learn that great science is done in many universities and research institutions throughout the world. In papers that use outmoded or unfamiliar terminology, or archaic gene symbols, we have substituted the modern equivalent because the use of a consistent terminology in the text and in the Connections makes the material more accessible to the student.

GENETICS ON THE WEB

The World Wide Web is a rich storehouse of information on all aspects of genetics. Many sites give nontechnical descriptions of human diseases, written at the level of a lay person, for people who have family members affected by a hereditary disease. Other sites give descriptions of ongoing research projects and explain why the research is important. At the most sophisticated level are databases of mutants, DNA and protein sequences, and other genetic information, which are designed for access by the professional geneticist.

To make the genetic information explosion on the Internet available to the beginning student, we have developed Internet Exercises, called GeNETics on the Web, which make use of Internet resources related to human genetics. One reason for developing these exercises is that genetic knowledge of human genetics is currently so vast that there can be no such thing as a comprehensive textbook. Detailed information must come from the Internet. The available information changes rapidly, too. Modern genetics is a dynamic science, and most of the key Internet resources are kept up to date. The addresses of the relevant genetic sites are not printed in the book. Instead, the sites are accessed through the use of key words that are highlighted in each exercise. The key words are maintained as hot links at the publisher's web site **(http://www.jbpub.com/genetics)** and are kept constantly up to date, tracking the address of each site if it should change. Each Internet Exercise includes a short assignment such as to draw a map, create a list, or write a short paper. Many instructors may wish to develop their own assignments.

LEVELS AND TYPES OF PROBLEMS

Each chapter provides numerous problems for solution, graded in difficulty, for the students to test their understanding. The problems are of two different types:

- **Essential Concepts** ask for genetic principles to be restated in the student's own words; some are matters of definition or call for the application of elementary principles.
- **Concepts in Action** are problems that require the student to reason using genetic concepts. The problems make use of a variety of formats, including true or false, multiple choice, matching, and traditional types of word problems. Many of the Concepts in Action require some numerical calculation. The level of mathematics is that of arithmetic and elementary probability as it pertains to genetics. None of the problems uses mathematics beyond elementary algebra.

SOLUTIONS STEP BY STEP

Each chapter contains a section entitled Solutions Step by Step that demonstrates problems worked in full, explaining step by step a path of logical reasoning that can be followed to analyze the problem. The Solutions Step by Step serve as another level of review of the important concepts used in working problems. The solutions also highlight some of the most common mistakes made by beginning students and give pointers on how the student can avoid falling into these conceptual traps.

ANSWERS TO PROBLEMS AND GLOSSARY

The answers to the even-numbered Concepts in Action are included in the answer section at the end of the book. There is also a Glossary of Key Words. The Answers are complete. They explain the logical foundation of the solution and lay out the methods. The answers to the rest of the Concepts in Action problems are available for the instructor in the Test Bank and Solutions Manual. We find that many of our students, like students everywhere, often sneak a look at the answer before attempting to solve a problem. This is a pity. Working backward from the answer should be a last resort. This is because problems are valuable opportunities to learn. Problems that the student cannot solve are usually more important than the ones that can be solved, because the sticklers usually identify trouble spots, areas of confusion, or gaps in understanding. So, forever in hope but against all experience, we urge our students to try answering each question before looking at the answer.

FURTHER READING

Each chapter also includes recommendations for Further Reading for the student who either wants more information or who needs an alternative explanation for the material presented in the book. Some additional "classic" papers and historical perspectives are included.

ILLUSTRATIONS

The art program is spectacular, thanks to the creative efforts of J/B Woolsey Associates, with special thanks to John Woolsey and Patrick Lane. Every chapter is richly illustrated with beautiful graphics in which color is used functionally to enhance the value of each illustration as a learning aid. The illustrations are also heavily annotated with "process labels" explaining step-by-step what is happening at each level of the illustration. These labels make the art inviting as well as informative. They also allow the illustrations to stand relatively independently of the text, enabling the student to review material without rereading the whole chapter.

The art program is used not only for its visual appeal but also to increase the pedagogical value of the book:

- Characteristic colors and shapes have been used consistently throughout the book to indicate different types of molecules—DNA, mRNA, tRNA, and so forth. For example, DNA is illustrated in any one of a number of ways, depending on the level of resolution necessary for the illustration, and each time a particular level of resolution is depicted, the DNA is shown in the same way. It avoids a great deal of potential confusion that DNA, RNA, and proteins are represented in the same manner in Chapter 13 as they are in Chapter 1.
- There are numerous full-color photographs of molecular models in three dimensions; these give a strong visual reinforcement of the concept of macromolecules as physical entities with defined three-dimensional shapes and charge distributions that serve as the basis of interaction with other macromolecules.
- The page design is clean, crisp, and uncluttered. As a result, the book is pleasant to look at and easy to read.

ADAPTABILITY AND FLEXIBILITY

There is no necessary reason to start at the beginning and proceed straight to the end. Each chapter is a self-contained unit that stands on its own. This feature gives the book the flexibility to be used in a variety of course formats. Throughout the book, we have integrated classical and molecular principles, so you can begin a course with almost any of the chapters. Most teachers will prefer starting with the overview in Chapter 1, possibly as suggested reading, because it brings every student to the same basic level of understanding. Teachers preferring the Mendel-early format should continue with Chapter 2; those preferring to teach the details of DNA early should continue with Chapter 6. Some teachers are partial to a chromosomes-early format, which would suggest continuing with Chapter 3, followed by Chapters 2 and 4. A novel approach would be a genomes-first format, which could be implemented by continuing with Chapter 10. Some teachers like to discuss mechanisms of mutation later in the course, and Chapter 7 can easily be assigned later. The writing and illustration program was designed to accommodate a variety of formats, and we encourage teachers to take advantage of this flexibility in order to meet their own special needs.

INSTRUCTOR AND STUDENT SUPPLEMENTS

An unprecedented offering of traditional and interactive multimedia supplements is available to assist instructors and aid students in mastering genetics. Additional information and review copies of any of the following items are available through your Jones and Bartlett Sales Representative.

For the Instructor

- **Printed Test Bank and Solutions Manual**—The Test Bank, prepared by Sarah C. Martinelli of Southern Connecticut State University with contributions from Michael Draper, Patrick McDermot, and Elena R. Lozovskaya, contains 700 test items, with 50 questions per chapter. There is a mix of factual, descriptive, analytical and quantitative question types. A typical chapter file contains 20 multiple choice objective questions, 15 fill-ins, and 15 quantitative. The Solutions Manual, authored by Elena R. Lozovskaya of Harvard University, contains worked solutions for all the end of chapter problems in the main text. Only solutions to even-numbered problems are provided in the back of the main text. This allows the instructor to control access to solutions for odd-numbered problems.

- **Instructor's ToolKit CD-ROM**—This MAC/IBM CD-ROM provides the instructor with a powerful set of five programs that can easily be integrated into your daily routine to help save time, while making classroom presentations more educational for students. The programs include:

  PowerPoint Slide Set—The PowerPoint slide set, authored by Sarah C. Martinelli of Southern Connecticut State University, provides outline summaries of each chapter and hyperlinks to selected key figures from the text. The slide set can be customized to meet your classroom needs.

  The Lecture Success Image Bank—The image bank is an easy to use multimedia tool containing over 300 figures from the text specially enhanced for classroom presentation. You select the images you need by chapter, topic, or figure number to create your own lecture aid.

  The WebCD—The CD contains key simulated web sites that allow you to bring the Internet into the classroom without the need for a live connection.

  The Computerized Test Bank and Solutions Manual—The Computerized Test Bank, using the ESATEST interface, allows the instructor to easily generate quizzes and tests from the complete set of over 700 questions. The solutions to all end-of-chapter problems are supplied as a Microsoft Word document.

- **Visual Genetics Plus: Tutorials and Laboratory Simulations. Faculty Version**—This Mac/IBM CD-ROM, created by Alan W. Day and Robert L. Dean of the University of Western Ontario and Harry Roy of Rensselaer Polytechnic Institute, is already in use at over 200 institutions worldwide. Visual Genetics 3.0 continues to provide a unique, dynamic presentation tool for viewing key genetic and molecular processes in the classroom. With this new, greatly expanded version of the Virtual Genetics Lab 2.0, instructors can now assign 18 comprehensive lab simulations. You can also bring the Lab into the classroom, as the program allows you to perform tasks on-screen—such as dragging of mutant colonies, using a pipette to make a dilution series, inoculating mutants to Petri dishes to test for response to growth factors—_ then to analyze and interpret the data. Through the testing feature and presentation capabilities you can offer a complete lab environment. Site Licenses and Instructor Copies are available.

- **An Electronic Companion to Genetics** © 1998, Cogito Learning Media, Inc.—This Mac/IBM CD-ROM, by Philip Anderson and Barry Ganetzky of the University of Wisconsin, Madison, reviews important genetics concepts using state of the art interactive multimedia. It consists of hundreds of animations, diagrams, and videos that dynamically explain difficult concepts to students.

- **Video Resource Library**—A full complement of quality videos are available to qualified adopters. Genetics related topics include: Origin and Evolution of Life, Human Gene Therapy, Biotechnology, the Human Genome Project, Oncogenes, and Science and Ethics.

For the Student

- **The Gist of Genetics: Guide to Learning and Review**—Written by Rowland H. Davis and Stephen G. Weller of the University of California, Irvine, this study aid uses illustrations, tables and text outlines to review all of the fundamental elements of genetics. It includes extensive practice problems and review questions with solutions for self-check. The Gist helps students formulate appropriate questions and generate hypothesis that can be tested with classical principles and modern genetic techniques.

- **GeNETics on the Web**—Corresponding to the end-of-chapter GeNETics on the Web exercises, this World Wide Web site offers genetics-related links, articles and monthly updates to other genetics sites on the web. Material for this site is carefully selected and updated by the authors, and Jones and Bartlett Publishers ensures that links for the site are regularly maintained. Visit the GeNETics on the Web site at **http://www.jbpub.com/genetics**.

- **An Electronic Companion to Genetics** © 1998, Cogito Learning Media, Inc.—This Mac/IBM CD-ROM, by Philip Anderson and Barry Ganetzky of the University of Wisconsin, Madison, reviews important genetics concepts covered in class using state of the art interactive multimedia. It consists of hundreds of animations, diagrams, and videos that dynamically explain difficult concepts to students. In addition, it contains over 400 interactive multiple choice, "drop and drag," true/false, and fill-in problems. These resources will prove invaluable to students in a self-study environment and to instructors as a lecture enhancement tool. This CD-ROM is available for packaging exclusively with Jones and Bartlett Publishers texts.

- **Visual Genetics Plus: Tutorials and Laboratory Simulations. Student Version**—This Mac/IBM CD-ROM is already in use at over 300 institutions worldwide. Visual Genetics 3.0 continues to provide a unique, dynamic multimedia review of key genetic and molecular processes. With this new, greatly expanded version of the Virtual Genetics Lab 2.0, students can now work on 18 comprehensive lab simulations. The Lab allows students to perform tasks on-screen—dragging of mutant colonies, using a pipette to make a dilution series, inoculating mutants to Petri dishes to test for response to growth factors—then to analyze and interpret the data. The Student Version is available for purchase and can be packaged with our text.

ACKNOWLEDGMENTS

We are indebted to the many colleagues whose advice and thoughts were immensely helpful throughout the preparation of this book. These colleagues range from specialists in various aspects of genetics who checked for accuracy or suggested improvement to instructors who evaluated the material for suitability in teaching or sent us comments on the text as they used it in their courses.

Jeremy C. Ahouse, Brandeis University
John C. Bauer, Stratagene, Inc., La Jolla, CA
Mary K. B. Berlyn, Yale University
Pierre Carol, Université Joseph Fourier, Grenoble, France
John W. Drake, National Institute of Environmental Health Sciences, Research Triangle Park, NC
Wolfgang Epstein, University of Chicago, Chicago, IL
Jeffrey C. Hall, Brandeis University
Steven Henikoff, Fred Hutchinson Cancer Research Center, Seattle, WA
Charles Hoffman, Boston College, Boston, MA
Joyce Katich, Monsanto, Inc., St. Louis, MO
Jeane M. Kennedy, Monsanto, Inc., St. Louis, MO
Jeffrey King, University of Berne, Switzerland
Yan B. Linhart, University of Colorado, Boulder
K. Brooks Low, Yale University
Gustavo Maroni, University of North Carolina
Jeffrey Mitton, University of Colorado, Boulder
Gisela Mosig, Vanderbilt University
Robert K. Mortimer, University of California, Berkeley

Steve O'Brien, National Cancer Institute, Frederick, MD
Ronald L. Phillips, University of Minnesota
Robert Pruitt, Harvard University
Pamela Reinagel, California Institute of Technology, Pasadena
Kenneth E. Rudd, National Library of Medicine
Leslie Smith, National Institute of Environmental Health Sciences, Research Triangle Park, NC
Johan H. Stuy, Florida State University
Jeanne Sullivan, West Virginia Wesleyan College, Buckhannon, WV
Irwin Tessman, Purdue University
David Ussery, Roanoke College, Salem, VA
Denise Wallack, Muhlenberg College, Allentown, PA
Kenneth E. Weber, University of Southern Maine

We would also like to thank the reviewers, listed below, who reviewed part or all of the material. Their comments and recommendations helped improve the content, organization, and presentation.

Laura Adamkewicz, George Mason University
Peter D. Ayling, University of Hull (UK)
Anna W. Berkovitz, Purdue University
John Celenza, Boston University
Stephen J. D'Surney, University of Mississippi
Kathleen Dunn, Boston College
David W. Francis, University of Delaware
Mark L. Hammond, Campbell University
Richard Imberski, University of Maryland
Sally A. MacKenzie, Purdue University
Kevin O'Hare, Imperial College (UK)
Peggy Redshaw, Austin College
Thomas F. Savage, Oregon State University
David Shepard, University of Delaware
Charles Staben, University of Kentucky
David T. Sullivan, Syracuse University
James H. Thomas, University of Washington

We also wish to acknowledge the superb art, production, and editorial staff who helped make this book possible: Mary Hill, Patrick Lane, Andrea Fincke, Judy Hauck, Bonnie Van Slyke, Sally Steele, John Woolsey, Brian McKean, Ivee Wong, Maggie Villiger, Kathryn Twombly, Rich Pirozzi, Mike Campbell, and Tom Walker. Much of the credit for the attractiveness and readability of the book should go to them. Thanks also to Jones and Bartlett, the publishers, for the high quality of the book production. We are also grateful to the many people, acknowledged in the legends of the illustrations, who contributed photographs, drawings, and micrographs from their own research and publications, especially those who provided color photographs for this edition. Every effort has been made to obtain permission to use copyrighted material and to make full disclosure of its source. We are grateful to the authors, journal editors, and publishers for their cooperation. Any errors or omissions are wholly inadvertant and will be corrected at the first opportunity. A special thanks goes to geneticist Elena R. Lozovskaya of Harvard University, who contributed substantially to the instructional materials at the end of each chapter.

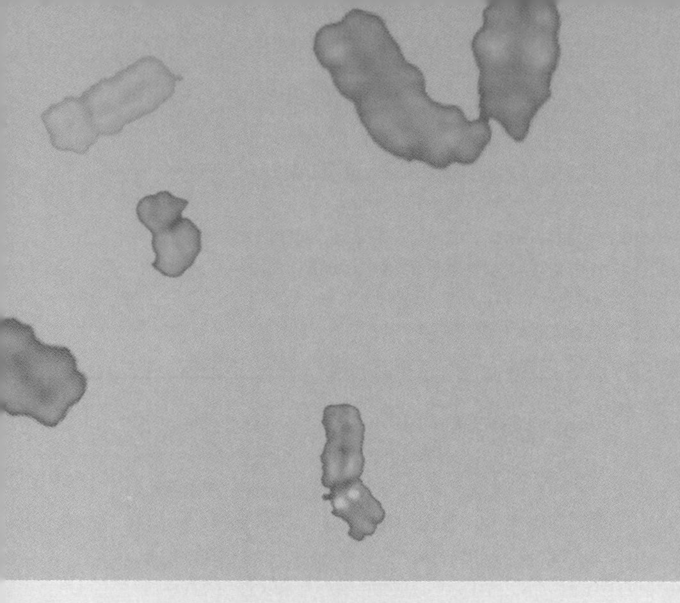

Key Concepts

- Inherited traits are affected by genes.
- Genes are composed of the chemical deoxyribonucleic acid (DNA).
- DNA replicates to form (usually identical) copies of itself.
- DNA contains a code specifying what types of enzymes and other proteins are made in cells.
- DNA occasionally mutates, and the mutant forms specify altered proteins.
- A mutant enzyme is an "inborn error of metabolism" that blocks one step in a biochemical pathway for the metabolism of small molecules.
- Traits are affected by environment as well as by genes.
- Organisms change genetically through generations in the process of biological evolution.

FRENCH SOCCER FANS celebrate at the end of the soccer World Cup '98 semifinal match in Saint Denis, France. What are some traits that differ from one person to the next? What are some traits that all human beings share? [© AP Photo/Michel Euler.]

CHAPTER

1

DNA: The Genetic Code

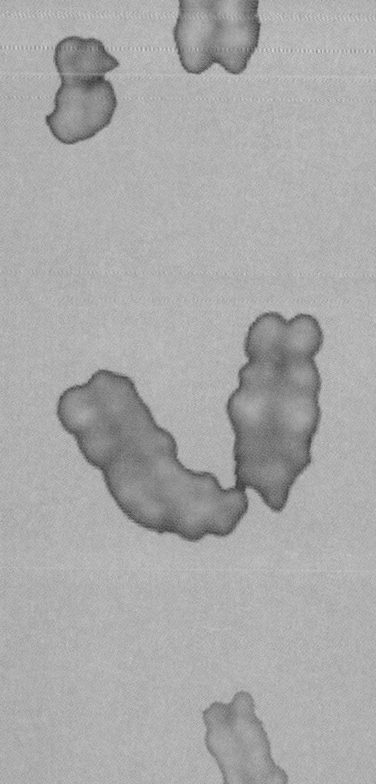

Any observer of living organisms soon notices similarities and differences among them. For example, all human beings are united by a common set of traits, or observable characteristics, that set us apart from all other species of organisms. Even small children have no difficulty distinguishing a human being from a chimpanzee or a gorilla, which are our closest living relatives. A human being habitually stands upright and has long legs, relatively little body hair, a large brain, and a flat face with a prominent nose, jutting chin, distinct lips, and small teeth. All of these traits are inherited. Some traits present in human beings we share with other animals to whom we are more distantly related. In common with other mammals, human beings are warm-blooded, and human mothers feed their young with milk secreted by mammary glands. In common with other vertebrates, human beings have a backbone and a spinal cord. Like the traits that distinguish us from other primates, the traits that define us as mammals, as vertebrates, and as belonging to other categories of animals are transmitted by biological inheritance from one generation of human beings to the next. Every normal human being exhibits these biological characteristics.

Though all human beings share many traits, human beings are by no means all identical. There is much variation within the human species. Many traits differ from one person to the next. Examples include hair color, eye color, skin color, height, weight, and personality traits. Some human traits are transmitted biologically, others culturally. The color of one's eyes results from biological inheritance; the native language one learns as a child results from cultural inheritance. Many traits are influenced jointly by biological inheritance and environmental factors. For example, one's weight is determined in part by inheritance but also in part by environment—how much food one eats, how much fat or carbohydrate the food contains, how much exercise one gets, and so forth. **Genetics** is the study of biologically inherited traits, including traits such as weight that are influenced in part by the environment.

The fundamental concept of genetics is that

> Inherited traits are determined by the elements of heredity that are transmitted from parents to offspring in reproduction; these elements of heredity are called **genes.**

The existence of genes and the rules governing the transmission of genes from generation to generation were discovered by Gregor Mendel in experiments with garden peas published in 1866. Mendel's experiments are among the most beautifully designed, carefully executed, and elegantly interpreted in the history of experimental science. Mendel's formulation of inheritance was in terms of the abstract rules by which hereditary elements are transmitted from parents to offspring. He was the first to study the precise statistical relationships between the different classes of progeny observed in successive generations. Mendel's experiments are the main subject of Chapter 2.

All in the family Cousins Sir Francis Galton (left) and Charles Darwin (right) share certain heritable characteristics, like heavy eyebrows and male pattern baldness. They share something else. Both made important contributions to 19th century genetics and evolutionary biology.

© Granger Collection

© Historical Picture Service, Chicago/PNI

In 1869, only 3 years after Mendel reported his experiments, Friedrich Miescher discovered a new type of weak acid, abundant in the nuclei of white blood cells, that turned out to be the chemical substance of which genes are made. Miescher's weak acid is now called **deoxyribonucleic acid,** or **DNA.** Nevertheless, even though the two main pieces of the puzzle of heredity—genes and DNA—had been discovered, the pieces were not put together until about the middle of the twentieth century when the chemical identity between genes and DNA was conclusively demonstrated. The next section shows how this connection was made.

1.1 DNA is the molecule of heredity.

The importance of the cell nucleus in inheritance became clear when the nuclei of the male and female reproductive cells were observed to fuse in the process of fertilization. This discovery was made in the 1870s. The next major advance was the discovery of **chromosomes,** thread-like objects inside the nucleus that become visible in the light microscope when stained with certain dyes. Chromosomes exhibit a characteristic "splitting" behavior, in which each daughter cell formed by cell division receives an identical complement of chromosomes. How this happens is taken up in Chapter 3. Further evidence of the importance of chromosomes was provided by the observation that, whereas the number of chromosomes in each cell differs from one biological species to the next, the number of chromosomes is nearly always constant within the cells of any one species. These features of chromosomes were well understood by about 1900, and they made it seem likely that chromosomes were the carriers of the genes.

By the 1920s, several lines of indirect evidence suggested a close relationship between chromosomes and DNA. Microscopic studies with special stains showed that DNA is present in chromosomes. Various types of proteins are present in chromosomes too. But whereas most of the DNA in cells of higher organisms is present in chromosomes, and the amount of DNA per cell is constant, the amount and kinds of proteins and other large molecules differ greatly from one cell type to another. The indirect evidence for DNA as the genetic material was unconvincing, because crude chemical analyses had suggested (erroneously, as it turned out) that DNA lacks the chemical diversity needed in a genetic substance. The favored candidate for the genetic material was protein, because proteins were known to be an exceedingly diverse collection of molecules. Proteins therefore became widely accepted as the genetic material, and the role of DNA was relegated to providing the structural framework of chromosomes. Any researcher who hoped to demonstrate that DNA was the genetic material had a double handicap. Such experiments had to demonstrate not only that DNA *is* the genetic material but also that proteins are *not* the genetic material. Two of the experimental approaches regarded as decisive are described in this section.

Genetic traits can be altered by treatment with pure DNA.

One type of bacterial pneumonia in mammals is caused by strains of *Streptococcus pneumoniae* able to synthesize a gelatinous capsule composed of polysaccharide (complex carbohydrate). This capsule surrounds the bacterium and protects it from the defense mechanisms of the infected animal; thus it enables the bacterium to cause disease. When a bacterial cell is grown on solid medium, it undergoes repeated cell divisions to form a visible clump of cells called a **colony.** The enveloping capsule makes the size of each colony large and gives it a glistening or smooth

(S) appearance. Certain strains of *S. pneumoniae* are unable to synthesize the capsular polysaccharide, and they form small colonies that have a rough (R) surface (Figure 1.1). The R strains do not cause pneumonia, because without the capsule, the bacteria are inactivated by the immune system of the host. Both types of bacteria "breed true" in the sense that the progeny formed by cell division have the capsular type of the parent, either S or R.

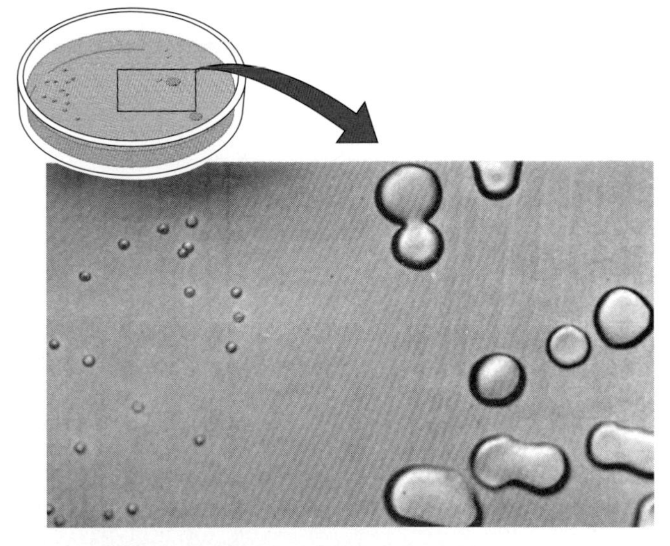

Figure 1.1 Colonies of rough (R, small colonies) and smooth (S, large colonies) strains of *Streptococcus pneumoniae.* The S colonies are larger because of the capsule on the S cells. (Photograph from O. T. Avery, C. M. MacLeod, and M. McCarty. 1944. *J. Exp. Med.* 79: 137.)

When mice are injected either with living R cells or with heat-killed S cells, they remain healthy. However, when mice are injected with a *mixture* of living R cells and heat-killed S cells, they often die of pneumonia. Bacteria isolated from blood samples of the dead mice produce S cultures with a capsule typical of the injected S cells, even though the injected S cells had been killed by heat. Evidently, the injected material from the dead S cells includes a substance that can give living R bacterial cells the ability to resist the immunological system of the mouse, multiply, and cause pneumonia. In other words, the R bacteria can be changed—or undergo **transformation**—into S bacteria, and the new characteristics are inherited by descendants of the transformed bacteria.

Transformation in *Streptococcus* had been discovered in 1928, but it was not until 1944 that the chemical substance responsible for changing the R cells into S cells was identified. In a milestone experiment, Oswald Avery, Colin MacLeod, and Maclyn McCarty showed that the substance causing the transformation of R cells into S cells was DNA. In preparation for the experiment, they had to develop chemical procedures for obtaining DNA in almost pure form from bacterial cells, which had not been done before. When they added DNA isolated from S cells to growing cultures of R cells, they observed that a few type-S cells were produced. Although the DNA preparations contained traces of protein and RNA (ribonucleic acid, an abundant cellular macromolecule chemically related to DNA), the transforming activity was not altered by treatments that destroy either protein or RNA. However, treatments that destroy DNA eliminated the transforming activity (Figure 1.2). These experiments implied that the substance responsible for genetic transformation was the DNA of the cell—and hence that DNA is the genetic material.

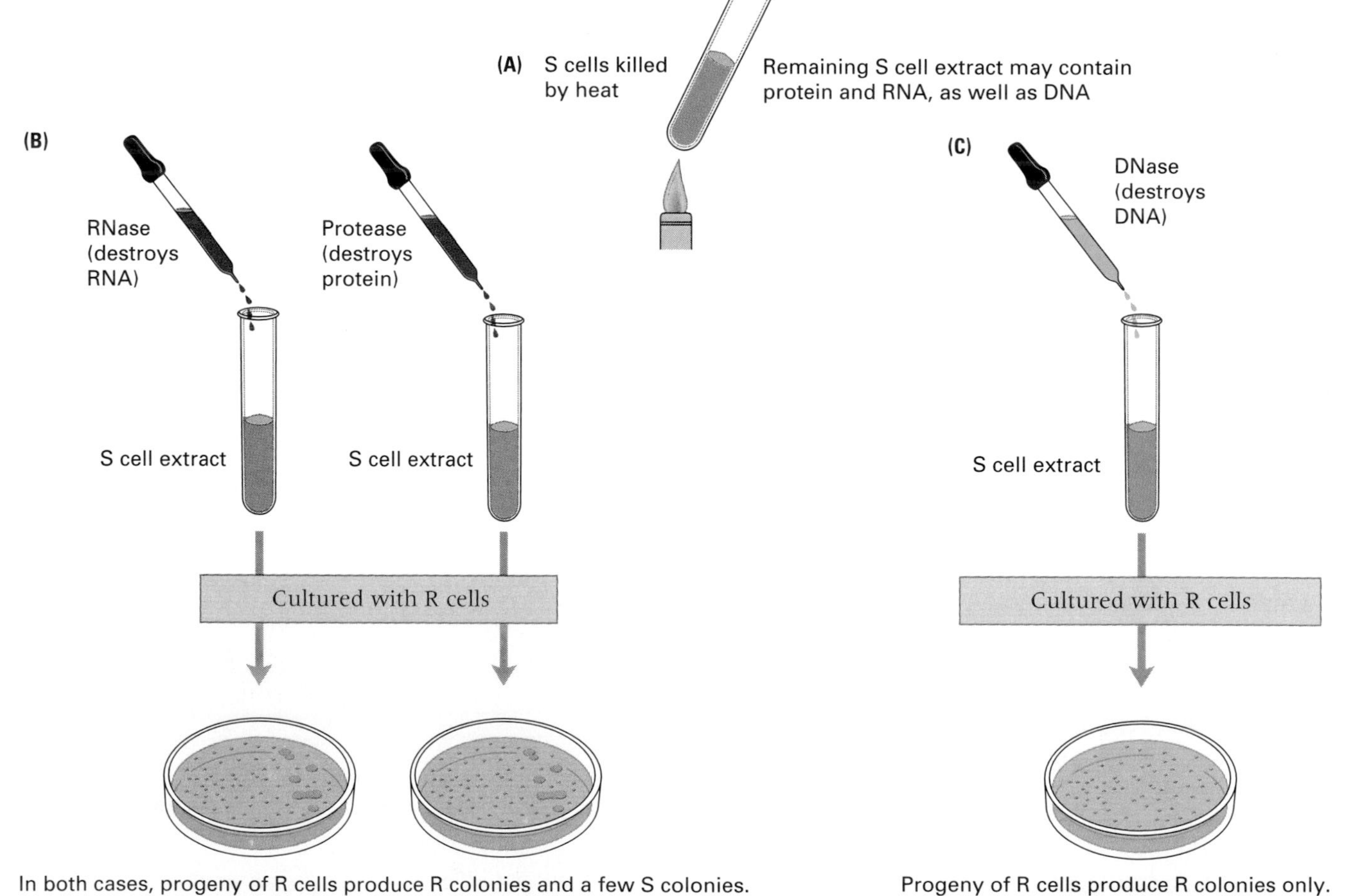

Figure 1.2 A diagram of the experiment that demonstrated that DNA is the active material in bacterial transformation. (A) Purified DNA extracted from heat-killed S cells can convert some living R cells into S cells, but the material may still contain undetectable traces of protein and/or RNA. (B) The transforming activity is not destroyed by either protease or RNase. (C) The transforming activity is destroyed by DNase and so probably consists of DNA.

Transmission of DNA is the link between generations.

A second pivotal finding was reported by Alfred Hershey and Martha Chase in 1952. They studied cells of the intestinal bacterium *Escherichia coli* after infection by the virus T2. A virus that attacks bacterial cells is called a **bacteriophage,** often shortened to **phage.** *Bacteriophage* means "bacteria-eater." The T2 particle is exceedingly small, yet it has a complex structure composed of head (which contains the phage DNA), collar, tail, and tail fibers. (The head of a human sperm is about 30–50 times larger in both length and width than the head of T2.) Hershey and Chase were already aware that T2 infection proceeds via the attachment of a phage particle by the tip of its tail to the bacterial cell wall, entry of phage material into the cell, multiplication of this material to form a hundred or more progeny phage, and release of the progeny phage by bursting (lysis) of the bacterial host cell. They also knew that T2 particles were composed of DNA and protein in approximately equal amounts.

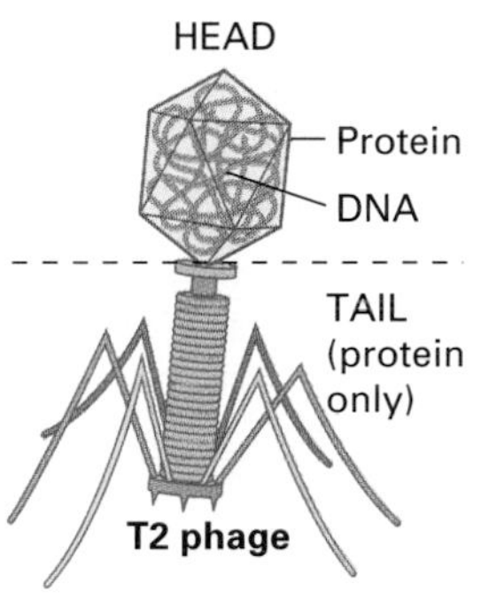

Because DNA contains phosphorus but no sulfur, whereas most proteins contain sulfur but no phosphorus, it is possible to label DNA and proteins differentially by the use of radioactive isotopes of the two elements. Hershey and Chase produced particles containing radioactive DNA by infecting *E. coli* cells that had been grown for several generations in a medium that included ^{32}P (a radioactive isotope of phosphorus) and then collecting the phage progeny. Other particles containing labeled proteins were obtained in the same way, using medium that included ^{35}S (a radioactive isotope of sulfur).

In the experiments summarized in Figure 1.3, nonradioactive *E. coli* cells were infected with phage labeled with *either* ^{32}P (part A) *or* ^{35}S (part B)

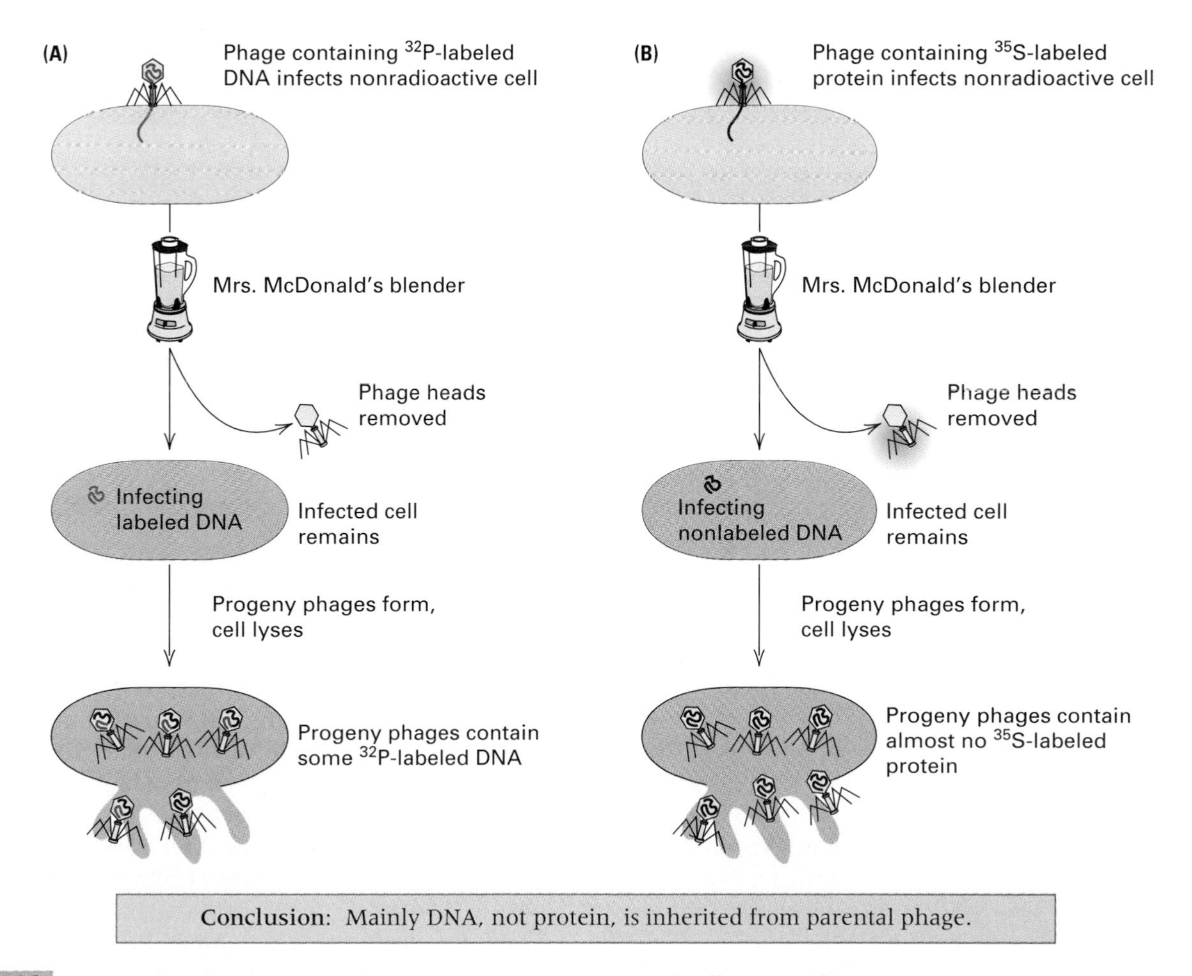

Figure 1.3 The Hershey–Chase ("blender") experiment, which demonstrated that DNA, not protein, is responsible for directing the reproduction of phage T2 in infected *E. coli* cells. (A) Radioactive DNA is transmitted to progeny phage in substantial amounts. (B) Radioactive protein is transmitted to progeny phage in negligible amounts.

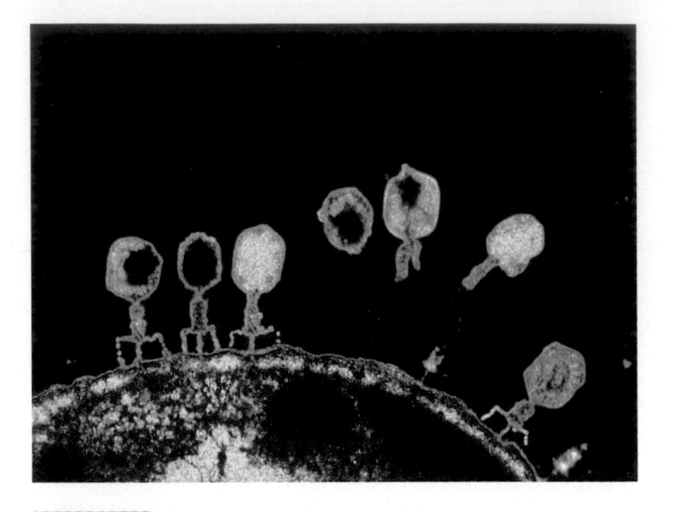

Figure 1.4 T2 phages infecting a cell of *E. coli*. Each phage attaches to the bacterial cell wall and injects its DNA into the host. The image has been color enhanced to show the phage DNA in green. [© Oliver Meckes/E.O.S./MPI Tubingen/Photo Researchers, Inc.]

in order to follow the DNA and proteins separately. Infected cells were separated from unattached phage particles by centrifugation, resuspended in fresh medium, and then swirled violently in a kitchen blender to shear attached phage material from the cell surfaces. This treatment was found to have no effect on the subsequent course of the infection, which implies that the genetic material must enter the infected cells very soon after phage attachment (Figure 1.4). The kitchen blender turned out to be the critical piece of equipment. Other methods had been tried to tear the phage heads from the bacterial cell surface, but nothing had worked reliably. Hershey later explained, "We tried various grinding arrangements, with results that weren't very encouraging. When Margaret McDonald loaned us her kitchen blender, the experiment promptly succeeded."

After the phage heads were removed by blending, the infected bacteria were examined. Most of the radioactivity from ^{32}P-labeled phage was found to be associated with the bacteria, whereas only a small fraction of the ^{35}S radioactivity was present in the infected cells. The retention of most of the labeled DNA, contrasted with the loss of most of the labeled protein, implied that a T2 phage transfers most of its DNA, but very little of its protein, to the cell it infects. The critical finding (Figure 1.3) was that about 50 percent of the transferred ^{32}P-labeled DNA, but less than 1 percent of the transferred ^{35}S-labeled protein, was inherited by the *progeny* phage particles. Hershey and Chase interpreted this result to mean that the genetic material in T2 phage is DNA.

The transformation experiment and the Hershey–Chase experiment are regarded as classics in the demonstration that genes consist of DNA. At the present time, the equivalent of the transformation experiment is carried out daily in many research laboratories throughout the world, usually with bacteria, yeast, or animal or plant cells grown in culture. These experiments indicate that DNA is the genetic material in these organisms as well as in phage T2.

> There are no known exceptions to the generalization that DNA is the genetic material in all cellular organisms.

It is worth noting, however, that in a few types of viruses, the genetic material consists of another kind of nucleic acid called RNA.

1.2 The structure of DNA is a double helix composed of two intertwined strands.

Even once it was known that genes consist of DNA, many questions remained. How is the DNA in a gene duplicated when a cell divides? How does the DNA in a gene control a hereditary trait? What happens to the DNA when a mutation (a change in the DNA) takes place in a gene? In the early 1950s, a number of researchers began to try to understand the detailed molecular structure of DNA in hopes that the structure alone would suggest answers to these questions. The first essentially correct three-dimensional structure of the DNA molecule was proposed in 1953 by James Watson and Francis Crick at Cambridge University. The structure was dazzling in its elegance and revolutionary in suggesting how DNA duplicates itself, controls hereditary traits, and undergoes mutation. Even while the tin sheet and wire model of the DNA molecule was still incomplete, Crick would visit his favorite pub and claim that "we have discovered the secret of life."

In the Watson–Crick structure, DNA consists of two long chains of subunits twisted around one another to form a double-stranded helix. The double helix is right-handed, which means that as one looks along the barrel, each chain follows a clockwise path as it progresses. You can see the right-handed coiling in Figure 1.5A if you imagine yourself looking up into the structure from the bottom: The dark spheres outline the "backbone" of each individual strand, and they coil in a clockwise direction. The subunits of each strand are **nucleotides,** each of which contains any one of

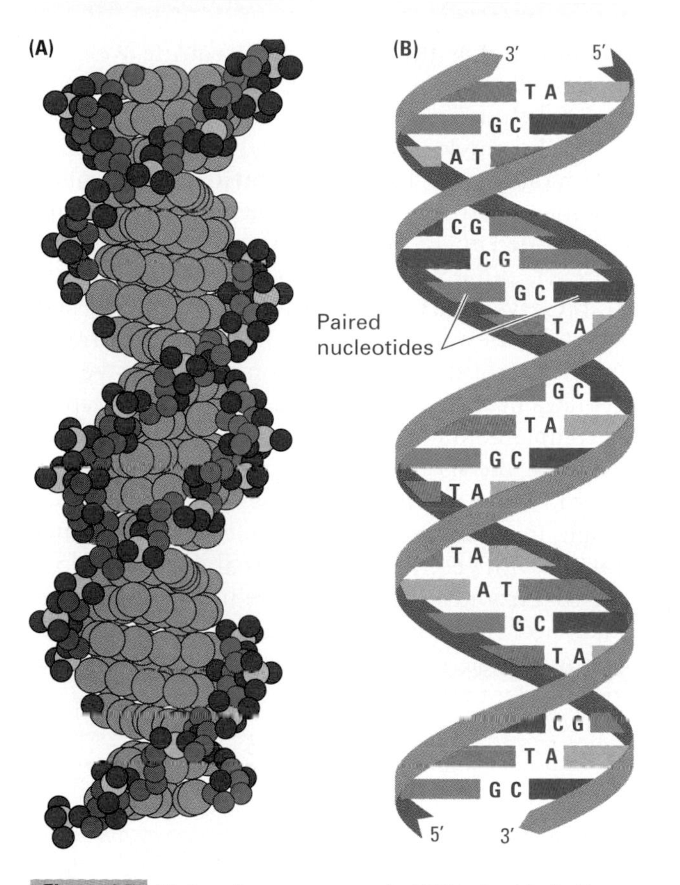

Figure 1.5 Molecular structure of a DNA double helix. (A) A "space-filling" model, in which each atom is depicted as a sphere. (B) A diagram highlighting the helical strands around the outside of the molecule and the A—T and G—C base pairs inside.

four chemical constituents called **bases.** The four bases in DNA are

- **Adenine (A)**
- **Guanine (G)**
- **Thymine (T)**
- **Cytosine (C)**

The chemical structures of the nucleotides and bases need not concern us at this point. They are examined in Chapter 6. A key point for our present purposes is that the bases in the double helix are paired as shown in Figure 1.5B. That is,

> At any position on the paired strands of a DNA molecule, if one strand has an A, then the partner strand has a T; and if one strand has a G, then the partner strand has a C.

The pairing between A and T and between G and C is said to be **complementary**; the complement of A is T, and the complement of G is C. The complementary pairing in the duplex molecule means that each base along one strand of the DNA is matched with a base in the opposite position on the other strand. Furthermore,

> Nothing restricts the sequence of bases in a single strand, so any sequence could be present along one strand.

This principle explains how only four bases in DNA can code for the huge amount of information needed to make an organism. It is the *sequence* of bases along the DNA that encodes the genetic information, and the sequence is completely unrestricted.

The complementary pairing is also called **Watson–Crick pairing.** In the three-dimensional structure (Figure 1.5A), the base pairs are represented by the lighter spheres filling the interior of the double helix. The base pairs lie almost flat, stacked on top of one another perpendicular to the long axis of the double helix, like pennies in a roll. When discussing a DNA molecule, biologists frequently refer to the individual strands as **single-stranded DNA** and to the double helix as **double-stranded DNA** or **duplex DNA.**

Each DNA strand has a **polarity,** or directionality, like a chain of circus elephants linked trunk to tail. In this analogy, each elephant corresponds to one nucleotide along the DNA strand. The polarity is determined by the direction in which the nucleotides are pointing. The "trunk" end of the strand is called the *5' end* of the strand, and the "tail" end is called the *3' end.* In double-stranded DNA, the paired strands are oriented in opposite directions, the 5' end of one strand aligned with the 3' end of the other. The molecular basis of the polarity, and the reason for the opposite orientation of the strands in duplex DNA, are explained in Chapter 6. In illustrating DNA molecules, we use an arrow-like ribbon to represent the backbone, and we use tabs jutting off the ribbon to represent the nucleotides. The polarity of a DNA strand is indicated by the direction of the arrow-like ribbon. The tail of the arrow represents the 5' end of the DNA strand, the head the 3' end.

Beyond the most optimistic hopes, knowledge of the structure of DNA immediately gave clues to its function:

1. The sequence of bases in DNA could be copied by using each of the separate "partner" strands as a pattern for the creation of a new partner strand with a complementary sequence of bases.
2. The DNA could contain genetic information in coded form in the sequence of bases, analogous to letters printed on a strip of paper.

3. Changes in genetic information (mutations) could result from errors in copying in which the base sequence of the DNA became altered.

In the remainder of this chapter, we discuss some of the implications of these clues.

1.3 In replication, each parental DNA strand directs the synthesis of a new partner strand.

In their first paper on the structure of DNA, Watson and Crick distinguished themselves as masters of understatement, admitting that "it has not escaped our notice that the specific base pairing we have postulated immediately suggests a copying mechanism for the genetic material." The copying process in which a single DNA molecule becomes two identical molecules is called **replication.** The replication mechanism that Watson and Crick had in mind is illustrated in Figure 1.6 The strands of the original (parent) duplex separate, and each individual strand serves as a pattern, or **template,** for the synthesis of a new strand (replica). The replica strands are synthesized by the addition of successive nucleotides in such a way that each base in the replica is complementary (in the Watson–Crick pairing sense) to the base across the way in the template strand. Although the mechanism in Figure 1.6 is simple in principle, it is a complex process that is fraught with geometric problems and requires a variety of enzymes and other proteins. The details are examined in Chapter 6. The end result of replication is that a single double-stranded molecule becomes replicated into two copies with identical sequences:

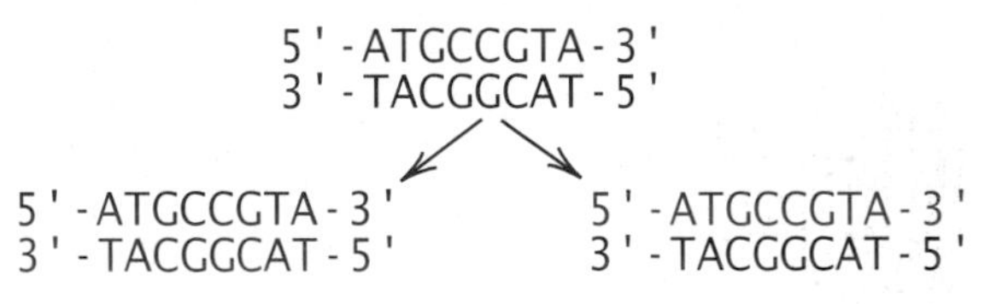

Here the bases in the newly synthesized strands are shown in red. In the duplex on the left, the top strand is the template from the parental molecule and the bottom strand is newly synthesized; in the duplex on the right, the bottom strand is the template from the parental molecule and the top strand is newly synthesized.

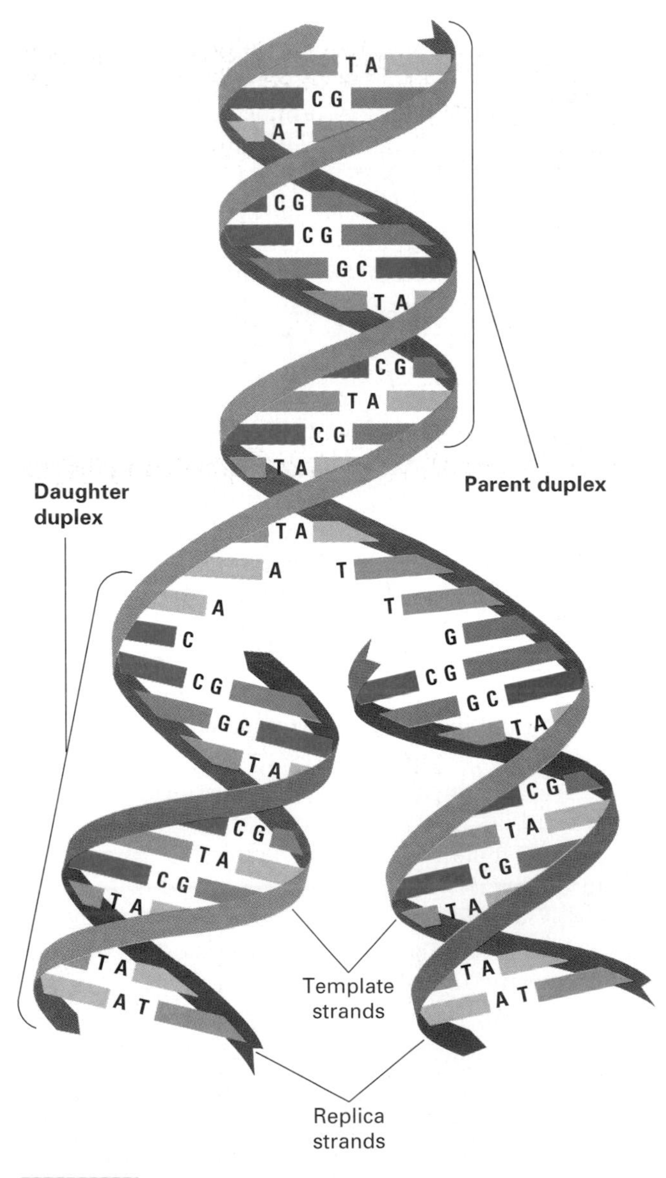

Figure 1.6 Replication in a long DNA duplex as originally proposed by Watson and Crick. The parental strands separate, and each parental strand serves as a template for the formation of a new daughter strand by means of A—T and G—C base pairing.

1.4 Genes code for proteins.

Long before DNA was shown to be the genetic material, it was already clear that proteins are responsible for most of the metabolic activities of cells. Proteins were known to be essential for the breakdown of organic molecules, generating the chemical energy needed for cellular activities. Proteins were also known to be essential for the assembly of small molecules into large molecules and complex cellular structures. In 1878 the term **enzyme** was introduced to refer to the biological catalysts that accelerate biochemical reactions in cells. By 1900, thanks largely to the genius of the German biochemist Emil Fischer, enzymes had been shown to be proteins. Other key components of cells also are proteins, including structural proteins that give the cell rigidity and mobility, proteins that form pores in the cell membrane to control the traffic of small molecules into and out of the cell, and receptor proteins that regulate cellular activities in response to molecular signals from the growth medium or from other cells.

The human connection — Black Urine

Archibald E. Garrod 1908
St. Bartholomew's Hospital,
London, England
Inborn Errors of Metabolism

Although he was a distinguished physician, Garrod's lectures on the relationship between heredity and congenital defects in metabolism had no impact when they were delivered. The important concept that one gene corresponds to one enzyme (the "one gene–one enzyme hypothesis") was developed independently in the 1940s by George W. Beadle and Edward L. Tatum, who used the bread mold *Neurospora crassa* as their experimental organism. When Beadle finally became aware of *Inborn Errors of Metabolism*, he was generous in praising it. This excerpt shows Garrod at his best, interweaving history, clinical medicine, heredity, and biochemistry in his account of alkaptonuria. The excerpt also illustrates how the severity of a genetic disease depends on its social context. Garrod writes as though alkaptonuria were a harmless curiosity. This is indeed largely true when the life expectancy is short. With today's longer life span, alkaptonuria patients accumulate the dark pigment in their cartilage and joints and eventually develop severe arthritis.

To students of heredity the inborn errors of metabolism offer a promising field of investigation. . . . It was pointed out [by others] that the mode of incidence of alkaptonuria finds a ready explanation if the anomaly be regarded as a rare recessive character in the Mendelian sense. . . . Of the cases of alkaptonuria a very large proportion have been in the children of first cousin marriages. . . . It is also noteworthy that, if one takes families with five or more children [with both parents normal and at least one child affected with alkaptonuria], the totals work out in strict conformity to Mendel's law, i.e. 57 [normal children] : 19 [affected children] in the proportions 3 : 1. . . . Of inborn errors of metabolism, alkaptonuria is that of which we know most. In itself it is a trifling matter, inconvenient rather than harmful. . . . Indications of the anomaly may be detected in early medical writings, such as that in 1584 of a schoolboy who, although he enjoyed good health, continuously excreted black urine; and that in 1609 of a monk who exhibited a similar peculiarity and stated that he had done so all his life. . . . There are no sufficient grounds [for doubting that the blackening substance in the urine originally called alkapton] is homogentisic acid, the excretion of which is the essential feature of the alkaptonuric. . . . Homogentisic acid is a product of normal metabolism. . . . The most likely sources of the benzene ring in homogentisic acid are phenylananine and tyrosine, [because when these amino acids are administered to an alkaptonuric] they cause a very conspicuous increase in the output of homogentisic acid. . . . Where the alkaptonuric differs from the normal individual is in having no power of destroying homogentisic acid when formed—in other words of breaking up the benzene ring of that compound. . . . We may further conceive that the splitting of the benzene ring in normal metabolism is the work of a special enzyme and that in congenital alkaptonuria this enzyme is wanting.

WE MAY FURTHER conceive that the splitting of the benzene ring in normal metabolism is the work of a special enzyme and that in congenital alkaptonuria this enzyme is wanting.

Source: Originally published in London, England, by the Oxford University Press. Excerpts from the reprinted edition in Harry Harris. 1963. *Garrod's Inborn Errors of Metabolism.* London, England: Oxford University Press.

Enzyme defects result in inborn errors of metabolism.

In 1908 the British physician Archibald Garrod gave a series of lectures in which he proposed a fundamental hypothesis about the relationship between enzymes and disease:

> Any hereditary disease in which cellular metabolism is abnormal results from an inherited defect in an enzyme.

Such diseases became known as **inborn errors of metabolism,** a term still in use today.

Garrod studied a number of inborn errors of metabolism in which the patients excreted abnormal substances in the urine. One of these was **alkaptonuria.** In this case, the abnormal substance excreted is **homogentisic acid:**

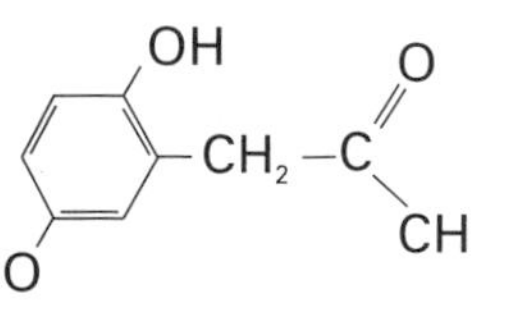

An early name for homogentisic acid was *alkapton*—hence the name *alkaptonuria.* Even though alkaptonuria is rare, with an incidence of about one in 200,000 people, it was well known

Figure 1.7 Urine from a person with alkaptonuria turns black because of the oxidation of the homogentisic acid that it contains. [Courtesy of Daniel De Aguiar.]

even before Garrod studied it. The disease itself is relatively mild, but it has one striking symptom: The urine of the patient turns black because of the oxidation of homogentisic acid (Figure 1.7). This is why alkaptonuria is also called *black urine disease.* The passing of black urine is likely to attract attention. One case was described in the year 1649.

> The patient was a boy who passed black urine and who, at the age of fourteen years, was submitted to a drastic course of treatment that had for its aim the subduing of the fiery heat of his viscera, which was supposed to bring about the condition in question by charring and blackening his bile. Among the measures prescribed were bleedings, purgation, baths, a cold and watery diet, and drugs galore. None of these had any obvious effect, and eventually the patient, who tired of the futile and superfluous therapy, resolved to let things take their natural course. None of the predicted evils ensued. He married, begat a large family, and lived a long and healthy life, always passing urine black as ink. (Quotation from Garrod, 1908.)

Garrod was primarily interested in the biochemistry of alkaptonuria, but he took note of family studies that indicated that the disease was inherited as though it were due to a defect in a single gene. As to the biochemistry, he deduced that the problem in alkaptonuria was the patients' inability to break down the benzene ring of six carbons that is present in homogentisic acid. Where does this ring come from? Most animals are unable to synthesize it. They obtain it from their diet. Garrod proposed that homogentisic acid originates as a breakdown product of two amino acids, phenylalanine and tyrosine, which also contain a benzene ring. An **amino acid** is one of the "building blocks" from which proteins are made. Phenylalanine and tyrosine are constituents of normal proteins. The scheme that

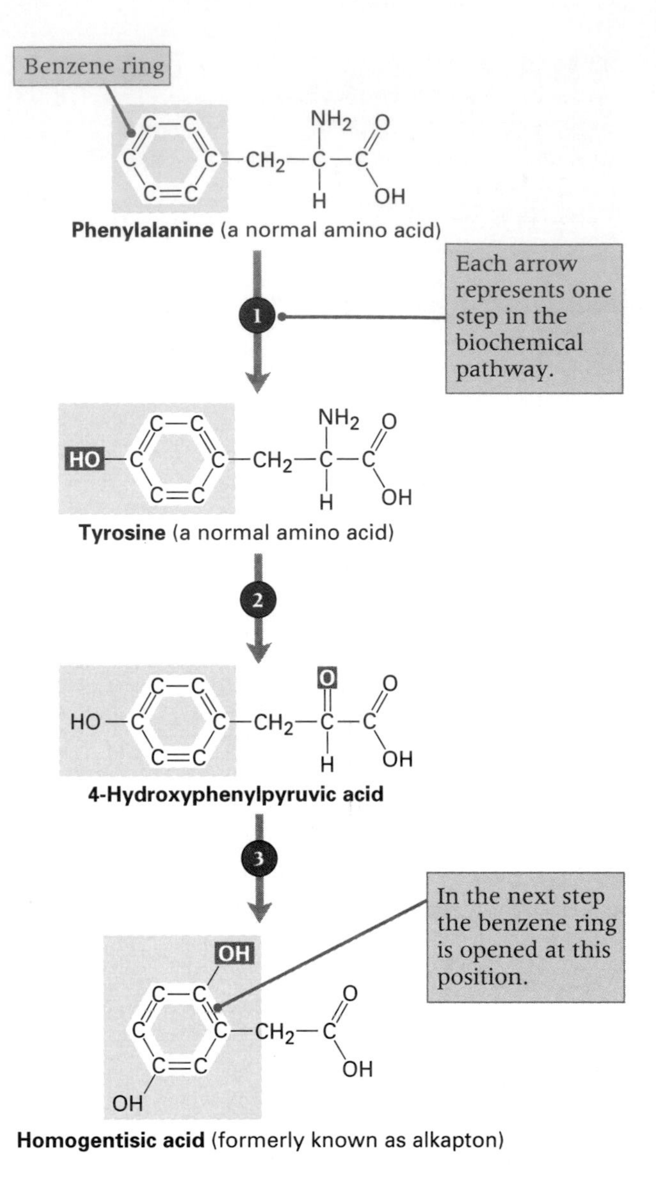

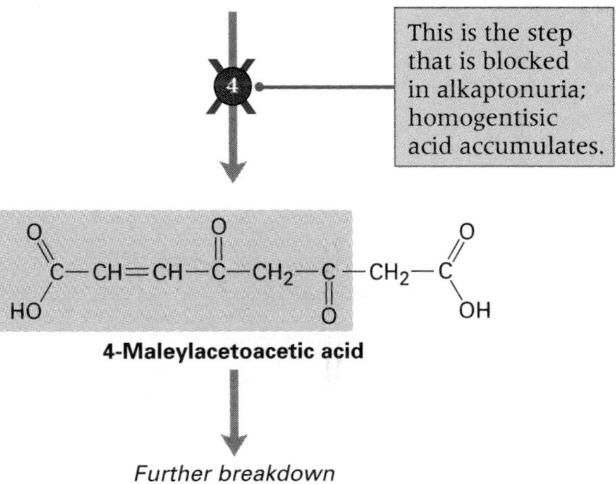

Figure 1.8 Metabolic pathway for the breakdown of phenylalanine and tyrosine. Each step in the pathway, represented by an arrow, requires a particular enzyme to catalyze the reaction. The key step in the breakdown of homogentisic acid is the breaking open of the benzene ring.

illustrates the relationship between the molecules is shown in Figure 1.8. Any such sequence of biochemical reactions is called a **biochemical pathway** or a **metabolic pathway.** Each arrow in the pathway represents a single step depicting the transition from the "input" or **substrate molecule,** shown at the tail of the arrow, to the "output" or **product molecule,** shown at the tip. Biochemical pathways are usually oriented either vertically with the arrows pointing down, as in Figure 1.8, or horizontally, with the arrows pointing from left to right. Garrod did not know all of the details of the pathway in Figure 1.8, but he did understand that the key step in the breakdown of homogentisic acid is the breaking open of the benzene ring and that the benzene ring in homogentisic acid comes from dietary phenylalanine and tyrosine.

What allows each step in a biochemical pathway to occur? Garrod's insight was to see that each step requires a specific enzyme to catalyze the reaction to allow the chemical transformation to take place. Persons with an inborn error of metabolism, such as alkaptonuria, have a defect in one step of a metabolic pathway because they lack a functional enzyme for that step. When an enzyme in a pathway is defective, the pathway is said to have a **block** at that step. One frequent result of a blocked pathway is that the substrate of the defective enzyme accumulates. Observing the accumulation of homogentisic acid in patients with alkaptonuria, Garrod proposed that there must be an enzyme whose function is to open the benzene ring of homogentisic acid and that this enzyme is missing in these patients. Discovery of all the enzymes in the pathway in Figure 1.7 took a long time. The enzyme that opens the benzene ring of homogentisic acid was not actually isolated until 50 years after Garrod's lectures. In normal people it is found in cells of the liver; just as Garrod predicted, the enzyme is defective in patients with alkaptonuria.

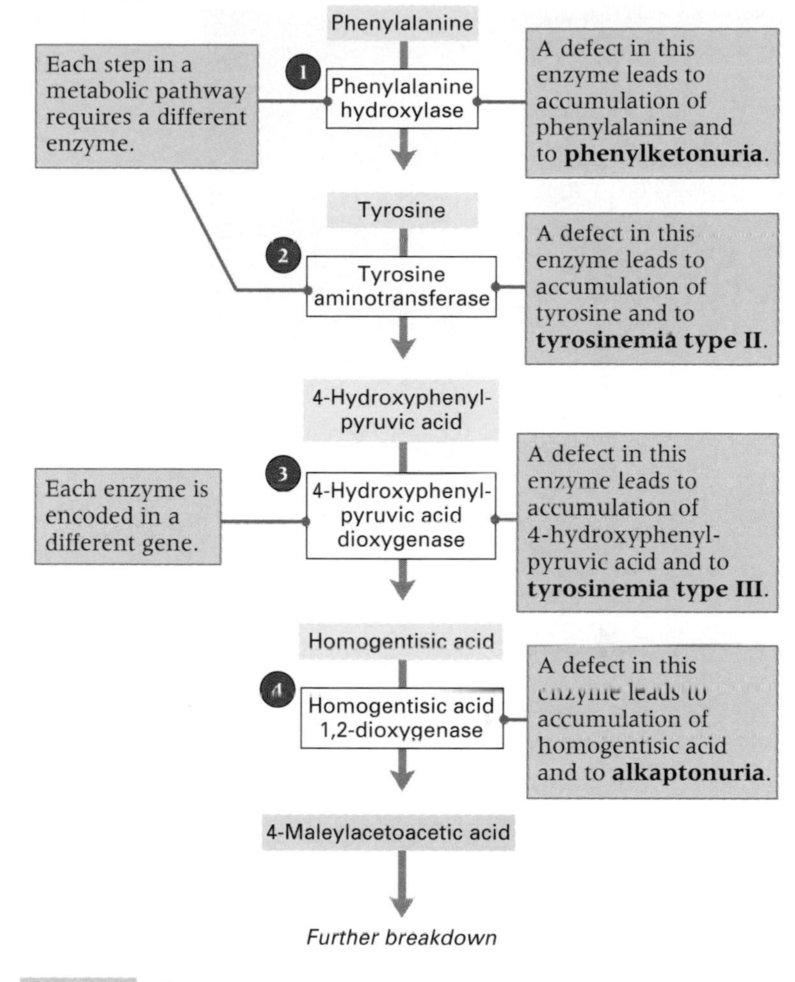

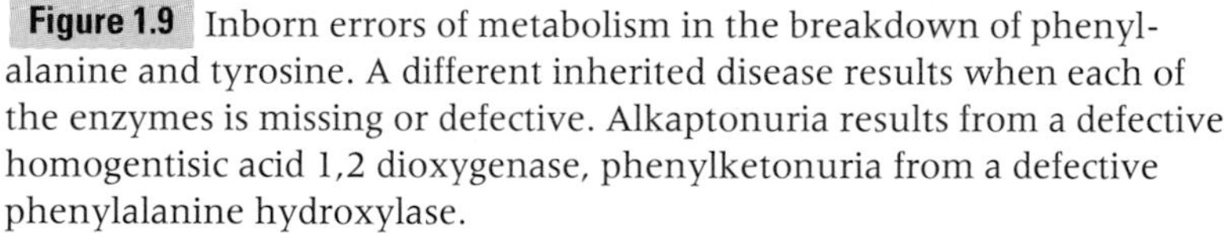

Figure 1.9 Inborn errors of metabolism in the breakdown of phenylalanine and tyrosine. A different inherited disease results when each of the enzymes is missing or defective. Alkaptonuria results from a defective homogentisic acid 1,2 dioxygenase, phenylketonuria from a defective phenylalanine hydroxylase.

The pathway for the breakdown of phenylalanine and tyrosine, as it is understood today, is shown in Figure 1.9. In this figure the emphasis is on the enzymes rather than on the structures of the **metabolites,** or small molecules, on which the enzymes act. As Garrod would have predicted, each step in the pathway requires the presence of a particular enzyme that catalyzes that step. Although Garrod knew only about alkaptonuria, in which the defective enzyme is homogentisic acid 1,2 dioxygenase, we now know the clinical consequences of defects in the other enzymes. Unlike alkaptonuria, which is a relatively benign inherited disease, the others are very serious. The condition known as **phenylketonuria (PKU)** results from the absence of (or a defect in) the enzyme **phenylalanine hydroxylase (PAH).** When this step in the pathway is blocked, phenylalanine accumulates. The excess phenylalanine is broken down into harmful metabolites that cause defects in myelin formation that damage a child's developing nervous system and lead to severe mental retardation. At the same time, the abnormally low levels of tyrosine in affected children often result in a reduced amount of skin and hair pigment, because the pigment molecule *melanin* is produced from tyrosine by an alternative to the breakdown pathway in Figure 1.8. The reduced pigmentation leads to the light complexion and hair color that are sometimes found in PKU. Because of modern treatment methods, this manifestation of the trait is rarely found today (Figure 1.10).

Figure 1.10 The women in this photograph are sisters. Both are homozygous for the same mutant *PAH* gene. The bride is the younger of the two. She was diagnosed just three days after birth and put on the PKU diet soon after. Her maid of honor, the older sister, was diagnosed too late to begin the diet. She is mentally retarded; she doesn't show the lighter skin and hair pigmentation often associated with phenylketonuria. [Courtesy of Charles R. Scriver.]

If PKU is diagnosed in children soon enough after birth, they can be placed on a specially formulated diet low in phenylalanine (Figure 1.10). The child is allowed only as much phenylalanine as can be used in the synthesis of proteins, so excess phenylalanine does not accumulate. The special diet is very strict. It excludes meat, poultry, fish, eggs, milk and milk products, legumes, nuts, and bakery goods manufactured with regular flour. These foods are replaced by a synthetic formula that is very expensive. With the special diet, however, the detrimental effects of excess phenylalanine on mental development can largely be avoided. Although it was once believed that the diet was needed only in childhood, it now appears that dietary treatment should continue for life, because termination of treatment is often associated with learning disabilities and behavior problems. In many countries, including the United States, all newborn babies have their blood tested for chemical signs of PKU. Routine screening is cost-effective because PKU is relatively common. In the United States, the incidence is about one in 8000 among Caucasian births. The disease is less common in other ethnic groups.

In the metabolic pathway in Figure 1.9, defects in the breakdown of tyrosine or of 4-hydroxyphenylpyruvic acid lead to types of tyrosinemia. These are also severe diseases. Type II is associated with skin lesions and mental retardation, Type III with severe liver dysfunction.

A defective enzyme results from a mutant gene.

It follows from Garrod's work that a defective enzyme results from a mutant gene. How does a mutant gene result in a defective enzyme? Garrod did not speculate. For all he knew, genes *were* enzymes. This would have been a logical hypothesis at the time. We now know that the relationship between genes and enzymes is somewhat indirect. With a few exceptions, each enzyme is *encoded* in a particular sequence of nucleotides present in a region of DNA. The DNA region that codes for the enzyme, as well as adjacent regions that regulate when and in which cells the enzyme is produced, make up the "gene" that encodes the enzyme.

The genes for the enzymes in the biochemical pathway in Figure 1.9 have all been identified and the nucleotide sequence of the DNA determined. In the following list, and throughout this book, we use the typographical convention that *genes* are written in *italic* type, whereas gene products are printed in regular type.

- The gene *PAH* on the long arm of chromosome 12 encodes phenylalanine hydroxylase (PAH).
- The gene *TAT* on the long arm of chromosome 16 encodes tyrosine aminotransferase (TAT).
- The gene *HPD* on the long arm of chromosome 12 encodes 4-hydroxyphenylpyruvic acid dioxygenase (HPD).
- The gene *HGD* on the long arm of chromosome 3 encodes homogentisic acid 1,2 dioxygenase (HGD).

Next we turn to the issue of *how* genes code for enzymes and other proteins.

One of the DNA strands directs the synthesis of a molecule of RNA.

Watson and Crick were correct in proposing that the genetic information in DNA is contained in the sequence of bases in a manner analogous to letters printed on a strip of paper. In a region of DNA that directs the synthesis of a protein, the genetic code for the protein is contained in only one strand, and it is decoded in a linear order. A typical protein is made up of one or more polypeptide chains; each **polypeptide chain** consists of a linear sequence of amino acids connected end to end. For example, the enzyme PAH consists of four identical polypeptide chains, each 452 amino acids in length (Figure 1.11). In the decoding of DNA, each successive "code word" in the DNA specifies the next amino

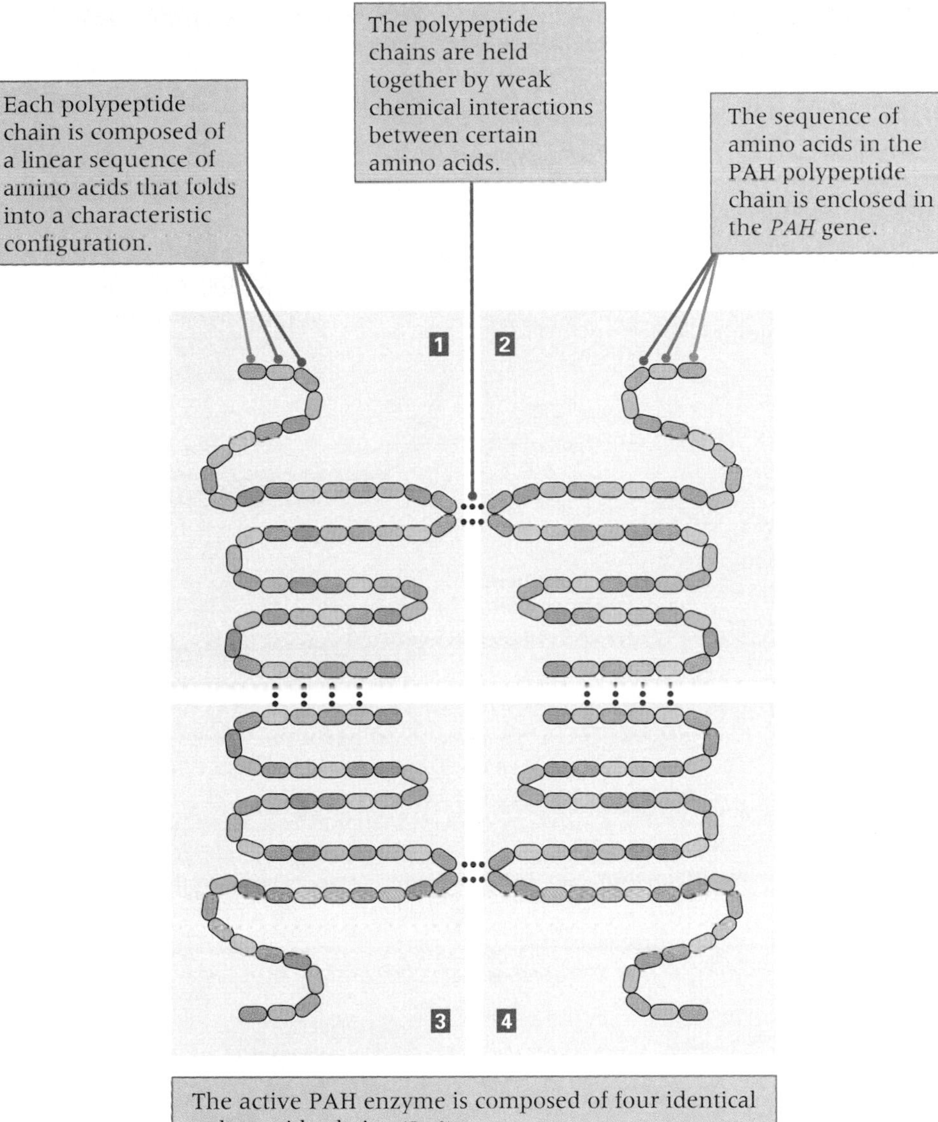

Figure 1.11 Many enzymes are composed of two or more polypeptide chains (there is usually an even number). Phenylalanine hydroxylase (PAH) is composed of four identical polypeptide chains encoded in the gene designated *PAH* (gene designations are conventionally set in italic type). The actual structure of the enzyme is more complex than that shown here.

acid to be added to the polypeptide chain as it is being made. The amount of DNA required to code for the polypeptide chain of PAH is therefore $452 \times 3 = 1356$ nucleotide pairs. The entire gene is very much longer—90,000 nucleotide pairs. Only 1.5 percent of the gene is devoted to coding for the amino acids. The noncoding part includes sequences that control the activity of the gene, but it is not known how much of the gene is involved in regulation.

There are 20 different amino acids. How can four bases code for 20 amino acids? Because each "word" in the genetic code consists of three adjacent bases. For example, the base sequence ATG specifies the amino acid methionine (Met), TCC specifies serine (Ser), ACT specifies threonine (Thr), and GCG specifies alanine (Ala). There are 64 possible three-base combinations but only 20 amino acids, because some combinations code for the same amino acid. For example, TCT, TCC, TCA, TCG, AGT, and AGC all code for serine (Ser), and CTT, CTC, CTA, CTG, TTA, and TTG all code for leucine (Leu). An example of the relationship between the base sequence in a DNA duplex and the amino acid sequence of the corresponding protein is shown in Figure 1.12.

This particular DNA duplex is the human sequence that codes for the first seven amino acids in the polypeptide chain of PAH.

The scheme outlined in Figure 1.12 indicates that DNA codes for protein not directly but indirectly through the processes of *transcription* and *translation.* The indirect route of information transfer,

$$DNA \rightarrow RNA \rightarrow Protein$$

is known as the **central dogma** of molecular genetics. The term *dogma* means "set of beliefs"; it dates from the time the idea was put forward first as a theory. Since then the "dogma" has been confirmed experimentally, but the term persists. The central dogma is shown in Figure 1.13. The main concept in the central dogma is that DNA does not code for protein directly but rather acts through an intermediary molecule of **ribonucleic acid (RNA).** The structure of RNA is similar to, but not identical with, that of DNA. There is a difference in the sugar (RNA contains the sugar **ribose** instead of deoxyribose), RNA is usually single-stranded (not a duplex), and RNA contains the base **uracil (U)** instead of thymine (T), which is present in DNA. Three types of RNA take part in the synthesis of proteins:

- A molecule of **messenger RNA (mRNA),** which carries the genetic information from DNA and is used as a template for polypeptide synthesis. In most mRNA molecules, there is a high proportion of nucleotides that actually code for amino acids. For example, the mRNA for PAH is 2400 nucleotides in length and codes for a polypeptide of 452 amino acids; in this case, more than 50 percent of the length of the mRNA codes for amino acids.

- Several types of **ribosomal RNA (rRNA),** which are major constituents of the cellular particles called **ribosomes** on which polypeptide synthesis takes place.

- A set of **transfer RNA (tRNA)** molecules, each of which carries a particular amino acid as well as a three-base recognition region that base-pairs with a group of three adjacent bases in the mRNA. As each tRNA participates in translation, its amino acid becomes the terminal subunit added to the length of the growing polypeptide chain. The tRNA that carries methionine is denoted $tRNA^{Met}$, that which carries serine is denoted $tRNA^{Ser}$, and so forth.

The central dogma is the fundamental principle of molecular genetics because it summarizes how the genetic information in DNA becomes

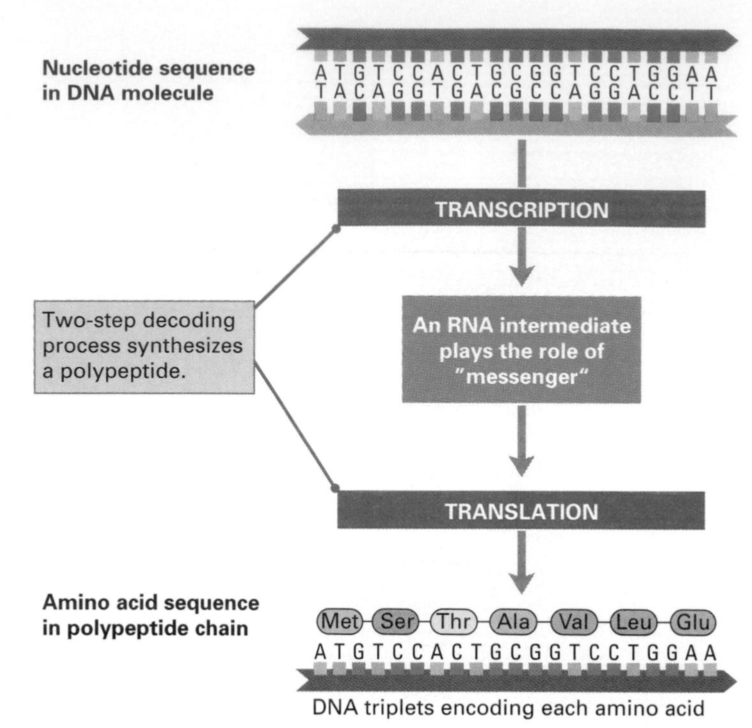

Figure 1.12 DNA sequence coding for the first seven amino acids in a polypeptide chain. The DNA sequence specifies the amino acid sequence through a molecule of RNA that serves as an intermediary "messenger." Although the decoding process is indirect, the net result is that each amino acid in the polypeptide chain is specified by a group of three adjacent bases in the DNA. In this example, the polypeptide chain is that of phenylalanine hydroxylase (PAH).

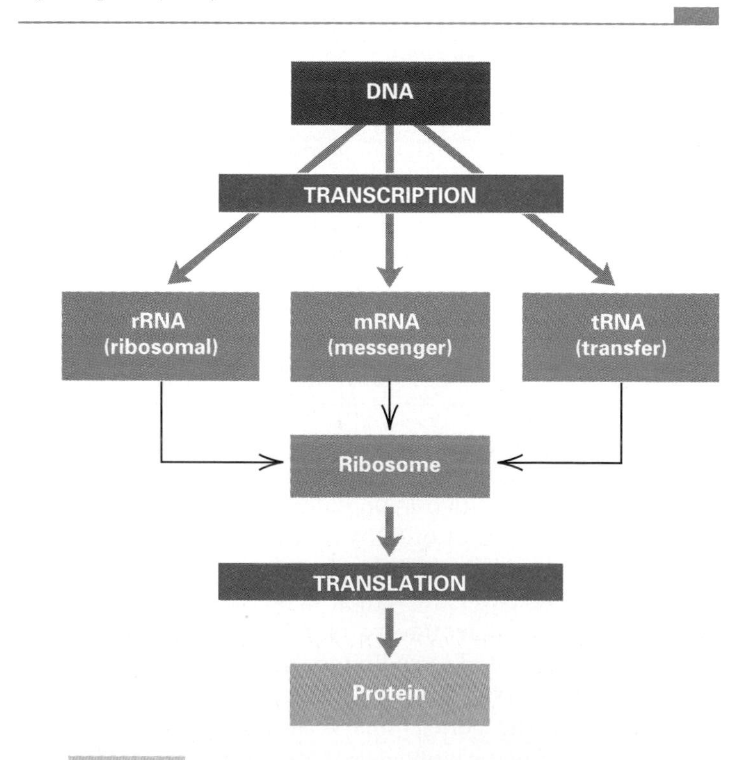

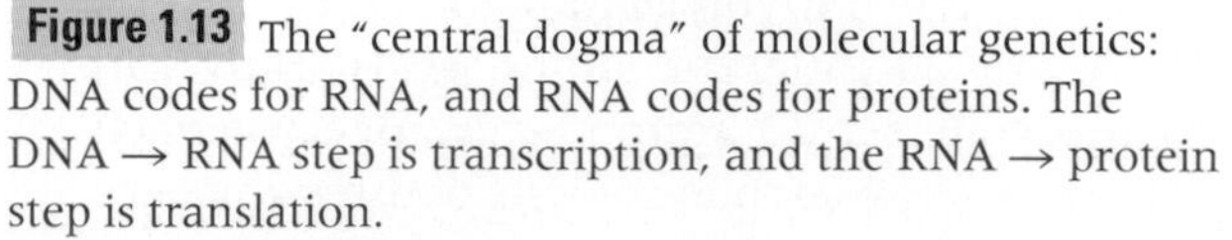

Figure 1.13 The "central dogma" of molecular genetics: DNA codes for RNA, and RNA codes for proteins. The DNA $\rightarrow$ RNA step is transcription, and the RNA $\rightarrow$ protein step is translation.

expressed in the amino acid sequence in a polypeptide chain:

> The sequence of nucleotides in a gene specifies the sequence of nucleotides in a molecule of messenger RNA; in turn, the sequence of nucleotides in the messenger RNA specifies the sequence of amino acids in the polypeptide chain.

Why should a process as conceptually simple as DNA coding for protein have the additional complexity of RNA intermediaries? Certain biochemical features of RNA suggest that RNA played a central role in the earliest forms of life and that RNA became locked into the processes of information transfer and protein synthesis. If this hypothesis is correct, then the participation of RNA in protein synthesis is a relic of the earliest stages of evolution—a "molecular fossil." The hypothesis that the first forms of life used RNA both for carrying information (in the base sequence) and as catalysts (accelerating chemical reactions) is supported by a variety of observations. For example, (1) DNA replication requires an RNA molecule in order to get started (Chapter 6), and (2) some RNA molecules act to catalyze biochemical reactions that are important in protein synthesis (Chapter 9). In the later evolution of the early life forms, additional complexity could have been added. The function of information storage and replication could have been transferred from RNA to DNA, and the function of RNA catalysis in metabolism could have been transferred from RNA to protein by the evolution of RNA-directed protein synthesis.

The manner in which genetic information is transferred from DNA to RNA is shown in Figure 1.14. The DNA opens up, and one of the strands is used as a template for the synthesis of a complementary strand of RNA. (How the template strand is chosen is discussed in Chapter 9.) The process of making an RNA strand from a DNA template is **transcription,** and the RNA molecule that is made is the **transcript.** The base sequence in the RNA is complementary (in the Watson–Crick pairing sense) to that in the DNA template, except that U (which pairs with A) is present in the RNA in place of T. The rules of base pairing between DNA and RNA are summarized in Figure 1.15. Like DNA, an RNA strand also exhibits polarity, its 5' and 3' ends determined by the orientation of the nucleotides. The 5' end of the RNA transcript is synthesized first, and in the RNA–DNA duplex formed in transcription, the polarity of the RNA strand is opposite to

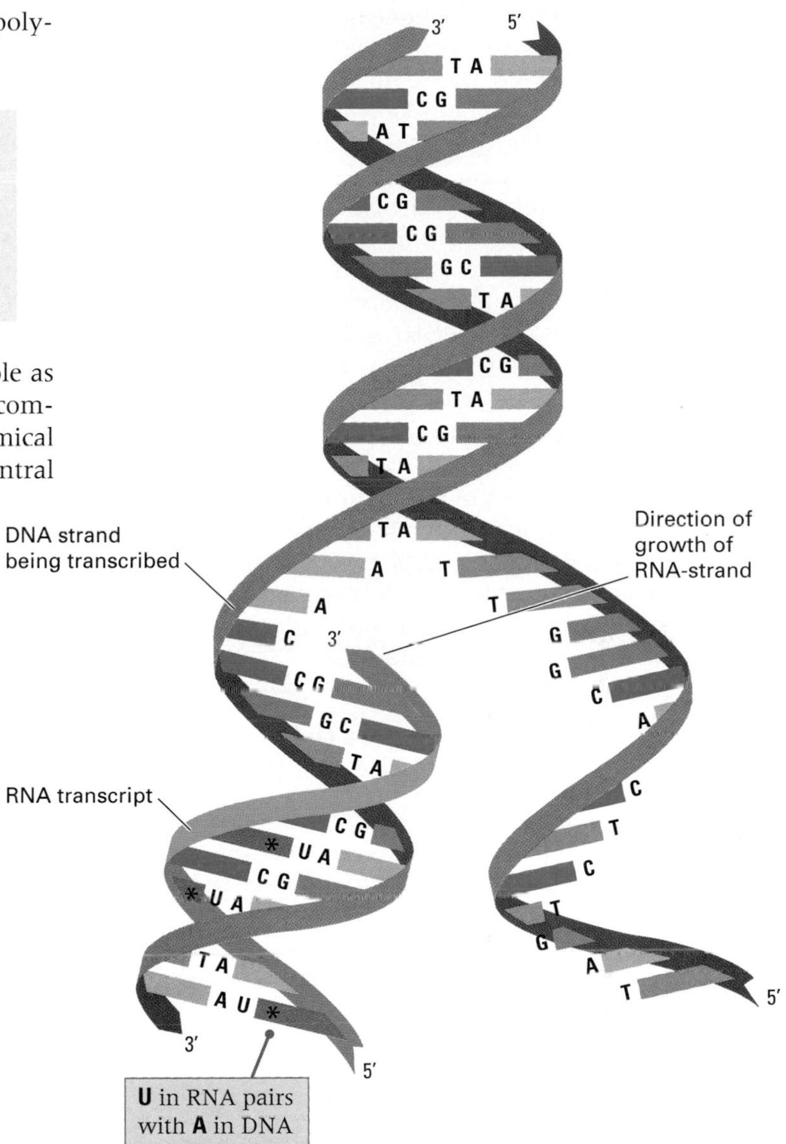

Figure 1.14 Transcription is the production of an RNA strand that is complementary in base sequence to a DNA strand. In this example, the DNA strand at the bottom is being transcribed into a strand of RNA. Note that in an RNA molecule, the base U (uracil) plays the role of T (thymine) in that it pairs with A (adenine). Each A—U pair is marked with an asterisk.

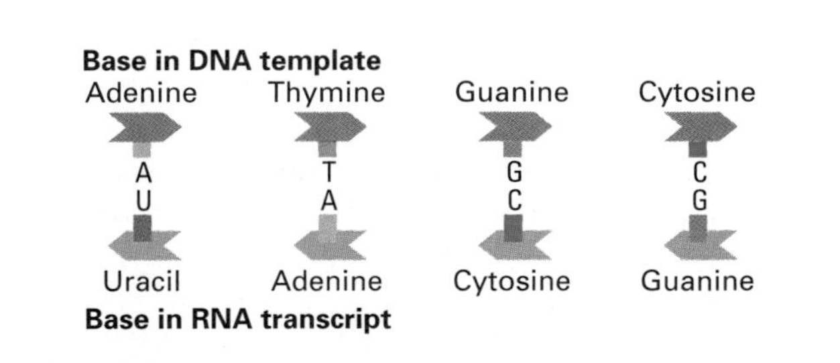

Figure 1.15 Pairing between bases in DNA and in RNA. The DNA bases A, T, G, and C pair with the RNA bases U, A, C, and G, respectively.

that of the DNA strand. Each gene includes particular nucleotide sequences that initiate and terminate transcription. The RNA transcript made from any gene begins at the initiation site in the template strand, which is located "upstream" from the amino-acid coding region, and ends at the terminate site, which is located "downstream" from the amino-acid coding region. For any gene, the length of the RNA transcript is very much smaller than the length of the DNA in the chromosome. For example, the transcript of the *PAH* gene for phenylalanine hydroxylase is 90,000 nucleotides in length, but the DNA in chromosome 12 is about 130,000,000 nucleotide pairs. In this case, the length of the *PAH* transcript is less than 0.1 percent of the length of the DNA in the chromosome. A different gene in chromosome 12 would be transcribed from a different region of the DNA molecule in chromosome 12, but the transcribed region would again be small in comparison with the total length of the DNA in the chromosome.

A molecule of RNA directs the synthesis of a polypeptide chain.

The synthesis of a polypeptide under the direction of an mRNA molecule is known as **translation.** Although the sequence of bases in the mRNA codes for the sequence of amino acids in a polypeptide, the molecules that actually do the "translating" are the tRNA molecules. The mRNA molecule is translated in nonoverlapping groups of three bases called **codons.** For each codon in the mRNA that specifies an amino acid, there is one tRNA molecule containing a complementary group of three adjacent bases that can pair with those in that codon. The correct amino acid is attached to the other end of the tRNA, and when the tRNA comes into line, the amino acid to which it is attached becomes the amino acid last added to the growing end of the polypeptide chain.

The role of tRNA in translation is illustrated in Figure 1.16 and can be described as follows:

> The mRNA is read codon by codon. Each codon that specifies an amino acid matches with a complementary group of three adjacent bases in a single tRNA molecule. One end of the tRNA is attached to the correct amino acid, so the correct amino acid is brought into line.

The tRNA molecules used in translation do not line up along the mRNA simultaneously as shown in Figure 1.16. The process of translation takes place on a ribosome, which combines with a single mRNA and moves along it from one end to the other in steps, three nucleotides at a time (codon by codon). As each new codon comes into place, the next tRNA binds with the ribosome. Here it seems logical to expect that the amino acid on the tRNA would be cleaved and added to the growing end of the polypeptide chain—but this is not what happens. Instead, it is the growing end of the polypeptide chain that becomes attached to the amino acid on the tRNA. In this way, each tRNA in turn serves temporarily to hold the polypeptide chain as it is being synthesized. As the polypeptide chain is transferred from each tRNA to the next in line, the tRNA that previously held the polypeptide is released from the ribosome. The polypeptide chain elongates one amino acid at a step until any one of three particular codons specifying "stop" is encountered. At this point, synthesis of the chain of amino acids is finished, and the polypeptide chain is re-

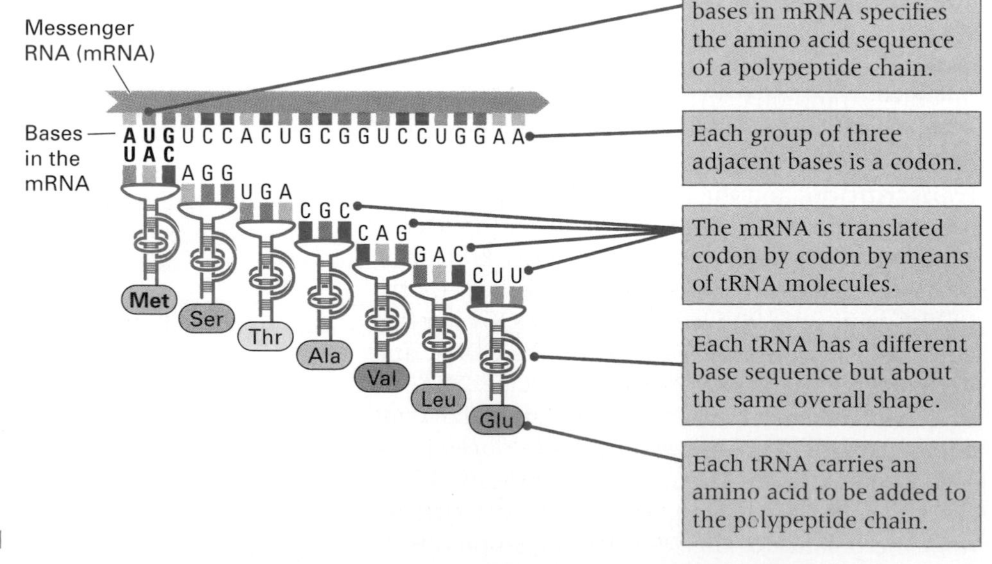

Figure 1.16 The role of messenger RNA in translation is to carry the information contained in a sequence of DNA bases to a ribosome, where it is translated into a polypeptide chain. Translation is mediated by transfer RNA (tRNA) molecules, each of which can base-pair with a group of three adjacent bases in the mRNA. Each tRNA also carries an amino acid, and when it is brought to the ribosome by base pairing, its amino acid becomes the growing end of the polypeptide chain.

Table 1.1
The Standard Genetic Code

First nucleotide in codon (5' end)	Second nucleotide in codon: U	C	A	G	Third nucleotide in codon (3' end)
U	UUU Phe F *Phenylalanine*	UCU Ser S *Serine*	UAU Tyr Y *Tyrosine*	UGU Cys C *Cysteine*	U
	UUC Phe F *Phenylalanine*	UCC Ser S *Serine*	UAC Tyr Y *Tyrosine*	UGC Cys C *Cysteine*	C
	UUA Leu L *Leucine*	UCA Ser S *Serine*	UAA Termination	UGA Termination	A
	UUG Leu L *Leucine*	UCG Ser S *Serine*	UAG Termination	UGG Trp W *Tryptophan*	G
C	CUU Leu L *Leucine*	CCU Pro P *Proline*	CAU His H *Histidine*	CGU Arg R *Arginine*	U
	CUC Leu L *Leucine*	CCC Pro P *Proline*	CAC His H *Histidine*	CGC Arg R *Arginine*	C
	CUA Leu L *Leucine*	CCA Pro P *Proline*	CAA Gln Q *Glutamine*	CGA Arg R *Arginine*	A
	CUG Leu L *Leucine*	CCG Pro P *Proline*	CAG Gln Q *Glutamine*	CGG Arg R *Arginine*	G
A	AUU Ile I *Isoleucine*	ACU Thr T *Threonine*	AAU Asn N *Asparagine*	AGU Ser S *Serine*	U
	AUC Ile I *Isoleucine*	ACC Thr T *Threonine*	AAC Asn N *Asparagine*	AGC Ser S *Serine*	C
	AUA Ile I *Isoleucine*	ACA Thr T *Threonine*	AAA Lys K *Lysine*	AGA Arg R *Arginine*	A
	AUG Met M *Methionine*	ACG Thr T *Threonine*	AAG Lys K *Lysine*	AGG Arg R *Arginine*	G
G	GUU Val V *Valine*	GCU Ala A *Alanine*	GAU Asp D *Aspartic acid*	GGU Gly G *Glycine*	U
	GUC Val V *Valine*	GCC Ala A *Alanine*	GAC Asp D *Aspartic acid*	GGC Gly G *Glycine*	C
	GUA Val V *Valine*	GCA Ala A *Alanine*	GAA Glu E *Glutamic acid*	GGA Gly G *Glycine*	A
	GUG Val V *Valine*	GCG Ala A *Alanine*	GAG Glu E *Glutamic acid*	GGG Gly G *Glycine*	G

Codon — Three-letter and single-letter abbreviations

leased from the ribosome. (This brief description of translation glosses over many of the details that are presented in Chapter 9.)

The genetic code is a triplet code.

Figure 1.16 indicates that the mRNA codon AUG specifies methionine (Met) in the polypeptide chain, UCC specifies Ser (serine), ACU specifies Thr (threonine), and so on. The complete decoding table is called the **genetic code,** and it is shown in Table 1.1. For any codon, the column on the left corresponds to the first nucleotide in the codon (reading from the 5' end), the row across the top corresponds to the second nucleotide, and the complete codon is given in the body of the table, along with the amino acid (or "stop") that the codon specifies. Each amino acid is designated by its full name and by a three-letter abbreviation and a single-letter abbreviation. Both types of abbreviations are used in molecular genetics. The code in Table 1.1 is the "standard" genetic code used in translation in the cells of nearly all organisms. In Chapter 9 we examine general features of the standard genetic code and the minor differences found in the genetic codes of certain organisms and cellular organelles. At this point, we are interested mainly in understanding how the genetic code is used to translate the codons in mRNA into the amino acids in a polypeptide chain.

In addition to the 61 codons that code only for amino acids, there are four codons that have specialized functions:

- The codon AUG, which specifies Met (methionine), is also the "start" codon for polypeptide synthesis. The positioning of a $tRNA^{Met}$ bound to AUG is one of the first steps in the initiation of polypeptide synthesis, so all polypeptide chains begin with Met. (Many polypeptides have the initial Met cleaved off after translation is complete.) In most organisms, the $tRNA^{Met}$ used for initiation of translation is the same $tRNA^{Met}$ used to specify methionine at internal positions in a polypeptide chain.
- The codons UAA, UAG, and UGA, each of which is a "stop," specify the termination of translation and result in release of the completed polypeptide chain from the ribosome. These codons do not have tRNA molecules that recognize them but are instead recognized by protein factors that terminate translation.

How the genetic code table is used to infer the amino acid sequence of a polypeptide chain may be illustrated using PAH again, in particular the DNA sequence coding for amino acid numbers 1 through 7. The DNA sequence is

5'-ATGTCCACTGCGGTCCTGGAA-3'
3'-TACAGGTGACGCCAGGACCTT-5'

This region is transcribed into RNA in a left-to-right direction, and because RNA grows by the addition of successive nucleotides to the 3' end (Figure 1.14), it is the bottom strand that is transcribed. The nucleotide sequence of the RNA is that of the top strand of the DNA, except that U replaces T, so the mRNA for amino acids 1 through 7 is

5'-AUGUCCACUGCGGUCCUGGAA-3'

The codons are read from left to right according to the genetic code shown in Table 1.1. Codon AUG codes for Met (methionine), UCC codes for Ser (serine), and so on. Altogether, the amino acid sequence of this region of the polypeptide is

5'-AUGUCCACUGCGGUCCUGGAA-3'
MetSerThrAlaValLeuGlu

or, in terms of the single-letter abbreviations,

5'-AUGUCCACUGCGGUCCUGGAA-3'
M S T A V L E

The full decoding operation for this region of the PH gene is shown in Figure 1.17. In this figure, the initiation codon AUG is highlighted because some patients with PKU have a mutation in this particular codon. As might be expected from the fact that methionine is the initiation codon for polypeptide synthesis, cells in patients with this particular mutation fail to produce any of the PAH polypeptide. Mutation and its consequences are considered next.

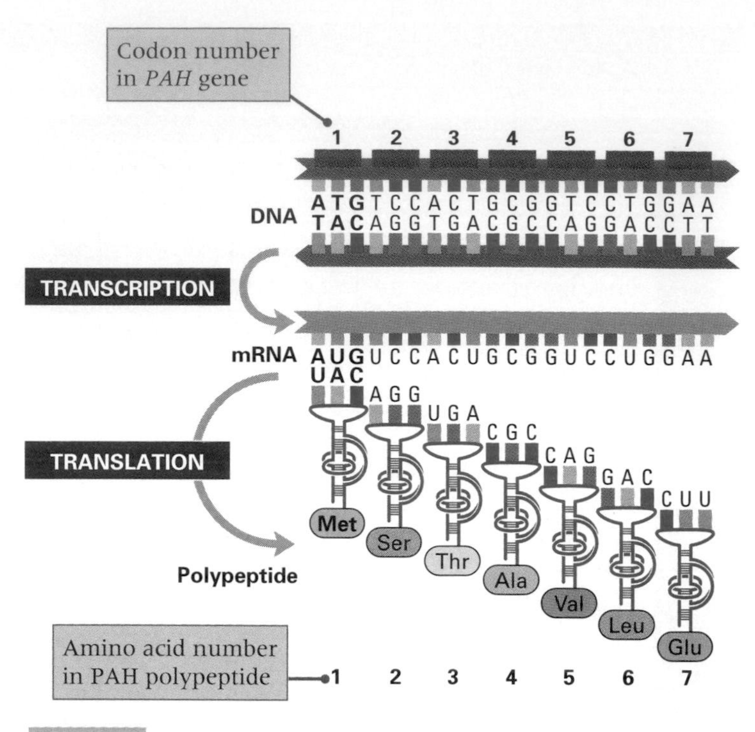

Figure 1.17 The central dogma in action. The DNA that encodes PAH serves as a template for the production of a messenger RNA, and the mRNA, in turn, serves to specify the sequence of amino acids in the PAH polypeptide chain through interactions with the tRNA molecules.

1.5 Genes change by mutation.

The term **mutation** refers to any heritable change in a gene (or, more generally, in the genetic material); the term also refers to the process by which such a change takes place. One type of mutation results in a change in the sequence of bases in DNA. The change may be simple, such as the substitution of one pair of bases in a duplex molecule for a different pair of bases. For example, a C—G pair in a duplex molecule may mutate to T—A, A—T, or G—C. The change in base sequence may also be more complex, such as the deletion or addition of base pairs. These and other types of mutations are considered in Chapter 7. Geneticists also use the term **mutant,** which refers to the result of a mutation. A mutation yields a mutant gene, which in turn produces a mutant mRNA, a mutant protein, and finally a mutant organism that exhibits the effects of the mutation—for example, an inborn error of metabolism.

DNA from patients from all over the world who have phenylketonuria has been studied to determine what types of mutations are responsible for the inborn error. There are a large variety of mutant types. More than 100 different mutations have been described. In some cases part of the gene is missing, so the genetic information to make a complete PAH enzyme is absent. In other cases the genetic defect is more subtle, but the result is still either the failure to produce a PAH protein or the production of a PAH protein that is inactive. In the mutation shown in Figure 1.18, substitution of a G—C base pair for the normal A—T base pair at the very first position in the coding sequence changes the normal codon AUG (Met) used for the initiation of translation into the codon GUG, which normally specifies valine (Val) and cannot be used as a "start" codon. The result is that translation of the PAH mRNA cannot occur, and so no PAH polypeptide is made. This mutant is designated M1V because the codon for M (methionine) at amino acid position 1 in the PAH polypeptide has been changed to a codon for V (valine). Although the M1V mutant is quite rare worldwide, it is common in some localities, such as in Quebec Province in Canada.

One PAH mutant that is quite common is designated R408W, which means that codon 408 in the PAH polypeptide chain has been changed from one

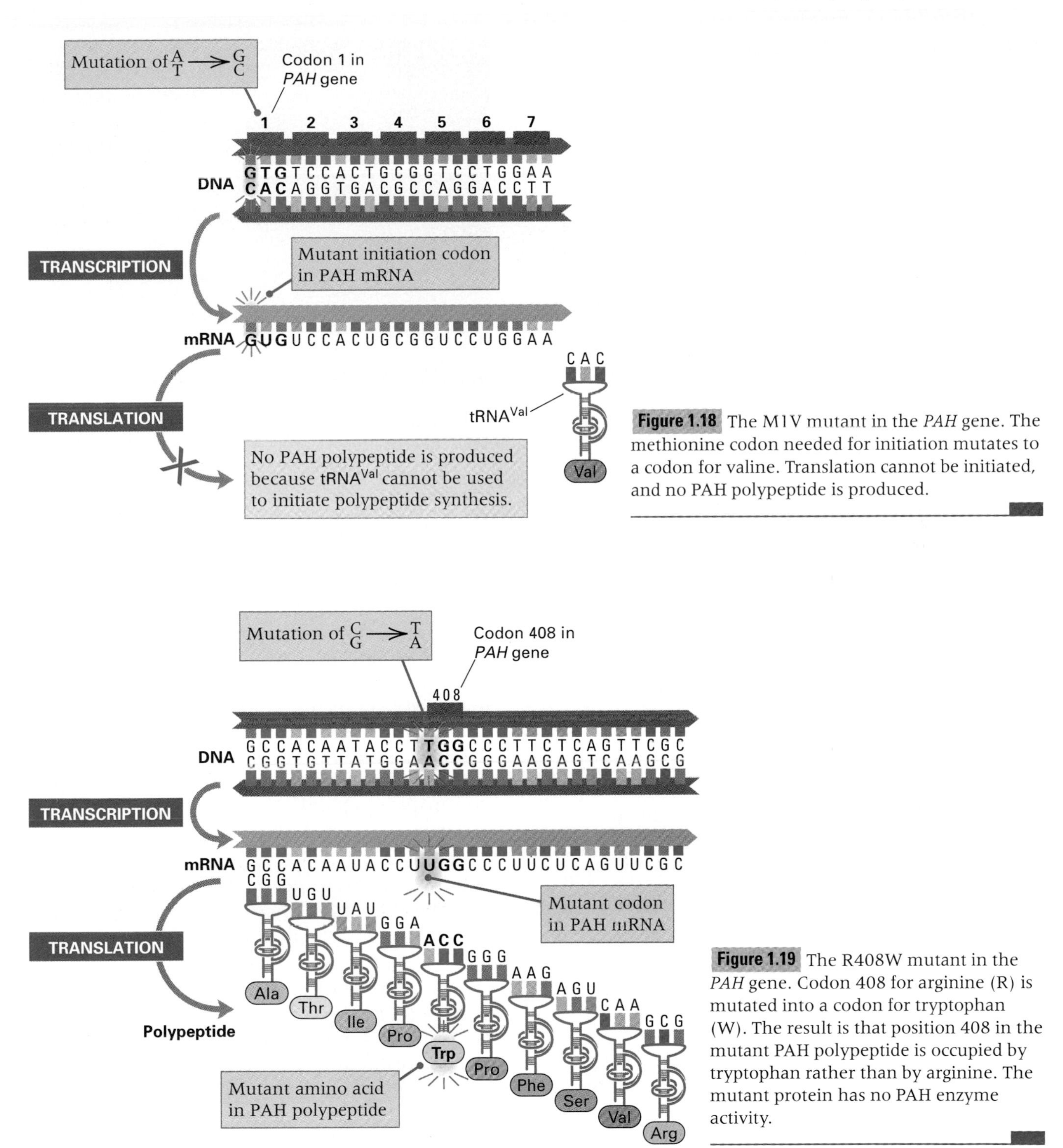

Figure 1.18 The M1V mutant in the *PAH* gene. The methionine codon needed for initiation mutates to a codon for valine. Translation cannot be initiated, and no PAH polypeptide is produced.

Figure 1.19 The R408W mutant in the *PAH* gene. Codon 408 for arginine (R) is mutated into a codon for tryptophan (W). The result is that position 408 in the mutant PAH polypeptide is occupied by tryptophan rather than by arginine. The mutant protein has no PAH enzyme activity.

coding for arginine (R) to one coding for tryptophan (W). This mutant is one of the four most common in cases of PKU among European Caucasians. The molecular basis of the mutant is shown in Figure 1.19. In this case, the first base pair in codon 408 is changed from a C—G base pair into a T—A base pair. The result is that the PAH mRNA has a mutant codon at position 408; specifically, it has UGG instead of CGG. Translation does occur in this mutant because everything else about the mRNA is normal, but the result is that the mutant PAH carries a tryptophan (Trp) instead of an arginine (Arg) at position 408 in the polypeptide chain. The consequence of the seemingly minor change of one amino acid is very drastic, because the mutant PAH has no enzyme activity and so is unable to catalyze its metabolic reaction. In other words, the mutant PAH protein is complete but inactive.

1.6 Traits are affected by environment as well as by genes.

Inborn errors of metabolism illustrate the general principle that genes code for proteins and that mutant genes code for mutant proteins. In cases such as PKU, mutant proteins cause such a drastic change in metabolism that a severe genetic defect results. But biology is not necessarily destiny. Organisms are also affected by the environment. PKU serves as an example of this principle, because patients who adhere to a diet restricted in the amount of phenylalanine develop mental capacities within the normal range. What is true in this example is true in general. Most traits are determined by the interaction of genes and environment.

It is also true that most traits are affected by multiple genes. No one knows how many genes are involved in the development and maturation of the brain and nervous system, but the number must be in the thousands. This number is in addition to the genes that are required in all cells to carry out metabolism and other basic life functions. It is easy to lose sight of the multiplicity of genes when considering extreme examples, such as PKU, in which a single mutation can have such a drastic effect on mental development. The situation is the same as that with any complex machine. An airplane can function if thousands of parts are working together in harmony, but it takes only one defective part, if it affects a vital system, to bring it down. Likewise, the development and functioning of every trait require a large number of genes working in harmony, but in some cases a single mutant gene can have catastrophic consequences.

In other words, the relationship between a gene and a trait is not necessarily a simple one. The biochemistry of organisms is a complex branching network in which different enzymes may share substrates, yield the same products, or be responsive to the same regulatory elements. The result is that most visible traits of organisms are the net result of many genes acting together and in combination with environmental factors. PKU affords examples of each of three principles governing these interactions:

1. One gene can affect more than one trait. We noted earlier that children with PKU tend also to have blond hair and reduced body pigment. This is because the absence of PAH is a metabolic block that prevents conversion of phenylalanine into tyrosine, which is the precursor of the pigment melanin. The relationship between severe mental retardation and decreased pigmentation in PKU makes sense only if one knows the metabolic connections among phenylalanine, tyrosine, and melanin. If these connections were not known, the traits would seem to be completely unrelated. PKU is not unusual in this regard. Many mutant genes affect multiple traits through their secondary or indirect effects. The various, sometimes seemingly unrelated, effects of a mutant gene are called **pleiotropic effects,** and the phenomenon itself is known as **pleiotropy.** PKU exemplifies pleiotropy because mental retardation and reduced pigmentation are pleiotropic effects of a mutant *PAH* gene.

Among cats with white fur and blue eyes, about 40% are born deaf. The reason for the defective hearing is not known, nor why it is so often associated with hair and eye color. This form of deafness can be regarded as a pleiotropic effect of white fur and blue eyes. [Courtesy of Janet Klein.]

2. One trait can be affected by more than one gene. We discussed this principle earlier in connection with the large number of genes that are required for the normal development and functioning of the brain and nervous system. Multiple genes affect even simpler traits. Take phenylalanine breakdown and excretion as an example. The metabolic pathway is illustrated in Figure 1.9. Four enzymes in the pathway are indicated, but even more enzymes are involved at the stage called "further breakdown." Because differences in the activity of any of these enzymes can affect the rate at which phenylalanine can be broken down and excreted, all of the genes for the enzymes in the pathway are important. The breakdown and excretion of phenylalanine are also affected, though less directly, by the genes required for normal liver function and those required for normal kidney function.

3. Many traits are affected by environmental factors as well as by genes. Here we come back to the low-phenylalanine diet. Children with PKU are not "doomed" to severe mental deficiency. Their capabilities can be brought into the normal range by dietary treatment. PKU serves as an example of what motivates geneticists to search to discover the molecular basis of inherited disease. The hope is that knowing the metabolic basis of the disease will eventually make it possible to develop methods for clinical intervention through diet, medication, or other treatments that will ameliorate the severity of the disease.

Maternal PKU illustrates the importance of genes and environment.

Some medical miracles have unforeseen and unhappy consequences. This was the case with the dietary treatment for PKU, which allowed a generation of children who otherwise would have been severely mentally handicapped to develop with mental capabilities in the normal range. What a shock it was to learn that women with PKU who became pregnant had a high frequency of spontaneous miscarriage and to find, among the babies born alive, a very high frequency of developmental abnormalities of the brain, slow fetal and postnatal growth, and congenital heart disease. These conditions were said to result from **maternal PKU.** It seemed likely that a high phenylalanine level in the mothers was affecting the growth of the fetus. This hypothesis was quickly confirmed with experiments in which mother monkeys were administered high levels of phenylalanine; the result was that their offspring had impaired learning ability. The developmental effects of excess maternal phenylalanine are exacerbated by the fact that amino acids are normally concentrated in the placenta, so the phenylalanine level is even higher in the placenta than in the mother.

The immediate response to the new findings was to alert each woman with PKU to the high risk of pregnancy. There were also studies that recorded the outcomes for women with PKU who returned to the low-phenylalanine diet prior to and during pregnancy. Under these conditions, the risk of severe developmental abnormalities was reduced but not completely eliminated, and even with the mother's phenylalanine level under control, the offspring showed some residual behavioral or developmental disabilities. At the present time, women with PKU are advised to return to the low-phenylalanine diet prior to and during pregnancy and to maintain their phenylalanine level at the lowest possible amount. In some cases surrogate motherhood is recommended, in which eggs from a woman with PKU are fertilized by her husband's sperm *in vitro* and then implanted into the womb of a surrogate mother.

Maternal PKU illustrates the critical role of the environment in normal development because, genetically, the offspring of PKU mothers have normal phenylalanine metabolism. The offspring receive a mutant *PAH* gene from their mother and a normal *PAH* gene from their father. Because only one copy of a normal *PAH* gene is enough to make sufficient PAH enzyme for normal development, the offspring would have been expected to be completely normal. This is the case in children born to parents with no family history of PKU, among whom about 1 in 45 carries a mutant *PAH* gene along with a normal one; these children are completely normal in their phenylalanine metabolism and development. Yet a fetus with this same genetic composition, when developing in the presence of high levels of phenylalanine, is severely compromised. Similarly, it is in the offspring in the monkey experiments, whose development was affected by administering phenylalanine to the mothers.

1.7 Evolution means continuity of life with change.

The pathway for the breakdown and excretion of phenylalanine is by no means unique to human beings. One of the remarkable generalizations to have emerged from molecular genetics is that organisms that are very distinct—for example, plants and animals—share many features in their genetics and biochemistry. These similarities indicate a fundamental "unity of life":

> All creatures on Earth share many features of the genetic apparatus, including genetic information encoded in the sequence of bases in DNA, transcription into RNA, and translation into protein on ribosomes with the use of transfer RNAs. All creatures also share certain characteristics in their biochemistry, including many enzymes and other proteins that are similar in amino acid sequence, three-dimensional structure, and function.

The molecular unity of life results from common ancestry.

Why is there unity of life? Because all creatures share a common origin. The process of **evolution** takes place whenever a population of organisms descended from a common ancestor gradually changes in genetic composition through time. From an evolutionary perspective, the unity of fundamental molecular processes is derived by inheritance from a distant common ancestor in which the mechanisms were already in place.

Not only the unity of life but also many other features of living organisms become comprehensible from an evolutionary perspective. For example, the interposition of an RNA intermediate in the

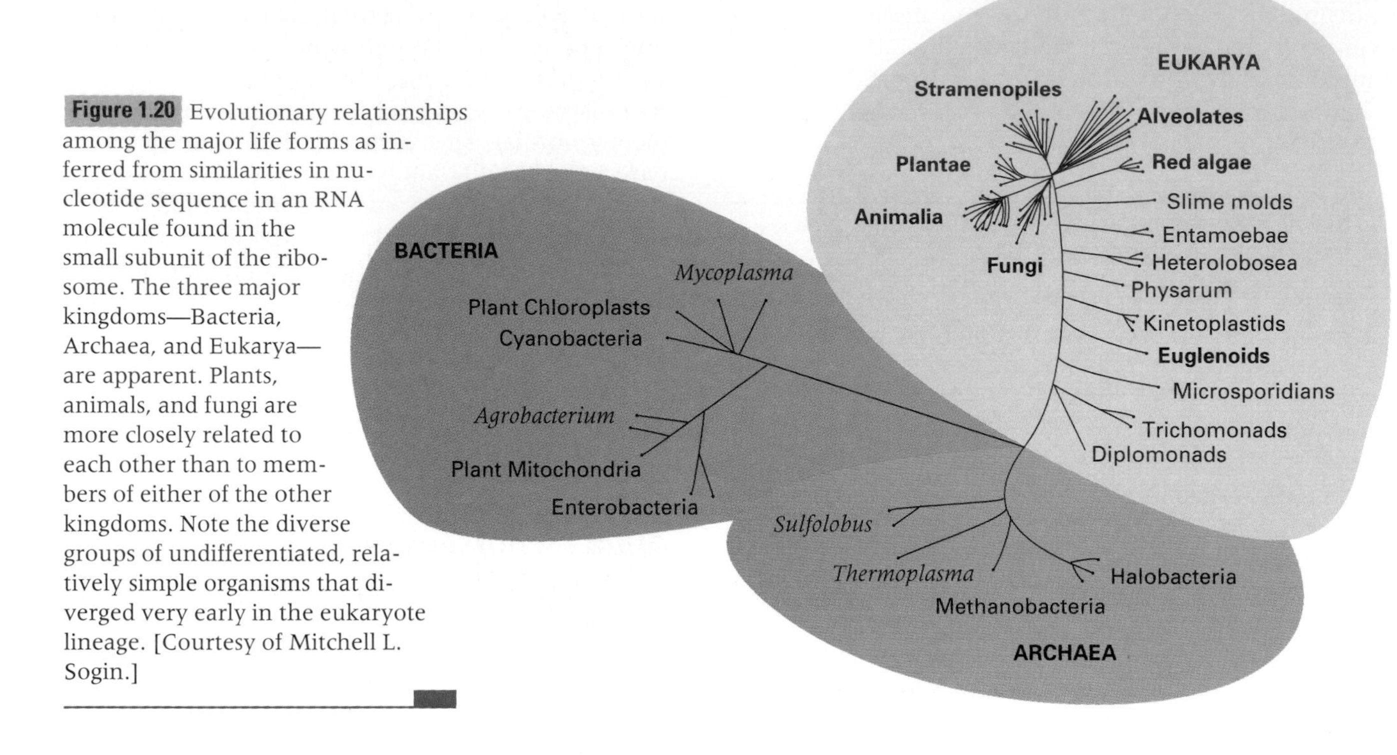

Figure 1.20 Evolutionary relationships among the major life forms as inferred from similarities in nucleotide sequence in an RNA molecule found in the small subunit of the ribosome. The three major kingdoms—Bacteria, Archaea, and Eukarya—are apparent. Plants, animals, and fungi are more closely related to each other than to members of either of the other kingdoms. Note the diverse groups of undifferentiated, relatively simple organisms that diverged very early in the eukaryote lineage. [Courtesy of Mitchell L. Sogin.]

basic flow of genetic information from DNA to RNA to protein makes sense if the earliest forms of life used RNA for both genetic information and enzyme catalysis. The importance of the evolutionary perspective in understanding aspects of biology that seem pointless or needlessly complex is summed up in a famous aphorism of the evolutionary biologist Theodosius Dobzhansky: "Nothing in biology makes sense except in the light of evolution."

One indication of the common ancestry among Earth's creatures is illustrated in Figure 1.20. The tree of relationships was inferred from similarities in nucleotide sequence in an RNA molecule found in the small subunit of the ribosome. Three major kingdoms of organisms are distinguished.

1. ***Bacteria*** This group includes most bacteria and cyanobacteria (formerly called blue-green algae). Cells of these organisms lack a membrane-bounded nucleus and mitochondria, are surrounded by a cell wall, and divide by binary fission.
2. ***Archaea*** This group was initially discovered among microorganisms that produce methane gas or that live in extreme environments, such as hot springs or high salt concentrations. They are widely distributed in more normal environments as well. Like those of Bacteria, the cells of Archaea lack internal membranes. DNA sequence analysis indicates that the machinery for DNA replication and transcription in Archaea resembles that of Eukarya, whereas their metabolism strongly resembles that of Bacteria. About half of the genes found in Archaea are unique to this group.
3. ***Eukarya*** This group includes all organisms whose cells contain an elaborate network of internal membranes, a membrane-bounded nucleus, and mitochondria. Their DNA is organized into true chromosomes, and cell division takes place by means of mitosis (discussed in Chapter 3). The eukaryotes include plants and animals as well as fungi and many single-celled organisms, such as amoebae and ciliated protozoa.

The members of the kingdoms Bacteria and Archaea are often grouped together into a larger assemblage called **prokaryotes,** which literally means "before [the evolution of] the nucleus." This terminology is convenient for designating prokaryotes as a group in contrast with eukaryotes, which literally means "good [well-formed] nucleus."

The diversity of life results mainly from natural selection.

Although Figure 1.20 illustrates the unity of life, it also illustrates life's diversity. Frogs are different from fungi, and beetles are different from bacteria. As a human being, it is sobering to consider that complex, multicellular organisms are relatively recent arrivals to the evolutionary scene of life on Earth. Animals came later still and primates very late indeed. What about human evolution? In the time scale of Earth history, human evolution is a matter of a few million years—barely a snap of the fingers.

If common ancestry is the source of the unity of life, what is the source of its diversity? Because differences among species are inherited, the original source of the differences must be mutation. However, mutations alone are not sufficient to explain why organisms are adapted to living in their environments—why ocean mammals have special adaptations for swimming and diving, why desert mammals have special adaptations to survive on minimal amounts of water. Mutations are chance events not directed toward any particular adaptive goal, such as longer fur among mammals living in the Arctic. The process that accounts for adaptation was described by Charles Darwin in his 1859 book *On the Origin of Species.* Darwin proposed that adaptation is the result of **natural selection:** Individual organisms carrying particular mutations or combinations of mutations that enable them to survive or reproduce more effectively in the prevailing environment leave more offspring than other organisms and so contribute their favorable genes disproportionately to future generations. If this process is repeated throughout the course of many generations, the entire species becomes genetically transformed—evolves—because a gradually increasing proportion of the population inherits the favorable mutations. The genetic basis of natural selection is discussed in Chapter 13.

Darwin's ideas about natural selection were strongly influenced by his observation that each species of finch had a beak whose length and shape matched the food sources and eating habits of the birds.

Chapter Summary

- **DNA is the molecule of heredity.**

  Genetic traits can be altered by treatment with pure DNA.

  Transmission of DNA is the link between generations.

Organisms of the same species have some traits (characteristics) in common but may differ from each other in innumerable other traits. Many of the differences between organisms result from genetic differences, the effects of the environment, or both. Genetics is the study of inherited traits, including those influenced in part by the environment. The elements of heredity consist of genes, which are transmitted from parents to offspring in reproduction. Although the sorting of genes in successive generations was first expressed numerically by Mendel, the chemical basis of genes was discovered by Miescher in the form of a weak acid—deoxyribonucleic acid (DNA). However, experimental proof that DNA is the genetic material did not come until about the middle of the twentieth century.

The first convincing evidence of the role of DNA in heredity came from the experiments of Avery, MacLeod, and McCarty, who showed that genetic characteristics in bacteria could be altered from one type to another by treatment with purified DNA. In studies of *Streptococcus pneumoniae,* they transformed mutant cells unable to cause pneumonia into cells that could do so by treating them with pure DNA from disease-causing forms. A second important line of evidence was the Hershey–Chase experiment. Hershey and Chase showed that the T2 bacterial virus injects primarily DNA into the host bacterium (*Escherichia coli*) and that a much higher proportion of parental DNA, compared with parental protein, is found among the progeny phage.

- **The structure of DNA is a double helix composed of two intertwined strands.**
- **In replication, each parental DNA strand directs the synthesis of a new partner strand.**

The three-dimensional structure of DNA, proposed in 1953 by Watson and Crick, gave many clues to the manner in which DNA functions as the genetic material. A molecule of DNA consists of two long chains of nucleotide subunits twisted around each other to form a right-handed helix. Each nucleotide subunit contains any one of four bases: A (adenine), T (thymine), G (guanine), or C (cytosine). The bases are paired in the two strands of a DNA molecule: Wherever one strand has an A, the partner strand has a T; and wherever one strand has a G, the partner strand has a C. The base pairing means that the two paired strands in a DNA duplex molecule have complementary base sequences along their lengths. The structure of the DNA molecule suggested that genetic information could be coded in DNA in the sequence of bases. Mutations—changes in the genetic material—could result from changes in the sequence of bases, such as the substitution of one nucleotide for another or the insertion or deletion of one or more nucleotides. The structure of DNA also suggested a mode of replication: The two strands of the parental DNA molecule separate, and each individual strand serves as a template for the synthesis of a new complementary strand.

- **Genes code for proteins.**

Enzyme defects result in inborn errors of metabolism.
A defective enzyme results from a mutant gene.
One of the DNA strands directs the synthesis of a molecule of RNA.
A molecule of RNA directs the synthesis of a polypeptide chain.
The genetic code is a triplet code.

Most genes code for proteins. More precisely stated, most genes specify the sequence of amino acids in a polypeptide chain. The transfer of genetic information from DNA into protein is a multistep process that includes several types of RNA (ribonucleic acid). Structurally, an RNA strand is similar to a DNA strand except that the "backbone" contains a different sugar (ribose instead of deoxyribose) and RNA contains the base uracil (U) instead of thymine (T). Also, RNA is usually present in cells in the form of single, unpaired strands. The initial step in gene expression is transcription, in which a molecule of RNA is synthesized that is complementary in base sequence to whichever DNA strand is being transcribed. In polypeptide synthesis, which takes place on a ribosome, the base sequence in the RNA transcript is translated in groups of three adjacent bases (codons). The codons are recognized by different types of transfer RNA (tRNA) through base pairing. Each type of tRNA is attached to a particular amino acid, and when a tRNA base-pairs with the proper codon on the mRNA, the growing end of the polypeptide chain is transferred to the amino acid on the tRNA. The table of all codons and the amino acids they specify is called the genetic code. Special codons specify the "start" (AUG, Met) and "stop" (UAA, UAG, and UGA) of polypeptide synthesis. The probable reason why various types of RNA are an intimate part of transcription and translation is that the earliest forms of life used RNA for both genetic information and enzyme catalysis.

- **Genes change by mutation.**

A mutation that alters one or more codons in a gene can change the amino acid sequence of the resulting polypeptide chain synthesized in the cell. Often the altered protein is functionally defective, so an inborn error of metabolism results. One of the first inborn errors of metabolism studied was alkaptonuria; it results from the absence of an enzyme for cleaving homogentisic acid, which accumulates and is excreted in the urine, turning black upon oxidation. Phenylketonuria (PKU) is an inborn error of metabolism that affects the same metabolic pathway. The enzyme defect in PKU results in an inability to convert phenylalanine to tyrosine. Phenylalanine accumulation has catastrophic effects on the development of the brain. Children with the disease have severe mental deficits unless they are treated with a special diet low in phenylalanine.

- **Traits are affected by environment as well as by genes.**

Maternal PKU illustrates the importance of genes and environment.

Most visible traits of organisms result from many genes acting together in combination with environmental factors. The relationship between genes and traits is often complex because (1) every gene potentially affects many traits (pleiotropy), (2) every trait is potentially affected by many genes, and (3) many traits are significantly affected by environmental factors as well as by genes. An example of environmental effects is maternal PKU, in which mothers with a defect in phenylalanine metabolism have children with severe brain and heart abnormalities, even though their children can metabolize phenylalanine normally.

- **Evolution means continuity of life with change.**

The molecular unity of life results from common ancestry.
The diversity of life results mainly from natural selection.

All living creatures are united by sharing many features of the genetic apparatus (for example, transcription and translation) and many aspects of metabolism. The unity of life results from all life being of common ancestry and provides evidence for evolution. There is also great diversity among living creatures. The three major kingdoms of organisms are the Bacteria (which lack a membrane-bounded nucleus), the Archaea (which share features with both Bacteria and Eukarya but form a distinct group), and Eukarya (all "higher" organisms whose cells have a membrane-bounded nucleus that contains DNA organized into discrete chromosomes). Members of the kingdoms Bacteria and Archaea collectively are often called prokaryotes.

The ultimate source of diversity among organisms is mutation, but natural selection is the process by which mutations that enhance survival and reproduction are retained and mutations that are harmful are eliminated. Natural selection, first proposed by Darwin, is therefore the primary mechanism by which organisms become progressively better adapted to their environments.

Key Terms

adenine (A)
alkaptonuria
amino acid
Archaea
Bacteria
bacteriophage
base (in DNA or RNA)
biochemical pathway
block (in a biochemical pathway)
central dogma
chromosome
codon
colony
complementary base sequence
cytosine (C)
deoxyribose
deoxyribonucleic acid (DNA)
double-stranded DNA
duplex DNA
enzyme
Eukarya
eukaryote
evolution
gene
genetic code

genetics
guanine (G)
homogentisic acid
inborn error of metabolism
maternal PKU
messenger RNA (mRNA)
metabolic pathway
metabolite
mutant
mutation
natural selection
nucleotide
phage
phenylalanine hydroxylase (PAH)
phenylketonuria (PKU)
pleiotropic effect
pleiotropy
polarity (of DNA or RNA)
polypeptide chain
product molecule
prokaryote
replication
ribonucleic acid (RNA)
ribose
ribosomal RNA (rRNA)
ribosome
single-stranded DNA
substrate molecule
template
thymine (T)
trait
transcript
transcription
transfer RNA (tRNA)
transformation
translation
uracil (U)
Watson–Crick base pairing

Concepts and Issues
Testing Your Knowledge

- What does it mean to say that DNA is the genetic material?
- What special feature of the structure of DNA allows each strand to be replicated separately?
- How does a strand of DNA specify the structure of a molecule of RNA?
- What types of RNA participate in protein synthesis, and what is the role of each type of RNA?
- What is meant by the phrase *the genetic code,* and how is the genetic code relevant to the translation of a polypeptide chain from a molecule of messenger RNA?
- How does the "central dogma" explain Garrod's discovery that nonfunctional enzymes result from mutant genes?
- In what way does phenylketonuria demonstrate the importance of the environment even for traits that are "determined" by genes?

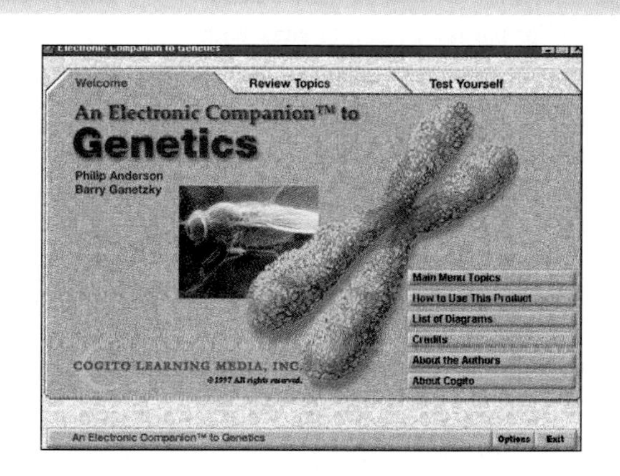

TO ENHANCE YOUR STUDY turn to *An Electronic Companion to Genetics*™, © 1998, Cogito Learning Media, Inc. This CD-ROM is a multimedia tool that provides interactive explanations of important genetic concepts through animations, diagrams, and videos. It also provides interactive test questions.

Solutions
Step by Step

Problem 1

In the human gene for the β chain of hemoglobin (the oxygen-carrying protein in the red blood cells), the first 30 nucleotides in the amino-acid–coding region have the sequence

3'-TACCACGTGGACTGAGGACTCCTCTTCAGA-5'

What is the sequence of the partner strand?

Solution The base pairing between the strands is A with T and G with C, but it is equally important that the strands in a DNA duplex have opposite polarity. The partner strand is therefore oriented with its 5' end at the left, and the base sequence is

5'-ATGGTGCACCTGACTCCTGAGGAGAAGTCT-3'

Problem 2

If the DNA duplex for the β chain of hemoglobin in Step-by-Step Problem 1 were transcribed from left to right, deduce the base sequence of the RNA in this coding region.

Solution To deduce the RNA sequence, we must apply three concepts. First, in the transcription of RNA, the base pairing is such that an A, T, G, or C in the DNA template strand is transcribed as U, A, C, or G, respectively, in the RNA strand. Second, the RNA transcript and the DNA template strand have opposite polarity. Third (and critically for this problem), the RNA transcript is always transcribed in the 5'-to-3' direction, so the 5' end of the RNA is the end synthesized first. This being the case, and considering the opposite polarity, the 3' end of the template strand must be transcribed first. Because we are told that transcription takes place from left to right, we can deduce that the transcribed strand is that in Problem 1. The RNA transcript therefore has the base sequence

5'-AUGGUGCACCUGACUCCUGAGGAGAAGUCU-3'

Problem 3

Given the RNA sequence coding for part of human β hemoglobin deduced in Step-by-Step Problem 2, what is the amino acid sequence in this part of the β polypeptide chain?

Solution The polypeptide chain is translated in successive groups of three nucleotides (each group constituting a codon), starting at the 5' end of the coding sequence and moving in the 5'-to-3' direction. The amino acid corresponding to each codon can be found in the genetic code table. The first ten amino acids in the polypeptide chain are therefore

5'-AUGGUGCACCUGACUCCUGAGGAGAAGUCU-3'
MetValHisLeuThrProGluGluLysSer

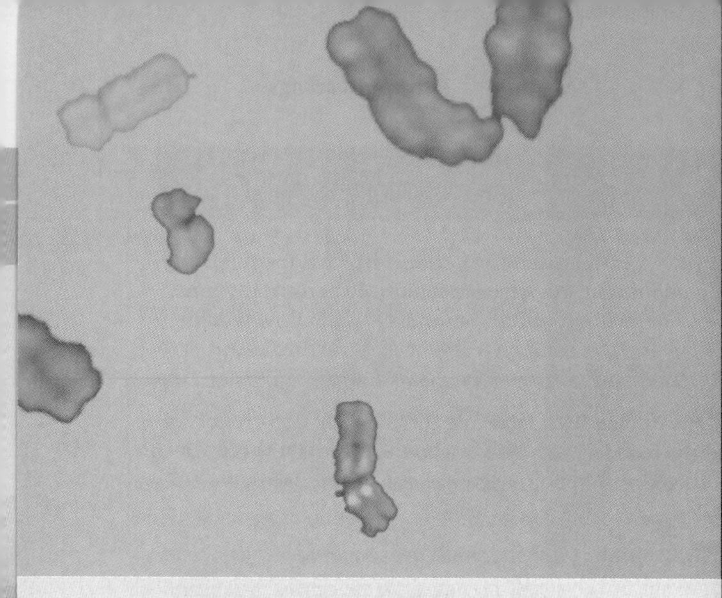

Key Concepts

- Inherited traits are determined by the genes present in the reproductive cells united in fertilization.
- Genes are usually inherited in pairs, one from the mother and one from the father.
- The genes in a pair may differ in DNA sequence and in their effect on the expression of a particular inherited trait.
- The maternally and paternally inherited genes are not changed by being together in the same organism.
- In the formation of reproductive cells, the paired genes separate again into different cells.
- Random combinations of reproductive cells containing different genes result in Mendel's ratios of traits appearing among the progeny.
- The ratios actually observed for any traits are determined by the types of dominance and gene interaction.

GREGOR MENDEL'S WORK with pea plants made a significant contribution to the field of genetics. His work was carried out in the years 1856–1863. In 1868 his fellow monks elected him Abbot, and thereafter he had little time to pursue experimental work. [© Richard Gross/Biological Photo.]

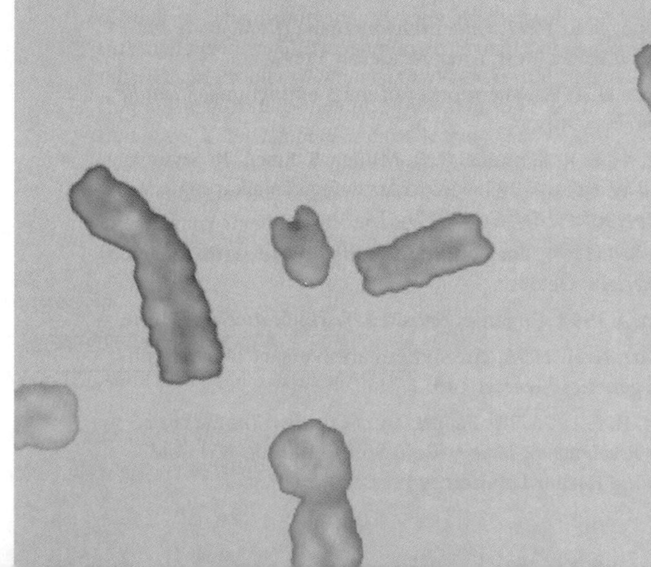

CHAPTER 2

Mendelian Genetics

Chapter Outline

THE STORY OF Gregor Mendel is one of the inspiring legends in the history of modern science. While serving as a monk at the distinguished monastery of St. Thomas in the town of Brno (Brünn), in what is now the Czech Republic, Mendel taught physics and natural history at a local secondary school and also carried out biological experiments. His teaching was said to be "clear, logical, and well suited to the needs of his students." His most important experiments were crosses of garden peas (*Pisum sativum*) carried out from 1856 to 1863 in a small garden plot nestled in a corner of the monastery grounds. He reported his experiments to a local natural history society, published the results and his interpretation in its scientific journal in 1866, and began exchanging letters with Carl Nägeli in Munich, one of the leading botanists of the time. No one understood the significance of the experiments. In 1868 Mendel was elected abbot of the monastery, and his scientific work effectively came to an end. Shortly before his death in 1884, Mendel is said to have remarked to one of the younger monks, "My scientific work has brought me a great deal of satisfaction, and I am convinced that it will be appreciated before long by the whole world." The prophecy was fulfilled 16 years later when Hugo de Vries, Carl Correns, and Erich von Tschermak, each working independently and in a different European country, published the results of experiments similar to Mendel's, drew attention to Mendel's paper, and attributed priority of discovery to him.

Although some modern historians of science disagree over Mendel's intentions in carrying out his work, everyone concedes that Mendel was a first-rate experimenter who performed careful and exceptionally well-documented experiments. His paper contains the first clear exposition of *transmission genetics*, or the statistical rules governing the transmission of hereditary elements from generation to generation. Whatever Mendel's intentions, the elegance of his experiments explains why they were embraced as the foundation of genetics. Indeed, the rules of hereditary transmission inferred from his results are often referred to as **Mendelian genetics.** Mendel's breakthrough experiments and concepts are the subject of this chapter.

GREGOR MENDEL [© Leslie Holzer/Photo Researcers, Inc./Science Source.]

2.1 Mendel was careful in his choice of traits.

The principal difference between Mendel's approach and that of other plant hybridizers of his era is that Mendel thought in quantitative terms about traits that could be classified into two contrasting categories, such as round seeds versus wrinkled

THE SMALL MONASTERY GARDEN in which Mendel grew and classified more than 33,500 pea plants.

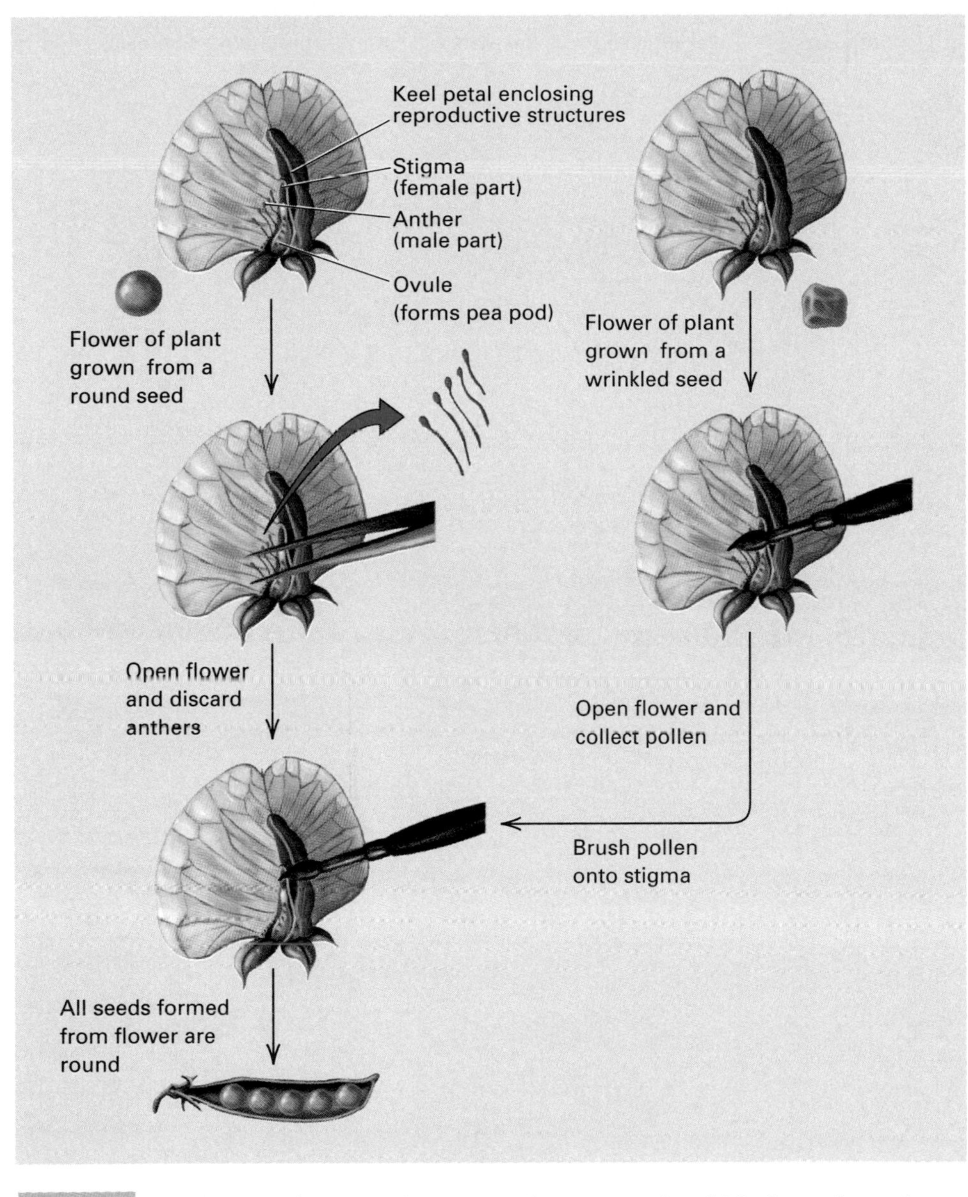

Figure 2.1 Crossing pea plants requires some minor surgery in which the anthers of a flower are removed before they produce pollen. The stigma, the female part of the flower, is not removed. It is fertilized by brushing with mature pollen grains taken from another plant.

seeds. He proceeded by carrying out quite simple crossing experiments and then looked for statistical regularities that might identify general rules. In his own words, he wanted to "determine the number of different forms in which hybrid progeny appear" and, among these, to "ascertain their numerical interrelationships."

Mendel selected peas for his experiments for two reasons. First, he had access to varieties that differed in observable alternative characteristics, such as round versus wrinkled seeds, or yellow versus green seeds. Second, his earlier studies had indicated that peas normally reproduce by self-fertilization, in which pollen produced in a flower is used to fertilize the eggs in the same flower. Left alone, pea flowers always self-fertilize. To carry out a cross between two different varieties, one must open the keel petal (which encloses the reproductive structures), remove the immature anthers (the pollen-producing structures) before they shed pollen, and dust the stigma (the female structure) with mature pollen taken from a flower on a different plant (Figure 2.1).

Mendel recognized the need to study inherited characteristics that were uniform within any given variety of peas but different between varieties. For this reason, at the beginning of his experiments, he established **true-breeding** varieties in which the

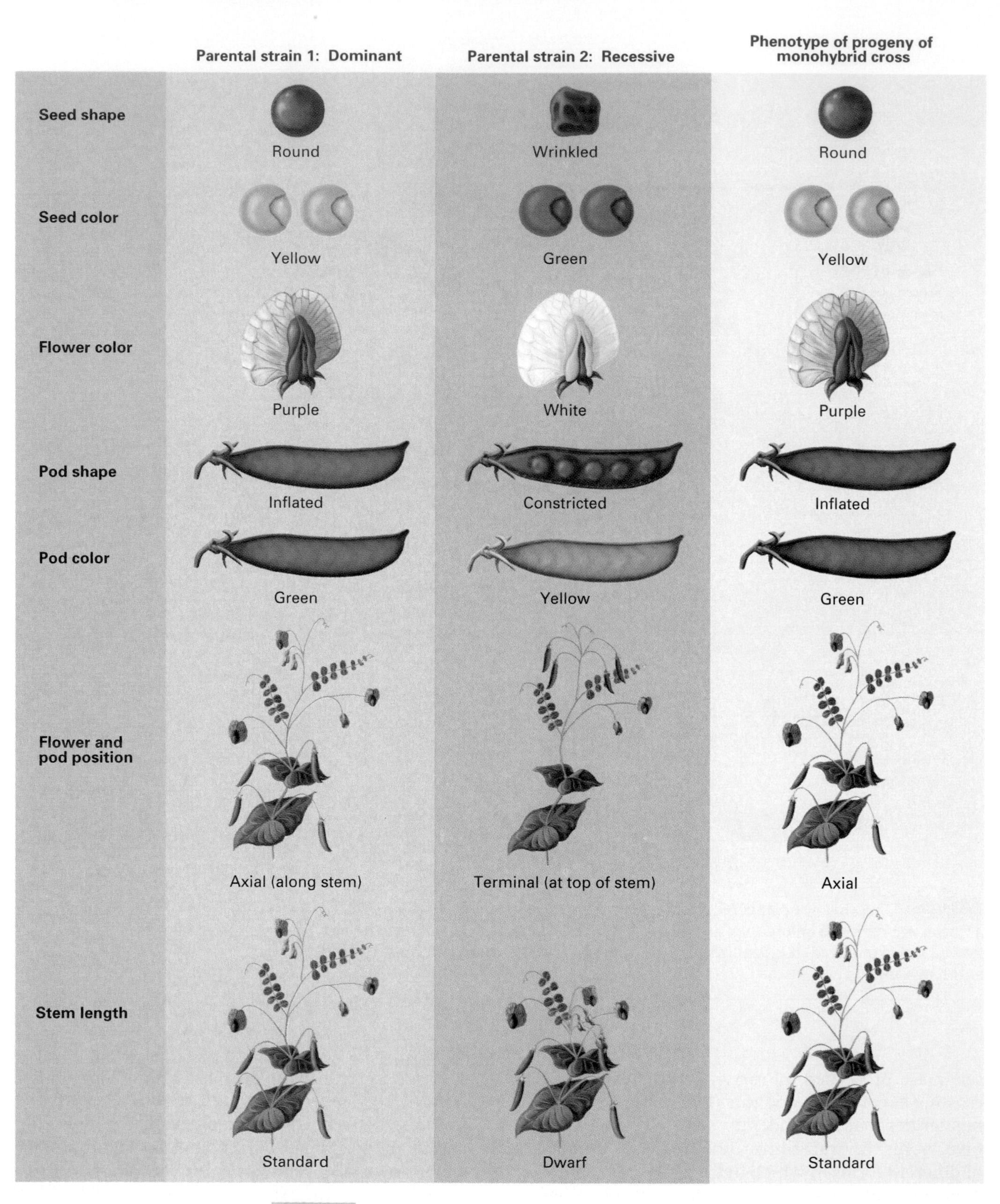

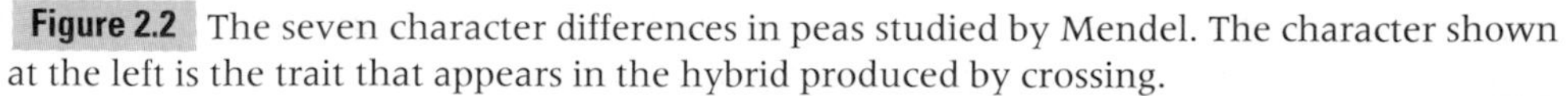

Figure 2.2 The seven character differences in peas studied by Mendel. The character shown at the left is the trait that appears in the hybrid produced by crossing.

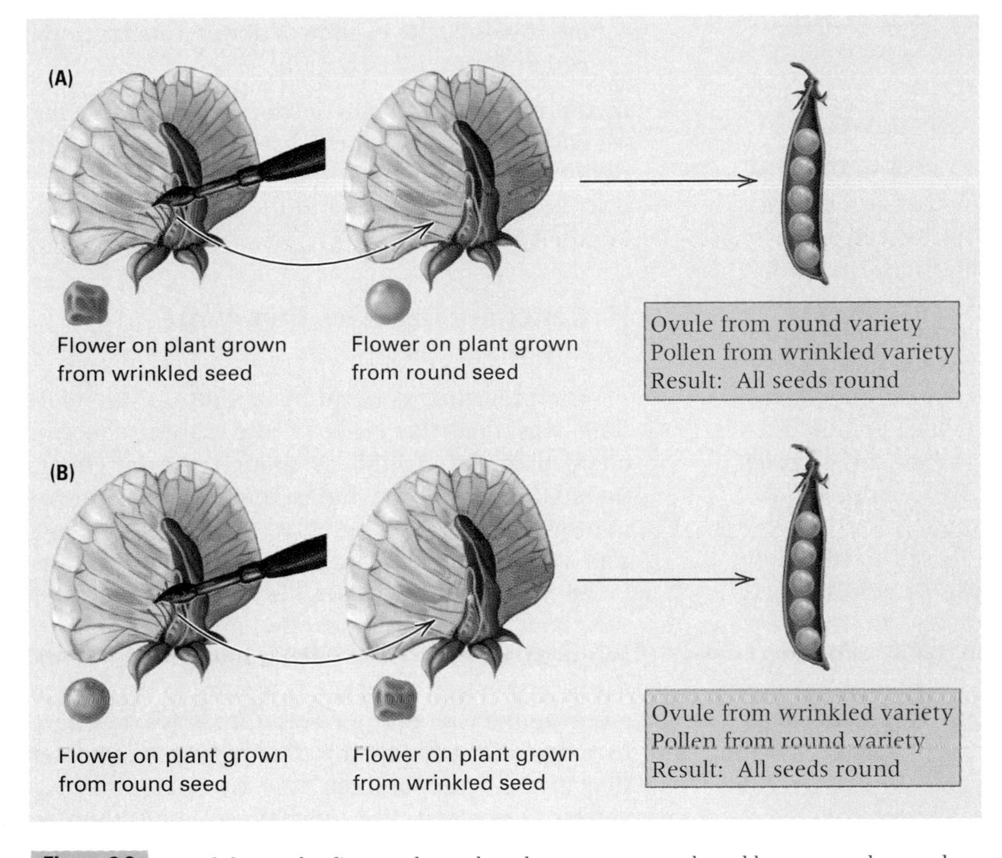

Figure 2.3 Mendel was the first to show that the progeny produced by a cross do not depend on which parent is the male and which the female. In this example, the seeds of the hybrid offspring are round whether the egg came from the round variety and the pollen from the wrinkled variety (A) or the other way around (B).

plants produced only progeny like themselves when allowed to self-fertilize. For example, one true-breeding variety always yielded round seeds, whereas another true-breeding variety always yielded wrinkled seeds. For his experiments, Mendel chose seven pairs of varieties, each of which was true-breeding for a different trait (Figure 2.2). The contrasting traits included seed shape (round versus wrinkled), seed color (yellow versus green), flower color (purple versus white), and pod shape (smooth versus constricted). When two varieties that differ in one or more traits are crossed, the progeny constitute a **hybrid** between the parental varieties. A **monohybrid** results from crossing parental varieties that differ in only one trait of interest.

It is worthwhile to examine of a few of Mendel's original experiments to learn what his methods were and how he interpreted his results. One pair of traits, or characters, that he studied was round versus wrinkled seeds. When pollen from a variety of plants with wrinkled seeds was used to cross-pollinate plants from a variety with round seeds, all of the resulting hybrid seeds were round (Figure 2.3A). Geneticists call the true-breeding parents the **P_1 generation** and the hybrid seeds or plants the **F_1 generation.** Mendel also performed the **reciprocal cross** (Figure 2.3B), in which plants from the variety with round seeds were used as the pollen parents and those from the variety with wrinkled seeds as the female parents. As before, all of the F_1 seeds were round. The reciprocal crosses in Figure 2.3 illustrate the principle that, with a few important exceptions to be discussed in later chapters,

> The outcome of a genetic cross does not depend on which trait is present in the male and which is present in the female; reciprocal crosses yield the same result.

Similar results were obtained when Mendel made crosses between plants that differed in any of the pairs of alternative characteristics. In each case, all of the F_1 progeny exhibited only one of the parental traits, and the other trait was absent. The trait expressed in the F_1 generation in each of the monohybrid crosses is shown on page 32 in Figure 2.2. The trait expressed in the hybrids Mendel called the **dominant** trait; the trait not expressed in the hybrids he called **recessive.**

2.2 Genes come in pairs, separate in gametes, and join randomly in fertilization.

Although the recessive trait is not expressed in the hybrid progeny of a monohybrid cross, it reappears in the next generation when the hybrid progeny are allowed to undergo self-fertilization. For example, when the round hybrid seeds from the round × wrinkled cross were grown into plants and allowed to undergo self-fertilization, some of the resulting seeds were round and others wrinkled. The two types were observed in definite numerical proportions. Mendel counted 5474 seeds that were round and 1850 that were wrinkled. He noted that this ratio was approximately 3 : 1.

The progeny seeds produced by self-fertilization of the F_1 generation constitute the **F_2 generation.** Mendel found that the dominant and recessive traits appear in the F_2 progeny in the proportions 3 round : 1 wrinkled. The results of crossing the round and wrinkled varieties are summarized in the following diagram:

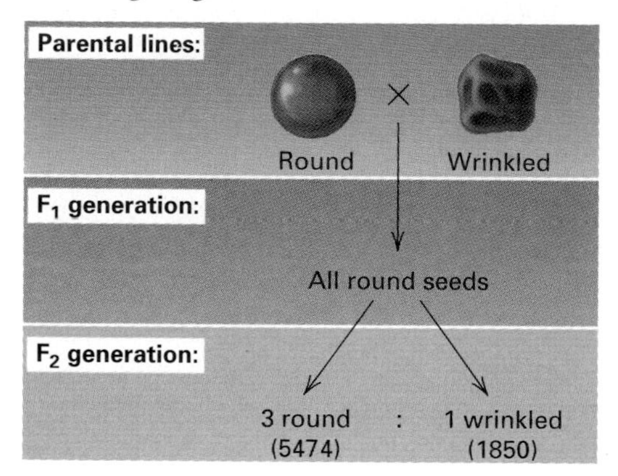

Similar results were obtained in the F_2 generation of crosses between plants that differed in any of the pairs of alternative characteristics (Table 2.1). Note that the first two traits (round versus wrinkled seeds and yellow versus green seeds) have many more observations than any of the other traits; this is because seed shape and color can be classified directly in the seeds, whereas the other traits can be classified only in the mature plants. The principal conclusions from the data in Table 2.1 were as follows:

- The F_1 hybrids express only the dominant trait.
- In the F_2 generation, plants with either the dominant or the recessive trait are present.
- In the F_2 generation, there are approximately three times as many plants with the dominant trait as plants with the recessive trait. In other words, the F_2 ratio of dominant : recessive equals approximately 3 : 1.

In the remainder of this section, we will see how Mendel followed up these basic observations and performed experiments that led to his concept of discrete genetic units and to the principles governing their inheritance.

Genes are particles that come in pairs.

The prevailing concept of heredity in Mendel's time was that the traits of the parents became blended in the hybrid, as though the hereditary material consisted of fluids that became permanently mixed when combined. The blending concept was directly contradicted by Mendel's results. In his monohybrid crosses, the recessive trait that seemingly disappeared in the F_1 generation reappeared again in the F_2 generation. Not only did the recessive trait reappear, it was in no way different from the trait present in the recessive P_1 plants. In a letter describing this finding, Mendel noted that in the F_2 generation, "the two parental traits appear, separated and unchanged, and there is nothing to indicate that one of them has either inherited or taken over anything from the other." From this finding, Mendel concluded that the hereditary determinants for the traits in the parental lines were transmitted as two different elements that retain their purity in the hybrids. In other words, the hereditary determinants do not "mix" or "contaminate" each other. The implication of this conclusion is that a plant with the dominant trait might carry, in unchanged form, a hereditary determinant for the recessive trait.

Mendel developed a genetic hypothesis to explain his results; it is outlined in Figure 2.4. The first element of the hypothesis is that each reproductive cell, or **gamete,** contains one representative of each kind of hereditary determinant in the plant. The hereditary determinant for round seeds he called *A*, and that for wrinkled seeds he called *a*. (It is customary to put gene symbols in italic type.) Mendel

Table 2.1
Results of Mendel's monohybrid experiments

Parental traits	F_1 trait	Number of F_2 progeny	F_2 ratio
round × wrinkled (seeds)	round	5474 round, 1850 wrinkled	2.96 : 1
yellow × green (seeds)	yellow	6022 yellow, 2001 green	3.01 : 1
purple × white (flowers)	purple	705 purple, 224 white	3.15 : 1
inflated × constricted (pods)	inflated	882 inflated, 299 constricted	2.95 : 1
green × yellow (unripe pods)	green	428 green, 152 yellow	2.82 : 1
axial × terminal (flower position)	axial	651 axial, 207 terminal	3.14 : 1
long × short (stems)	long	787 long, 277 short	2.84 : 1

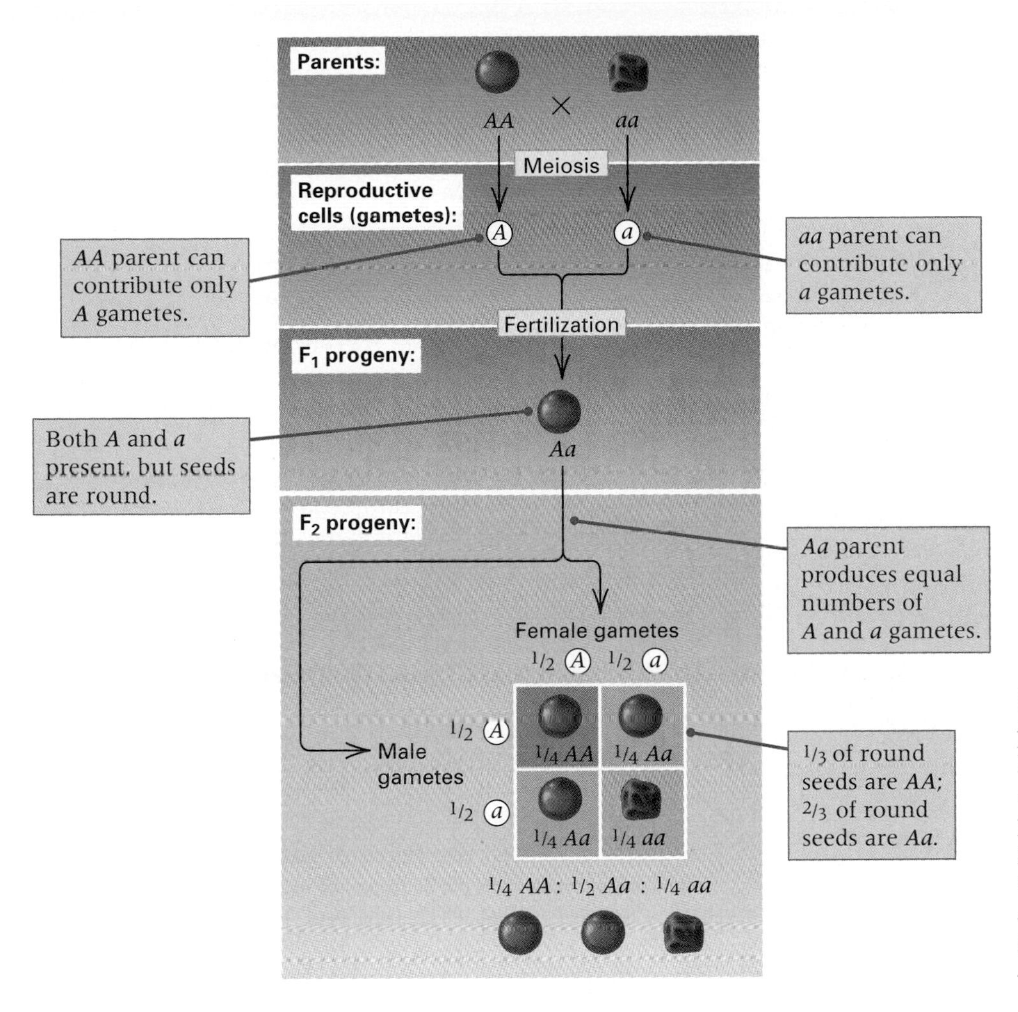

Figure 2.4 A diagrammatic explanation of Mendel's genetic hypothesis to explain the 3 : 1 ratio of dominant : recessive phenotypes observed in the F_2 generation of a monohybrid cross. Note that the ratio of *AA* : *Aa* : *aa* genetic types in the F_2 generation is 1 : 2 : 1.

proposed that in the true-breeding variety with round seeds, all of the reproductive cells would contain *A* and that in the true-breeding variety with wrinkled seeds, all of the reproductive cells would contain *a*. When the varieties are crossed, the F_1 hybrid should receive one of each of *A* and *a* and so have the genetic constitution *Aa* (Figure 2.4). Because *A* is dominant to *a*, the presence of *a* in the F_1 seeds is concealed, and so the seeds are round.

The paired genes separate (segregate) in the formation of reproductive cells.

The second element of Mendel's hypothesis in Figure 2.4 is that when an F_1 plant is self-fertilized, the *A* and *a* determinants separate from one another and are included in the gametes in equal numbers. The separation of the hereditary elements is the heart of Mendelian genetics. The principle is called **segregation.**

> **The Principle of Segregation:** In the formation of gametes, the paired hereditary determinants separate (segregate) in such a way that each gamete is equally likely to contain either member of the pair.

The principle of segregation implies not only that the hereditary determinants separate in the formation of gametes but also that, when separated, the hereditary determinants are completely unaltered by their having been paired in the previous generation. In Mendel's words, neither of them has "inherited or taken over anything from the other."

Gametes unite at random in fertilization.

The third element in Mendel's hypothesis (Figure 2.4) is that the gametes produced by segregation should come together in pairs at random to yield the progeny of the next generation. The assumption of random fertilization means that the result of self-fertilization of the F_1 plants can be deduced by cross-multiplication in a square grid as shown in Figure 2.4. Across the top are the female gametes, which occur in the proportions 1/2 *A* and 1/2 *a* because of segregation. Along the left edge are the male gametes, again in the proportions 1/2 *A* and 1/2 *a* because of segregation. Each square within the grid represents the result of fertilization of one type of pollen with one type of egg. When an *A*-bearing pollen fertilizes an *A*-bearing egg, the result is an *AA* fertilized egg, which is also called a

zygote. Similarly, when an *A*-bearing pollen fertilizes an *a*-bearing egg, the result is an *Aa* zygote. The consequences of fertilization by *a*-bearing pollen are either an *Aa* zygote (if the egg carries *A*) or an *aa* zygote (if the egg carries *a*). These possibilities are shown in the grid in Figure 2.4. The critical point is that with random fertilization, the probability of each gene combination in the zygotes is given by the product of the frequencies of the gametes along the margins. In this case, each zygote combination has a probability of 1/4. Because, in an *Aa* seed, it does not matter whether the *A* came through the pollen or the egg, random combinations of the gametes result in an F_2 generation with the genetic composition 1/4 *AA*, 1/2 *Aa*, and 1/4 *aa*, as indicated at the bottom of Figure 2.4. The *AA* and *Aa* types should have round seeds, and the *aa* types should have wrinkled seeds, so the predicted ratio of round : wrinkled seeds is 3 : 1.

Genotype means genetic endowment; phenotype means observed trait.

The genetic hypothesis in Figure 2.4 also illustrates another of Mendel's important deductions: Two plants with the same outward appearance—for example, with round seeds—might nevertheless differ in their hereditary makeup as revealed by the types of progeny observed when they are crossed. One of the handicaps under which Mendel wrote was the absence of an established vocabulary of terms suitable for describing his concepts. Hence he made a number of seemingly elementary mistakes, such as occasionally confusing the outward appearance of an organism with its hereditary constitution. The necessary vocabulary was developed only after Mendel's work was rediscovered, and it includes the following essential terms.

1. A hereditary determinant of a trait is called a **gene.**
2. The different forms of a particular gene are called **alleles.** In Figure 2.4, the alleles of the gene for seed shape are *A* for round seeds and *a* for wrinkled seeds. *A* and *a* are alleles because they are alternative forms of the gene for seed shape. Alternative alleles are typically represented by the same letter, or combination of letters, distinguished either by upper case versus lower case or by means of superscripts or subscripts or some other typographic identifier.
3. The **genotype** is the genetic constitution of an organism or cell. With respect to seed shape in peas, *AA, Aa,* and *aa* are examples of the possible genotypes for the *A* and *a* alleles. Because gametes contain only one allele of each gene, *A* and *a* are examples of genotypes of gametes.
4. A genotype in which the members of a pair of alleles are different, as in the *Aa* hybrids in Figure 2.4, is said to be **heterozygous;** a genotype in which the two alleles are alike is said to be **homozygous.** A homozygous organism may be homozygous dominant (*AA*) or homozygous recessive (*aa*). The terms *homozygous* and *heterozygous* cannot apply to gametes, because gametes contain only one allele of each gene.
5. The observable properties of an organism constitute its **phenotype.** Round seeds and wrinkled seeds are phenotypes. So are yellow seeds and green seeds. The phenotype of an organism does not necessarily imply anything about its genotype. For example, a seed with the phenotype "round" could have either the genotype *AA* or the genotype *Aa.*

The progeny of the F_2 generation support Mendel's hypothesis.

But could Mendel's genetic hypothesis in Figure 2.4 be tested? Mendel realized that a key prediction concerned the genetic composition of the round seeds in the F_2 generation. If his hypothesis were correct, then one-third of the round seeds should have the genetic composition *AA* and two-thirds of the round seeds should have the genetic composition *Aa.* The reason for the 1 : 2 ratio is shown in Figure 2.5. The ratio of *AA* : *Aa* : *aa* in the F_2 generation is 1 : 2 : 1, but if we disregard the *aa* seeds, then the ratio of *AA* : *Aa* is 1 : 2. In other words, 1/3 of the seeds are *AA* and 2/3 are *Aa.* Upon self-fertilization, the *AA* types should be true-breeding for round seeds, whereas the *Aa* types should yield round and wrinkled seeds in the ratio 3 : 1. Furthermore, among the wrinkled seeds in the F_2 gen-

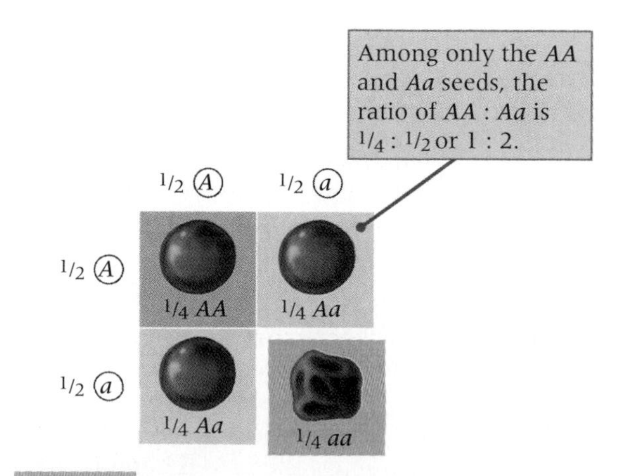

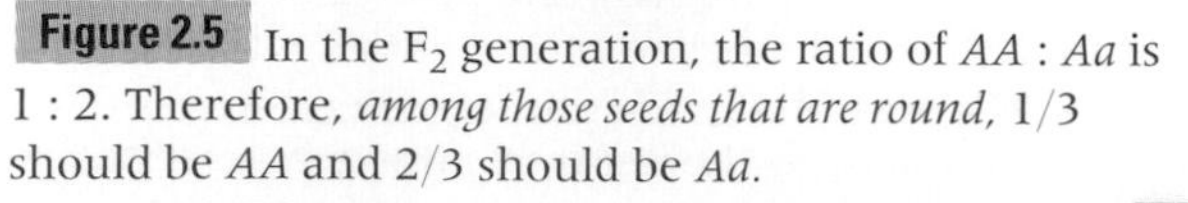
Figure 2.5 In the F_2 generation, the ratio of *AA* : *Aa* is 1 : 2. Therefore, *among those seeds that are round,* 1/3 should be *AA* and 2/3 should be *Aa.*

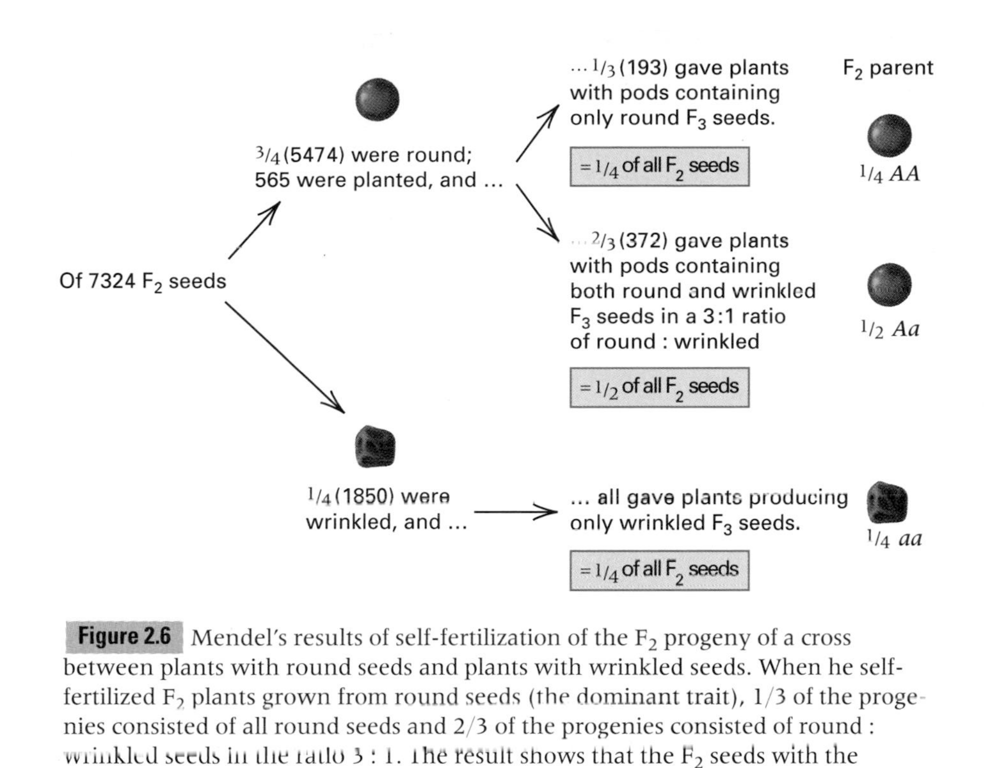

Figure 2.6 Mendel's results of self-fertilization of the F_2 progeny of a cross between plants with round seeds and plants with wrinkled seeds. When he self-fertilized F_2 plants grown from round seeds (the dominant trait), 1/3 of the progenies consisted of all round seeds and 2/3 of the progenies consisted of round : wrinkled seeds in the ratio 3 : 1. The result shows that the F_2 seeds with the dominant trait (round) include two genetic types, *AA* and *Aa*, in a ratio of 1 : 2.

eration, all should have the genetic composition *aa*, and so, upon self-fertilization, they should be true-breeding for wrinkled seeds.

For several of his traits, Mendel carried out self-fertilization of the F_2 plants in order to test these predictions. His results for round versus wrinkled seeds are summarized in Figure 2.6. As predicted from Mendel's genetic hypothesis, the plants grown from F_2 wrinkled seeds were true-breeding for wrinkled seeds. They produced only wrinkled seeds in the F_3 generation. Moreover, among 565 plants grown from F_2 round seeds, 193 were true-breeding, producing only round seeds in the F_3 generation, whereas the other 372 plants produced both round and wrinkled seeds in a proportion very close to 3 : 1. The ratio 193 : 372 equals 1 : 1.93, which is very close to the ratio 1 : 2 of *AA* : *Aa* types predicted theoretically from the genetic hypothesis in Figure 2.4. Overall, taking all of the F_2 plants into account, the ratio of genotypes observed was very close to the predicted 1 : 2 : 1 of *AA* : *Aa* : *aa* expected from Figure 2.4.

The progeny of testcrosses also support Mendel's hypothesis.

A second way in which Mendel tested the genetic hypothesis in Figure 2.4 was by crossing the F_1 heterozygous genotypes with plants that were homozygous recessive. Such a cross, between an organism that is heterozygous for one or more genes (for example, *Aa*) and an organism that is homozygous for the recessive alleles (for example, *aa*), is called a **testcross.** The result of such a testcross is shown in Figure 2.7. Because of segregation, the heterozygous parent is expected to produce *A* and *a* gametes in equal numbers. When these gametes combine at random with the *a*-bearing gametes produced by the homozygous

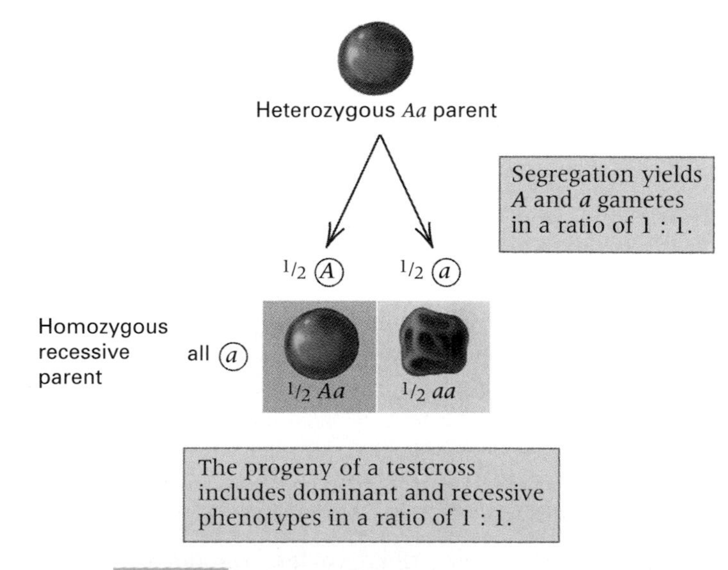

Figure 2.7 A testcross shows the result of segregation in the phenotypes of the progeny. This example illustrates a testcross of an *Aa* heterozygous parent with an *aa* homozygous recessive. The progeny are expected to have genotypes *Aa* and *aa* in the ratio of 1 : 1.

Table 2.2
Results of Mendel's experiments

Testcross (F_1 heterozygote × homozygous recessive)	Progeny from testcross	Ratio
Round × wrinkled seeds	193 round 192 wrinkled	1.01 : 1
Yellow × green seeds	196 yellow 189 green	1.04 : 1
Purple × white flowers	85 purple 81 white	1.05 : 1
Long × short stems	85 long 79 short	1.01 : 1

recessive parent, the expected progeny are 1/2 with the genotype *Aa* and 1/2 with the genotype *aa*. The former have the dominant phenotype (because *A* is dominant to *a*), whereas the latter have the recessive phenotype, and so the expected ratio of dominant : recessive phenotypes is 1 : 1. This is why a testcross is often extremely useful in genetic analysis.

> In a testcross, the relative frequencies of the different gametes produced by the heterozygous parent can be observed directly in the phenotypes of the progeny, because the recessive parent contributes only recessive alleles.

Mendel carried out a series of testcrosses with the genes for round versus wrinkled seeds, yellow versus green seeds, purple versus white flowers, and long versus short stems. The results are shown in Table 2.2. In all cases, the ratio of phenotypes among the progeny is very close to the 1 : 1 ratio expected from segregation of the alleles in the heterozygous parent.

Another valuable type of cross is a **backcross,** in which hybrid organisms are crossed with one of the parental genotypes. Backcrosses are commonly used by geneticists and by plant and animal breeders, as we will see in later chapters. Note that the testcrosses in Table 2.2 are also backcrosses, because in each case, the F_1 heterozygous parent came from a cross between the homozygous dominant and the homozygous recessive.

2.3 The alleles of different genes segregate independently.

In the experiments described so far, Mendel was concerned with successive generations of progeny from parents that differed in a single contrasting trait, such as round seeds versus wrinkled seeds. In each case, he observed the result of segregation of the pair of alleles determining the trait. Mendel also carried out experiments in which he examined the inheritance of two or more traits simultaneously to determine whether the same pattern of inheritance applied to each pair of alleles separately even though two allelic pairs were segregating in the hybrids. For example, plants from a true-breeding variety with round and yellow seeds were crossed with plants from a variety with wrinkled and green seeds. The F_1 progeny were hybrid for both characteristics, or **dihybrid**, and the phenotype of the seeds was round and yellow. The F_1 phenotype was round and yellow because round is dominant to wrinkled, and yellow is dominant to green (Figure 2.2). Then Mendel self-fertilized the F_1 progeny to obtain seeds in the F_2 generation. He observed four types of seed phenotypes in the progeny and, in counting the seeds, he obtained the following numbers:

round, yellow	315
round, green	108
wrinkled, yellow	101
wrinkled, green	32
Total	556

In these data, Mendel noted the presence of the expected monohybrid 3 : 1 ratio for each trait separately. With respect to each trait, the progeny were

round : wrinkled
= (315 + 108) : (101 + 32)
= 423 : 133
= 3.18 : 1

yellow : green
= (315 + 101) : (108 + 32)
= 416 : 140
= 2.97 : 1

Furthermore, in the F_2 progeny of the dihybrid cross, the separate 3 : 1 ratios for the two traits were combined at random, as shown in Figure 2.8. When the phenotypes of two traits are combined at random, then among the 3/4 of the progeny that are round, 3/4 will be yellow and 1/4 green; similarly, among the 1/4 of the progeny that are wrinkled, 3/4 will be yellow and 1/4 green. The overall proportions of round–yellow to round–green to wrinkled–yellow to wrinkled–green are therefore expected to be 3/4 × 3/4 to 3/4 × 1/4 to 1/4 × 3/4 to 1/4 × 1/4 or

$$9/16 : 3/16 : 3/16 : 1/16$$

The observed ratio of 315 : 108 : 101 : 32 equals 9.84 : 3.38 : 3.16 : 1, which is reasonably close to the 9 : 3 : 3 : 1 ratio expected from the cross-multiplication of the separate 3 : 1 ratios in Figure 2.8.

Figure 2.8 The 3 : 1 ratio of round : wrinkled, when combined at random with the 3 : 1 ratio of yellow : green, yields the 9 : 3 : 3 : 1 ratio that Mendel observed in the F_2 progeny of the dihybrid cross.

The F_2 genotypes in a dihybrid cross conform to Mendel's prediction.

Mendel carried out similar experiments with other combinations of traits and, for each pair of traits he examined, consistently observed the 9 : 3 : 3 : 1 ratio. He also deduced the biological reason for the observation. To illustrate his explanation using the dihybrid round × wrinkled cross, we can represent the dominant and recessive alleles of the pair affecting seed shape as *W* and *w*, respectively, and the allelic pair affecting seed color as *G* and *g*. Mendel proposed that the underlying reason for the 9 : 3 : 3 : 1 ratio in the F_2 generation is that the segregation of the alleles *W* and *w* for round or wrinkled seeds has no effect on the segregation of the alleles *G* and *g* for yellow or green seeds. Each pair of alleles undergoes segregation into the gametes independently of the segregation of the other pair of alleles. In the P_1 generation, the parental genotypes are *WW GG* (round, yellow seeds) and *ww gg* (wrinkled, green seeds). Then, the genotype of the F_1 is the double heterozygote *Ww Gg*.

Gametes unite at random in fertilization.

The result of independent segregation in the F_1 plants is that the *W* allele is just as likely to be included in a gamete with *G* as with *g*, and the *w* allele is just as likely to be included in a gamete with *G* as with *g*. The independent segregation is illustrated in Figure 2.9. The independent segregation of the *W*, *w* and the *G*, *g* allele pairs implies that the gametes produced by the double heterozygote *Ww Gg* are

$$1/4\ WG \quad 1/4\ Wg \quad 1/4\ wG \quad 1/4\ wg$$

When the four types of gametes combine at random to form the zygotes of the next generation, the result of independent assortment is as shown in Figure 2.10. The cross-multiplication-like format, which is used to show how the F_1 female and male

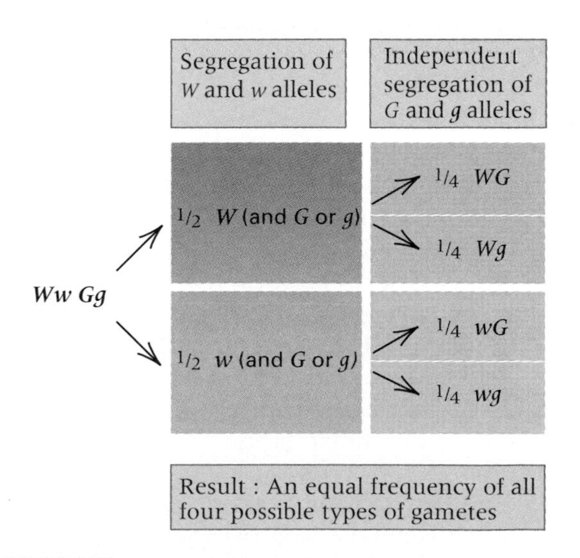

Figure 2.9 Independent segregation of the *Ww* and *Gg* allele pairs means that among each of the *W* and *w* classes, the ratio of *G* : *g* is 1 : 1. Likewise, among each of the *G* and *g* classes, the ratio of *W* : *w* is 1 : 1.

gametes combine at random to produce the F_2 genotypes, is called a **Punnett square.** In the Punnett square, the combinations of seed shape and color phenotypes of the F_2 progeny are indicated. Note that the ratio of phenotypes is 9 : 3 : 3 : 1 for round yellow : wrinkled yellow : round green : wrinkled green.

The Punnett square in Figure 2.10 also shows that the ratio of *genotypes* in the F_2 generation is not 9 : 3 : 3 : 1. With independent assortment, the ratio of genotypes in the F_2 generation is

1 : 2 : 1 : 2 : 4 : 2 : 1 : 2 : 1

The reason for this ratio is shown in Figure 2.11. Among seeds with the *WW* genotype, the ratio of *GG* : *Gg* : *gg* is 1 : 2 : 1. Among seeds with the *Ww* genotype, the ratio is 2 : 4 : 2 (the 1 : 2 : 1 is multiplied by 2 because there are twice as many *Ww* genotypes as either *WW* or *ww*). And among seeds with the *ww* genotype, the ratio of *GG* : *Gg* : *gg* is 1 : 2 : 1. The phenotypes of the seeds are shown beneath the genotypes. The combined ratio of phenotypes is 9 : 3 : 3 : 1. Figure 2.10 also shows that among seeds that are *GG*, the ratio of *WW* : *Ww* : *ww* is 1 : 2 : 1, among seeds that are *Gg* it is 2 : 4 : 2, and among seeds that are *gg* it is 1 : 2 : 1. Therefore, the independent segregation means that, among each of the possible genotypes formed by the allele pair, the ratio of homozygous dominant : heterozygous : homozygous recessive for the other allele pair is 1 : 2 : 1.

Mendel tested the hypothesis of independent segregation by ascertaining whether the predicted genotypes were actually present in the expected proportions. He did the tests by growing plants from the F_2 seeds and obtaining F_3 progeny by self-pollination. To illustrate the tests, consider one series of crosses in which he grew plants from F_2 seeds that were round, green. Note in Figures 2.10 and 2.11 that round, green F_2 seeds are expected to have either the genotype *Ww gg* or the genotype *WW gg* in the ratio 2 : 1. Mendel grew 102 plants from such seeds and found that 67 of them produced both round, green and wrinkled, green seeds (indicating that the parental plants must have been *Ww gg*) and 35 of them produced only round, green seeds (indicating that the parental genotype was *WW gg*). The ratio 67 : 35 is in good agreement with the expected 2 : 1 ratio of genotypes. Similar good agreement with the predicted relative frequencies of the different genotypes was found when plants were grown from round, yellow or from wrinkled, yellow F_2 seeds. (As expected, plants

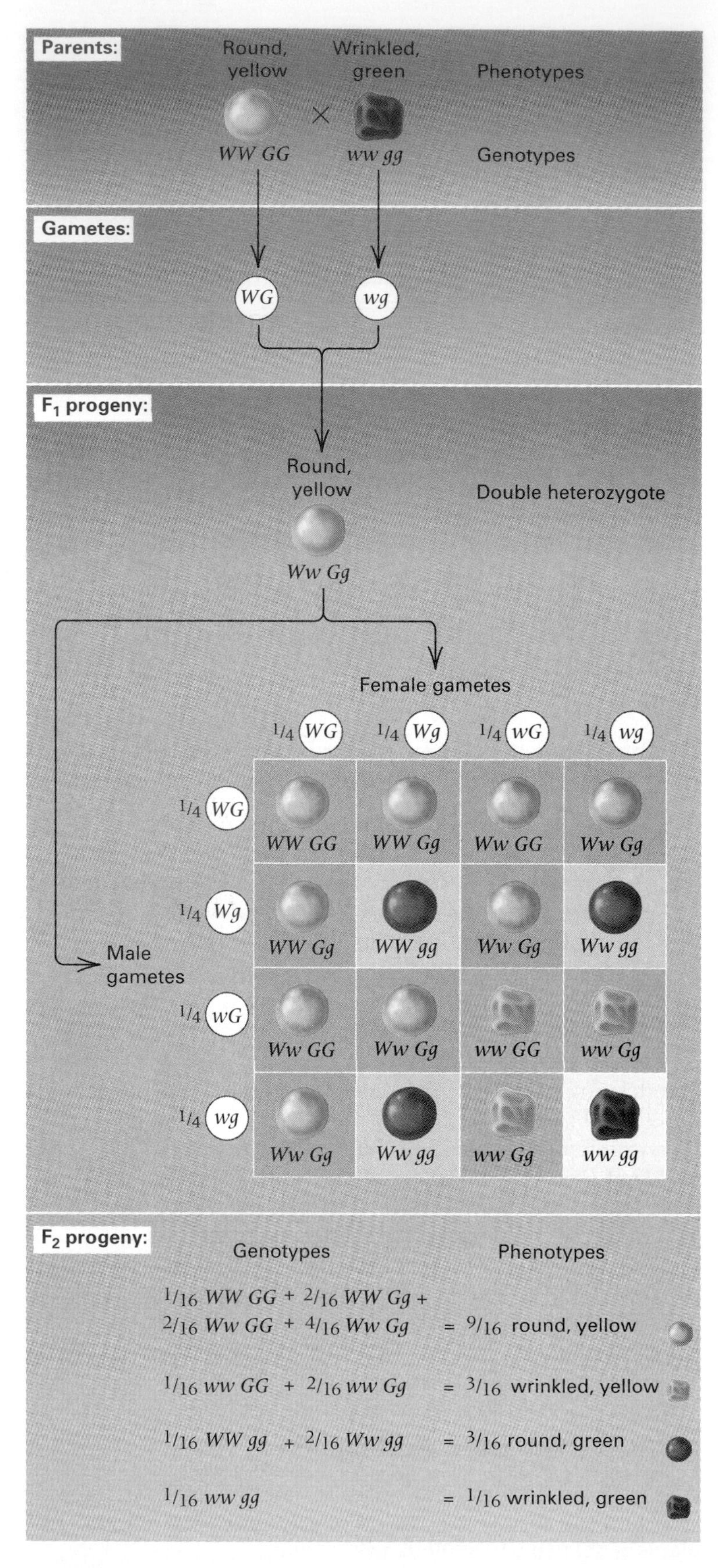

Figure 2.10 Diagram showing the basis for the 9 : 3 : 3 : 1 ratio of F_2 phenotypes resulting from a cross in which the parents differ in two traits determined by genes that undergo independent assortment.

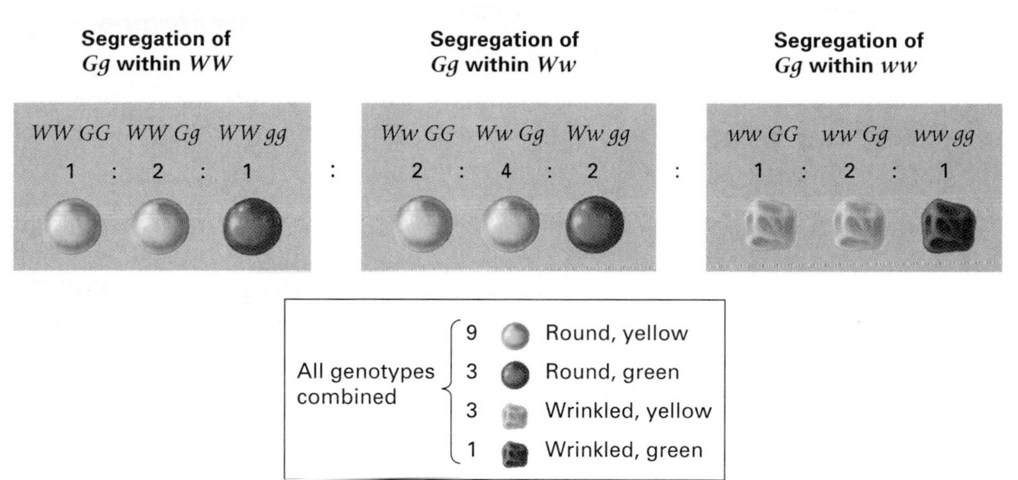

Figure 2.11 In the F_2 progeny of the dihybrid cross for seed shape and seed color, in each of the genotypes for one of the allele pairs, the ratio of homozygous dominant, heterozygous, and homozygous recessive genotypes for the other allele pair is 1 : 2 : 1.

grown from the wrinkled, green seeds, which have the predicted homozygous recessive genotype *ww gg*, produced only wrinkled, green seeds.)

Mendel's observation of independent segregation of two pairs of alleles has come to be known as the principle of **independent assortment:**

> **The Principle of Independent Assortment**: Segregation of the members of any pair of alleles is independent of the segregation of other pairs in the formation of reproductive cells.

Although the principle of independent assortment is of fundamental importance in Mendelian genetics, in later chapters we will see that there are important exceptions.

The progeny of testcrosses show the result of independent segregation.

A second way in which Mendel tested the hypothesis of independent assortment was by carrying out a testcross with the F_1 genotypes that were heterozygous for both genes (*Ww Gg*). In a testcross, one parental genotype is always multiple homozygous recessive—in this case, *ww gg*. As shown in Figure 2.12, the double heterozygotes produce four types of gametes—*W G*, *W g*, *w G*, and *w g*—in equal frequencies, whereas the *ww gg* plants produce only *w g* gametes. Thus the progeny phenotypes are expected to consist of round yellow, round green, wrinkled yellow, and wrinkled green in a ratio of 1 : 1 : 1 : 1. As in a testcross of a monohybrid, the ratio of phenotypes in the progeny is a direct demonstration of the ratio of

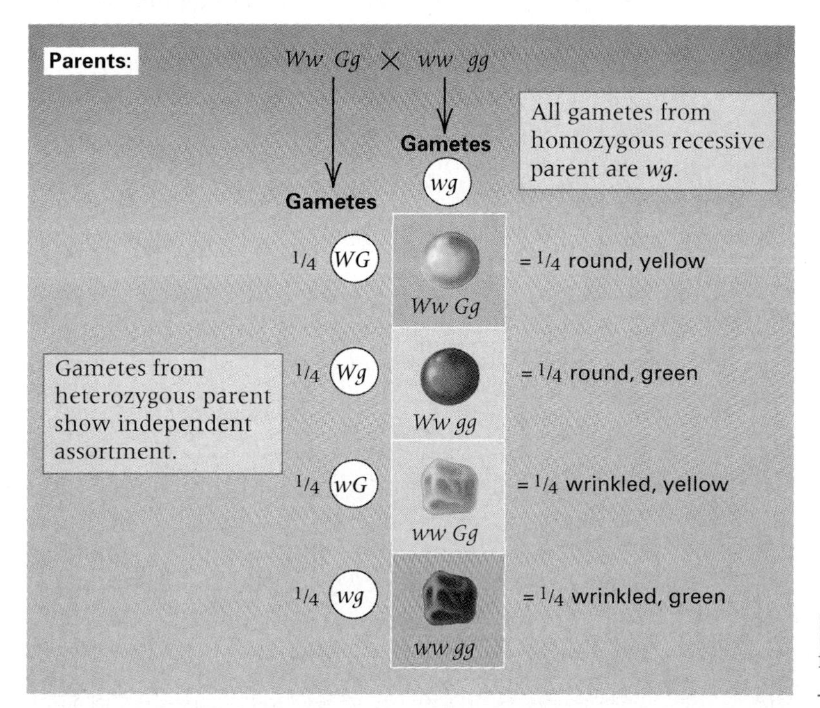

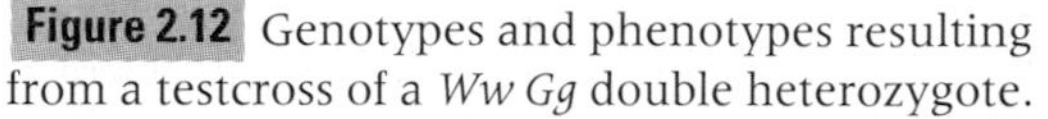

Figure 2.12 Genotypes and phenotypes resulting from a testcross of a *Ww Gg* double heterozygote.

gametes produced by the heterozygous parent, because no dominant alleles are contributed by the homozygous recessive parent to obscure the results. In the actual cross, Mendel obtained 55 round yellow, 51 round green, 49 wrinkled yellow, and 53 wrinkled green, which is in good agreement with the predicted 1 : 1 : 1 : 1 ratio. The results were the same in the reciprocal cross with *Ww Gg* as the female parent and *ww gg* as the male parent. This observation confirmed Mendel's assumption that the gametes of both sexes included each possible genotype in approximately equal proportions.

An interesting historical note: Mendel's paper does not explicitly state either the principle of segregation (sometimes called Mendel's first law) or the principle of independent assortment (sometimes called Mendel's second law). On this basis, one could argue that Mendel did not discover Mendel's laws! On the other hand, Mendel did seem to have a pretty clear idea of what was going on. Six times in his relatively short paper, he repeated what he evidently thought was the main message: "Pea hybrids form germinal and pollen cells that in their composition correspond in equal numbers to all the constant forms resulting from the combination of traits united through fertilization." One could not draw this conclusion without invoking both segregation and independent assortment.

2.4 Chance plays a central role in Mendelian genetics.

Chance plays a central role in Mendelian genetics because the union of gametes in fertilization is a random process. In each fertilization, the particular combination of dominant and recessive alleles that occurs is random and subject to chance variation. In a genetic cross, the proportions of the different types of offspring obtained are the cumulative result of numerous individual events of fertilization. It is for this reason that a working knowledge of the rules of probability is basic to understanding the transmission of hereditary characteristics.

In the analysis of genetic crosses, the probability of an event may be considered equivalent to the proportion of times that the event is expected to be realized in numerous repeated trials. Likewise, the proportion of times that an event is expected to be realized in numerous repeated trials is equivalent to the probability that it is realized in a single trial. For example, in the F_2 generation of the hybrid between pea varieties with round seeds and those with wrinkled seeds, Mendel observed 5474 round seeds and 1850 wrinkled seeds (Table 2.1). In this case, the proportion of wrinkled seeds was $1850/(1850 + 5474) = 1/3.96$, or very nearly 1/4. We may therefore regard 1/4 as the approximate proportion of wrinkled seeds to be expected among a large number of progeny from this cross. Completely equivalently, we can regard 1/4 as the probability that any particular seed chosen at random will be wrinkled.

Evaluating the probability of a genetic event usually requires an understanding of the mechanism of inheritance and knowledge of the particular cross. For example, in evaluating the probability of obtaining a round seed from a particular cross, one needs to know that there are two alleles, *W* and *w*, with *W* dominant to *w*; one also needs to know the particular cross, because the probability of round seeds is determined by whether the cross is

WW × *ww*, in which all the progeny seeds are expected to be round,

Ww × *Ww*, in which 3/4 of the progeny seeds are expected to be round, or

Ww × *ww*, in which 1/2 of the progeny seeds are expected to be round.

In many genetic crosses, the possible outcomes of fertilization are equally likely. Suppose that there are n possible outcomes, each equally likely, and that in m of these a particular outcome of interest is realized. Then the probability of the outcome of interest is m/n. In the language of probability, an outcome of interest is typically called an *event*. As an example, consider the progeny produced by self-fertilization of an *Aa* plant. Four equally likely progeny genotypes (outcomes) are possible: *AA*, *Aa*, *aA*, and *aa*. Two of the four possible outcomes are heterozygous, so the probability of a heterozygous genotype, *Aa*, is 2/4, or 1/2.

The addition rule applies to mutually exclusive events.

Sometimes an outcome of interest can be expressed in terms of two or more possibilities. For example, a seed with the phenotype "round" may have either of two genotypes, *WW* or *Ww*. A seed that is round cannot have both genotypes at the same time. Only one event such as the formation of the *WW* or the *Ww* genotype can be realized in any one organism, and the realization of one such event in an organism precludes the realization of others in the same organism. In this example, the realization of the genotype *WW* in a seed precludes the realization of the genotype *Ww* in the same seed, and the other way around. Events that exclude each other in this manner are said to be *mutually exclusive*. When events are mutually exclusive, their probabilities are combined according to the addition rule.

> **Addition Rule**: The probability of the realization of one or the other of two mutually exclusive events, A or B, is the sum of their separate probabilities.

In symbols, where Prob is used to mean *probability,* the addition rule is written

$$\text{Prob}\{\text{A or B}\} = \text{Prob}\{\text{A}\} + \text{Prob}\{\text{B}\}$$

The addition rule can be applied to determine the proportion of round seeds expected from the cross *Ww* × *Ww*, which is illustrated in Figure 2.5. The round-seed phenotype results from the expression of either of two genotypes, *WW* and *Ww*, and these events are mutually exclusive. In any particular progeny seed, the probability of genotype *WW* is 1/4 and that of *Ww* is 1/2. Hence the overall probability of either *WW* or *Ww* is

$$\text{Prob}\{WW \text{ or } Ww\} = \text{Prob}\{WW\} + \text{Prob}\{Ww\}$$
$$= 1/4 + 1/2 = 3/4$$

Because 3/4 is the probability of an individual seed being round, it is also the expected proportion of round seeds among a large number of progeny.

The multiplication rule applies to independent events.

Events that are not mutually exclusive may be *independent,* which means that the realization of one event has no influence on the possible realization of any others. For example, in Mendel's crosses for seed shape and color, the two traits are independent, and the ratio of phenotypes in the F_2 generation is expected to be 9/16 round yellow, 3/16 round green, 3/16 wrinkled yellow, and 1/16 wrinkled green. These proportions can be obtained by considering the traits separately, because they are independent. Considering only seed shape, we can expect the F_2 generation to consist of 3/4 round and 1/4 wrinkled seeds. Considering only seed color, we can expect the F_2 generation to consist of 3/4 yellow and 1/4 green. Because the traits are inherited independently, among the 3/4 of the seeds that are round, there should be 3/4 that are yellow, and so the overall proportion of round yellow seeds is expected to be 3/4 × 3/4 = 9/16 (Figure 2.8). Likewise, among the 3/4 of the seeds that are round, there should be 1/4 green, yielding 3/4 × 1/4 = 3/16 as the expected proportion of round green seeds. The proportions of the other phenotypic classes can be deduced in a similar way using the cross-multiplication method illustrated in Figure 2.8. The principle is that when events are independent, the probability that they are realized together is obtained by multiplication.

Successive offspring from a cross are also independent events, which means that the genotypes of early progeny have no influence on the relative proportions of genotypes in later progeny. The independence of successive offspring contradicts the widespread belief that in each human family, the ratio of girls to boys must "even out" at approximately 1 : 1 such that if a family already has, say, four girls, then they are somehow more likely to have a boy the next time around. But this belief is not supported by theory, and it is also contradicted by actual data on the sex ratios in human sibships. (The term **sibship** refers to a group of offspring from the same parents.) The data indicate that a human family is no more likely to have a girl on the next birth if it already has five boys than if it already has five girls. The statistical reason is that although the sex ratios tend to balance out when they are averaged across a large number of sibships, they do not need to balance within individual sibships. Thus, among families in which there are five children, the sibships consisting of five boys balance those consisting of five girls, for an overall sex ratio of 1 : 1. However, both of these sibships are unusual in their sex distribution.

A SIBSHIP CONSISTING of four girls has a probability given by the multiplication rule as 1/2 × 1/2 × 1/2 × 1/2 = 1/16.

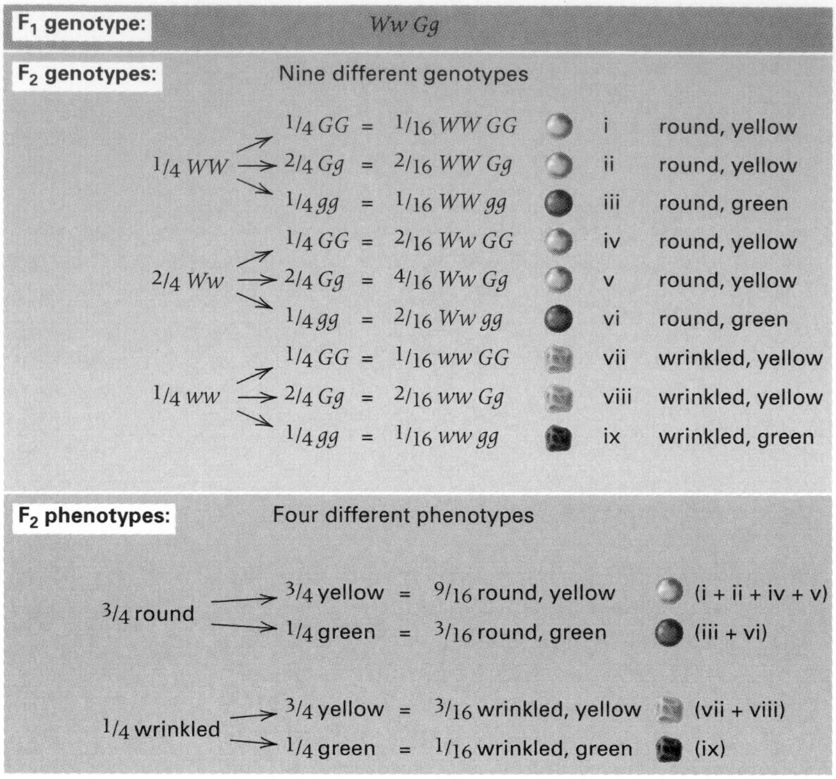

F1 genotype: Ww Gg
F2 genotypes: Nine different genotypes
1/4 WW
1/4 GG = 1/16 WW GG i round, yellow
2/4 Gg = 2/16 WW Gg ii round, yellow
1/4 gg = 1/16 WW gg iii round, green
2/4 Ww
1/4 GG = 2/16 Ww GG iv round, yellow
2/4 Gg = 4/16 Ww Gg v round, yellow
1/4 gg = 2/16 Ww gg vi round, green
1/4 ww
1/4 GG = 1/16 ww GG vii wrinkled, yellow
2/4 Gg = 2/16 ww Gg viii wrinkled, yellow
1/4 gg = 1/16 ww gg ix wrinkled, green
F2 phenotypes: Four different phenotypes
3/4 round
3/4 yellow = 9/16 round, yellow (i + ii + iv + v)
1/4 green = 3/16 round, green (iii + vi)
1/4 wrinkled
3/4 yellow = 3/16 wrinkled, yellow (vii + viii)
1/4 green = 1/16 wrinkled, green (ix)

Figure 2.13 Example of the use of the addition and multiplication rules to determine the probabilities of the nine genotypes and four phenotypes in the F_2 progeny obtained from self-pollination of a dihybrid F_1. The roman numerals are arbitrary labels identifying the F_2 genotypes.

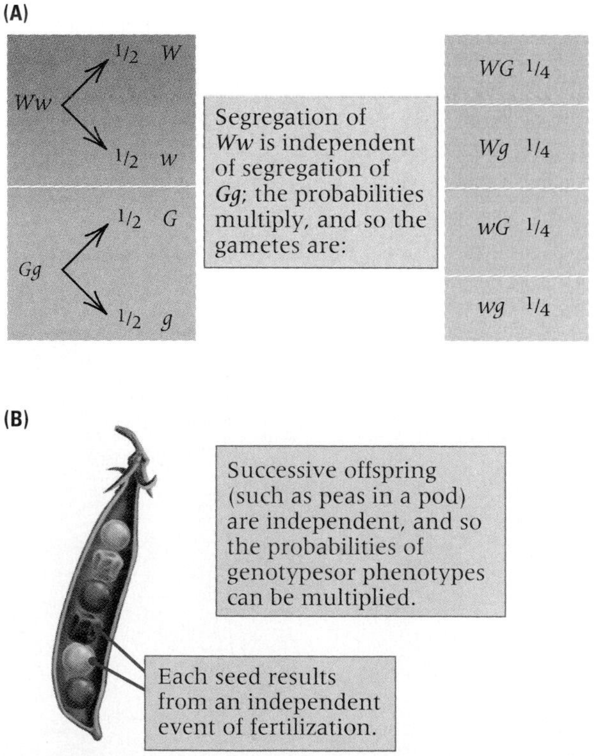

(A)
Ww
1/2 W
1/2 w
Gg
1/2 G
1/2 g
Segregation of Ww is independent of segregation of Gg; the probabilities multiply, and so the gametes are:
WG 1/4
Wg 1/4
wG 1/4
wg 1/4
(B)
Successive offspring (such as peas in a pod) are independent, and so the probabilities of genotypesor phenotypes can be multiplied.
Each seed results from an independent event of fertilization.

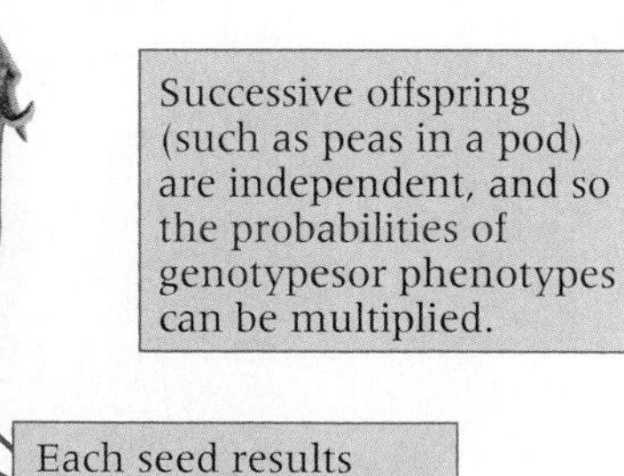

Figure 2.14 In genetics, two important types of independence are independent segregation of alleles that show independent assortment (A), and independent fertilizations resulting in successive offspring (B). In these cases, the probabilities of each of the individual outcomes of segregation or fertilization are multiplied to obtain the overall probability.

When events are independent (such as independent traits or successive offspring from a cross), the probabilities are combined by means of the multiplication rule.

> **Multiplication Rule**: The probability of two independent events, A and B, being realized simultaneously is given by the product of their separate probabilities.

In symbols, the multiplication rule is

$$\text{Prob}\{\text{A and B}\} = \text{Prob}\{\text{A}\} \cdot \text{Prob}\{\text{B}\}$$

The multiplication rule can be used to answer questions like the following: Of two offspring from the mating $Aa \times Aa$, what is the probability that both have the dominant phenotype? Because the mating is $Aa \times Aa$, the probability that any particular offspring has the dominant phenotype equals 3/4. Using the multiplication rule, the probability that both of two offspring have the dominant phenotype is $3/4 \times 3/4 = 9/16$.

Here is a typical genetic question that can be answered by using the addition and multiplication rules together: Of two offspring from the mating $Aa \times Aa$, what is the probability of one dominant phenotype and one recessive? Sibships of one dominant phenotype and one recessive can come about in two different ways, with the dominant born first or with the dominant born second, and these outcomes are mutually exclusive. The probability of the first case is $3/4 \times 1/4$ and that of the second is $1/4 \times 3/4$; because the events are mutually exclusive, the probabilities are added. The answer is therefore $(3/4 \times 1/4) + (1/4 \times 3/4) = 2(3/4)(1/4) = 3/8$.

The addition and multiplication rules are very powerful tools for calculating the probabilities of genetic events. Figure 2.13 shows how the rules are applied to determine the expected proportions of the nine different genotypes possible among the F_2 progeny produced by self-pollination of a *Ww Gg* dihybrid.

In genetics, independence applies not only to the successive offspring formed by a mating but also to genes that segregate according to the principle of independent assortment (Figure 2.14). The independence means that the multiplication rule can be used to determine the probability of the various types of progeny from a cross in which there is independent assortment among numerous pairs of alleles. This principle is the theoretical basis for the expected progeny types from the dihybrid cross shown in Figure 2.13. One can also use the multiplication rule to calculate the probability of a specific genotype among the progeny of a cross. For example, if a quadruple heterozygote of genotype *Aa Bb Cc Dd* is self-fertilized, the probability of a quadruple heterozygote *Aa Bb Cc Dd* offspring is $(1/2)(1/2)(1/2)(1/2) = (1/2)^4$, or 1/16, assuming independent assortment of all four pairs of alleles.

2.5 The results of segregation can be observed in human pedigrees.

Determination of the genetic basis of a trait from the kinds of crosses that we have considered requires that we control matings between organisms and obtain large numbers of offspring to classify with regard to phenotype. The analysis of segregation by this method is not possible in human beings, and it is not usually feasible for traits in large domestic animals. However, the mode of inheritance of a trait can sometimes be determined by examining the appearance of the phenotypes that reflect the segregation of alleles in several generations of related individuals. This is typically done with a family tree that shows the phenotype of each individual; such a diagram is called a **pedigree.** An important application of probability in genetics is its use in pedigree analysis.

Figure 2.15 depicts most of the standard symbols used in drawing a human pedigree. Females are represented by circles and males by squares. (A diamond is used if the sex of an individual is unknown.) Persons with the phenotype of interest are indicated by colored or shaded symbols. For recessive alleles, heterozygous carriers are depicted with half-filled symbols. A mating between a female and a male is indicated by joining their symbols with a horizontal line, which is connected vertically to a second horizontal line, below, that connects the symbols for their offspring. The off-

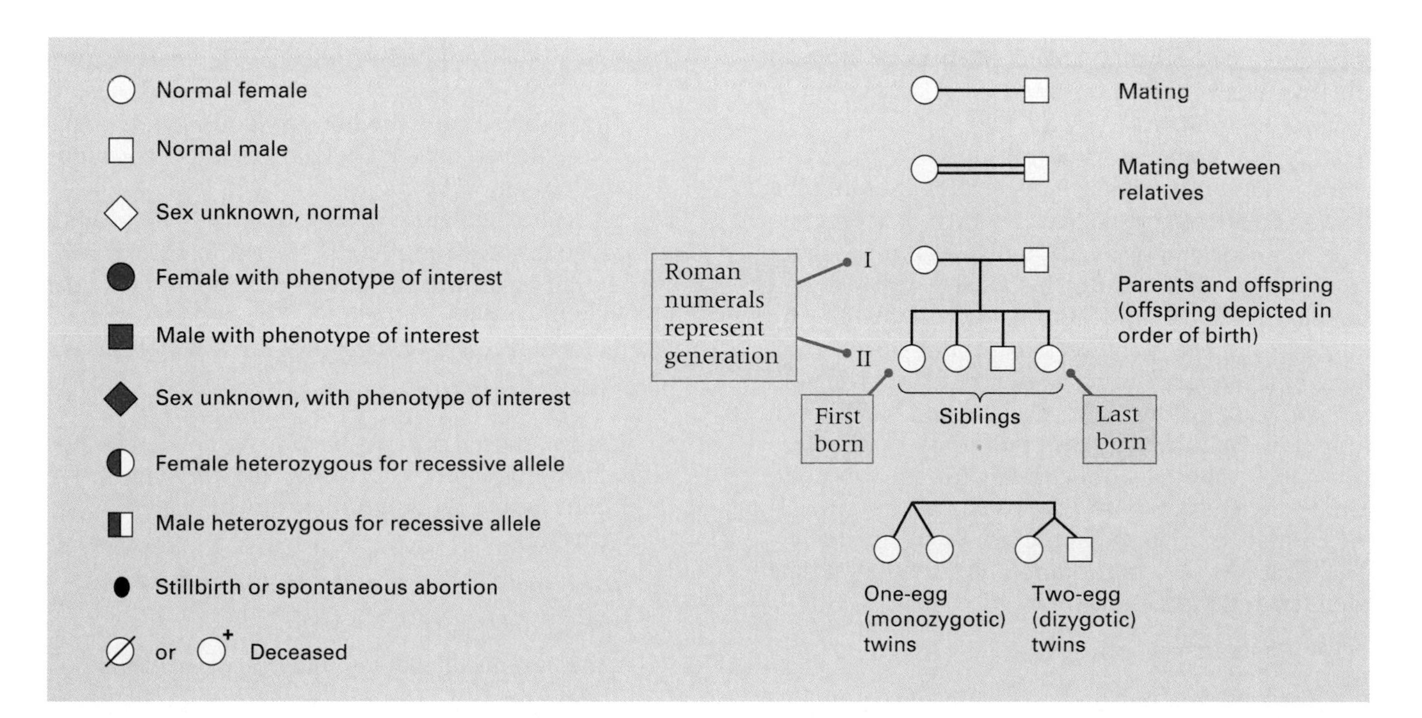

Figure 2.15 Conventional symbols used in depicting human pedigrees.

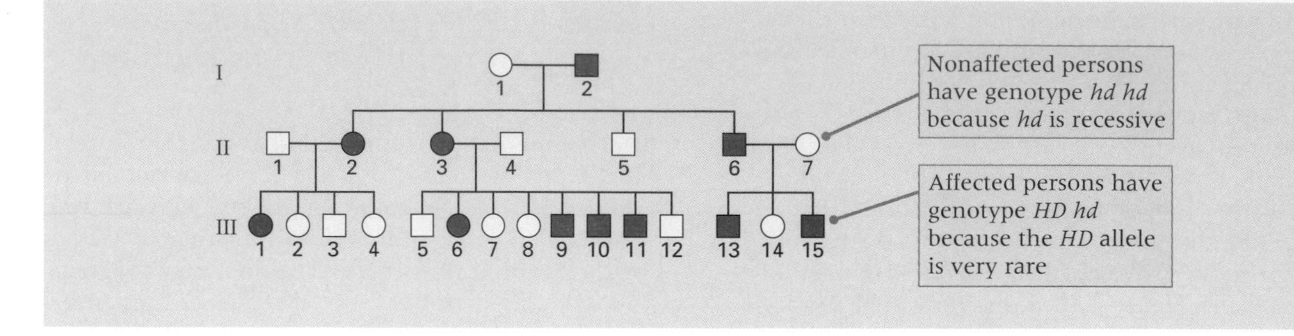

Figure 2.16 Pedigree of a human family showing the inheritance of the dominant gene for Huntington disease. Females and males are represented by circles and squares, respectively. Red symbols indicate persons affected with the disease.

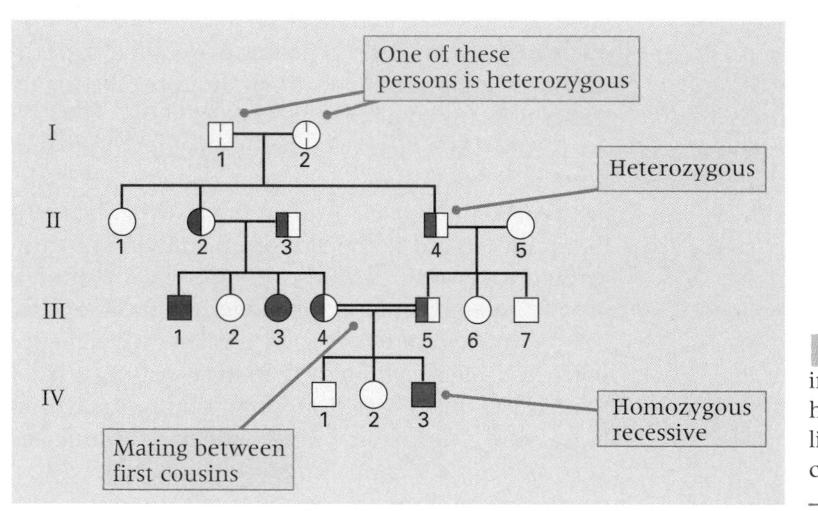

Figure 2.17 Pedigree of albinism. With recessive inheritance, affected persons (red symbols) often have unaffected parents. The double horizontal line indicates a mating between relatives—in this case, first cousins.

spring within a sibship, called **siblings** or **sibs** regardless of sex, are represented from left to right in order of their birth.

A typical pedigree for a trait due to a dominant allele is shown in Figure 2.16. In this example the trait is **Huntington disease,** which is a progressive nerve degeneration, usually beginning about middle age, that results in severe physical and mental disability and ultimately in death. The numbers in the pedigree are added for convenience in referring to particular persons. The successive generations are designated by Roman numerals. Within any generation, all of the persons are numbered consecutively from left to right. The pedigree starts with the woman I-1 and the affected man I-2. The pedigree shows the characteristic features of inheritance due to a simple Mendelian dominant allele:

- The trait affects both sexes.
- Every affected person has an affected parent.
- Approximately 1/2 of the offspring of affected persons are affected.

Because the dominant allele, *HD*, that causes Huntington disease is very rare, all affected persons in the pedigree have the heterozygous genotype *HD hd*. Nonaffected persons have the homozygous normal genotype *hd hd*.

A typical pedigree pattern for a trait due to a homozygous recessive allele is shown in Figure 2.17. The trait is **albinism**, absence of pigment in the skin, hair, and iris of the eyes. The pedigree characteristics of recessive inheritance are as follows:

- The trait affects both sexes.
- Most affected persons have parents who are not themselves affected; the parents are heterozygous for the recessive allele and are called **carriers.**
- Approximately 1/4 of the children of heterozygous parents are affected.
- The parents of affected individuals are often relatives.

The reason for the 1/4 ratio is that in a mating between carriers ($Aa \times Aa$), each offspring has a 1/4 chance of being homozygous *aa* and hence being affected. The reason why mating between relatives is important, particularly with traits due to rare recessive alleles, is that when a recessive allele is rare, it is more likely to become homozygous through inheritance from a common ancestor than from parents who are completely unrelated. When a common ancestor of an individual's parents is a carrier, the recessive allele may, by chance, be transmitted down both sides of the pedigree to the parents of the individual. That allele then has a 1/4 chance of becoming homozygous when the relatives mate. Mating between relatives constitutes *inbreeding;* the consequences of inbreeding are discussed further in Chapter 13.

2.6 The alleles of some genes do not show complete dominance.

In Mendel's experiments, all traits had clear dominant–recessive patterns. This was fortunate, because otherwise he might not have made his discoveries. Departures from strict dominance are also frequently observed. In fact, even for such a classic trait as round versus wrinkled seeds in peas, it is an oversimplification to say that round is dominant. At the level of whether a seed is round or wrinkled, round is dominant in the sense that the genotypes *WW* and *Ww* cannot be distinguished by the outward appearance of the seeds. However, as we noted in Chapter 1, every gene potentially affects many traits. It often happens that the same pair of alleles show complete dominance for one trait but not complete dominance for another trait. For example, in the case of round versus wrinkled seeds, the biochemical defect in wrinkled seeds is the absence of an active form of an enzyme called starch-branching enzyme I (SBEI), which is needed for the synthesis of a branched-chain form of starch known as amylopectin. Seeds that are heterozygous *Ww* have only half as much SBEI as homozygous *WW* seeds, and seeds that are homozygous *ww* have virtually none (Figure 2.18A). Homozygous *WW* peas contain large, well-rounded starch grains. As a result, the seeds retain water and shrink uniformly as they ripen, and so they do not become wrinkled. In homozygous *ww* seeds, the starch grains lack amylopectin; they are irregular in shape. When these seeds ripen, they lose water too rapidly and shrink unevenly, resulting in the wrinkled phenotype (Figure 2.18B and C).

The *w* allele also affects the shape of the starch grains in *Ww* heterozygotes. In heterozygous seeds, the starch grains are intermediate in shape (Figure 2.18B). Nevertheless, their amylopectin content is high enough to result in uniform shrinking of the seeds and no wrinkling (Figure 2.18C). Thus there is an apparent paradox of dominance. If we consider

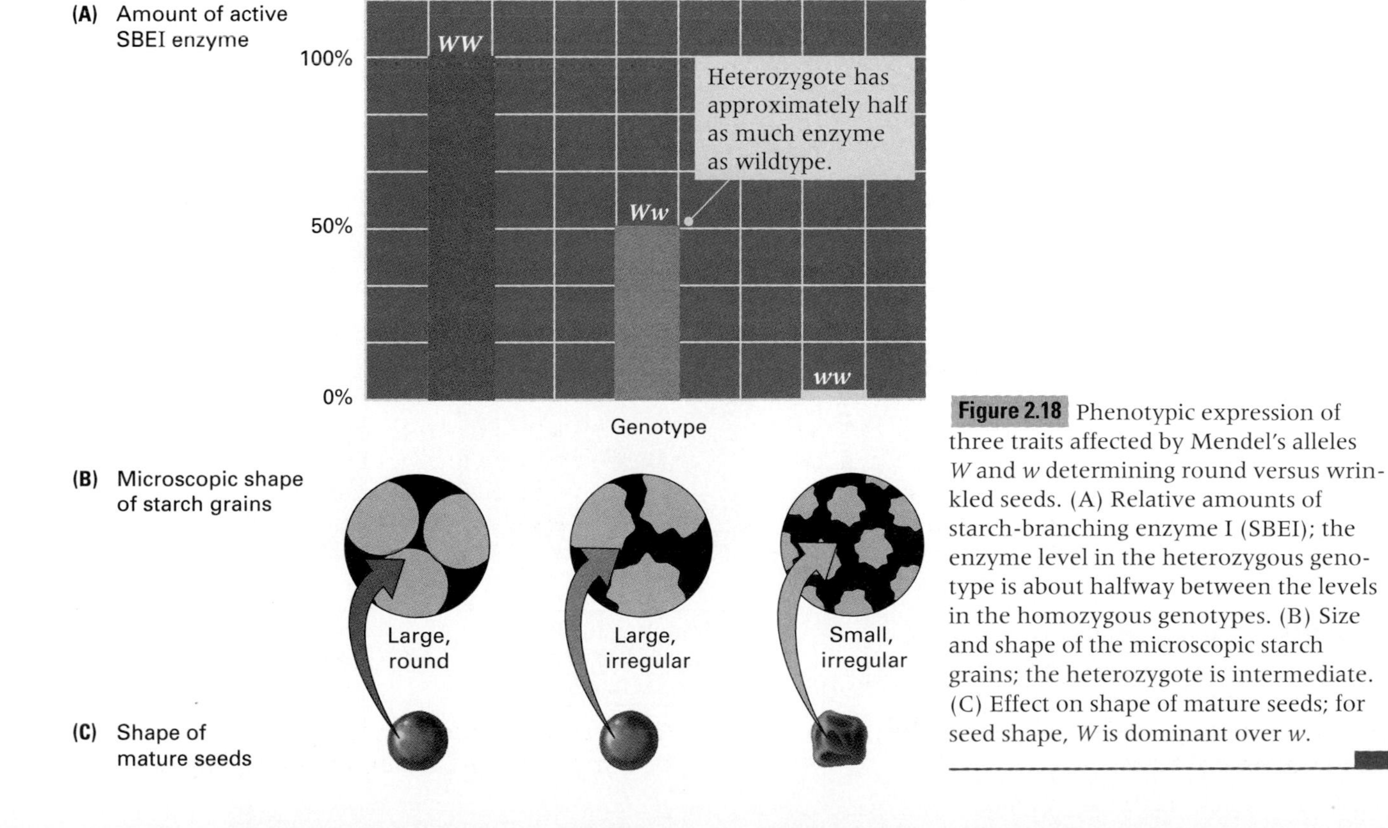

Figure 2.18 Phenotypic expression of three traits affected by Mendel's alleles *W* and *w* determining round versus wrinkled seeds. (A) Relative amounts of starch-branching enzyme I (SBEI); the enzyme level in the heterozygous genotype is about halfway between the levels in the homozygous genotypes. (B) Size and shape of the microscopic starch grains; the heterozygote is intermediate. (C) Effect on shape of mature seeds; for seed shape, *W* is dominant over *w*.

only the overall shape of the seeds, round is dominant over wrinkled. There are only two phenotypes. If we examine the shape of the starch grains with a microscope, all three genotypes can be distinguished from each other: large rounded starch grains in *WW*, large irregular grains in *Ww*, and small irregular grains in *ww*. If we consider the amount of the SBEI enzyme, the *Ww* genotype has an amount about halfway between the amounts in *WW* and *ww*.

The round-wrinkled pea example in Figure 2.18 makes it clear that "dominance" is not simply a property of a particular pair of alleles no matter how the resulting phenotypes are observed. When a gene affects multiple traits (as most genes do), a particular pair of alleles might show simple dominance for some traits but not others. The general principle illustrated in Figure 2.18 is that

> The phenotype consists of many different physical and biochemical attributes, and dominance may be observed for some of these attributes and not for others; thus dominance is a property of a pair of alleles in relation to a particular attribute of phenotype.

The phenotype of a heterozygous genotype is often intermediate.

When the phenotype of the heterozygous genotype lies in the range between the phenotypes of the homozygous genotypes, there is said to be **incomplete dominance.** Most genes code for enzymes, and each allele in a genotype often makes its own contribution to the total level of the enzyme in the cell or organism. In such cases, the phenotype of the heterozygote falls in the range between the phenotypes of the corresponding homozygotes, as illustrated in Figure 2.19. The terms *incomplete dominance, partial dominance,* and *semidominance* are all used to describe the situation.

A classic example of incomplete dominance concerns flower color in the snapdragon *Antirrhinum* (Figure 2.20). In wildtype flowers, a red type of anthocyanin pigment is formed by a sequence of enzymatic reactions. A wildtype enzyme, encoded by the *I* allele, is limiting to the rate of the overall reaction, so the amount of red pigment is determined by the amount of enzyme that expression of the *I* allele produces. The alternative *i* allele codes for an inactive enzyme, and *ii* flowers are ivory in color. Because the amount of the critical enzyme is reduced in *Ii* heterozygotes, the amount of red pigment in the flowers is reduced also, and the effect of the dilution is to make the flowers pink.

The result of Mendelian segregation is observed directly when snapdragons that differ in flower color are crossed. For example, a cross between plants from a true-breeding red-flowered variety and a true-breeding ivory-flowered variety results in F_1 plants with pink flowers. In the F_2 progeny obtained by self-fertilization of the F_1 hybrids (self-fertilization is denoted by the encircled cross sign), one experiment resulted in 22 plants with red flowers, 52 with pink flowers, and 23 with ivory flowers. These numbers agree fairly well with the Mendelian ratio of 1 dominant homozygote : 2 heterozygotes : 1 recessive homozygote. In agreement with the predictions from simple Mendelian inheritance, when self-fertilized, the red-flowered F_2 plants produced only red-flowered progeny; the ivory-flowered plants produced only

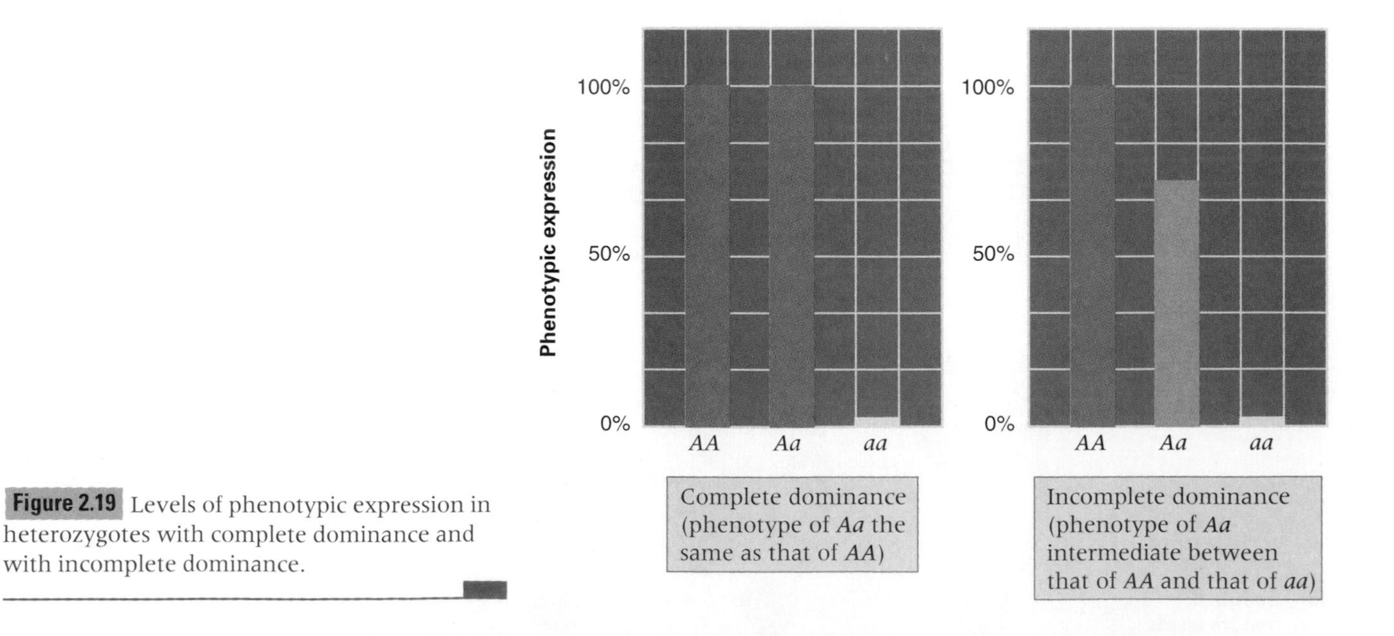

Figure 2.19 Levels of phenotypic expression in heterozygotes with complete dominance and with incomplete dominance.

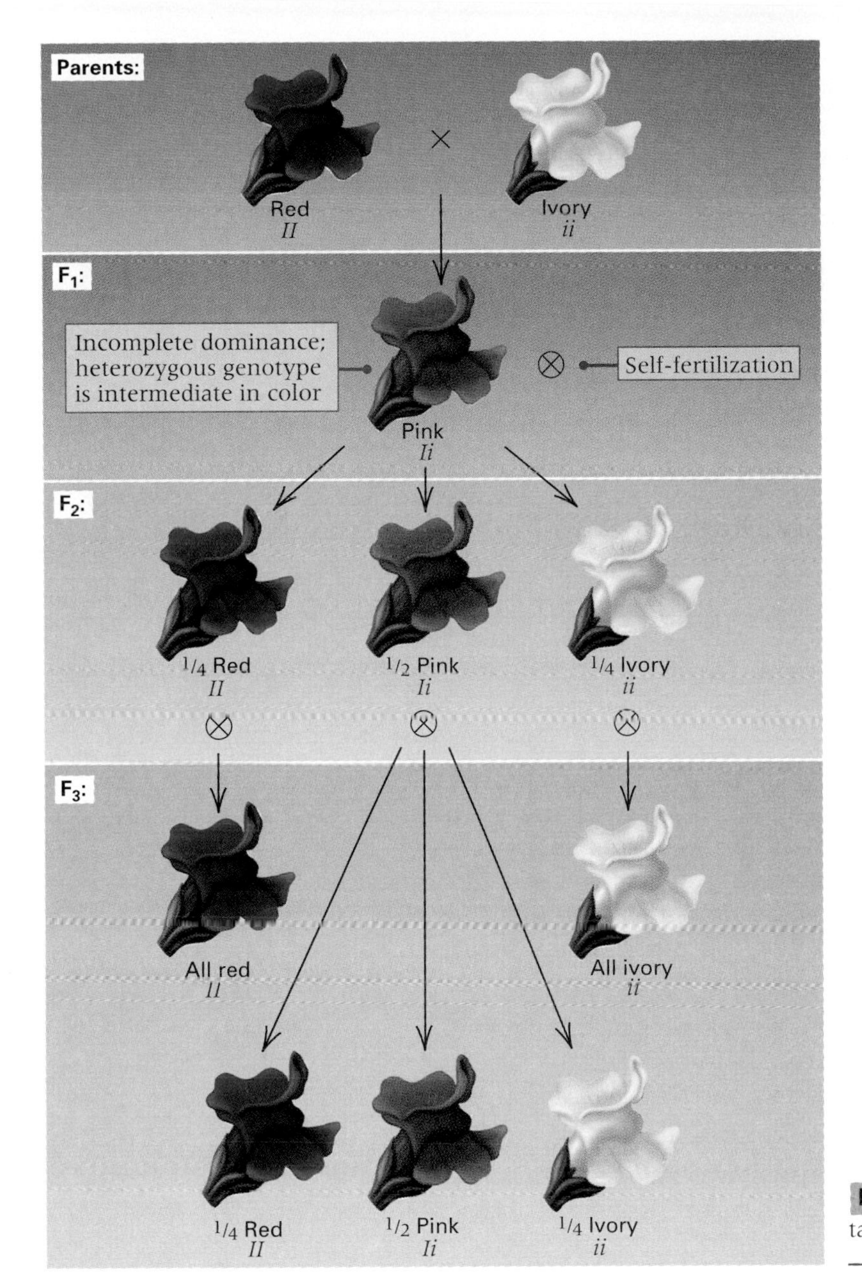

Figure 2.20 Absence of dominance in the inheritance of flower color in snapdragons.

ivory-flowered progeny; and the pink-flowered plants produced red, pink, and ivory progeny in the proportions 1/4 red : 1/2 pink : 1/4 ivory.

Incomplete dominance is often observed when the phenotype is quantitative rather than discrete. A trait that is *quantitative* can be measured on a continuous scale; examples include height, weight, number of eggs laid by a hen, time of flowering of a plant, and amount of enzyme in a cell or organism. A trait that is *discrete* is all-or-nothing; examples include round versus wrinkled seeds, and yellow versus green seeds. With a phenotype that is quantitative, the measured value of a heterozygote usually falls in the range between the homozygotes, and thus there is incomplete dominance.

Biochemical tests often reveal the products of both alleles in heterozygotes.

A special term, **codominance**, refers to a situation in which the phenotype of a heterozygous genotype is a mixture of the phenotypes of both of the corresponding homozygous genotypes. In such cases, the heterozygous phenotype is not intermediate between the homozygous genotypes (like pink snapdragons) but rather has the characteristics of both homozygous genotypes.

What we mean by "having the characteristics of both homozygous genotypes" is illustrated by one of the classic examples of codominance. These are the alleles that determine the A, B, AB, and O human blood groups. Blood type is determined by the

The human connection

Blood Drive

Karl Landsteiner 1901
Anatomical Institute,
Vienna, Austria
On Agglutination Phenomena in Normal Blood

Early blood transfusions sometimes gave very unsatisfactory results. The patient receiving the blood often went into shock and even died. This outcome was caused by massive clumping (agglutination) of red blood cells in the recipient, leading to blockage of the oxygen supply to many vital organs. In this paper, Landsteiner demonstrates that the clumping reaction can be observed in the test tube and that blood cells from each person can be classified as type A, type B, or type O, according to whether the cells are agglutinated by blood sera from other persons. The blood samples were taken from volunteers at the Institute at which Landsteiner worked ("Dr. St.", "Dr. Plecn.", and so forth; the abbreviation "Landst." refers to the author himself). In this excerpt, we have preserved the terms *agglutinin* (antibody) and *corpuscle* (red blood cell) as in the original but have replaced the blood type that Landsteiner called type C with its modern equivalent, type O. (Blood type AB was not found in these experiments, because the number of persons tested was too small.) Landsteiner's discovery led quickly to the matching of donor and recipient for the ABO blood groups in blood transfusions, and the disastrous incompatibility reactions were almost completely eliminated. As an interesting exercise, you may wish to deduce the blood type of "Landst."

	Blood corpuscles of					
Sera	**Dr. St.**	**Dr. Plecn.**	**Dr. Sturl.**	**Dr. Erdh.**	**Zar.**	**Landst.**
Dr. St.	−	+	+	+	+	−
Dr. Plecn.	−	−	+	+	−	−
Dr. Sturl.	−	+	−	−	+	−
Dr. Erdh.	−	+	−	−	+	−
Zar.	−	−	+	+	−	−
Landst.	−	+	+	+	+	−

Some time ago I observed and reported that blood serum of normal human beings is often capable of agglutinating red blood corpuscles of other healthy individuals. . . . I will mention in the following the results obtained in some recent experiments. . . . The tables [only one is shown] are self-explanatory. About equal amounts of serum and approximately 5 percent blood suspension were mixed in 0.6 percent saline solution and observed in test tubes. The plus sign denotes agglutination.

The experiment demonstrates that my data require no correction. All examined sera [22 altogether] from healthy persons gave the reaction. The result obviously would have been different had I not used a number of different corpuscles for the test. . . . In several cases (group A) the serum reacted on the corpuscles of another group (B), but not on those of group A, whereas the A corpuscles are again influenced in the same manner by serum B. In the third group (O) the serum aggregates the corpuscles of A and B, while the O corpuscles are not affected by sera of A and B. In ordinary speech, it can be said that in these cases at least two different kinds of agglutinins are present: some in A, others in B, and both together in O. The corpuscles are naturally to be considered as insensitive for the agglutinins which are present in the same serum. . . . I also did the agglutination successfully with blood which had been dried on linen and preserved for 14 days. Thus the reaction may possibly be suitable for forensic purposes of identification in some cases. . . . Finally, it must be mentioned that the reported observations allow us to explain the variable results in therapeutic transfusions of human blood.

THE REACTION MAY possibly be suitable for forensic purposes of identification in some cases.

Source: *Weiner Klinische Wochenschrift* 14: 1132–1134. Original in German. Excerpt from translation in S. H. Boyer, IV. 1963. *Papers on Human Genetics.* Englewood Cliffs, NJ: Prentice-Hall, pp. 27–31.

types of polysaccharides (polymers of sugars) present on the surface of red blood cells. Two different polysaccharides, A and B, can be formed. Both are formed from a precursor substance that is modified by the enzyme product of either the I^A or the I^B allele. The gene products are transferase enzymes that attach either of two types of sugar units to the precursor (Figure 2.21). People of genotype I^AI^A produce red blood cells having only the A polysaccharide and are said to have blood type A. Those of genotype I^BI^B have red blood cells with only the B polysaccharide and have blood type B. Hetero-

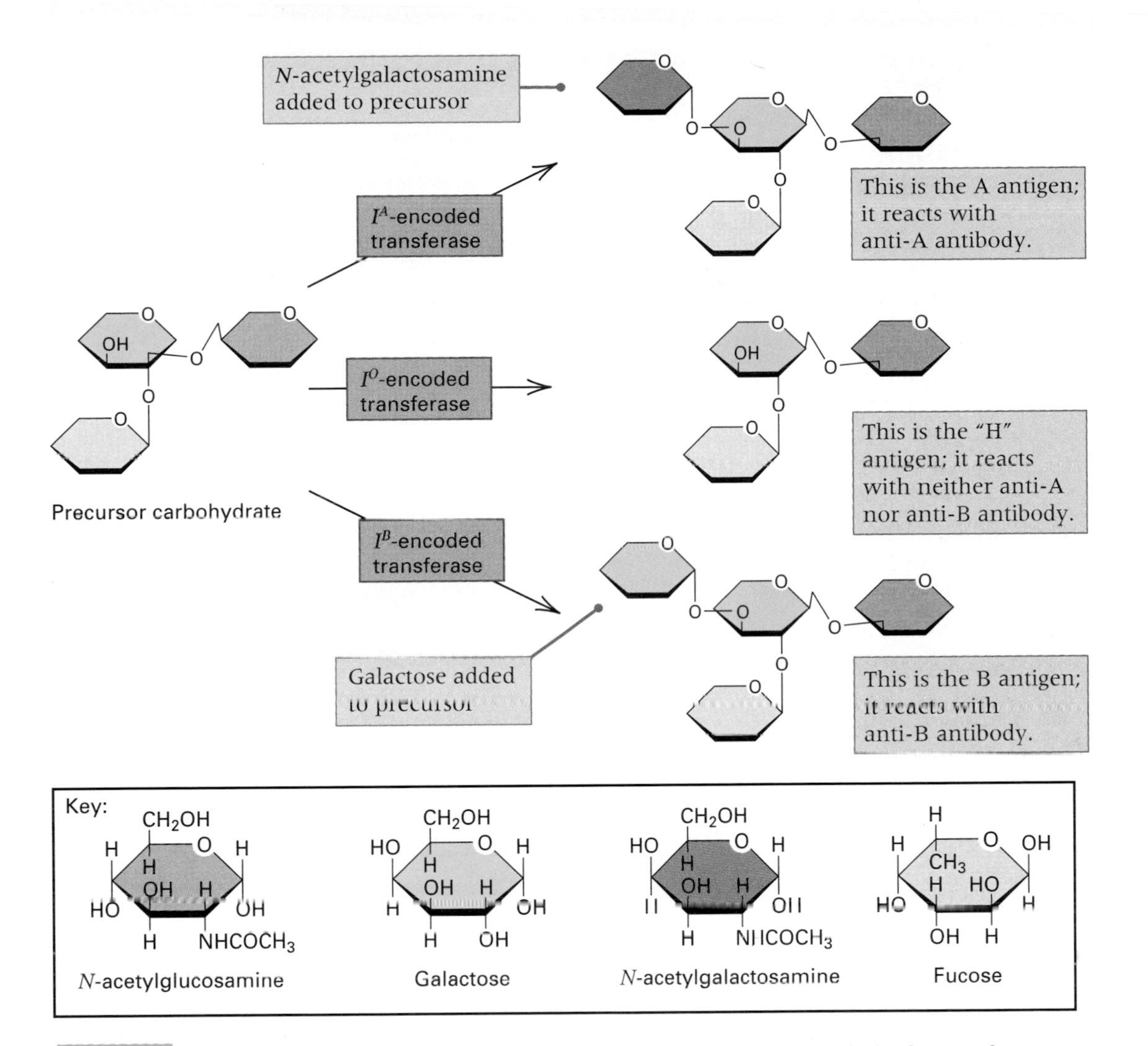

Figure 2.21 The ABO antigens on the surface of human red blood cells are carbohydrates. They are formed from a precursor carbohydrate by the action of transferase enzymes encoded by alleles of the *I* gene. Allele I^O codes for an inactive enzyme and leaves the precursor unmodified. The unmodified form is called the H substance. The I^A allele encodes an enzyme that adds *N*-acetylgalactosamine (purple) to the precursor. The I^B allele encodes an enzyme that adds galactose (green) to the precursor. The other colored sugar units are *N*-acetylglucosamine (orange) and fucose (yellow). The sugar rings also have side groups attached to one or more of their carbon atoms; these are shown in the detailed structures inside the box.

zygous I^AI^B people have red cells with both the A and the B polysaccharide and have blood type AB. The I^AI^B genotype illustrates codominance, because the heterozygous genotype has the characteristic of both homozygous genotypes—in this case, the presence of both the A and the B carbohydrate on the red blood cells.

The third allele, I^O, does not show codominance. It encodes a defective enzyme that leaves the precursor unchanged; neither the A nor the B type of polysaccharide is produced. Homozygous I^OI^O persons therefore lack both the A and the B polysaccharide; they are said to have blood type O. In I^AI^O heterozygotes, presence of the I^A allele results in production of the A polysaccharide; and in I^BI^O heterozygotes, presence of the I^B allele results in production of the B polysaccharide. The result is that I^AI^O persons have blood type A and I^BI^O persons have blood type B, and so I^O is recessive to both I^A and I^B. The genotypes and phenotypes of the ABO blood group system are summarized in the first three columns of Table 2.3.

The ABO blood groups are critical in medicine because of the frequent need for blood transfusions. An important feature of the ABO system is that most human blood contains antibodies to either the A or the B polysaccharide. An **antibody** is a protein that is made by the immune system in response to a stimulating molecule called an **antigen** and is capable of binding to the antigen. An antibody is usually

Table 2.3
Genetic control of the human ABO blood groups

Genotype	Antigens present on red blood cells	ABO blood group phenotype	Antibodies present in blood fluid	Blood types that can be tolerated in transfusion	Blood types that can accept blood for transfusion
I^AI^A	A	Type A	Anti-B	A & O	A & AB
I^AI^O	A	Type A	Anti-B	A & O	A & AB
I^BI^B	B	Type B	Anti-A	B & O	B & AB
I^BI^O	B	Type B	Anti-A	B & O	B & AB
I^AI^B	A & B	Type AB	Neither anti-A nor anti-B	A, B, AB & O	AB only
I^OI^O	Neither A nor B	Type O	Anti-A & anti-B	O only	A, B, AB & O

specific in that it recognizes only one antigen. Some antibodies combine with antigen and form large molecular aggregates that may precipitate.

Antibodies act in the body's defense against invading viruses and bacteria, as well as other cells, and help remove such invaders from the body. Although antibodies do not normally form without prior stimulation by the antigen, people capable of producing anti-A and anti-B antibodies do produce them. Production of these antibodies may be stimulated by antigens similar to polysaccharides A and B present on the surfaces of many common bacteria. However, a mechanism called *tolerance* prevents an organism from producing antibodies against its own antigens. This mechanism ensures that A-antigen or B-antigen elicits antibody production only in people whose own red blood cells do not contain A or B, respectively. The end result:

> People of blood type O make both anti-A and anti-B antibodies; those of blood type A make anti-B antibodies; those of blood type B make anti-A antibodies; and those of blood type AB make neither type of antibody.

The antibodies found in the blood fluid of people with each of the ABO blood types are shown in the fourth column in Table 2.3. The clinical significance of the ABO blood groups is that transfusion of blood containing A or B red-cell antigens into persons who make antibodies against these antigens results in an agglutination reaction in which the donor red blood cells are clumped. In this reaction, the anti-A antibody will agglutinate red blood cells of either blood type A or blood type AB, because both carry the A antigen (Figure 2.22). Similarly, anti-B antibody will agglutinate red blood cells of either blood type B or blood type AB. When the blood cells agglutinate, many blood vessels are blocked, and the recipient of the transfusion goes into shock and may die. Incompatibility in the other direction, in which the donor blood contains antibodies against the recipient's red blood cells, is usually acceptable because the donor's antibodies are diluted so rapidly that clumping is avoided. The types of compatible blood transfusions are shown in the last two columns of Table 2.3. Note that a person of blood type AB can receive blood from a person of any other ABO type; type AB is called a *universal recipient*. Conversely, a person of blood type O can donate blood to a person of any ABO type; type O is called a *universal donor.*

A mutant gene is not always expressed in exactly the same way.

Monohybrid Mendelian ratios, such as 3 : 1 (or 1 : 2 : 1 when the heterozygote is intermediate), are not always observed even when a trait is determined by the action of a single recessive allele. Regular ratios such as these indicate that organisms with the same genotype also exhibit the same phenotype. Although the phenotypes of organisms with a particular genotype are often very similar, this is not always the case—particularly in natural populations in which neither the matings nor the environmental conditions are under an experimenter's control. Variation in the phenotypic expression of a particular genotype may happen because other genes modify the phenotype or because the biological processes that produce the phenotype are sensitive to environmental conditions.

The types of variable gene expression are usually grouped into two categories:

- **Variable expressivity** refers to genes that are expressed to different degrees in different organisms. For example, inherited genetic diseases in human beings are often variable in expression from one person to the next. One patient may be very sick, whereas another with the same disease is less severely affected. Variable expressivity means that the same mutant gene can result in a severe form of the disease in one person but

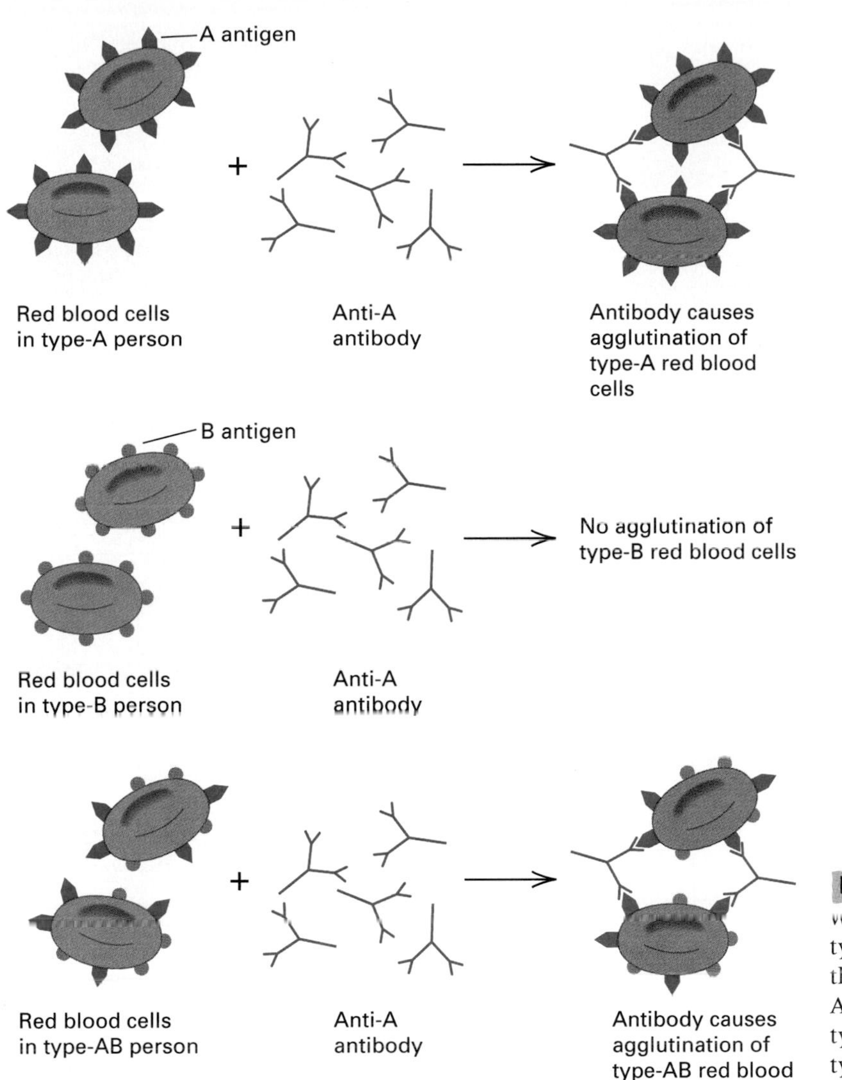

Figure 2.22 Antibody against type-A antigen will agglutinate red blood cells carrying the type-A antigen whether or not they also carry the type-B antigen. Blood fluid containing anti-A antibody will agglutinate red blood cells of type A and type AB, but not red blood cells of type B or type O.

a mild form in another. The different degrees of expression often form a continuous series from full expression to almost no expression of the expected phenotypic characteristics.

- **Incomplete penetrance** means that the phenotype expected from a particular genotype is not always expressed. For example, a person with a genetic predisposition to lung cancer may not get the disease if he or she does not smoke tobacco. A lack of gene expression may result from environmental conditions, such as in the example of not smoking, or from the effects of other genes. Incomplete penetrance is but an extreme of variable expressivity in which the expressed phenotype is so mild as to be undetectable. The proportion of organisms whose phenotype matches their genotype for a given character is called the **penetrance** of the genotype. A genotype that is always expressed has a penetrance of 100 percent.

2.7 Epistasis can affect the observed ratios of phenotypes.

In Chapter 1 we saw that the products of several genes may be necessary to carry out all the steps in a biochemical pathway. In genetic crosses in which two mutations that affect different steps in a single pathway are both segregating, the typical F_2 dihybrid ratio of 9 : 3 : 3 : 1 is not observed. One example is found in the interaction of two recessive mutations, each in a different gene, that affect flower coloration in peas. Plants of genotypes *CC* and *Cc* have purple flower color, which is the normal or **wildtype** expression of the trait, whereas homozygous *cc* plants have white flowers. For the other gene, plants of genotype *PP* and *Pp* have wildtype purple flowers, whereas homozygous *pp* plants have white flowers. Geneticists often use a dash to indicate an allele whose identity is not specified; for example, the symbol $C-$ means that in this genotype, one allele is known to be C and the other (unspecified) allele, indicated by the dash, may be either C or c. The symbol $C-$ is therefore a shorthand designation meaning "either CC or Cc." Using this type of symbolism, we could say that genotypes $C-$ and $P-$ have wildtype purple flowers, whereas genotypes cc and pp have white flowers. Homozygous recessive cc or pp plants have white flowers regardless of the genotype of the other gene.

Figure 2.23 shows a cross between plants of genotype $CC\ pp$ and plants of genotype $cc\ PP$. Both plants have white flowers because they are homozygous for either pp or cc. Even though both parental plants have white flowers, the flower-color phenotype of the plants in the F_1 generation is purple because the genotype of the F_1 progeny is $Cc\ Pp$ and heterozygous for both recessive alleles. Self-fertilization of the F_1 plants results in the F_2 progeny genotypes shown in the Punnett square. Because only the progeny with at least one C allele ($C-$) and at least one P allele ($P-$) have purple flowers, and all the rest have white flowers, the ratio of purple flowers to white flowers in the F_2 generation is 9 : 7.

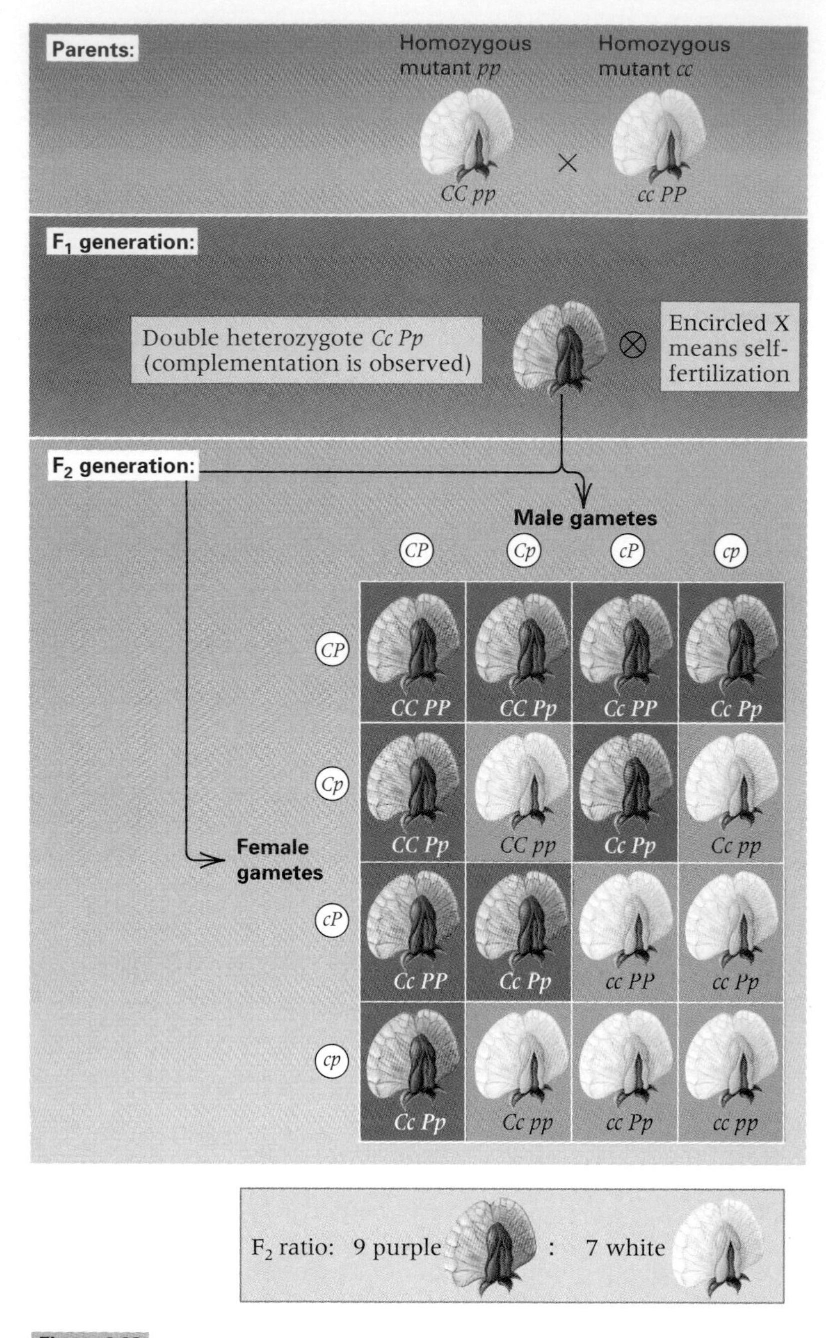

Figure 2.23 A cross showing epistasis in the determination of flower color in sweet peas. Formation of the purple pigment requires the dominant allele of both the C and P genes. With this type of epistasis, the dihybrid F_2 ratio is modified to 9 purple : 7 white.

The 9 : 7 ratio of purple : white flowers is a modified form of the 9 : 3 : 3 : 1 ratio in which the "9" class has purple flowers and the "3 : 3 : 1" classes all have white flowers. This is an example of **epistasis,** a term that refers to any type of gene interaction that results in the F_2 dihybrid ratio of 9 : 3 : 3 : 1 being modified into some other ratio. For a trait determined by the interaction of two genes, each with a dominant allele, there are only a limited number of ways in which the 9 : 3 : 3 : 1 dihybrid ratio can be modified. The possibilities are illustrated in Figure 2.24. Across the top are the genotypes produced in the F_2 generation by independent assortment and the ratios in which these genotypes occur. In the absence of epistasis, the F_2 ratio of phenotypes is 9 : 3 : 3 : 1. The possible modified ratios are shown in the lower part of the figure. In each row, the color coding indicates phenotypes that are indistinguishable because of epistasis, and the resulting modified ratio is given. For example, in the modified ratio at the bottom, the phenotypes of the "3 : 3 : 1" classes are indistinguishable, resulting in a 9 : 7 ratio. This is the ratio observed in the segregation of the C, c and P, p alleles in Figure 2.23, and the 9 : 7 ratio is the ratio of purple flowers to white flowers.

Taking all the possible modified ratios in Figure 2.24 together, there are nine possible dihybrid ratios when both genes show complete dominance. Examples are known of each of the modified ratios. However, the most frequently encountered modified ratios are 9 : 7, 12 : 3 : 1, 13 : 3 , 9 : 4 : 3, and 9 : 6 : 1. The types of epistasis that result in these modified ratios are illustrated in the following examples, which are taken from a variety of organisms. Other examples can be found in the problems at the end of the chapter.

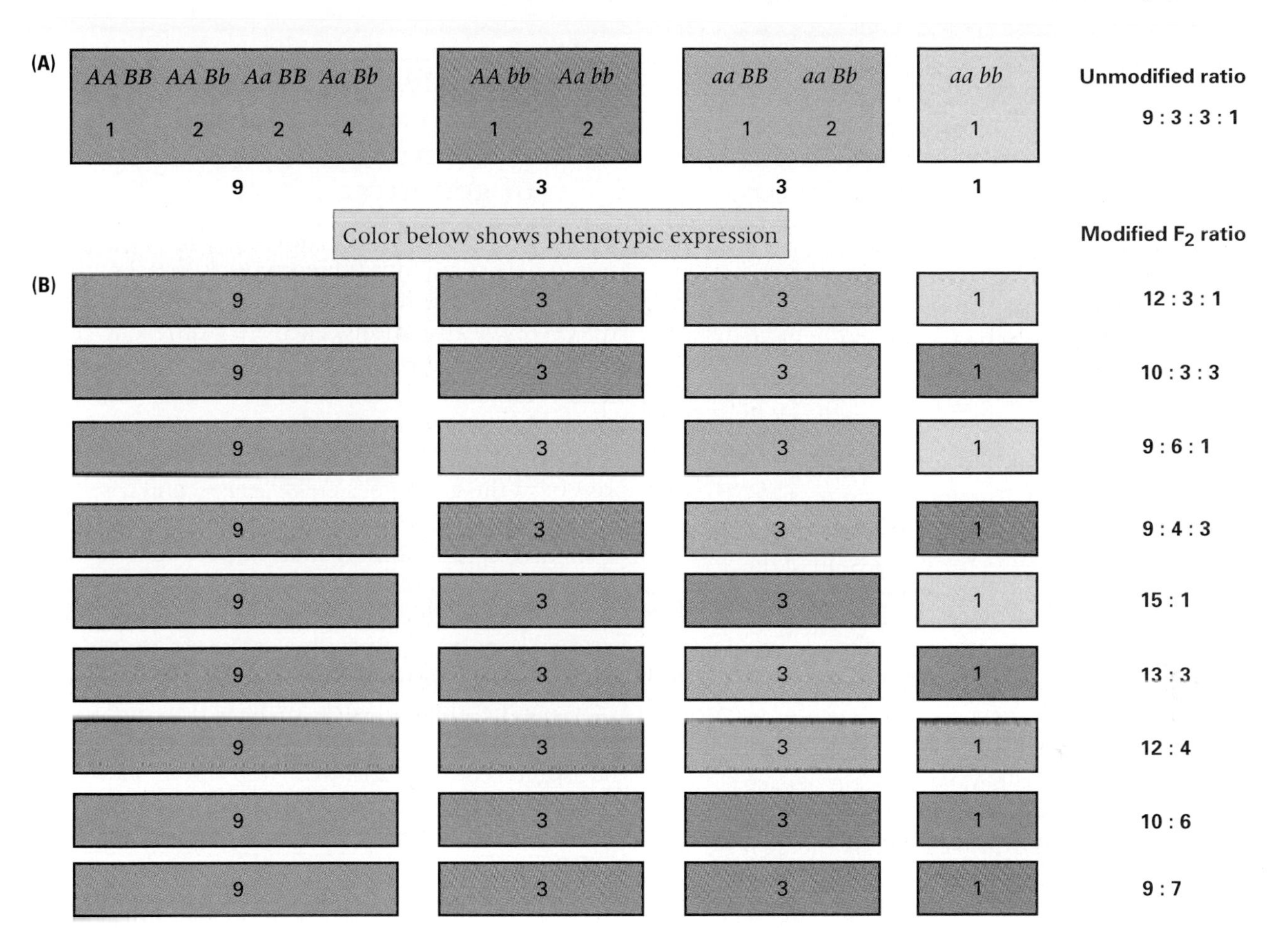

Figure 2.24 Modified F_2 dihybrid ratios. The F_2 genotypes of two independently assorting genes with complete dominance result in a 9 : 3 : 3 : 1 ratio of phenotypes if there is no interaction between the genes (epistasis). If there is epistasis that renders two or more of the phenotypes indistinguishable (indicated by the colors), then the F_2 ratio is modified. The most frequently encountered modified ratios are 9 : 7, 12 : 3 : 1, 13 : 3, 9 : 4 : 3, and 9 : 6 : 1.

9 : 7 This is the ratio observed when a homozygous recessive mutation in either or both of two different genes results in the same mutant phenotype. It is exemplified by the segregation of purple and white flowers in Figure 2.23. Genotypes that are *C*— for the *C* gene and *P*— for the *P* gene have purple flowers; all other genotypes have white flowers. Recall that the dash in *C*— means that the unspecified allele could be either *C* or *c*, and so *C*— means "either *CC* or *Cc*." Similarly, the dash in *P*— means that the unspecified allele could be either *P* or *p*.

12 : 3 : 1 A modified dihybrid ratio of the 12 : 3 : 1 variety results when the presence of a dominant allele of one gene masks the genotype of a different gene. For example, if the *A*— genotype renders the *B*— and *bb* genotypes indistinguishable, then the dihybrid ratio is 12 : 3 : 1 because the *A*— *B*— and *A*— *bb* genotypes are expressed as the same phenotype.

In a genetic study of the color of the hull in oat seeds, a variety having white hulls was crossed with a variety having black hulls. The F_1 hybrid seeds had black hulls. Among 560 progeny in the F_2 generation produced by self-fertilization of the F_1, the following seed phenotypes were observed in the indicated numbers:

418 black hulls 106 gray hulls 36 white hulls

Note that the observed ratio of phenotypes is 11.6 : 2.9 : 1, or very nearly 12 : 3 : 1. These results can be explained by a genetic hypothesis in which the black-hull phenotype results from the presence of a dominant allele (say, *A*) and the gray-hull phenotype results from another dominant allele (say, *B*) whose effect is apparent only in the *aa* homozygotes. On the basis of this hypothesis, the original true-breeding varieties must have had genotypes *aa bb* (white) and *AA BB* (black). The F_1 has genotype *Aa Bb* (black). If the *A*, *a* allele pair and the *B*, *b* allele pair undergo independent assortment, then the F_2 generation is expected to have the following composition of genotypes:

9/16	*A*— *B*—	(black hull)
3/16	*A*— *bb*	(black hull)
3/16	*aa B*—	(gray hull)
1/16	*aa bb*	(white hull)

This type of epistasis accounts for the 12 : 3 : 1 ratio.

13 : 3 This type of epistasis is illustrated by the difference between White Leghorn chickens (genotype *CC II*) and White Wyandotte chickens (genotype *cc ii*). Both breeds have white feathers because the *C* allele is necessary for colored feathers but the *I* allele in White Leghorns is a dominant inhibitor of feather coloration. The F_1 generation of a dihybrid cross between these breeds has the genotype *Cc Ii*, which results in the presence of white feathers because of the inhibitory effects of the *I* allele. In the F_2 generation, only the *C*— *ii* genotype has colored feathers; hence there is a 13 : 3 ratio of white : colored.

9 : 4 : 3 This dihybrid ratio (often stated as 9 : 3 : 4) is observed when homozygosity for a recessive allele with respect to one gene masks the expression of the genotype of a different gene. For example, if the *aa* genotype has the same phenotype regardless of whether the genotype is *B*— or *bb*, then the 9 : 4 : 3 ratio results.

In the mouse, the grayish coat color called agouti is produced by the presence of a horizontal band of yellow pigment just beneath the tip of each hair. The agouti pattern results from the presence of a dominant allele *A*, and in *aa* animals the coat color is black. A second dominant allele, *C*, is necessary for the formation of hair pigments of any kind, and *cc* animals are albino (white fur). In a cross of *AA CC* (agouti) × *aa cc* (albino), the F_1 progeny are *Aa Cc* and agouti. Crosses between F_1 males and females produce F_2 progeny in the following proportions:

9/16	*A*— *C*—	(agouti)
3/16	*A*— *cc*	(albino)
3/16	*aa C*—	(black)
1/16	*aa cc*	(albino)

The dihybrid ratio is therefore 9 agouti : 4 albino : 3 black.

9 : 6 : 1 This dihybrid ratio is observed when homozygosity for a recessive allele of either of two genes results in the same phenotype but the phenotype of the double homozygote is distinct. For example, red coat color in Duroc-Jersey pigs requires the presence of two dominant alleles *R* and *S*. Pigs of genotype *R*— *ss* and *rr S*— have sandy-colored coats, and *rr ss* pigs are white. The F_2 dihybrid ratio is therefore

9/16	*R*— *S*—	(red)
3/16	*R*— *ss*	(sandy)
3/16	*rr S*—	(sandy)
1/16	*rr ss*	(white)

The 9 : 6 : 1 ratio results from the fact that both single recessives have the same phenotype.

2.8 Complementation between mutations of different genes is a fundamental principle of genetics.

A 9 : 7 ratio in the F_2 generation results from epistasis when a mutant phenotype is present in a genotype that is homozygous for either (or both) of two recessive mutations, each in a different gene. The situation is illustrated for purple (wildtype) and white (mutant) flower color in Figure 2.23, in which the recessive mutations in different genes are *p* and *c*. In this example, white flower color is present in the genotypes *C*— *pp*, *cc P*—, and *cc pp*.

Suppose that a pea geneticist isolates yet another recessive mutation that causes white flower color. Such a mutant gene might be recovered in an experiment in which peas are exposed to radiation or to a chemical agent known to cause mutations, and the progeny screened to identify new mutants with white flowers. Such experiments are done routinely in genetics to identify mutations in genes that affect each of the key steps in a biological process—in this example, the production of color pigment in flowers; an experiment of this type is called a **mutant screen.**

Having discovered a new recessive mutation that causes white flowers, the geneticist must first address the issue of whether the new mutant gene has a genetic defect in the *C* gene, a genetic defect in the *P* gene, or a genetic defect in a third flower-color gene not previously identified. It could easily be mutant in either the *C* or the *P* gene, because mutations in the same gene can occur more than once. If the new mutation is in the *C* gene, it is an allele of *c*; even so, it need not be identical in DNA sequence with the mutant *c* allele. We saw in Chapter 1 that genes typically consist of thousands of base pairs, many of which are so critical to function that a single-base change results in a visible mutant phenotype. Mutations can also be more drastic changes, such as a deletion of part of the gene. Because there are so many different ways in which a gene can mutate to impair or eliminate its function, it is very unlikely that two independent mutations are identical in DNA sequence, even if they are alleles of the same gene. This is the basis for saying that if the new mutation is in the *C* gene, it is unlikely to be identical with *c*. Likewise, if the mutation is in the *P* gene, it is unlikely to be identical with *p*. The various possible forms of a gene are called **multiple alleles.** (It is important to note that the concept of multiple alleles includes not only mutant alleles. In natural populations of organisms there are usually also multiple *wildtype* alleles, even though we often refer to wildtype as "the" wildtype allele. The multiple wildtype alleles differ in nucleotide sequence, but they all encode a functional gene product.)

How could our geneticist find out whether the new mutant gene is an allele of *c*, an allele of *p*, or neither? One possibility is to purify DNA from the mutant strain, isolate the *C* and *P* genes, and determine their DNA sequences. This approach has numerous practical difficulties. First, the sequences of the *C* and *P* genes are not known at present, so there is no basis for comparison of the sequences. Second, even if the sequences of *C* and *P* were known, isolating the genes from the mutant strain and sequencing them would be expensive and could require many months of effort. The third problem is even worse. As we have noted, in many populations the wildtype gene has multiple alleles, so it is altogether possible that the *C* and *P* alleles in the new mutant differ in sequence from other *C* and *P* alleles and yet are both wildtype in their function. Here we have an example in which knowledge of the complete DNA sequence of the *C* gene and the *P* gene affords no practical help in determining whether a new mutant allele results from a mutation in either of these genes!

Is there another approach to determining the allelism of the new mutant gene? Yes, there is, and it is illustrated in Figure 2.23. The key observation is that the cross of *CC pp* × *cc PP* yields F_1 progeny whose flowers are wildtype. The reason for the wildtype phenotype is that the genotype of the F_1 progeny is *Cc Pp*. The F_1 progeny are heterozygous for a mutation in each of two different genes. Each mutation is recessive, and its presence is concealed by the dominant allele in the F_1 progeny. When a cross between two homozygous recessive mutants results in wildtype F_1 progeny, the mutant alleles are said to *complement* one another or to exhibit **complementation.** The finding of complementation between two recessive mutations means that the mutations are in different genes, as shown for the genes *C* and *P* in Figure 2.23.

Lack of complementation means that two mutations are alleles of the same gene.

Complementation between mutant alleles in different genes is illustrated in Figure 2.25A for mutant strains designated 1 and 2. Mutant 1 is known to be homozygous recessive *pp*, and mutant 2 is "unknown." The finding of complementation (a wildtype phenotype in the F1 generation) indicates that the mutant gene in strain 2 is not an allele of the mutant *p* in strain 1. In other words, the mutation in strain 2 is not in the wildtype *P* gene, because a recessive mutation in *P* would yield an allele of *p*.

Figure 2.25B shows the result that is expected in a cross between two mutants when the mutations are in the same gene. Mutant strain 1 is again the "known" (*pp*), and mutant strain 3 is the "unknown." In this case, the F_1 progeny have a mutant phenotype, so there is no complementation between the mutant alleles in strains 1 and 3. This finding means that the mutant allele in strain 3 must be in the same gene as the mutant allele in strain 1. It must be mutated in the same gene, because if it were mutated in a different gene, then complementation would have been observed, as in part A. The lack of complementation means that

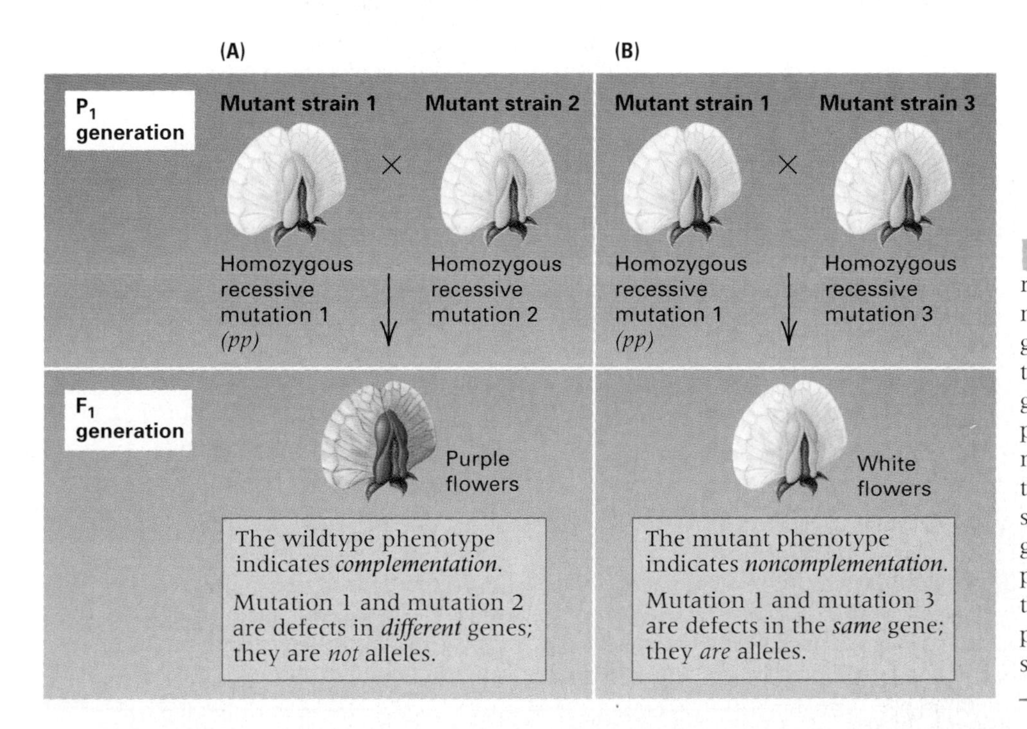

Figure 2.25 Complementation reveals whether two recessive mutations are alleles of different genes. To test for complementation, homozygous recessive genotypes are crossed. If the phenotype of the F_1 progeny is nonmutant (A), it means that the mutations in the parental strains are alleles of different genes. If the phenotype of the F_1 progeny is mutant (B), it means that the mutations in the parental strains are alleles of the same gene.

the genotype of the F_1 progeny can be written as $p\ p_3$, where p_3 denotes the recessive mutant allele in strain 3. We assign the mutant allele the symbol "p" because the lack of complementation with p demonstrates that it is an allele of p, but we must distinguish the mutant allele from p itself because, in all likelihood, it has a different nucleotide sequence from p. The distinguishing mark is often a subscript, and in this case we use the subscript 3 (p_3) because the mutant allele is present in strain 3.

The complementation test enables us to group mutants into allelic classes.

The kind of cross illustrated in Figure 2.25 is a **complementation test.** As we have seen, it is used to determine whether recessive mutations in each of two different strains are alleles of the same gene. Because the result indicates presence or absence of allelism, the complementation test is one of the key experimental operations in genetics. To illustrate the application of the test in practice, suppose a mutant screen were carried out to isolate new mutants with white flowers in peas. Starting with a true-breeding strain with purple flowers, we treat pollen with x rays and use the irradiated pollen to fertilize ovules to obtain seeds. The F_1 seeds are grown and the resulting plants allowed to self-fertilize, after which the F_2 plants are grown. A few of the F_1 seeds may contain a new mutation that causes white flowers, but because the white phenotype is recessive, the flower will be purple. However, the resulting F_1 plant will be heterozygous for the new white mutation so self-fertilization will result in the formation of F_2 plants with a 3 : 1 ratio of purple : white flowers. Because mutations that result in a particular phenotype are quite rare, even when induced by radiation, among many thousands of self-fertilized plants only a few will be found to be a new mutant with white flowers. Let us suppose that we were lucky enough to obtain four new mutants, in addition to the p and c mutants already identified.

How are we going to name these four new mutants? We can make no assumptions about the number of genes represented. All four could be recurrences of either p or c. On the other hand, each

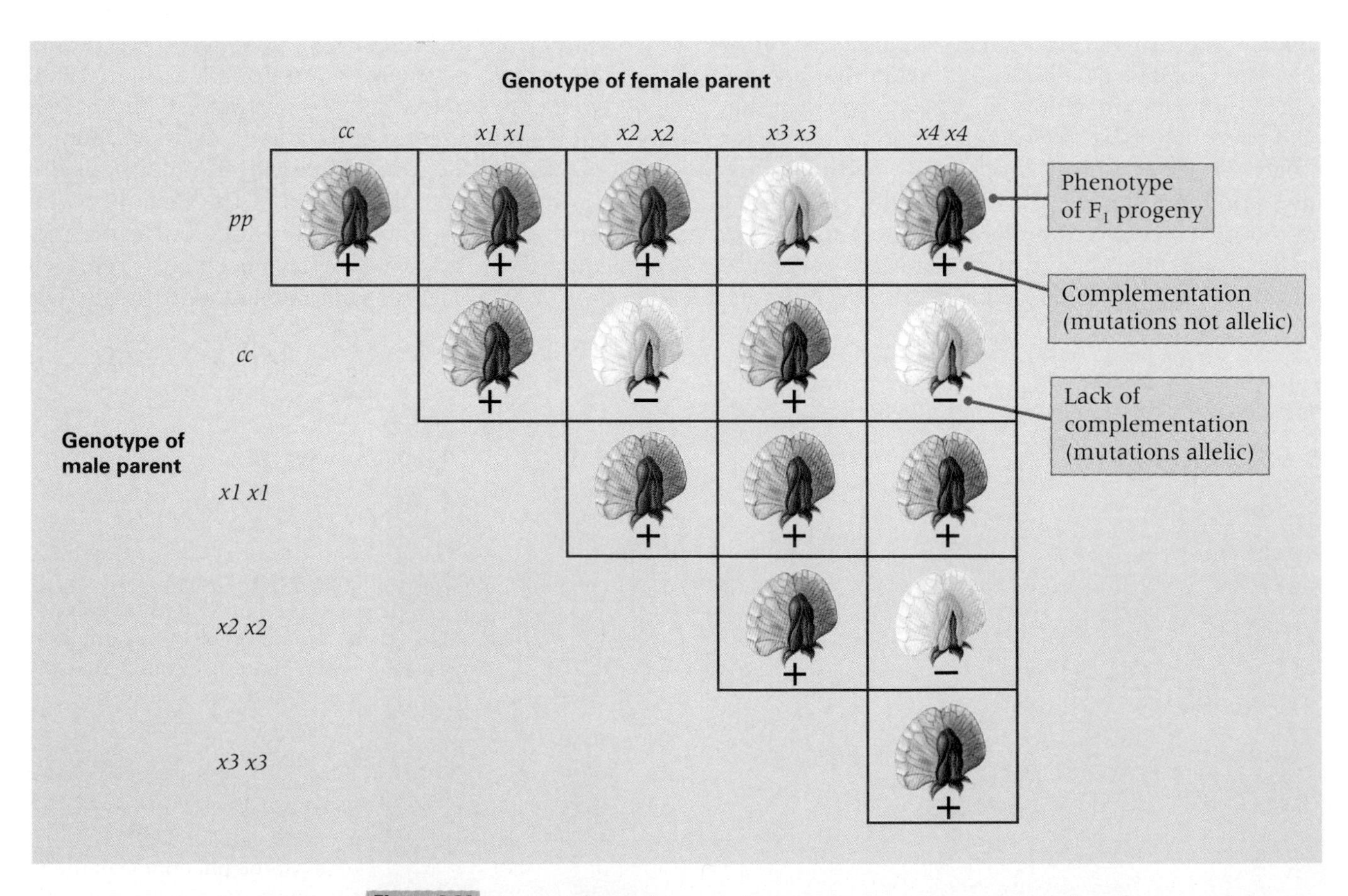

Figure 2.26 Results of complementation tests among six mutant strains of peas, each homozygous for a recessive allele resulting in white flowers. Each box gives the phenotype of the F_1 progeny of a cross between the male parent whose genotype is indicated in the far left column and the female parent whose genotype is indicated in the top row.

of the four could be the result of a new mutation in a different gene needed for flower color. For the moment, let us call the new mutant genes *x1*, *x2*, *x3*, and *x4*, where the *x* does not imply a gene but rather indicates that the mutant gene was obtained with x irradiation. Each mutant gene is recessive and was identified through the white flowers of the homozygous recessive F_2 seeds (for example, *x1 x1*).

Now the complementation test is used to classify the "*x*" mutants into groups. Figure 2.26 shows that the results of a complementation test are reported in a triangular array of + and − signs. The crosses that yield F_1 progeny with the wildtype phenotype (in this case, purple flowers) are denoted with a + in the box where imaginary lines from the male parent and the female parent intersect. The crosses that yield F_1 progeny with the mutant phenotype (white flowers) are denoted with a − sign. The + signs indicate complementation between the mutant alleles in the parents; the − signs indicate lack of complementation. The bottom half of the triangle is unnecessary, because the reciprocal of each cross produces F_1 progeny with the same genotype and phenotype as the cross that is shown. The diagonal elements are also unnecessary, because a cross between any two organisms that carry the same mutant gene—for example, *x1 x1* × *x1 x1*—must yield homozygous recessive *x1 x1* progeny, which will have the mutant phenotype. As we saw in Figure 2.25, complementation in a cross means that the parental strains are mutant for different genes. Lack of complementation means that the parental strains are mutant for the same gene. The principle underlying the complementation test is as follows:

> **The Principle of Complementation**: If two recessive mutations are alleles of the same gene, then the phenotype of an organism that contains both mutations is mutant; if they are alleles of different genes, then the phenotype of an organism that contains both mutations is wildtype (nonmutant).

In the interpretation of complementation data such as those in Figure 2.26, the principle is actually applied the other way around. Examination of the phenotype of the F_1 progeny of each possible cross reveals which of the mutations are alleles of the same gene.

> In a complementation test, if the combination of two recessive mutations results in a mutant phenotype, then the mutations are regarded as alleles of the same gene; if the combination results in a wildtype phenotype, then the mutations are regarded as alleles of different genes.

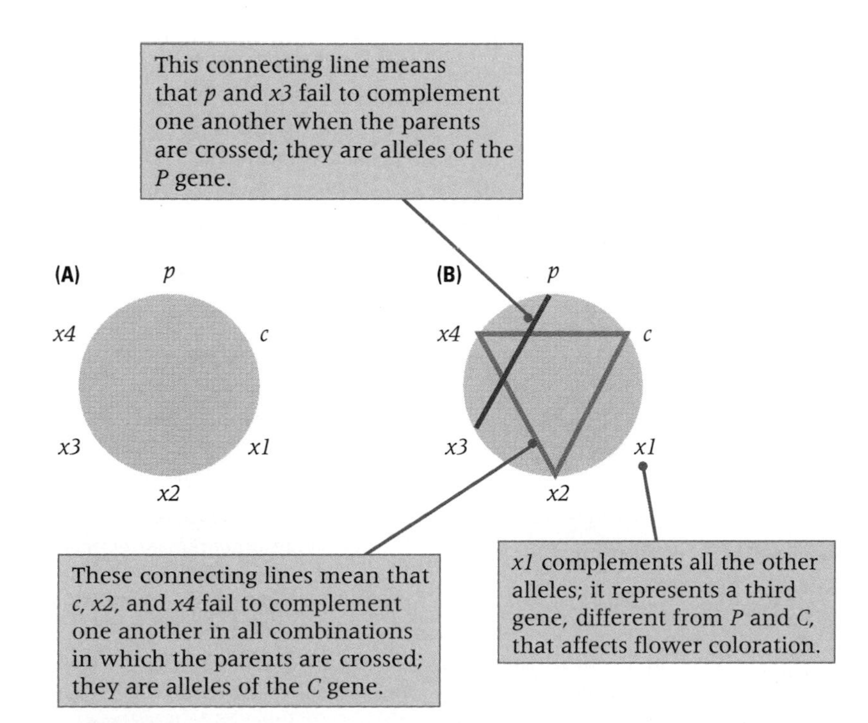

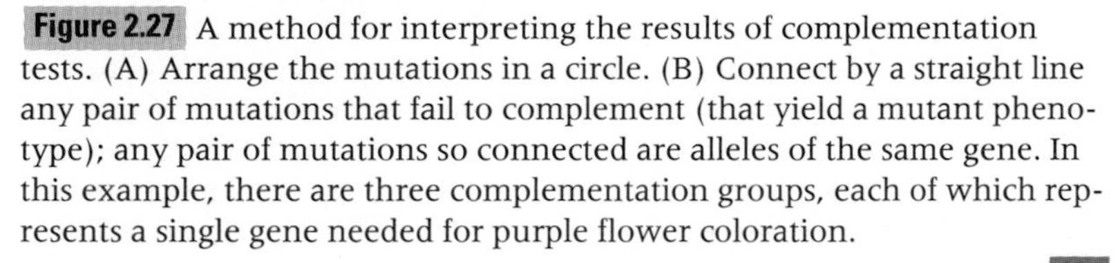
Figure 2.27 A method for interpreting the results of complementation tests. (A) Arrange the mutations in a circle. (B) Connect by a straight line any pair of mutations that fail to complement (that yield a mutant phenotype); any pair of mutations so connected are alleles of the same gene. In this example, there are three complementation groups, each of which represents a single gene needed for purple flower coloration.

A convenient way to analyze the data in Figure 2.26 is to arrange the alleles in a circle as shown in Figure 2.27A. Then, for each possible pair of mutants, connect the pair by a straight line if the mutant genes *fail* to complement (Figure 2.27B). According to the principle of complementation, the lines must connect mutant genes that are alleles of each other, because in a complementation test, lack of complementation means that the mutant genes are alleles. In this example, mutant *x3* is an allele of *p*, so *x3* and *p* are different mutant alleles of the gene *P*. Similarly, the mutants *x2*, *x4*, and *c* are different mutant alleles of the gene *C*. The mutant *x1* complements all of the other alleles. It represents a third gene, different from *P* and *C*, that affects flower coloration.

In an analysis like that in Figure 2.27, each of the groups of noncomplementing mutants is called a **complementation group.** As we have seen, each complementation group defines a gene.

> A gene is defined experimentally as a set of mutant alleles that make up one complementation group. Any pair of mutant alleles in such a group fail to complement one another and result in an organism with an observable mutant phenotype.

The mutants in Figure 2.27 therefore represent three genes, a mutation in any one of which results in white flowers. The gene *P* is represented by the alleles *p* and *x3;* the gene *C* is represented by the alleles *c, x2,* and *x4;* and the allele *x1* represents a third gene different from either *P* or *C.* Each gene coincides with one of the complementation groups.

At this point in a genetic analysis, it is possible to rename the mutant genes to indicate which ones are true alleles. Because the *p* allele already had its name before the mutant screen was carried out to obtain more flower-color mutants, the new allele of *p* that was previously called *x3* should be renamed to reflect its allelism with *p.* We did this in the discussion of Figure 2.25, calling the new allele p_3 to reflect its identity with the mutant gene *x3.* For similar reasons, we might rename the *x2* and *x4* mutant genes c_2 and c_4 to reflect their allelism with the original *c* allele and to convey their independent origins. The mutant *x1* represents an allele of a new gene to which we can assign a name arbitrarily. For example, we might call the mutant allele *albus* (Latin for "white") and assign the *x1* allele the new name *alb.* The wildtype dominant allele of *alb,* which is necessary for purple coloration, would then be symbolized as *Alb* or as alb^+. The procedure of sorting new mutants into complementation groups and renaming them according to their allelism is an example of how geneticists identify genes and name alleles. Such renaming of alleles is the typical manner in which genetic terminology evolves as knowledge advances.

Chapter Summary

- **Mendel was careful in his choice of traits.**

- **Genes come in pairs, separate in gametes, and join randomly in fertilization.**

> Genes are particles that come in pairs.
> The paired genes separate (segregate) in the formation of reproductive cells.
> Gametes unite at random in fertilization.
> Genotype means genetic endowment; phenotype means observed trait.
> The progeny of the F_2 generation support Mendel's hypothesis.
> The progeny of testcrosses also support Mendel's hypothesis.

Inherited traits are determined by particulate elements called genes. In a higher plant or animal, the genes are present in pairs. One member of each gene pair is inherited from the maternal parent, the other member from the paternal parent.

A gene can have different forms that result from differences in DNA sequence. The different forms of a gene are called alleles. The particular combination of alleles present in an organism constitutes its genotype. The observable characteristics of an organism constitute its phenotype. In an organism, if the two alleles of a gene pair are the same (for example, *AA* or *aa*), the genotype is homozygous for the *A* or *a* allele; if the alleles are different (*Aa*), the genotype is heterozygous. When the phenotype of a heterozygote is the same as that of one of the homozygous genotypes, the allele that is expressed is called dominant and the hidden allele is called recessive.

- **The alleles of different genes segregate independently.**

> The F_2 genotypes in a dihybrid cross conform to Mendel's prediction.
> The progeny of testcrosses show the result of independent segregation.

In genetic studies, the organisms produced by a mating constitute the F_1 generation. Matings between members of the F_1 generation produce the F_2 generation. In a cross such as $AA \times aa$, in which only one gene is considered (a monohybrid cross), the ratio of genotypes in the F_2 generation is 1 dominant homozygote (*AA*) : 2 heterozygotes (*Aa*) : 1 recessive homozygote (*aa*). The phenotypes in the F_2 generation appear in the ratio 3 dominant : 1 recessive. The Mendelian ratios of genotypes and phenotypes result from segregation in gamete formation (when the members of each allelic pair segregate into different gametes) and random union of gametes in fertilization.

- **Chance plays a central role in Mendelian genetics.**

> The addition rule applies to mutually exclusive events.
> The multiplication rule applies to independent events.

- **The results of segregation can be observed in human pedigrees.**

The processes of segregation, independent assortment, and random union of gametes follow the rules of probability, which provide the basis for predicting outcomes of genetic crosses. Two basic rules for combining probabilities are the addition rule and the multiplication rule. The addition rule applies to mutually exclusive events; it states that the probability of the realization of either one or the other (or both) of two events equals the sum of the respective probabilities. The multiplication rule applies to independent events; it states that the probability of the simultaneous realization of both of two events is equal to the product of the respective probabilities. In some organisms—for example, human beings—it is not possible to perform controlled crosses, and genetic analysis is accomplished through the study of several generations of a family tree, called a pedigree. Pedigree analysis is determination of the possible genotypes of the family members in a pedigree and of the probability that an individual member has a particular genotype.

- **The alleles of some genes do not show complete dominance.**

The phenotype of a heterozygous genotype is often intermediate.

Biochemical tests often reveal the products of both alleles in heterozygotes.

A mutant gene is not always expressed in exactly the same way.

In heterozygous genotypes, complete dominance of one allele over the other is not always observed. In most cases, a heterozygote for a wildtype allele and a mutant allele that encodes a defective gene product produces less gene product than does the wildtype homozygote. If the phenotype is determined by the amount of wildtype gene product rather than by its mere presence, the heterozygote will have an intermediate phenotype. This situation is called incomplete dominance. Codominance means that both alleles in a heterozygote are expressed, and so the heterozygous genotype exhibits the phenotypic characteristics of both homozygous genotypes. Codominance is exemplified by the I^A and I^B alleles in persons with blood group AB. Codominance is often observed for proteins when each alternative allele codes for a different amino acid replacement, because the alternative forms of the protein may be able to be distinguished by chemical or physical means. Genes are not always expressed to the same extent in different organisms; this phenomenon is called variable expressivity. A genotype that is not expressed at all in some organisms is said to have incomplete penetrance.

- **Epistasis can affect the observed ratios of phenotypes.**

Dihybrid crosses differ in two genes—for example, *AA BB* × *aa bb*. The phenotypic ratios in the dihybrid F_2 are 9 : 3 : 3 : 1, provided that both the *A* and the *B* allele are dominant and that the genes undergo independent assortment. The 9 : 3 : 3 : 1 ratio can be modified in various ways by interaction between the genes (epistasis). Different types of epistasis may result in dihybrid ratios such as 9 : 7, 12 : 3 : 1, 13 : 3, 9 : 4 : 3, and 9 : 6 : 1.

- **Complementation between mutations of different genes is a fundamental principle of genetics.**

Lack of complementation means that two mutations are alleles of the same gene.

The complementation test enables us to group mutants into allelic classes.

The complementation test is the functional definition of a gene. Two recessive mutations are considered alleles of different genes if a cross between the homozygous recessives results in nonmutant progeny. Such alleles are said to complement each other. On the other hand, two recessive mutations are considered alleles of the same gene if a cross between the homozygous recessives results in mutant progeny. Such alleles are said to fail to complement. For any group of recessive mutant alleles, a complete complementation test entails crossing the homozygous recessives in all pairwise combinations.

Key Terms

addition rule	F_1 generation	multiplication rule	true-breeding
albinism	F_2 generation	mutant screen	variable expressivity
allele	gamete	P_1 generation	wildtype
antibody	gene	pedigree	zygote
antigen	genotype	penetrance	
backcross	heterozygous	phenotype	
carrier	homozygous	Punnett square	
codominance	Huntington disease	recessive	
complementation	hybrid	reciprocal cross	
complementation group	incomplete penetrance	segregation	
complementation test	independent assortment	sib	
dihybrid	Mendelian genetics	sibling	
dominant	monohybrid	sibship	
epistasis	multiple alleles	testcross	

Concepts and Issues
Testing Your Knowledge

- What constitutes the genotype of an organism? What constitutes the phenotype? Why is it important in genetics that genotype and phenotype be distinguished?
- What is the difference between a "gene" and an "allele"? How can a gene have more than two alleles? Give an example of multiple alleles of a gene.
- What is the "principle of segregation," and how is this principle demonstrated in the results of Mendel's monohybrid crosses?
- What is the "principle of independent assortment," and how is this principle demonstrated in the results of Mendel's dihybrid crosses?

- Explain why random union of male and female gametes is necessary for Mendelian segregation and independent assortment to occur.
- What is the difference between mutually exclusive events and independent events? How are the probabilities of these two types of events combined? Give two examples of genetic events that are mutually exclusive and two examples of genetic events that are independent.
- When two pairs of alleles show independent assortment, under what conditions will a 9 : 3 : 3 : 1 ratio of phenotypes in the F_2 generation *not* be observed?
- Explain this statement: In genetics, a "gene" is identified experimentally by a set of mutant alleles that fail to show complementation.

Solutions Step by Step

Problem 1

In garden peas, an allele *T* for axial flowers (positioned along the stem) is dominant to an allele *t* for terminal flowers (positioned at the tips of the branches).

(a) In the F_2 generation of a monohybrid cross, what is the expected ratio of axial : terminal?
(b) Among the F_2 progeny, what proportion are heterozygous?
(c) Among the F_2 progeny with axial flowers, what proportion are heterozygous?
(d) In a testcross of the F_1 progeny of a monohybrid cross, what is the expected ratio of axial : terminal?

Solution **(a)** The monohybrid cross is between the genotypes *TT* × *tt*, which yields F_1 plants with genotype *Tt*. Crossing these among themselves (*Tt* × *Tt*) yields F_2 plants given by the accompanying Punnett square. Therefore, the expected ratio of axial (*TT* or *Tt*) to terminal (*tt*) is 1/4 + 1/2 = 3/4 to 1/4, or 3 : 1.

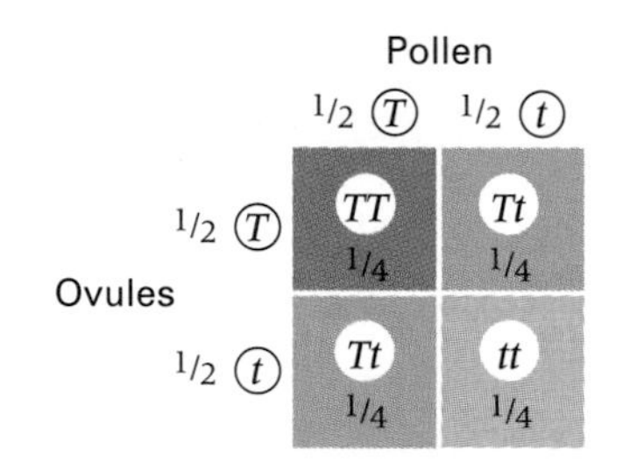

(b) The proportion of heterozygous *Tt* progeny in the F_2 generation is evident from the Punnett square as 1/4 + 1/4 = 1/2. **(c)** The critical phrase in this question is *with axial flowers*, because it means that in considering the Punnett square, we are supposed to disregard the *tt* plants with terminal flowers. When these are disregarded, there remain 1/3 *TT* plants and 2/3 *Tt* plants, so the proportion of F_2 plants *with axial flowers* that are heterozygous is 2/3.
(d) By definition, the testcross is with a homozygous recessive parent, so the parents in the testcross are *Tt* × *tt*. The Punnett square is indicated, and it shows that the expected ratio of axial (genotype *Tt*) to terminal (genotype *tt*) plants is 1/2 : 1/2, or 1 : 1.

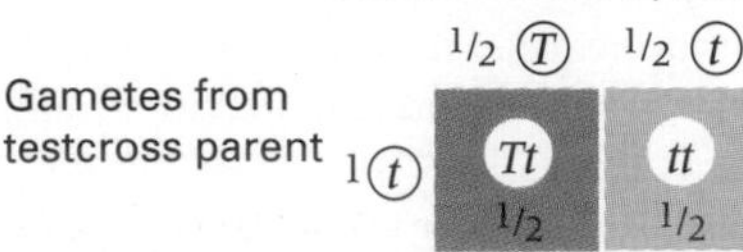

Problem 2

In human genetics, an allele *B* for early baldness ("pattern baldness") is dominant to its allele *b* in males but recessive in females. (Pattern baldness in females is also expressed at a later age.) In the accompanying pedigree, identify the genotype of each person or, if more than one genotype is possible, the possible genotypes and their respective probabilities.

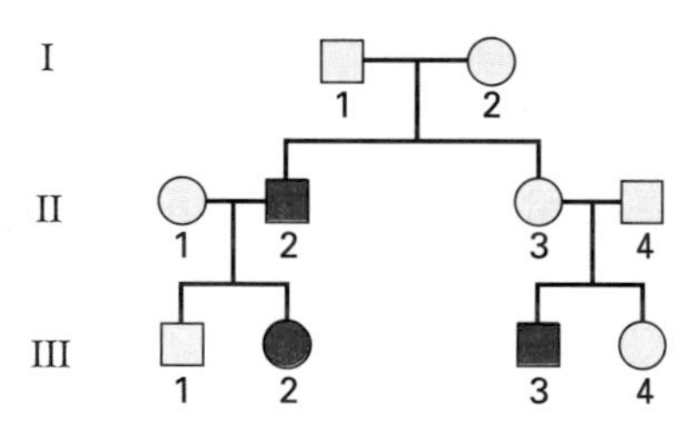

Solution Because *B* is dominant in males, all nonbald males must have the genotype *bb*. This includes male I-1. But because the mating in generation I produces a bald son, the mother must carry *B*. She is not bald, so her genotype must be *Bb* (if she were *BB*, she would be bald). Because the bald male II-2 comes from the mating *Bb* × *bb*, his genotype must be *Bb*. His daughter, III-2, is bald and so must have the genotype *BB*. This implies that the female II-1 has genotype *Bb*, because (1) she transmitted a *B* allele to III-2 but (2) she herself is not bald. The nonbald female II-3 has a bald son, so II-3 has genotype *Bb*. The cross II-3 × II-4 is therefore *Bb* × *bb*, so the bald male III-3 has genotype *Bb*. The nonbald female III-4 could be either *Bb* or *bb*, each with the probability 1/2. These deductions are summarized in the accompanying pedigree.

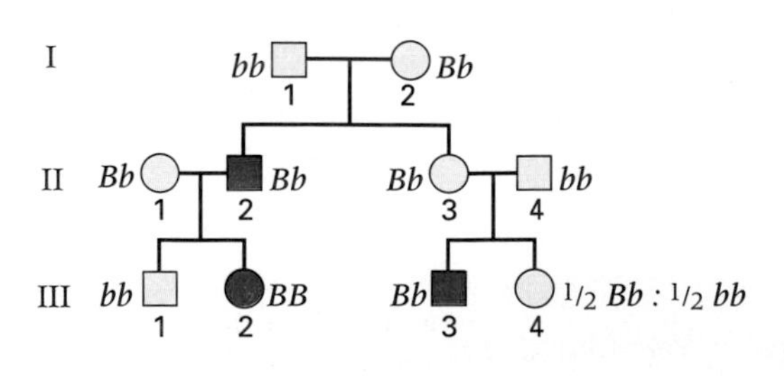

Problem 3

In some people, the carbohydrate A and B antigens of the ABO blood groups not only are found on the surface of the red blood cells but are also secreted into the saliva and other body fluids. Secretion of the antigens is due to a dominant allele of another genetic locus called *secretor*. People of genotype *Se Se* or *Se se* do secrete the A and B antigens, but those

geNETics on the web

www.jbpub.com/genetics

These GeNETics on the Web exercises will introduce some of the most useful sites for finding genetic information on the World Wide Web. Genetic sites provide access to a rich storehouse of information on all aspects of genetics. These range from sites written in nontechnical language for the lay person to sites with sophisticated databases designed for the professional geneticist. In carrying out these exercises, you will get a taste of what the Internet can offer a student in genetics.

The keywords shown in color in the following exercises are available on the Jones and Bartlett Publishers' web site as hyperlinks to various genetic sites. To complete the exercises, visit the GeNETics on the Web home page at

http://www.jbpub.com/genetics

Select the link to Essential geNETics on the Web. Then choose a chapter. You will find a list of keywords that correspond to the exercises that follow. Select a keyword to link to a web site containing the genetic information necessary to complete the exercise. Each exercise includes a specific assignment that makes use of the information available at the site.

Exercises

- Huntington disease is a devastating degeneration of the brain that begins in middle life. This disease affects about 30,000 Americans, and because of its dominant mode of inheritance and complete penetrance, each of their 150,000 siblings and children has a 50–50 chance of developing it. Named after George Huntington, a Long Island physician who first described the disease in 1872, its principal symptom is an involuntary, jerky motion of the head, trunk, and limbs called *chorea*, after the Greek word for "dance." At this keyword site you can learn about the molecular basis of Huntington disease and genetic testing programs to detect it. Write a short report defining *presymptomatic testing,* and explain the basis of the test. Do you think that everyone at risk for this disease should necessarily be tested? If you were at risk, would you?
- The MendelWeb is a treasure trove of information about Mendel, including his famous paper, essays, commentary, and a collection of images—all richly linked to additional Internet resources. At this keyword site you should find the MendelWeb Timeline, a listing of some of the events that occurred during Mendel's lifetime (1822–1884), and these are linked to additional Internet resources that put his life into historical perspective. From the timeline, you will learn that Mendel had two disastrous academic experiences, one in 1850 and another in 1855. Write a brief report on what these experiences were, where they took place, and how they happened. How soon after Mendel's second disaster did he begin his pea-breeding experiments?

Pic Site

The Pic Site showcases some of the most visually appealing genetics sites on the World Wide Web. To visit the showcase genetics site, select the Pic Site for Chapter 2.

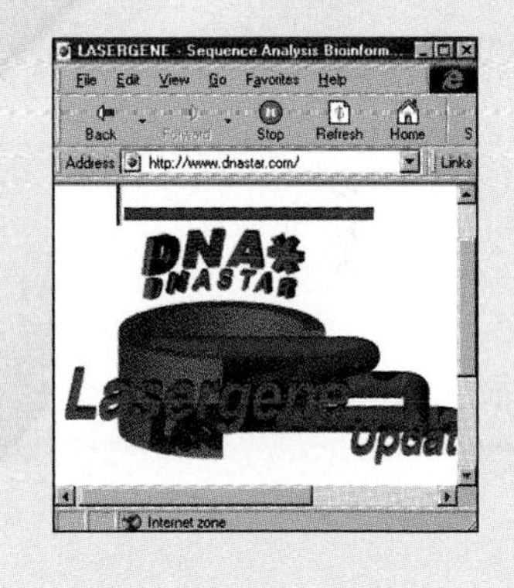

of genotype *se se* do not. There is epistasis between the genetic systems, because people of blood group O are nonsecretors regardless of their *secretor* genotype. The ABO and *secretor* loci are on different chromosomes and therefore show independent assortment. In crosses of the following type, what is the expected ratio of secretors to nonsecretors?

$$I^A I^O\ Se\ se \times I^B I^O\ Se\ se$$

Solution One way to work this problem is to make a 4 × 4 Punnett square with the gametes I^A *Se*, I^A *se*, I^O *Se*, and I^O *se* (1/4 for each) along one axis and I^B *Se*, I^B *se*, I^O *Se*, and I^O *se* (again 1/4 for each) along the other; then fill in all the boxes and tabulate which are secretors and which are not. But because the genes segregate independently, the approach can be simplified by considering each gene separately. The ABO segregation produces the genotypes 1/4 $I^A I^B$, 1/4 $I^A I^O$, 1/4 $I^B I^O$, and 1/4 $I^O I^O$. The first three are potential secretors, depending on the *secretor* locus, whereas the $I^O I^O$ genotype is a nonsecretor no matter what. Segregation of the *secretor* locus produces 3/4 *Se*— (where the dash indicates *either Se or se*) and 1/4 *se se*. The former are potential secretors, depending on the ABO genotype, whereas the *se se* genotype is a nonsecretor no matter what. Because these segregations are independent, we can make the sort of Punnett square illustrated, which shows that the ratio of secretor : nonsecretor is 9/16 : 7/16, or 9 : 7.

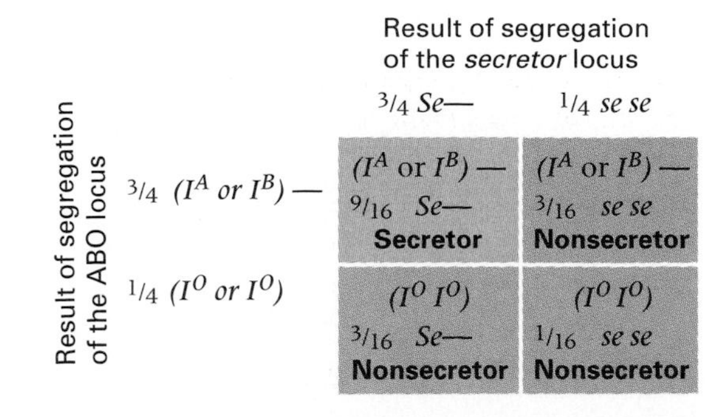

Problem 4

Mutant genes *a* through *f*, all recessive, are identified in a mutant screen for genes affecting pharynx development in the nematode worm *Caenorhabditis elegans*. Complementation tests are carried out between the homozygous mutant strains, with the results shown in the accompanying matrix. A + sign indicates complementation, and a − sign indicates lack of complementation. Configure the mutant genes in the form of a circle, and use straight lines to connect the alleles that are in the same complementation group. How many different genes affecting pharynx development do these data indicate?

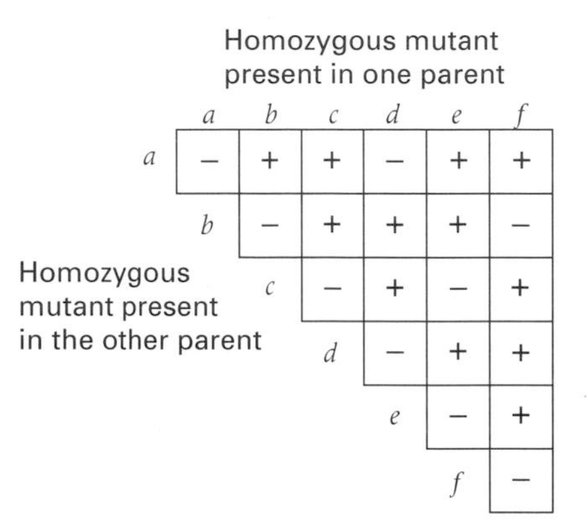

Homozygous mutant present in one parent (columns); Homozygous mutant present in the other parent (rows)

	a	*b*	*c*	*d*	*e*	*f*
a	−	+	+	−	+	+
b		−	+	+	+	−
c			−	+	−	+
d				−	+	+
e					−	+
f						−

Solution This problem uses the key concept of a *complementation test,* which is a cross between two genotypes, each homozygous for an independently identified mutant gene. The mutations may or may not be alleles of the same gene. If they *are* alleles of the same gene, then all of the progeny will have the mutant phenotype. If they are *not* alleles of the same gene, then all of the progeny will have the nonmutant phenotype. A set of recessive mutations in which all pairwise crosses between the homozygous mutants yield progeny with the mutant phenotype constitutes a *complementation group,* which is the geneticist's experimental definition of a *gene.* The circular pattern called for in the problem is indicated. The lines connect mutants that, when crossed, yield only mutant progeny (− sign in the complementation matrix). The circular analysis of the complementation data indicates that mutants *a* and *d* are alleles of one gene affecting pharynx development, *b* and *f* alleles of a different gene affecting pharynx development, and *c* and *e* are alleles of a third such gene.

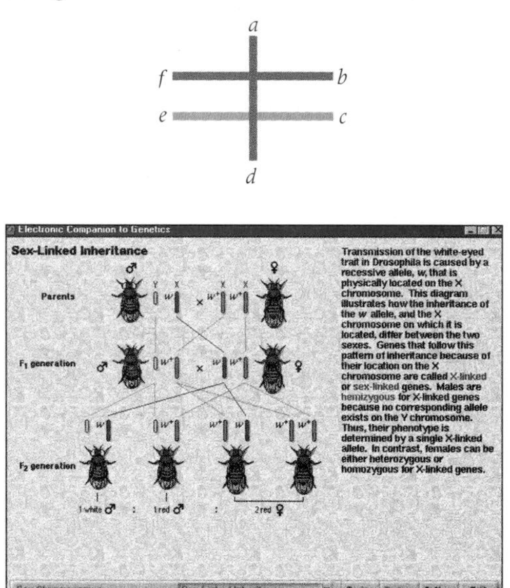

To enhance your study turn to *An Electronic Companion to Genetics*™, © 1998, Cogito Learning Media, Inc. This CD-ROM is a multimedia tool that provides interactive explanations of important genetic concepts through animations, diagrams, and videos. It also provides interactive test questions.

Concepts in Action

Problems for Solution

2.1 How many different alleles of a particular gene may exist in the population? How many alleles of a particular gene can be present in single diploid organism?

2.2 Complementation tests of the recessive mutant genes *a* through *f* produced the data in the accompanying matrix. The circles represent missing data. Assuming that all of the missing mutant combinations would yield data consistent with the entries that are known, complete the table by filling each circle with a + or − as needed.

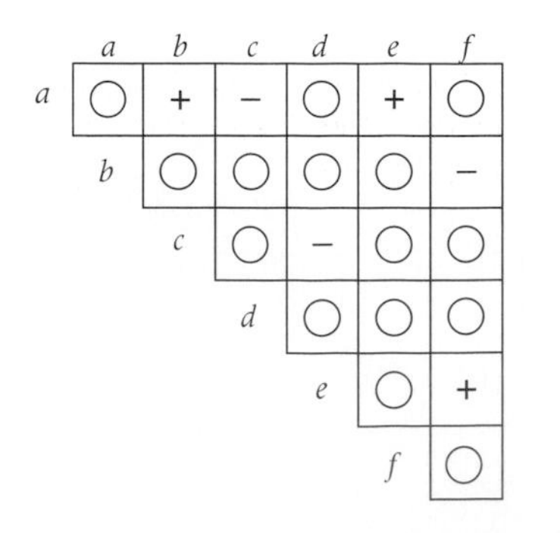

	a	*b*	*c*	*d*	*e*	*f*
a	○	+	−	○	+	○
b		○	○	○	○	−
c			○	−	○	○
d				○	○	○
e					○	+
f						○

2.3 In the shepherd's purse, *Capsella bursapastoris,* the capsule containing the seeds can be either triangular or ovoid. A cross between certain true-breeding strains with triangular capsules yielded an F_1 with triangular capsules. The observed F_2 ratio was 15 triangular : 1 ovoid. What genetic hypothesis can explain these results? What crosses would you carry out to test this hypothesis?

2.4 The dominant allele *Cy* (*Curly*) in *Drosophila* results in curly wings. The cross *Cy*/+ × *Cy*/+ (where + represents the wildtype allele of *Cy*) results in a ratio of 2 curly : 1 wildtype F_1 progeny. The cross between curly F_1 progeny also gives a ratio of 2 curly : 1 wildtype F_2 progeny. How can this result be explained?

2.5 White Leghorn chickens are homozygous for a dominant allele, *C,* of a gene responsible for colored feathers, and also for a dominant allele, *I,* of an independently segregating gene that prevents the expression of *C.* The White Wyandotte breed is homozygous recessive for both genes *c*/*c*; *i*/*i*. What proportion of the F_2 progeny obtained from mating White Leghorn × White Wyandotte F_1 hybrids would be expected to have colored feathers?

2.6 A trihybrid cross $A/A; B/B; r/r \times a/a; b/b; R/R$ is made in a plant species in which A and B are dominant to their respective alleles but there is no dominance between R and r. Assume independent assortment, and consider the F_2 progeny from this cross.

(a) How many phenotypic classes are expected?
(b) What is the probability of the parental $a/a; b/b; R/R$ genotype?
(c) What proportion of the progeny would be expected to be homozygous for all three genes?

2.7 *Meiotic drive* is a phenomenon observed occasionally in which a heterozygous genotype does not produce a 1 : 1 proportion of functional gametes, usually because one of the gametic types is not formed or fails to function. Suppose that an allele A shows meiotic drive such that heterozygous Aa genotypes form 3/4 A-bearing and 1/4 a-bearing functional gametes. What is the expected ratio of genotypes in the F_2 generation of a monohybrid cross:

(a) If the meiotic drive occurs equally in both sexes?
(b) If the meiotic drive occurs only in females?

2.8 Huntington disease is a rare degenerative human disease determined by a dominant allele, *HD*. The disorder is usually manifested after the age of 45. A young man has learned that his father has developed the disease.

(a) What is the probability that the young man will later develop the disorder?
(b) What is the probability that a child of the young man carries the *HD* allele?

2.9 Consider human families with four children, and assume that each birth is equally likely to result in a boy or a girl.

(a) What proportion will include at least one boy?
(b) What fraction will have the gender order FMFM?

2.10 In plants, certain mutant genes are known that affect the ability of gametes to participate in fertilization. Suppose that an allele A is such a mutation and that pollen cells bearing the A allele are only half as likely to survive and participate in fertilization as pollen cells bearing the a allele. What is the expected ratio of AA : Aa : aa plants in the F_2 generation in a monohybrid cross?

2.11 Assume that the trait in the accompanying pedigree is due to simple Mendelian inheritance.

(a) Is it likely to be due to a dominant allele or a recessive allele? Explain.
(b) What is the meaning of the double horizontal line connecting III-1 with III-2?
(c) What is the biological relationship between III-1 and III-2?
(d) If the allele responsible for the condition is rare, what are the most likely genotypes of all of the persons in the pedigree in generations I, II, and III? (Use A and a for the dominant and recessive alleles, respectively.)

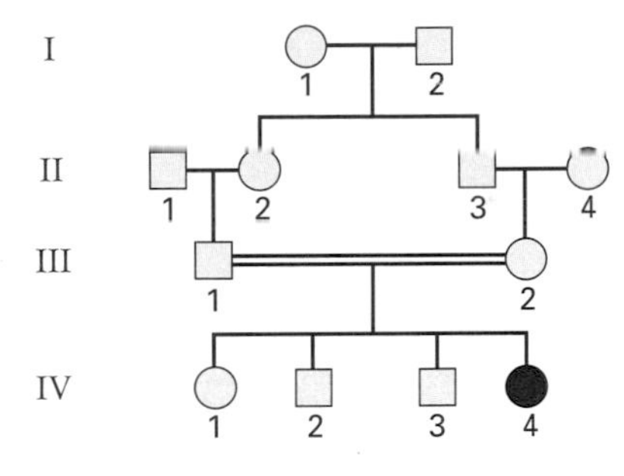

2.12 A certain man and a certain woman both have a 50 percent chance of being a carrier (heterozygous) for a recessive allele associated with a genetic disease. If they have one child, what is the chance that the child will be homozygous recessive?

Further Readings

Ashley, C. T., and S. T. Warren. 1995. Trinucleotide repeat expansion and human disease. *Annual Review of Genetics* 29: 703.

Bowler, P. J. 1989. *The Mendelian Revolution.* Baltimore, MD: Johns Hopkins University Press.

Carlson, E. A. 1987. *The Gene: A Critical History.* 2d ed. Philadelphia: Saunders.

Dunn, L. C. 1965. *A Short History of Genetics.* New York: McGraw-Hill.

Hartl, D. L., and V. Orel. 1992. What did Gregor Mendel think he discovered? *Genetics* 131: 245.

Judson, H. F. 1996. *The Eighth Day of Creation: The Makers of the Revolution in Biology.* Cold Spring Harbor, NY: Cold Spring Harbor Laboratory Press.

Mendel, G. 1866. Experiments in plant hybridization. (Translation.) In *The Origins of Genetics: A Mendel Source Book*, ed. C. Stern and E. Sherwood. 1966. New York: Freeman.

Olby, R. C. 1966. *Origins of Mendelism.* London: Constable.

Orel, V. 1996. *Gregor Mendel: The First Geneticist.* Oxford, England: Oxford University Press.

Orel, V., and D. L. Hartl. 1994. Controversies in the interpretation of Mendel's discovery. *History and Philosophy of the Life Sciences* 16: 423.

Stern, C., and E. Sherwood. 1966. *The Origins of Genetics: A Mendel Source Book.* New York: Freeman.

Sturtevant, A. H. 1965. *A Short History of Genetics.* New York: Harper & Row.

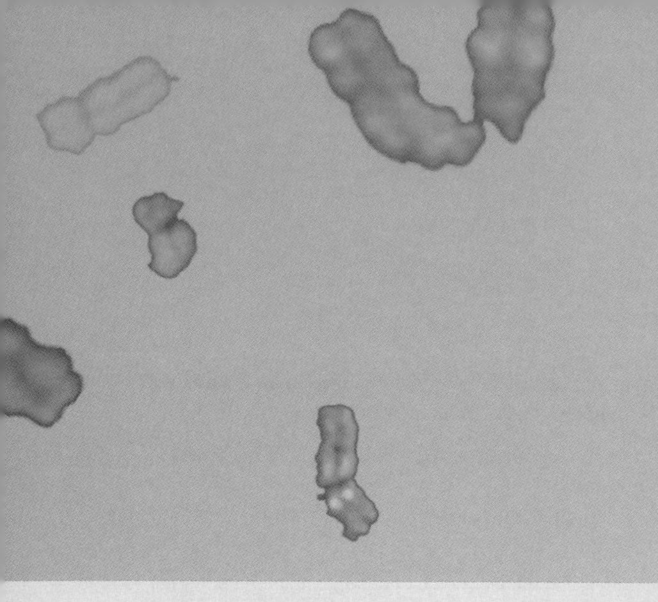

Key Concepts

- Chromosomes in eukaryotic cells are usually present in pairs.
- The chromosomes of each pair separate in meiosis, one going to each gamete.
- In meiosis, the chromosomes of different pairs undergo independent assortment.
- Chromosomes consist largely of DNA combined with histone proteins.
- In many animals, sex is determined by a special pair of chromosomes, the X and Y.
- Irregularities in the inheritance of an X-linked gene in *Drosophila* gave experimental proof of the chromosomal theory of heredity.
- The progeny of genetic crosses follow the binomial probability formula.
- The chi-square statistical test is used to determine how well observed genetic data agree with expectations from a hypothesis.

THE RATE OF CHROMOSOME evolution in cats is relatively conservative compared with that in some other mammals. The domestic cat has 19 pairs of chromosomes. [© Ron Kimball. All rights reserved.]

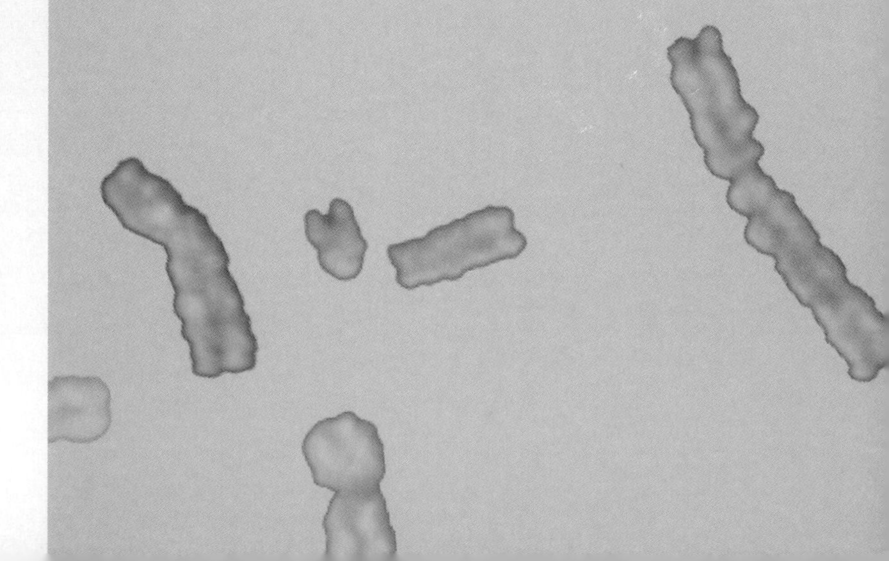

CHAPTER 3

The Chromosomal Basis of Heredity

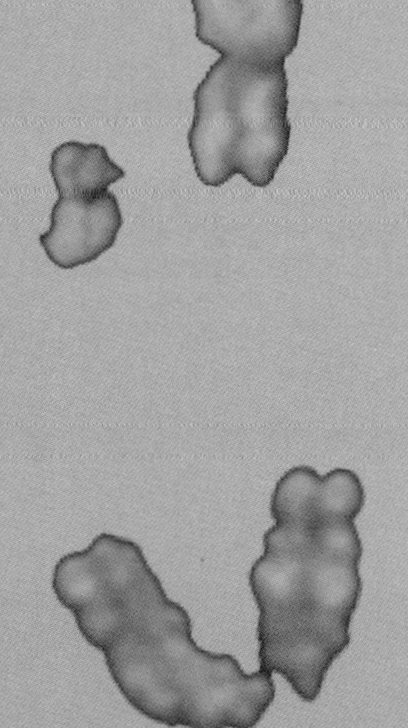

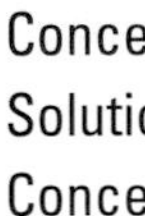

Gregor Mendel's experiments made it clear that in heterozygous genotypes, neither allele is altered by the presence of the other. The hereditary units remain stable and unchanged in passing from one generation to the next. Mendel emphasized this finding in a long letter to Carl Nägeli sent on April 18, 1867: "I have never observed gradual transitions between the parental traits or a progressive approach toward one of them. . . . In each generation, the two parental traits appear, separated and unchanged, and there is nothing to indicate that one of them has either inherited or taken over anything from the other." Nevertheless, at the time, the biological basis of the transmission of the hereditary factors from one generation to the next was quite mysterious. Neither the role of the nucleus in reproduction nor the details of cell division had been discovered. Once these phenomena were understood, and when microscopy had improved enough that the chromosomes could be observed and were finally realized to be the bearers of the genes, new understanding came at a rapid pace. This chapter examines the mechanism of chromosome segregation in cell division and the relationship between DNA and chromosomes.

3.1 Each species has a characteristic set of chromosomes.

The importance of the cell nucleus and its contents was suggested as early as the 1840s when Carl Nägeli observed that in dividing cells, the nucleus divided first. This was the same man with whom Mendel would later correspond. Nägeli failed to understand Mendel's findings, and he also failed to see the importance of his own discovery of nuclear division. He regarded the cells in which he saw nuclear division as aberrant. Nevertheless, by the 1870s it was recognized that nuclear division is a universal attribute of cell division. The importance of the nucleus in inheritance was reinforced by the nearly simultaneous discovery that the nuclei of two gametes fuse in the process of fertilization. The next major advance came a decade later with the discovery of **chromosomes,** which had been made visible by light microscopy when stained with basic dyes. A few years later, chromosomes were found to segregate by an orderly process into the daughter cells formed by cell division, as well as into the gametes formed by the division of reproductive cells. Finally, three important regularities were observed about the **chromosome complement** (the complete set of chromosomes) of plants and animals:

1. The nucleus of each **somatic cell** (a cell of the body, in contrast with a **germ cell,** or gamete) contains a fixed number of chromosomes typical of the particular species. However, the numbers vary tremendously among species and have little relationship to the complexity of the organism (Table 3.1).
2. The chromosomes in the nuclei of somatic cells are usually present in pairs. For example, the 46 chromosomes of human beings consist of 23 pairs (Figure 3.1). Similarly, the 14 chromosomes of peas consist of 7 pairs. Cells with nuclei of this sort, containing two similar sets of chromosomes, are called **diploid.** The chromosomes are present in pairs because one chromosome of each pair derives from the maternal parent of the organism and the other from its paternal parent.

Table 3.1
Somatic chromosome numbers of some plant and animal species

Organism	Chromosome number	Organism	Chromosome number
Field horsetail	216	Yeast (*Saccharomyces cerevisiae*)	32
Bracken fern	116	Fruit fly (*Drosophila melanogaster*)	8
Giant sequoia	22	Nematode (*Caenorhabditis elegans*)	11 ♂, 12 ♀
Macaroni wheat	28	House fly	12
Bread wheat	42	Scorpion	4
Fava bean	12	Geometrid moth	224
Garden pea	14	Common toad	22
Wall cress (*Arabidopsis thaliana*)	10	Chicken	78
Corn (*Zea mays*)	20	Mouse	40
Lily	12	Gibbon	44
Snapdragon	16	Human being	46

3. The germ cells, or gametes, contain only one set of chromosomes, consisting of one member of each of the pairs. The gamete nuclei are said to be **haploid.** The haploid gametes unite in fertilization to produce the diploid state of somatic cells.

In a multicellular organism, which develops from a single fertilized egg, the presence of the diploid chromosome number in somatic cells and the haploid chromosome number in germ cells indicates that there are *two* different processes of nuclear division. One of these, mitosis, maintains chromosome number, whereas the other, meiosis, halves the number. These two processes are examined in the following sections.

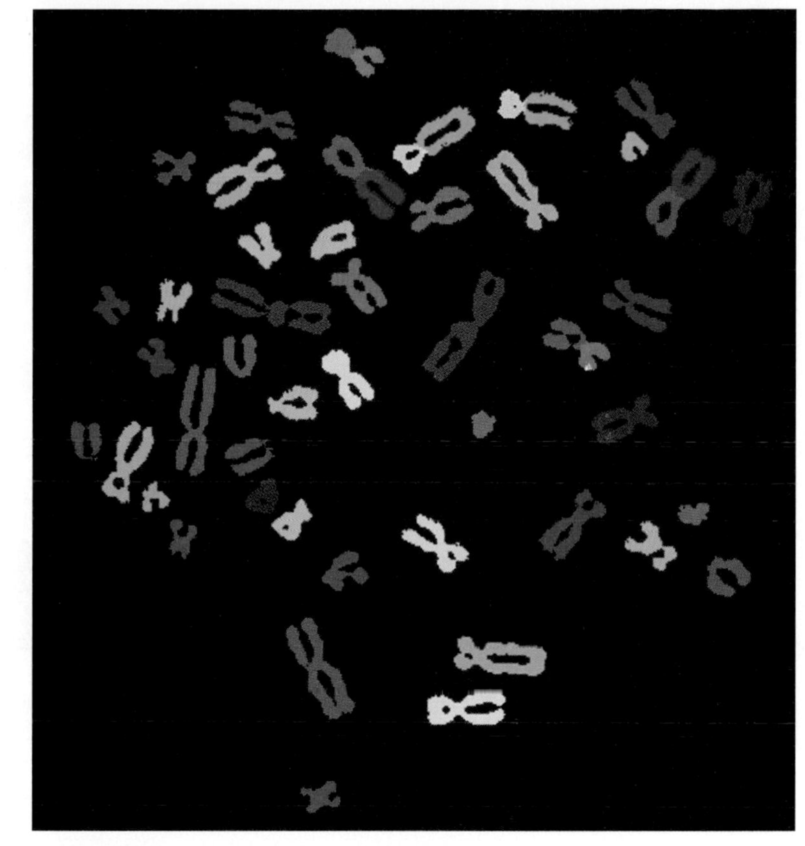

Figure 3.1 Chromosome complement of a human male. There are 46 chromosomes, present in 23 pairs. At the stage of the division cycle in which these chromosomes were observed, each chromosome consists of two identical halves lying side by side longitudinally. Except for the members of one chromosome pair (the pair that determines sex), the members of all of the chromosome pairs are the same color because they contain DNA molecules that were labeled with the same mixture of fluorescent dyes. The colors differ from one pair to the next because the dye mixtures differ in color. In some cases, the long and the short arm have been labeled with a different color. [Courtesy of David C. Ward and Michael R. Speicher.]

3.2 The daughter cells of mitosis have identical chromosomes.

Mitosis is a precise process of nuclear division that ensures that each of two daughter cells receives a diploid complement of chromosomes identical with the diploid complement of the parent cell. Mitosis is usually accompanied by **cytokinesis,** the process in which the cell itself divides to yield two daughter cells. The essential details of mitosis are the same in all organisms, and the basic process is remarkably uniform:

1. Each chromosome is already present as a duplicated structure at the beginning of nuclear division. (The duplication of each chromosome coincides with the replication of the DNA molecule contained within it.)
2. Each chromosome divides longitudinally into identical halves that become separated from each other.
3. The separated chromosome halves move in opposite directions, and each becomes included in one of the two daughter nuclei that are formed.

In a cell that is not undergoing mitosis, the chromosomes are not visible with a light microscope. This stage of the cell cycle is called **interphase.** In preparation for mitosis, the genetic material (DNA) in the chromosomes is replicated during a period of late interphase called **S** (Figure 3.2). (The S stands for *synthesis* of DNA.) DNA replication is accompanied by chromosome duplication. Before and after S, there are periods, called $\mathbf{G_1}$ and $\mathbf{G_2}$, respectively, in which DNA replication does not take place. The **cell cycle,** or the life cycle of a cell, is commonly described in terms of these three interphase periods followed by mitosis, **M.** The order of events is therefore $G_1 \rightarrow S \rightarrow G_2 \rightarrow M$, as shown in Figure 3.2. In this representation, the M period includes cytokinesis, which is the division of the cytoplasm into two approximately equal parts, each containing one daughter nucleus. The length of time required for a complete life cycle varies with cell type. In higher eukaryotes, the majority of cells require from 18 to 24 hours. The relative duration of the different periods in the cycle also varies considerably with cell type. Mitosis is usually the shortest period, requiring from 1/2 hour to 2 hours.

The cell cycle itself is under genetic control. The mechanisms of control appear to be essentially identical in all eukaryotes. There are two critical transitions: from G_1 into S and from G_2 into M

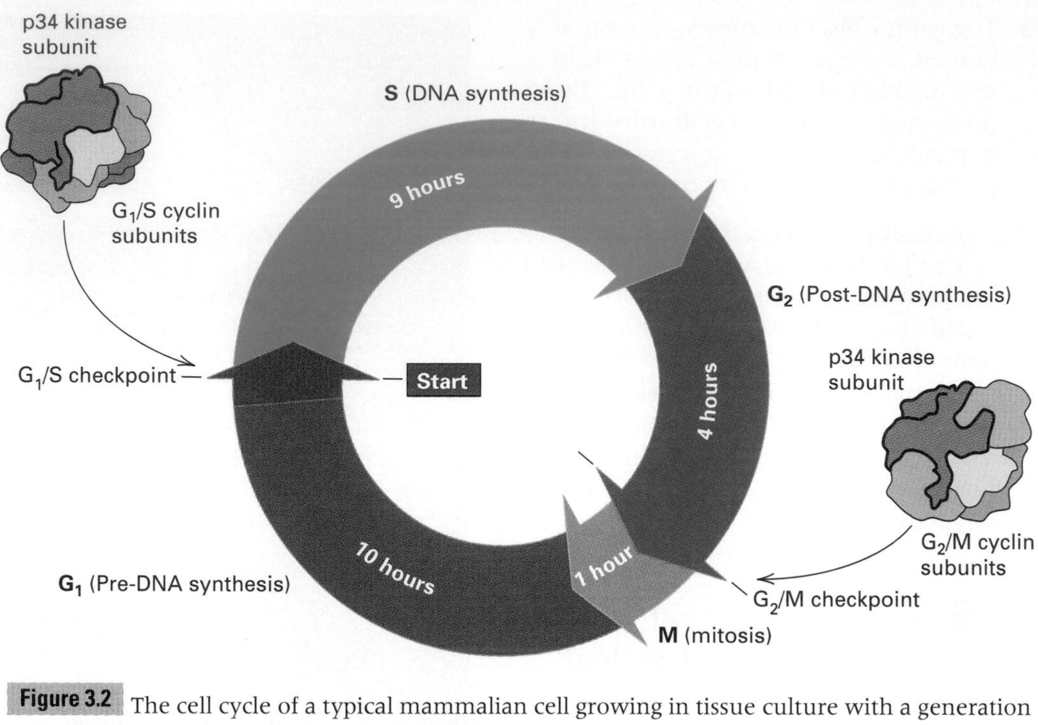

Figure 3.2 The cell cycle of a typical mammalian cell growing in tissue culture with a generation time of 24 hours. The critical control points for the G_1/S and G_2/M transitions are governed by a p34 kinase that is activated by stage-specific cyclins and that regulates the activity of its target proteins through phosphorylation.

(Figure 3.2). The G_1/S and G_2/M transitions are called "checkpoints" because the transitions are delayed unless key processes have been completed. For example, the G_1/S checkpoint requires either that sufficient time should have elapsed since the preceding mitosis (in some cell types) or that the cell should have attained sufficient size (in other cell types) for DNA replication to be initiated. Similarly, the G_2/M checkpoint requires that DNA replication and repair of any DNA damage be completed for the M phase to commence. Both major control points are regulated in a similar manner and make use of a specialized protein kinase (called the p34 kinase subunit in Figure 3.2) that regulates the activity of target proteins by phosphorylation (transfer of phosphate groups). At the G_1/S control point, one set of protein subunits combines with the p34 polypeptide to yield the active kinase that triggers DNA replication and other events of the S period. The protein subunits are called **cyclins** because their abundance and activity cycles in synchrony with the cell cycle. Similarly, at the G_2/M control point, a second set of cyclins combines with the p34 polypeptide to yield the active kinase that initiates condensation of the chromosomes, breakdown of the nuclear envelope, and reorganization of the cytoskeleton in preparation for cytokinesis.

The essential features of chromosome behavior in mitosis are illustrated in Figure 3.3. Mitosis is conventionally divided into four stages: **prophase, metaphase, anaphase,** and **telophase.** The stages have the following characteristics:

1. Prophase In interphase, the chromosomes have the form of extended filaments and cannot be seen as discrete bodies with a light microscope. Except for the presence of one or more conspicuous

Figure 3.3 Diagram of mitosis in an organism with two pairs of chromosomes. At each stage, the smaller circle represents the entire cell, and the larger circle is an exploded view showing the chromosomes at that stage. Interphase is usually not considered part of mitosis proper; it is typically much longer than the rest of the cell cycle, and the chromosomes are not yet visible. In early prophase, the chromosomes first become visible as fine strands, and the nuclear envelope and one or more nucleoli are intact. As prophase progresses, the chromosomes condense, and each chromosome can be seen to consist of two sister chromatids; the nuclear envelope and nucleoli disappear. In metaphase, the chromosomes are highly condensed and aligned on the central plane of the spindle, which forms at the end of prophase. In anaphase, the centromeres split longitudinally, and the sister chromatids of each chromosome move to opposite poles of the spindle. In telophase, the separation of sister chromatids is complete, the spindle breaks down, new nuclear envelopes are formed around each group of chromosomes, the condensation process of prophase is reversed, and the cell cycles back into interphase.

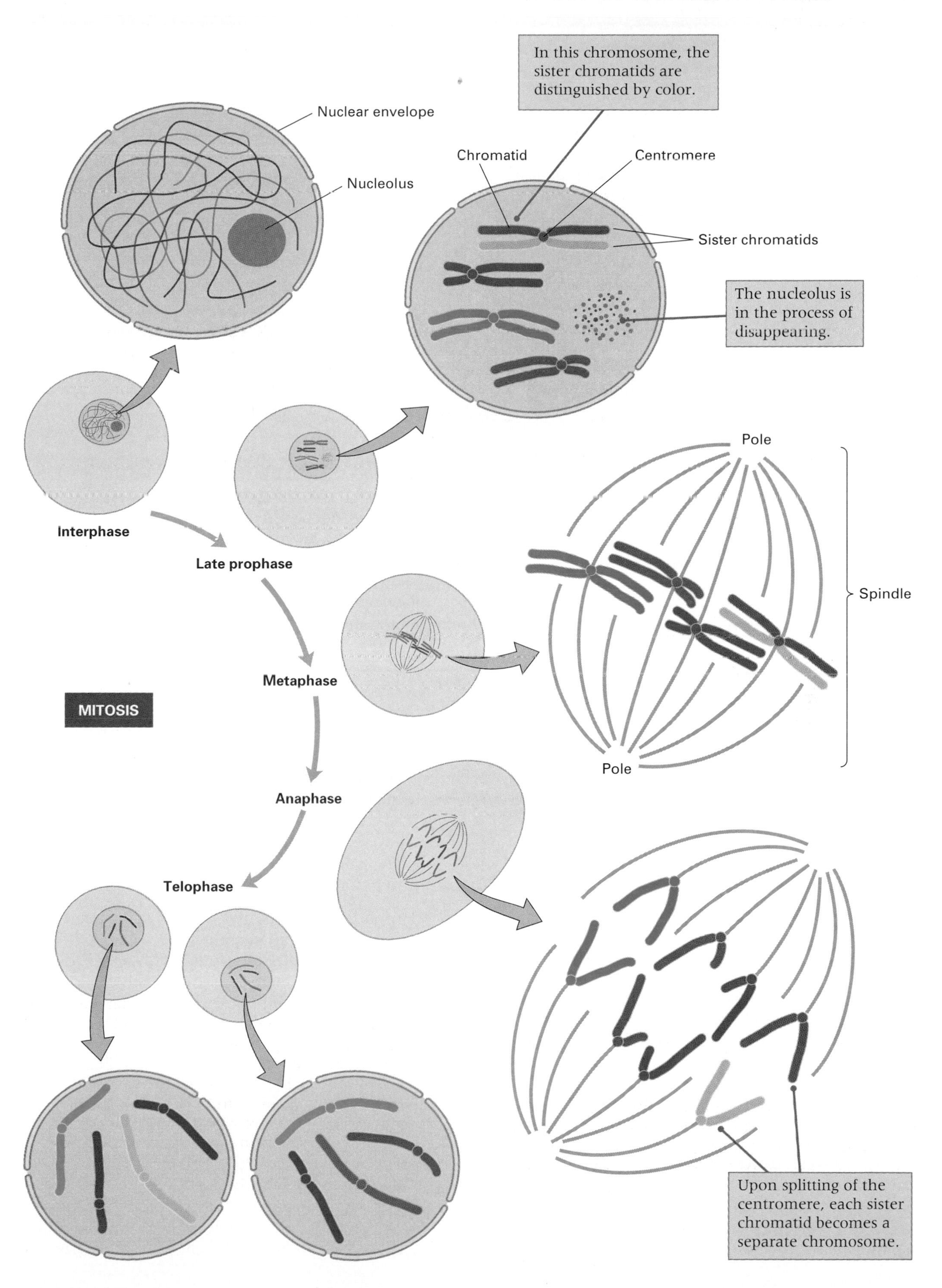

In this chromosome, the sister chromatids are distinguished by color.
Nuclear envelope
Nucleolus
Chromatid
Centromere
Sister chromatids
The nucleolus is in the process of disappearing.
Interphase
Late prophase
Pole
Spindle
Metaphase
MITOSIS
Pole
Anaphase
Telophase
Upon splitting of the centromere, each sister chromatid becomes a separate chromosome.

dark bodies, each called a **nucleolus,** the nucleus has a diffuse, granular appearance. The beginning of prophase is marked by the condensation of chromosomes to form visibly distinct, thin threads within the nucleus. Each chromosome is already longitudinally double, consisting of two closely associated subunits called **chromatids.** The longitudinally bipartite nature of each chromosome is readily seen later in prophase. Each pair of chromatids is the product of the duplication of one chromosome in the S period of interphase. The chromatids in a pair are held together at a specific region of the chromosome called the **centromere.** As prophase progresses, the chromosomes become shorter and thicker, as a result of further coiling. At the end of prophase, the nucleoli disappear and the nuclear envelope, a membrane surrounding the nucleus, abruptly disintegrates.

2. Metaphase At the beginning of metaphase, the mitotic **spindle** forms. The spindle is a bipolar structure consisting of fiber-like bundles of microtubules that extend through the cell between the poles of the spindle. Each chromosome becomes attached to several spindle fibers in the region of the centromere. The structure associated with the centromere to which the spindle fibers attach is technically known as the **kinetochore.** After the chromosomes are attached to spindle fibers, they move toward the center of the cell until all the kinetochores lie on an imaginary plane equidistant from the spindle poles. This imaginary plane is called the **metaphase plate.** Aligned on the metaphase plate, the chromosomes reach their maximum contraction and are easiest to count and examine for differences in morphology. Proper chromosome alignment is an important checkpoint for controlling the cell cycle at metaphase in both mitosis and meiosis; when all of the kinetochores are aligned on the metaphase plate, the metaphase checkpoint is passed and the cell continues the process of division.

3. Anaphase In anaphase, the centromeres divide longitudinally, and the two **sister chromatids** of each chromosome move toward opposite poles of the spindle. Once the centromeres divide, each sister chromatid is regarded as a separate chromosome in its own right. Chromosome movement results in part from progressive shortening of the spindle fibers attached to the centromeres, which pulls the chromosomes in opposite directions toward the poles. At the completion of anaphase, the chromosomes lie in two groups near opposite poles of the spindle. Each group contains the same number of chromosomes that was present in the original interphase nucleus.

4. Telophase In telophase, a nuclear envelope forms around each compact group of chromosomes, nucleoli are formed, and the spindle disappears. The chromosomes undergo a reversal of condensation until they are no longer visible as discrete entities. The two daughter nuclei slowly assume a typical interphase appearance as the cytoplasm of the cell divides in two by means of a gradually deepening furrow around the periphery. (In plants, a new cell wall is synthesized between the daughter cells and separates them.)

3.3 Meiosis results in gametes that differ genetically.

Meiosis is a mode of cell division in which cells are created that contain only one member of each pair of chromosomes present in the premeiotic cell. When a diploid cell with two sets of chromosomes undergoes meiosis, the result is four daughter cells, each genetically different and each containing one haploid set of chromosomes.

Meiosis consists of two successive nuclear divisions. The essentials of chromosome behavior during meiosis are outlined in Figure 3.4. This outline affords an overview of meiosis as well as an introduction to the process as it takes place in a cellular context.

1. Prior to the first nuclear division, the members of each pair of chromosomes become closely associated along their length (part A). The chromosomes that pair with each other are said to be **homologous** chromosomes. Because each member of a pair of homologous chromosomes is already replicated, it consists of a duplex of two sister chromatids joined at the centromere. The pairing of the homologous chromosomes therefore produces a four-stranded structure.
2. In the first nuclear division, the homologous chromosomes are separated from each other, one member of each pair going to opposite poles of the spindle (part B). Two nuclei are formed, each containing a haploid set of duplex chromosomes (part C).
3. The second nuclear division loosely resembles a mitotic division, *but there is no chromosome replication.* At metaphase, the chromosomes align on the metaphase plate, and at anaphase, the chromatids of each chromosome are separated into opposite daughter nuclei (part D). The net effect of the two divisions in meiosis is the creation of four haploid daughter nuclei, each containing the equivalent of a single sister chromatid from each pair of homologous chromosomes (part E).

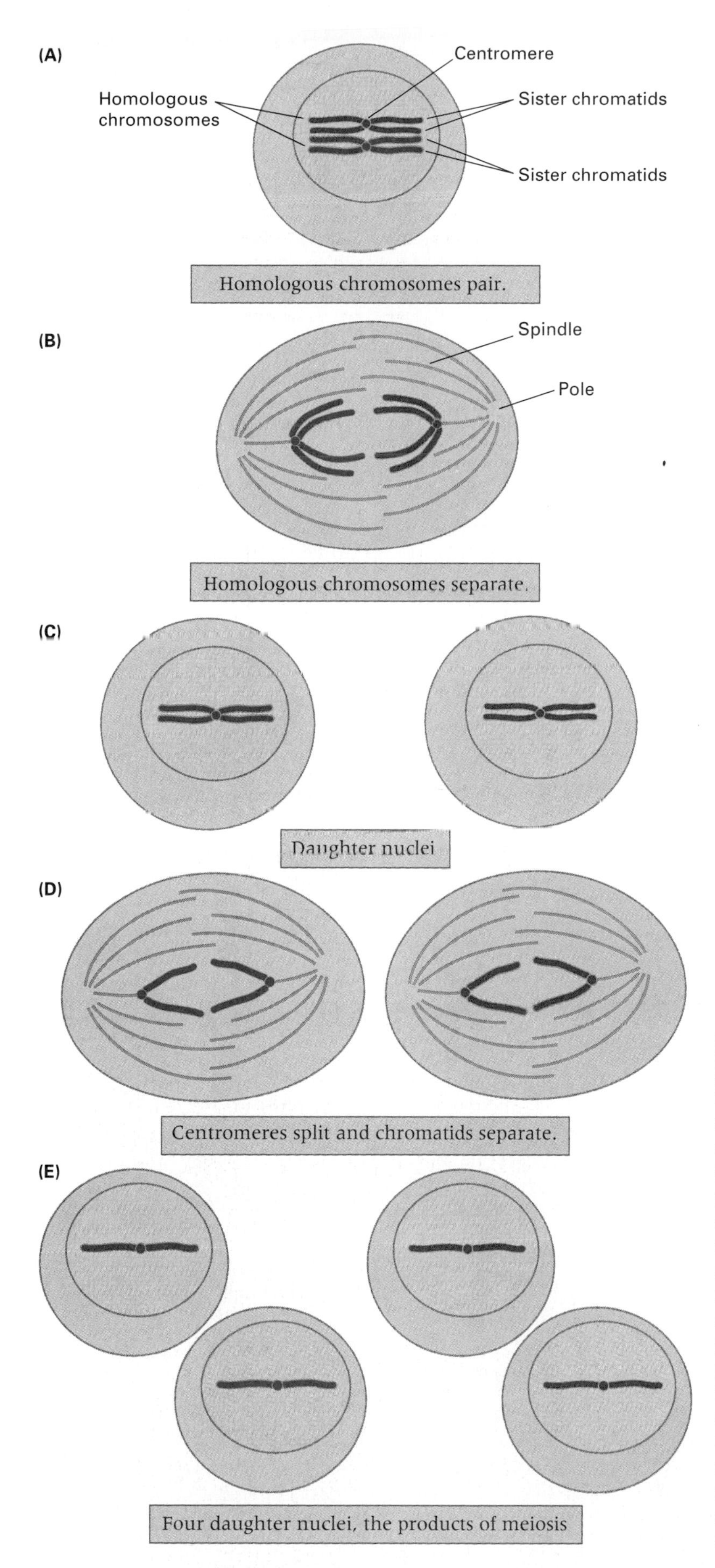

Figure 3.4 Overview of the behavior of a single pair of homologous chromosomes in meiosis. (A) The homologous chromosomes form a pair by coming together; each chromosome already consists of two chromatids joined at a single centromere. (B) The members of each homologous pair separate. (C) At the end of the first meiotic division, each daughter nucleus carries one or the other of the homologous chromosomes. (D) In the second meiotic division, in each of the daughter nuclei formed in meiosis I, the sister chromatids separate. (E) The end result is four products of meiosis, each containing one of each pair of homologous chromosomes. For clarity, this diagram does not incorporate crossing-over, an interchange of chromosome segments that takes place at the stage depicted in part A. If crossing-over were included, each chromatid would consist of one or more segments of red and one of more segments of blue. (Crossing-over is depicted in Figure 3.7.)

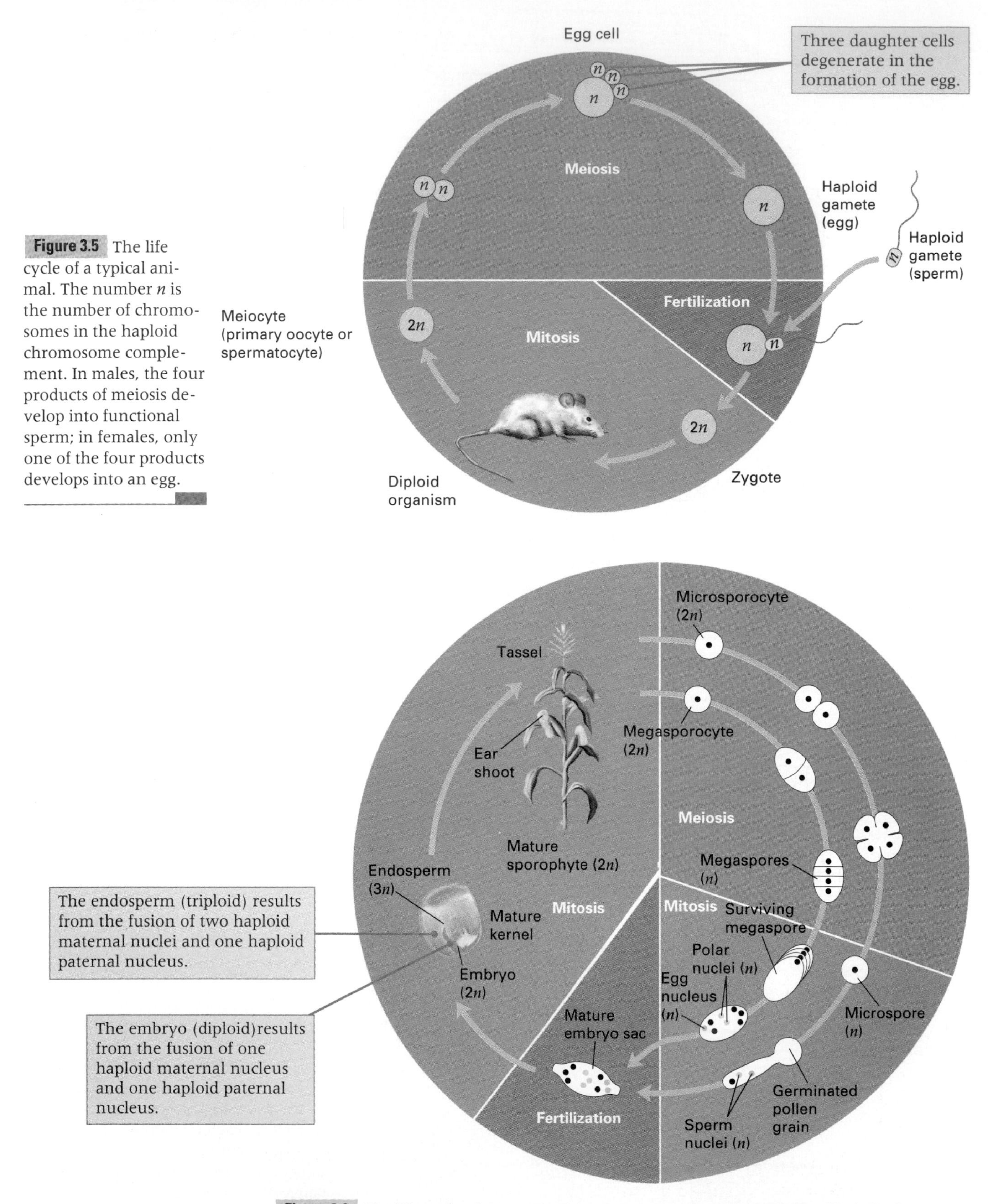

Figure 3.5 The life cycle of a typical animal. The number n is the number of chromosomes in the haploid chromosome complement. In males, the four products of meiosis develop into functional sperm; in females, only one of the four products develops into an egg.

Figure 3.6 The life cycle of corn, *Zea mays*. As is typical in higher plants, the diploid spore-producing (sporophyte) generation is conspicuous, whereas the gamete-producing (gametophyte) generation is microscopic. The egg-producing spore is the *megaspore* and the sperm-producing spore is the *microspore*. Nuclei participating in meiosis and fertilization are shown in yellow and green.

Figure 3.4 does not show that at the time of chromosome pairing, the homologous chromosomes can exchange genes. The exchanges result in the formation of chromosomes that consist of segments from one homologous chromosome intermixed with segments from the other. In Figure 3.4, the exchanged chromosomes would be depicted as segments of alternating color. The exchange process is one of the critical feature of meiosis, and it will be examined in the next section.

In animals, meiosis takes place in specific cells called **meiocytes,** a general term for the primary oocytes and spermatocytes in the gamete-forming tissues (Figure 3.5). The oocytes form egg cells and the spermatocytes form sperm cells. Although the process of meiosis is similar in all sexually reproducing organisms, in the female of both animals and plants, only one of the four products develops into a functional cell (the other three disintegrate). In animals, the products of meiosis form gametes (sperm or eggs).

In plants, the situation is slightly more complicated:

1. The products of meiosis typically form **spores,** which undergo one or more mitotic divisions to produce a haploid **gametophyte** organism. The gametophyte produces gametes by mitotic division of a haploid nucleus (Figure 3.6).
2. Fusion of haploid gametes creates a diploid zygote that develops into the **sporophyte** plant, which undergoes meiosis to produce spores and so restarts the cycle.

Meiosis is a more complex and considerably longer process than mitosis and usually requires days or even weeks. The entire process of meiosis is illustrated in its cellular context in Figure 3.7. The essence is that *meiosis consists of two divisions of the nucleus but only one duplication of the chromosomes.* The nuclear divisions—called the **first meiotic division** and the **second meiotic division**—can be separated into a sequence of stages similar to those used to describe mitosis. The distinctive events of this important process occur during the first division of the nucleus; these events are described in the following section.

The first meiotic division reduces the chromosome number by half.

The first meiotic division (meiosis I) is sometimes called the **reductional division** because it divides the chromosome number in half. By analogy with mitosis, the first meiotic division can be split into the four stages of **prophase I, metaphase I, anaphase I,** and **telophase I.** These stages are generally more complex than their counterparts in mitosis. The stages and substages can be visualized with reference to Figures 3.7 and 3.8.

1. Prophase I This long stage lasts several days in most higher organisms and is commonly divided into five substages: *leptotene, zygotene, pachytene, diplotene,* and *diakinesis.* These are descriptive terms that indicate the appearance of the chromosomes at each substage.

In **leptotene,** which literally means "thin thread," the chromosomes first become visible as long, thread-like structures. The pairs of sister chromatids can be distinguished by electron microscopy. In this initial phase of condensation of the chromosomes, numerous dense granules appear at irregular intervals along their length. These localized contractions, called **chromomeres,** have a characteristic number, size, and position in a given chromosome (Figure 3.8A).

The **zygotene** ("paired thread") period is marked by the lateral pairing, or **synapsis,** of homologous chromosomes, beginning at the chromosome tips. As the pairing process proceeds along the length of the chromosomes, it results in a precise chromomere-by-chromomere association (Figure 3.8B and F). Each pair of synapsed homologous chromosomes is referred to as a **bivalent.**

During **pachytene** (Figure 3.8C and D), condensation of the chromosomes continues. The term literally means "thick thread," and throughout this period, the chromosomes continue to shorten and thicken (Figure 3.6). By late pachytene, it can sometimes be seen that each bivalent (that is, each set of paired chromosomes) actually consists of a **tetrad** of four chromatids, but the two sister chromatids of each chromosome are usually juxtaposed very tightly. The important event of genetic exchange, which is called **crossing-over,** takes place during pachytene, but crossing-over does not become apparent until the transition to diplotene. In Figure 3.7, the sites of exchange are indicated by the points where chromatids of different color cross over each other.

At the onset of **diplotene** ("double thread"), the synapsed chromosomes begin to separate, and the diplotene chromosomes are clearly double (Figure 3.8E). However, the homologous chromosomes remain held together at intervals along their length by cross-connections resulting from crossing-over. Each cross-connection, called a **chiasma** (plural, *chiasmata*), is formed by a breakage and rejoining between nonsister chromatids. As shown in the chromosome and diagram in Figure 3.9, *a chiasma results from physical exchange between chromatids of homologous chromosomes.* In

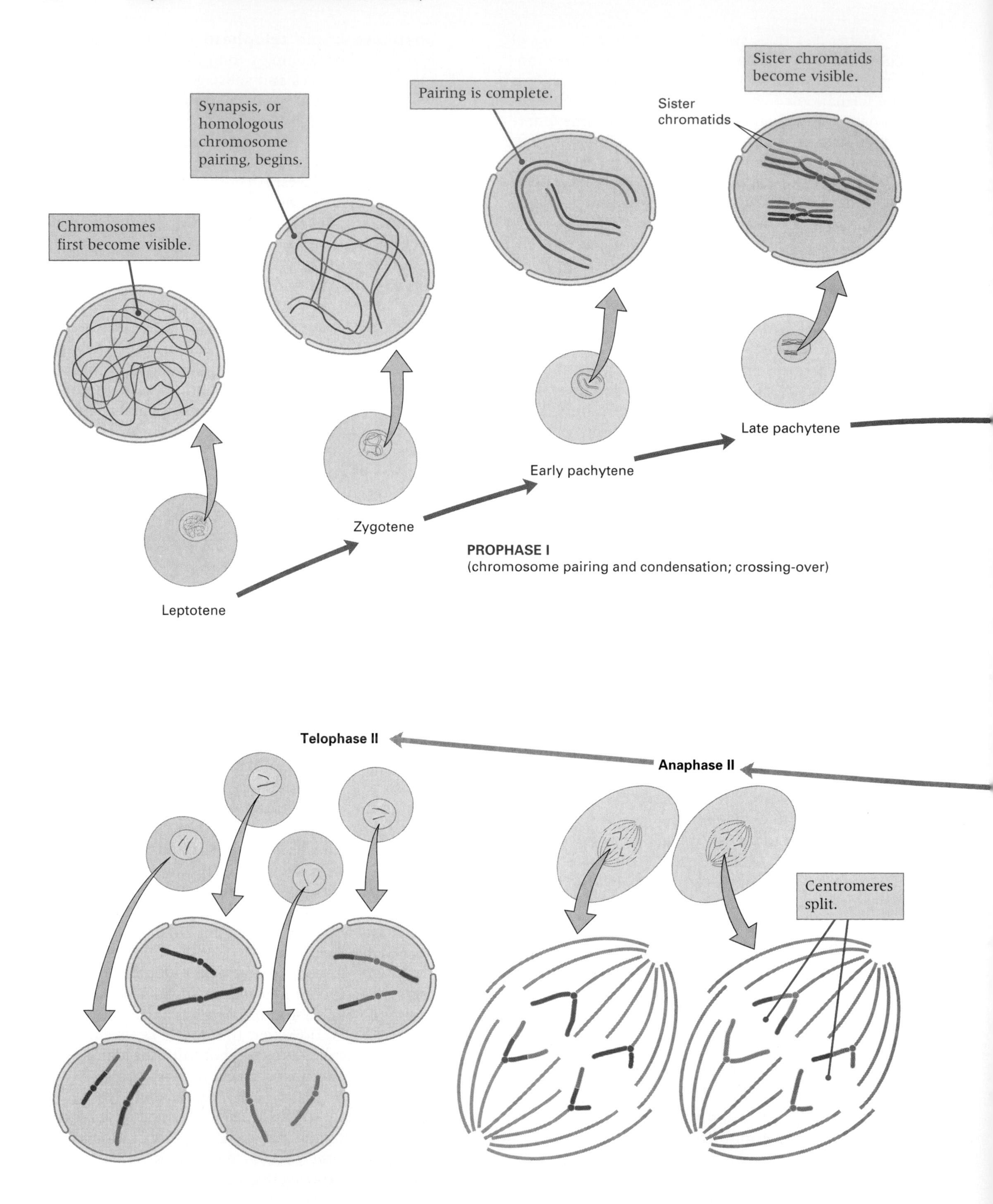
Chromosomes first become visible.
Synapsis, or homologous chromosome pairing, begins.
Pairing is complete.
Sister chromatids become visible.
Sister chromatids
Leptotene
Zygotene
Early pachytene
Late pachytene
PROPHASE I
(chromosome pairing and condensation; crossing-over)
Telophase II
Anaphase II
Centromeres split.

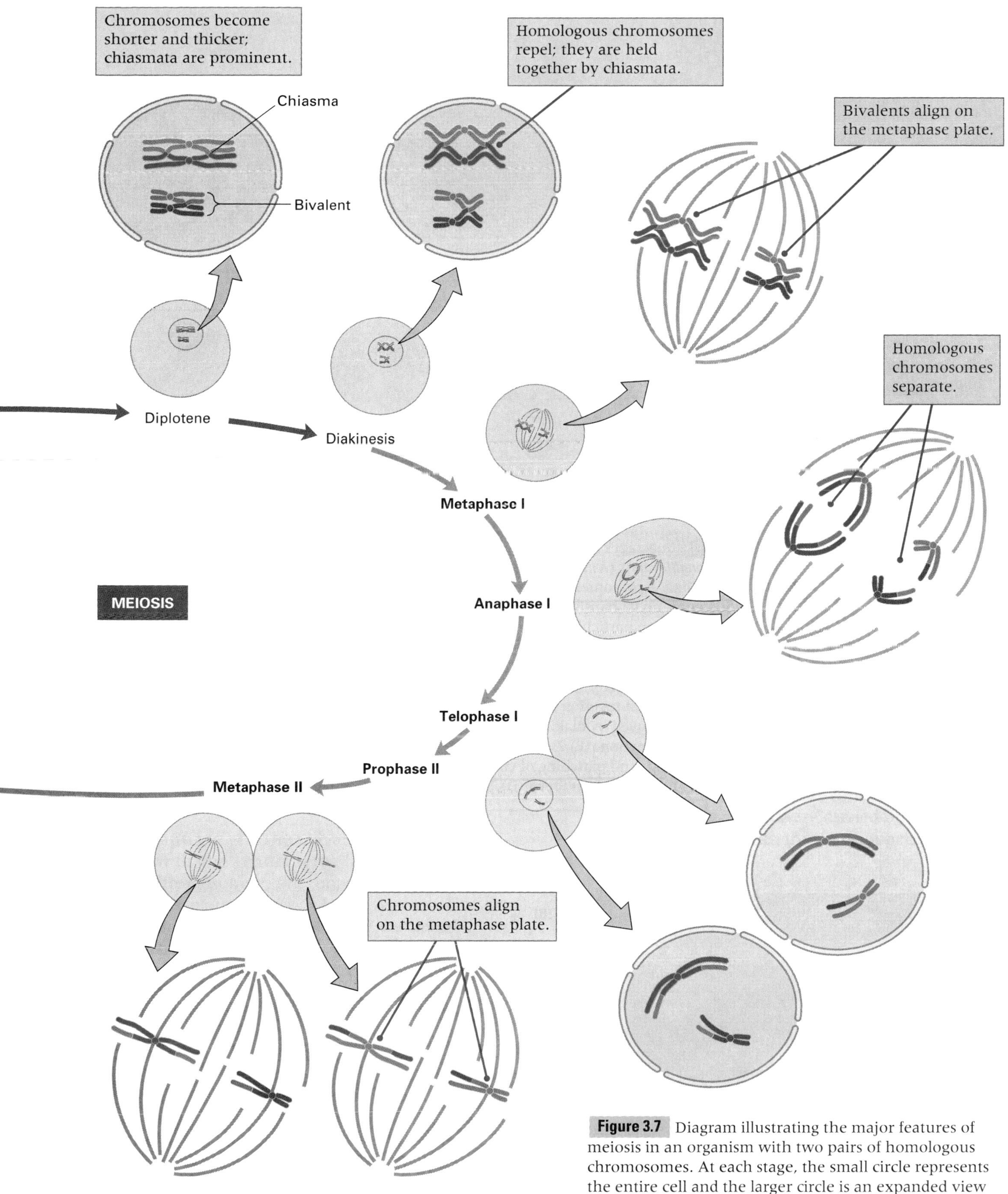

Figure 3.7 Diagram illustrating the major features of meiosis in an organism with two pairs of homologous chromosomes. At each stage, the small circle represents the entire cell and the larger circle is an expanded view of the chromosomes at that stage.

recessive phenotype. This is the "expected" Mendelian ratio. In this case, $n = 8$, $s = 6$, $t = 2$, and the probability of this combination of events is

$$\frac{8!}{6!2!}p^6q^2 = \frac{6! \times 7 \times 8}{6! \times 2!}(3/4)^6(1/4)^2 = 0.31$$

That is, in only 31 percent of the families with eight children would the offspring exhibit the expected 3 : 1 phenotypic ratio; the other sibships would deviate in one direction or the other because of chance variation. The importance of this example is in demonstrating that although a 3 : 1 ratio is the "expected" outcome (and also the single most probable outcome), the majority of the families (69 percent) actually have a distribution of offspring different from 3 : 1.

Chi-square tests goodness of fit of observed to expected numbers.

Geneticists often need to decide whether an observed ratio is in satisfactory agreement with a theoretical prediction. Mere inspection of the data is unsatisfactory because different investigators may disagree. Suppose, for example, that we crossed a plant having purple flowers with a plant having white flowers and, among the progeny, observed 14 plants with purple flowers and 6 with white flowers. Is this result close enough to be accepted as a 1 : 1 ratio? What if we observed 15 plants with purple flowers and 5 with white flowers? Is this result consistent with a 1 : 1 ratio? There is bound to be statistical variation in the observed results from one experiment to the next. Who is to say what results are consistent with a particular genetic hypothesis? In this section, we describe a test of whether observed results deviate too far from a theoretical expectation. The test is called a test for **goodness of fit,** where the word *fit* means how closely the observed results "fit", or agree with, the expected results.

A conventional measure of goodness of fit is a value called **chi-square** (symbol, χ^2), which is calculated from the number of progeny observed in each of various classes, compared with the number expected in each of the classes on the basis of some genetic hypothesis. For example, in a cross between plants with purple flowers and those with white flowers, we may be interested in testing the hypothesis that the parent with purple flowers is heterozygous for a pair of alleles determining flower color and that the parent with white flowers is homozygous recessive. Suppose further that we examine 20 progeny plants from the mating and find that 14 are purple and 6 are white. The procedure to be followed in testing this genetic hypothesis (or any other genetic hypothesis) by means of the chi-square method is as follows:

1. *State the genetic hypothesis in detail, specifying the genotypes and phenotypes of the parents and the possible progeny.* In the example using flower color, the genetic hypothesis implies that the genotypes in the cross purple × white could be symbolized as *Pp* × *pp*. The possible progeny genotypes are either *Pp* or *pp*.

2. *Use the rules of probability to make explicit predictions of the types and proportions of progeny that should be observed if the genetic hypothesis is true. Convert the proportions to numbers of progeny (percentages are not allowed in a χ^2 test).* If the hypothesis about the flower-color cross is true, then we expect the progeny genotypes *Pp* and *pp* in a ratio of 1 : 1. Because the hypothesis is that *Pp* flowers are purple and *pp* flowers are white, we expect the phenotypes of the progeny to be purple or white in the ratio 1 : 1. Among 20 progeny, the expected numbers are 10 purple and 10 white.

3. *For each class of progeny in turn, subtract the expected number from the observed number. Square this difference and divide the result by the expected number.* In our example, the calculation for the purple progeny is $(14 - 10)^2/10 = 1.6$, and that for the white progeny is $(6 - 10)^2/10 = 1.6$.

4. *Sum the result of the numbers calculated in step 3 for all classes of progeny. The summation is the value of χ^2 for these data.* The sum for the purple and white classes of progeny is $1.6 + 1.6 = 3.2$, and this is the value of χ^2 for the experiment, calculated on the assumption that our genetic hypothesis is correct.

In symbols, the calculation of χ^2 can be represented by the expression

$$\chi^2 = \sum\frac{(Observed - Expected)^2}{Expected}$$

in which Σ means the summation over all the classes of progeny. Note that χ^2 is calculated using the observed and expected *numbers,* not the proportions, ratios, or percentages. Using something other than the actual numbers is the beginner's most common mistake in applying the χ^2 method. The χ^2 value is reasonable as a measure of goodness of fit, because the closer the observed numbers are to the expected numbers, the smaller the value of χ^2. A value of $\chi^2 = 0$ means that the observed numbers fit the expected numbers perfectly.

Table 3.3
Calculation of χ^2 for a monohybrid ratio

Phenotype (class)	Observed number	Expected number	Deviation from expected	$\frac{(\text{Deviation})^2}{\text{expected number}}$
Wildtype	99	108	−9	0.75
Mutant	45	36	+9	2.25
Total	144	144		$\chi^2 = 3.00$

As another example of the calculation of χ^2, suppose that the progeny of an $F_1 \times F_1$ cross include two contrasting phenotypes observed in the numbers 99 and 45. The genetic hypothesis might be that the trait is determined by a pair of alleles of a single gene, in which case the expected ratio of dominant : recessive phenotypes among the F_2 progeny is 3 : 1. Considering the data, the question is whether the observed ratio of 99 : 45 is in satisfactory agreement with the expected 3 : 1. Calculation of the value of χ^2 is illustrated in Table 3.3. The total number of progeny is 99 + 45 = 144. The *expected* numbers in the two classes, on the basis of the genetic hypothesis that the true ratio is 3 : 1, are calculated as $(3/4) \times 144 = 108$ and $(1/4) \times 144 = 36$. Because there are two classes of data, there are two terms in the χ^2:

$$\chi^2 = \frac{(99 - 108)^2}{108} + \frac{(45 - 36)^2}{36}$$
$$= 0.75 + 2.25 = 3.00$$

Once the χ^2 value has been calculated, the next step is to interpret whether this value represents a good fit or a bad fit to the expected numbers. This assessment is done with the aid of the graphs in Figure 3.34. The *x*-axis gives the χ^2 values measuring goodness of fit, and the *y*-axis gives the probability *P* that a worse fit (or one equally bad) would be obtained by chance, assuming that the genetic hypothesis is true. If the genetic hypothesis is true, then the observed numbers should be reasonably close to the expected numbers. Suppose that the observed χ^2 is so large that the probability of a fit as bad or worse is very small. Then the observed results do *not* fit the theoretical expectations. This means that the genetic hypothesis used to calculate the expected numbers of progeny must be rejected, because the observed numbers of progeny deviate too much from the expected numbers.

In practice, the critical values of *P* are conventionally chosen as 0.05 (the 5 percent level) and 0.01 (the 1 percent level). For *P* values ranging from 0.01 to 0.05, the probability that chance alone would lead to a fit as bad or worse is between 1 in 20 experiments and 1 in 100. This is the purple region in Figure 3.34; if the *P* value falls in this range, the correctness of the genetic hypothesis is considered very doubtful. The result is said to be **statistically significant** at the 5 percent level. For *P* values smaller than 0.01, the probability that chance alone would lead to a fit as bad or worse is less than 1 in 100 experiments. This is the green region in Figure 3.34; in this case, the result is said to be **statistically highly significant** at the 1 percent level, and the genetic hypothesis is rejected outright. If the terminology of statistical significance seems backwards, it is because the term *significant* refers to the magnitude of the difference between the observed and the expected numbers; in a result that is statistically significant, there is a large ("significant") difference between what is observed and what is expected.

To use Figure 3.34 to determine the *P* value corresponding to a calculated χ^2, we need the number of **degrees of freedom** of the particular χ^2 test. For the type of χ^2 test illustrated in Table 3.3, the number of degrees of freedom equals the number of classes of data minus 1. Table 3.3 contains two classes of data (wildtype and mutant), so the number of degrees of freedom is 2 − 1 = 1. The reason for subtracting 1 is that, in calculating the expected numbers of progeny, we make sure that the total number of progeny is the same as that actually observed. For this reason, one of the classes of data is not really "free" to contain any number we might specify; because the expected number in one class must be adjusted to make the total come out correctly, one "degree of freedom" is lost. Analogous χ^2 tests with three classes of data have 2 degrees of freedom, and those with four classes of data have 3 degrees of freedom.

Once we have determined the appropriate number of degrees of freedom, we can interpret the χ^2 value in Table 3.3. Refer to Figure 3.34, and observe that each curve is labeled with its degrees of freedom. To determine the *P* value for the data in Table 3.3, in which the χ^2 value is 3.00, first find the location of $\chi^2 = 3.00$ along the *x*-axis in Figure 3.34. Trace vertically from 3.00 until you intersect the curve with 1 degree of freedom. Then trace horizontally to the left until you intersect the *y*-axis and read the *P* value—in this case, $P = 0.08$. This means that chance alone would produce a χ^2 value as great as or greater than 3 in about 8 percent of

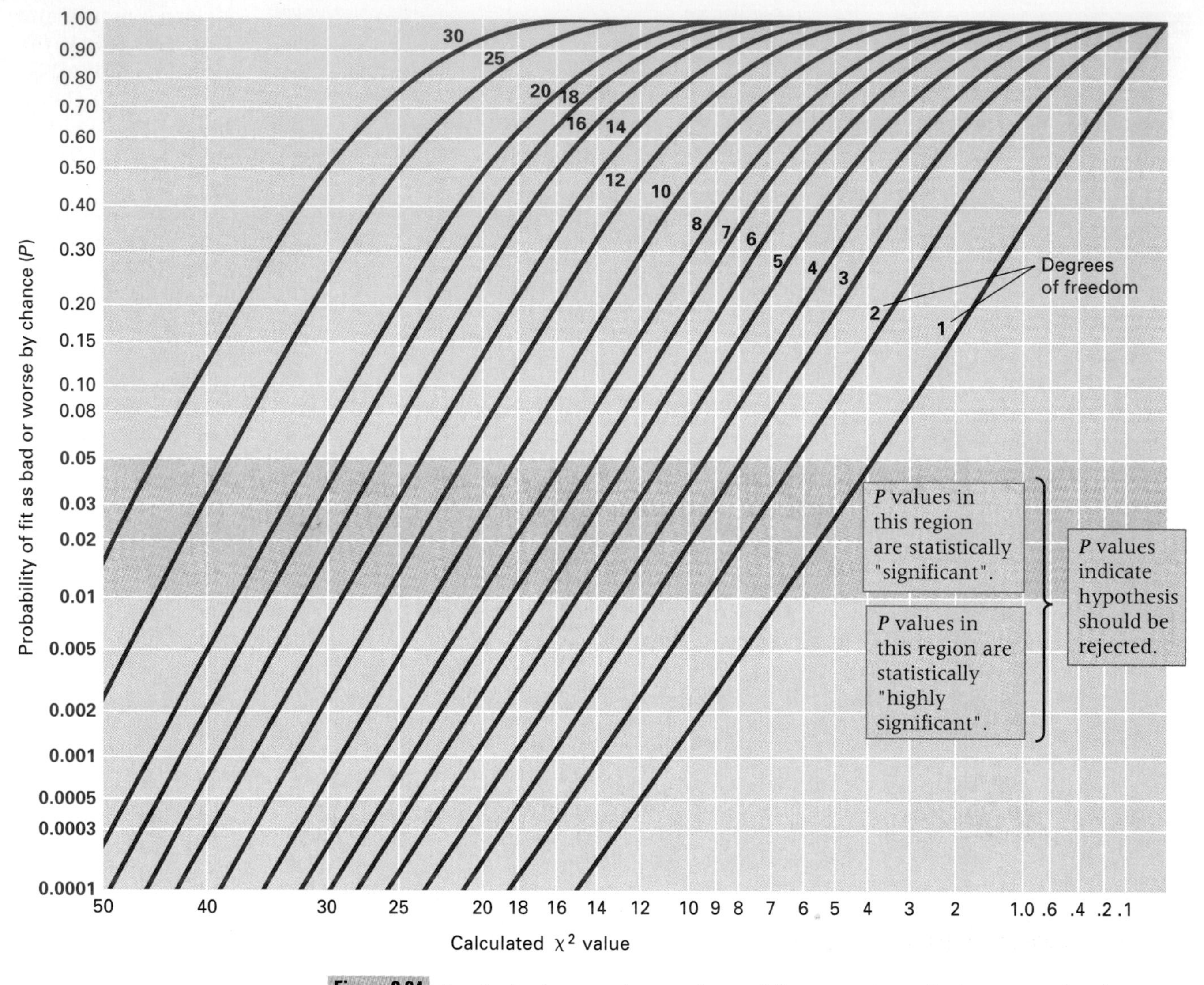

Figure 3.34 Graphs for interpreting goodness of fit to genetic predictions using the chi-square test. For any calculated value of χ^2 along the x-axis, the y-axis gives the probability P that chance alone would produce a result as bad as or worse than that actually observed, when the genetic predictions are correct. Tests with P in the purple region (less than 5 percent) or in the green region (less than 1 percent) are regarded as statistically significant and normally require rejection of the genetic hypothesis leading to the prediction. Each χ^2 test has a number of degrees of freedom associated with it. In the types of tests illustrated in this chapter, the number of degrees of freedom equals the number of classes in the data minus 1.

experiments of the type in Table 3.3; and because the P value is in the blue region, the goodness of fit to the hypothesis of a 3 : 1 ratio of wildtype : mutant is judged to be satisfactory.

As a second illustration of the χ^2 test, we will determine the goodness of fit of Mendel's round versus wrinkled data to the expected 3 : 1 ratio. Among the 7324 seeds that he observed, 5474 were round and 1850 were wrinkled. The expected numbers are $(3/4) \times 7324 = 5493$ round and $(1/4) \times 7324 = 1831$ wrinkled. The χ^2 value is calculated as

$$\chi^2 = \frac{(5474 - 5493)^2}{5493} + \frac{(1850 - 1831)^2}{1831} = 0.26$$

The fact that the χ^2 is less than 1 already implies that the fit is very good. To find out how good, note that the number of degrees of freedom equals $2 - 1 = 1$ because there are two classes of data (round and wrinkled). From Figure 3.34, the P value for $\chi^2 = 0.26$ with 1 degree of freedom is approximately 0.65. This means that in about 65 percent of all experiments of this type, a fit as bad or worse would be expected simply because of chance; only about 35 percent of all experiments would yield a better fit.

Are Mendel's data a little too good?

Many of Mendel's experimental results are very close to the expected values. For the ratios listed in Table 2.1 in Chapter 2, the χ^2 values are 0.26 (round versus wrinkled seeds), 0.01 (yellow versus green seeds), 0.39 (purple versus white flowers), 0.06 (inflated versus constricted pods), 0.45 (green versus yellow pods), 0.35 (axial versus terminal flowers), and 0.61 (long versus short stems). (As an exercise in χ^2, you should confirm these calculations for yourself.) All of the χ^2 tests have P values of 0.45 or greater (Figure 3.34), which means that the reported results are in excellent agreement with the theoretical expectations.

The statistician Ronald Fisher pointed out in 1936 that Mendel's results are *suspiciously* close to the theoretical expectations. In a large number of experiments, some experiments can be expected to yield fits that appear doubtful simply because of chance variation from one experiment to the next. In Mendel's data, the doubtful values that are to be expected by chance appear to be missing. Figure 3.35 shows the observed deviations in Mendel's experiments compared with the deviations expected to occur by chance. (The measure of deviation is the square root of the χ^2, which is assigned either a plus or a minus sign according to whether the dominant or the recessive phenotypic class was in excess of the expected number.) For each magnitude of deviation, the height of the yellow bar gives the number of experiments that Mendel observed with that magnitude of deviation, and the orange bar gives the number of experiments expected to deviate by this amount as a result of chance alone. There are clearly too few experiments with deviations smaller than -1 or larger than $+1$. This type of discrepancy could be explained if Mendel discarded or repeated a few experiments with large deviations that made him suspect that the results were not to be trusted.

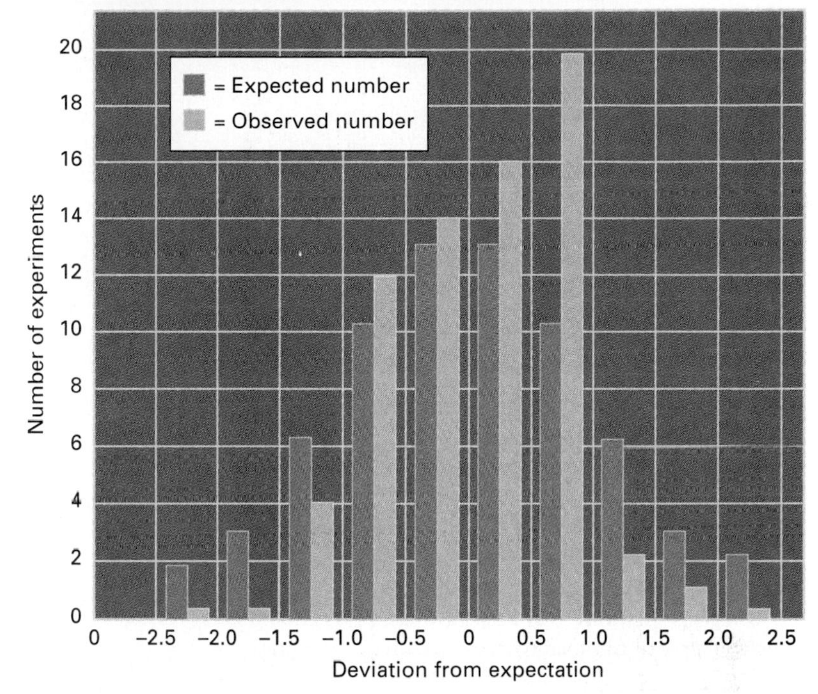

Figure 3.35 Distribution of deviations observed in 69 of Mendel's experiments (yellow bars) compared with expected values (orange bars). There is no suggestion that the data in the middle have been adjusted to improve the fit. However, several experiments with large deviations may have been discarded or repeated, because there are not so many experiments with large deviations as might be expected.

Did Mendel cheat? Did he deliberately falsify his data to make them appear better than they were? Mendel's paper reports extremely deviant ratios from individual plants, as well as experiments repeated a second time when the first results were doubtful. These are not the kinds of things that a dishonest person would admit. Only a small bias is necessary to explain the excessive goodness of fit in Figure 3.35. In a count of seeds or individual plants, only about 2 phenotypes per 1000 would need to be assigned to the wrong category to account for the bias in the 91 percent of the data related to the testing of monohybrid ratios. The excessive fit could also be explained if three or four entire experiments were discarded or repeated because deviant results were attributed to pollen contamination or other accident. After careful reexamination of Mendel's data in 1966, the evolutionary geneticist Sewall Wright concluded,

> Mendel was the first to count segregants at all. It is rather too much to expect that he would be aware of the precautions now known to be necessary for completely objective data. . . . Checking of counts that one does not like, but not of others, can lead to systematic bias toward agreement. I doubt whether there are many geneticists even now whose data, if extensive, would stand up wholly satisfactorily under the χ^2 test. . . . Taking everything into account, I am confident that there was no deliberate effort at falsification.

Mendel's data are some of the most extensive and complete "raw data" ever published in genetics. Additional examinations of the data will surely be carried out as new statistical approaches are developed. However, the point to be emphasized is that up to the present time, no reputable statistician has alleged that Mendel knowingly and deliberately adjusted his data in favor of the theoretical expectation.

Chapter Summary

- **Each species has a characteristic set of chromosomes.**

The chromosomes in somatic cells of higher plants and animals are present in pairs. The members of each pair are homologous chromosomes, and each member is a homolog. Pairs of homologs are usually identical in appearance, whereas nonhomologous chromosomes often show differences in size and structural detail that make them visibly distinct from each other. A cell whose nucleus contains two sets of homologous chromosomes is diploid. One set of chromosomes comes from the maternal parent and the other from the paternal parent. Gametes are haploid. A gamete contains only one set of chromosomes, consisting of one member of each pair of homologs.

- **The daughter cells of mitosis have identical chromosomes.**

Mitosis is the process of nuclear division that maintains the chromosome number when a somatic cell divides. Before mitosis, each chromosome replicates, forming a two-part structure that consists of two sister chromatids joined at the centromere (kinetochore). At the onset of mitosis, the chromosomes become visible and, at metaphase, become aligned on the metaphase plate perpendicular to the spindle. At anaphase, the centromere of each chromosome divides, and the sister chromatids are pulled by spindle fibers to opposite poles of the cell. The separated sets of chromosomes present in telophase nuclei are genetically identical.

- **Meiosis results in gametes that differ genetically.**

The first meiotic division reduces the chromosome number by half.
The second meiotic division is equational.

Meiosis is the type of nuclear division that takes place in germ cells, and it reduces the diploid number of chromosomes to the haploid number. The genetic material is replicated before the onset of meiosis, so each chromosome consists of two sister chromatids. The first meiotic division is the reduction division, which reduces the chromosome number by half. The homologous chromosomes first pair (synapsis) and then, at anaphase I, separate. The resulting products contain chromosomes that still consist of two chromatids attached to a common centromere. However, as a result of crossing-over, which takes place in prophase I, the chromatids may not be genetically identical along their entire length. In the second meiotic division, the centromeres divide and the homologous chromatids separate. The end result of meiosis is the formation of four genetically different haploid nuclei.

A distinctive feature of meiosis is the synapsis, or side-by-side pairing, of homologous chromosomes in the zygotene substage of prophase I. During the pachytene substage, the paired chromosomes become connected by chiasmata (the physical manifestations of crossing-over), and they do not separate until anaphase I. This separation is called disjunction (unjoining), and failure of chromosomes to separate is called nondisjunction. Nondisjunction results in a gamete that contains either two copies or no copies of a particular chromosome. Meiosis is the physical basis of the segregation and independent assortment of genes. In *Drosophila*, an unexpected pattern of inheritance of the *white* gene was shown to be accompanied by nondisjunction of the X chromosome; these observations gave experimental proof to the chromosome theory of heredity.

- **Eukaryotic chromosomes are highly coiled complexes of DNA and protein.**

Chromosome-sized DNA molecules can be separated by electrophoresis.
The nucleosome is the basic structural unit of chromatin.
Chromatin fibers are formed of coiled coil.
Heterochromatin is rich in satellite DNA and low in gene content.

In eukaryotes, the DNA is compacted into chromosomes, which contain several proteins and which are thick enough to be visible by light microscopy during the mitotic phase of the cell cycle. The DNA–protein complex of eukaryotic chromosomes is called chromatin. The protein component of chromatin consists primarily of five distinct proteins: histones H1, H2A, H2B, H3, and H4. The last four histones aggregate to form an octameric protein containing two molecules of each. DNA is wrapped around the histone octamer, forming a particle called a nucleosome. This wrapping is the first level of compaction of the DNA in chromosomes. Each nucleosome unit contains about 200 nucleotide pairs of which about 145 are in contact with the protein. The remaining 55 nucleotide pairs link adjacent nucleosomes. Histone H1 binds to the linker segment and draws the nucleosomes nearer to one another. The DNA in its nucleosome form is further compacted into a helical fiber, the 30-nm fiber. In forming a visible chromosome, this unit undergoes several additional levels of folding, producing a highly compact visible chromosome. The result is that a eukaryotic DNA molecule, whose length and width are about 50,000 and 0.002 μm, respectively, is folded to form a chromosome with a length of about 5 μm and a width of about 0.5 μm.

- **The centromere and telomeres are essential parts of chromosomes.**

The centromere is essential for chromosome segregation.
The telomere is essential for the stability of the chromosome tips.

Centromeres and telomeres are regions of eukaryotic chromosomes specialized for spindle-fiber attachment and stabilization of the tips, respectively. The centromeres of most higher eukaryotes are associated with localized, highly repeated, satellite DNA sequences. Telomeres are formed by a telomerase enzyme that contains a guide RNA that serves as a template for the addition of nucleotides to the 3' end of a telomerase addition site. The complementary strand is synthesized by the normal DNA replication enzymes. In mammals and other vertebrates, the 3' strand of the telomere terminates in tandem repeats of the simple sequence 5'-TTAGGG-3'. Relatively few copies of this sequence are needed to prime the telomerase.

- **Genes are located in chromosomes.**

  Special chromosomes determine sex in many organisms.
  X-linked genes are inherited according to sex.
  Experimental proof of the chromosome theory came from nondisjunction.
  Sex in *Drosophila* is determined by differential gene expression.

The X and Y sex chromosomes differ from other chromosome pairs in that they are visibly different and contain different genes. In mammals and in many insects and other animals, as well as in some flowering plants, the female contains two X chromosomes (XX) and hence is homogametic, and the male contains one X chromosome and one Y chromosome (XY) and hence is heterogametic. In birds, moths, butterflies, and some reptiles, the situation is the reverse: Females are the heterogametic sex (WZ) and males the homogametic sex (WW). The Y chromosome in many species contains only a few genes. In human beings and other mammals, the Y chromosome includes a male-determining factor. In *Drosophila,* sex is determined by a male-specific or female-specific pattern of gene expression that is regulated by the ratio of the number of X chromosomes to the number of sets of autosomes. In most organisms, the X chromosome contains many genes unrelated to sexual differentiation. These X-linked genes show a characteristic pattern of inheritance that is due to their location in the X chromosome.

- **Genetic data analysis makes use of probability and statistics.**

  Progeny of crosses are predicted by the binomial probability formula.
  Chi-square tests goodness of fit of observed to expected numbers.
  Are Mendel's data a little too good?

The progeny of genetic crosses often conform to the theoretical predictions of the binomial probability formula. The degree to which the observed numbers of different genetic classes of progeny fit theoretically expected numbers is usually ascertained with a chi-square (χ^2) test. On the basis of the criterion of the χ^2 test, Mendel's data fit the expectations somewhat more closely than chance would dictate. However, the bias in the data is relatively small and is unlikely to be due to anything more than his recounting or repeating certain experiments whose results were regarded as unsatisfactory.

Key Terms

alpha satellite
anaphase
anaphase I
anaphase II
autosomes
bivalent
cell cycle
centromere
chiasma
chi-square
chromatid
chromatin
chromomere
chromosome
chromosome complement
chromosome theory of heredity
crossing-over
conserved sequence
core particle
cyclin
cytokinesis
degrees of freedom
diakinesis
diploid
diplotene
equational division
euchromatin
factorial
first meiotic division
gametophyte
G_1 period
G_2 period
germ cell
goodness of fit
guide RNA
haploid
hemophilia A
heterochromatin
histone
homologous
interphase
kilobase pair (kb)
kinetochore
leptotene
linker DNA
M period
megabase pair (Mb)
meiocyte
meiosis
metaphase
metaphase I
metaphase II
metaphase plate
mitosis
nucleosome
nondisjunction
nucleolus
pachytene
Pascal's triangle
prophase
prophase I
prophase II
reductional division
satellite DNA
S period
scaffold
second meiotic division
sex chromosome
sister chromatid
somatic cell
spindle
spore
sporophyte
statistically highly significant
statistically significant
synapsis
telomerase
telomere
telophase
telophase I
telophase II
tetrad
30-nm fiber
X chromosome
X-linked gene
Y chromosome
zygotene

Concepts and Issues Testing Your Knowledge

- What is the genetic significance of the fact that gametes contain half the chromosome complement of somatic cells?
- The term *mitosis* derives from the Greek *mitos,* which means "thread." The term *meiosis* derives from the Greek *meioun,* which means "to make smaller." What feature, or features, of these types of nuclear division might have led to the choice of these terms?
- Explain the meaning of the terms *reductional division* and *equational division.* What is "reduced" or "kept equal"? To which nuclear divisions do the terms refer?

- How is independent assortment of genes on different chromosomes related physically to the process of chromosome alignment on the metaphase plate in meiosis I?
- What are some of the important differences between the first meiotic division and the second meiotic division?
- Why is X-linked inheritance often called "criss-cross inheritance"? How can this term be misleading in regard to the genetic transmission of the X chromosome?
- In what ways is the inheritance of Y-linked genes different from that of X-linked genes?
- How did nondisjunction "prove" the chromosome theory of heredity?
- Why is a statistical test necessary to determine whether an observed set of data yields an acceptable fit to the result expected from a particular genetic hypothesis? What statistical test is conventionally used for this purpose?

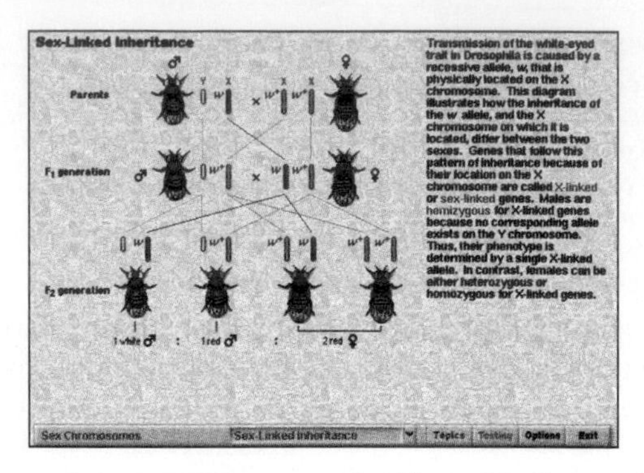

TO ENHANCE YOUR STUDY turn to *An Electronic Companion to Genetics*™, © 1998, Cogito Learning Media, Inc. This CD-ROM is a multimedia tool that provides interactive explanations of important genetic concepts through animations, diagrams, and videos. It also provides interactive test questions.

Solutions
Step by Step

Problem 1

The accompanying diagrams show the appearance of a pair of homologous chromosomes in prophase I of meiosis. Arrange the diagrams in chronological order, and identify each stage as leptotene, zygotene, early pachytene, late pachytene, or diakinesis.

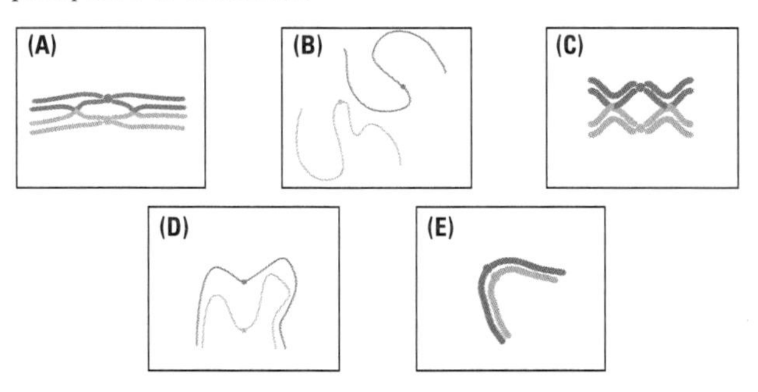

Solution First recall what distinguishes one stage of prophase I from the next. *Leptotene* literally means "thin thread," when each chromosome is in an extended, thread-like condition prior to synapsis; this stage corresponds to diagram B. *Zygotene* means "paired threads," and the pairing begins at the chromosome tips; this is configuration D. *Pachytene* means "thick thread"; it commences when pairing is completed and the homologous chromosomes still appear to be single, which corresponds to diagram E. By late pachytene each homologous chromosome clearly consists of two sister chromatids, and chiasmata are apparent, which is shown in diagram A. *Diakinesis* means "moving apart," and in this stage the synapsed homologous chromosomes begin to repel one another, being held together by the chiasmata, producing configuration C. Therefore, the order of the stages is B–D–E–A–C: leptotene (B), zygotene (D), early pachytene (E), late pachytene (A), and diakinesis (C).

Problem 2

Most color blindness in people is due to relatively common X-linked recessive alleles. A woman with normal color vision whose father was color blind marries a normal man. What types of color vision are expected in the offspring, and in what frequencies?

Solution In these kinds of problems it is helpful to draw a pedigree, showing the information given, and to identify the genotypes of the persons in the pedigree insofar as possible. In this case the pedigree is as indicated, the woman whose father was color blind being number II-1. Her father's genotype must be as shown, because he was color blind. We are told nothing about the mother's genotype, but because II-1 has normal color vision, the mother I-1 must have at least one nonmutant allele (designated cb^+). The normal male II-2 must have a nonmutant allele in his X chromosome, as shown.

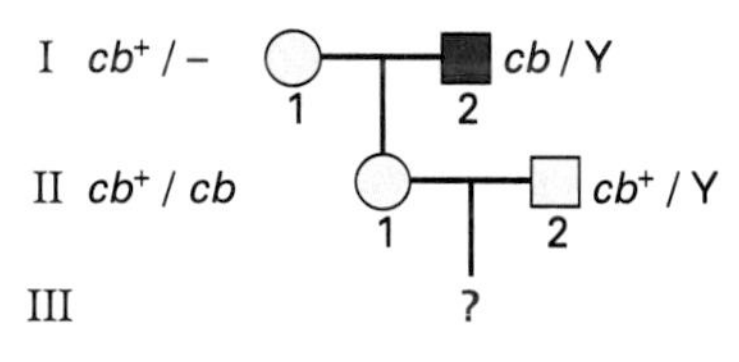

The progeny in question are those in generation III, and their expected composition is shown in the Punnett square that follows. The expected offspring are 1/2 normal females, 1/4 normal males, and 1/4 color-blind males. Note that half of the female offspring are carriers of the recessive allele (heterozygous).

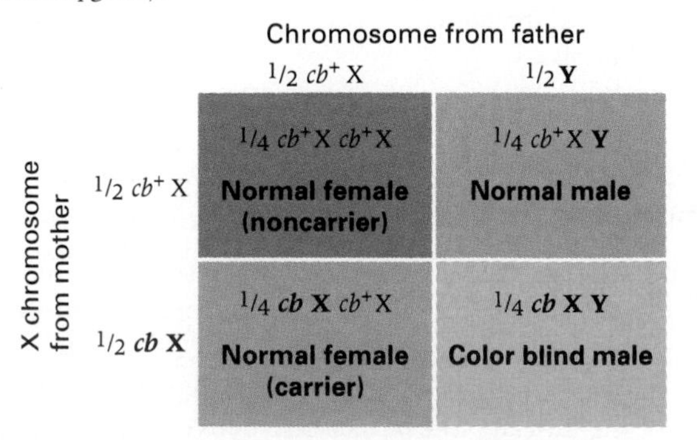

X chromosome from mother	Chromosome from father: 1/2 cb^+ X	Chromosome from father: 1/2 Y
1/2 cb^+ X	1/4 cb^+X cb^+X Normal female (noncarrier)	1/4 cb^+X Y Normal male
1/2 cb X	1/4 cb X cb^+X Normal female (carrier)	1/4 cb X Y Color blind male

geNETics on the web

www.jbpub.com/genetics

These GeNETics on the Web exercises will introduce some of the most useful sites for finding genetic information on the World Wide Web. Genetic sites provide access to a rich storehouse of information on all aspects of genetics. These range from sites written in nontechnical language for the lay person to sites with sophisticated databases designed for the professional geneticist. In carrying out these exercises, you will get a taste of what the Internet can offer a student in genetics.

The keywords shown in color in the following exercises are available on the Jones and Bartlett Publishers' web site as hyperlinks to various genetic sites. To complete the exercises, visit the GeNETics on the Web home page at

http://www.jbpub.com/genetics

Select the link to Essential geNETics on the web. Then choose a chapter. You find a list of keywords that correspond to the exercises below. Select a keyword to link to a web site containing the genetic information necessary to complete the exercise. Each exercise includes a specific assignment that makes use of the information available at the site.

Exercises

- The skeletons of six adults and three children were found in a shallow, unmarked grave near Ekaterinburg, Russia, in 1979, but they remained hidden until July 1991, when they were revealed under the openness policy (glasnost) of Mikhail Gorbachev. The remains have been positively identified as those of Tsar Nicholas II, Tsarina Alexandra, and their daughters Olga, Tatiana, and Anastasia, who had been killed by a firing squad in the basement of a merchant's house and their bodies dumped in a pit north of the city. The other skeletons were those of Anna Demidova, the Tsarina's lady-in-waiting; Eugene Bodkin, the family's physician; Ivan Kharitonov, their cook; and Alouzy Trupp, the Tsar's valet. The remains of Marie and Alexei, afflicted with hemophilia, were never found and are believed to have been burned. At the keyword site Last Tsar you can learn what genetic methods were used to identify the skeletons. Write a short description of the studies that were carried out, and specify what very unusual genetic condition the Tsar was found to have.

- About one in 15,000 people in the United States has X-linked hemophilia, but there are actually two X-linked forms, called hemophilia A and hemophilia B, which differ in the particular blood-clotting factor that is mutant. The "Royal hemophilia" transmitted by Queen Victoria is hemophilia A. Much more common than either X-linked hemophilia is another bleeding disorder known as von Willebrand disease, after the physician who first described it in 1931. (He called it "pseudohemophilia.") At this keyword site you can learn more about this condition, which is due to an autosomal mutation present in 2.5 to 5 percent of the population. Use this keyword site to prepare an outline of a report about von Willebrand disease, including its cause, symptoms, severity, and treatment and some of the complications of treatment. Include a topic sentence for the key paragraph under each main heading in the outline.

Pic Site

The Pic Site showcases some of the most visually appealing genetics sites on the World Wide Web. This site is operated by the Estacion Experimental de Aula Dei near Zaragoza, Spain, where Joe Hin Tijo carried out much of his pioneering work in human cytogenetics. To visit this site, select the Pic Site for Chapter 3.

Problem 3

In a cell undergoing meiosis in a normal human male, nondisjunction of the sex chromosomes takes place. Determine what chromosome constitution would result in a zygote formed from a normal egg and either of the abnormal gametes resulting from nondisjunction, under the following conditions:

(a) The nondisjunction takes place in meiosis I.
(b) Nondisjunction happens to the X chromosome in meiosis II.
(c) Nondisjunction happens to the Y chromosome in meiosis II.

Solution In approaching problems like this, it is extremely helpful to draw diagrams of the meiotic divisions, showing the nondisjunctions that are postulated to happen. The consequences then become quite clear. The accompanying diagrams illustrate the normal situation, along with the three types of nondisjunction stipulated in the problem. The raised dot in the symbols X · X and Y · Y serves to indicate that at this stage, each chromosome consists of two chromatids attached to a single centromere. The consequences of the nondisjunction events are clear from the diagrams.
(a) The abnormal gametes resulting from XY nondisjunction in meiosis I carry either no sex chromosome ("nullo-X") or

both an X and a Y. A zygote from a nullo-X gamete will have 45 chromosomes, with a missing X (this chromosome constitution is designated 45,X); a zygote from an XY gamete will have 47 chromosomes with an XXY sex-chromosome constitution (designated 47,XXY). **(b)** The abnormal gametes resulting from X nondisjunction in meiosis II are either nullo-X or XX. The nullo-X gamete yields a 45,X zygote, and the XX gamete yields a 47,XXX zygote (47 chromosomes total, with three X chromosomes). **(c)** The abnormal gametes resulting from Y nondisjunction in meiosis II are either nullo-Y or YY. The nullo-Y gamete yields a 45,X zygote, and the YY gamete yields a 47,XYY zygote (47 chromosomes total, with an XYY sex-chromosome constitution).

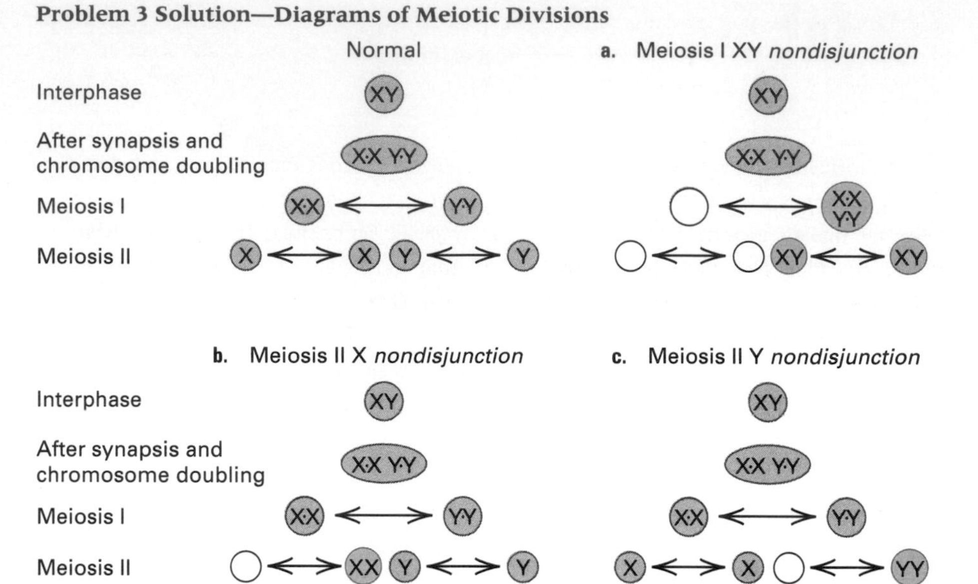

Problem 4

Certain people are able to taste the chemical phenylthiocarbamide (PTC) when it is present in wet filter paper. These people are called "tasters." Others are unable to detect PTC and have the "nontaster" phenotype. PTC tasting is an example of variation in sensory perception, and the genetic basis is thought to be rather simple. The ability to taste PTC is attributed to a dominant allele, denoted *T*, located in chromosome 7. The recessive allele is designated *t*. The genotypes *TT* and *Tt* are tasters of PTC, and *tt* genotypes are nontasters. From 204 matings of heterozygous tasters with nontasters, 259 taster and 278 nontaster progeny were observed. Use a chi-square test to determine whether these numbers give a satisfactory fit to the Mendelian expectation.

Solution First you need to determine the Mendelian expectation. The problem specifies that the matings are heterozygous tasters with nontasters, or *Tt* × *tt*. The Mendelian expectation is therefore 1/2 *Tt* (taster) : 1/2 *tt* (nontaster) progeny. The total number of progeny observed was 259 + 278 = 537, so the expected number of tasters : nontasters is 268.5 : 268.5. The chi-square value is given by Σ *(observed − expected)*^{2}*/expected*, where the sum is over all classes of data. In this case,

$$\chi^2 = \frac{(259 - 268.5)^2}{268.5} + \frac{(278 - 268.5)^2}{268.5}$$

$$= 0.336 + 0.336$$

$$= 0.672$$

This chi-square has 1 degree of freedom, because there are two classes of data. The corresponding *P* value from Figure 3.34 is about 0.47, which means that there is about a 47 percent chance of obtaining a fit as bad or worse than 259 : 278. This means that there is no reason, on the basis of these data, to reject the hypothesis of simple Mendelian inheritance of tasting ability.

Concepts in Action
Problems for Solution

3.1 A cytogeneticist examining cells in *Tradescantia* stamen hairs is trying to determine the length of the various stages in mitosis and the cell cycle. She examines 2000 cells and finds 320 cells in prophase, 150 cells in metaphase, 80 cells in anaphase, and 120 cells in telophase. What conclusion can be drawn about the relative length of each stage of the cell cycle, including the time spent in interphase? Express each answer as a percentage of the total cell-cycle time.

3.2 The diagrams shown here depict anaphase in cell division in a cell of a hypothetical organism with one pair of chromosomes. Identify the panels as being anaphase of mitosis, anaphase I of meiosis, or anaphase II of meiosis, stating on what basis you reached your conclusions.

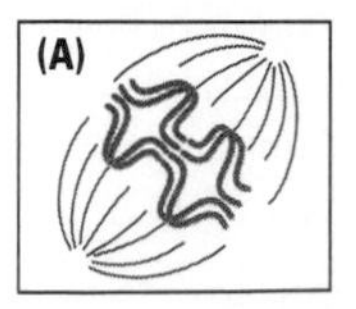

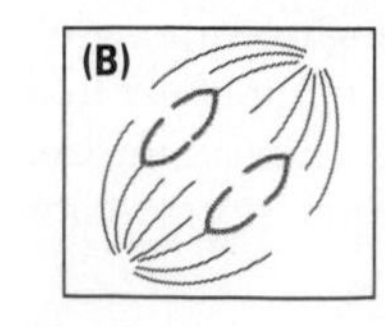

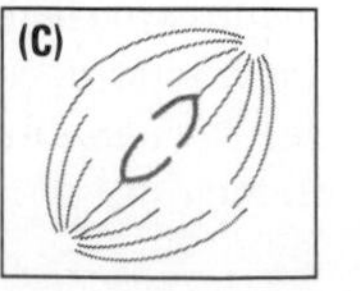

3.3 The most common form of color blindness in human beings results from X-linked recessive alleles. One type of allele, call it cb^r, results in defective red perception, whereas another type of allele, call it cb^g, results in defective green perception. A woman who is heterozygous cb^r/cb^g and a normal male produce a son whose chromosome constitution is XXY. What are the possible genotypes of this boy if:

(a) The nondisjunction took place in meiosis I?
(b) The nondisjunction took place in the cb^r-bearing chromosome in meiosis II?
(c) The nondisjunction took place in the chromosome in meiosis II?

3.4 In *Drosophila*, sex is determined by the ratio of X chromosomes to autosomes. In the following X/A combinations, A indicates one complete haploid set of autosomes. What is the expected sex of each of the following flies?

(a) X/AA **(b)** XX/AA
(c) XXX/AA **(d)** XX/AAA

3.5 People with the chromosome constitution 47,XXY are phenotypically males. A normal woman whose father had hemophilia mates with a normal man and produces an XXY son who also has hemophilia. What kind of nondisjunction can explain this result?

3.6 Most color blindness in human beings is due to relatively common X-linked recessive alleles, and premature baldness ("pattern baldness") is due to an allele, *B*, that is dominant in males but recessive in females. A woman whose father was color blind and bald mates with a normal male and they have a son. Assuming that the *B* allele is present only in the maternal grandfather, what is the probability that the son will be color blind and bald:

(a) If the boy's maternal grandfather was homozygous *BB*?
(b) If the boy's maternal grandfather was heterozygous *Bb*?

3.7 Duchenne-type muscular dystrophy is an inherited disease of muscle due to a mutant form of a protein called dystrophin. The pattern of inheritance of the disease has these characteristics: (1) affected males have unaffected children, (2) the unaffected sisters of affected males often have affected sons, and (3) the unaffected brothers of affected males have unaffected children. What type of inheritance do these findings suggest? Explain your reasoning.

3.8 Mendel studied the inheritance of phenotypic characters determined by seven pairs of alleles. It is an interesting coincidence that the pea plant also has seven pairs of chromosomes. What is the probability that no two of the traits studied by Mendel were determined by genes located on the same pair of chromosomes?

3.9 A recessive mutation of an X-linked gene in human beings results in hemophilia, marked by a prolonged increase in the time needed for blood clotting. Suppose that a phenotypically normal couple produces two normal daughters and a son affected with hemophilia.

(a) What is the probability that both of the daughters are heterozygous carriers?
(b) If one of the daughters and a normal man produce a son, what is the probability that the son will be affected?

3.10 Tall, red-flowered hibiscus is mated with short, white-flowered hibiscus. Both varieties are true-breeding. All the F_1 plants are backcrossed with the short, white-flowered variety. This backcross yields 188 tall red, 203 tall white, 175 short red, and 178 short white plants. Does the observed result fit the genetic hypothesis of 1 : 1 : 1 : 1 segregation as assessed by a chi-square test?

3.11 What are the values of chi-square that yield *P* values of 5 percent (statistically significant) when there are 1, 2, 3, 4, and 5 degrees of freedom? For the types of chi-square tests illustrated in this chapter, how many classes of data do these degrees of freedom represent? Because the significant chi-square values increase with the number of degrees of freedom, does this mean that it becomes increasingly "hard" (less likely) to obtain a statistically significant chi-square value when the genetic hypothesis is true?

3.12 Are the observed progeny numbers of 11, 11, 22, and 22 consistent with a genetic hypothesis that predicts a 1 : 1 : 1 : 1 ratio?

Further Readings

Allshire, R. C. 1997. Centromeres, checkpoints and chromatid cohesion. *Current Opinion in Genetics & Development* 7: 264.

Chandley, A. C. 1988. Meiosis in man. *Trends in Genetics* 4: 79.

Cohen, J. S., and M. E. Hogan. 1994. The new genetic medicine. *Scientific American*, December.

Ingber, D. E. 1998. The architecture of life. *Scientific American*, January.

McIntosh, J. R., and K. L. McDonald. 1989. The mitotic spindle. *Scientific American*, October.

McKusick, V. A. 1965. The royal hemophilia. *Scientific American*, August.

Miller, O. J. 1995. The fifties and the renaissance of human and mammalian genetics. *Genetics* 139: 484.

Page, A. W. and T. L. Orr-Weaver. 1997. Stopping and starting the meiotic cell cycle. *Current Opinion in Genetics & Development* 7: 23.

Sokal, R. R., and F. J. Rohlf. 1969. *Biometry*. New York: Freeman.

Sturtevant, A. H. 1965. *A Short History of Genetics*. New York: Harper & Row.

Voeller, B. R., ed. 1968. *The Chromosome Theory of Inheritance: Classical Papers in Development and Heredity*. New York: Appleton-Century-Crofts.

Welsh, M. J., and A. E. Smith. 1995. Cystic fibrosis. *Scientific American*, December.

Zielenski, J., and L. C. Tsui. 1995. Cystic fibrosis: Genotypic and phenotypic variations. *Annual Review of Genetics* 29: 777.

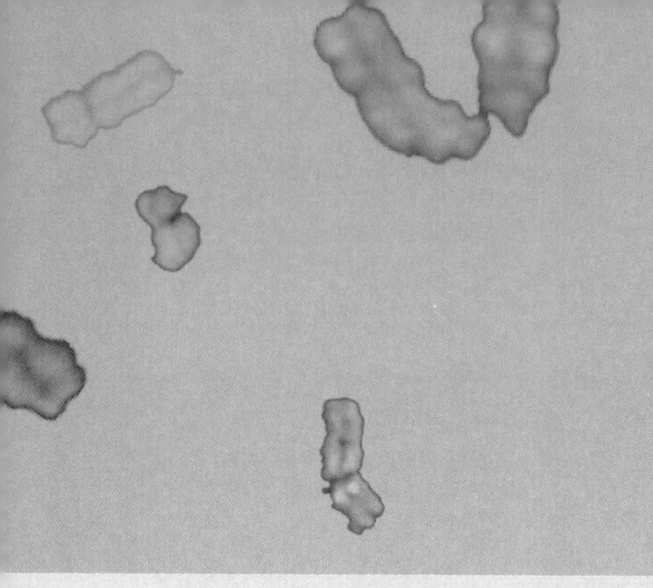

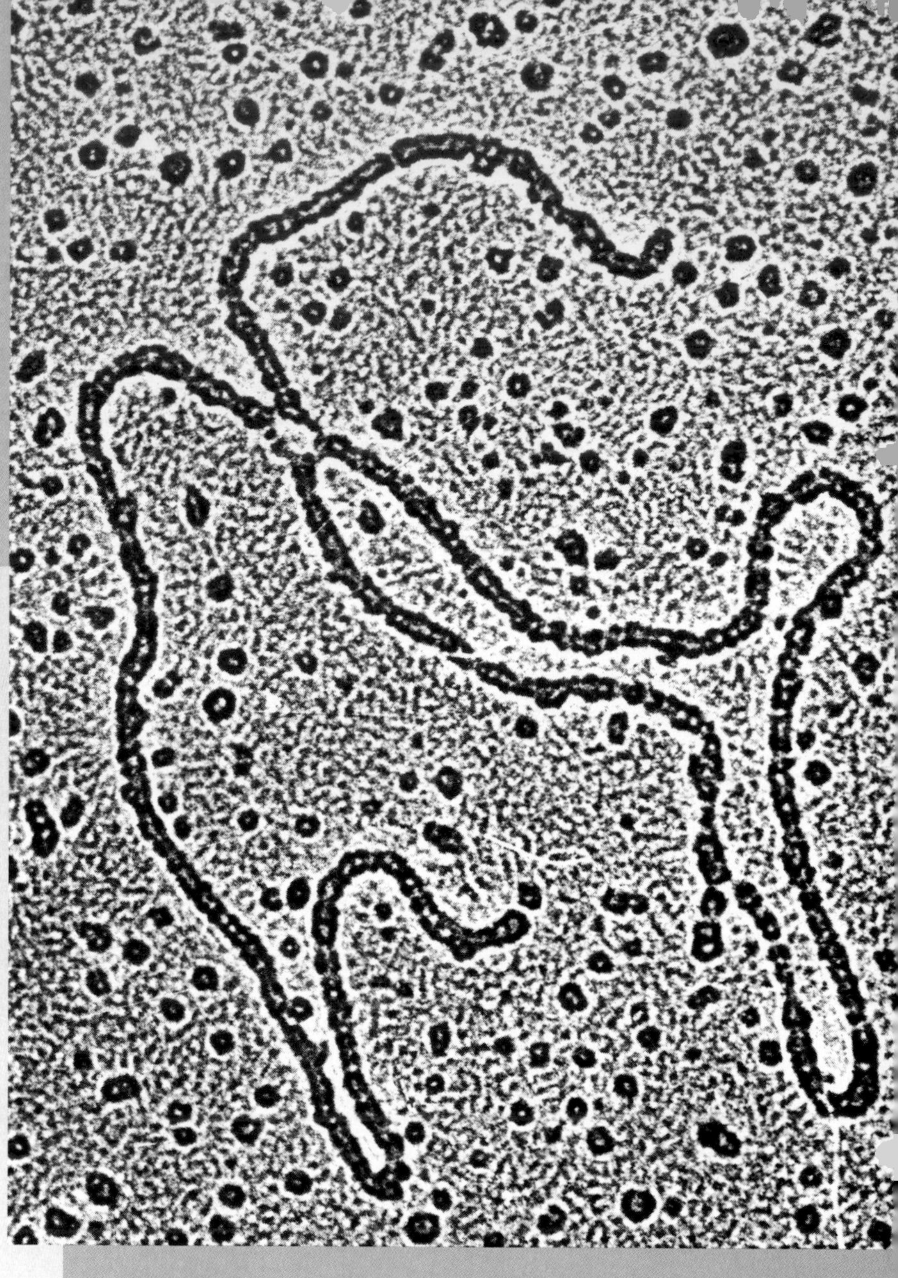

AN ELECTRON MICROGRAPH of two duplex DNA molecules caught in the act of recombination.

Key Concepts

- Genes that are located in the same chromosome and that do not show independent assortment are said to be linked.
- The alleles of linked genes present together in the same chromosome tend to be inherited as a group.
- Crossing-over between homologous chromosomes results in recombination, which breaks up combinations of linked alleles.
- A genetic map depicts the relative positions of genes along a chromosome.
- The map distance between genes in a genetic map is related to the rate of recombination between the genes.
- Physical distance along a chromosome is often, but not always, correlated with map distance.
- Tetrads are sensitive indicators of linkage because they include all the products of meiosis.
- At the DNA level, recombination can be initiated with the interchange of a single strand between two duplex DNA molecules; this creates a cross-shaped "Holliday" structure that is resolved into two separate duplexes.

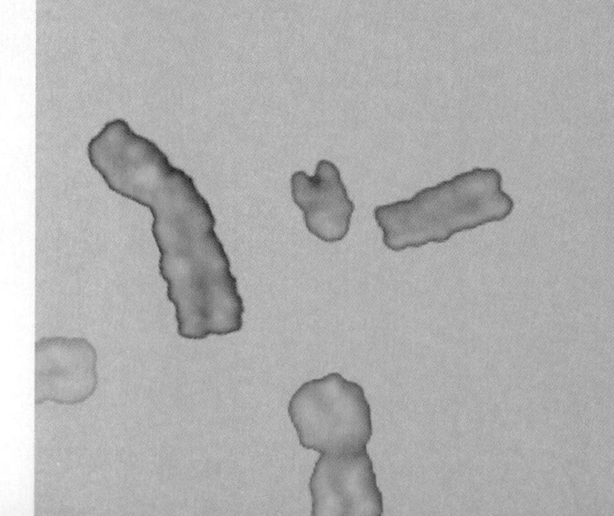

CHAPTER

4

Gene Linkage and Genetic Mapping

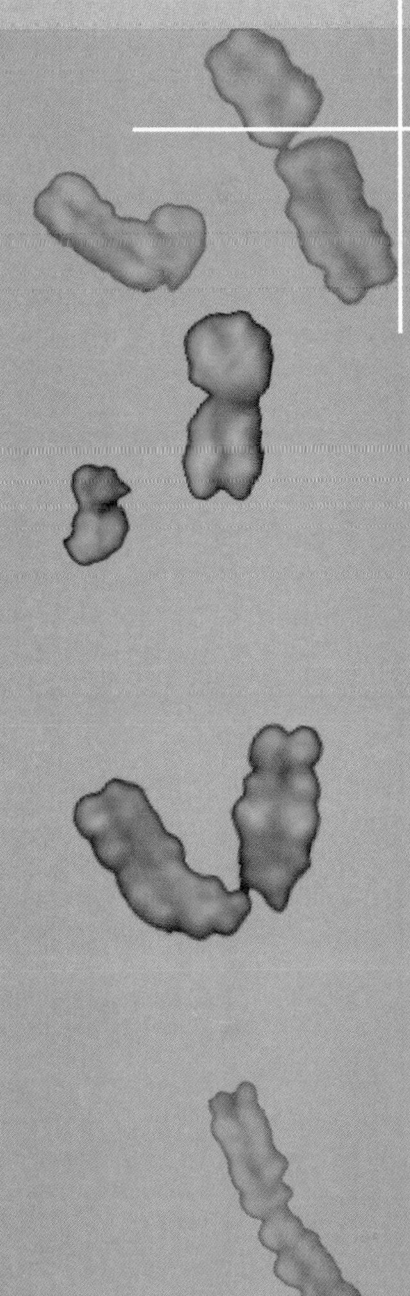

Genetic mapping means determining the relative positions of genes along a chromosome. It is one of the main experimental tools in genetics. This may seem odd in an era when the DNA sequences of some genomes are completely known and the sequencing of other genomes is well advanced. If every gene in an organism is already known, then what is the point of genetic mapping? The answer is that a gene's sequence does not always reveal its function, nor does a genomic DNA sequence reveal which genes interact in a complex biological process. The genetic analysis of a biological process usually begins with a mutant screen, in which many mutants affecting the process are identified through their effects on phenotype. This approach was examined in Chapter 2, in which we used flower color as an example. Once the mutants are identified, they are sorted into groups by complementation tests. Each complementation group consists of the mutant alleles of one gene. Each gene so identified is mapped genetically to determine the chromosome in which it is located and its position in the chromosome. It is at this point that the genome sequence, if known, becomes useful, because in some cases the position of the mutant gene coincides with a gene whose sequence suggests a role in the biological process being investigated. For example, in the case of flower color, one of the mutations may map to a position where there is a gene whose sequence suggests that it encodes an enzyme in anthocyanin synthesis. But the function of a gene is not always revealed by its DNA sequence, and so in some cases further genetic or molecular analysis is necessary to sort out which one of the genes in a sequenced region corresponds to a mutant gene mapped to that region.

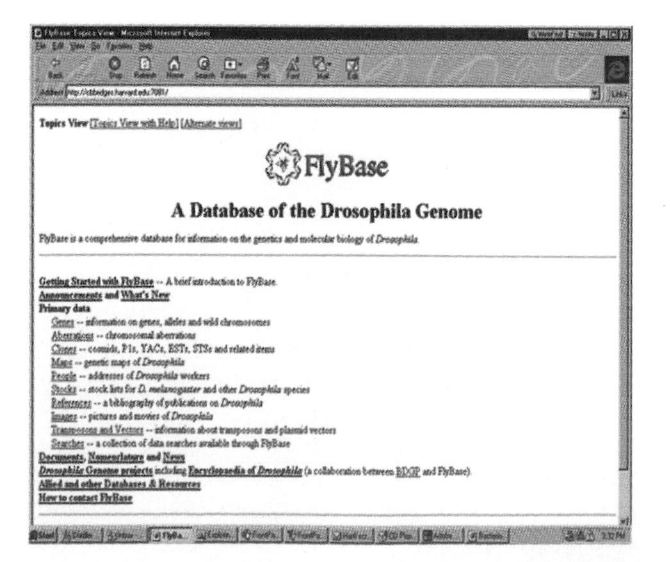

FLYBASE, A MAJOR repository of information about the genetics of *Drosophila.*

Genetic mapping is an essential tool in genetic analysis because it reveals which regions of the genome contain genes that are relevant to the biological process being studied. In human genetics, genetic mapping also makes possible the identification of genes associated with hereditary diseases, such as those that predispose to breast cancer, because a mutant human gene is usually identified only after its map position is known.

As we saw in Chapter 3, homologous chromosomes form pairs in prophase I of meiosis by undergoing synapsis, and the individual members of each pair separate from one another at anaphase I. The observation that homologous chromosomes behave as complete units when they separate at anaphase led to the expectation that genes located in the same chromosome might not undergo independent assortment. Genes in the same chromosome might always be transmitted together and so exhibit complete **linkage.** Thomas Hunt Morgan examined this issue using two genes that he knew were both present in the X chromosome of *Drosophila.* One was a mutation for white eyes, the other a mutation for miniature wings. Although Morgan did observe linkage, it was incomplete. The *white* and *miniature* alleles present in each X chromosome of a female did tend to remain together in inheritance, but some X chromosomes were produced that had new combinations of the *white* and *miniature* alleles. Morgan's observation of incomplete linkage is the rule for genes present in the same chromosome. Linkage is incomplete because the homologous chromosomes, when they are paired, can undergo an exchange of segments. An exchange event between homologous chromosomes (crossing-over), results in the **recombination** of genes in the homologous chromosomes. The probability of crossing-over between any two genes serves as a measure of genetic distance between the genes and allows the construction of a **genetic map,** which is a diagram of a chromosome showing the relative positions of the genes.

4.1 Linked alleles tend to stay together in meiosis.

A direct test of independent assortment is to carry out a testcross between an F_1 double heterozygote (*Aa Bb*) and the double recessive homozygote (*aa bb*). When the genes are on different chromosomes, the expected gametes from the *Aa Bb* parent are as shown in Figure 4.1. Because the pairs of homologous chromosomes segregate independently of each other in meiosis, the double heterozygote produces all four possible types of gametes—*A B, A b, a B,* and *a b*—in equal proportions.

Independent assortment takes place in the *Aa Bb* genotype whether the genotype was produced by the cross *AA BB* × *aa bb* or by the cross *AA bb* × *aa BB*. The four products of meiosis are still expected in equal proportions. An expected 50 percent of the testcross progeny result from gametes with the same combination of alleles present in the parents of the double heterozygote (**parental combinations**), and 50 percent result from gametes with new combinations of the alleles (**recombinants**). For example, if the double heterozygote came from the mating *AA BB* × *aa bb*, then the *A B* and *a b* gametes would be parental and the *A b* and *a B* gametes recombinant. On the other hand, if the double heterozygote came from the mating *AA bb* × *aa BB*, then the *A b* and *a B* gametes would be parental and the *a b* and *A B* gametes recombinant. In either case, with independent assortment, the genotypes of the testcross progeny are expected in the ratio of 1 : 1 : 1 : 1.

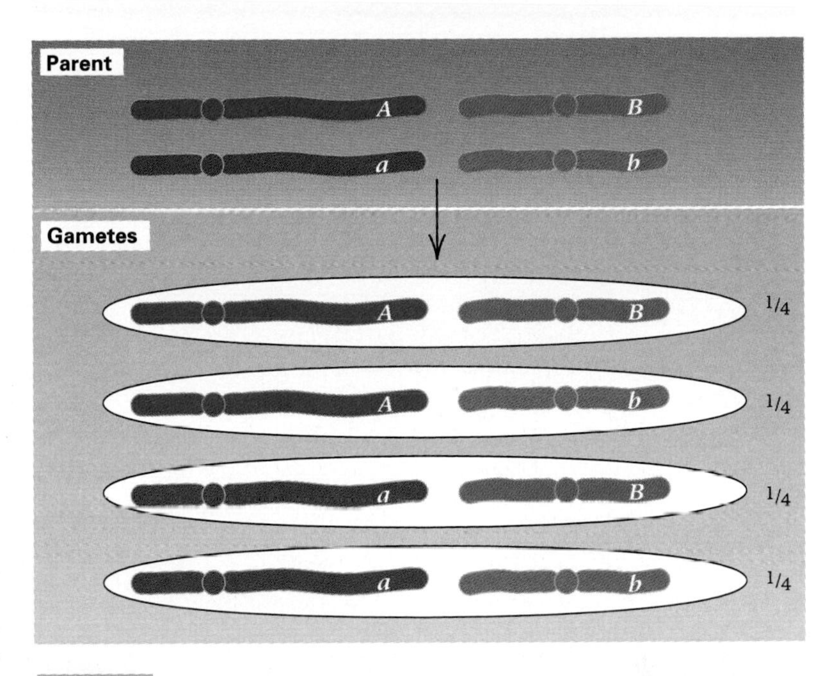

Figure 4.1 Alleles of genes in different chromosomes undergo independent assortment. The pairs of homologous chromosomes segregate at random with respect to one another, so an *A*-bearing chromosome is as likely to go to the same anaphase pole with a *B*-bearing chromosome as with a *b*-bearing chromosome. The result is that each possible combination of chromatids is equally likely among the gametes: 1/4 each for *A B*, *A b*, *a B*, and *a b*.

In dealing with linked genes, it is necessary to distinguish which alleles are present together on the chromosomes. This is done by means of a slash ("/", also called a virgule). The alleles in one chromosome are depicted to the left of the slash, and those in the homologous chromosome are depicted to the right of the slash. For example, in the cross *AA BB* × *aa bb*, the genotype of the heterozygous progeny is denoted *A B*/*a b* because the *A* and *B* alleles were inherited in one parental chromosome and the alleles *a* and *b* were inherited in the other parental chromosome. Similarly, in the cross *AA bb* × *aa BB*, the genotype of the heterozygous progeny is denoted *A b*/*A B* because the *A* and *b* alleles are inherited from one parent and the *a* and *B* alleles from the other.

In his early experiments with *Drosophila*, Morgan found mutations in each of several X-linked genes that provided ideal materials for studying the inheritance of genes in the same chromosome. One of these genes, with alleles w^+ and *w*, determines normal red eye color versus white eyes, as discussed in Chapter 3; another such gene, with the alleles m^+ and *m*, determines whether the size of the wings is normal or miniature. The initial cross is shown as Cross 1 in Figure 4.2. It was a cross between females with white eyes and normal wings and males with red eyes and miniature wings:

$$\frac{w\,m^+}{w\,m^+}\,♀♀ \times \frac{w^+\,m}{Y}\,♂♂$$

In this way of writing the genotypes, the horizontal line replaces the slash. Alleles written above the line are present in one chromosome, and those written below the line are present in the homologous chromosome. In the females, both X chromosomes carry *w* and m^+. In males, the X chromosome carries the alleles w^+ and *m*. (The Y written below the line denotes the Y chromosome in the male.) Figure 4.2 illustrates a simplified symbolism, commonly used in *Drosophila* genetics, in which a wild-type allele is denoted by a + sign in the appropriate position. The + symbolism is unambiguous because the linked genes in a chromosome are always written in the same order. Using the + notation,

$$\frac{w\,+}{w\,+} \quad \text{means} \quad \frac{w\,m^+}{w\,m^+}$$

and

$$\frac{+\,m}{Y} \quad \text{means} \quad \frac{w^+\,m}{Y}$$

The resulting F_1 female progeny from Cross 1 have the genotype *w* +/+ *m* (or, equivalently, *w* m^+/w^+ *m*). In this genotype, the mutant alleles are in different chromosomes. When these females were mated with *w m*/Y males, the offspring denoted as Progeny 1 in Figure 4.2 were obtained. In each class of progeny, the gamete from the female parent is shown in the column at the left, and the gamete from the male parent carries either *w m* or

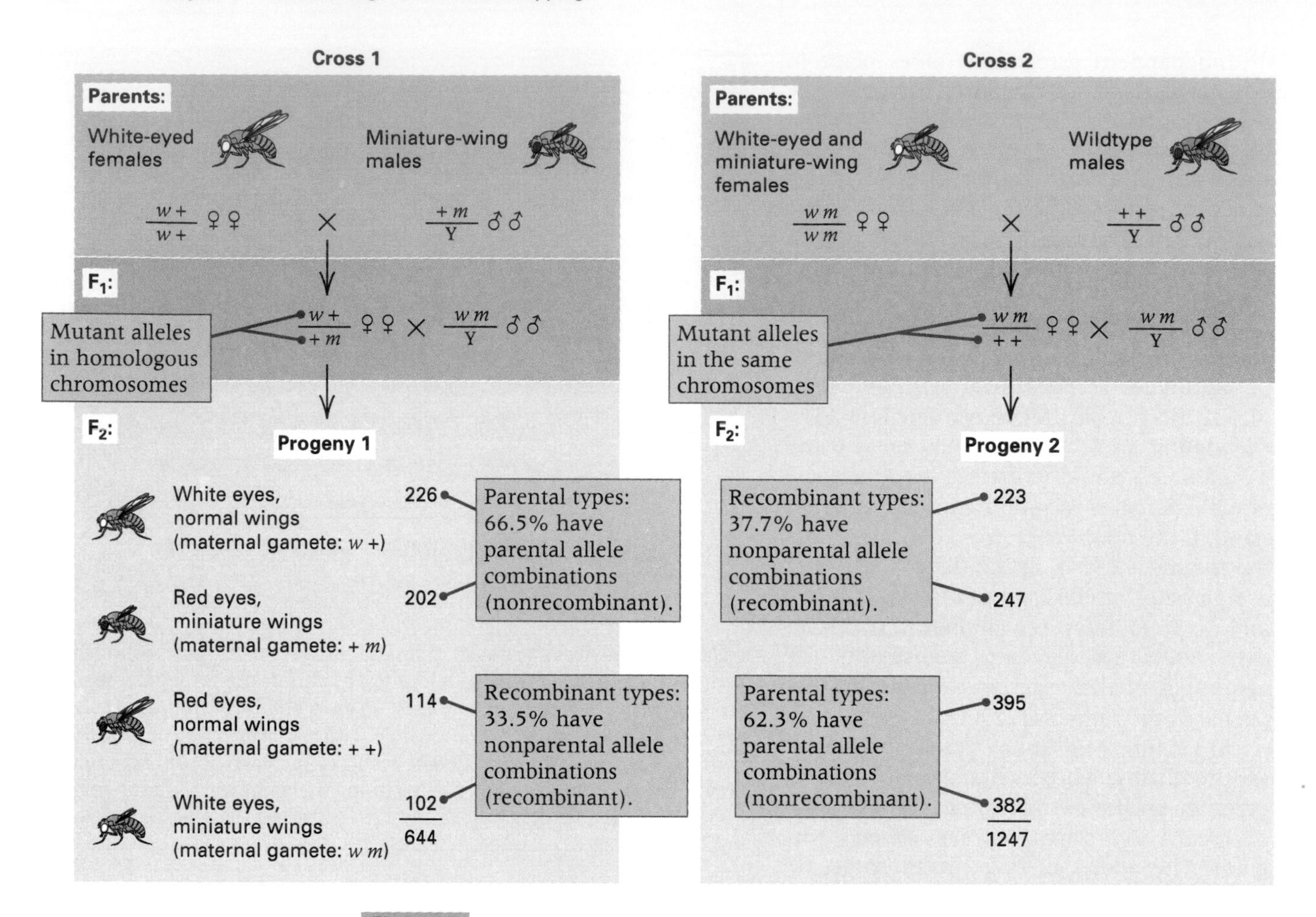

Figure 4.2 An experiment demonstrating that the frequency of recombination between two mutant alleles is independent of whether they are present in the same chromosome or in homologous chromosomes. (A) Cross 1 produces F_1 females with the genotype $w\,+/+\,m$, and the $w-m$ recombination frequency is 33.5 percent. (B) Cross 2 produces F_1 females with the genotype $w\,m/+\,+$, and the $w-m$ recombination frequency is 37.7 percent. These values are within the range of variation expected to occur by chance.

the Y chromosome. The cross is equivalent to a test-cross, and so the phenotype of each class of progeny reveals the alleles present in the gamete from the mother.

The results of Cross 1 show a great departure from the 1 : 1 : 1 : 1 ratio of the four male phenotypes expected with independent assortment. If genes in the same chromosome tended to remain together in inheritance but were not completely linked, this pattern of deviation might be observed. In this case, the combinations of phenotypic traits in the parents of the original cross (parental phenotypes) were present in 428/644 (66.5 percent) of the F_2 males, and nonparental combinations (recombinant phenotypes) of the traits were present in 216/644 (33.5 percent). The 33.5 percent recombinant X chromosomes is called the **frequency of recombination,** and it should be contrasted with the 50 percent recombination expected with independent assortment.

The recombinant X chromosomes $w^+\ m^+$ and $w\ m$ result from crossing-over in meiosis in F_1 females. In this example, the frequency of recombination between the linked *w* and *m* genes was 33.5 percent. With other pairs of linked genes, the frequency of recombination ranges from near 0 to 50 percent. Even genes in the same chromosome can undergo independent assortment (frequency of recombination equal to 50 percent) if they are sufficiently far apart. This implies the following principle:

> Genes with recombination frequencies smaller than 50 percent are present in the same chromosome (linked). Two genes that undergo independent assortment, indicated by a recombination frequency equal to 50 percent, either are in nonhomologous chromosomes or are located far apart in a single chromosome.

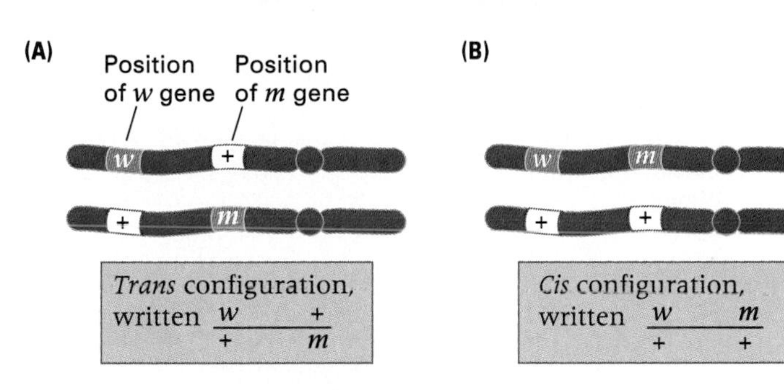

Figure 4.3 Two configurations of the mutant alleles are possible in a genotype that is heterozygous for both mutations. (A) The *trans,* or repulsion, configuration has the mutant alleles on opposite chromosomes. (B) The *cis,* or coupling, configuration has the mutant alleles on the same chromosome.

The frequency of recombination is the same for *cis* and *trans* heterozygotes.

A genotype that is heterozygous for each of two linked genes can have the alleles in either of two possible configurations, as shown in Figure 4.3 for the *w* and *m* alleles. In one configuration, called the ***trans,*** or **repulsion,** configuration, the mutant alleles are in opposite chromosomes, and the genotype is written as *w* +/+ *m*. In the alternative configuration, called the ***cis,*** or **coupling,** configuration, the mutant alleles are present in the same chromosomes, and the genotype is written as *w m*/+ +.

Morgan's study of linkage between the *white* and *miniature* alleles began with the *trans* configuration, diagrammed as Cross 1 in Figure 4.2. He also studied progeny from the *cis* configuration of the *w* and *m* alleles, which results from the mating designated as Cross 2 in Figure 4.2. In this case, the original parents had the genotypes:

$$\frac{w\ m}{w\ m} ♀♀ \times \frac{+\ +}{Y} ♂♂$$

The resulting F_1 female progeny from Cross 2 have the genotype *w m*/+ + (equivalently, $w\ m/w^+\ m^+$). In this case the mutant alleles are in the same chromosomes, in the *cis* (coupling) configuration. When these F_1 female progeny were crossed with *w m*/Y males, they yielded the types of progeny tabulated as Progeny 2 in Figure 4.2.

Because the alleles in Cross 2 are in the *cis* configuration, the parental-type gametes carry either *w m* or + +, and the recombinant gametes carry either *w* + or + *m*. The types of gametes are the same as those observed in Cross 1, but the parental and recombinant types are interchanged. Yet the frequency of recombination is approximately the same: 37.7 percent versus 33.5 percent. The difference is within the range expected to result from random variation from experiment to experiment. The consistent finding of equal recombination frequencies in experiments in which the mutant alleles are in the *trans* or the *cis* configuration leads to the following conclusion:

> Recombination between linked genes takes place with the same frequency whether the alleles of the genes are in the *trans* configuration or in the *cis* configuration; it is the same no matter how the alleles are arranged.

The frequency of recombination differs from one gene pair to the next.

The principle that the frequency of recombination depends on the genes may be illustrated using the recessive allele *y* of another X-linked gene in *Drosophila,* which results in yellow body color instead of the usual gray color determined by the y^+ allele. The *yellow body (y)* and *white eye (w)* genes are linked. The frequency of recombination between the genes is as shown in the data in Figure 4.4. The layout of the crosses is like that in Figure 4.2. In Cross 1, the female has *y* and *w* in the *trans* configuration (+ *w*/*y* +); in Cross 2, the alleles are in the *cis* configuration (*y w*/+ +). The *y* and *w* genes exhibit a much lower frequency of recombination than that observed with *w* and *m* in Figure 4.2. To put it another way, the genes *y* and *w* are more closely linked than are *w* and *m*. In Cross 1, the recombinant progeny are + + and *y w,* and they account for 130/9027 = 1.4 percent of the total. In Cross 2, the recombinant progeny are + *w* and *y* +, and they account for 94/7838 = 1.2 percent of the total. Once again, the parental and recombinant gametes are reversed in Crosses 1 and 2, because the configuration of alleles in the female parent is *trans* in Cross 1 but *cis* in Cross 2, yet the frequency of recombination between the genes is within experimental error.

The results of these and other experiments give support to two general principles of recombination:

- The recombination frequency is a characteristic of a particular pair of genes.
- Recombination frequencies are the same in *cis* (coupling) and *trans* (repulsion) heterozygotes.

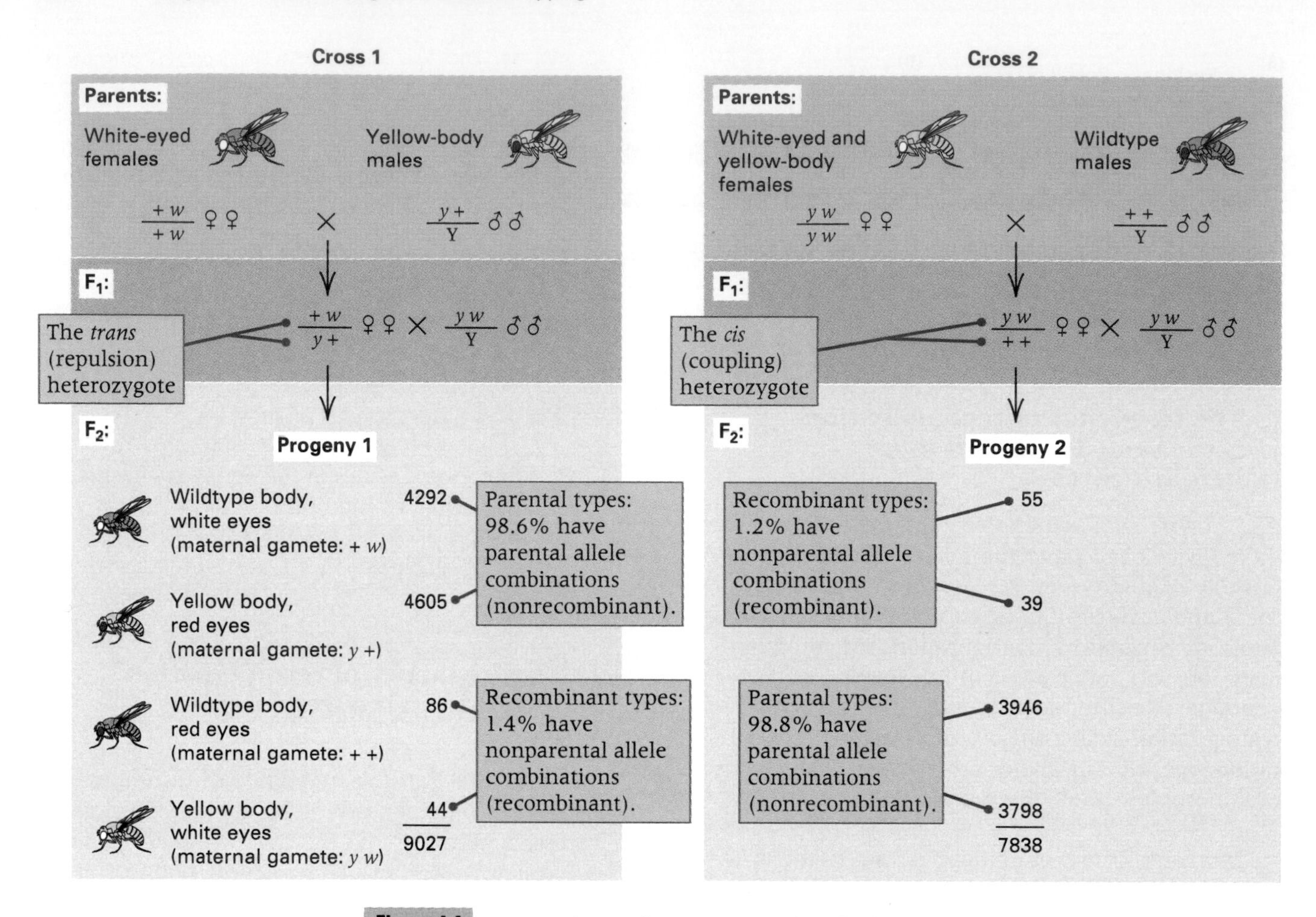

Figure 4.4 An experiment demonstrating that the frequency of recombination between two genes depends on the genes. The frequency of recombination between *w* and *y* is much less than that between *w* and *m* in Figure 4.2. The *y*−*w* experiment also confirms the equal frequency of recombination in *trans* and *cis* heterozygous genotypes. (A) The *trans* heterozygous females, + *w*/*y* +, yield 1.4 percent recombination. (B) The *cis* heterozygous females, *y w*/+ +, yield 1.2 percent recombination.

Recombination does not occur in *Drosophila* males.

Early experiments in *Drosophila* genetics also indicated that the organism is unusual in that recombination does not take place in males. Although it is not known how (or why) crossing-over is prevented in males, the result of the absence of recombination in *Drosophila* males is that all alleles located in a particular chromosome show complete linkage in the male. For example, the genes *cn* (cinnabar eyes) and *bw* (brown eyes) are both in chromosome 2, but they are so far apart that, in females, there is 50 percent recombination. Because the genes exhibit 50 percent recombination, the cross

$$\frac{cn\ bw}{+\ +} ♀♀ \times \frac{cn\ bw}{cn\ bw} ♂♂$$

yields progeny of genotype *cn bw*/*cn bw* and + +/*cn bw* (the nonrecombinant types) as well as *cn* +/*cn bw* and + *bw*/*cn bw* (the recombinant types) in the proportions 1 : 1 : 1 : 1. The outcome of the reciprocal cross is completely different. Because there is no crossing-over in males, the reciprocal cross

$$\frac{cn\ bw}{cn\ bw} ♀♀ \times \frac{cn\ bw}{+\ +} ♂♂$$

yields progeny only of the nonrecombinant genotypes *cn bw*/*cn bw* and + +/*cn bw* in equal proportions. The absence of recombination in *Drosophila* males is a convenience often exploited in experimental design; as shown in the case of *cn* and *bw*, all the alleles present in any chromosome in a male must be transmitted as a group, without being recombined with alleles present in the homologous chromosome. The absence of crossing-over in *Drosophila* males is atypical; in most other animals and plants, recombination takes place in both sexes.

4.2 Recombination results from crossing-over between linked alleles.

The linkage of the genes in a chromosome can be represented in the form of a *genetic map*, which shows the linear order of the genes along the chromosome with the distances between adjacent genes proportional to the frequency of recombination between them. A genetic map is also called a **linkage map** or a **chromosome map.** The concept of genetic mapping was first developed by Morgan's student Alfred H. Sturtevant in 1913. The early geneticists understood that recombination between genes takes place by an exchange of segments between homologous chromosomes in the process now called crossing-over. Each crossing-over is manifested physically as a chiasma, or cross-shaped configuration, between homologous chromosomes; chiasmata are observed in prophase I of meiosis (Chapter 3). Each chiasma results from the breaking and rejoining of chromatids during synapsis, with the result that there is an exchange of corresponding segments between them. The theory of crossing-over is that each chiasma results in a new association of genetic markers. This process is illustrated in Figure 4.5. When there is no crossing-over (part A), the alleles present in each homologous chromosome remain in the same combination. When a crossing-over does take place (part B), the outermost alleles in two of the chromatids are interchanged (recombined).

The unit of distance in a genetic map is called a **map unit;** one map unit is equal to 1 percent recombination. For example, two genes that recombine with a frequency of 3.1 percent are said to be located 3.1 map units apart. One map unit is also called a **centimorgan,** abbreviated cM, in honor of T. H. Morgan. A distance of 3.1 map units therefore equals 3.1 centimorgans and indicates 3.1 percent recombination between the genes. An example is shown in Figure 4.6A, which deals with the *Drosophila* mutants *w* for white eyes and *dm (diminutive)* for small body. The female parent in the testcross is the *trans* heterozygote, but as we have seen, this configuration is equivalent in frequency of recombination to the *cis* heterozygote. Among 1000 progeny there are 31 recombinants. Using this estimate, we can express the genetic distance between *w* and *dm* in four completely equivalent ways:

- As the *frequency of recombination*—in this case 0.035
- As the *percent recombination,* or 3.5 percent

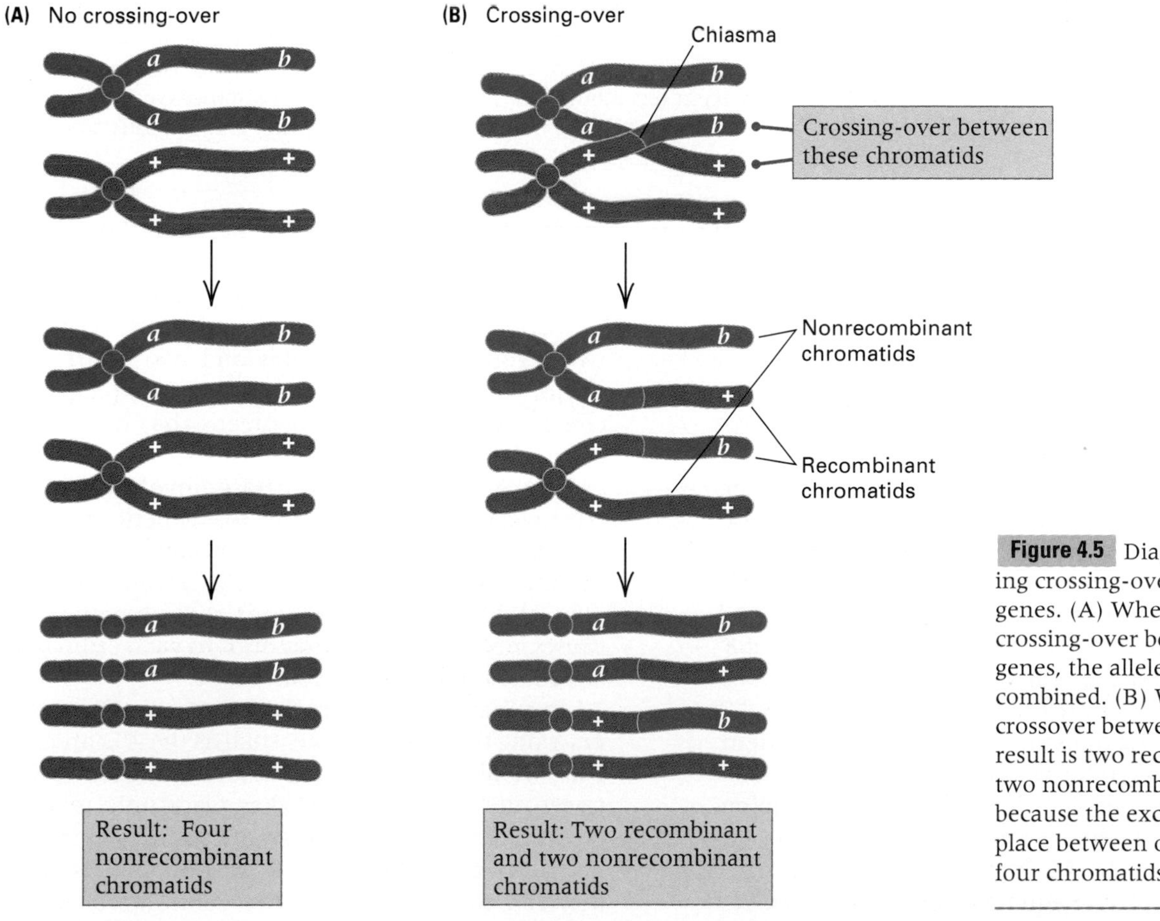

Figure 4.5 Diagram illustrating crossing-over between two genes. (A) When there is no crossing-over between two genes, the alleles are not recombined. (B) When there is a crossover between them, the result is two recombinant and two nonrecombinant products, because the exchange takes place between only two of the four chromatids.

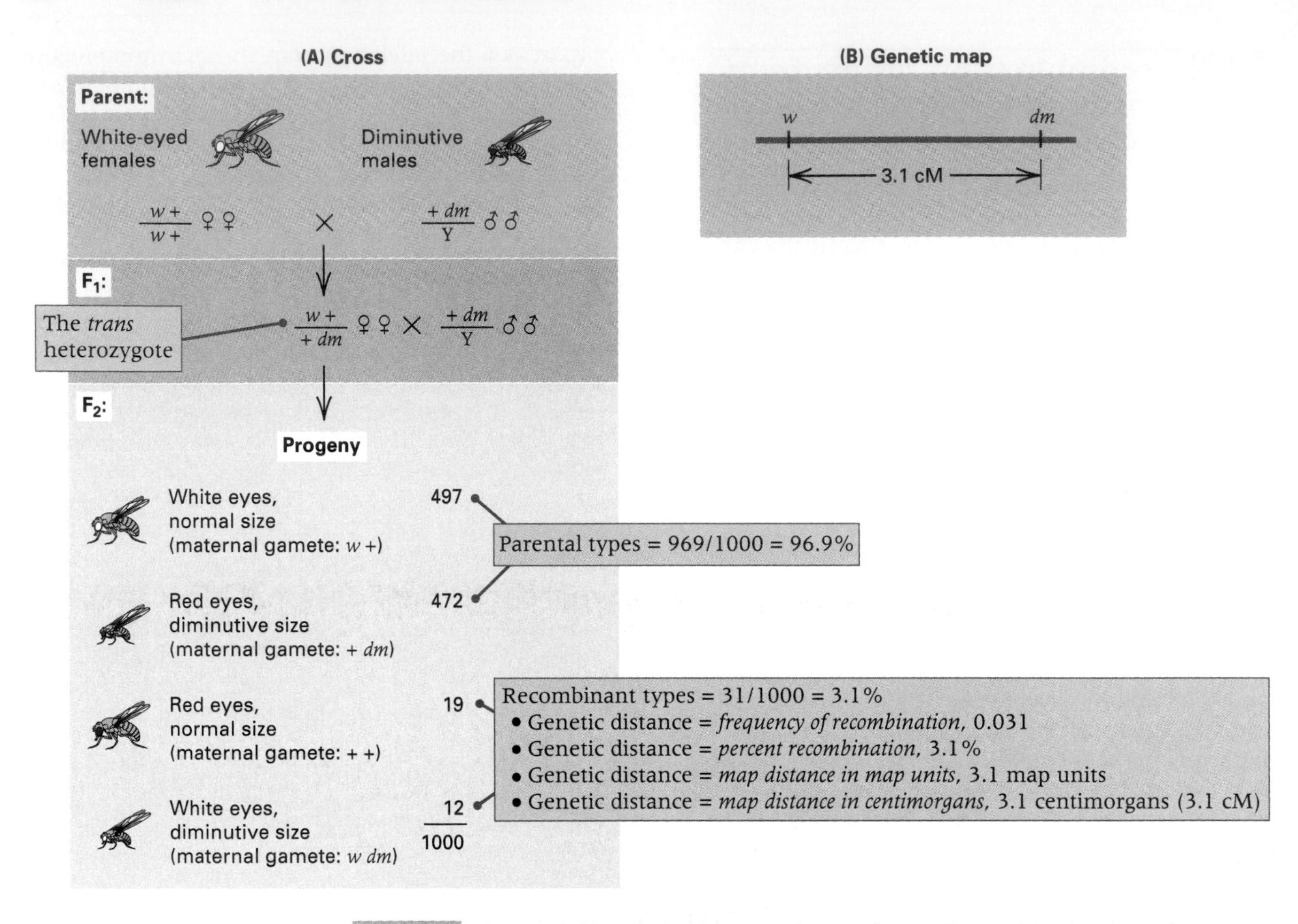

Figure 4.6 An experiment illustrating how the frequency of recombination is used to construct a genetic map. (A) There is 3.1 percent recombination between the genes *w* and *dm*. (B) A genetic map with *w* and *dm* positioned 3.1 map units (3.1 centimorgans, cM) apart, corresponding to 3.1 percent recombination. We shall see that map distance equals frequency of recombination only when the frequency of recombination is sufficiently small.

- As the distance in *map units*—in this example, 3.5 map units
- As the distance in *centimorgans,* or 3.5 centimorgans (3.5 cM)

A genetic map based on these data is shown in Figure 4.6B. The chromosome is represented as a horizontal line, and each gene is assigned a position on the line according to its genetic distance from other genes. In this example, there are only two genes, *w* and *dm,* and they are separated by a distance of 3.1 centimorgans (3.1 cM), or 3.1 map units. Genetic maps are usually truncated to show only the genes of interest. The full genetic map of the *Drosophila* X chromosome extends considerably farther in both directions than indicated in this figure.

Physically, one map unit corresponds to a length of the chromosome in which, on the average, one crossover is formed in every 50 cells undergoing meioses. This principle is illustrated in Figure 4.7. If one meiotic cell in 50 has a crossover, the frequency of *crossing-over* equals 1/50, or 2 percent. Yet the frequency of *recombination* between the genes is 1 percent. The correspondence of 1 percent recombination with 2 percent crossing-over is a little confusing until you consider that a crossover results in two recombinant chromatids and two nonrecombinant chromatids (Figure 4.7). A frequency of crossing-over of 2 percent means that of the 200 chromosomes that result from meiosis in 50 cells, exactly 2 chromosomes (those involved in the crossover) are recombinant for genetic markers spanning the particular chromosome segment. To put the matter in another way, 2 percent crossing-over corresponds to 1 percent recombination because only half of the chromatids in each cell with a crossover are actually recombinant.

In situations in which there are **genetic markers** along the chromosome, such as the *A, a* and *B, b* pairs of alleles in Figure 4.7, recombination between the marker genes takes place only when a crossing-over occurs *between* the genes. Figure 4.8 illustrates a case in which a crossover takes place between the gene *A* and the centromere, rather

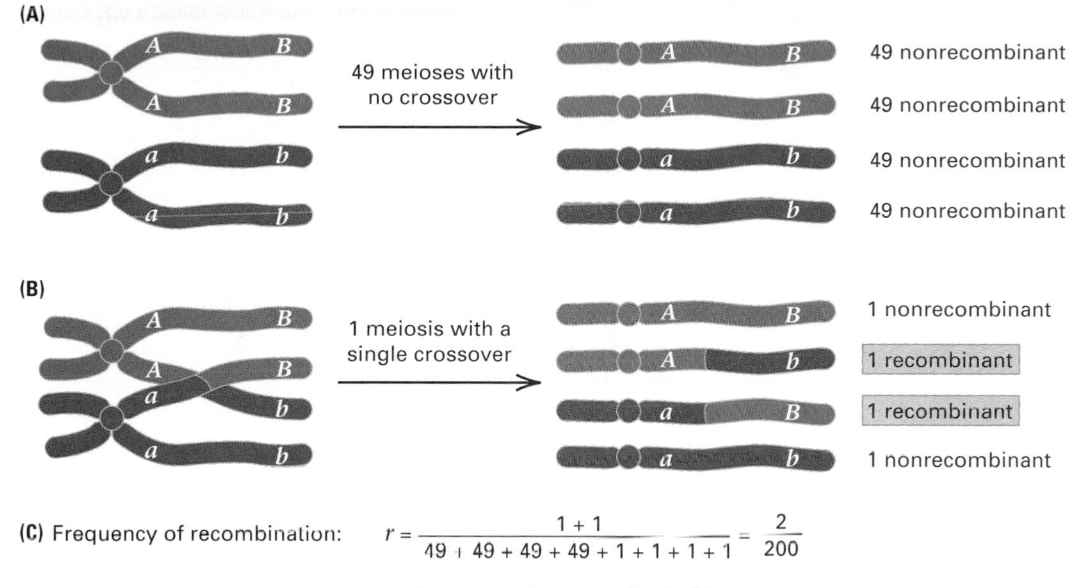

(C) Frequency of recombination: $r = \frac{1+1}{49+49+49+49+1+1+1+1} = \frac{2}{200}$

= 1 percent = 1 map unit = 1 cM

Figure 4.7 Diagram of chromosomal configurations in 50 meiotic cells, in which one has a crossover between two genes. (A) The 49 cells without a crossover result in 98 *A B* and 98 *a b* chromosomes; these are all nonrecombinant. (B) The cell with a crossover yields chromosomes that are *A B*, *A b*, *a B*, and *a b*, of which the middle two types are recombinant chromosomes. (C) The recombination frequency equals 2/200 or 1 percent, also called 1 map unit or 1 cM. Hence, one percent recombination means that 1 meiotic cell in 50 has a crossing-over in the region between the genes.

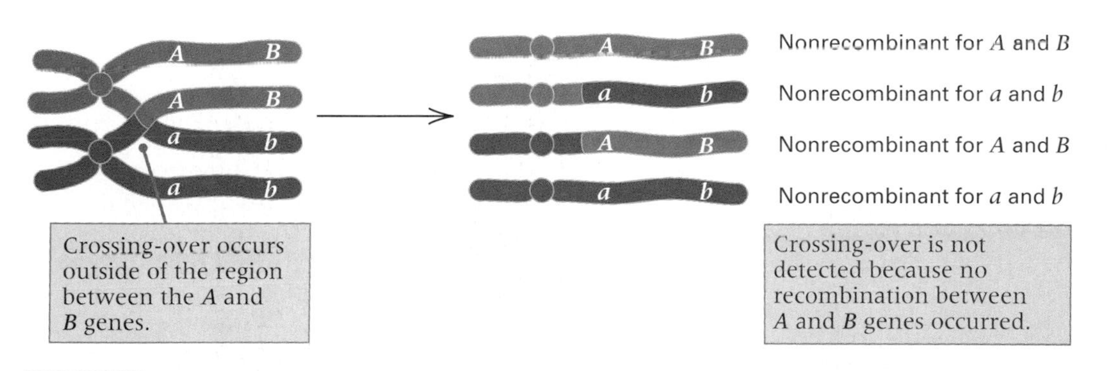

Figure 4.8 Crossing-over outside the region between two genes is not detectable through recombination. Although a segment of chromosome is exchanged, the genetic markers outside the region of the crossovers stay in the nonrecombinant configurations, in this case *A B* and *a b*.

than between the genes *A* and *B*. The crossover does result in the physical exchange of segments between the innermost chromatids. However, because it is located outside the region between *A* and *B*, all of the resulting gametes must carry either the *A B* or the *a b* allele combination. These are nonrecombinant chromosomes. The presence of the crossover is undetected because it is not in the region between the genetic markers.

In some cases, the region between genetic markers is large enough that two (or even more) crossovers can be formed in a single meiotic cell. One possible configuration for two crossovers is shown in Figure 4.9. In this example, both crossovers are between the same pair of chromatids. The result is that there is a physical exchange of a segment of chromosome between the marker genes, but the double crossover remains undetected because the markers themselves are not recombined. The absence of recombination results from the fact that the second crossover reverses the effect of the first, insofar as recombination between *A* and *B* is concerned. The resulting chromosomes are either *A B* or *a b*, both of which are nonrecombinant.

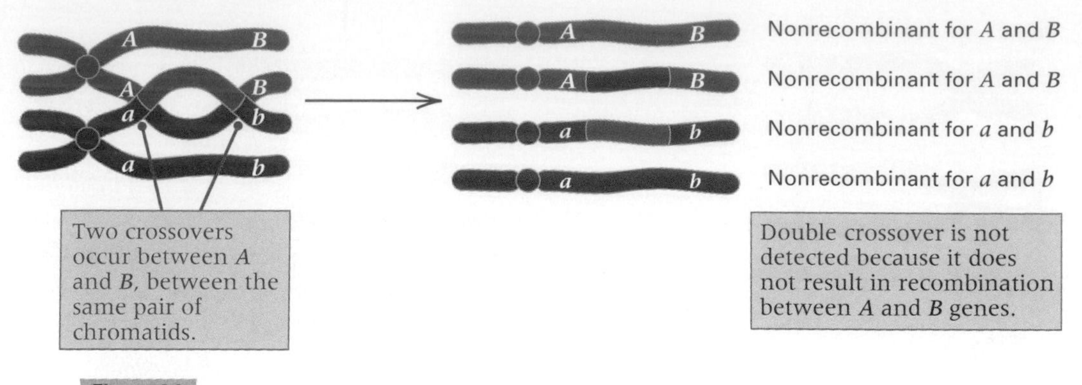

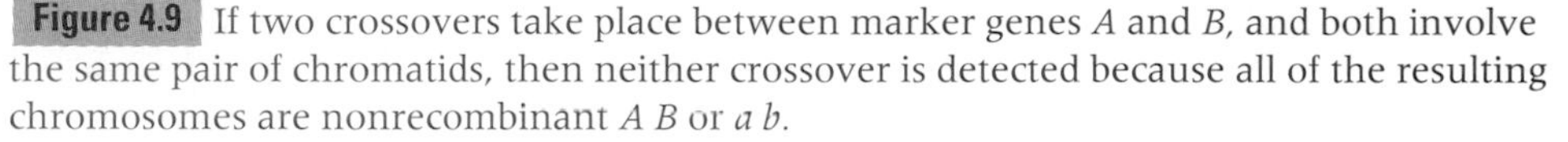

Figure 4.9 If two crossovers take place between marker genes *A* and *B*, and both involve the same pair of chromatids, then neither crossover is detected because all of the resulting chromosomes are nonrecombinant *A B* or *a b*.

Because double crossing-over in a region between two genes can remain undetected to the extent that it does not result in recombinant chromosomes, there is an important distinction between the distance between two genes as measured by the recombination frequency and as measured in map units:

- The *map distance* between two genes equals one-half of the average number of crossovers that take place in the region per meiotic cell; it is a measure of crossing-over.
- The *recombination frequency* between two genes indicates how much recombination is actually observed in a particular experiment; it is a measure of recombination.

FOUR GENERATIONS OF a family. Pattern baldness in human beings is due to an allele that is dominant in males but recessive in females. Affected females also have a later age of onset. [Courtesy of Judith Hauck.]

The difference between map distance and recombination frequency arises because double crossovers that do not yield recombinant gametes, like the one depicted in Figure 4.9, *do* contribute to the map distance but *do not* contribute to the recombination frequency. The distinction is important only when the region in question is large enough for double crossing-over to occur. If the region between the genes is short enough that no more than one crossover can be formed in the region in any one meiosis, then map units and recombination frequencies are the same (because there are no multiple crossovers that can undo each other). This is the basis for defining a map unit as being equal to 1 percent recombination:

> Over an interval so short that multiple crossovers are precluded (typically yielding 10 percent recombination or less), the map distance equals the recombination frequency because all crossovers result in recombinant gametes.

Furthermore, when adjacent chromosome regions separating linked genes are so short that multiple crossovers are not formed, the recombination frequencies (and hence the map distances) between the genes are additive. This important feature of recombination, and also the logic used in genetic mapping, is illustrated by the example in Figure 4.10. The genes are located in the X chromosome of *Drosophila*—*y* for yellow body, *rb* for ruby eye color, and *cv* for shortened wing crossvein. The experimentally measured recombination frequency between genes *y* and *rb* is 7.5 percent and that between *rb* and *cv* is 6.2 percent. The genetic map might be any one of three possibilities, depending on which gene is in the middle (*y*, *cv*, or *rb*). Map C, which has *y* in the middle, can be excluded because

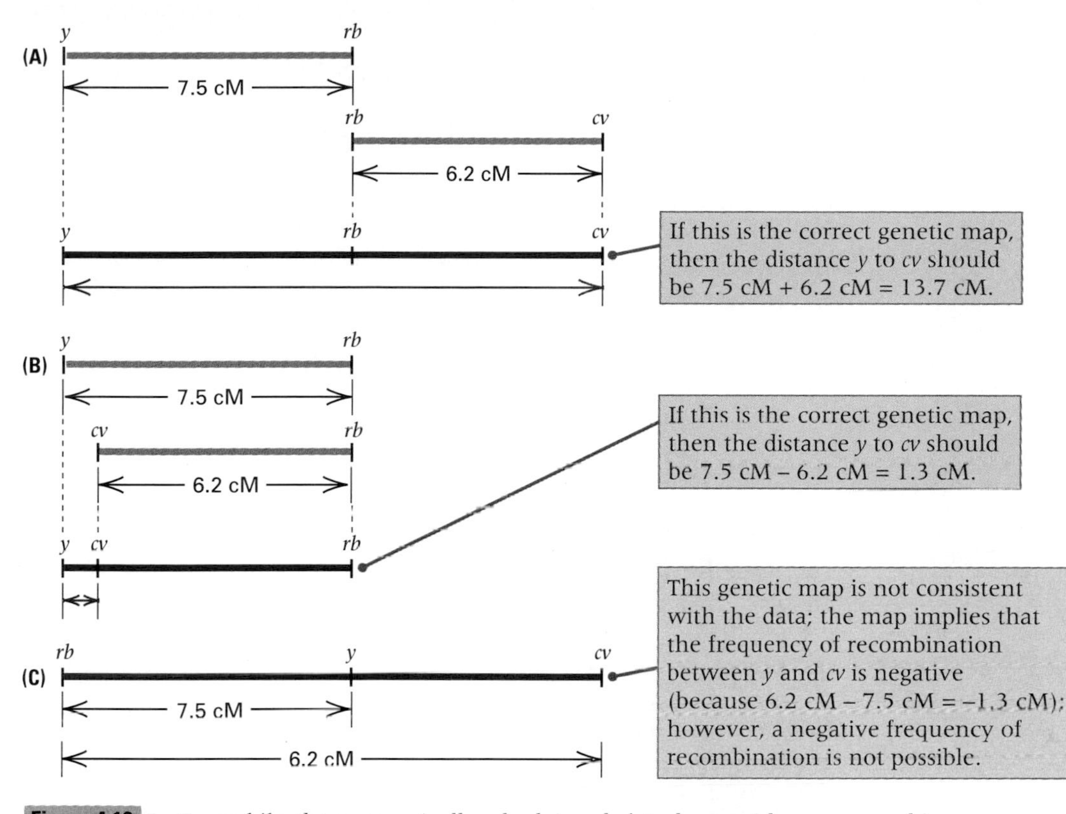

Figure 4.10 In *Drosophila,* the genes *y* (yellow body) and *rb* (ruby eyes) have a recombination frequency of 7.5 percent, and *rb* and *cv* (shortened wing crossvein) have a recombination frequency of 6.2 percent. There are three possible genetic maps, depending on whether *rb* is in the middle (part A), *cv* is in the middle (part B), or *y* is in the middle (part C). Map (C) can be excluded because it implies that *rb* and *y* should be closer than *rb* and *cv,* whereas the observed recombination frequency between *rb* and *y* is actually greater than that between *rb* and *cv.* Maps (B) and (C) are compatible with the data given.

it implies that the recombination frequency between *rb* and *cv* should be greater than that between *rb* and *y,* and this contradicts the observed data. In other words, map C can be excluded because it implies that the frequency of recombination between *y* and *cv* must be negative.

Maps A and B are both consistent with the observed recombination frequencies. They differ in their predictions regarding the recombination frequency between *y* and *cv.* Using the principle of additivity of map distances, the predicted *y*–*cv* map distance in A is 13.7 map units, whereas the predicted *y*–*cv* map distance in B is 1.3 map units. In fact, the observed recombination frequency between *y* and *cv* is 13.3 percent. Map A is therefore correct. However, there are actually two genetic maps corresponding to map A. They differ only in whether *y* is placed at the left or at the right. One map is *y*–*rb*–*cv,* which is the one shown in Figure 4.10; the other is *cv*–*rb*–*y.* The two ways of depicting the genetic map are completely equivalent because there is no way of knowing from the recombination data whether *y* or *cv* is closer to the telomere. (Other data indicate that *y* is, in fact, near the telomere.)

A genetic map can be expanded by this type of reasoning to include all of the known genes in a chromosome; these genes constitute a **linkage group.** The number of linkage groups is the same as the haploid number of chromosomes of the species. For example, cultivated corn (*Zea mays*) has ten pairs of chromosomes and ten linkage groups. A partial genetic map of chromosome 10 is shown in Figure 4.11, along with the dramatic phenotypes caused by some of the mutations. The ears of corn in photographs parts C and F demonstrate the result of Mendelian segregation. The ear in part C shows a 3 : 1 segregation of yellow : orange kernels produced by the recessive *orange pericarp-2 (orp-2)* allele in a cross between two heterozygous genotypes; the ear in part F shows a 1 : 1 segregation of marbled : white kernels produced by the dominant allele *R1-mb* in a cross between a heterozygous genotype and a homozygous wildtype.

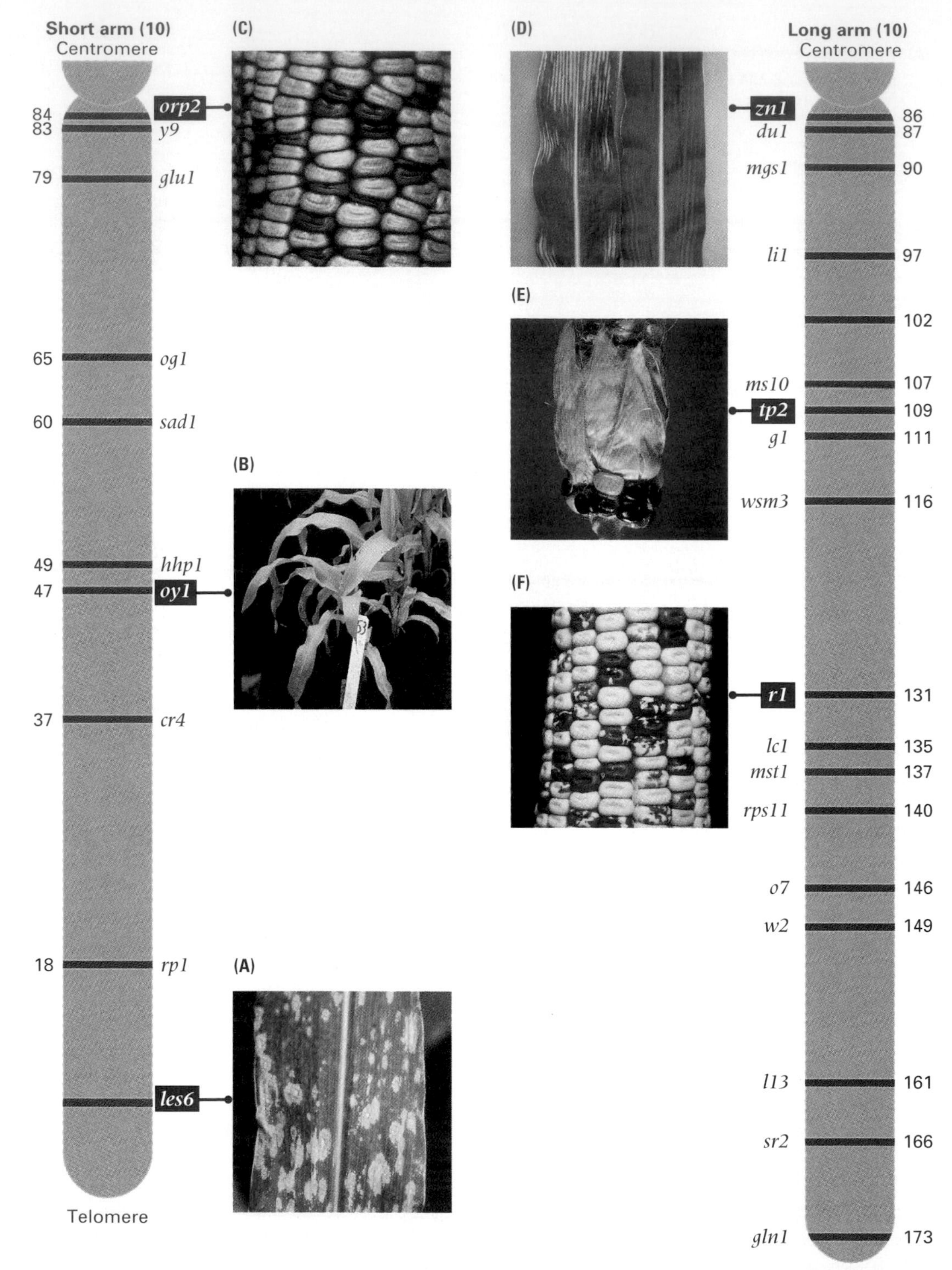

Figure 4.11 Genetic map of chromosome 10 of corn, *Zea mays.* The map distance to each gene is given in map units (centimorgans) relative to a position 0 for the telomere of the short arm (lower left). (A) Mutations in the gene *lesion-6 (les6)* result in many small to medium-sized, irregularly spaced, discolored spots on the leaf blade and sheath. (B) Mutations in the gene *oil yellow-1 (oy1)* result in a yellow-green plant; in the photograph, the plant in front shows the mutant phenotype; behind it is a normal plant. (C) The *orp2* allele is a recessive expressed as orange pericarp, a maternal tissue that surrounds the kernels; the photograph shows segregation in the F_2, yielding a 3 : 1 ratio of yellow : orange seeds. (D) In *zn1* (*zebra necrotic-1*) mutants, stripes of leaf tissue die; in the photograph, the left leaf is homozygous *zn1*, and the right is wildtype. (E) Mutants for the gene *teopod-2* (*tp2*) have small, partially podded ears and a simple tassle. (F) The mutation *R1-mb* is an allele of the *r1* gene resulting in red or purple color in the aleurone layer of the seed; the photograph shows the marbled color in kernels of an ear segregating for *R1-mb.* [Photographs courtesy of M. G. Neuffer; genetic map courtesy of E. H. Coe.]

Physical distance is often—but not always—correlated with map distance.

Generally speaking, the greater the physical separation between genes along a chromosome, the greater the map distance between them. Physical distance and genetic map distance are usually correlated, because a greater distance between genetic markers affords a greater chance for a crossover to take place; crossing-over is a physical exchange between the chromatids of paired homologous chromosomes.

On the other hand, the general correlation between physical distance and genetic map distance is by no means absolute. We have already noted that the frequency of recombination between genes may differ in males and females. An unequal frequency of recombination means that the sexes can have different map distances in their genetic maps, although the physical chromosomes of the two sexes are the same and the genes must have the same linear order. For example, because there is no recombination in male *Drosophila*, the map distance between any pair of genes located in the same chromosome, when measured in the male, is 0. (On the other hand, genes on different chromosomes do undergo independent assortment in males.)

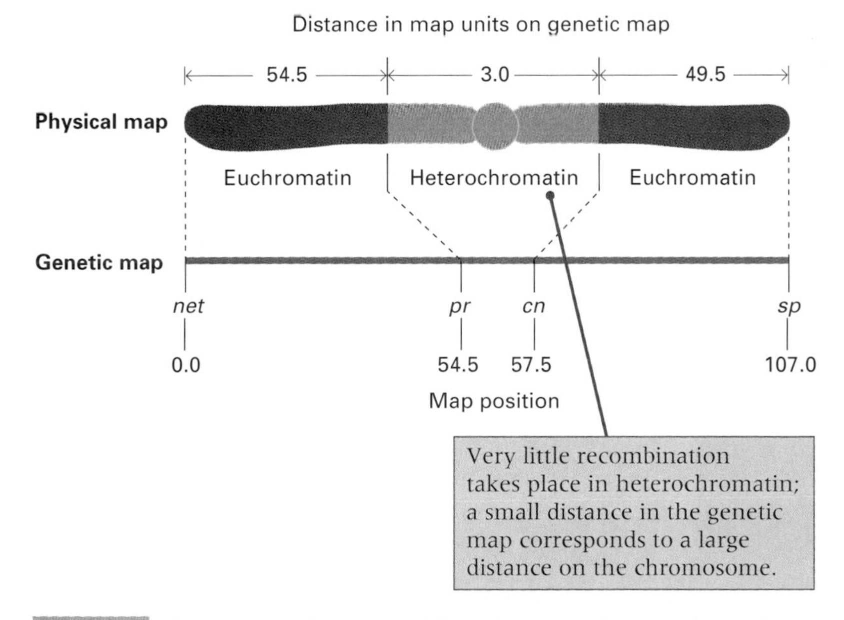

Figure 4.12 Chromosome 2 in *Drosophila* as it appears in metaphase of mitosis (physical map, top) and in the genetic map (bottom). Heterochromatin and euchromatin are in contrasting colors. The genes indicated on the map are *net* (net wing veins), *pr* (purple eye color), *cn* (cinnabar eye color), and *sp* (speck of wing pigment). The genes *pr* and *cn* are actually in euchromatin but are located near the junction with heterochromatin. The total map length is 54.5 + 49.5 + 3.0 = 107.0 map units. The heterochromatin accounts for 3.0/107.0 = 2.8 percent of the total map length but constitutes approximately 25 percent of the physical length of the metaphase chromosome.

The general correlation between physical distance and genetic map distance can even break down in a single chromosome. For example, crossing-over is much less frequent in certain regions of the chromosome than in other regions. The term **heterochromatin** refers to certain regions of the chromosome that have a dense, compact structure in interphase; these regions take up many of the standard dyes used to make chromosomes visible. The rest of the chromatin, which becomes visible only after chromosome condensation in mitosis or meiosis, is called **euchromatin.** In most organisms, the major heterochromatic regions are adjacent to the centromere; smaller blocks are present at the ends of the chromosome arms (the telomeres) and interspersed with the euchromatin. In general, crossing-over is much less frequent in regions of heterochromatin than in regions of euchromatin.

Because there is less crossing-over in heterochromatin, a given length of heterochromatin will appear much shorter in the genetic map than an equal length of euchromatin. In heterochromatic regions, therefore, the genetic map gives a distorted picture of the physical map. An example of such distortion is illustrated in Figure 4.12, which compares the physical map and the genetic map of chromosome 2 in *Drosophila*. The physical map depicts the appearance of the chromosome in metaphase of mitosis. Two genes near the tips and two near the euchromatin–heterochromatin junction are indicated in the genetic map. The map distances across the euchromatic arms are 54.5 and 49.5 map units, respectively, for a total euchromatic map distance of 104.0 map units. However, the heterochromatin, which constitutes approximately 25 percent of the entire chromosome, has a genetic length in map units of only 3.0 percent. The distorted length of the heterochromatin in the genetic map results from the reduced frequency of crossing-over in the heterochromatin. In spite of the distortion of the genetic map across the heterochromatin, in the regions of euchromatin there is a good correlation between the physical distance between genes and their distance, in map units, in the genetic map.

Crossing-over is reciprocal and takes place at the four-strand stage.

The orderly arrangement of genes represented by a genetic map is consistent with the conclusion that each gene occupies a well-defined site, or **locus,** in the chromosome, with the alleles of a gene in a heterozygote occupying corresponding locations in the pair of homologous chromosomes. Crossing-over, which is brought about by a physical exchange of segments that results in a new association of genes in the same chromosome, has the following features:

1. The exchange of segments between parental chromatids takes place in the first meiotic prophase, *after the chromosomes have duplicated.* The group of four chromatids (strands) of a pair of homologous chromosomes are closely synapsed at this stage. Crossing-over is a physical exchange between chromatids in a pair of homologous chromosomes.

2. The exchange process consists of the breaking and rejoining of the two chromatids, resulting in the *reciprocal* exchange of equal and corresponding segments between them (see Figure 4.5).

3. The sites of crossing-over are more or less random along the length of a chromosome pair. Hence the probability of crossing-over between two genes increases as the physical distance between the genes along the chromosome becomes larger. This principle is the basis of genetic mapping.

So far we have asserted, without citing experimental evidence, that crossing-over takes place in meiosis after the chromosomes have duplicated, at the stage when each bivalent has four chromatid strands. One experimental proof that crossing-over takes place after the chromosomes have duplicated came from a study of laboratory stocks of *D. melanogaster* in which the two X chromosomes in a female are joined to a common centromere to form an aberrant chromosome called an **attached-X** chromosome. The normal X chromosome in *Drosophila* has a centromere almost at the end of the chromosome, and the attachment of two of these chromosomes to a single centromere results in a chromosome with two equal arms, each consisting of a virtually complete X (Figure 4.13B). Females with an attached-X chromosome usually contain a Y chromosome as well, and they produce two classes of viable offspring (Figure 4.13D): females with the maternal attached-X chromosome plus a paternal Y chromosome, and males with the maternal Y chromosome along with a paternal X chromosome. Attached-X chromosomes are frequently used to study X-linked genes in *Drosophila* because a male carrying any X-linked mutant allele, when crossed with an attached-X female, produces sons who carry the mutant allele and daughters who carry the attached-X chromosome. In matings with attached-X females, therefore, an X-linked gene in the male is transmitted from father to son to grandson and so on, which is the opposite of the usual pattern of X-linked inheritance.

In an attached-X chromosome in which one X carries a recessive allele and the other carries the wildtype nonmutant allele, crossing-over between the X-chromosome arms can yield attached-X products in which the recessive allele is present in both arms of the attached-X chromosome (Figure 4.14). Hence attached-X females that are heterozygous can produce some female progeny that are homozygous for the recessive allele. The frequency with which homozygosity is observed increases with increasing map distance of the gene from the centromere. The diagrams in Figure

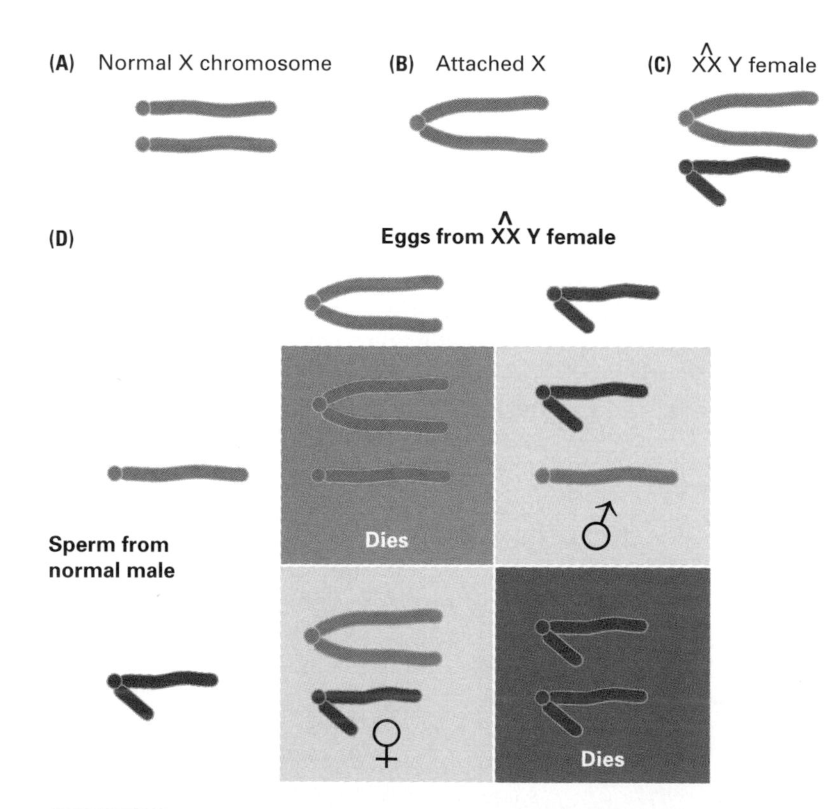

Figure 4.13 Attached-X chromosomes in *Drosophila.* (A) Structurally normal X chromosomes in a female. (B) An attached-X chromosome, with the long arms of two normal X chromosomes attached to a common centromere. (C) Typical attached-X females also contain a Y chromosome. (D) The outcome of a cross between an attached-X female and a normal male. The eggs contain either the attached-X or the Y chromosome; they combine at random with X-bearing or Y-bearing sperm. Genotypes with either three X chromosomes or no X chromosomes are lethal. Note that a male fly receives its X chromosome from its father and its Y chromosome from its mother, which is the opposite of the usual situation in *Drosophila.*

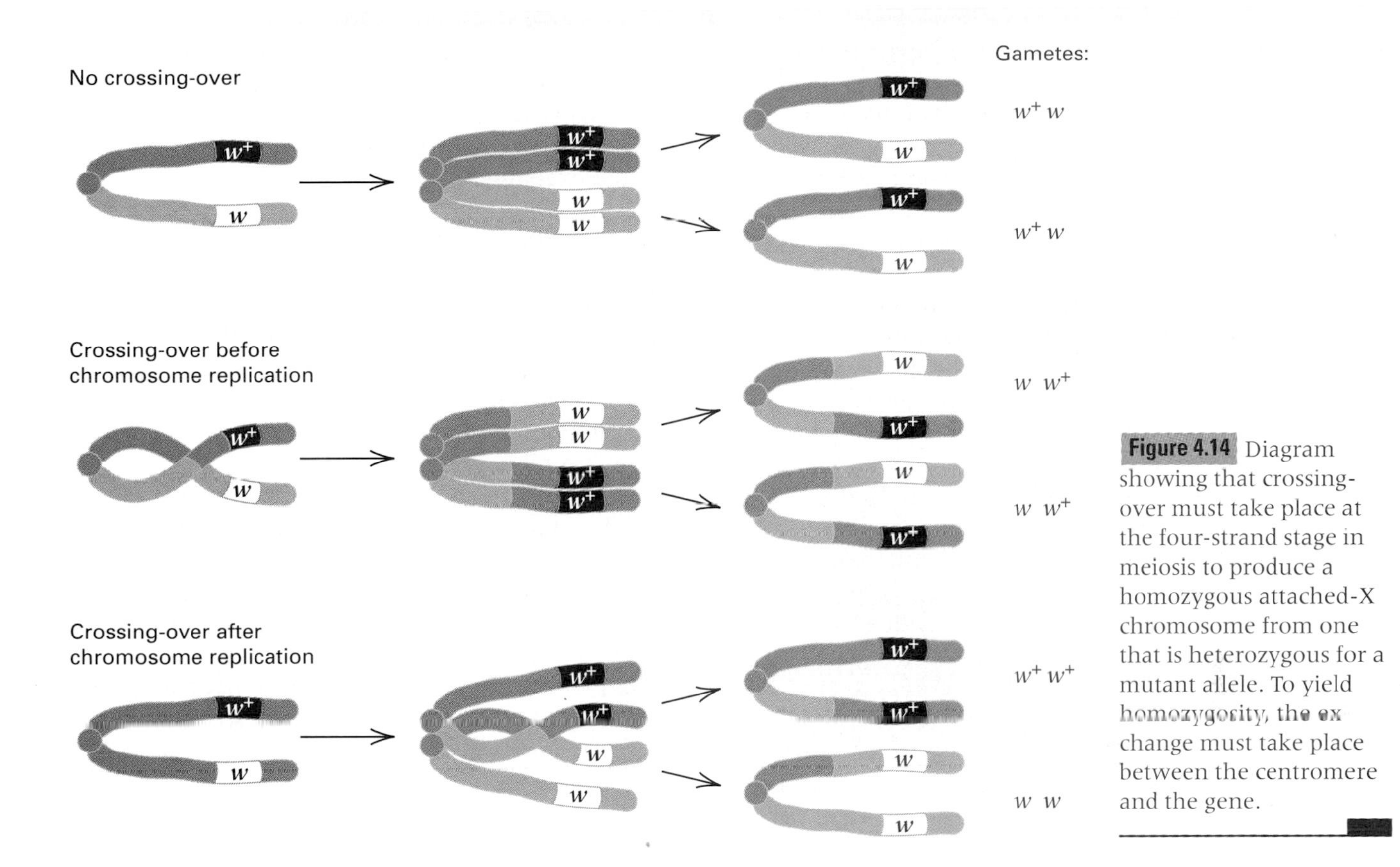

Figure 4.14 Diagram showing that crossing-over must take place at the four-strand stage in meiosis to produce a homozygous attached-X chromosome from one that is heterozygous for a mutant allele. To yield homozygosity, the exchange must take place between the centromere and the gene.

4.14 show that homozygosity can result only if the crossover between the gene and the centromere takes place *after* the chromosome has replicated. The implication of finding homozygous attached-X female progeny is therefore that crossing-over takes place at the *four-strand* stage of meiosis. If this were not the case, and crossing-over happened before replication of the chromosome (at the two-strand stage), it would result only in a swap of the alleles between the chromosome arms and would never yield the homozygous products that are actually observed.

One crossover can undo the effects of another.

When two genes are located far apart along a chromosome, more than one crossover can be formed between them in a single meiosis, and this complicates the interpretation of recombination data. The probability of multiple crossovers increases with the distance between the genes. Multiple crossing-over complicates genetic mapping because map distance is based on the number of physical exchanges that are formed, and some of the multiple exchanges between two genes do not result in recombination of the genes and hence are not detected. As we saw in Figure 4.9, the effect of one crossover can be canceled by another crossover further along the way. If two exchanges between the same two chromatids take place between the genes *A* and *B*, then their net effect will be that all chromosomes are nonrecombinant, either *A B* or *a b*. Two of the products of this meiosis have an interchange of their middle segments, but the chromosomes are not recombinant for the genetic markers and so are genetically indistinguishable from noncrossover chromosomes. The possibility of such canceling events means that the observed recombination value is an *underestimate* of the true exchange frequency and the map distance between the genes. In higher organisms, double crossing-over is effectively precluded in chromosome segments that are sufficiently short, usually about 10 map units or less. Therefore, multiple crossovers that cancel each other's effects can be avoided by using recombination data for closely linked genes to build up genetic linkage maps.

The minimum recombination frequency between two genes is 0. The recombination frequency also has a maximum:

> No matter how far apart two genes may be, the maximum frequency of recombination between any two genes is 50 percent.

Fifty percent recombination is the same value that would be observed if the genes were on nonhomologous chromosomes and assorted independently. The maximum frequency of recombination is observed when the genes are so far apart in the chromosome that at least one crossover is almost always formed between them. In Figure 4.7B, it can be seen that a single exchange in every meiosis would result in half of the products having parental combinations and the other half having recombinant combinations of the genes. The occurrence of two exchanges between two genes has the same effect, as shown in Figure 4.15. Part A shows a two-strand double crossover, in which the same chromatids participate in both exchanges; no recombination of the marker genes is detectable. When the two exchanges have one chromatid in common (three-strand double crossover, parts B and C), the result is indistinguishable from that of a single exchange; two products with parental combinations and two with recombinant combinations are produced. Note that there are two types of three-strand doubles, depending on which three chromatids participate. The final possibility is that the second exchange connects the chromatids that did not participate in the first exchange (four-strand double crossover, part D), in which case all four products are recombinant.

In most organisms, when double crossovers are formed, the chromatids that take part in the two exchange events are selected at random. In this case, the expected proportions of the three types of double exchanges are 1/4 four-strand doubles, 1/2 three-strand doubles, and 1/4 two-strand doubles. This means that on the average, $(1/4)(0) + (1/2)(2) + (1/4)(4) = 2$ recombinant chromatids will be found among the 4 chromatids produced from meioses with two exchanges between a pair of genes. This is the same proportion obtained with a single exchange between the genes. Moreover, a maximum of 50 percent recombination is obtained for any number of exchanges.

Double crossing-over is detectable in recombination experiments that employ **three-point crosses,** which include three pairs of alleles. If a third pair of alleles, c^+ and c, is located between the two with which we have been concerned (the outermost genetic markers), double exchanges in the region can be detected when the crossovers flank the c gene (Figure 4.16). The two crossovers, which in this example take place between the same pair of chromatids, would result in a reciprocal exchange

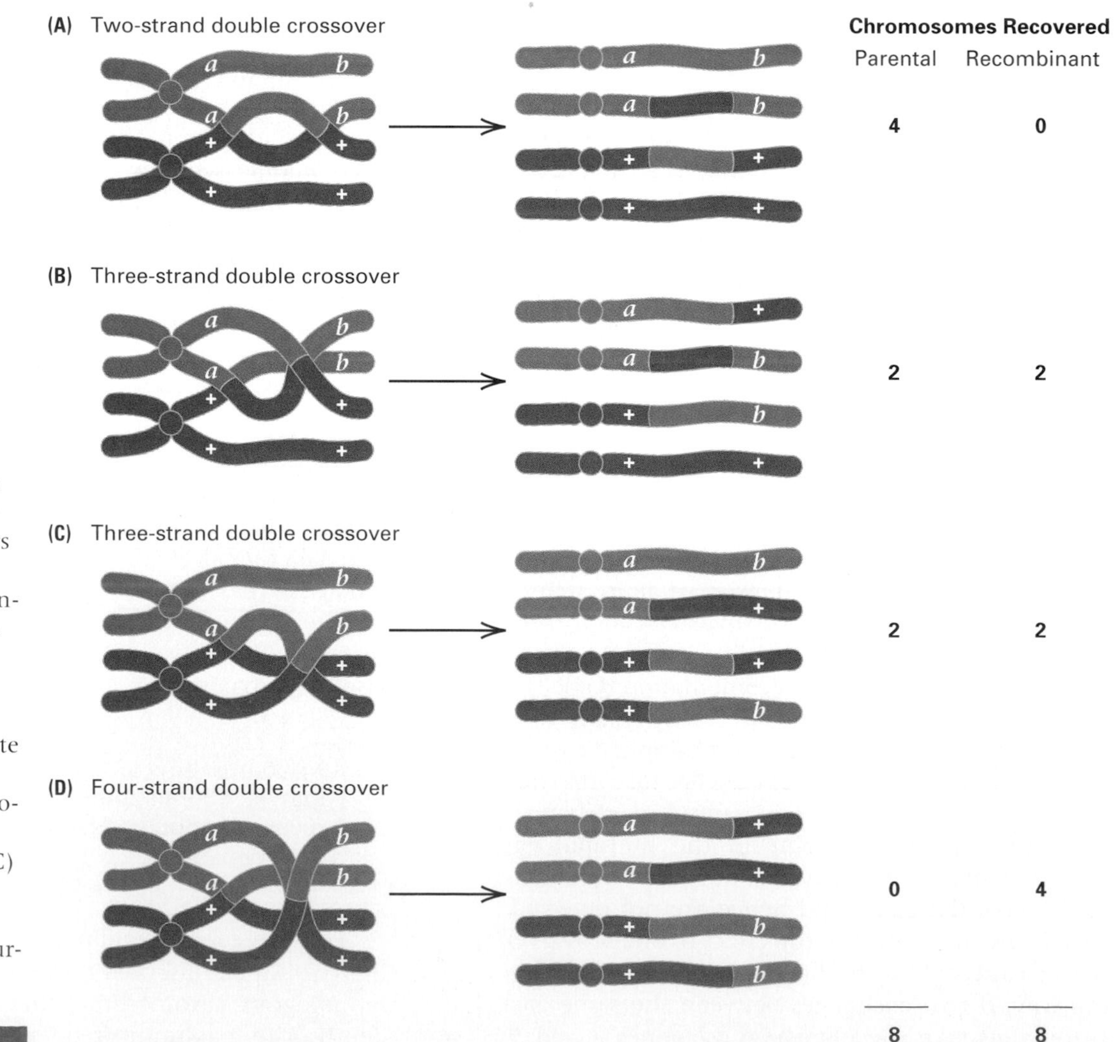

Figure 4.15 Diagram showing that the result of two crossovers in the interval between two genes is indistinguishable from independent assortment of the genes, provided that the chromatids participate at random in the crossovers. (A) A two-strand double crossover. (B) and (C) The two types of three-strand double crossovers. (D) A four-strand double crossover.

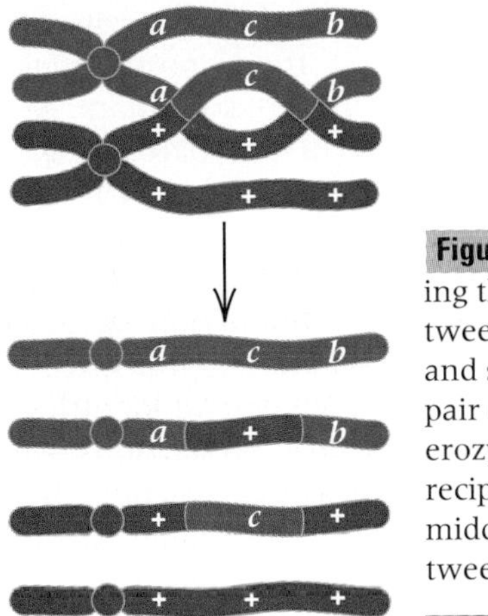

Figure 4.16 Diagram showing that two crossovers between the same chromatids and spanning the middle pair of alleles in a triple heterozygote will result in a reciprocal exchange of the middle pair of alleles between the two chromatids.

of the c^+ and c alleles between the chromatids. A three-point cross is an efficient way to obtain recombination data; it is also a simple method for determining the order of the three genes, as we will see in the next section.

4.3 Double crossovers are revealed in three-point crosses.

The data in Table 4.1 result from a testcross in corn with three genes in a single chromosome. The analysis illustrates the approach to interpreting a three-point cross. The recessive alleles of the genes in this cross are *lz* (for lazy or prostrate growth habit), *gl* (for glossy leaf), and *su* (for sugary endosperm), and the multiply heterozygous parent in the cross had the genotype

$$\frac{Lz\ Gl\ Su}{lz\ gl\ su}$$

where each symbol with an initial capital letter represents the dominant allele. (This type of symbolism is usual in corn genetics.) The two classes of progeny that inherit noncrossover (parental-type) gametes are therefore the wildtype plants and those with the lazy-glossy-sugary phenotype. The number of progeny in these classes is far larger than the number in any of the crossover classes. Because the frequency of recombination is never larger than 50 percent, the very fact that these progeny are the most numerous indicates that the gametes that gave rise to them have the parental allele configurations, in this case *Lz Gl Su* and *lz gl su*. Using this principle, we could have inferred the genotype of the heterozygous parent even if the genotype had not been stated. This is a point important enough to state more generally:

> In any genetic cross, no matter how complex, the two most frequent types of gametes with respect to any pair of genes are *nonrecombinant;* these provide the linkage phase (*cis* versus *trans*) of the alleles of the genes in the multiply heterozygous parent.

In mapping experiments, the gene sequence is usually not known. In this example, the order in which the three genes are shown is entirely arbitrary. However, there is an easy way to determine the correct order from three-point data. The gene order can be deduced by identifying the genotypes

Table 4.1
Interpreting a Three-Point Cross

Phenotype of Testcross Progeny	Genotype of Gamete from Hybrid Parent	Number of Progeny
Wildtype	*Lz Gl Su*	286
Lazy	*lz Gl Su*	33
Glossy	*Lz gl Su*	59
Sugary	*Lz Gl su*	4
Lazy, glossy	*lz gl Su*	2
Lazy, sugary	*lz Gl su*	44
Glossy, sugary	*Lz gl su*	40
Lazy, glossy, sugary	*lz gl su*	272
		740

of the double-crossover gametes produced by the heterozygous parent and comparing these with the nonrecombinant gametes. Because the probability of two simultaneous exchanges is considerably smaller than that of either single exchange, the double-crossover gametes will be the least frequent types. It is clear in Table 4.1 that the classes composed of four plants with the sugary phenotype and two plants with the lazy-glossy phenotype (products of the *Lz Gl su* and *lz gl Su* gametes, respectively) are the least frequent and therefore constitute the double-crossover progeny. Now we apply another principle:

> The effect of double crossing-over is to interchange the members of the *middle* pair of alleles between the chromosomes.

This principle is illustrated in Figure 4.17. With three genes there are three possible orders, depending on which gene is in the middle. If *gl* were in the middle (part A), the double-recombinant gametes would be *Lz gl Su* and *lz Gl su,* which is inconsistent with the data. Likewise, if *lz* were in the middle (part C), the double-recombinant gametes would be *Gl lz Su* and *gl Lz su,* which is also inconsistent with the data.

The correct order of the genes, *lz*−*su*−*gl,* is given in part B, because in this case, the double-recombinant gametes are *Lz su Gl* and *lz Su gl,* which Table 4.1 indicates is actually the case. Although one can always infer which gene is in the middle by going through all three possibilities, there is a shortcut. Each double-recombinant gamete will always match one of the parental gametes in two of the alleles. In Table 4.1, for example, the double-recombinant gamete *Lz Gl su* matches the parental gamete *Lz Gl Su* except for the allele *su.* Similarly, the double-recombinant gamete *lz gl Su* matches the parental gamete *lz gl su* except for the allele *Su.* The middle gene can be identified because the "odd man out" in the comparisons—in this case, the alleles of *Su*—is always the gene in the middle. The reason is that only the middle pair of alleles is interchanged by double crossing-over.

Taking the correct gene order into account, the genotype of the heterozygous parent in the cross yielding the progeny in Table 4.1 should be written as

$$\frac{Lz\ Su\ Gl}{lz\ su\ gl}$$

The consequences of single crossing-over in this genotype are shown in Figure 4.18. A single crossover in the *lz*−*su* region (part A) yields the reciprocal recombinants *Lz su gl* and *lz Gl Su,* and a single crossover in the *su*−*gl* region (part B) yields the reciprocal recombinants *Lz Su gl* and *lz gl Su.* The consequences of double crossing-over are illustrated in Figure 4.19. There are four different types of double crossing-over: a two-strand double (part A), two types of three-strand doubles (parts B

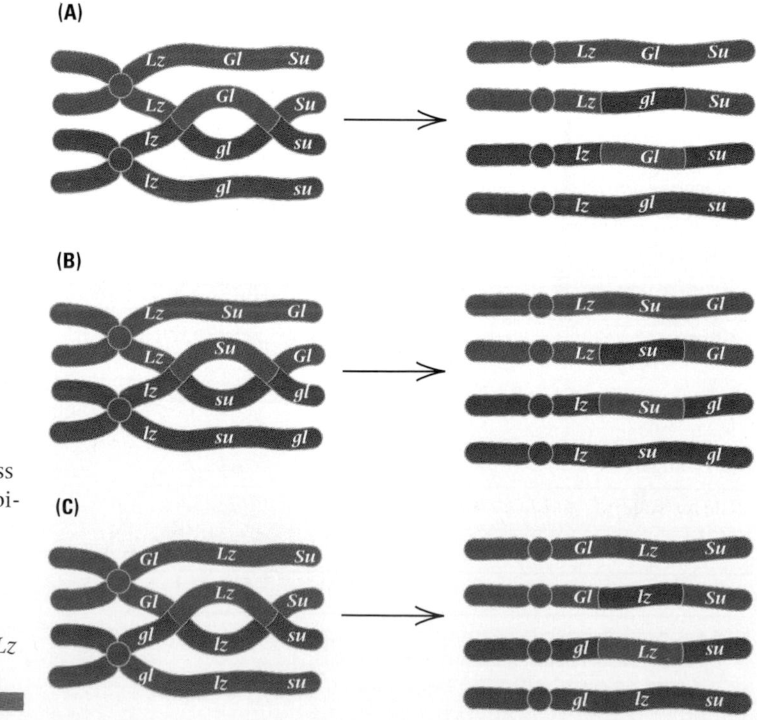

Figure 4.17 The order of genes in a three-point testcross may be deduced from the principle that double recombination interchanges the middle pair of alleles. For the genes *Lu, Gl,* and *Su,* there are three possible orders (parts A, B, and C), each of which predicts a different pair of gametes as the result of double recombination. Only the order in B is consistent with the finding that *Lz Gl su* and *lz gl Su* are the double-recombinant gametes.

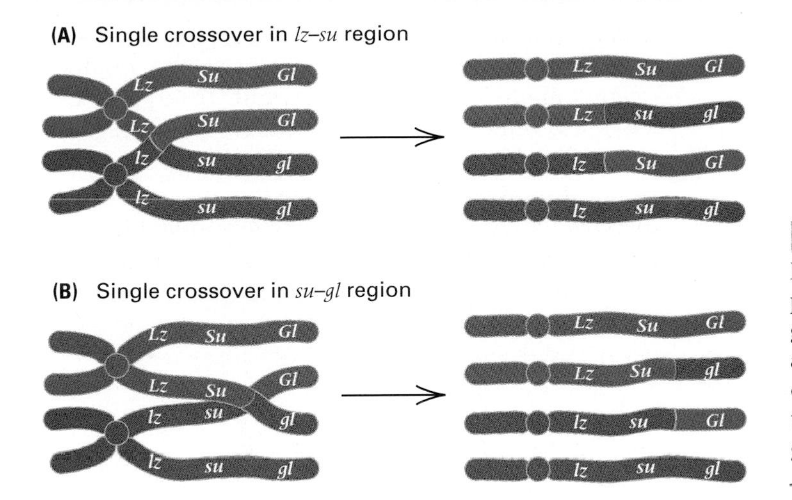

Figure 4.18 Result of single crossing-over in a triple heterozygote, using the *Lz*–*Su*–*Gl* region as an example. (A) A crossover between *Lz* and *Su* results in two gametes that show recombination between *Lz* and *Su* and two gametes that are nonrecombinant. (B) A crossover between *Su* and *Gl* results in two gametes that show recombination between *Su* and *Gl* and two gametes that are nonrecombinant.

and C), and a four-strand double (part D). These types were illustrated earlier in Figure 4.15, where the main point was that with two genetic markers flanking the crossovers, the occurrence of double crossing-over cannot be detected genetically. The difference in the present case is that, here, the genetic marker *su* is located in the middle between the two crossovers, so some of the double crossovers can be detected genetically. On the right in Figure 4.19, the asterisks mark the sites of crossing-over between nonsister chromatids. In terms of recombination, the result is that

- A two-strand double crossover (part A) yields the reciprocal double-recombinant products *Lz su Gl* and *lz Su gl*.

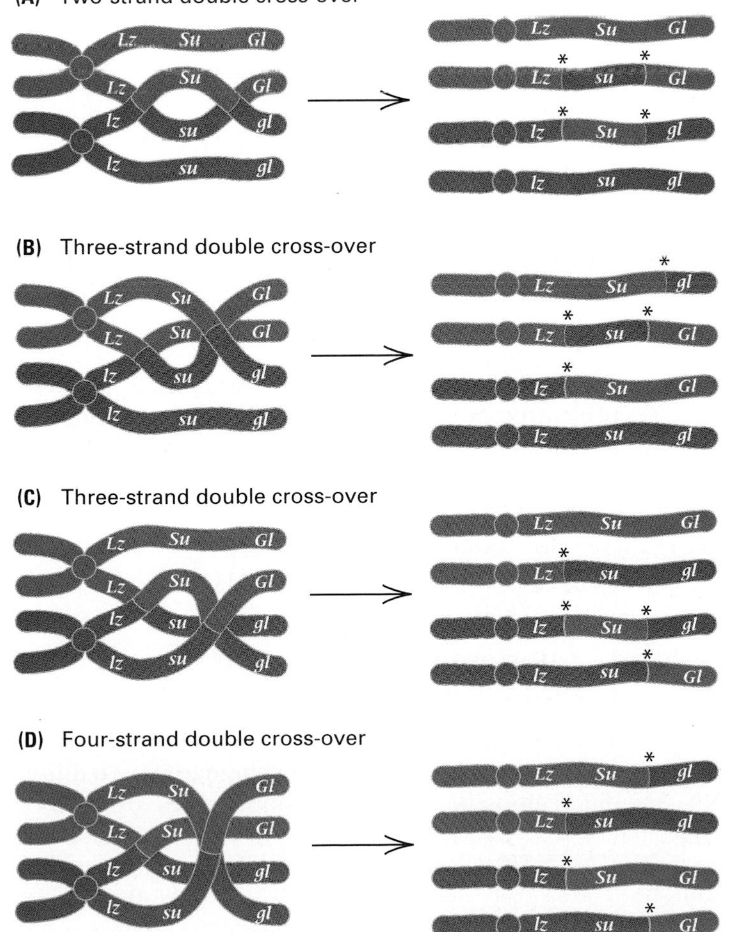

Figure 4.19 Result of double crossing-over in a triple heterozygote, using the *Lz*–*Su*–*Gl* region as an example. Note that chromosomes showing double recombination derive from two-strand double crossover (A) or from either type of three-strand double crossover (B and C). The four-strand double crossover (D) results only in single-recombinant chromosomes.

- One three-strand double crossover (part B) yields the double-recombinant product *Lz su Gl* and two single-recombinant products, *Lz Su gl* and *lz Su Gl.*
- The other three-strand double crossover (part C) yields the double-recombinant product *lz Su gl* and two single-recombinant products, *Lz su gl* and *lz su Gl.*
- The four-strand double crossover (part C) yields reciprocal single recombinants in the *lz–su* region, namely *Lz su gl* and *lz Su Gl,* and reciprocal single recombinants in the *su–gl* region, namely *Lz Su gl* and *lz su Gl.*

Note that the products of recombination in the three-strand double crossovers (parts B and C) are the reciprocals of each other. Because these two types of double crossovers are equally frequent, the reciprocal products of recombination are expected to appear in equal numbers.

We can now summarize the data in Table 4.1 in a more informative way by writing the genes in the correct order and grouping reciprocal gametic genotypes together. This grouping is shown in Table 4.2. Note that each class of single recombinants consists of two reciprocal products and that these are found in approximately equal frequencies (40 versus 33 and 59 versus 44). This observation illustrates an important principle:

> The two reciprocal products resulting from any crossover, or any combination of crossovers, are expected to appear in approximately equal frequencies among the progeny.

Table 4.2
Comparing Reciprocal Products in a Three-Point Cross

Genotype of Gamete from Hybrid Parent	Number of Progeny	Intervals Showing Recombination
Lz Su Gl	286	
lz su gl	272	
Lz su gl	40	*lz–su*
lz Su Gl	33	
Lz Su gl	59	*su–gl*
lz su Gl	44	
Lz su Gl	4	*lz–su* + *su–gl*
lz Su gl	2	
	740	

Total number of recombinants in *lz–su* region: 40 + 33 + 4 + 2 = 79

Total number of recombinants in *su–gl* region: 59 + 44 + 4 + 2 = 109

In calculating the frequency of recombination from the data, remember that the double-recombinant chromosomes result from *two* exchanges, one in each of the chromosome regions defined by the three genes. Therefore, chromosomes that are recombinant between *lz* and *su* are represented by the following chromosome types:

Lz su gl	40
lz Su Gl	33
Lz su Gl	4
lz Su gl	2
	79

The total implies that 79/740, or 10.7 percent, of the chromosomes recovered in the progeny are recombinant between the *lz* and *su* genes, so the map distance between these genes is 10.7 map units, or 10.7 centimorgans. Similarly, the chromosomes that are recombinant between *su* and *gl* are represented by

Lz Su gl	59
lz su Gl	44
Lz su Gl	4
lz Su gl	2
	109

In this case the recombination frequency between *su* and *gl* is 109/740, or 14.8 percent, so the map distance between these genes is 14.8 map units, or 14.8 centimorgans. The genetic map of the chromosome segment in which the three genes are located is therefore

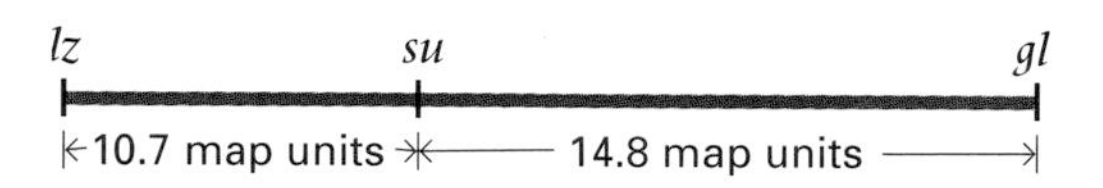

The most common error in learning how to interpret three-point crosses is to forget to include the double recombinants when calculating the recombination frequency between adjacent genes. You can keep from falling into this trap by remembering that the double-recombinant chromosomes have single recombination in *both* regions.

Interference decreases the chance of multiple crossing-over.

The detection of double crossing-over makes it possible to determine whether exchanges in two different regions of a pair of chromosomes are formed independently of each other. Using the information from the example with corn, we know from the recombination frequencies that the probability of recombination is 0.107

The human connection

Count Your Blessings

Joe Hin Tijo[1] and Albert Levan[2] 1956
[1]Estacion Experimental de Aula Dei, Zaragoza, Spain
[2]Institute of Genetics, Lund, Sweden
The Chromosome Number in Man

This paper marks the beginning of modern human cytogenetics. Tijo and Levan made a technical improvement for spreading chromosomes on a microscope slide, which for the first time allowed accurate chromosome counts to be made. Previous methods ran up against the problem that human chromosomes are small, relatively numerous, and bunched together at metaphase. Until 1956 it was widely believed that the human chromosome number was 48, as it is in the chimpanzee and gorilla. The number 48 became dogma, so much so that other counts were disbelieved, as illustrated in this excerpt by the reference to a previous researcher who repeatedly obtained a chromosome number of 46 in liver cells but abandoned the study because she was unable to find the two "missing" chromosomes. What worked for Tijo and Levan was the simple trick of soaking the cells in a hypotonic solution (a solution with a lower concentration of charged inorganic ions than of the cells themselves). When cells are bathed in a hypotonic solution, water surges through the cell membrane into the cells and causes them to swell. The nucleus becomes considerably enlarged, thereby spreading out the chromosomes. Once this technique was in use, the discovery of many human chromosomal abnormalities followed quickly, such as the finding of three copies of chromosome 21 in Down syndrome. (See "The Human Connection" in Chapter 5.)

While staying last summer at the Sloan-Kettering Institute, New York, one of us tried out hypotonic treatment on various human tissue cultures. . . . The results were promising inasmuch as some fairly satisfactory chromosome analyses were obtained. . . . The treatment [had] a tendency to make the chromosome outlines somewhat blurred and vague. We consequently tried to abbreviate the treatment to a minimum, hoping to induce the scattering of the chromosomes without unfavorable effects on the chromosome surface. . . . Treatment with hypotonic solution for only one or two minutes gave good results. . . . Ordinary squash preparations were made. For chromosome counts the squashing was made very mild in order to keep the chromosomes in the metaphase group. For studies of chromosome mophology a more thorough squashing was preferable. In many cases single cells were squashed under the microscope by a slight pressure of a needle. In such cases it was directly observed that no chromosomes escaped. . . . We were surprised to find that [among 261 cells] the chromosome number 46 predominated in the tissue cultures from embryonic cells. . . . Lower numbers were frequent, of course, but always in cells that seemed damaged. These were consequently disregarded. . . . The chromosomes are easily arranged in pairs, but only certain of these pairs are individually distinguishable. . . . The almost exclusive occurrence of the chromosome number 46 in embryonic cell cultures is a very unexpected finding. . . . After the conclusion had been drawn that the tissue studied by us had 46 as a chromosome number, Dr. Eva Hansen-Melander kindly informed us that during last Spring she had studied the chromosomes of embryonic liver mitosis. This study, however, was temporarily discontinued because the workers were unable to find all the 48 human chromosomes in their material; as a matter of fact, the number 46 was repeatedly counted in their slides. This finding suggests that 46 may be the correct chromosome number for human liver tissue, too. . . . We do not wish to generalize our present findings into a statement that the chromosome number of human beings is $2n = 46$, but it is hard to avoid the conclusion that this would be the most natural explanation of our data.

WE WERE SURPRISED to find that the chromosome number 46 predominated

Source: Hereditas 42: 1–6.

between *lz* and *su* and 0.148 between *su* and *gl*. If recombination is independent in the two regions (which means that the formation of one crossover does not alter the probability of the second crossover), the probability of a single recombination in both regions is the product of these separate probabilities, or $0.107 \times 0.148 = 0.0158$ (1.58 percent). This implies that in a sample of 740 gametes, the expected number of double recombinants would be 740×0.0158, or 12, whereas the number actually observed was only 6 (Table 4.2). Such deficiencies in the observed number of double recombinants are common; they reflect a phenomenon called chromosome **interference,** in which a crossover in one region of a chromosome reduces the probability of a second crossover in a nearby region. Over sufficiently short genetic distances, chromosome interference is nearly complete.

The **coefficient of coincidence** is the observed number of double-recombinant chromosomes divided by the expected number. Its value provides a quantitative measure of the degree of interference, which is defined as

$$i = \text{Interference}$$
$$= 1 - (\text{Coefficient of coincidence})$$

From the data in the corn example, the coefficient of coincidence is calculated as follows:

- Observed frequency of double recombinants = 6
- Expected frequency of double recombinants = $0.107 \times 0.148 \times 740 = 12$
- Coefficient of coincidence = $6/12 = 0.50$

The 0.50 means that the observed number of double recombinants was only 50 percent of the number expected if crossing-over in the two regions were independent. The value of the interference depends on the distance between the genetic markers and on the species. In some species, the interference increases as the distance between the two outside markers becomes smaller, until a point is reached at which double crossing-over is eliminated; that is, no double recombinants are found, and the coefficient of coincidence equals 0 (or, to say the same thing, the interference equals 1). In *Drosophila* this distance is about 10 map units. In yeast, by contrast, interference is incomplete even over short distances. For markers separated by 3 map units, the interference is in the range 0.3 to 0.6; and for those separated by 7 map units, it is in the range 0.1 to 0.3. In most organisms, when the total distance between the genetic markers is greater than about 30 map units, interference essentially disappears and the coefficient of coincidence approaches 1.

The effect of interference on the relationship between genetic map distance and the frequency of recombination is illustrated in Figure 4.20. Each curve is an example of a **mapping function,** which is the mathematical relation between the genetic distance across an interval in map units (centimorgans) and the observed frequency of recombination across the interval. In other words, a mapping function tells one how to convert a *map distance* between genetic markers into a *recombination frequency* between the markers. As we have seen, when the map distance between the markers is small, the recombination frequency equals the map distance. This principle is reflected in the curves in Figure 4.20 in the region in which the map distance is smaller than about 10 cM. At less than this distance, all of the curves are nearly straight lines, which means that map distance and recombination frequency are equal; 1 map unit equals 1 percent recombination, and 10 map units equals 10 percent recombination. For distances greater than 10 map units, the recombination frequency becomes smaller than the map distance according to the pattern of interference along the chromosome. Each pattern of interference yields a different mapping function, as shown by the three examples in Figure 4.20.

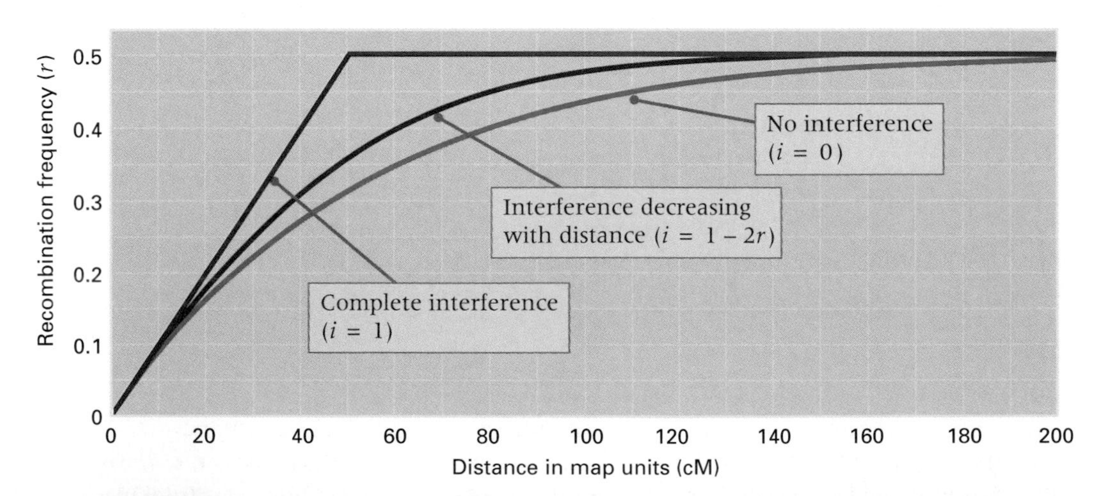

Figure 4.20 A mapping function is the relation between genetic map distance across an interval and the observed frequency of recombination across the interval. Map distance is defined as one-half the average number of crossovers converted into a percentage. The three mapping functions correspond to different assumptions about interference, *i*. The mapping function in the middle is based on the assumption that *i* decreases as a linear function of distance.

4.4 Polymorphic DNA sequences are used in human genetic mapping.

Until quite recently, mapping genes in human beings was very tedious and slow. Numerous practical obstacles complicated genetic mapping in human pedigrees:

1. Most genes that cause genetic diseases are rare, so they are observed in only a small number of families.
2. Many mutant genes of interest in human genetics are recessive, so they are not detected in heterozygous genotypes.
3. The number of offspring per human family is relatively small, so segregation cannot usually be detected in single sibships.
4. The human geneticist cannot perform testcrosses or backcrosses, because human matings are not dictated by an experimenter.

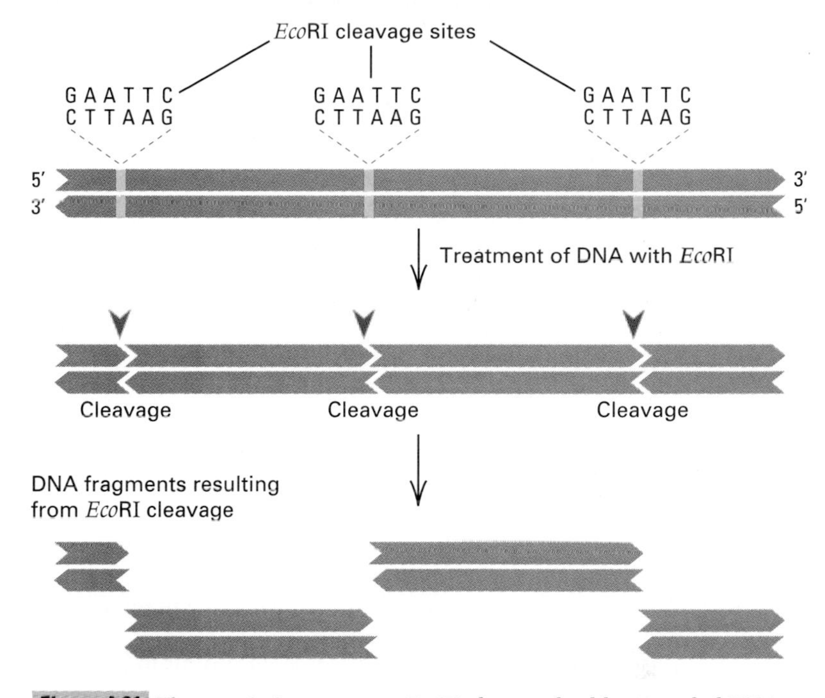

Figure 4.21 The restriction enzyme *Eco*RI cleaves double-stranded DNA wherever the sequence 5'-GAATTC-3' is present. In the example shown here, the DNA molecule contains three *Eco*RI cleavage sites, and it is cleaved at each site, producing a number of fragments.

In recent years, the use of techniques for manipulating DNA have enabled investigators to carry out genetic mapping in human pedigrees primarily by using genetic markers present in the DNA itself, rather than through the phenotypes produced by mutant genes. There are many minor differences in DNA sequence from one person to the next. On the average, the DNA sequences at corresponding positions in any two chromosomes, taken from any two people, differ at approximately one in every thousand base pairs. A genetic difference that is relatively common in a population is called a **polymorphism.** Most polymorphisms in DNA sequence are not associated with any inherited disease or disability; many occur in DNA sequences that do not code for proteins. Nevertheless, each of the polymorphisms serves as a convenient genetic marker, and those genetically linked to genes that cause hereditary diseases are particularly important. Some polymorphisms in DNA sequence are detected by means of a type of enzyme called a **restriction endonuclease,** which cleaves double-stranded DNA molecules wherever a particular, short sequence of bases is present. For example, the restriction enzyme *Eco*RI cleaves DNA wherever the sequence GAATTC appears in either strand, as illustrated in Figure 4.21. Restriction enzymes are considered in detail in Chapter 6. For present purposes, their significance is related to the fact that a difference in DNA sequence that eliminates a cleavage site can be detected because the region lacking the cleavage site will be cleaved into one larger fragment instead of two smaller ones (Figure 4.22). More rarely, a mutation in the DNA sequence will create a new site rather than destroy one already present. The main point is that any difference in DNA sequence that alters a cleavage site also changes the length of the DNA fragments produced by cleavage with the corresponding restriction enzyme. The different DNA fragments can be separated by size by an electric field in a supporting gel and detected by various means. Differences in DNA fragment length produced by presence or absence of the cleavage sites in DNA molecules are known as **restriction fragment length polymorphisms (RFLPs).**

RFLPs are typically formed in one of two ways. A mutation that changes a base sequence may result in loss or gain of a cleavage site that is recognized by the restriction endonuclease in use. Figure 4.23A gives an example. On the left is shown the relevant region in the homologous DNA molecules in a person who is heterozygous for such a sequence polymorphism. The homologous chromosomes in the person are distinguished by the letters *a* and *b*. In the region of interest, chromosome *a* contains two cleavage sites and chromosome *b* contains three. On the right is shown the position of the DNA fragments produced by cleavage after separation in an electric field. Each fragment appears

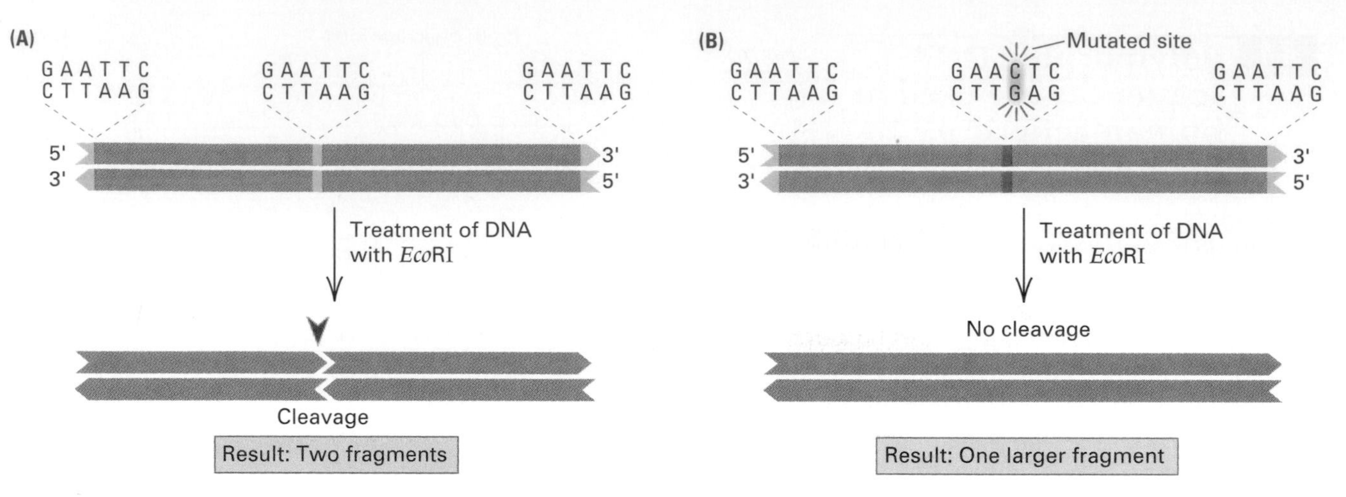

Figure 4.22 A minor difference in the DNA sequence of two molecules can be detected if the difference eliminates a restriction site. (A) This molecule contains three restriction sites for *Eco*RI, including one at each end. It is cleaved into two fragments by the enzyme. (B) This molecule has a mutant base sequence in the *Eco*RI site in the middle. It changes 5'-GAATTC-3' into 5'-GAACTC-3', which is no longer cleaved by *Eco*RI. Treatment of this molecule with *Eco*RI results in one larger fragment.

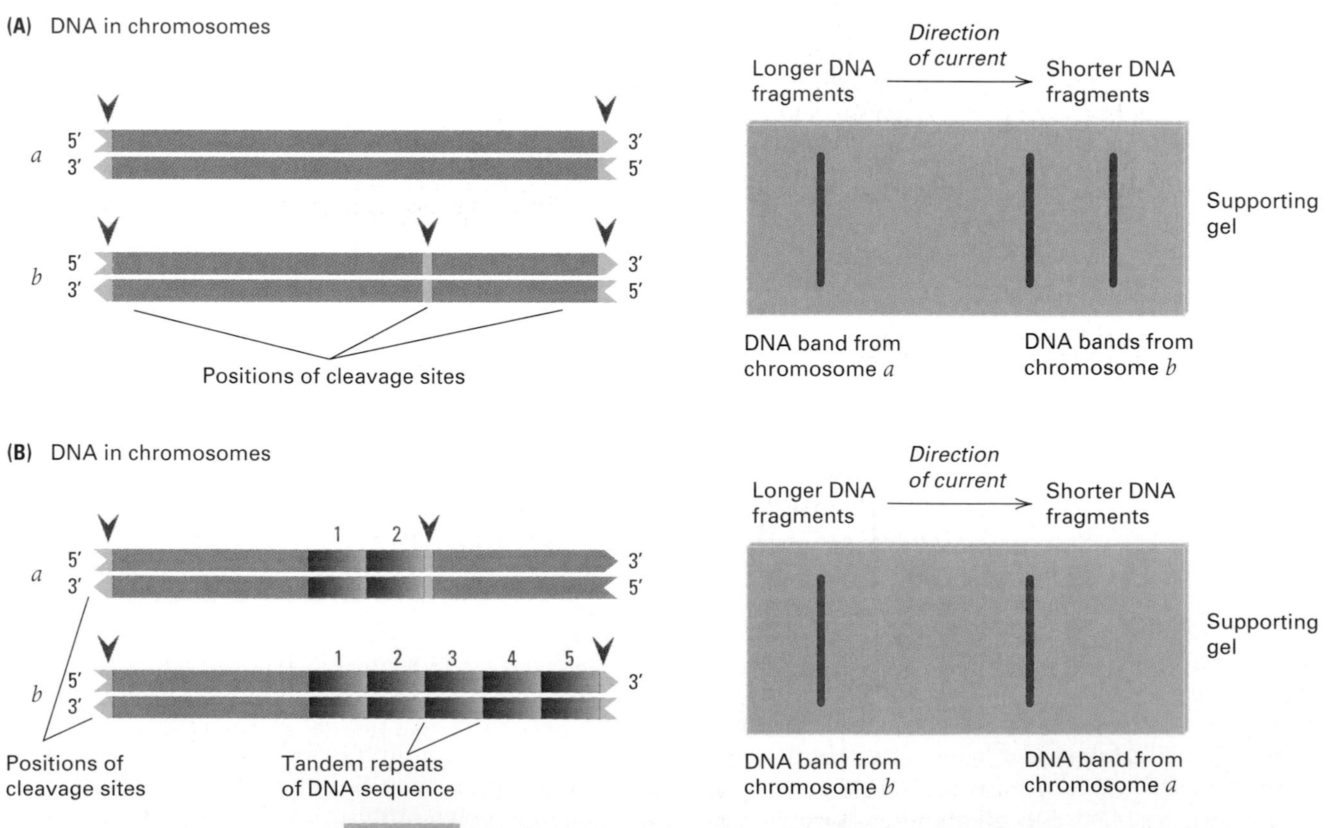

Figure 4.23 Two types of genetic variation that are widespread in most natural populations of animals and plants. (A) RFLP (restriction fragment length polymorphism), in which alleles differ in the presence or absence of a cleavage site in the DNA; the different alleles yield different fragment lengths (shown in the gel pattern at the right) when the molecules are cleaved with a restriction enzyme. (B) STRP (simple tandem repeat polymorphism), in which alleles differ in the number of repeating units present between two cleavage sites.

as a discrete band in the gel. The fragment from chromosome *a* migrates more slowly than those from chromosome *b* because it is longer, and longer fragments move more slowly through the gel. In this example, DNA from a person heterozygous for the *a* and *b* types of chromosomes (genotype *ab*) would yield three bands in a gel. Similarly, DNA from homozygous *aa* would yield one band, and that from homozygous *bb* would yield two bands.

A second type of DNA polymorphism results from differences in the number of copies of a short DNA sequence that may be repeated many times in tandem at a particular site in a chromosome (Figure 4.23B). In a particular chromosome, the tandem repeats may contain any number of copies, typically ranging from ten to a few hundred. When a DNA molecule is cleaved with a restriction endonuclease that cleaves at sites flanking the tandem repeat, the size of the DNA fragment produced is determined by the number of repeats present in the molecule. Figure 4.23B illustrates homologous DNA sequences in a heterozygous person containing one chromosome *a* with two copies of the repeat and another chromosome *b* with five copies of the repeat. When cleaved and separated in a gel, chromosome *a* yields a shorter fragment than that from chromosome *b*, because *a* contains fewer copies of the repeat. A genetic polymorphism resulting from a tandemly repeated short DNA sequence is called a **simple tandem repeat polymorphism (STRP).** An example of a STRP is the repeating sequence

5'-...TGTGTGTGTGTG...-3'

and the polymorphism consists of differences in the number of TG repeats. A particular "allele" of the STRP is defined by the number of TG repeats it includes.

The utility of STRPs in human genetic mapping derives from the very large number of alleles that may be present in the human population. The large number of alleles also implies that most people will be heterozygous, and so their DNA will yield two bands upon cleavage with the appropriate restriction endonuclease. Because of their high degree of variation among people, DNA polymorphisms are also widely used in DNA typing in criminal investigations (Chapter 13).

In genetic mapping, the phenotype of a person with respect to a DNA polymorphism is a pattern of bands in a gel. As with any other type of genetic marker, the genotype of a person with respect to the polymorphism is inferred, insofar as it is possible, from the phenotype. Linkage between different polymorphic loci is detected through lack of independent assortment of the alleles in pedigrees, and recombination and genetic mapping are carried out using the same principles as apply in other organisms, except that in human beings, because of the small family size, different pedigrees are pooled together for analysis. Primarily through the use of DNA polymorphisms, genetic mapping in humans has progressed rapidly.

To give an example of the type of data used in human genetic mapping, a three-generation pedigree of a family segregating for several alleles of an STRP is illustrated in Figure 4.24. In this example, each of the parents is heterozygous, as are all of the children. Yet every person can be assigned his or her genotype because the STRP alleles are codominant. At present, DNA polymorphisms are the principal types of genetic markers used in genetic mapping in human pedigrees. Such polymorphisms are prevalent, are located in virtually all regions of the chromosome set, and have multiple alleles and so yield a high proportion of heterozygous genotypes. Furthermore, only a small amount of biological material is needed to perform the necessary tests.

The present human genetic map is based on more than 5000 genetic markers, primarily STRPs, each heterozygous in an average of 70 percent of people tested. In human beings, there is about 60

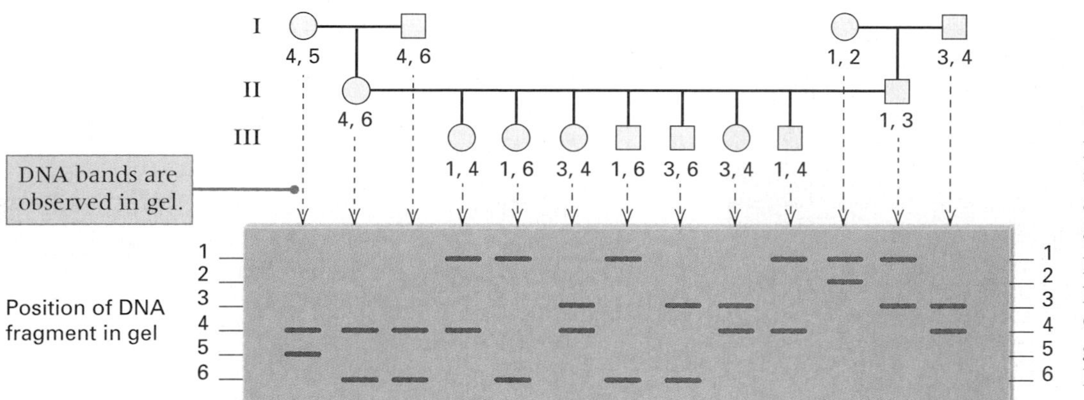

Figure 4.24 Human pedigree showing segregation of STRP alleles. Six alleles (1–6) are present in the pedigree, but any one person can have only one allele (if homozygous) or two alleles (if heterozygous).

percent more recombination in females than in males, so the female and male genetic maps differ in length. The female map is about 4400 cM, the male map about 2700 cM. Averaged over both sexes, the length of the human genetic map for all 23 pairs of chromosomes is about 3500 cM. Because the total DNA content per haploid set of chromosomes is 3154 million base pairs, there is, very roughly, 1 cM per million base pairs in the human genome.

Single-nucleotide polymorphisms (SNPs) are detected with DNA chips.

The most prevalent type of polymorphism in the human genome, as well as in those of other organisms, is the **simple-nucleotide polymorphism (SNP),** in which a single base pair at a particular nucleotide site may differ from one individual to the next. Rare mutants of virtually every nucleotide can probably be found, if one searches hard enough. But rare variants are not usually useful for family studies of heritable variation in susceptibility to disease or to reconstruct historical patterns of human migration. For this reason, in order for a difference in nucleotide sequence to be considered as an SNP, the less-frequent base must have a frequency of greater than 1 percent in the human population. By this definition, the density of SNPs in the human genome is about 1 SNP per 750 nucleotide pairs. This implies a total of about 4 million SNPs, or approximately 1 SNP every 0.001 cM.

A RESEARCH SCIENTIST carrying out gel electrophoresis. Samples containing various sized fragments are separated within a gel using an electric current. Such techniques are used during DNA fingerprint analysis and when isolating particular genes. [© Simon Fraser/Photo Researchers, Inc./Science Photo library.]

SNPs are not as amenable to identification by electrophoretic methods as are RFLPs or STRPs, but they are more abundant. The current method for identifying the alternative bases at a site makes use of a special type of **DNA chip,** a small piece of glass on which an array of a large number of synthetic, single-stranded DNA fragments about 25 nucleotides in length are immobilized at very high density. In each synthetic strand of DNA, the central base is either A, T, G, or C (to detect the SNP), whereas the flanking bases perfectly match the nonpolymorphic bases surrounding the SNP in the human genome. After it is suitably prepared, single-stranded human DNA, labeled with a fluorescent molecule, is spread upon the chip. The labeled DNA sticks to the perfectly matched 25-nucleotide fragment because it forms a stable duplex; the same DNA forms an unstable duplex with any mismatched fragment and washes off the chip later in the procedure. So, for example, DNA from a person heterozygous for an SNP that has an A—T base pair in one strand and a G—C pair in the other will form duplexes with the A-bearing synthetic fragment and the G-bearing synthetic fragment for one strand and with the T-bearing fragment and the C-bearing fragment for the other strand. The positions of the labeled DNA strands that stick to the chip are determined using a high-resolution fluorescent scanner, and in this way the genotype for each SNP detected by the chip is identified. Genotyping chips that are capable of simultaneously detecting 100 SNPs are already in use, and chips capable of detecting 2000 SNPs are expected soon. Already more than 2000 SNPs have been assigned genetic positions in the human genetic map, with an average spacing of 2 cM.

4.5 Tetrads contain all four products of meiosis.

In some species of fungi, each meiotic tetrad is contained in a sac-like structure, called an **ascus,** and can be recovered as an intact group. Each product of meiosis is included in a reproductive cell called an **ascospore,** and all of the ascospores formed from one meiotic cell remain together in the ascus (Figure 4.25). The advantage of these organisms for the study of recombination is the potential for analyzing all of the products from each meiotic division. Two other features of the organisms are especially useful for genetic analysis: (1) They are haploid, so dominance is not a complicating factor because the genotype is expressed directly in the phenotype. (2) They produce very large numbers of progeny, making it possible to detect rare events and to estimate their frequencies accurately.

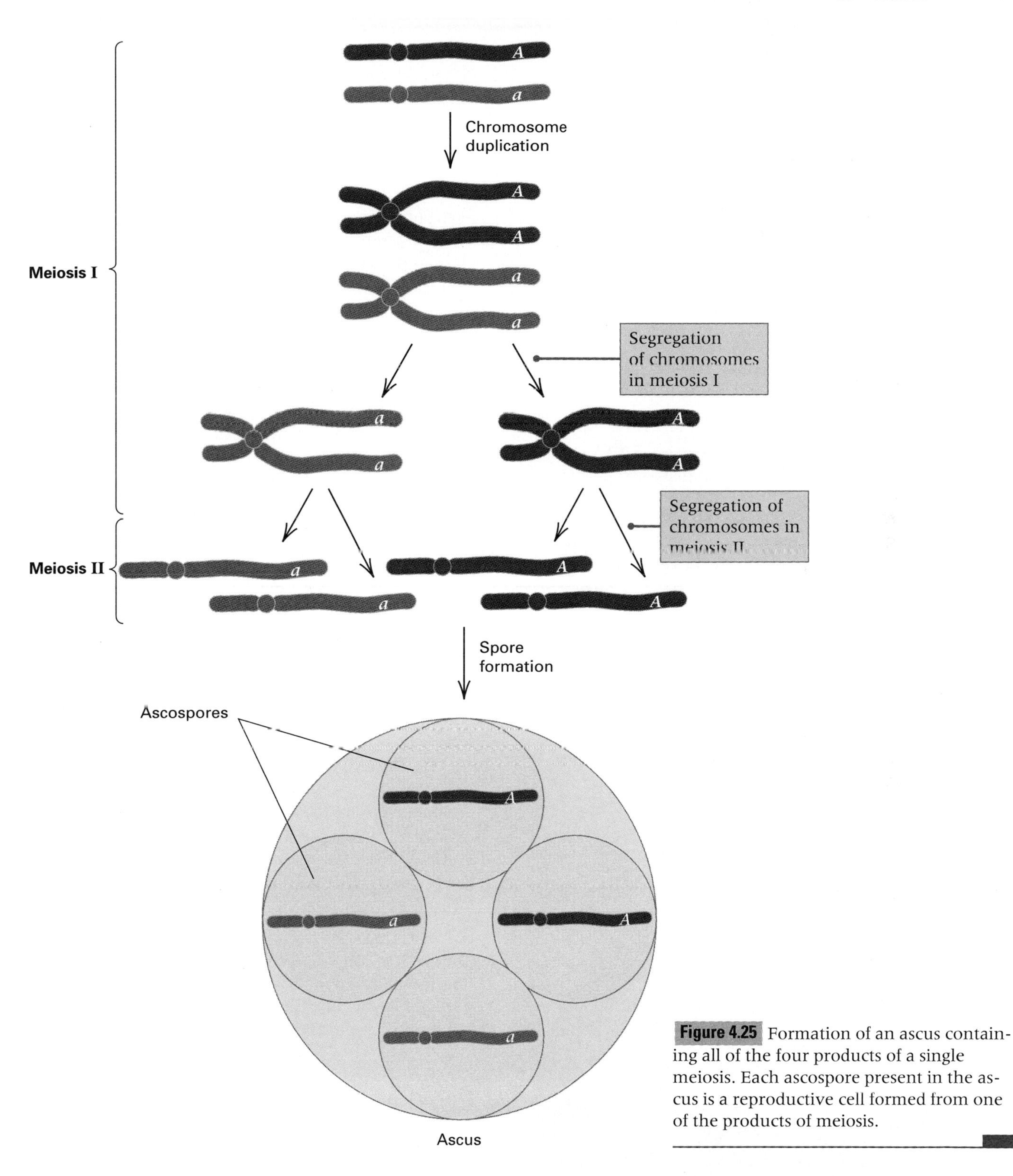

Figure 4.25 Formation of an ascus containing all of the four products of a single meiosis. Each ascospore present in the ascus is a reproductive cell formed from one of the products of meiosis.

The life cycles of these organisms tend to be short. The only diploid stage is the zygote, which undergoes meiosis soon after it is formed; the resulting haploid meiotic products (which form the ascospores) germinate to regenerate the vegetative stage (Figure 4.26). In some species, each of the four products of meiosis subsequently undergoes a mitotic division, with the result that each member of the tetrad yields a *pair* of genetically identical ascospores. In most of the organisms, the meiotic products, or their derivatives, are not arranged in any particular order in the ascus. However, bread molds of the genus *Neurospora* and related organisms have the useful characteristic that the meiotic products are arranged in a definite order directly related to the planes of the meiotic divisions. We will examine the ordered system after first looking at unordered tetrads.

Unordered tetrads have no relation to the geometry of meiosis.

In the tetrads when two pairs of alleles are segregating, three patterns of segregation are possible. For example, in the cross $A\ B \times a\ b$, the three types of tetrads are

(*AB*)(*AB*)(*ab*)(*ab*) referred to as **parental ditype,** or **PD.** Only two genotypes are represented, and their alleles have the same combinations found in the parents.

(*Ab*)(*Ab*)(*aB*)(*aB*) referred to as **nonparental ditype,** or **NPD.** Only two genotypes are represented, but their alleles have nonparental combinations.

(*AB*)(*Ab*)(*aB*)(*ab*) referred to as **tetratype,** or **TT.** All four of the possible genotypes are present.

Tetrad analysis is an effective way to determine whether two genes are linked, because of the following principle:

> When genes are *unlinked,* the parental ditype tetrads and the nonparental ditype tetrads are expected in equal frequencies (PD = NPD).

The reason for the equality PD = NPD for unlinked genes is shown in Figure 4.27A for two pairs of alleles, *A a* and *B b,* located in different chromosomes. In the absence of crossing-over between either gene and its centromere, the two chromosomal configurations are equally likely at metaphase I, and so PD = NPD. When there is crossing-over between either gene and its centromere (Figure 4.27B), a tetratype tetrad results, but this does not change the fact that PD = NPD.

In contrast, when genes are linked, parental ditypes are far more frequent than nonparental ditypes. To see why, assume that the genes are linked and consider the events required for the production of the three types of tetrads. Figure 4.28 shows that when no crossing-over takes place between the genes, a PD tetrad is formed. Single crossover between the genes results in a TT tetrad. The formation of a two-strand, three-strand, or four-strand double crossover results in a PD, TT, or NPD tetrad, respectively. With linked genes, meiotic cells with no crossovers always outnumber those with four-strand double crossovers. Therefore,

> Linkage is indicated when nonparental ditype tetrads appear with a much lower frequency than parental ditype tetrads (NPD << PD).

The relative frequencies of the different types of tetrads can be used to determine the map distance between two linked genes. The simplest case is one in which the genes are sufficiently close that double and higher levels of crossing-over can be neglected. In this case, tetratype tetrads arise only from meiotic cells in which a single crossing-over occurs between the genes (Figure 4.28 A and B). As we saw in Figure 4.7, the genetic map distance across an interval is defined as one-half the proportion of cells with a crossover in the interval, so the map distance implied by the tetrads is given by

$$\text{Map distance} = \frac{1}{2} \times \frac{\text{Number of tetratype tetrads}}{\text{Total number of tetrads}} \times 100 \qquad (4.1)$$

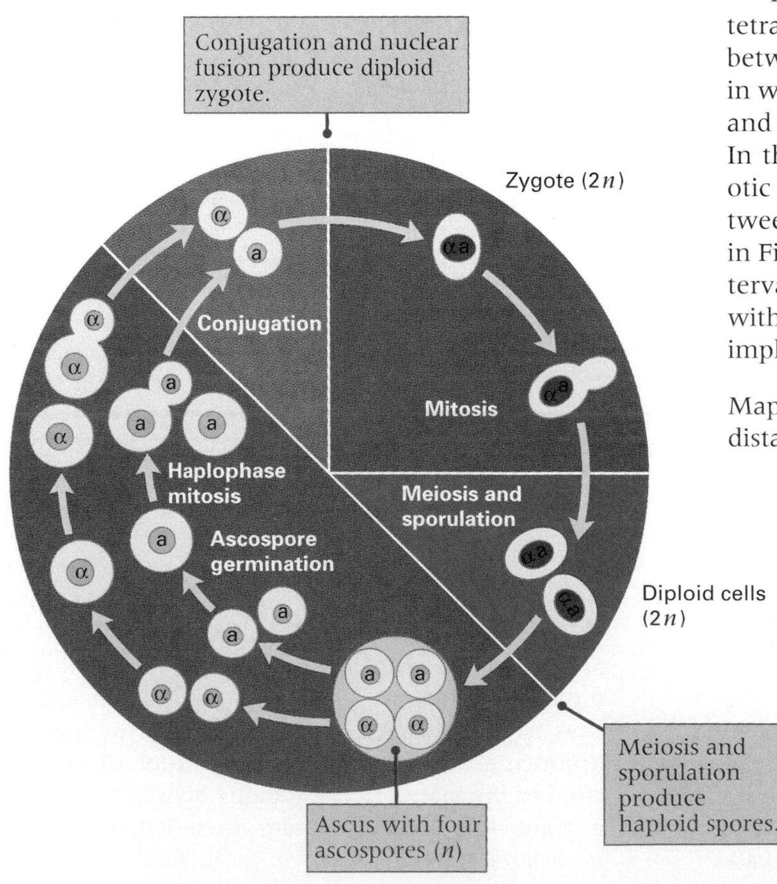

Figure 4.26 Life cycle of the yeast *Saccharomyces cerevisiae.* Mating type is determined by the alleles **a** and α. Both haploid and diploid cells normally multiply by mitosis (budding). Depletion of nutrients in the growth medium induces meiosis and sporulation of cells in the diploid state. Diploid nuclei are red; haploid nuclei are yellow.

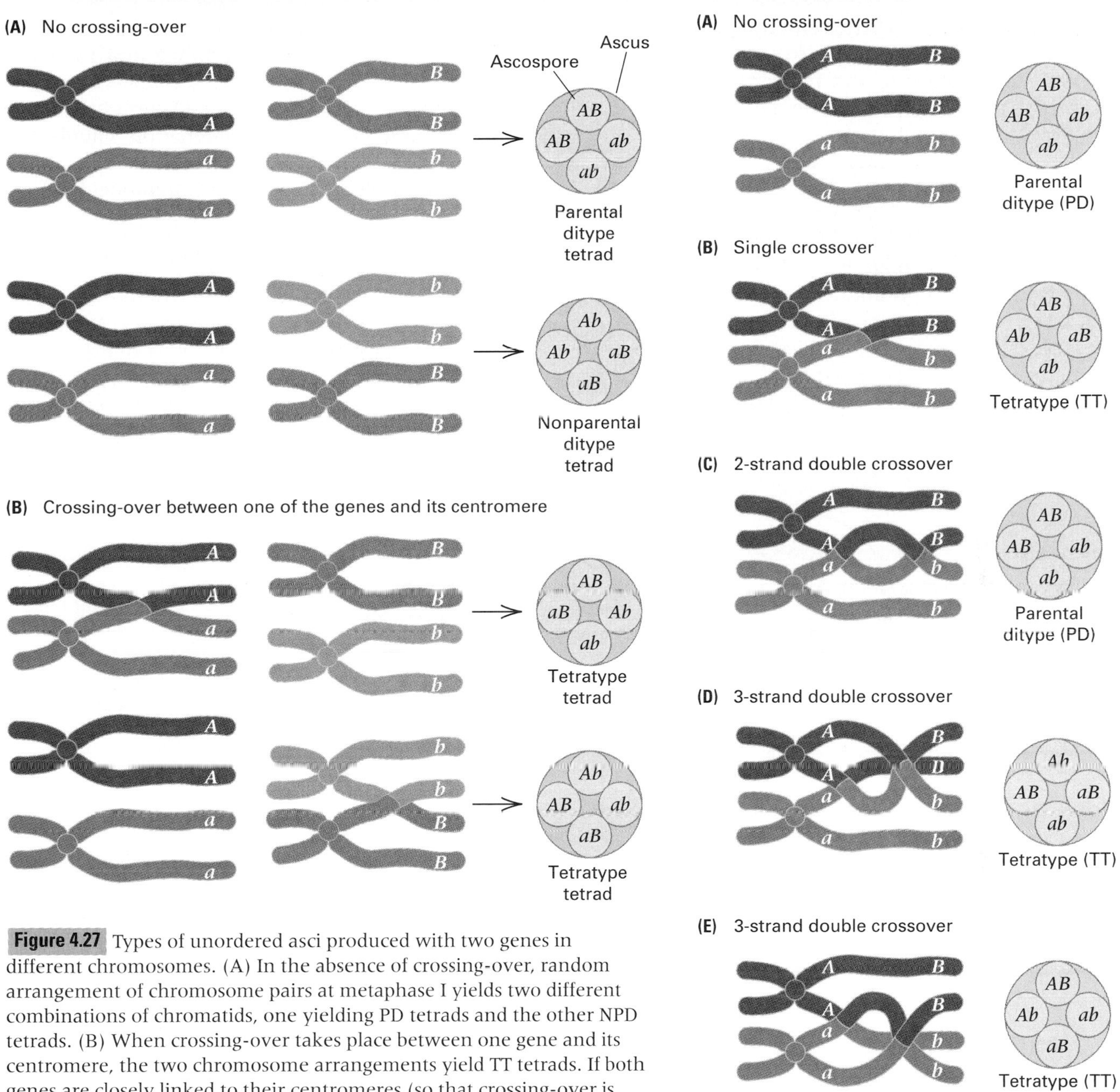

Figure 4.27 Types of unordered asci produced with two genes in different chromosomes. (A) In the absence of crossing-over, random arrangement of chromosome pairs at metaphase I yields two different combinations of chromatids, one yielding PD tetrads and the other NPD tetrads. (B) When crossing-over takes place between one gene and its centromere, the two chromosome arrangements yield TT tetrads. If both genes are closely linked to their centromeres (so that crossing-over is rare), few TT tetrads are produced.

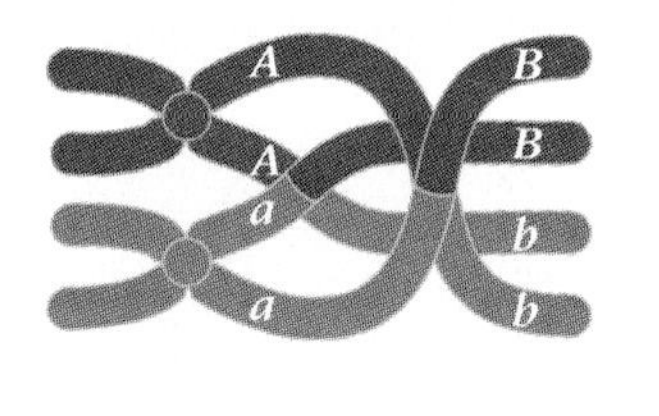
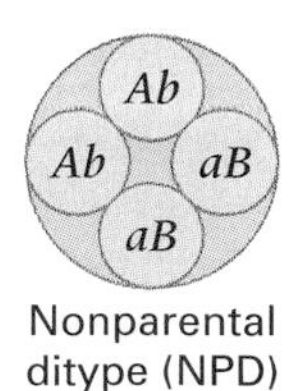

Figure 4.28 Types of tetrads produced with two linked genes. (A) In the absence of crossing-over, a PD tetrad is produced. (B) With a single crossover between the genes, a TT tetrad is produced. (C–F) Among the four possible types of double crossovers between the genes, only the four-strand double crossover in part F yields an NPD tetrad.

To take a specific example, suppose 100 tetrads are analyzed from the cross $A\,B \times a\,b$, and the result is that 91 are PD and 9 TT. The finding that NPD $<<$ PD means that the genes are linked, and the fact that NPD $= 0$ means that the genes are so closely linked that double crossing-over does not occur between them. The map distance between A and B is calculated as follows:

$$\text{Map distance} = \frac{1}{2} \times \frac{9}{100} \times 100 = 4.5\text{cM}$$

We must emphasize that Equation (4.1) is valid only when NPD $= 0$, so that interference across the region prevents the occurrence of double crossing-over. When double crossovers do take place in the interval, then NPD ≠ 0, and the formula for map distance has to be modified to take the double crossovers into account.

The mapping procedure using tetrads differs from that presented earlier in the chapter in that the map distance is not calculated directly from the number of recombinant and nonrecombinant chromatids. Instead, the map distance is calculated directly from the tetrads and the inferred crossovers that give rise to each type of tetrad. However, it is not necessary to carry out a full tetrad analysis for estimating linkage. The alternative is to examine spores chosen at random after allowing the tetrads to break open and disseminate their spores. This procedure is called **random-spore analysis,** and the linkage relationships are determined exactly as described earlier for *Drosophila* and corn. In particular, the frequency of recombination equals the number of spores that are recombinant for the genetic markers divided by the total number of spores.

The geometry of meiosis is revealed in ordered tetrads.

In the bread mold *Neurospora crassa,* a species used extensively in genetic investigations, the products of meiosis are contained in an *ordered* array of ascospores (Figure 4.29). A zygote nucleus, con-

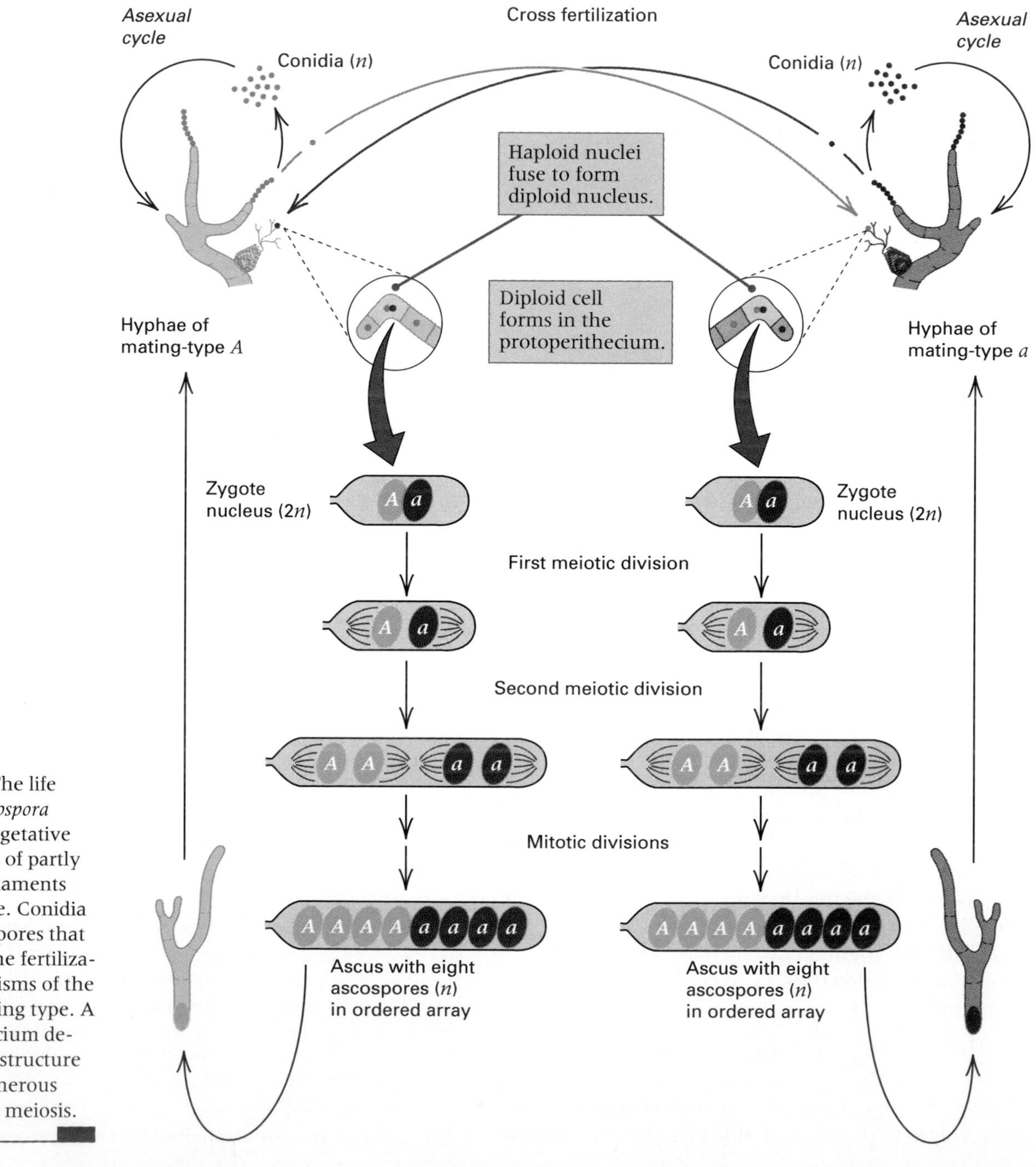

Figure 4.29 The life cycle of *Neurospora crassa.* The vegetative body consists of partly segmented filaments called hyphae. Conidia are asexual spores that function in the fertilization of organisms of the opposite mating type. A protoperithecium develops into a structure in which numerous cells undergo meiosis.

tained in a sac-like ascus, undergoes meiosis almost immediately after it is formed. The four nuclei produced by meiosis are in a linear, ordered sequence in the ascus, and each of them undergoes a mitotic division to form two genetically identical and adjacent ascospores. Each mature ascus contains eight ascospores arranged in four pairs, each pair derived from one of the products of meiosis. The ascospores can be removed one by one from an ascus and each germinated in a culture tube to determine its genotypes.

Ordered asci also can be classified as PD, NPD, or TT with respect to two pairs of alleles, which makes it possible to assess the degree of linkage between the genes. The fact that the arrangement of meiotic products is ordered also makes it possible to determine the recombination frequency between any particular gene and its centromere. The logic of the mapping technique is based on the feature of meiosis shown in Figure 4.30.

> Homologous centromeres of parental chromosomes separate at the first meiotic division; the centromeres of sister chromatids separate at the second meiotic division.

Thus, in the absence of crossing-over between a gene and its centromere, the alleles of the gene (for example, *A* and *a*) must separate in the first meiotic division; this separation is called **first-division segregation.** If, instead, a crossover is formed between the gene and its centromere, the *A* and *a* alleles do not become separated until the second meiotic division; this separation is called **second-division segregation.** The distinction between first-division and second-division segregation is shown in Figure 4.30. As shown in part A, only two possible arrangements of the products of meiosis can yield first-division segregation—*A A a a* or *a a A A*. However, four patterns of second-division segregation are possible because of the random arrangement of homologous chromosomes at metaphase I and of the chromatids at metaphase II. These four arrangements, shown in part B, are *A a A a*, *a A a A*, *A a a A*, and *a A A a*.

The percentage of asci with second-division segregation patterns for a gene can be used to map the gene with respect to its centromere. For example, let us assume that 30 percent of a sample of asci from a cross have a second-division segregation pattern for the *A* and *a* alleles. This means that 30 percent of the cells undergoing meiosis had a crossover between the *A* gene and its centromere. Because the map distance between two genes is, by definition, equal to one-half times the proportion of cells with a crossover between the genes, the map distance between a gene and its centromere is given by the equation

$$\text{Map distance} = \frac{1}{2} \times \frac{\text{Number of asci with second division segregation}}{\text{Total number of asci}} \times 100 \qquad (4.2)$$

Equation (4.2) is valid as long as the gene is close enough to the centromere for us to neglect multiple crossing-over. Reliable linkage values are best determined for genes that are near the centromere. The location of more distant genes is then accomplished by mapping these genes relative to genes nearer the centromere.

If a gene is far from its centromere, crossing-over between the gene and its centromere will be so frequent that the *A* and *a* alleles become randomized with respect to the four chromatids. The result is that the six possible spore arrangements shown in Figure 4.27 are all equally frequent. Therefore, in the absence of chromatid interference:

> The maximum frequency of second-division segregation asci is 2/3.

Gene conversion suggests a molecular mechanism of recombination.

Genetic recombination may be regarded as a process of breakage and repair between two DNA molecules. In eukaryotes, the process takes place early in meiosis after each molecule has replicated, and with respect to genetic markers, it results in two molecules of the parental type and two recombinants (Chapter 3). For genetic studies of recombination, fungi such as yeast or *Neurospora* are particularly useful because all four products of any meiosis can be recovered in a four-spore (yeast) or eight-spore (*Neurospora*) ascus. Most asci from heterozygous *Aa* diploids contain ratios of

2 *A* : 2 *a* in four-spored asci, or
4 *A* : 4 *a* in eight-spored asci

because of normal Mendelian segregation. Occasionally, however, aberrant ratios are also found, such as

3 *A* : 1 *a* or 1 *A* : 3 *a* in four-spored asci, and
5 *A* : 3 *a* or 3 *A* : 5 *a* in eight-spored asci

Different types of aberrant ratios can also occur. The aberrant asci are said to result from **gene conversion** because it appears as if one allele has "converted" the other allele into a form like itself. Gene

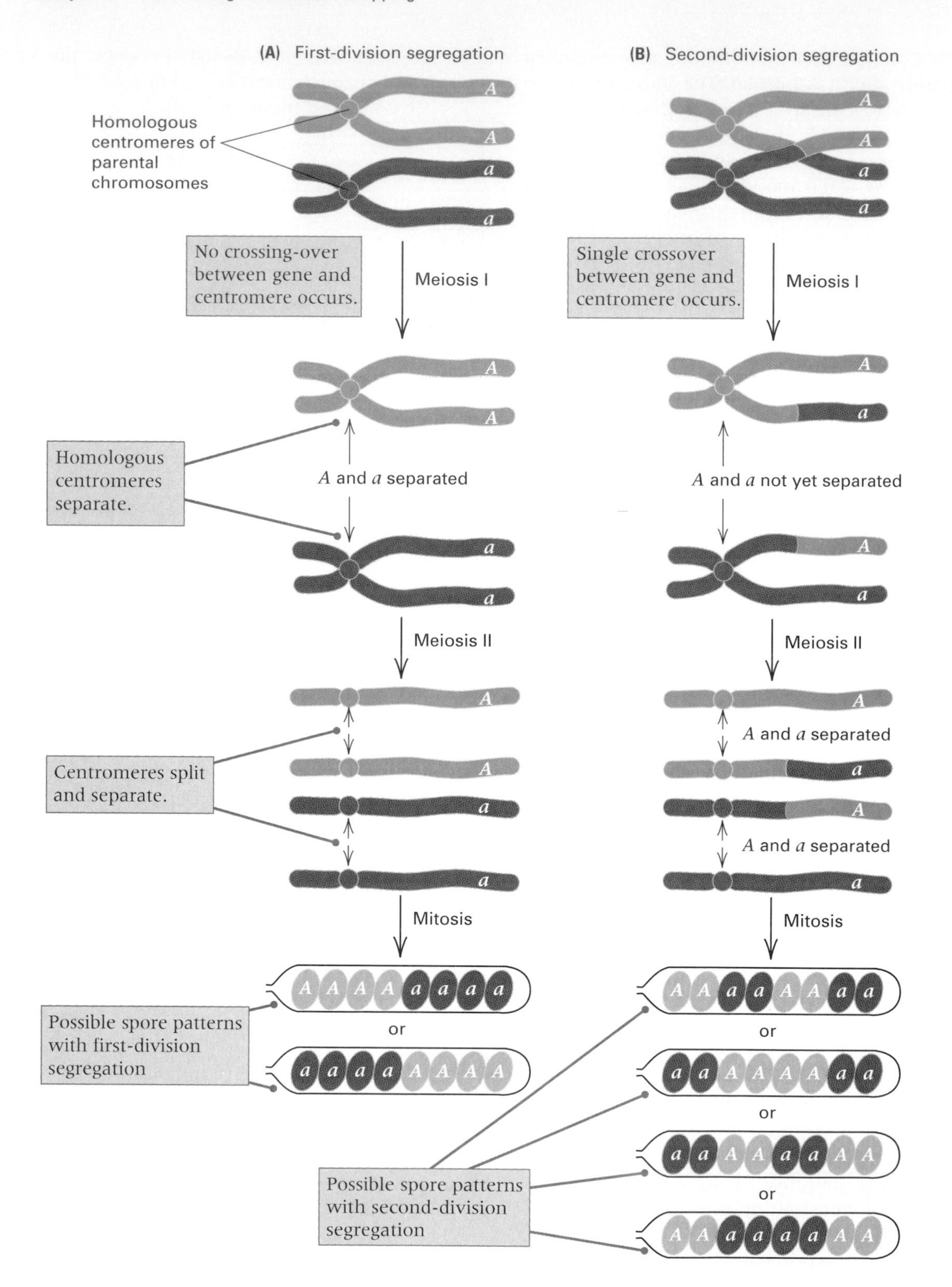

Figure 4.30 First- and second-division segregation in *Neurospora.* (A) First-division segregation patterns are found in the ascus when crossing-over between the gene and centromere does not take place. The alleles separate (segregate) in meiosis I. Two spore patterns are possible, depending on the orientation of the pair of chromosomes on the first-division spindle. The orientation shown results in the pattern in the upper ascus. (B) Second-division segregation patterns are found in the ascus when crossing-over between the gene and the centromere delays separation of *A* from *a* until meiosis II. Four patterns of spores are possible, depending on the orientation of the pair of chromosomes on the first-division spindle and that of the chromatids of each chromosome on the second-division spindle. The orientation shown results in the pattern in the top ascus.

conversion is frequently accompanied by recombination between genetic markers on either side of the conversion event, even when the flanking markers are tightly linked. This implies that gene conversion can be one consequence of the recombination process.

Gene conversion results from a normal DNA repair process in the cell known as **mismatch repair.** In this process, an enzyme recognizes any base pair in a DNA duplex in which the paired bases are mismatched—for example, G paired with T, or A paired with C. When such a mismatch is found in a molecule of duplex DNA, a small segment of one strand is excised and replaced with a new segment synthesized using the remaining strand as a template. In this manner the mismatched base pair is replaced. Figure 4.31 shows an example in which a mismatched G—T pair is being repaired. The strand that is excised could be either the strand containing T or the one containing G, and the newly synthesized (repaired) segment, shown in red, would contain either a C or an A, respectively. The two possible products of repair differ in DNA sequence.

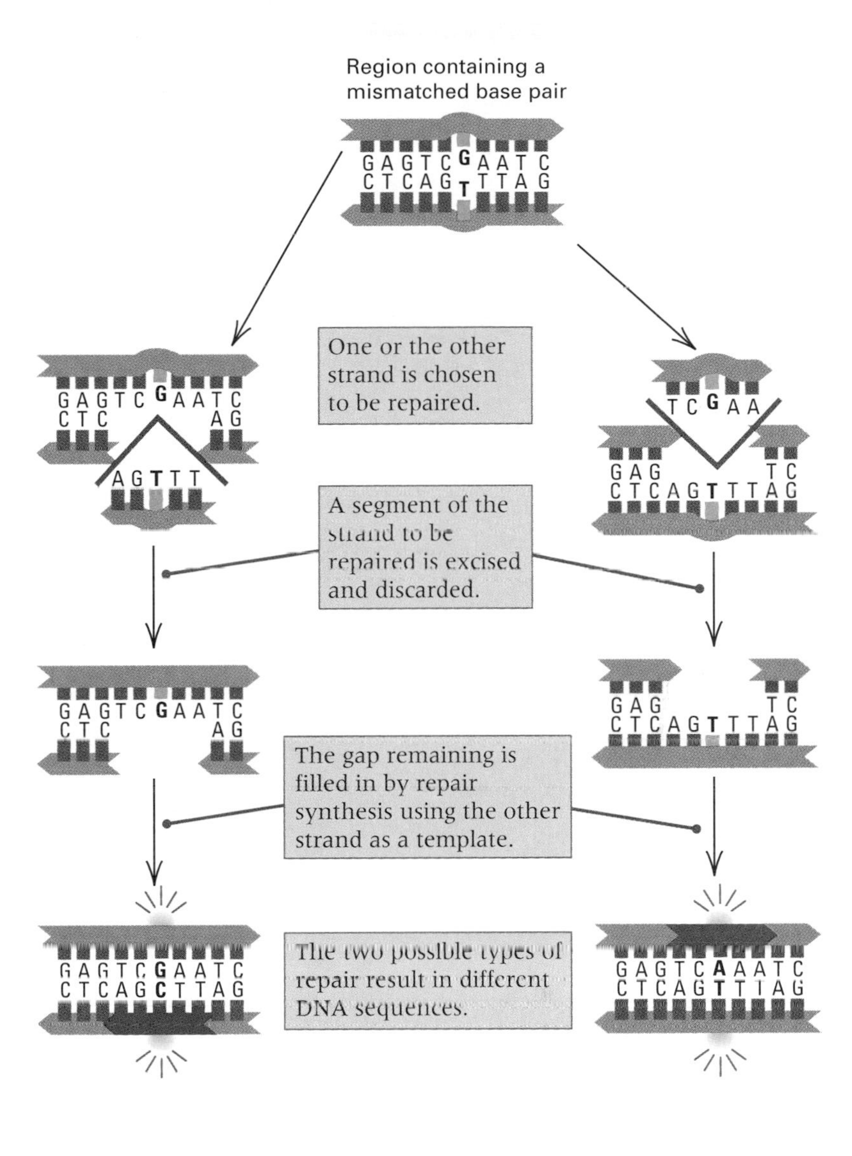

Figure 4.31 Mismatch repair consists of the excision of a segment of a DNA strand containing a base mismatch followed by repair synthesis. Either strand can be excised and corrected. In this example, the G—T mismatch is corrected to either G—C (left) or A—T (right). The length of the excised strand is typically 13 nucleotides in prokaryotes and 29 nucleotides in eukaryotes.

The role of mismatch repair in gene conversion is illustrated in Figure 4.32. The row of DNA duplexes across the top represents the DNA molecules in two pairs of sister chromatids in pachytene of meiosis, immediately prior to recombination. One pair of sister chromatids carries the *A* allele, the sequence of which contains the G—C base pair highlighted in color, whereas the other pair of sister chromatids contains the *a* allele, in which the same base pair is A—T. In the process of recombination, the participating DNA duplexes at first exchange a single strand, each of which pairs with the nonexchanged strand in the other duplex. The second row shows the situation after the middle two duplexes have exchanged the bottom strand. In each molecule, the exchange of pairing partners creates a **heteroduplex** region in which any bases that are not identical in the parental duplexes become mismatched. In this example, one heteroduplex contains a G—T base pair and the other contains an A—C base pair. At this point the mismatch repair system comes into play and corrects the mismatches. Each mismatch can be repaired in either of two ways, so there are four possible ways in which the mismatches can be repaired. Two of the possibilities are shown. In one, the G—T mismatch is repaired to G—C and the A—C mismatch is repaired to G—C also. Because the G—C pair defines the *A* allele, the result is that both middle duplexes carry *A*. After meiosis, when each sister chromatid is segregated to a different ascospore, the ascus will contain 3 *A* : 1 *a* and show an apparent $a \rightarrow A$ gene conversion. The other possibility for mismatch repair shown in Figure 4.32 is one in which both mismatches are repaired to A—T, which defines the *a* allele. The ascus that results from this type of correction will contain 1 *A* : 3 *a* and show an apparent $A \rightarrow a$ gene conversion.

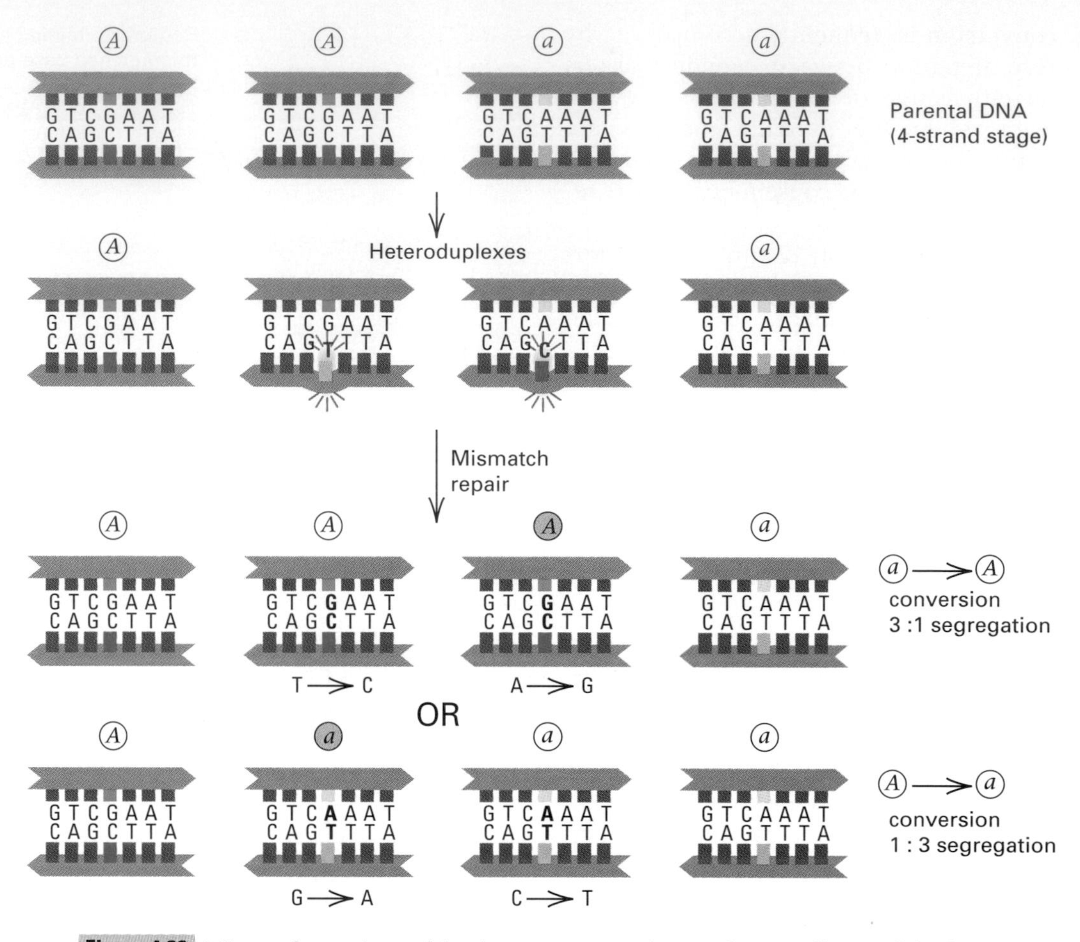

Figure 4.32 Mismatch repair resulting in gene conversion. Only a small part of the heteroduplex region is shown. Each heteroduplex containing a mismatched base pair can be repaired to give an *A* allele or an *a* allele. The patterns of repair leading to 3 *A* : 1 *a* and 1 *A* : 3 *a* segregation are shown.

4.6 Recombination results from breakage and reunion of DNA molecules.

By what process are the heteroduplexes in Figure 4.32 formed as an intermediate in recombination? The first molecular model for recombination was suggested by Robin Holliday in 1964. This model is now known to be somewhat oversimplified, but it illustrates why gene conversion in a small region is so frequently associated with the recombination of genetic markers flanking the region.

In the **Holliday model** of recombination, each of the participating DNA duplexes (Figure 4.33A) initially undergoes a symmetrically positioned single-stranded nick in strands of the same polarity (Figure 4.33B). The nicked strands unwind in the region of the nicks, switch pairing partners, and form a heteroduplex region in each molecule (Figure 4.33C). The juxtaposed free ends of the DNA strands are joined together (Figure 4.33D), and further unwinding of the DNA duplexes and exchange of pairing partners increases the length of the heteroduplex region (this process is called *branch migration*). Two more nicks and rejoinings are necessary to resolve the structure in part D. As the configuration is drawn, these breaks can be in the inner strands, resulting in molecules that are *nonrecombinant* for markers flanking the heteroduplex region, or they can be in the outer strands, resulting in molecules that are *recombinant* for outside markers (Figure 4.33E).

Of the two possibilities for resolution of the structure in Figure 4.33D, one results in recombination of the outside markers and the other does not. These outcomes are equally likely, because there is no topological difference between the inner strands and the outer strands. To understand why, note that an end-for-end rotation of the lower duplex in Figure 4.33D gives the configura-

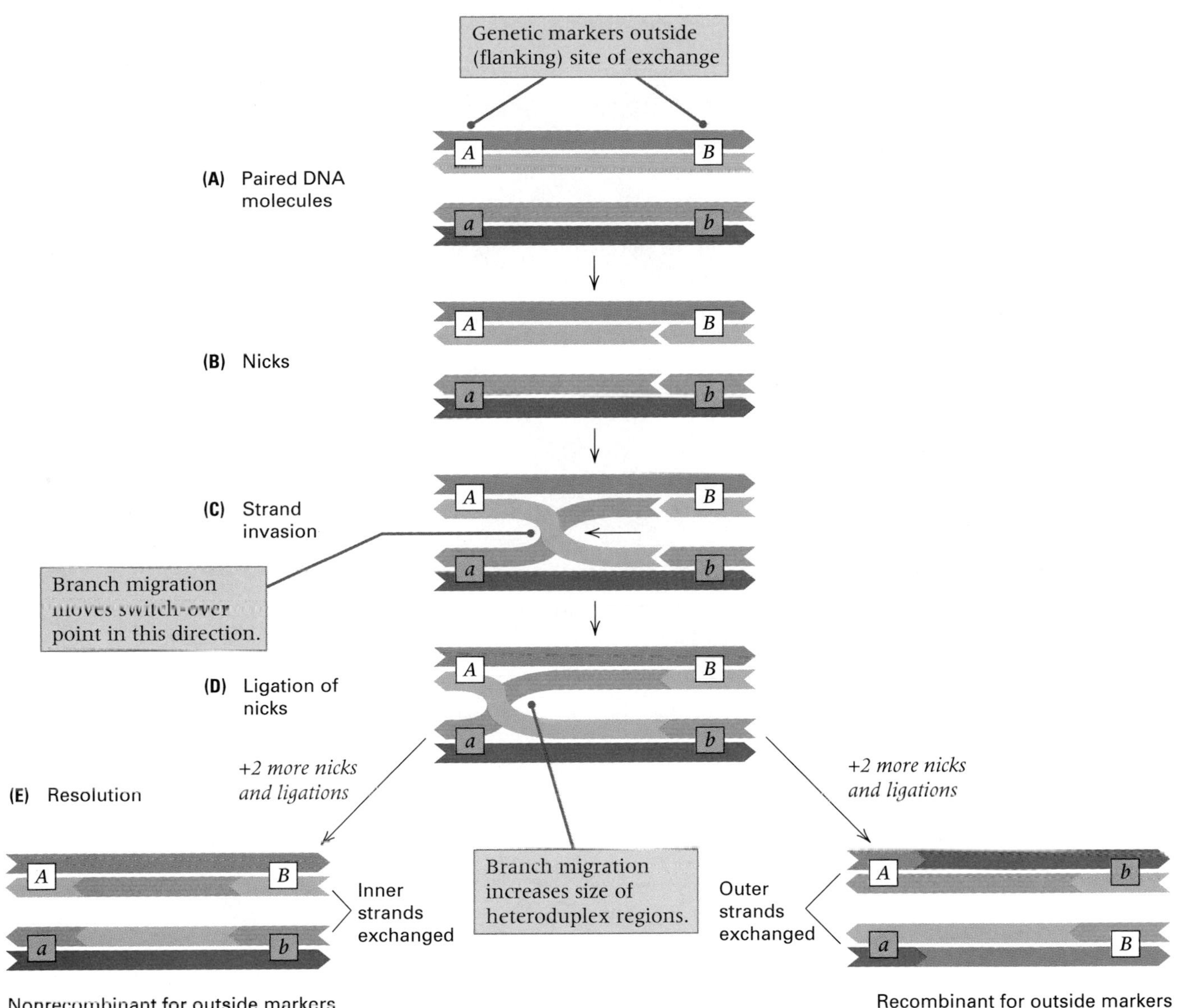

Figure 4.33 Holliday model of recombination. (A) Paired DNA molecules. (B) The exchange process is initiated by single-stranded breaks in strands of the same polarity. (C) Strands exchange pairing partners ("invasion"). (D) Ligation and branch migration. (E) Exchange is resolved either by the breaking and rejoining of the inner strands (resulting in molecules that are nonrecombinant for outside markers) or by the breaking and rejoining of the outer strands (resulting in recombinant molecules). Either type of resolution gives a heteroduplex region of the same length in each participating molecule.

tion in Figure 4.34. This type of configuration is known as a **Holliday structure.** Drawing it in this manner makes it clear that breakage and rejoining of the east-west strands and of the north-south strands are topologically equivalent. However, north-south resolution results in recombination of the outside markers (*A b* and *a B* gametes), whereas east-west resolution gives nonrecombinants (*A B* and *a b* gametes).

An electron micrograph of a Holliday structure formed between two DNA duplexes is shown in Figure 4.35A. This is the physical structure of DNA that corresponds to the diagram in Figure 4.34. An interpretation of the structure that preserves the double-helical structure of duplex DNA is shown in Figure 4.35B, in which the colors match those at the top in Figure 4.34. The locations of the genetic markers (*A a* and *B b*) flanking the heteroduplex region are also indicated. The key feature of the Holliday structure is that there are four short, single-stranded regions where the duplexes come together. The Holliday structure can be resolved

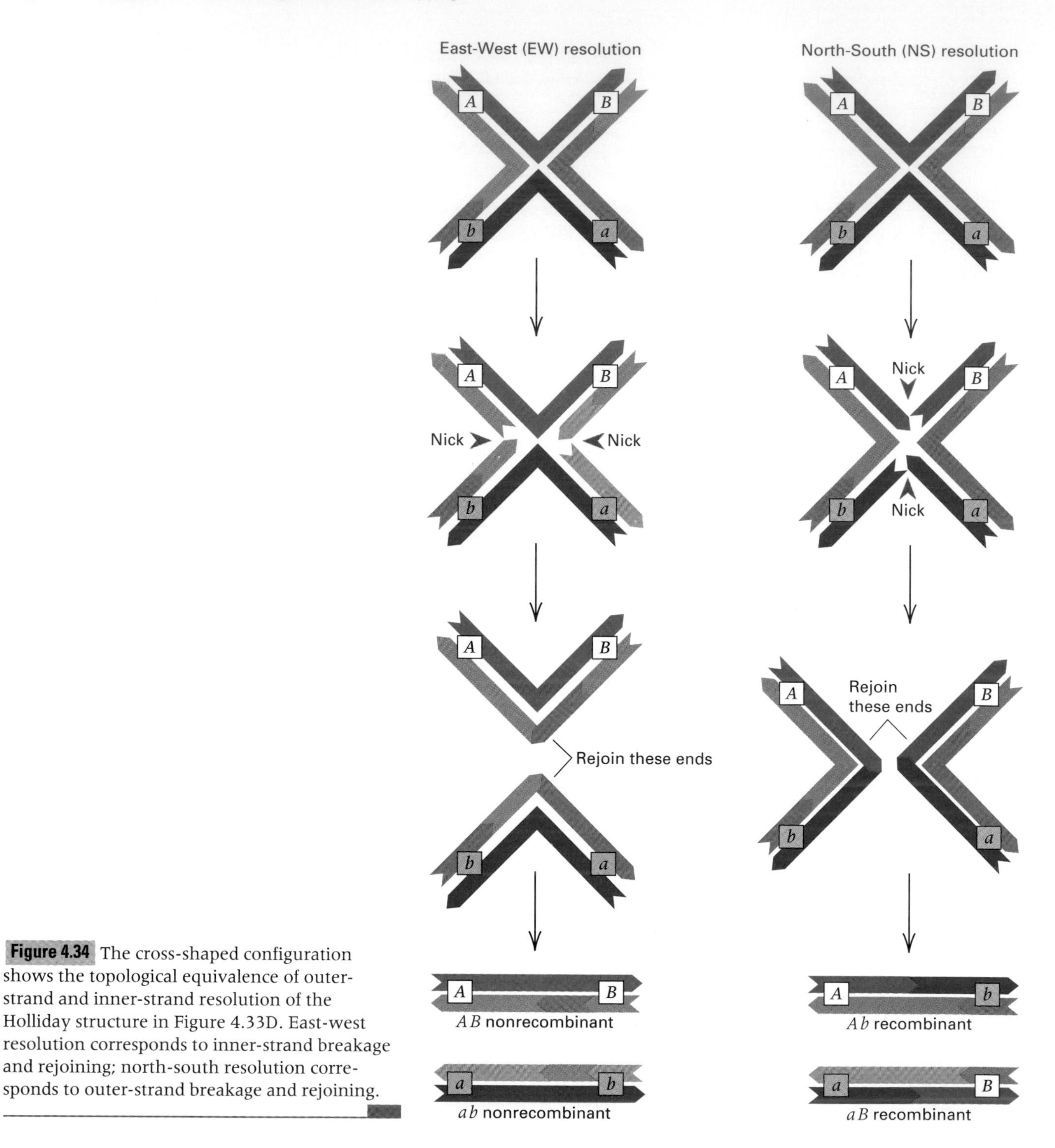

Figure 4.34 The cross-shaped configuration shows the topological equivalence of outer-strand and inner-strand resolution of the Holliday structure in Figure 4.33D. East-west resolution corresponds to inner-strand breakage and rejoining; north-south resolution corresponds to outer-strand breakage and rejoining.

either by breakage and reunion in the single-stranded regions on the top and bottom (called NS in Figure 4.34) or by breakage and reunion in the single-stranded regions on the left and right (called EW in Figure 4.34). The top-bottom breakage and rejoining result in two duplexes that are nonrecombinant for the outside markers (*A B* and *a b*), whereas the right-left breakage and rejoining result in two duplexes that are recombinant for the outside markers (*A b* and *a B*).

Several types of enzymes are required for recombination. In *E. coli,* the exchange of strands is mediated by the recA protein. Another enzyme, called the recBCD nuclease because it contains three protein subunits (B, C, and D), has been implicated in strand exchange and branch migration. The resolution of the Holliday structure is also an enzymatic function, which is carried out by the **Holliday junction-resolving enzyme,** denoted RuvC.

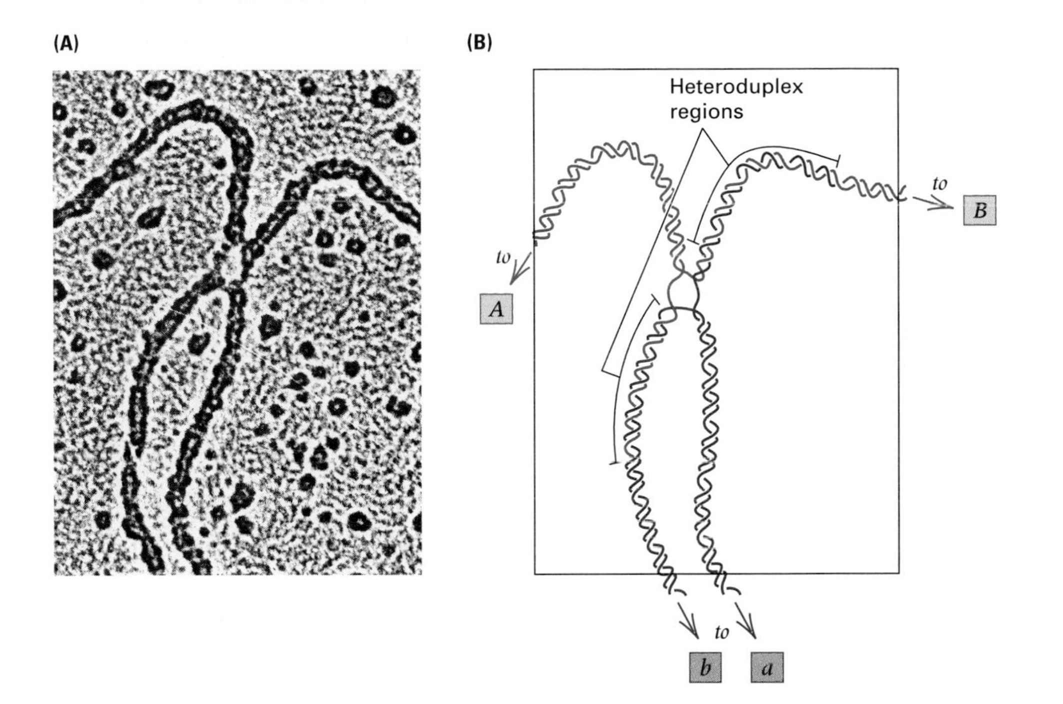

Figure 4.35 (A) Electron micrograph of a Holliday structure. Note the short, single-stranded region of DNA on each side of the oval-shaped structure where the duplexes come together. (B) Interpretative drawing of a Holliday structure in terms of double-helical molecules. Note the single-stranded regions where the duplexes come together, as well as the short heteroduplex region in two of the duplexes. For ease of comparison with Figure 4.34, the positions of the alleles *A, a* and *B, b* are indicated. Breakage and rejoining of the pair of single strands at the left and right yield duplexes that are nonrecombinant for the outside markers (*A B* and *a b*). Breakage and rejoining of the pair of single strands at the top and bottom yield duplexes that are recombinant for the outside markers (*A b* and *a B*).

Chapter Summary

- **Linked alleles tend to stay together in meiosis.**

  The frequency of recombination is the same for *cis* and *trans* heterozygotes.
  The frequency of recombination differs from one gene pair to the next.
  Recombination does not occur in *Drosophila* males.

Nonallelic genes located in the same chromosome tend to remain together in meiosis rather than undergoing independent assortment. This phenomenon is called linkage. The indication of linkage is deviation from the 1 : 1 : 1 : 1 ratio of phenotypes in the progeny of a mating of the form *Aa Bb* × *aa bb*. When alleles of two linked genes segregate, more than 50 percent of the gametes produced have parental combinations of the segregating alleles and fewer than 50 percent have nonparental (recombinant) combinations of the alleles. The recombination of linked genes results from crossing-over, a process in which nonsister chromatids of the homologous chromosomes exchange corresponding segments in the first meiotic prophase.

- **Recombination results from crossing-over between linked alleles.**

  Physical distance is often—but not always—correlated with map distance.
  Crossing-over is reciprocal and takes place at the four-strand stage.
  One crossover can undo the effects of another.

The frequencies of crossing-over between different genes can be used to determine the relative order and locations of the genes in chromosomes. This is called genetic mapping. Distance between adjacent genes in such a map (a genetic or linkage map) is defined in map units. Across regions in which multiple crossing-over does not take place, the map distance between two genes is proportional to the frequency of recombination between them. In this case, one unit of map distance corresponds to 1 percent recombination. Across longer regions in which multiple crossovers are possible, map distance equals one-half the average number of crossovers expressed as a percentage. One map unit therefore corresponds to a physical length of the chromosome in which a crossover event occurs, on the average, once in every 50 meioses. For short distances, map units are additive. (For example, for three genes with order *a b c*, if the map distances $a-b$ and $b-c$ are 2 and 3 map units, respectively, then the map distance $a-c$ is $2 + 3 = 5$ map units.) The recombination frequency underestimates the map distance between genes if the length of the region is too great. This discrepancy results from multiple crossovers, which yield either no recombinants or the same number produced by a single event. For example, two crossovers in the region between two genes may yield no recombinants, and three crossover events may yield recombinants of the same type as that from a single crossover.

- **Double crossovers are revealed in three-point crosses.**

  Interference decreases the chance of multiple crossing-over.

When many genes are mapped in a particular species, they form linkage groups equal in number to the haploid chromosome number of the species. The maximum frequency of recombination between any two genes in a mating is 50 percent; this happens when the genes are in nonhomologous chromosomes and assort independently or when the genes are sufficiently far apart in the same chromosome that at least one crossing-over is formed between them in every meiosis.

- **Polymorphic DNA sequences are used in human genetic mapping.**

In many organisms, some of the most useful genetic markers are polymorphisms in nucleotide sequence that are not associated with any phenotypic abnormalities. Prominent among these are nucleotide substitutions that create or destroy a particular cleavage site recognized by a restriction endonuclease. Such mutations can be detected because different chromosomes yield restriction fragments differing in size according to the positions of the cleavage sites. Genetic variation of this type is called restriction fragment length polymorphism (RFLP). Most species also have considerable genetic variation in which one allele differs from the next according to the number of copies of a tandemly repeated DNA sequence it contains (simple tandem repeat polymorphism, or STRP).

- **Tetrads contain all four products of meiosis.**

  Unordered tetrads have no relation to the geometry of meiosis.

  The geometry of meiosis is revealed in ordered tetrads.

  Gene conversion suggests a molecular mechanism of recombination.

The four haploid products of individual meiotic divisions can be used to analyze linkage and recombination in some species of fungi and unicellular algae. The method is called tetrad analysis. In *Neurospora* and related fungi, the meiotic tetrads are contained in a tubular sac, or ascus, in a linear order, making it possible to determine whether a pair of alleles segregated in the first or the second meiotic division. Linkage analysis in unordered tetrads is based on the frequencies of parental ditype (PD), nonparental ditype (NPD), and tetratype (TT) tetrads. The observation that NPD < PD is a sensitive indicator of linkage between two genetic markers. Ordered tetrads are convenient for genetic analysis because the distance between a gene and its centromere is related to the frequency of asci showing first-division segregation.

- **Recombination results from breakage and reunion of DNA molecules.**

Genetic recombination is intimately connected with DNA repair because the process always includes breakage and rejoining of DNA strands. In gene conversion, one allele becomes converted into a homologous allele, which is detected by aberrant segregation in fungal asci, yielding ratios of alleles such as 3 : 1 or 1 : 3. Gene conversion is the outcome of mismatch repair in heteroduplexes. At the DNA level, the process of recombination includes the creation of heteroduplexes in the region of the exchange, so gene conversion is often accompanied by recombination of genetic markers flanking the conversion event. The Holliday model assumes that recombination is initiated by single-strand nicks that occur at homologous positions in the participating duplexes.

Key Terms

ascospore
ascus
attached-X chromosome
centimorgan
chromosome map
cis configuration
coefficient of coincidence
coupling
DNA chip
euchromatin
first-division segregation
frequency of recombination
gene conversion
genetic map
genetic marker
heterochromatin
heteroduplex
Holliday junction-resolving enzyme
Holliday model
Holliday structure
interference
linkage
linkage group
linkage map
locus
mapping function
map unit
mismatch repair
nonparental ditype (NPD)
parental combination
parental ditype (PD)
polymorphism
random-spore analysis
recombinant
recombination
repulsion
restriction endonuclease
restriction fragment length polymorphism (RFLP)
second-division segregation
simple tandem repeat polymorphism (STRP)
single-nucleotide polymorphism
tetratype (TT)
three-point cross
trans configuration

Concepts and Issues
Testing Your Knowledge

- Distinguish between genetic recombination and genetic complementation. How is it possible for two mutant genes to show complementation but not recombination? How is it possible for two mutant genes to show recombination but not complementation?
- In genetic analysis, why is it important to know the position of a gene along a chromosome?
- What is the maximum frequency of recombination between two genes? Is there a maximum map distance between two genes?
- Why is the frequency of recombination over a long interval of a chromosome always smaller than the map distance over the same interval?
- What is meant by the term *chromosome interference?*
- What is special about the inheritance of an attached-X chromosome, as compared with the inheritance of unattached X chromosomes?
- How was an attached X chromosome used to show that crossing-over takes place at the four-strand stage of meiosis, after the chromosomes have replicated?
- In human genetics, why are molecular variations in DNA sequence, rather than phenotypes such as eye color or blood-group differences, used for genetic analysis?
- In genetic analysis, what is so special about the ability to examine tetrads in certain fungi?

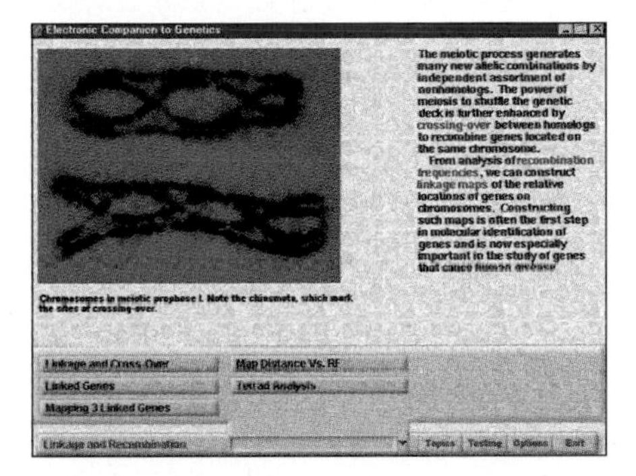

TO ENHANCE YOUR STUDY turn to *An Electronic Companion to Genetics*™, © 1998, Cogito Learning Media, Inc. This CD-ROM is a multimedia tool that provides interactive explanations of important genetic concepts through animations, diagrams, and videos. It also provides interactive test questions.

Solutions
Step by Step

Problem 1

The mutations *cinnabar* (*cn*, bright red eyes) and *vestigial* (*vg*, malformed wings) are linked in the second chromosome of *Drosophila*. Among 1000 progeny of the cross *cn vg* / + + females × *cn vg* / *cn vg* males, the following genotypes of progeny were observed. From these data, estimate the frequency of recombination between the *cn* and *vg* genes.

Genotype	Number
cn vg / cn vg	455
cn + / cn vg	35
+ vg / cn vg	45
+ + / cn vg	465

Solution In linkage problems of this type, you must first classify the progeny in terms of whether they are recombinant or nonrecombinant. Because the parental genotype is given as *cn vg* / + +, the nonrecombinant progeny are *cn vg* / *cn vg* (455) and + + / *cn vg* (465), and the recombinant progeny are *cn* + / *cn vg* (35) and + *vg* / *cn vg* (45). If the parental genotypes had not been given, it would still have been possible to classify the progeny using the principle that when there is linkage, the recombinant progeny are always less frequent than the nonrecombinant progeny. Because the frequency of recombination is determined by the frequency of progeny carrying recombinant chromosomes, the frequency of recombination between *cn* and *vg* is estimated as $r = (35 + 45)/1000 = 0.08$. The frequency of recombination can also be expressed as a percentage. For recombination frequencies smaller than about 10 percent, the percent recombination equals the number of map units between the genes; in this case, $r = 8$ percent recombination, or 8 map units.

Problem 2

In addition to *cn* and *vg*, the genes *curved* (*c*, curved wings) and *plexus* (*px*, extra wing veins) are linked in the second chromosome of *Drosophila*. In a cross of *cn c px* / + + + females × *cn c px* / *cn c px* males, the following progeny were counted:

Genotype	Number
cn c px / cn c px	296
cn c + / cn c px	63
cn + + / cn c px	119
cn + px / cn c px	10
+ c px / cn c px	86
+ c + / cn c px	15
+ + + / cn c px	329
+ + px / cn c px	82
Total	1000

(a) What is the frequency of recombination between *cn* and *c*?
(b) What is the frequency of recombination between *c* and *px*?
(c) What is the frequency of recombination between *cn* and *px*?
(d) Why is the frequency of recombination between *cn* and *px* smaller than the sum of that between *cn* and *c* and that between *c* and *px*?
(e) What is the coefficient of coincidence across this region? What is the value of the interference?
(f) Draw a genetic map of the region, showing the locations of *cn*, *c*, and *px* and the map distances between the genes.
(g) From the data in Step-by-Step Problem 1, where in this map would you put the gene *vg*?

Solution Do not try to hurry through linkage problems! You will be rewarded by taking time to organize the information in the optimal manner. First, group the progeny types into reciprocal pairs—*cn c px* with + + +, *cn c* + with + + *px*, and so forth—and make a new list organized as shown here. (Ignore the *cn c px* chromosome from the father because it contributes no information about recombination.)

cn c px	296	625
+ + +	329	
cn c +	63	145
+ + px	82	
cn + +	119	205
+ c px	86	
cn + px	10	25
+ c +	15	
Total		1000

In this tabulation, a space has been inserted between the pairs of reciprocal products in order to keep the groups separate. The number next to each brace is the total number of chromosomes in the group. The most numerous group of reciprocal chromosomes (in this case, *cn c px* and + + +) consists of the nonrecombinants, and the least numerous group of reciprocal chromosomes (in this case, *cn* + *px* and + *c* +) consists of the double recombinants. Rearrange the order of the groups, if necessary, so that the nonrecombinants are at the top of the list and the double recombinants are at the bottom. (In the present example, rearrangement is not necessary.) At this point, also make sure that the order of the genes is correct as given, by comparing the genotypes of the double recombinants with those of the nonrecombinants. If the gene order is correct, then it will require two recombination events (one in each interval) to derive the double-recombinant chromosomes from the nonrecombinants. If this is not the case, rearrange the order of the genes. (The "odd man out" in comparing the double recombinants with the nonrecombinants is always the gene in the middle.) In this particular example, the gene order is correct as given. Finally, with this preliminary bookkeeping done, we can proceed to tackle the questions. **(a)** The frequency of recombination between *cn* and *c* is given by the totals of all classes of progeny showing recombination in the *cn*–*c* interval, in this case $(205 + 25)/1000 = 0.23$. **(b)** The frequency of recombination between *c* and *px* equals $(145 + 25)/1000 = 0.17$. **(c)** The frequency of recombination between *cn* and *px* equals $(145 + 205)/1000 = 0.35$. (Note that the double recombinants are not included in this total, because the double recombinants are not recombined for *cn* and *px*; their allele combinations for *cn* and *px* are the same as in the nonrecombinants.) **(d)** The frequency of recombination between *cn* and *px* (0.35) is smaller than the sum of that between *cn* and *c* and that between *c* and *px* ($0.23 + 0.17 = 0.40$) because of double recombination. **(e)** The coefficient of coincidence equals the observed number of double recombinants divided by the expected number. The observed number is 25 and the expected number is $0.23 \times 0.17 \times 1000 = 39.1$; the coefficient of coincidence therefore equals $25/39.1 = 0.64$. The interference equals 1 − *coefficient of coincidence*, so the interference equals $1 - 0.64 = 0.36$.

(f) The genetic map is shown in the accompanying diagram. The distances are in map units (centimorgans). However, the map distances of 23 and 17 map units are based on the 23 percent and 17 percent recombination observed between *cn* and *c* and between *c* and *px*, respectively; the actual distances in map units are probably a little greater than these estimates because of a small amount of double recombination within each of the intervals. **(g)** The position of *vg* is located 10 map units from *cn* because there is 10 percent recombination observed between *vg* and *cn*. However, there is no way of knowing from the information given whether *vg* is to the left of *cn* or the right, so both possible positions for *vg* are indicated. (In fact, *vg* is located between *cn* and *c*.)

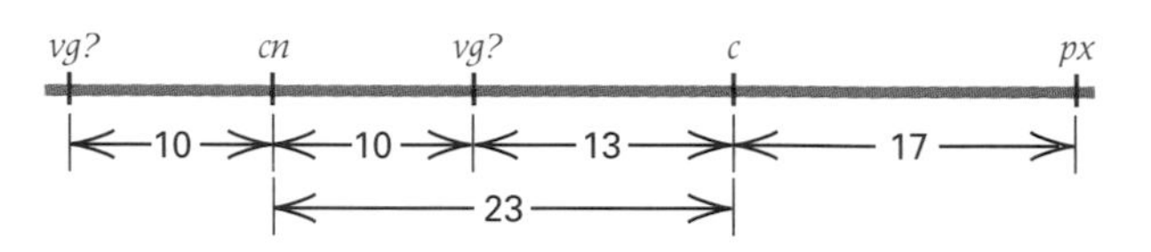

Problem 3

Genes called spore killers are relatively frequent in natural populations of *Neurospora crassa*. In asci produced by a heterozygous spore-killer genotype, all spores that do not carry the spore-killer allele are killed, and they are unable to be germinated. For example, in a cross of *Sk* × *sk*, where *Sk* represents the spore-killer allele, all spores carrying *sk* are killed. For one spore-killer allele, a cross of *Sk* × *sk* produced 125 asci. All of the asci had four dead ascospores and four live ascospores. In 95 of the asci, the four dead spores were all adjacent at either the top end of the ascus or the bottom end. What is the map distance between the *Sk* allele and its centromere?

Solution Centromere mapping in *Neurospora* is based on the relative frequencies of first-division and second-division segregation. One feature that distinguishes these two types of asci is that in first-division segregation, the four spores carrying a particular allele are at one end of the ascus or the other. In the *Sk* × *sk* cross, the first-division asci would have one of the configurations shown in the diagram. Because the *sk*-bearing asci die, in these asci (and only in these) the dead ascospores are all either at the top or the bottom of the ascus. The observation is that 95/125 asci are of this type, which implies that $95/125 = 76$ percent of the asci show first-division segregation. Because the distance between a gene and its centromere is estimated as 1/2 times the frequency of *second*-division segregation, this spore-killer gene is estimated to be at a distance of $(1 - 0.76)/2 = 12$ map units from its centromere.

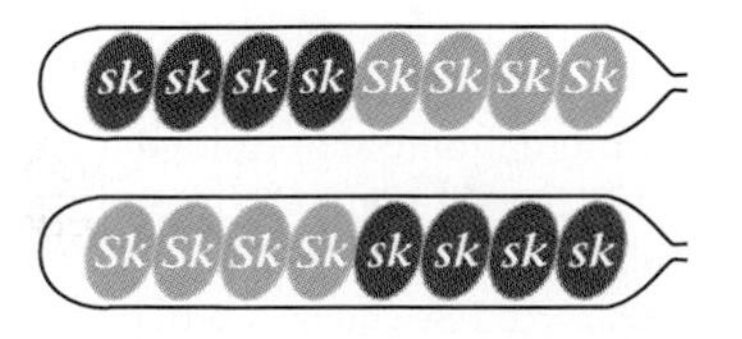

geNETics on the web

www.jbpub.com/genetics

These GeNETics on the Web exercises will introduce some of the most useful sites for finding genetic information on the World Wide Web. Genetic sites provide access to a rich storehouse of information on all aspects of genetics. These range from sites written in nontechnical language for the lay person to sites with sophisticated databases designed for the professional geneticist. In carrying out these exercises, you will get a taste of what the Internet can offer a student in genetics.

The keywords shown in color in the following exercises are available on the Jones and Bartlett Publishers' web site as hyperlinks to various genetic sites. To complete the exercises, visit the GeNETics on the Web home page at

http://www.jbpub.com/genetics

Select the link to Essential geNETics on the Web. Then choose a chapter. You will find a list of keywords that correspond to the exercises that follow. Select a keyword to link to a web site containing the genetic information necessary to complete the exercise. Each exercise includes a specific assignment that makes use of the information available at the site.

Exercises

- One of the first detected genetic linkages for autosomal genes in human beings was between the gene *ABO* for the ABO blood groups and the gene *NPS1* for the nail-patella syndrome. Another gene, *AK1*, coding for the enzyme adenylate kinase 1, was later found to be linked to these genes as well. All of them are located near the tip of the long arm of human chromosome 9. Use the search engine at the keyword site OMIM (it stands for *Online Mendelian Inheritance in Man*) to search for both *NP1* and *AK1*, and read the report of the genetic linkages. Draw a genetic map of this region based on the given data.
- An animated model of the process of recombination can be found at this keyword site. The animation shows the Holliday structure in the process of formation and resolution. Make a series of diagrams showing key stages in the recombination process and then appropriately label using the colors in the animation. Make a set of similar diagrams showing how the Holliday structure is resolved without an exchange of genetic markers flanking the heteroduplex region.

Pic Site

The Pic Site showcases some of the most visually appealing genetics sites on the World Wide Web. This site is sponsored by one of the leading suppliers of DNA chips. To visit this showcase genetics site, select the Pic Site for Chapter 4.

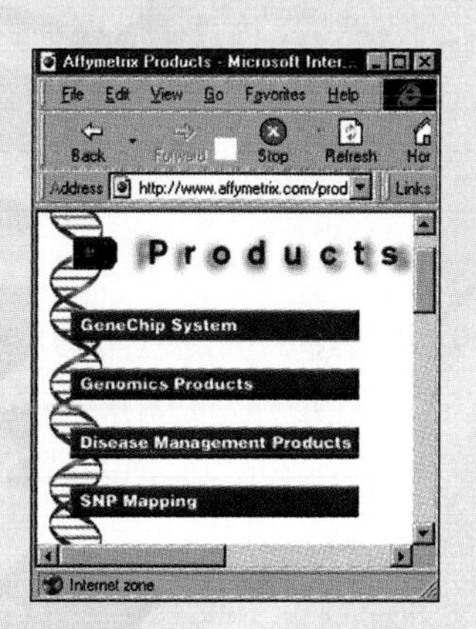

Concepts in Action
Problems for Solution

4.1 The genetic map of the X chromosome of *Drosophila melanogaster* has a length of 73.1 map units. The X chromosome of the related species *Drosophila virilis* is much longer (170.5 map units). In view of the fact that the recombination rate between two genes cannot exceed 50 percent, how is it possible for the map distance between genes at opposite ends of a chromosome to exceed 50 map units?

4.2 Classify each of the following statements as true or false. A coefficient of coincidence of 0.25 means that:

(a) The frequency of double crossovers was 1/4.
(b) The frequency of double crossovers was 1/4 of the number that would be expected if there were no interference.
(c) There were four times as many single crossovers as double crossovers.
(d) There were four times as many single crossovers in one region as there were in an adjacent region.
(e) There were four times as many parental as recombinant progeny.

4.3 In the case of independently assorting genes in fungi, why are parental ditype (PD) and nonparental ditype (NPD) tetrads formed in equal numbers? Why are NPD tetrads rare for linked genes?

4.4 In *Drosophila*, the eye-color mutation scarlet (*st*) and the bristle mutation spineless (*ss*) are located in chromosome 3 at a distance of 14 map units. What phenotypes, and in what proportions, would you expect in the progeny from the mating of $st^+ \ ss^+ / st \ ss$ females with $st \ ss / st \ ss$ males?

4.5 If the recombination frequency between the two genes *A* and *B* is 6.2 percent, what is the distance, in map units, between the genes in the linkage map?

4.6 In corn, the genes *v* (virescent seedlings), *pr* (red aleurone), and *bm* (brown midrib) are all on chromosome 5, but not necessarily in the order given. The cross

$$v^+ \ pr \ m \ / \ v \ pr^+ \ bm^+ \times v \ pr \ bm \ / \ v \ pr \ bm$$

produces 1000 progeny with the following phenotypes:

v^+	pr	bm	209
v	pr^+	bm^+	213
v^+	pr	bm^+	175
v	pr^+	bm	181
v^+	pr^+	bm	69
v	pr	bm^+	76
v^+	pr^+	bm^+	36
v	pr	bm	41

(a) Determine the gene order, the recombination frequencies between adjacent genes, the coefficient of coincidence, and the interference.
(b) Explain why, in this example, the recombination frequencies are not good estimates of map distance.

4.7 In the accompanying human pedigree, the parents I-1 and I-2 are both heterozygous for a recessive allele for cystic fibrosis (*cf;* the wildtype allele is *CF*) and heterozygous for a restriction fragment length polymorphism (RFLP) with phenotypes shown in the diagram of the electrophoresis gel. DNA fragments *A* and *a* are the result of a restriction-enzyme cleavage site that differs between alleles *A* and *a*. The RFLP is linked to the *CF* gene. I-1 and I-2 are first cousins, and a molecular analysis of their parents and siblings indicates that both I-1 and I-2 have the genotype *CF A* / *cf a*. The frequency of recombination in females is 16 percent, and in males is 10 percent. None of the children II-1 through II-5 is affected with cystic fibrosis.

(a) What is the probability that the child II-6 is affected?
(b) What is the probability that the child II-1 is a carrier?

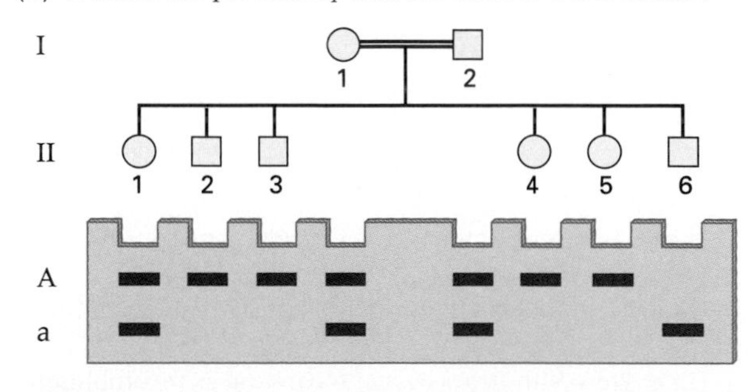

4.8 A human geneticist discovers the molecular variation in DNA sequence illustrated in the accompanying diagrams of electrophoresis gels. In the human population as a whole, she finds any of four phenotypes, shown in panel A. She believes that this may be a simple genetic polymorphism with three alleles, like the ABO blood groups. There are two alleles that yield DNA fragments of different sizes, fast (*F*) or slow (*S*) migration, and a "null" allele (*O*) in which the DNA fragment is deleted. The genotypes in panel A would therefore be, from left to right, *FF* or *FO*, *SS* or *SO*, *FS*, and *OO*. In the population as a whole, the putative *OO* genotype is extremely common, and the *FS* genotype is quite rare. To investigate this hypothesis further, the geneticist studies offspring of matings between parents who have the putative *FS* genotype (panel B). The types of progeny, and their numbers, are shown in panel C.

(a) What result would be expected from the three-allele hypothesis?
(b) Are the observed data consistent with this result? Why or why not?
(c) Suggest a genetic hypothesis that can explain the data in panel C.
(d) Are the data consistent with your hypothesis?

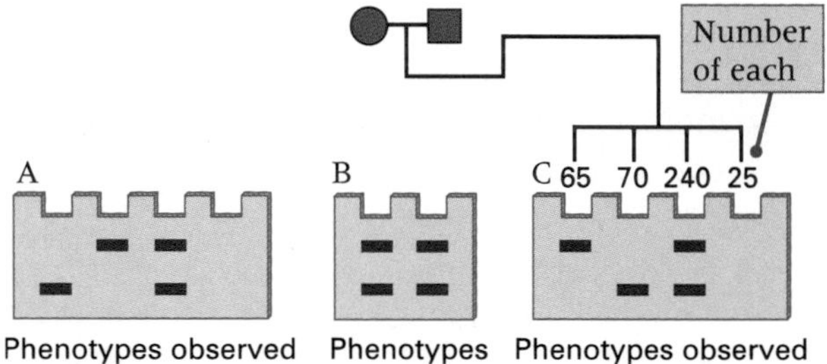

4.9 The following classes and frequencies of ordered tetrads were obtained from the cross $a^+ \ b^+ \times a \ b$ in *Neurospora*. (Only one member of each pair of spores is shown.) What is the order of the genes in relation to the centromere?

Spore Pair				Number of
1–2	**3–4**	**5–6**	**7–8**	**Asci**
$a^+ \ b^+$	$a^+ \ b^+$	$a \ b$	$a \ b$	1766
$a^+ \ b^+$	$a \ b$	$a^+ \ b^+$	$a \ b$	220
$a^+ \ b^+$	$a \ b^+$	$a^+ \ b$	$a \ b$	14

4.10 A portion of the linkage map of chromosome 2 in the tomato is illustrated here. The oblate phenotype has flattened fruit, peach results in hairy fruit (like a peach), and compound inflorescence means clustered flowers.

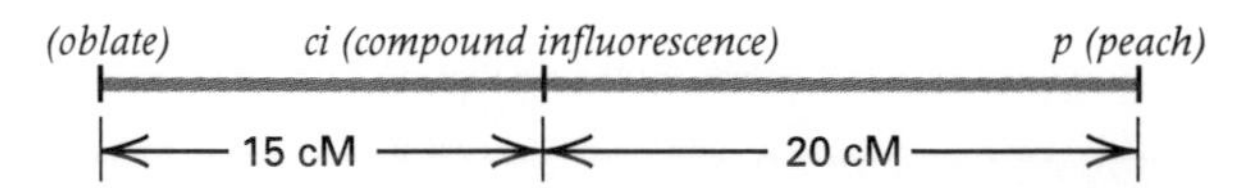

Among 1000 gametes produced by a plant of genotype $o \ ci + / + + p$, what types of gametes would be expected, and what number would be expected of each? Assume that the chromosome interference across this region is 80 percent.

4.11 The yeast *Saccharomyces cerevisiae* has unordered tetrads. In a cross made to study the linkage relationships among three genes, the tetrads in the table at the top of page 155 were obtained. The cross was between a strain of genotype $+ \ b \ c$ and one of genotype $a + +$.

(a) From these data determine which, if any, of the genes are linked.
(b) For any linked genes, determine the map distances.

Problem 4-11—Tetrads

Tetrad Type	Genotypes of Spores in Tetrads				Number of Tetrads
1	*a* + +	*a* + +	+ *b c*	+ *b c*	132
2	*a b* +	*a b* +	+ + *c*	+ + *c*	124
3	*a* + +	*a* + *c*	+ *b* +	+ *b c*	64
4	*a b* +	*a b c*	+ + +	+ + *c*	80
Total					400

4.12 A small portion of the genetic map of *Neurospora crassa* chromosome VI is illustrated here. The *cys-1* mutation blocks cysteine synthesis, and the *pan-2* mutation blocks pantothenic acid synthesis. Assuming complete chromosome interference, determine the expected frequencies of the following types of asci in a cross of *cys-1 pan-2* × *CYS-1 PAN-2*.

(a) First-division segregation of *cys-1* and first-division segregation of *pan-2*.
(b) First-division segregation of *cys-1* and second-division segregation of *pan-2*.
(c) Second-division segregation of *cys-1* and first-division segregation of *pan-2*.
(d) Second-division segregation of *cys-1* and second-division segregation of *pan-2*.
(e) Parental ditype, tetratype, and nonparental ditype tetrads.

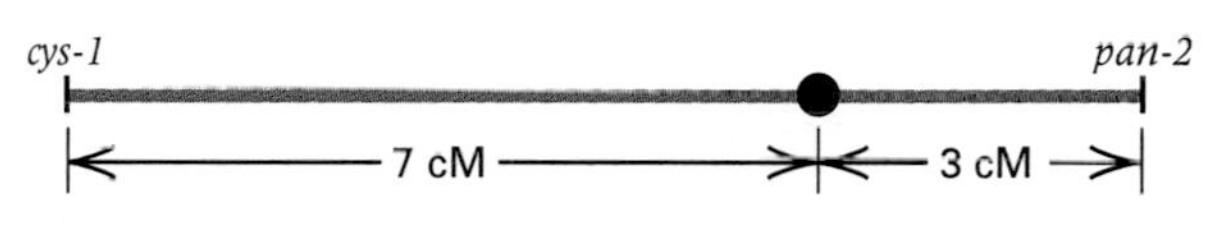

Further Readings

Carlson, E. A. 1987. *The Gene: A Critical History.* 2d ed. Philadelphia: Saunders.

Creighton, H. S., and B. McClintock. 1931. A correlation of cytological and genetical crossing over in *Zea mays. Proceedings of the National Academy of Sciences, USA* 17: 492.

Fincham, J. R. S., P. R. Day, and A. Radford. 1979. *Fungal Genetics.* Oxford, England: Blackwell.

Green, M. M. 1996. The "Genesis of the White-Eyed Mutant" in *Drosophila melanogaster*: A reappraisal. *Genetics* 142: 329.

Kohler, R. E. 1994. *Lords of the Fly.* University of Chicago Press.

Levine, L. 1971. *Papers on Genetics.* St. Louis, MO: Mosby.

Lewis, E. B. 1995. Remembering Sturtevant. *Genetics* 141: 1227.

Morton, N. E. 1995. LODs past and present. *Genetics* 140: 7.

Stewart, G. D., T. J. Hassold, and D. M. Kurnit. 1988. Trisomy 21: Molecular and cytogenetic studies of nondisjunction. *Advances in Human Genetics* 17: 99.

Sturtevant, A. H. 1965. *A History of Genetics.* New York: Harper & Row.

Sturtevant, A. H., and G. W. Beadle. 1962. *An Introduction to Genetics.* New York: Dover.

Voeller, B. R., ed. 1968. *The Chromosome Theory of Inheritance: Classical Papers in Development and Heredity.* New York: Appleton-Century-Crofts.

Wang, D. G., J.-B. Fan, C.-J. Siao, A. Berno, P. Young, R. Sapolsky et al. 1998. Large-scale identification, mapping and genotyping of single-nucleotide polymorphisms in the human genome. *Science* 280: 1077.

White, R., and J.-M. Lalouel. 1988. Chromosome mapping with DNA markers. *Scientific American,* February.

Key Concepts

- Duplication of the entire chromosome complement that is present in a species, or in a hybrid between species, is a major process in the evolution of higher plants.
- The genetic imbalance caused by a single chromosome that is extra or missing may have a more serious phenotypic effect than an entire extra set of chromosomes.
- Chromosome abnormalities are an important cause of human genetic disease and are a major factor in spontaneous abortions.
- Aneuploid (unbalanced) chromosome rearrangements usually have greater phenotypic effects than euploid (balanced) chromosome rearrangements.
- Reciprocal translocations result in abnormal gametes because they upset segregation.
- Some types of cancer are associated with particular chromosome rearrangements.
- Transposable elements are DNA sequences that are able to change their location within a chromosome or to move between chromosomes.

ORANGUTANS, LIKE GORILLAS and chimpanzees, have 24 pairs of chromosomes. In the human lineage, two chromosomes underwent end-to-end fusion and produced human chromosome number 2, reducing the number of pairs of chromosomes to 23. [© Masterfile.]

CHAPTER 5

Variation in Chromosome Number and Structure

In all species, an occasional organism is found that has an extra chromosome or that lacks a particular chromosome. Such a deviation from the norm is an abnormality in chromosome *number.* Other organisms, usually rare, are found to have alterations in the arrangement of genes in the genome, such as by having a chromosome with a particular segment missing, reversed in orientation, or attached to a different chromosome. These variations are abnormalities in chromosome *structure.* This chapter deals with the genetic effects of both numerical and structural chromosome abnormalities. We shall see that animals are much less tolerant of such changes than are plants. Furthermore, in animals, numerical alterations often produce greater effects on phenotype than do structural alterations.

5.1 Chromosomes may differ in size and in position of the centromere.

The centromere is required for a chromosome to segregate properly in cell division. When a cell divides, spindle fibers attach to the centromere of each chromosome and pull the sister chromatids to opposite poles. Occasionally, a chromosome arises that has an abnormal number of centromeres, as diagrammed in Figure 5.1A. The chromosome at the left has two centromeres and is said to be **dicentric.** A dicentric chromosome is genetically unstable, which means it is not transmitted in a predictable fashion. The dicentric chromosome is frequently lost from a cell when the two centromeres proceed to opposite poles in the course of cell division; in this case, the chromosome is stretched and forms a *bridge* between the daughter cells. This bridge may not be included in either daughter nucleus, or it may break such that each daughter nucleus receives a broken chromosome. The chromosome at the right in Figure 5.1A is an **acentric** chromosome, which lacks a centromere. Acentric chromosomes also are genetically unstable because they cannot be maneuvered properly during cell division and tend to be lost.

In eukaryotic organisms, virtually all chromosomes have a single centromere and are rod-shaped. Monocentric rod chromosomes are often classified according to the relative position of their centromeres. A chromosome with its centromere approximately in the middle is a **metacentric chromosome;** the arms are of approximately equal length and form a V shape at anaphase (Figure 5.1B). When the centromere is somewhat off center, the chromosome is a **submetacentric chromosome,** and the arms form a J shape at anaphase. A chromosome with the centromere very close to one end appears I-shaped at anaphase because the arms are grossly unequal in length; such a chromosome is **acrocentric.**

The distinction among metacentric, submetacentric, and acrocentric chromosomes is useful because it draws attention to the chromosome arms. In the evolution of chromosomes, often the number of chromosome *arms* is conserved without conservation of the individual *chromosomes.* For example, *Drosophila melanogaster* has two large metacentric autosomes, but many other *Drosophila* species have four acrocen-

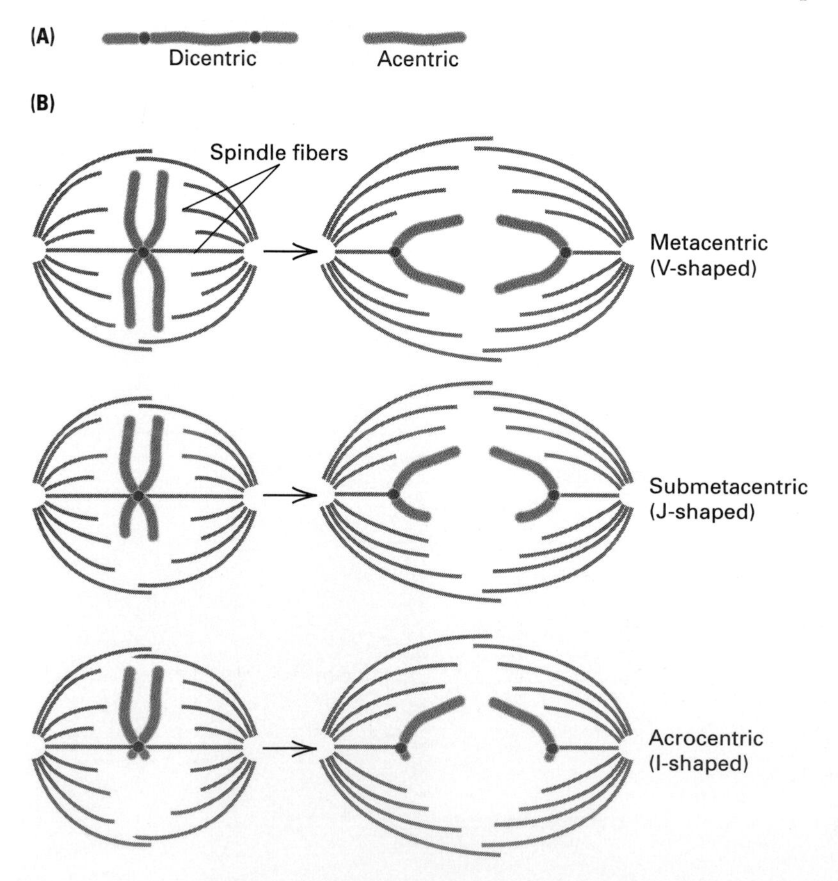

Figure 5.1 (A) Diagram of a chromosome that is dicentric (two centromeres) and one that is acentric (no centromere). Dicentric and acentric chromosomes are frequently lost in cell division, the former because the two centromeres may bridge between the daughter cells, the latter because the chromosome cannot attach to the spindle fibers. (B) Three possible shapes of monocentric chromosomes in anaphase as determined by the position of the centromere. The centromeres are shown in dark blue.

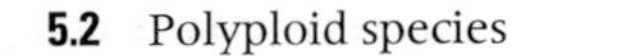

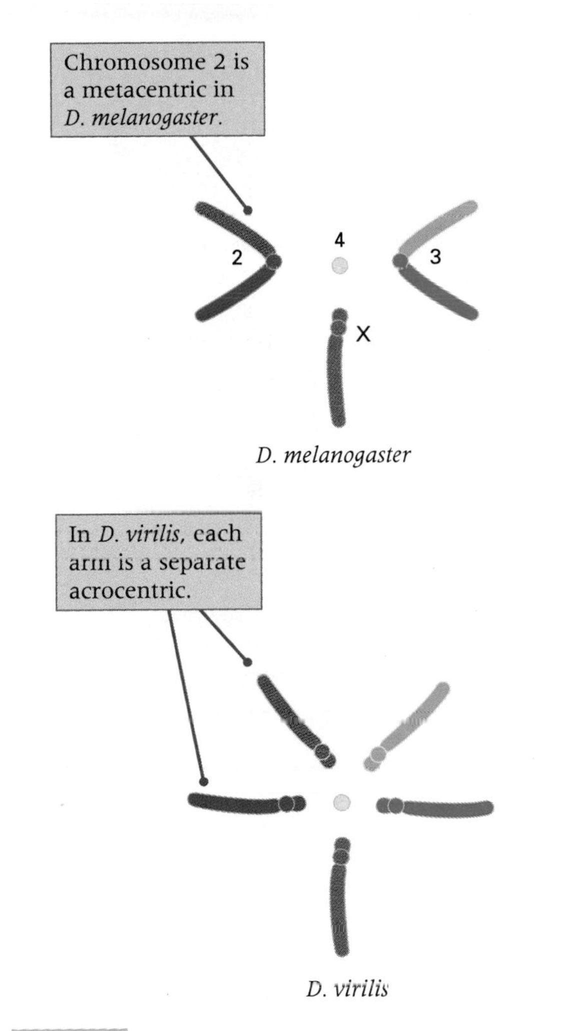

Figure 5.2 The haploid chromosome complement of two species of *Drosophila*. Color indicates homology of the chromosome arms. The large metacentric chromosomes of *Drosophila melanogaster* (chromosomes 2 and 3) correspond arm for arm with the four large acrocentric autosomes of *Drosophila virilis*.

tric autosomes instead of the two metacentrics. Detailed comparison of the genetic maps of these species reveals that the acrocentric chromosomes in the other species correspond, arm for arm, with the large metacentrics in *D. melanogaster* (Figure 5.2).

Among higher primates, chimpanzees and human beings have 22 pairs of chromosomes that are morphologically similar, but chimpanzees have two pairs of acrocentrics not found in human beings, and human beings have one pair of metacentrics not found in chimpanzees. In this case, the human metacentric chromosome was formed by fusion of the telomeres between the short arms of the chromosomes that, in chimpanzees, remain acrocentrics (Figure 5.3). The fusion was accompanied by the inactivation, by an unknown mechanism, of one of the centromeres, so that the fused chromosome has only a single functional centromere. The metaphase chromosome resulting from the fusion is human chromosome 2, and it reduces the chromosome number from 48, which is characteristic of the great apes (chimpanzee, gorilla, and orangutan), to the number 46 present in the human genome.

5.2 Polyploid species have multiple sets of chromosomes.

The genus *Chrysanthemum* illustrates **polyploidy,** an important phenomenon found frequently in higher plants. In polyploidy, a species has a genome composed of multiple complete sets of chromosomes. One *Chrysanthemum* species, a diploid species, has 18 chromosomes. A closely related species has 36 chromosomes. However, comparison of chromosome morphology indicates that the

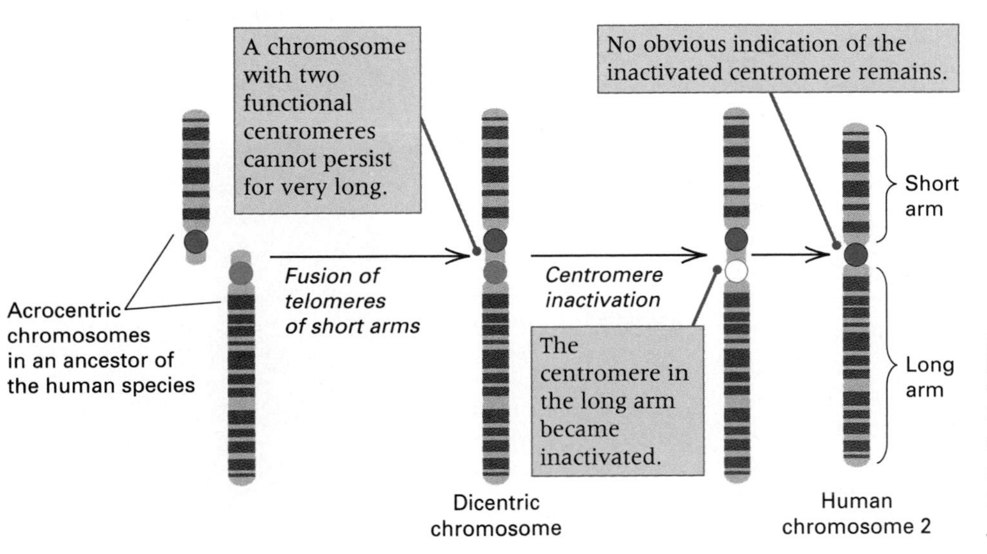

Figure 5.3 Human ancestors had 48 pairs of chromosomes rather than 46. In the evolution of the human genome, two acrocentric chromosomes fused to create human chromosome 2.

36-chromosome species has two complete sets of the chromosomes found in the 18-chromosome species (Figure 5.4). The basic chromosome set in the group, from which all the other genomes are formed, is called the **monoploid** chromosome set. In *Chrysanthemum*, the monoploid chromosome number is 9. The diploid species has two complete copies of the monoploid set, or 18 chromosomes altogether. The 36-chromosome species has four copies of the monoploid set ($4 \times 9 = 36$) and is a **tetraploid.** Other species of *Chrysanthemum* have 54 chromosomes (6×9, constituting the **hexaploid**), 72 chromosomes (8×9, constituting the *octoploid*), and 90 chromosomes (10×9, constituting the *decaploid*).

In meiosis, the chromosomes of all *Chrysanthemum* species synapse normally in pairs to form bivalents (Section 3.3). The 18-chromosome species forms 9 bivalents, the 36-chromosome species forms 18 bivalents, the 54-chromosome species forms 27 bivalents, and so forth. Gametes receive one chromosome from each bivalent, so the number of chromosomes in the gametes of any species is exactly half the number of chromosomes in its somatic cells. The chromosomes present in the gametes of a species constitute the **haploid** set of chromosomes. In the species of *Chrysanthemum* with 90 chromosomes, for example, the haploid chromosome number is 45; in meiosis, 45 bivalents are formed, and so each gamete contains 45 chromosomes. When two such gametes come together in fertilization, the complete set of 90 chromosomes in the species is restored. Thus the gametes of a polyploid organism are not always monoploid, as they are in a diploid organism; for example, a tetraploid organism has diploid gametes.

The distinction between the term *monoploid* and the term *haploid* is subtle:

- The *monoploid* chromosome set is the basic set of chromosomes that is multiplied in a polyploid series of species, such as *Chrysanthemum*.
- The *haploid* chromosome set is the set of chromosomes present in a gamete, irrespective of the chromosome number in the species.

Confusion can arise because of diploid organisms, in which the monoploid chromosome set and the haploid chromosome set are the same. It helps to clarify the difference by considering the tetraploid, which contains four monoploid chromosome sets and in which the haploid gametes are diploid.

Polyploidy is widespread in certain plant groups. Among flowering plants, from 30 to 35 percent of existing species are thought to have originated as some form of polyploid. Valuable

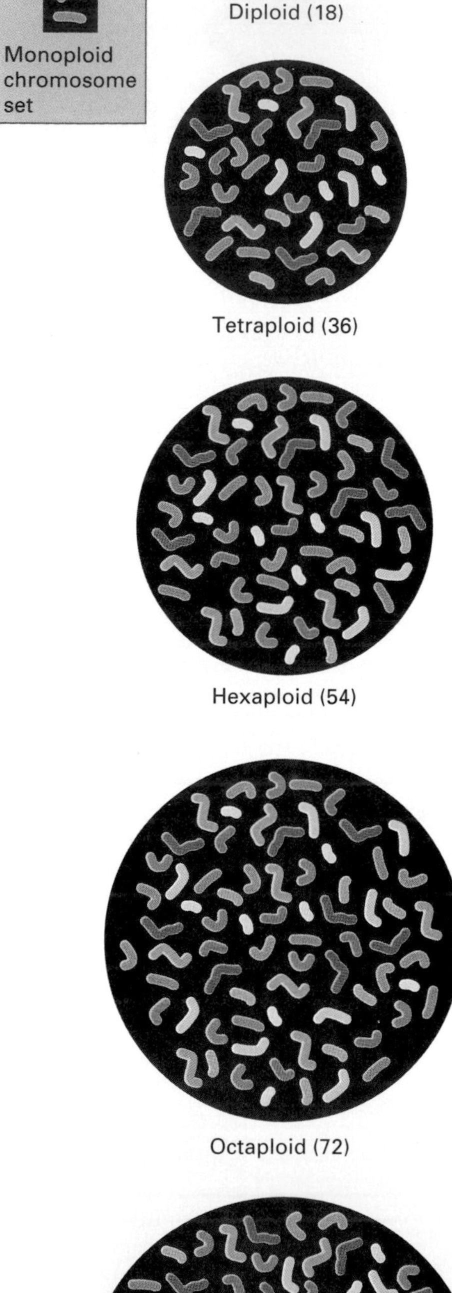

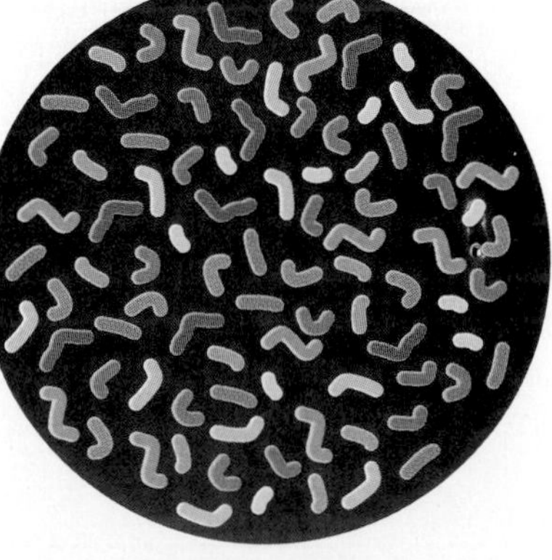

Figure 5.4 Chromosome numbers in diploid and polyploid species of *Chrysanthemum*. Each set of homologous chromosomes is depicted in a different color.

agricultural crops that are polyploid include wheat, oats, cotton, potatoes, bananas, coffee, and sugar cane. Polyploidy often leads to an increase in the size of individual cells, and polyploid plants are often larger and more vigorous than their diploid ancestors; however, there are many exceptions to these generalizations. Polyploidy is rare in vertebrate animals, but it is found in a few groups of invertebrates. One reason why polyploidy is rare in animals is the difficulty in regular segregation of the sex chromosomes. For example, a tetraploid animal with XXXX females and XXYY males would produce XX eggs and XY sperm (if all chromosomes paired to form bivalents), so the progeny would be exclusively XXXY and unlike either of the parents.

Polyploid plants found in nature almost always have an even number of sets of chromosomes because organisms with an odd number have low fertility. Organisms with three monoploid sets of chromosomes are known as **triploids.** As far as growth is concerned, a triploid is quite normal because the triploid condition does not interfere with mitosis; in mitosis in triploids (or any other type of polyploid), each chromosome replicates and divides just as in a diploid. However, because each chromosome has more than one pairing partner, chromosome segregation is severely upset in meiosis, and most gametes are defective. Unless the organism can perpetuate itself by means of asexual reproduction, it will eventually become extinct.

The infertility of triploids is sometimes of commercial benefit. For example, the seeds in commercial bananas are small and edible because the plant is triploid and most of the seeds fail to develop to full size. In oysters, triploids are produced by treating fertilized diploid eggs with a chemical that causes the second polar body of the egg to be retained. The triploid oysters are sterile and do not spawn. In Florida and in certain other states, weed control in waterways is aided by the release of weed-eating fish (the grass carp), which do not become overpopulated, and hence a problem themselves, because the released fish are sterile triploids.

Tetraploid organisms can be produced in several ways. The simplest mechanism is a failure of chromosome separation in either mitosis or meiosis, which instantly doubles the chromosome number. Chromosome doubling through an abortive cell division is called **endoreduplication** (Figure 5.5). In a plant species that can undergo self-fertilization,

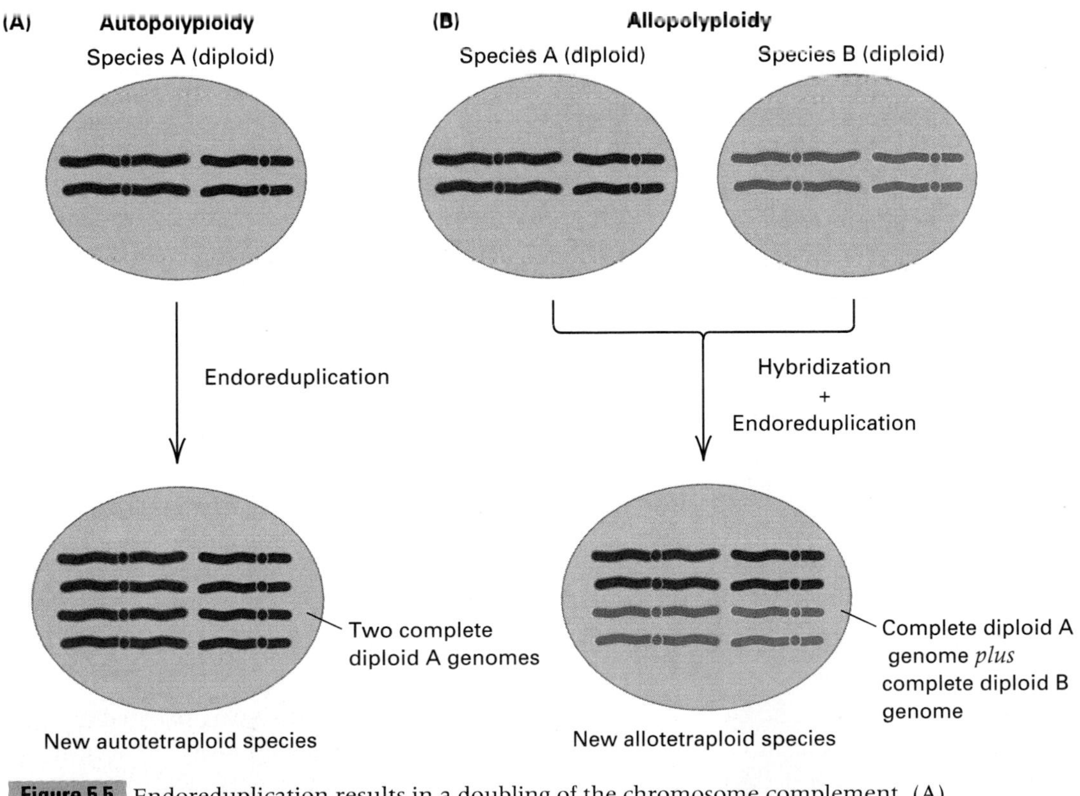

Figure 5.5 Endoreduplication results in a doubling of the chromosome complement. (A) When it takes place in a diploid species, it results in the formation of an autotetraploid species. (B) When endoreduplication takes place in the hybrid formed by cross-fertilization between different species, it results in the formation of an allotetraploid species that has a complete diploid set of chromosomes from each parental species.

endoreduplication creates a new, genetically stable species, because the chromosomes in the tetraploid can pair two by two in meiosis and therefore segregate regularly, each gamete receiving a full diploid set of chromosomes. Self-fertilization of the tetraploid restores the chromosome number, so the tetraploid condition can be perpetuated. The genetics of tetraploid species, like that of other polyploids, is more complex than that of diploid species because the organism carries more than two alleles of any gene. With two alleles in a diploid, only three genotypes are possible: *AA, Aa,* and *aa.* In a tetraploid, by contrast, five genotypes are possible: *AAAA, AAAa, AAaa, Aaaa,* and *aaaa.* Among these genotypes, the middle three represent different types of tetraploid heterozygotes.

An octoploid species (eight sets of chromosomes) can be generated by failure of chromosome separation in mitosis in a tetraploid. If only bivalents form in meiosis, an octoploid organism can be perpetuated sexually by self-fertilization or through crosses with other octoploids. Furthermore, cross-fertilization between an octoploid and a tetraploid results in a hexaploid (six sets of chromosomes). Repeated episodes of polyploidization and cross-fertilization may ultimately produce an entire polyploid series of closely related organisms that differ in chromosome number, as exemplified in *Chrysanthemum.*

Chrysanthemum represents a type of polyploidy, known as **autopolyploidy,** in which all chromosomes in the polyploid species derive from a single diploid ancestral species (Figure 5.5A). In many cases of polyploidy, the polyploid species have complete sets of chromosomes from two or more *different* ancestral species. Such polyploids are known as **allopolyploids** (Figure 5.5B). They derive from occasional hybridization between different diploid species when pollen from one species germinates on the stigma of another species and sexually fertilizes the ovule, followed by endoreduplication in the zygote to yield a hybrid plant in which each chromosome has a pairing partner in meiosis. The pollen may be carried to the wrong flower by wind, insects, or other pollinators. Figure 5.5B illustrates hybridization between species A and B in which endoreduplication leads to the formation of an allopolyploid (in this case, an *allotetraploid*), which carries a complete diploid genome from each of its two ancestral species. The formation of allopolyploids through hybridization and endoreduplication is an extremely important process in plant evolution and plant breeding. At least half of all naturally occurring polyploids are allopolyploids. Cultivated wheat provides an excellent example of allopolyploidy. Cultivated bread wheat is a hexaploid with 42 chromosomes constituting a complete diploid genome of 14 chromosomes from each of three ancestral species. The 42-chromosome allopolyploid is thought to have originated by the series of hybridizations and endoreduplications outlined in Figure 5.6.

Flower of one of the autoploid species of Chrysanthemum [© Beth Maynor/ Photo Researchers, Inc.]

The ancestral origin of the chromosome sets in an allopolyploid can often be revealed by the technique of **chromosome painting,** in which chromosomes are "painted" different colors by hybridization with DNA strands labeled with fluorescent dyes. DNA from each of the putative ancestral species is isolated, denatured, and labeled with a different fluorescent dye. Then the labeled single strands are spread on a microscope slide and allowed to renature with homologous strands present in the chromosomes of the allopolyploid species.

An example of chromosome painting is shown in Figure 5.7. The flower is from a variety of crocus called Golden Yellow. Its genome contains seven pairs of chromosomes, shown painted in yellow and green. Golden Yellow was thought to be an allopolyploid formed by hybridization of two closely related species followed by endoreduplication of the chromosomes in the hybrid. The putative ancestral species are *Crocus flavus,* which has four pairs of chromosomes, and *Crocus angustifolius,* which has three pairs of chromosomes. To paint the chromosomes of Golden Yellow, DNA from *C. flavus* was isolated and labeled with a fluorescent green dye, and that from *C. angustifolius* was isolated and labeled with a fluorescent yellow dye. The result of the chromosome painting is very clear: Three pairs of chromosomes hybridize with the green-labeled DNA from *C. flavus,* and four pairs of chromosomes hybridize with the yellow-labeled DNA from *C. angustifolius.* This pattern of hybridization strongly supports the hypothesis of the allopolyploid origin of Golden Yellow diagrammed in Figure 5.8.

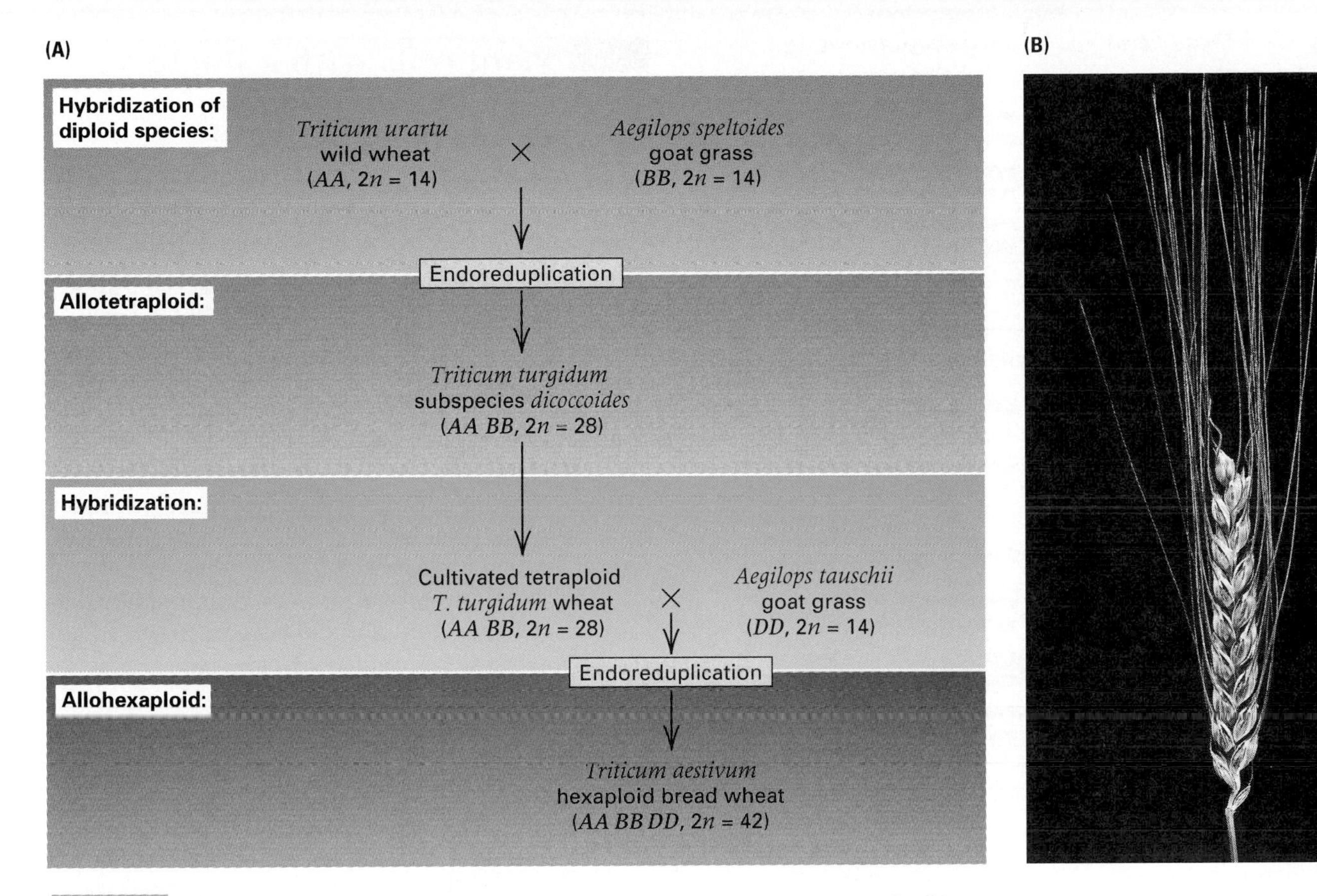

Figure 5.6 (A) Repeated hybridization and endoreduplication in the ancestry of cultivated bread wheat (*Triticum aestivum*), which is an allohexaploid containing complete diploid genomes (*AA, BB, DD*) from three ancestral species. (B) The spike of *T. turgidum*, an allotetraploid species. The large grains of this species made it attractive to early hunter-gatherer societies in the Middle East. One of the earliest cultivated wheats, *T. turgidum* is the progenitor of commercial macaroni wheat. [Photograph courtesy of Gordon Kimber.]

Figure 5.7 Flower of the crocus, variety Golden Yellow, and chromosome painting that reveals its origin as an allopolyploid. Its seven pairs of chromosomes are shown at the right. The chromosomes in green hybridized with DNA from *C. angustifolius,* which has three pairs of chromosomes, and those in yellow hybridized with DNA from *C. flavus,* which has four pairs of chromosomes. [Courtesy of J. S. Heslop-Harrison, John Innes Centre, Norwich, UK. With permission of the *Annals of Botany.*]

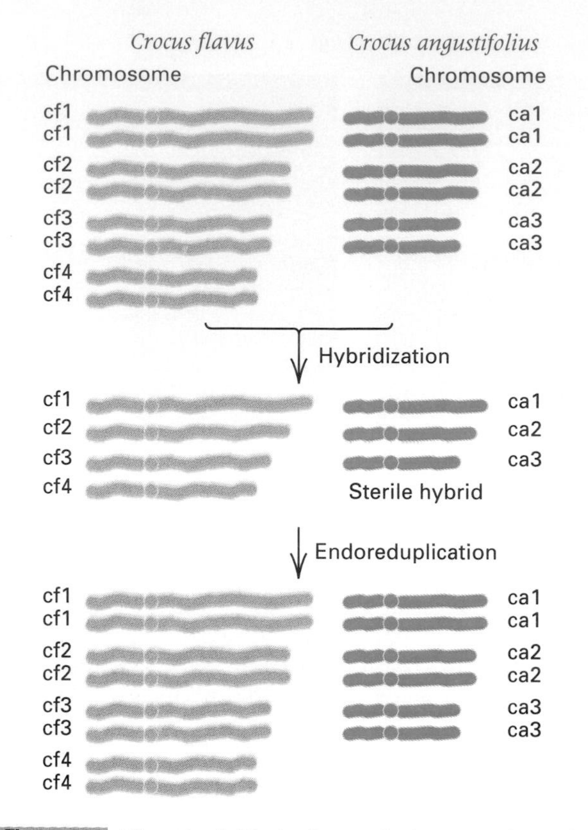

Figure 5.8 Allopolyploidy in the evolutionary origin of the Golden Yellow crocus. The original hybrid between *C. flavus* and *C. angustifolius* produced a sterile monoploid with one copy of each chromosome from each species. Chromosome endoreduplication in the monoploid results in a fertile allotetraploid with a complete diploid genome from each species.

CERTAIN INSECT SPECIES, such as ants and bees, produce males that are derived from unfertilized eggs. These organisms are known as monoploids and are quite rare. [© Robert Brons/BPS/Tony Stone Images]

5.3 Plant cells with a single chromosome set can be cultured.

As we have said, the monoploid chromosome set is the set of chromosomes multiplied in polyploid species. An organism is monoploid if it develops from a monoploid cell. Meiosis cannot take place normally in the germ cells of a monoploid, because each chromosome lacks a pairing partner, and hence monoploids are usually sterile. Monoploid organisms are quite rare, but they occur naturally in certain insect species (ants, bees) in which males are derived from unfertilized eggs. These monoploid males are fertile because the gametes are produced by a modified meiosis in which chromosomes do not separate in meiosis I.

Monoploids are important in plant breeding, because in the selection of diploid organisms with desired properties, favorable recessive alleles may be masked by heterozygosity. This problem can be avoided by studying monoploids, provided that their sterility can be overcome. In many plants, the production of monoploids capable of reproducing can be stimulated by conditions that yield aberrant cell divisions. Two techniques make this possible.

With some diploid plants, monoploids can be derived from cells in the anthers (the pollen-bearing structures). Extreme chilling of the anthers causes some of the haploid cells destined to become pollen grains to begin to divide. These cells are monoploid as well as haploid. If the cold-shocked cells are placed on an agar surface containing suitable nutrients and certain plant hormones, a small dividing mass of cells called an **embryoid** forms. A subsequent change of plant hormones in the growth medium causes the embryoid to form a small plant with roots and leaves that can be potted in soil and allowed to grow normally. Because monoploid cells have only a single set of chromosomes, their genotypes can be identified without regard to the dominance or recessiveness of individual alleles. A plant breeder can then select a monoploid plant with the desired traits. In some cases, the desired genes are present in the original diploid plant and are merely sorted out and selected in the monoploids. In other cases, the anthers are treated with mutagenic agents in the hope of producing the desired traits.

When a desired mutation is isolated in a monoploid, it is necessary to convert the monoploid into a homozygous diploid because the monoploid plant is sterile and does not produce seeds. The monoploid is converted into a diploid by treatment of the meristematic tissue (the growing point of a stem or branch) with the substance **colchicine.** This chemical is an inhibitor of the formation of the mitotic spindle. When the treated cells in the monoploid meristem begin mitosis, the chromosomes replicate normally; however, the colchicine blocks metaphase and anaphase, so the result is endoreduplication (doubling of each chromosome in a given cell). Many of the cells are killed by colchicine, but fortunately for the plant breeder, a few of the monoploid cells are converted into the diploid state (Figure 5.9). The colchicine is removed to allow continued cell multiplication, and many of the now-diploid cells multiply to form a small sector of tissue that can be recognized microscopically. If placed on a nutrient-agar surface, this tissue will develop into a complete plant. Such plants, which are completely homozygous, are fertile and produce normal seeds.

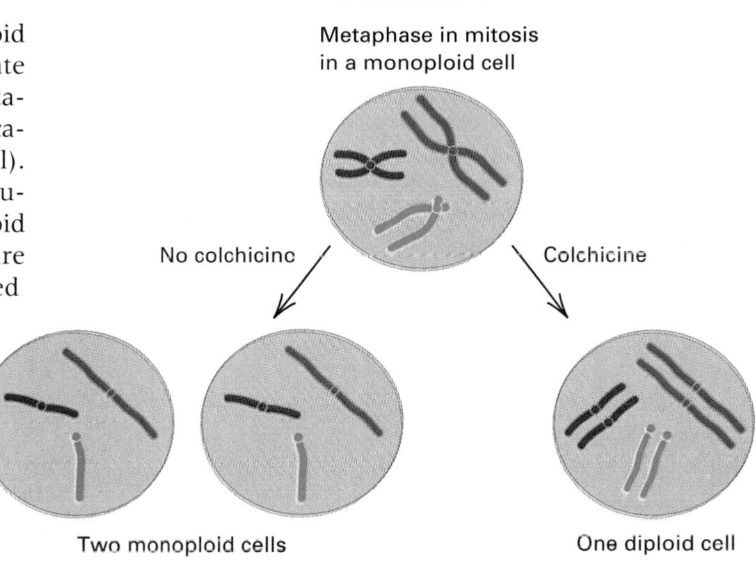

Figure 5.9 Production of a diploid from a monoploid by treatment with colchicine. The colchicine results in endoreduplication because it disrupts the spindle and thereby prevents separation of the chromatids after the centromeres (circles) divide.

5.4 Polysomic organisms have extra or missing chromosomes.

Organisms are occasionally found that have extra copies of individual chromosomes rather than extra entire sets of chromosomes. This situation is called **polysomy.** In contrast with polyploids, which in plants are often healthy and in some cases more vigorous than the diploid, polysomics are usually less vigorous than the diploid and have abnormal phenotypes. In most plants, a single extra chromosome (or a missing chromosome) has a more severe effect on phenotype than the presence of a complete extra set of chromosomes. Each chromosome that is extra or missing results in a characteristic phenotype. For example, Figure 5.10 shows the seed capsule of the normal diploid form of the Jimson weed *Datura stramonium,* along with the capsule of a series of strains, each having an extra copy of a different chromosome. The seed capsule of each of the strains is distinctive.

An otherwise diploid organism that has an extra copy of an individual chromosome is called a **trisomic.** In a trisomic organism, the segregation of chromosomes in meiosis is upset because the trisomic chromosome has two pairing partners instead of one. The behavior of the chromosomes in meiosis depends on the manner in which the homologous chromosome arms pair and the chiasmata are formed between them. In some cells, the three chromosomes form a **trivalent,** in which distinct

Figure 5.10 Seed capsules of the normal diploid *Datura stramonium* (Jimson weed), which has a haploid number of 12 chromosomes, and 4 of the 12 possible trisomics. The phenotype of the seed capsule in each trisomic differs according to the chromosome that is trisomic.

parts of one chromosome are paired with homologous parts of each of the others (Figure 5.11A). In metaphase, the trivalent usually orients with two centromeres pointing toward one pole and the other centromere toward the other so that, at the end of both meiotic divisions, one pair of gametes contains two copies of the trisomic chromosome and the other pair of gametes contains only a single copy. Alternatively, the trisomic chromosome can form one normal bivalent and one **univalent,** or unpaired, chromosome as shown in Figure 5.11B. In anaphase I, the bivalent disjoins normally and the univalent usually proceeds randomly to one pole or the other. Again, the end result is the formation of two products of meiosis that contain two copies of the trisomic chromosome and two products of meiosis that contain one copy. To state the matter in another way, a trisomic organism with three copies of a chromosome (say, *C C C*) will produce gametes among which half contain *C C* and half contain *C*. The unequal segregation of trisomic chromosomes is also the cause of the infertility in triploids, because in a triploid, each chromosome behaves independently as though it were trisomic. For each of the homologous chromosomes in a triploid, a gamete can receive either one copy or two copies.

Polysomy generally results in more severe phenotypic effects than does polyploidy. The usual explanation is that the greater harmful phenotypic effects in trisomics are related to the imbalance in the number of copies of different genes. A polyploid organism has a "balanced" genome in the sense that the ratio of the number of copies of any pair of genes is the same as in the diploid. For example, in a tetraploid, each gene is present in twice as many copies as in a diploid, so no gene or group of genes is out of balance with the others. Balanced chromosome abnormalities, which retain equality in the number of copies of each gene, are said to be **euploid.** In contrast, gene equality is upset in a trisomic because three copies of the genes located in the trisomic chromosome are present, whereas two copies of the genes in the other chromosomes are

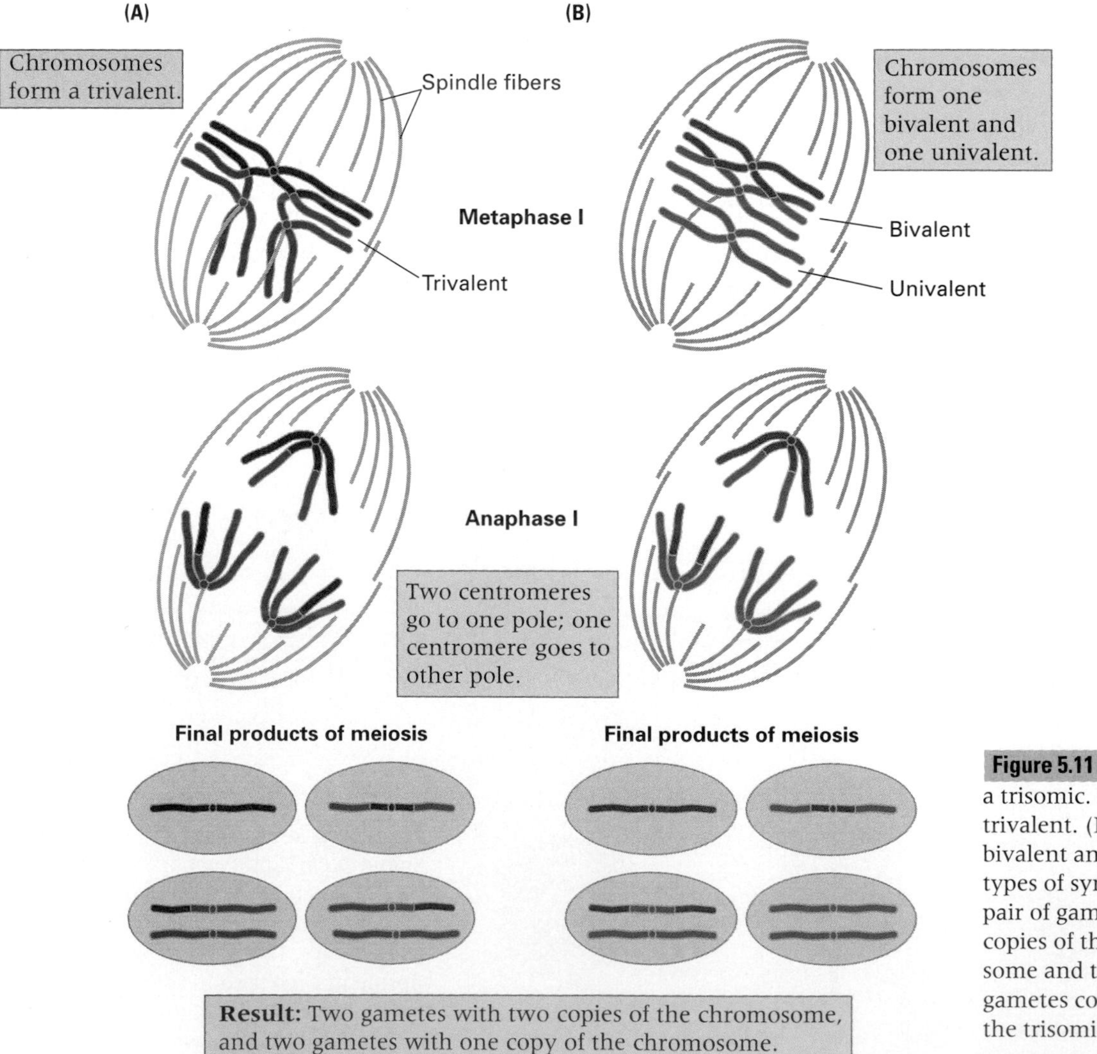

Figure 5.11 Meiotic synapsis in a trisomic. (A) Formation or a trivalent. (B) Formation of a bivalent and a univalent. Both types of synapsis result in one pair of gametes containing two copies of the trisomic chromosome and the other pair of gametes containing one copy of the trisomic chromosome.

present. Such unbalanced chromosome complements are said to be **aneuploid.** In general, aneuploid abnormalities are usually more severe than euploid abnormalities. For example, in *Drosophila,* triploid females are viable, fertile, and nearly normal in morphology, whereas trisomy for either of the two large autosomes is invariably lethal (the larvae die at an early stage).

Just as an occasional organism may have an extra chromosome, a chromosome may be missing. Such an individual is said to be **monosomic.** In general, a missing copy of a chromosome results in more harmful effects than an extra copy of the same chromosome, and monosomy is often lethal. If the monosomic organism can survive to sexual maturity, chromosomal segregation in meiosis is unequal because the monosomic chromosome forms a univalent. Half the gametes will carry one copy of the monosomic chromosome, and the other half will contain no copies.

5.5 Human beings have 46 chromosomes in 23 pairs.

The chromosome complement of a normal human male is illustrated in Figure 5.12. The chromosomes have been treated with a staining reagent called Giemsa, which causes the chromosomes to exhibit transverse bands that are specific for each pair of homologs. These bands permit the chromosome pairs to be identified individually. By convention, the autosome pairs are arranged and numbered from longest to shortest and, on the basis of size and centromere position, separated into seven groups designated by the letters A through G. This conventional representation of chromosomes is called a **karyotype,** and it is obtained by cutting each chromosome out of a photograph taken at metaphase and pasting it into place. In a karyotype of a normal human female, the autosomes would not differ from those of a male and hence would be identical with those in Figure 5.12; however, there would be two X chromosomes instead of an X and a Y. Abnormalities in chromosome number and morphology are made evident by a karyotype.

A diagrammatic representation of the banding patterns in human chromosomes is shown in Figure 5.13. For each chromosome, the short arm is designated with the letter *p,* which stands for *petite,* and the long arm by the letter *q,* which stands for "not-*p*." Within each arm, the regions are numbered consecutively from the centromere toward the telomere; within each region, the second number indicates the next-smaller division. For example, the designation 1p34 indicates chromosome 1, short arm, division 34. For each chromosome in Figure 5.13, the number in parentheses indicates the number of genes presently assigned a position on the chromosome. These numbers do not include the approximately 5000 DNA markers used in

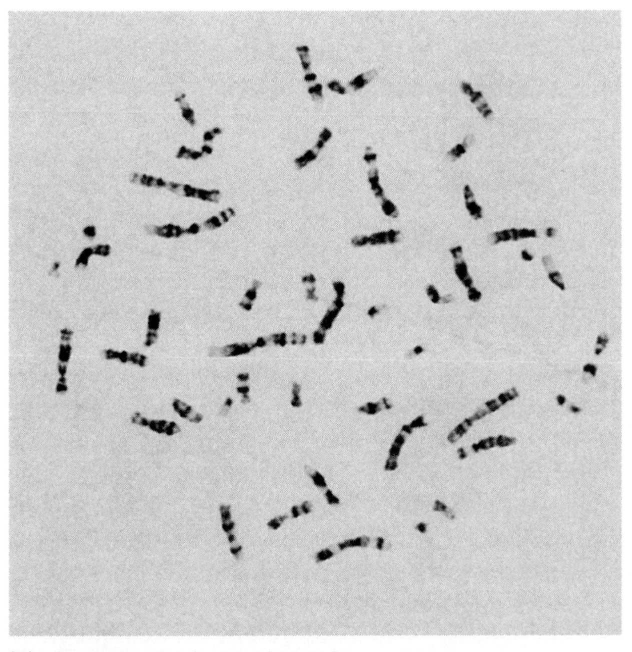

(A) Photograph of metaphase chromosomes

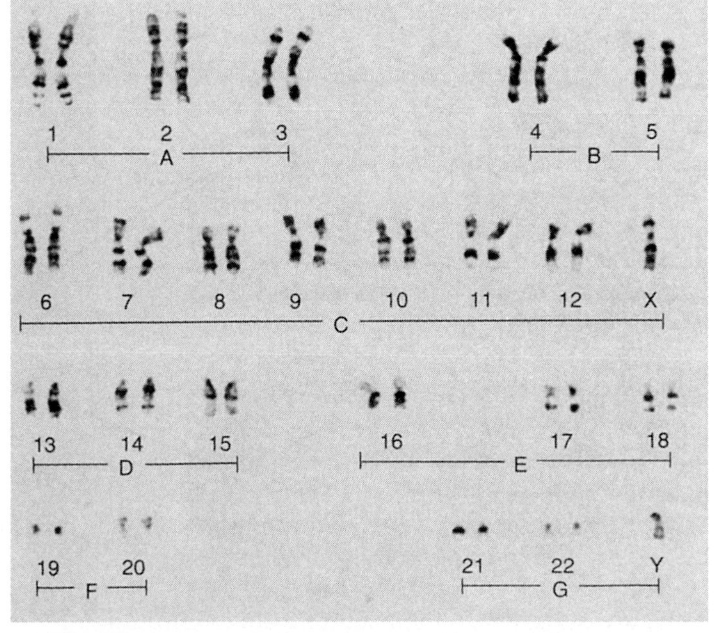

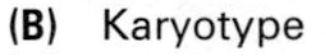
(B) Karyotype

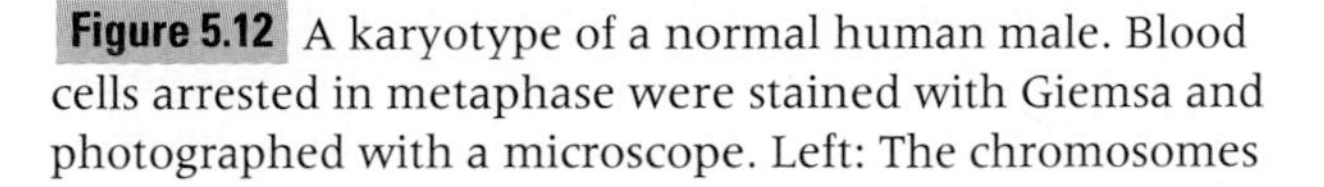
Figure 5.12 A karyotype of a normal human male. Blood cells arrested in metaphase were stained with Giemsa and photographed with a microscope. Left: The chromosomes as seen in the cell by microscopy. Right: The chromosomes have been cut out of the photograph and paired with their homologs. [Courtesy of Patricia Jacobs.]

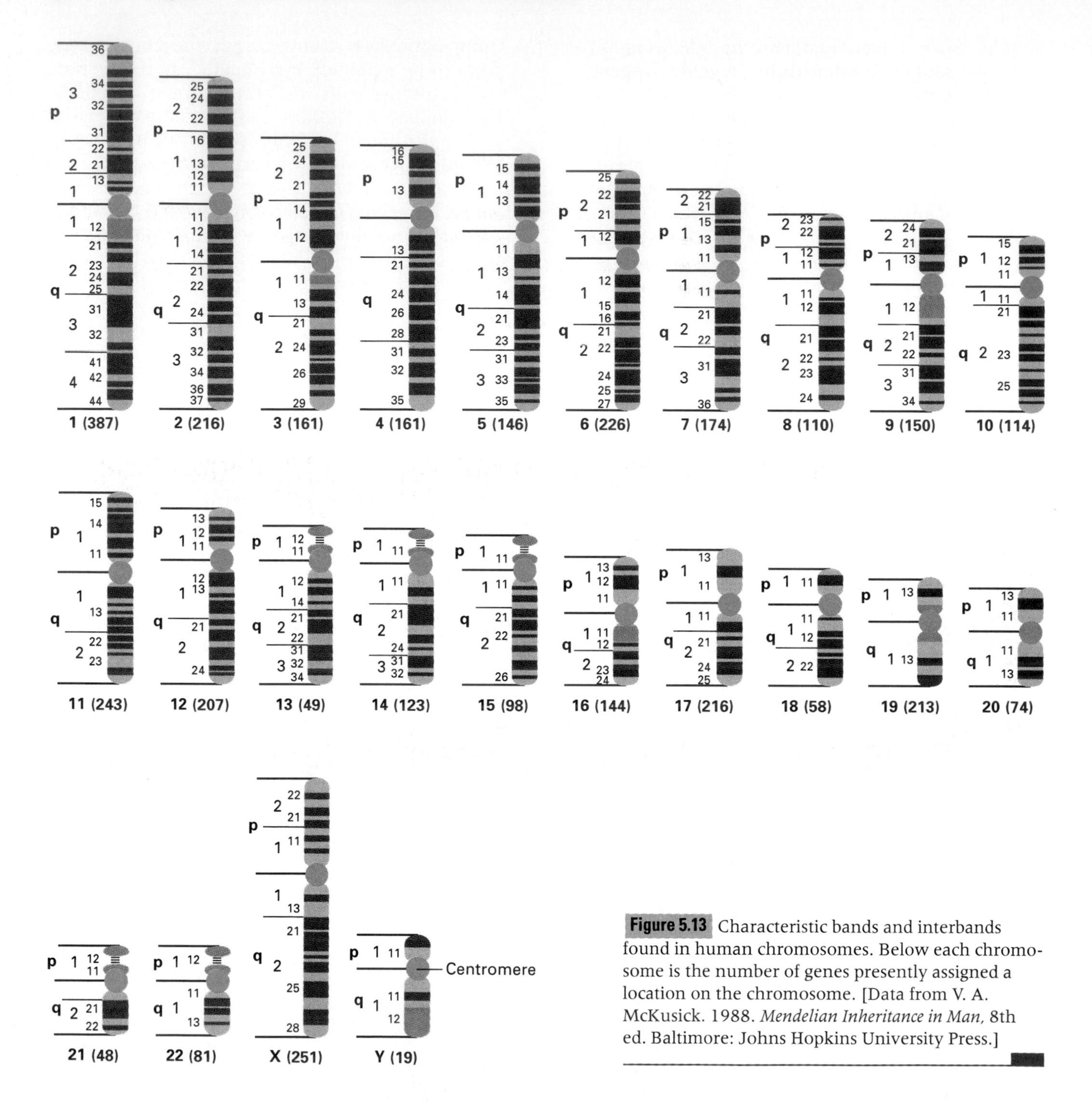

Figure 5.13 Characteristic bands and interbands found in human chromosomes. Below each chromosome is the number of genes presently assigned a location on the chromosome. [Data from V. A. McKusick. 1988. *Mendelian Inheritance in Man,* 8th ed. Baltimore: Johns Hopkins University Press.]

developing the human genetic map. Many of the genes enumerated in Figure 5.13 are associated with inherited diseases. There are about 3500 genes assigned a position on one of the autosomes, 250 assigned a position on the X chromosome, and 20 assigned a position on the Y chromosome.

The technique of chromosome painting shown in Figure 5.7 has been applied to human chromosomes to achieve spectacular effects. An example is shown in the metaphase spread and karyotype in Figure 5.14. To produce this effect, the DNA inside each chromosome was separated into single strands by chemical treatment and then bathed in a solution containing complementary DNA strands labeled with fluorescent dyes. The fluorescent strands form duplexes with the complementary chromosomal DNA, a process called *DNA hybridization,* and are retained on the microscope slide when the excess fluorescent DNA is washed off. In this example, the DNA of each chromosome was hybridized simultaneously with 27 different fluorescent DNA samples, each specific for hybridization with a single chromosome or chromosome arm. The labeled DNA was obtained by microdissection of metaphase nuclei to isolate DNA from each individual chromosome or chromosome arm. After hybridization,

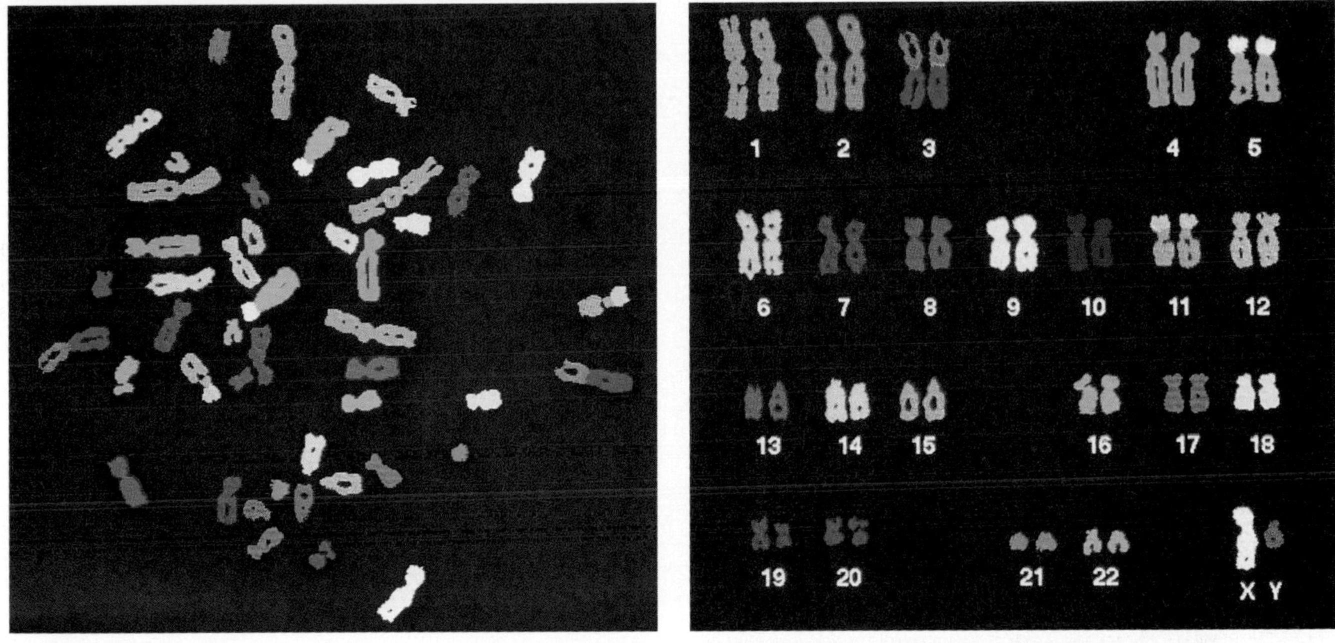

(A) Metaphase spread (B) Karyotype

Figure 5.14 Metaphase spread (A) and karyotype (B) in which human chromosomes have been "painted" in 27 different colors according to their hybridization with fluorescent probes specific to individual chromosomes or chromosome arms. [Courtesy of M. R. Speicher and D. C. Ward. See M. R. Speicher, S. G. Ballard, and D. C. Ward. 1996. *Nature Genetics* 12: 368.]

each chromosome was scanned along its length, and at each point, the fluorescent signal was converted to a specific color in the visible spectrum. The chromosomes in Figure 5.14 are therefore "painted" in 27 different colors. This technique is of considerable utility in human cytogenetics because even complex chromosome rearrangements can be detected rapidly and easily. The painting technique makes it possible to decipher some chromosome rearrangements, particularly those involving small pieces of chromosome, that are not amenable to analysis by conventional banding procedures.

Down syndrome results from three copies of chromosome 21.

Monosomy or trisomy of most human autosomes is usually incompatible with life. Most zygotes with missing chromosomes or extra chromosomes either fail to begin embryonic development or undergo spontaneous abortion at an early stage. There are a few exceptions. One exception is **Down syndrome,** which is caused by trisomy of chromosome 21. Down syndrome affects about 1 in 750 liveborn children. Its major symptom is mental retardation, but there can be multiple physical abnormalities as well, such as serious heart defects.

Most cases of Down syndrome are caused by nondisjunction, which means the failure of homologous chromosomes to separate in meiosis, as explained in Chapter 3. The result of chromosome-21 nondisjunction is one gamete that contains two copies of chromosome 21 and one that contains none. If the gamete with two copies participates in fertilization, a zygote with trisomy 21 is produced. The gamete with one copy may also participate in fertilization, but zygotes with monosomy 21 do not survive even through the first few days or weeks of pregnancy. For unknown reasons, nondisjunction of chromosome 21 is more likely to happen in oogenesis than in spermatogenesis, so the abnormal gamete in Down syndrome is usually the egg. (Students already familiar with trisomy 21 may know it as *Down's* syndrome. This text follows current practice in human genetics in avoiding the possessive form of proper names used to designate syndromes. Down's syndrome thereby becomes "Down syndrome" or "the Down syndrome.")

The risk of nondisjunction of chromosome 21 increases dramatically with the age of the mother, and the incidence of Down syndrome reaches 6 percent in mothers of age forty-five and older (Figure 5.15). It is for this reason that many physicians recommend that older women who are pregnant have cells from the fetus tested in order to detect Down syndrome prenatally. Detection can be done from 15 to 16 weeks after fertilization by **amniocentesis,** in which cells of a developing fetus are obtained by insertion of a fine needle through the wall of the uterus and into the sac of fluid (the

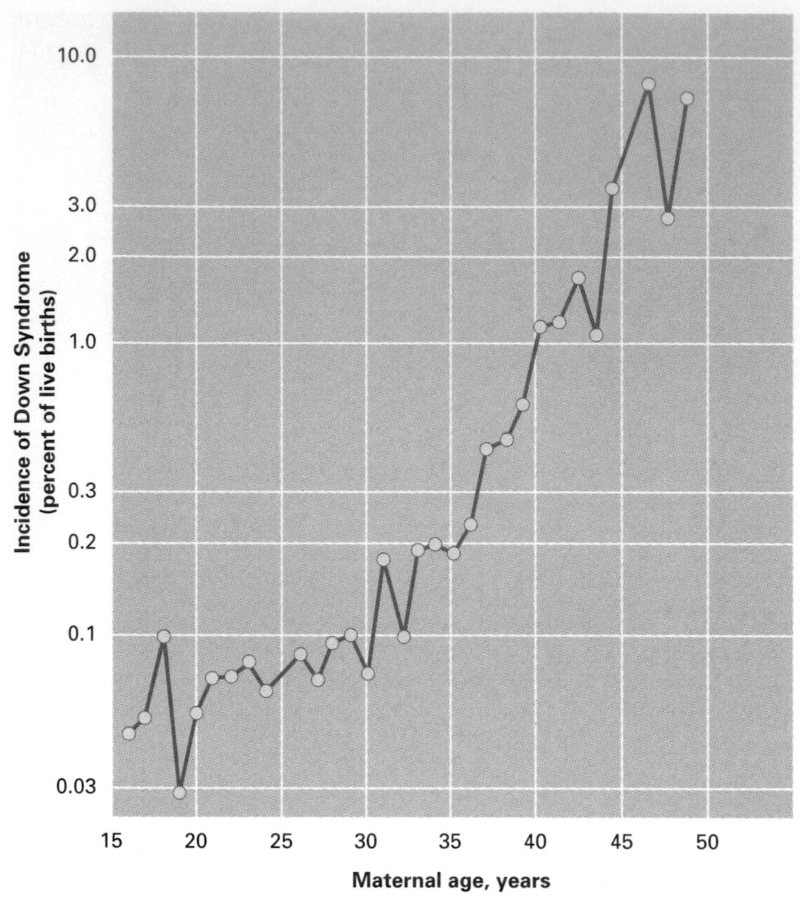

Figure 5.15 Frequency of Down syndrome (number of cases per 100 live births) related to age of mother. The graph is based on 438 Down-syndrome births (among 330,859 total births) in Sweden in the period 1968 to 1970. [Data from E. B. Hook and A. Lindsjö. 1978. *Am. J. Human Genet.* 30:19.]

amnion) containing the fetus, or even earlier in pregnancy by sampling cells from another of the embryonic membranes (the *chorion*). In about 3 percent of families with a Down syndrome child, the risk of another affected child is very high—up to 20 percent of births. This high risk is caused by a chromosome abnormality called a translocation (Section 5.6).

Dosage compensation adjusts the activity of X-linked genes in females.

Abnormal numbers of sex chromosomes usually produce less severe phenotypic effects than do abnormal numbers of autosomes. For example, the effects of extra Y chromosomes are relatively mild. This is in part because the Y chromosome in mammals is largely heterochromatic. In human beings, there is a region at the tip of the short arm of the Y that is homologous with a corresponding region at the tip of the short arm of the X chromosome. It is in this region of homology that the X and Y chromosomes synapse in spermatogenesis, and an obligatory crossover in the region holds the chromosomes together and ensures proper separation during anaphase I. The crossover is said to be obligatory because it takes place somewhere in this region in every meiotic division. The shared X-Y homology defines the **pseudoautosomal** region. Genes within the pseudoautosomal region show a pattern of inheritance very similar to ordinary autosomal inheritance because they are not completely linked to either the X chromosome or the Y chromosome but can exchange between the sex chromosomes by crossing-over. Near the pseudoautosomal region, but not within it, the Y chromosome contains the master sex-controller gene *SRY,* which encodes a protein transcription factor, the **testis-determining factor (TDF),** that triggers male embryonic development by inducing the undifferentiated embryonic genital ridge, the precursor of the gonad, to develop as a testis. A transcription factor, as the name implies, is a component that, together with other factors, stimulates transcription of its target genes. Beyond the pseudoautosomal pairing region and *SRY,* the Y chromosome contains very few genes, so extra Y chromosomes are milder in their effects on phenotype than are extra autosomes.

Extra X chromosomes have milder effects than extra autosomes, because in mammals, all X chromosomes except one are genetically inactivated very early in embryonic development. The inactivation tends to minimize the phenotypic effects of extra X chromosomes, but there are still some effects due to a block of genes near the tip of the short arm that are *not* inactivated.

In female mammals, X-chromosome inactivation is a normal process in embryonic development. In human beings, at an early stage of embryonic development, one of the two X chromosomes is inactivated in each somatic cell; different tissues undergo X inactivation at different times. The X chromosome that is inactivated in a particular somatic cell is selected at random, but once the decision is made, the same X chromosome remains inactive in all of the descendants of the cell.

X-chromosome inactivation has two consequences. First, it equalizes the number of active copies of X-linked genes in females and males. A female has two X chromosomes and a male has only one, but because of inactivation of one X chromosome in each of the somatic cells of the female, the number of *active* X chromosomes in both sexes is one. In effect, gene dosage is equalized except for the block of genes in the short arm of the X that escapes inactivation. The equalization in dosage of active genes is called **dosage compensation.** The mammalian method of dosage compensation by means of X-inactivation was originally proposed by Mary Lyon and is called the **single-active-X principle.**

The second consequence of X-chromosome inactivation is that a normal female is a **mosaic** for X-linked genes (Figure 5.16A). That is, each somatic cell expresses the genes in only one X chromosome, but the X chromosome that is active genetically differs from one cell to the next. This mosaicism has been observed directly in females that are heterozygous for X-linked alleles that determine different forms of an enzyme, A and B; when cells from the heterozygous female are individually cultured in the laboratory, half of the clones are found to produce only the A form of the enzyme and the other half to produce only the B form. Mosaicism can be observed directly in women who are heterozygous for an X-linked recessive mutation resulting in the absence of sweat glands; these women exhibit patches of skin in which sweat glands are present (these patches are derived from embryonic cells in which the normal X chromosome remained active and the mutant X was inactivated) and other patches of skin in which sweat glands are absent (these patches are derived from embryonic cells in which the normal X chromosome was inactivated and the mutant X remained active.)

In certain cell types, the inactive X chromosome in females can be observed microscopically as a densely staining body in the nucleus of interphase cells. This is called a **Barr body** (upper left in Figure 5.16B). Although cells of normal females have one Barr body, cells of normal males have none. Persons with two or more X chromosomes have all but one X chromosome per cell inactivated, and the number of Barr bodies equals the number of inactivated X chromosomes.

The calico cat shows visible evidence of X-chromosome inactivation.

In some cases, the result of random X inactivation in females can be observed in the external phenotype. One example is the "calico" pattern of coat

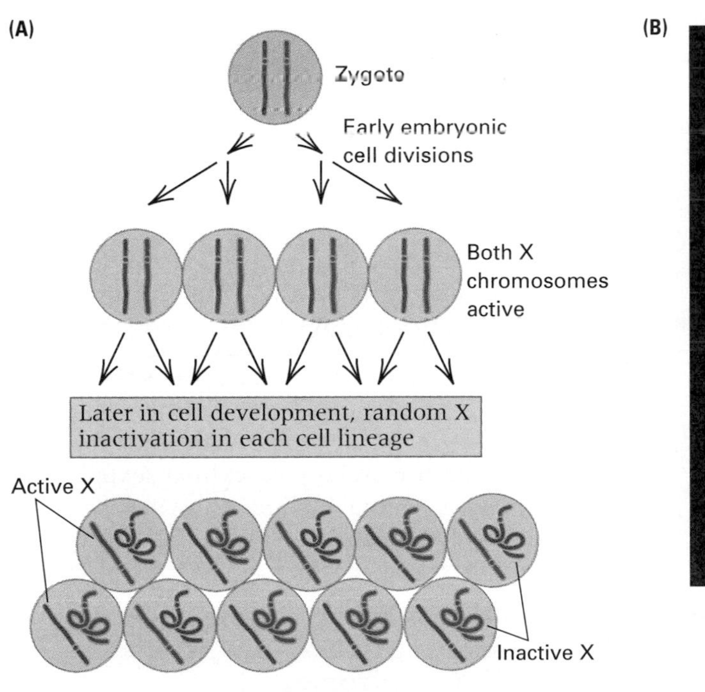

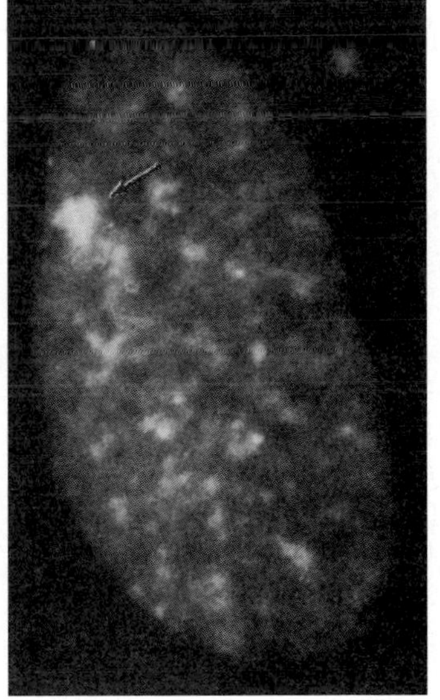

Figure 5.16 (A) Schematic diagram of somatic cells of a normal female showing that the female is a mosaic for X-linked genes. The two X chromosomes are shown in red and blue. An active X is depicted as a straight chromosome, an inactive X as a tangle. Each cell has just one active X, but the particular X that remains active is a matter of chance. In human beings, the inactivation includes all but a few genes in the tip of the short arm. (B) Fluorescence micrograph of a human cell showing a Barr body (bright spot at the upper left; see arrow). This cell is from a normal human female, and it has one Barr body. [Micrograph courtesy of A. J. R. de Jonge.]

Figure 5.17 A female cat heterozygous for the orange and black coat-color alleles and showing the classic "calico" pattern of patches of orange, black, and white.

coloration in female cats. Two alleles affecting coat color are present in the X chromosome in cats. One allele results in an orange coat color, the other in a black coat color. Because he has only one X chromosome, a normal male has either the orange or the black allele. A female can be heterozygous for orange and black, and in this case the coat color is "calico"—a mosaic of orange and black patches mixed with patches of white. Figure 5.17 is a photograph of a female cat with the classic calico pattern. The orange and black patches result from X-chromosome inactivation. In cell lineages in which the X chromosome bearing the orange allele is inactivated, the X chromosome with the black allele is active and so the fur is black. In cell lineages in which the X chromosome with the black allele is inactivated, the orange allele in the active X chromosome results in orange fur.

The white patches have a completely different explanation than the orange and black patches. The white patches are due to an autosomal gene *S* for white spotting, which prevents pigment formation in the cell lineages in which it is expressed. Why the *S* gene is expressed in some cell lineages and not others is not known. Homozygous *S*/*S* cats have more white than heterozygous *S*/*s* cats. This female is homozygous.

An extra X or Y chromosome has a relatively mild effect.

Many types of sex-chromosomal abnormalities have been observed. As we have noted, they are usually less severe in their phenotypic effects than abnormal numbers of autosomes. The four most common types are

- 47,XXX This condition is often called **trisomy-X.** The number 47 in the chromosome designation refers to the total number of chromosomes, and XXX indicates that the person has three X chromosomes. People with the karyotype 47,XXX are female. Many are phenotypically normal or nearly normal, though the frequency of mild mental retardation is somewhat greater than it is among 46,XX females.
- 47,XYY This condition is often called **double-Y.** These people are male and tend to be tall, but they are otherwise phenotypically normal. At one time it was thought that 47,XYY males developed severe personality disorders and were at a high risk of committing crimes of violence, a belief based on an elevated incidence of 47,XYY among violent criminals. Further study indicated that most 47,XYY males have slightly impaired mental function and that, although their rate of criminality is higher than that of normal males, the crimes are mainly nonviolent petty crimes such as theft. The majority of 47,XYY males are phenotypically and psychologically normal, have mental capabilities in the normal range, and have no criminal convictions.
- XXY This condition is called **Klinefelter syndrome.** Affected persons are male. They tend to be tall, do not undergo normal sexual maturation, are sterile, and in some cases have enlargement of the breasts. Mild mental impairment is common.
- 45,X Monosomy of the X chromosome in females is called **Turner syndrome.** Affected persons are phenotypically female but are short in stature and do not exhibit sexual maturation. Mental abilities are typically within the normal range.

Amplification of a simple DNA sequence is associated with the fragile-X syndrome.

An important form of inherited mental retardation is associated with a class of X chromosomes containing a site toward the end of the long arm (Xq27−Xq28) that tends to break in cultured cells that are starved for DNA precursors, such as the nucleotides. The X chromosomes containing this site are called **fragile-X** chromosomes, and the associated form of mental retardation is the **fragile-X syndrome.** The fragile-X syndrome affects about one in 2500 children. It accounts for about half of all cases of X-linked mental retardation and is second only to Down syndrome as a cause of inherited mental impairment. The fragile-X syndrome has an unusual pattern of inheritance in which approximately 1 in 5 males with the fragile-X chromosome are phenotypically normal and also have phenotypically normal children. However, the heterozygous daughters of such a "transmitting male" often have affected sons, and about 1 in 3 of their heterozygous daughters are also affected. This pattern is illustrated in Figure 5.18. The transmitting male denoted I-2 is not affected, but the X chromosome that he transmits to his daughters (II-2 and II-5) somehow becomes altered in the female germ line in such a way that sons and daughters in the next generation (III) are affected. Affected and normal granddaughters of the transmitting male sometimes have affected progeny (generation IV). Males that are affected are severely retarded and do not reproduce. Among females, there is substantial variation in severity of expression (variable expressivity). In general, females are less severely affected than males, and some heterozygous females are not affected (incomplete penetrance).

The molecular basis of the fragile-X chromosome has been traced to a **trinucleotide repeat** of the form $(CCG)_n$ present in the DNA at the site where the breakage takes place. Normal X chromosomes have from 6 to 54 tandem copies of the repeating unit (the average is about 30), and affected persons have more than 230 copies of the repeat. Transmitting males have an intermediate number, between 52 and 230 copies, which is called the fragile-X "premutation." The unprecedented feature of the premutation is that in females, it invariably increases in copy number to reach a level of 230 copies or greater, at which stage the chromosome causes mental retardation. The amplification in the number of copies present in the germ line of daughters of transmitting males is related to the cycle of X-chromosome inactivation and reactivation. Chromosomes with the premutation that go through the X-inactivation cycle become permanently altered. In these chromosomes, the inactivation of a gene designated *FMR1* (fragile-site mental retardation-1) is irreversible, and the gene is unable to be reactivated in oogenesis. The variable expressivity of the fragile-X syndrome in affected females also is related to X-chromosome inactivation: More severely affected females have a higher proportion of cells in which the normal X chromosome is inactivated.

Many spontaneous abortions result from chromosome abnormalities.

Approximately 15 percent of all recognized pregnancies in human beings end in spontaneous abortion, and in about half of them the fetus has a major chromosome abnormality. Table 5.1 summarizes the average numbers of chromosome abnormality found per 100,000 recognized pregnancies in several studies. Many of the spontaneously aborted fetuses have trisomy of one of the autosomes. Triploids and tetraploids also are common in spontaneous abortions. Note that the majority of trisomy-21 fetuses are spontaneously aborted, as are the vast majority of 45,X fetuses. If all trisomy-21 fetuses survived to birth, the incidence of Down syndrome would rise to 1 in 250, approximately a threefold increase over the incidence observed.

Although many autosomal trisomies are found in spontaneous abortions, autosomal monosomies are not found. Monosomic embryos do exist. They are probably created in greater numbers than the trisomic fetuses because chromosome loss, leading to monosomy, is usually much more frequent than chromosome gain, leading to trisomy. Autosomal monosomies do not show up in spontaneously aborted fetuses, doubtless because these embryos abort so early in development that the pregnancy goes unrecognized. The spontaneous abortions summarized in Table 5.1, although they represent a huge fetal wastage, serve the important biological function of eliminating many fetuses that are grossly abnormal in their development because of major chromosome abnormalities.

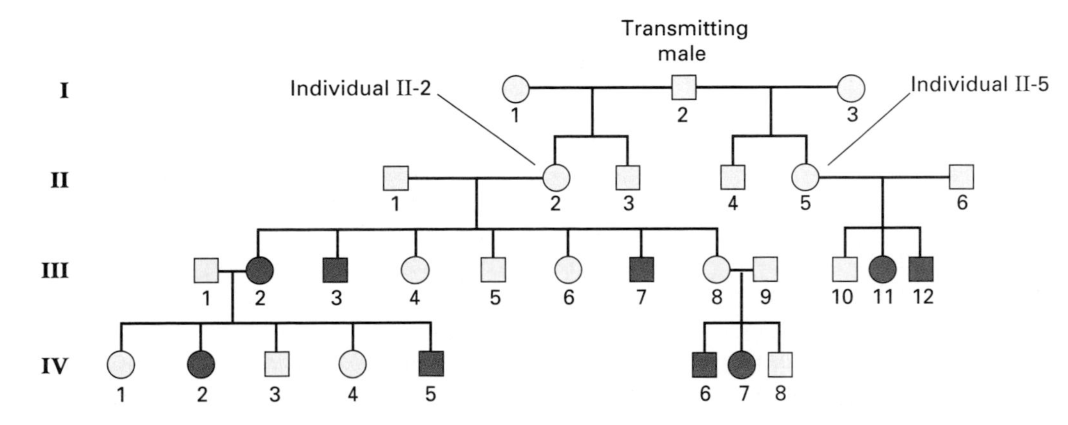

Figure 5.18 Pedigree showing transmission of the fragile-X syndrome. Male I-2 is not affected, but his daughters (II-2 and II-5) have affected children and grandchildren. [After C. D. Laird. 1987. *Genetics*, 117: 587.]

Table 5.1
Chromosome abnormalities per 100,000 recognized human pregnancies

Chromosome constitution	Number among spontaneously aborted fetuses	Number among live births
Normal	7,500	84,450
Trisomy		
13	128	17
18	223	13
21	350	113
Other autosomes	3,176	0
Sex chromosomes		
47,XYY	4	46
47,XXY	4	44
45,X	1,350	8
47,XXX	21	44
Translocations		
Balanced (euploid)	14	164
Unbalanced (aneuploid)	225	52
Polyploid		
Triploid	1,275	0
Tetraploid	450	0
Others (mosaics, etc.)	280	49
Total	15,000	85,000

5.6 Chromosome rearrangements can have important genetic effects.

Thus far, abnormalities in chromosome *number* have been described. This section deals with abnormalities in chromosome *structure*. There are several principal types of structural aberrations, each of which has characteristic genetic effects. Chromosome aberrations were initially discovered through their genetic effects, which, though confusing at first, were eventually understood as resulting from abnormal chromosome structure. This was later confirmed directly by microscopic observations.

A chromosome with a deletion has genes missing.

A chromosome sometimes arises in which a segment is missing. Such a chromosome is said to have a **deletion** or a **deficiency.** Deletions are generally harmful to the organism, and the usual rule is the larger the deletion, the greater the harm. Very large deletions are usually lethal, even when heterozygous with a normal chromosome. Small deletions are often viable when they are heterozygous with a structurally normal homolog, because the normal homolog supplies gene products that are necessary for survival. However, even small deletions are usually homozygous-lethal (when both members of a pair of homologous chromosomes carry the deletion).

Deletions can be detected genetically by making use of the fact that a chromosome with a deletion no longer carries the wildtype alleles of the genes that have been eliminated. For example, in *Drosophila,* many *Notch* deletions are large enough to remove the nearby wildtype allele of *white* also. When these deleted chromosomes are heterozygous with a structurally normal chromosome carrying the recessive *w* allele, the fly has white eyes because the wildtype w^+ allele is no longer present in the deleted *Notch* chromosome. This "uncovering" of the recessive allele implies that the corresponding wildtype allele of *white* has also been deleted. Once a deletion has been identified, its size can be assessed genetically by determining which recessive mutations in the region are uncovered by the deletion. This method is illustrated in Figure 5.19.

Rearrangements are apparent in giant polytene chromosomes.

In the nuclei of cells in the larval salivary glands and certain other tissues of *Drosophila* and other two-winged (dipteran) flies, there are giant chromosomes, called **polytene chromosomes,** that contain about 1000 DNA molecules laterally aligned (Figure 5.20). Each of these chromosomes has a volume many times greater than that of the corresponding chromosome at mitotic metaphase in ordinary somatic cells, as well as a constant and distinctive pattern of transverse banding. The polytene structures are formed by repeated replication of the DNA in a closely synapsed pair of homologous chromosomes without separation of the replicated chromatin strands or of the two chromosomes. Polytene chromosomes are atypical chromosomes and are formed in "terminal" cells; that is, the larval cells containing them do not divide further during the development of the fly and are later eliminated in the formation of the pupa. However, the polytene chromosomes have been especially valuable in the genetics of *Drosophila* and are ideal for the study of chromosome rearrangements.

About 5000 darkly staining transverse bands have been identified in the polytene chromosomes of *D. melanogaster* (Figure 5.20). The linear array of bands, which has a pattern that is constant and characteristic for each species, provides a finely detailed **cytological map** of the chromosomes. The banding pattern is such that short regions in any of the chromosomes can be identified. Because of their large size and finely detailed morphology, polytene chromosomes are exceedingly useful for the study of deletions and other chromosome aberrations. For example, all the *Notch* deletions

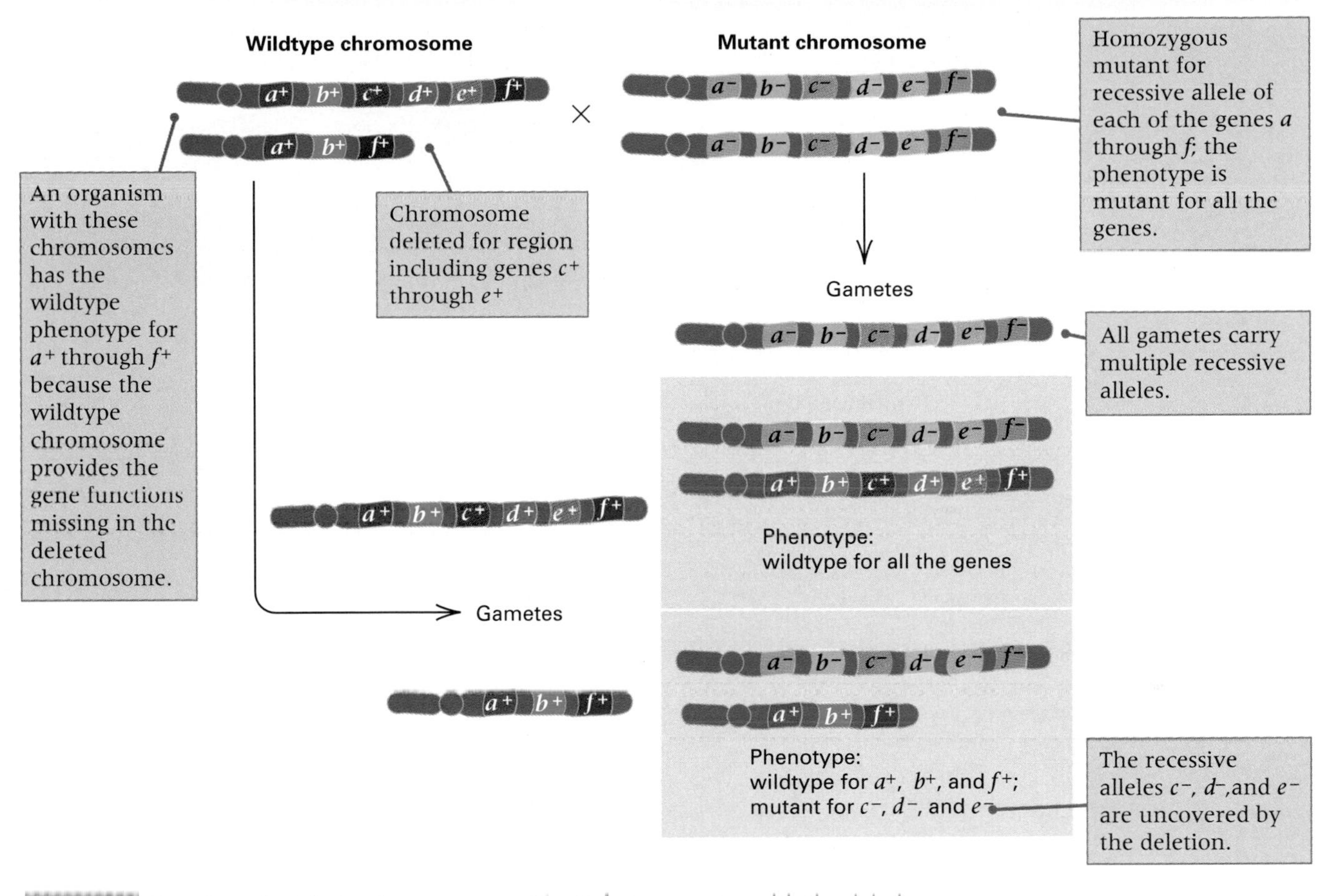

Figure 5.19 Mapping of a deletion by testcrosses. The F_1 heterozygotes with the deletion express the recessive phenotype of all deleted genes. The expressed recessive alleles are said to be "uncovered" by the deletion.

cause particular bands to be missing in the salivary chromosomes. Physical mapping of deletions also allows particular genes, otherwise known only from genetic studies, to be assigned to specific bands or regions in the salivary chromosomes.

Physical mapping of genes in part of the *Drosophila* X chromosome is illustrated in Figure 5.21. The banded chromosome is shown, and beneath it are the designations of the individual bands. On the average, each band contains about 20 kb of DNA, but there is considerable variation in DNA content from band to band. The mutant X chromosomes labeled I through VI in the figure have deletions. The deleted part of each chromosome is shown in red. These deletions define regions along the chromosome, some of which correspond to specific bands. For example, the deleted region in both chromosomes I and II that is present in all the other chromosomes consists of band 3A3. In crosses, only deletions I and II uncover the mutation *zeste (z)*, so the *z* gene must be in band 3A3, as indicated at the top. Similarly, the recessive-lethal mutation *zw2* is uncovered by all

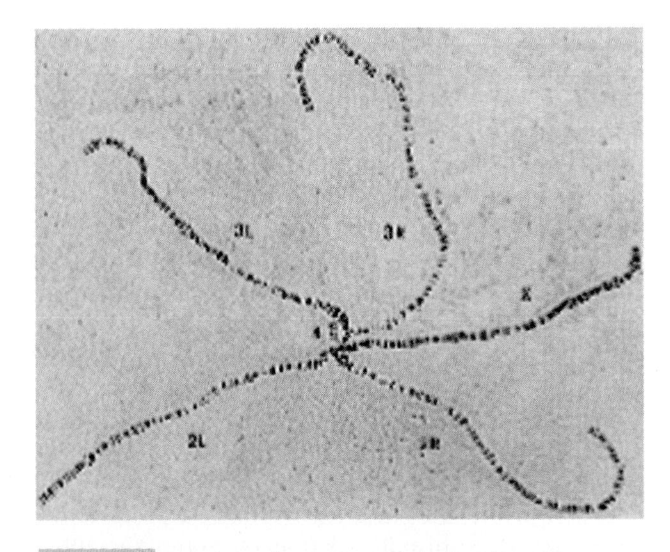

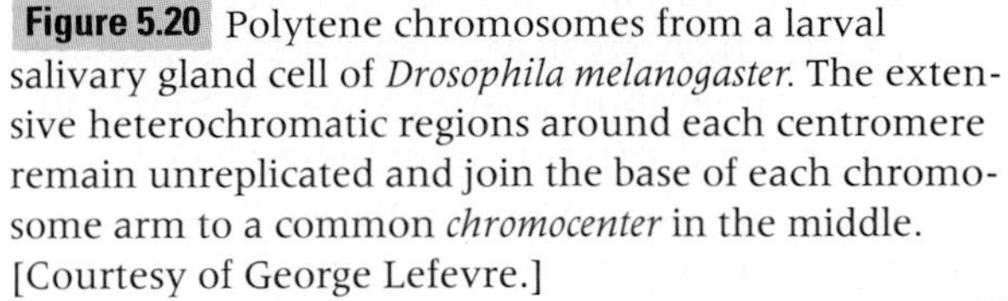

Figure 5.20 Polytene chromosomes from a larval salivary gland cell of *Drosophila melanogaster.* The extensive heterochromatic regions around each centromere remain unreplicated and join the base of each chromosome arm to a common *chromocenter* in the middle. [Courtesy of George Lefevre.]

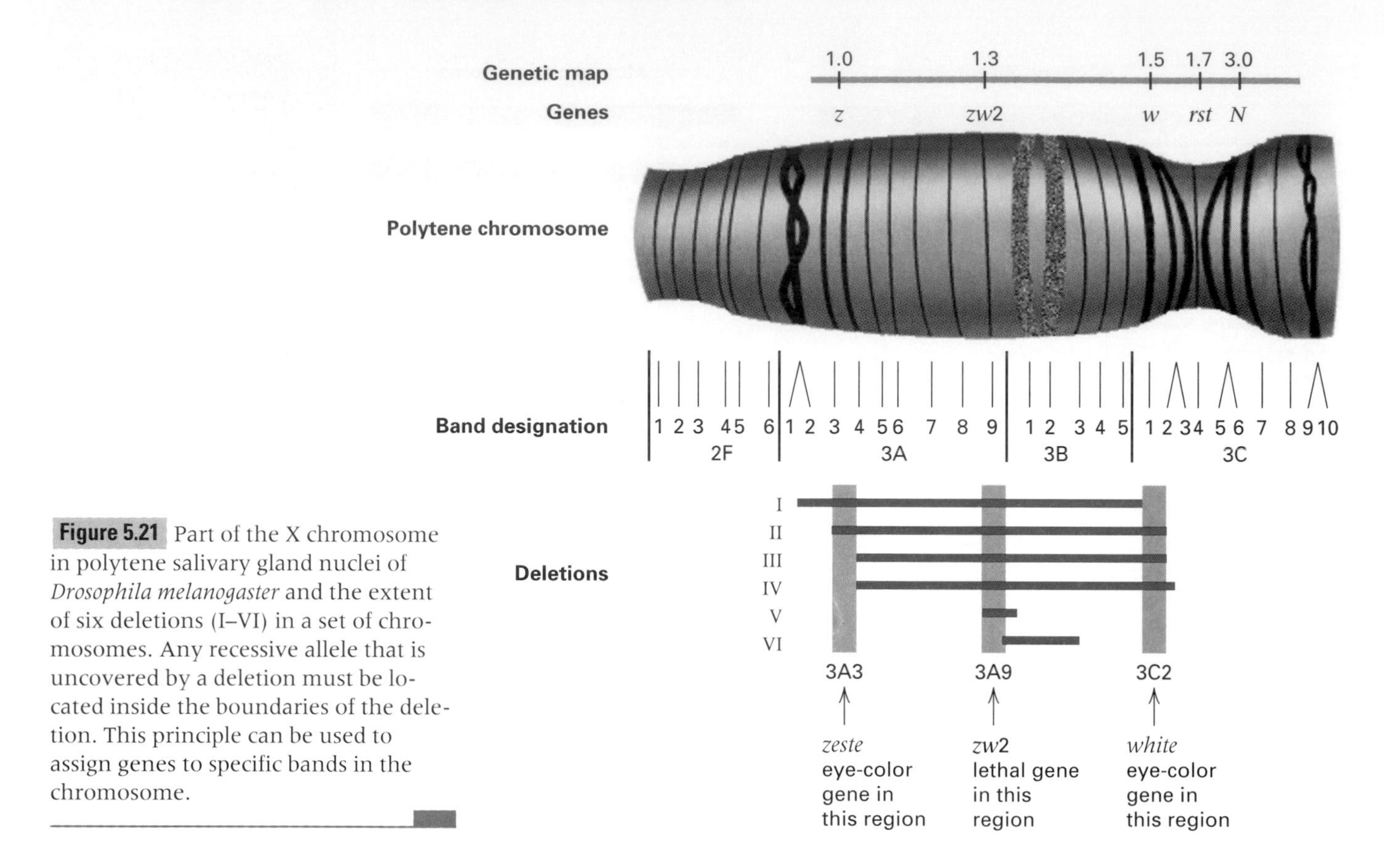

Figure 5.21 Part of the X chromosome in polytene salivary gland nuclei of *Drosophila melanogaster* and the extent of six deletions (I–VI) in a set of chromosomes. Any recessive allele that is uncovered by a deletion must be located inside the boundaries of the deletion. This principle can be used to assign genes to specific bands in the chromosome.

deletions except VI; therefore, the *zw2* gene must be in band 3A9. As a final example, the *w* mutation is uncovered only by deletions II, III, and IV; thus the *w* gene must be in band 3C2. The *rst* (rough eye texture) and *N* (notched wing margin) genes are not uncovered by any of the deletions. These genes were localized by a similar analysis of overlapping deletions in regions 3C5 to 3C10.

A chromosome with a duplication has extra genes.

Some abnormal chromosomes have a region that is present twice. These chromosomes are said to have a **duplication.** A **tandem duplication** is one in which the duplicated segment is present in the same orientation immediately adjacent to the normal region in the chromosome. Tandem duplications are able to produce even more copies of the duplicated region by means of a process called **unequal crossing-over.** Figure 5.22A illustrates the chromosomes in meiosis of an organism that is homozygous for a tandem duplication (brown region). When they undergo synapsis, these chromosomes can mispair with each other, as illustrated in part B. A crossover within the mispaired part of the duplication (part C) will thereby produce a chromatid carrying three copies of the region as well as a reciprocal product containing a single copy (part D).

Human color-blindness mutations result from unequal crossing-over.

Human color vision is mediated by three light-sensitive protein pigments present in the cone cells of the retina. Each of the pigments is related to *rhodopsin,* the pigment found in the rod cells that mediates vision in dim light. The light sensitivities of the cone pigments are toward blue, red, and green. These are our primary colors. We perceive all other colors as mixtures of these primaries. The gene for the blue-sensitive pigment is in chromosome 7, whereas the genes for the red and green pigments are in the X chromosome near the tip of the long arm, separated by less than 5 cM (roughly 5 Mb of DNA). Because the red and green pigments arose from the duplication of a single ancestral pigment gene and are still 96 percent identical in amino acid sequence, the genes are similar enough that they can pair and undergo unequal crossing-over. The process of unequal crossing-over is the genetic basis of red-green color blindness.

Almost everyone is familiar with **red-green color blindness;** it is one of the most common inherited conditions in human beings. Approximately 5 percent of males have some form of red-green color blindness. The preponderance of affected males immediately suggests X-linked inheritance, which is confirmed by pedigree studies (Section 3.6). Affected males have normal sons and

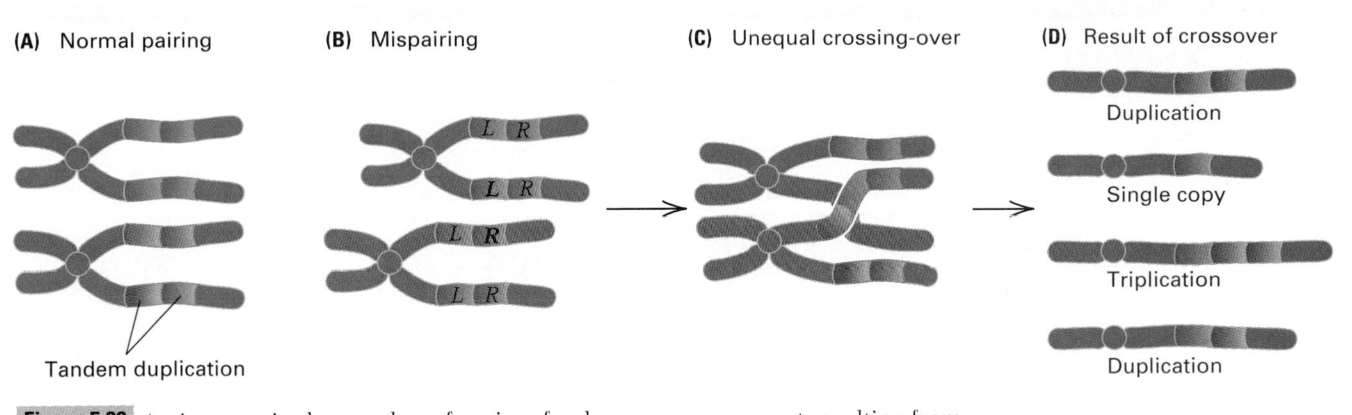

Figure 5.22 An increase in the number of copies of a chromosome segment resulting from unequal crossing-over of tandem duplications (brown). (A) Normal synapsis of chromosomes with a tandem duplication. (B) Mispairing. The right-hand element of the lower chromosome is paired with the left-hand element of the upper chromosome. (C) Crossing-over within the mispaired duplication, which is called unequal crossing-over. (D) The outcome of unequal crossing-over. One product contains a single copy of the duplicated region, another chromosome contains a triplication, and the two strands that do not participate in the crossover retain the duplication.

carrier daughters, and the carrier daughters have 50 percent affected sons and 50 percent carrier daughters.

There are several distinct varieties of red-green color blindness. Defects in red vision go by the names of *protanopia,* an inability to perceive red, and *protanomaly,* an impaired ability to perceive red. The comparable defects in green perception are called *deuteranopia* and *deuteranomaly,* respectively. Isolation of the red and green pigment genes and study of their organization in people with normal and defective color vision have indicated quite clearly how the "-opias" and "-omalies" differ but also explain why the frequency of color blindness is so relatively high.

The organization of the red and green pigment genes in men with normal vision is illustrated in Figure 5.23A. Unexpectedly, a significant proportion of normal X chromosomes contain two or three green-pigment genes. How these arise by unequal crossing-over is shown in part B. The red-pigment and green-pigment genes pair, and the crossover takes place in the region of homology between the genes. The result is a duplication of the green-pigment gene in one chromosome and a deletion of the green-pigment gene in the other.

The recombinational origin of the defects in color vision are illustrated in Figure 5.24. The top chromosome in part A is the result of deletion of the green-pigment gene shown in Figure 5.23B. Males with such an X chromosome have deuteranopia, or "green blindness." Other types of abnormal pigments result when crossing-over takes place within mispaired red-pigment and green-pigment

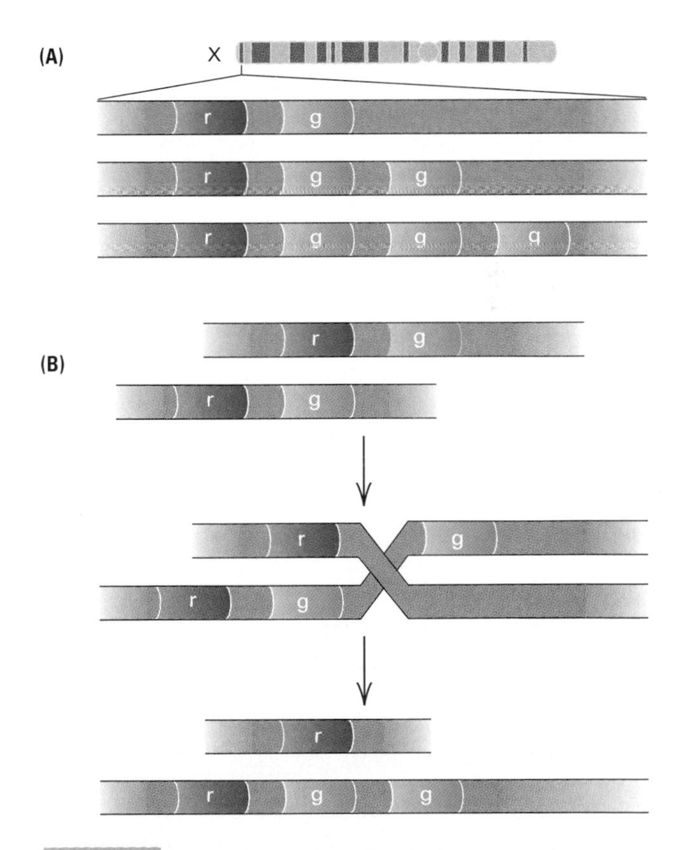

Figure 5.23 (A) Organization of red-pigment and green-pigment genes in normal X chromosomes. Some chromosomes contain one copy of the green-pigment gene, others two, still others three. (B) Origin of multiple green-pigment genes by unequal crossing-over in the region of DNA homology between the genes. Note that one product of unequal crossing-over is a chromosome that contains a red-pigment gene but no green-pigment gene.

The human connection

Catch 21

Jerome Lejeune, Marthe Gautier, and Raymond Turpin 1959
National Center for Scientific Research, Paris, France
Study of the Somatic Chromosomes of Nine Down Syndrome Children
(original in French)

Prior to this study, Down syndrome was one of the greatest mysteries in human genetics. One of the most common forms of mental retardation, the syndrome did not follow any pattern of Mendelian inheritance. Yet some families had two or more children with Down syndrome. (Many of these cases are now known to be due to a translocation involving chromosome 21.) This paper marked a turning point in human genetics by demonstrating that Down syndrome actually results from the presence of an extra chromosome. It was the first chromosomal disorder to be identified. The excerpt uses the term *telocentric,* which means a chromosome that has its centromere very near one end. In the human genome, the smallest chromosomes are three very small telocentric chromosomes. These are chromosomes 21 and 22, and the Y chromosome. A normal male has five small telocentrics (21, 21, 22, 22, and Y); a normal female has four (21, 21, 22, and 22). (The X is a medium-sized chromosome with its centromere somewhat off center.) In the accompanying table, note the variation in chromosome counts in the "doubtful" cells. The methods for counting chromosomes were very difficult then, and investigators made many errors, either by counting two nearby chromosomes as one or by including in the count of one nucleus a chromosome that actually belonged to a nearby nucleus. Lejeune and collaborators wisely chose to ignore these doubtful counts and based their conclusion only on the "perfect" cells. Sometimes good science is a matter of knowing which data to ignore.

	Number of chromosomes					
	"Doubtful" cells			"Perfect" cells		
	46	47	48	46	47	48
Boys 1	6	10	2	–	11	–
2	–	2	1	–	9	–
3	–	1	1	–	7	–
4	–	3	–	–	1	–
5	–	–	–	–	8	–
Girls 1	1	6	1	–	5	–
2	1	2	–	–	8	–
3	1	2	1	–	4	–
4	1	1	2	–	4	–

The culture of fibroblast cells from nine Down syndrome children reveals the presence of 47 chromosomes, the supernumerary chromosome being a small telocentric one. The hypothesis of the chromosomal determination of Down syndrome is considered. . . . The observations made in these nine cases (five boys and four girls) are recorded in the table above.

The number of cells counted in each case may seem relatively small. This is due to the fact that only the pictures [of the spread chromosomes] that claim a minimum of interpretation have been retained in this table. The apparent variation in the chromosome number in the "doubtful" cells, that is to say, cells in which each chromosome cannot be noted individually with certainty, has been pointed out by several authors. It does not seem to us that this phenomenon represents a cytological reality, but merely reflects the difficulties of a delicate technique. It therefore seems logical to prefer a small number of absolutely certain counts ("perfect" cells in the table) to a mass of doubtful observations, the statistical variance of which rests solely on the lack of precision of the observations. Analysis of the chromosome set of the "perfect" cells reveals the presence in Down syndrome boys of 6 small telocentric chromosomes (instead of 5 in the normal man) and 5 small telocentric ones in Down syndrome girls (instead of 4 in the normal woman). . . . It therefore seems legitimate to conclude that there exists in Down syndrome children a small supernumerary telocentric chromosome, accounting for the abnormal figure of 47. To explain these observations, the hypothesis of nondisjunction of a pair of small telocentric chromosomes at the time of meiosis can be considered. . . . It is, however, not possible to say that the supernumerary small telocentric chromosome is indeed a normal chromosome and at the present time the possibility cannot be discarded that a fragment resulting from another type of aberration is involved.

THERE EXISTS IN Down syndrome children a small supernumerary telocentric chromosome.

Source: Comptes rendus des séances de l'Académie des Sciences 248: 1721–1722.

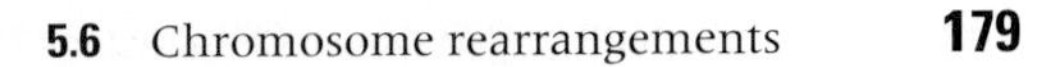

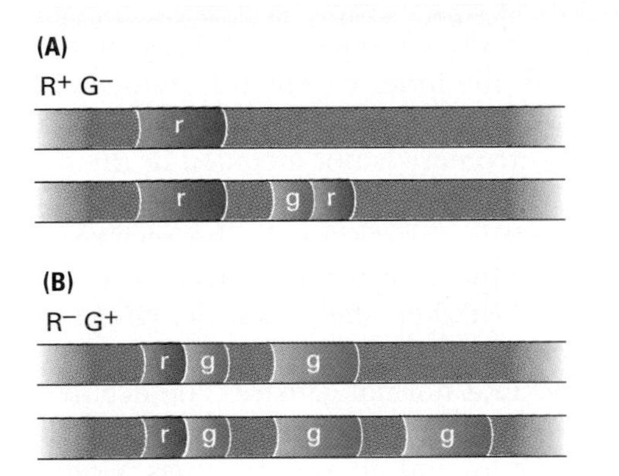

Figure 5.24 Genetic basis of absent or impaired red-green color vision. (A) Defects in green vision result from unequal crossing-over between mispaired red-pigment and green-pigment genes, yielding a green-red chimeric gene. If the green-pigment gene is missing or the chimeric gene is largely "red" in its sequence, deuteranopia is the result. If the chimeric gene is largely "green" in its sequence, deuteranomaly is the result. (B) Defects in red vision result from unequal crossing-over between mispaired red-pigment and green-pigment genes, again yielding a green-red chimeric gene. If the chimeric gene is largely "green," protanopia results; if it is largely "red," protanomaly results. Note that the red gene cannot be eliminated altogether (as the green gene can), because the red gene is at the end of the region of homology between the chromosomes.

genes. Crossing-over between the genes yields a **chimeric gene,** which is a composite gene, part of one joined with part of the other. The chimeric gene in Figure 5.24A joins the 5' end of the green-pigment gene with the 3' end of the red-pigment gene. If the crossover point is toward the 5' end of the gene, the resulting chimeric gene is mostly "red" in sequence, and hence the chromosome will cause deuteranopia or "green blindness." However, if the crossover point is near the 3' end of the gene, most of the green-pigment gene remains intact, and the chromosome will cause deuteranomaly.

Chromosomes associated with defects in red vision are illustrated in Figure 5.24B. The chimeric genes are the reciprocal products of the unequal crossovers that yield defects in green vision. In this case, the chimeric gene consists of the red-pigment gene at the 5' end and the green-pigment gene at the 3' end. If the crossover point is near the 5' end, most of the red-pigment gene is replaced with the green-pigment gene. The result is protanopia, or "red blindness." The same is true of the other chromosome indicated in Figure 5.24B. However, if the crossover point is near the 3' end, then most of the red-pigment gene remains intact and the result is protanomaly.

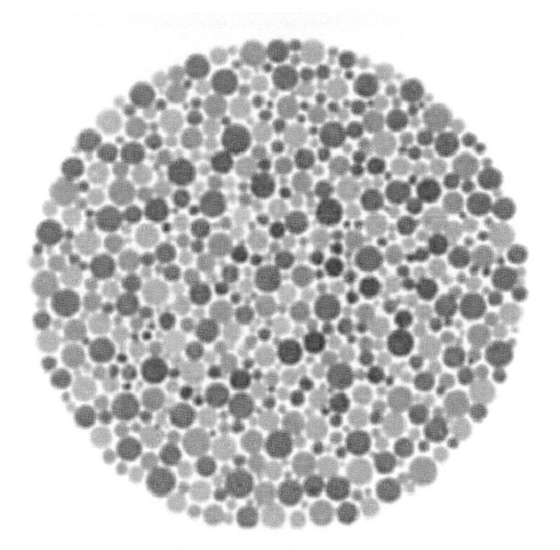

A STANDARD COLOR CHART used in initial testing for color blindness. The pattern tests for an inability to distinguish red from green. Those with red-green color blindness will not be able to distinguish the green dots from red and therefore will not see the green path.

A chromosome with an inversion has some genes in reverse order.

Another important type of chromosome abnormality is an **inversion,** a segment of a chromosome in which the order of the genes is the reverse of the normal order. In an organism that is heterozygous for an inversion, one chromosome is structurally normal (wildtype), and the other carries an inversion (Figure 5.25). These chromosomes pass through mitosis without difficulty because each chromosome duplicates and its chromatids are separated into the daughter cells without regard to the other chromosome. There is a problem in meiosis, however. The problem is that the chromosomes are attracted gene-for-gene in the process of synapsis, as shown in Figure 5.25. In an inversion heterozygote, in order for gene-for-gene pairing to take place everywhere along the length of the chromosome, one or the other of the chromosomes must twist into a loop in the region in which the gene order is inverted. In Figure 5.25 the structurally normal chromosome is shown as looped, but in other cells it may be the inverted chromosome that is looped. In either case, the loop is called an **inversion loop.**

The loop itself does not create a problem. The looping apparently takes place without difficulty and can be observed through the microscope. As long as there is no crossing over within the inversion, the homologous chromosomes can separate normally at anaphase I. When there is crossing-over within the inversion loop, then the chromatids involved in the crossing-over become physically

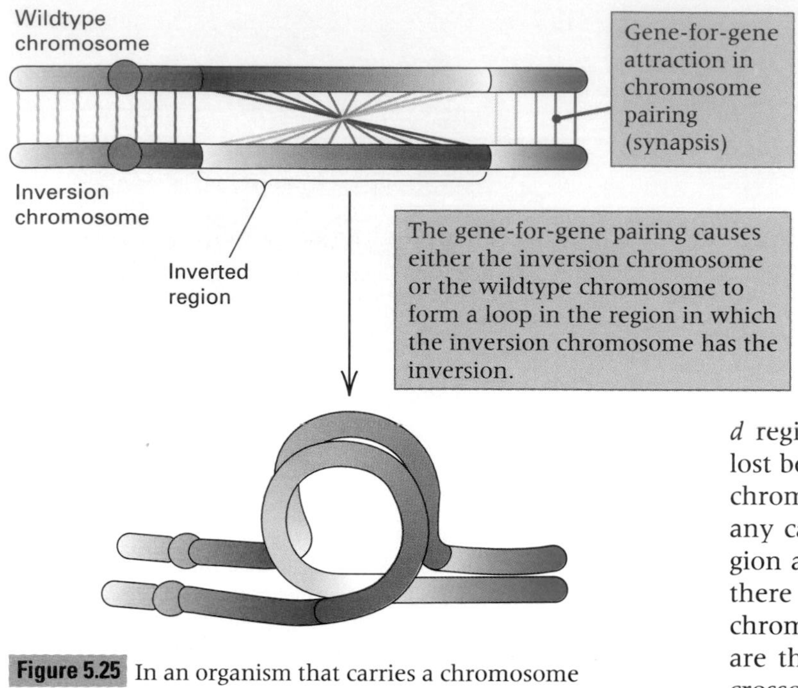

Figure 5.25 In an organism that carries a chromosome that is structurally normal along with a homologous chromosome with an inversion, the gene-for-gene attraction between the chromosomes during synapsis causes one of the chromosomes to form into a loop in the region in which the gene order is inverted. In this example, the structurally normal chromosome forms the loop.

joined, and the result is the formation of chromosomes containing large duplications and deletions. The products of the crossing-over can be deduced from Figure 5.26 by tracing along the chromatids in part A. The outer chromatids are the ones that do not participate in the crossover. One of these contains the inverted sequence and the other the normal sequence, as shown in part B. Because of the crossover, the inner chromatids, which did participate in the crossover, are connected. If the centromere is not included in the inversion loop, as is the case here, the result is a dicentric chromosome. The reciprocal product of the crossover is an acentric chromosome. Neither the dicentric chromosome nor the acentric chromosome can be included in a normal gamete. The acentric chromosome is usually lost because it lacks a centromere and, in any case, has a deletion of the *a* region and a duplication of the *d* region. The dicentric chromosome is also often lost because it is held on the meiotic spindle by the chromatid bridging between the centromeres; in any case, this chromosome is deleted for the *d* region and duplicated for the *a* region. Hence, when there is a crossover in the inversion loop, the only chromatids that can be recovered in the gametes are the chromatids that did not participate in the crossover. One of these carries the inversion and the other does not.

The inversion in Figure 5.26, in which the centromere is not included in the inverted region, is known as a **paracentric inversion,** which means inverted "beside" (*para-*) the centromere. As seen in the figure, the products of crossing-over include a dicentric and an acentric chromosome.

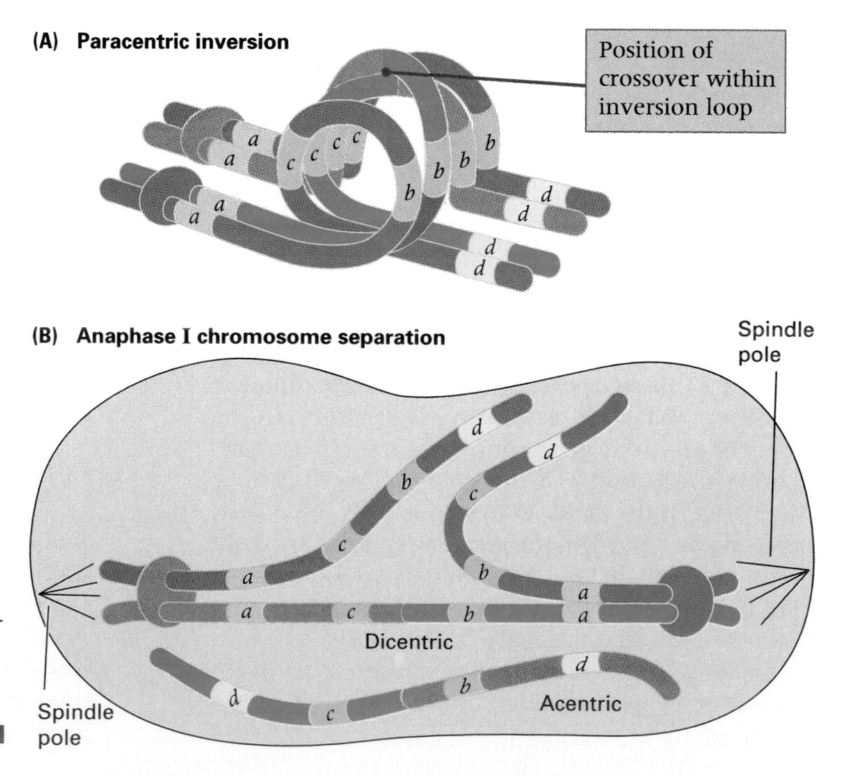

Figure 5.26 (A) Synapsis between homologous chromosomes, one of which contains an inversion. There is a crossover within the inversion loop. (B) Anaphase I configuration resulting from the crossover. Because the centromere is not included in the inverted region, one of the crossover products is a dicentric chromosome, and the reciprocal product is an acentric chromosome. Among the two chromatids not involved in the crossover, one carries the inversion and the other the normal gene sequence.

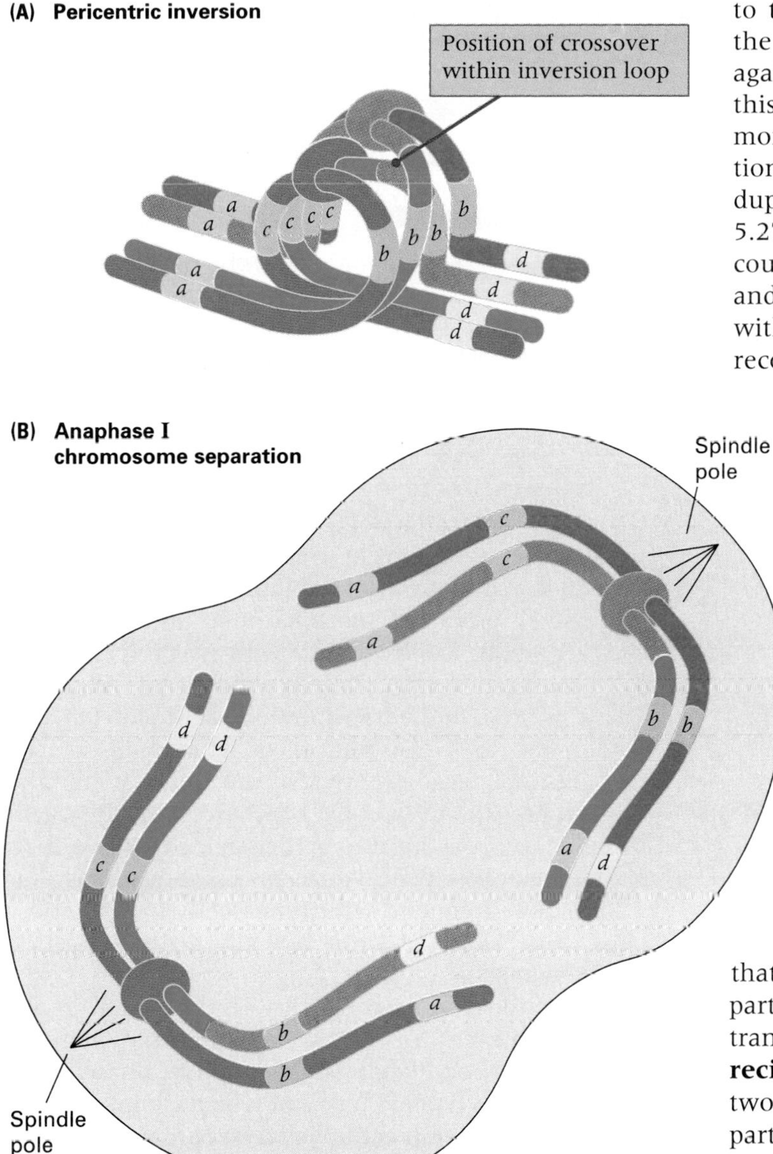

Figure 5.27 (A) Synapsis between homologous chromosomes, one of which carries an inversion that includes the centromere. A crossover within the inversion loop is shown. (B) Anaphase I configuration resulting from the crossover. One of the crossover products is duplicated for *a* and deficient for *d*; the other is duplicated for *d* and deficient for *a*. Among the two chromatids not involved in the crossover, one carries the inversion and the other is normal.

When the inversion does include the centromere, it is called a **pericentric inversion,** which means "around" (*peri-*) the centromere. Chromatids with duplications and deficiencies are also created by crossing-over within the inversion loop of a pericentric inversion, but in this case the crossover products are monocentric. The situation is illustrated in Figure 5.27A. The diagram is identical to that in Figure 5.26 except for the position of the centromere. The products of crossing-over can again be deduced by tracing the chromatids. In this case, both products of the crossover are monocentric, but one chromatid carries a duplication of *a* and a deletion of *d* and the other carries a duplication of *d* and a deletion of *a* (Figure 5.27B). Although either of these chromosomes could be included in a gamete, the duplication and deficiency usually cause inviability. Thus, as with the paracentric inversion, the products of recombination are not recovered, but for a different reason. Among the chromatids that do not participate in the crossing-over in Figure 5.27A, one carries the pericentric inversion and the other has the normal sequence.

Reciprocal translocations interchange parts between chromosomes.

A chromosomal aberration resulting from the interchange of parts between nonhomologous chromosomes is called a **translocation.** In Figure 5.28, organism A is homozygous for two pairs of structurally normal chromosomes. Organism B contains one structurally normal pair of chromosomes and another pair of chromosomes that have undergone an interchange of terminal parts. This organism is said to be *heterozygous* for the translocation. The translocation is properly called a **reciprocal translocation** because it consists of two reciprocally interchanged parts. As indicated in part C, an organism can also be homozygous for a translocation if both pairs of homologous chromosomes undergo an interchange of parts.

An organism that is heterozygous for a reciprocal translocation usually produces only about half as many offspring as normal, which is called **semisterility.** The reason for the semisterility is difficulty in chromosome segregation in meiosis. When meiosis takes place in a translocation heterozygote, the normal and translocated chromosomes must undergo synapsis as shown in Figure 5.29. Ordinarily, there would also be chiasmata between nonsister chromatids in the arms of the homologous chromosomes, but these are not shown, as if the translocation were present in an organism with no crossing-over, such as a male *Drosophila.* Segregation from this configuration can take place in any of three ways. In the list that follows, the notation $1 + 2 \longleftrightarrow 3 + 4$ means that at the first meiotic anaphase, the chromosomes in Figure 5.29 labeled 1 and 2 go to one pole and those labeled 3 and 4 go to the opposite pole. The red numbers 1 and 4

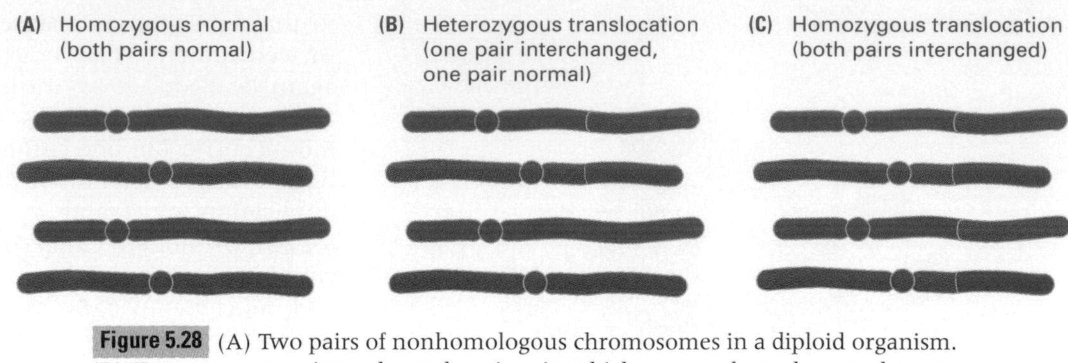

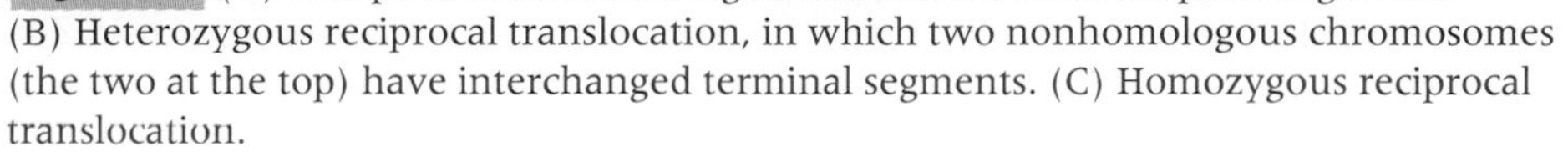

Figure 5.28 (A) Two pairs of nonhomologous chromosomes in a diploid organism. (B) Heterozygous reciprocal translocation, in which two nonhomologous chromosomes (the two at the top) have interchanged terminal segments. (C) Homozygous reciprocal translocation.

indicate the two parts of the reciprocal translocation. The three types of segregation are

- 1 + 2 ⟷ 3 + 4 This mode is called **adjacent-1 segregation.** Homologous centromeres go to opposite poles, but each normal chromosome goes with one part of the reciprocal translocation. All gametes formed from adjacent-1 segregation have a large duplication and deficiency for the distal part of the translocated chromosomes. (The *distal* part of a chromosome is the part farthest from the centromere.) The pair of gametes that originate from the 1 + 2 pole are duplicated for the distal part of the blue chromosome and deficient for the distal part of the red chromosome; the pair of gametes from the 3 + 4 pole have the reciprocal deficiency and duplication.
- 1 + 3 ⟷ 2 + 4 This mode is **adjacent-2 segregation,** in which homologous centromeres go to the same pole at anaphase I. In this case, all gametes have a large duplication and deficiency of the proximal part of the translocated chromosome. (The *proximal* part of a chromosome is the part closest to the centromere.) The pair of gametes from the 1 + 3 pole have a duplication of the proximal part of the red chromosome and a deficiency of the proximal part of the blue chromosome; the pair of gametes from the 2 + 4 have the complementary deficiency and duplication.
- 1 + 4 ⟷ 2 + 3 In this type of segregation, which is called **alternate segregation,** the gametes are all balanced (euploid), which means that none has a duplication or deficiency. The gametes from the 1 + 4 pole have both parts of the reciprocal translocation; those from the 2 + 3 have both normal chromosomes.

The semisterility of genotypes that are heterozygous for a reciprocal translocation results from lethality due to the duplication and deficiency gametes produced by adjacent-1 and adjacent-2 segregation. The frequency with which these types of segregation take place is strongly influenced by the position of the translocation breakpoints, by the number and distribution of chiasmata in the interstitial region between the centromere and each breakpoint, and by whether the quadrivalent tends to open out into a ring-shaped structure on the metaphase plate. Adjacent-1 segregation is usually quite frequent, which means that semisterility is to be expected from virtually all translocation heterozygotes.

Translocation semisterility is manifested in different life-history stages in plants and animals. Plants have an elaborate gametophyte phase of the life cycle—a haploid phase in which complex metabolic and developmental processes are necessary. In plants, large duplications and deficiencies are usually lethal in the gametophyte stage. Because the gametophyte produces the gametes, in higher plants the semisterility is manifested as pollen or seed lethality. In animals, by contrast, minimal gene activity is necessary in the gametes, which function in spite of very large duplications and deficiencies. In animals, therefore, the semisterility is usually manifested as zygotic lethality.

A special type of *non*reciprocal translocation is a **Robertsonian translocation,** in which the centromeric regions of two nonhomologous acrocentric chromosomes become fused to form a single centromere (Figure 5.30). This kind of fusion happened in human evolution to create the present metacentric chromosome 2 from two acrocentric chromosomes in an ancient human ancestor (Figure 5.3). Robertsonian translocations are an important risk factor to be considered in Down syndrome. When chromosome 21 is one of the

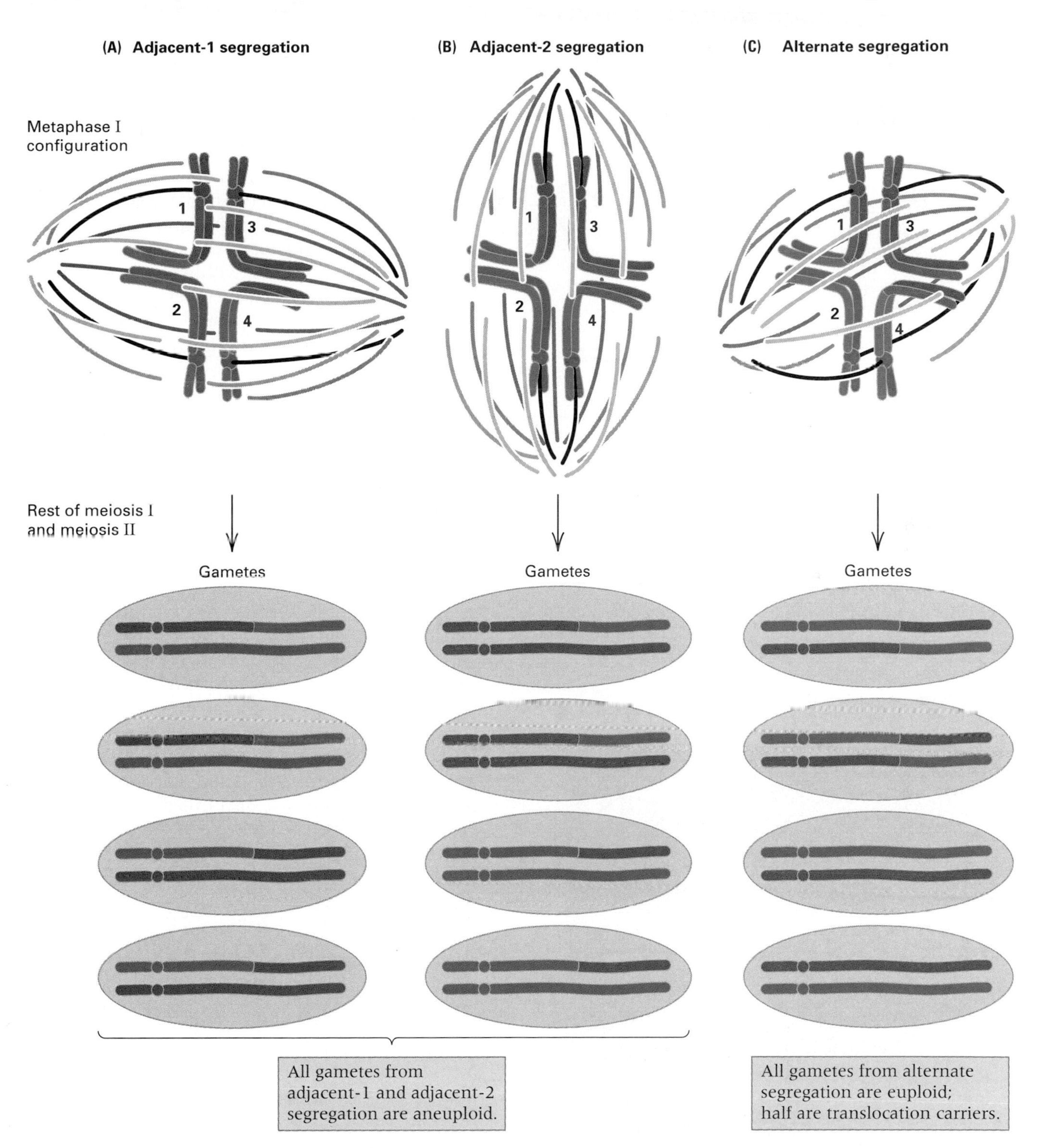

Figure 5.29 A quadrivalent formed in the synapsis of a heterozygous reciprocal translocation. The translocated chromosomes are numbered in red, their normal homologs in black. No chiasmata are shown. (A) Adjacent-1 segregation, in which homologous centromeres separate at anaphase I; all of the resulting gametes have a duplication of one terminal segment and a deficiency of the other. (B) Adjacent-2 segregation, in which homologous centromeres go together at anaphase I; all of the resulting gametes have a duplication of one basal segment and a deficiency of the other. (C) Alternate segregation, in which half of the gametes receive both parts of the reciprocal translocation and the other half receive both normal chromosomes.

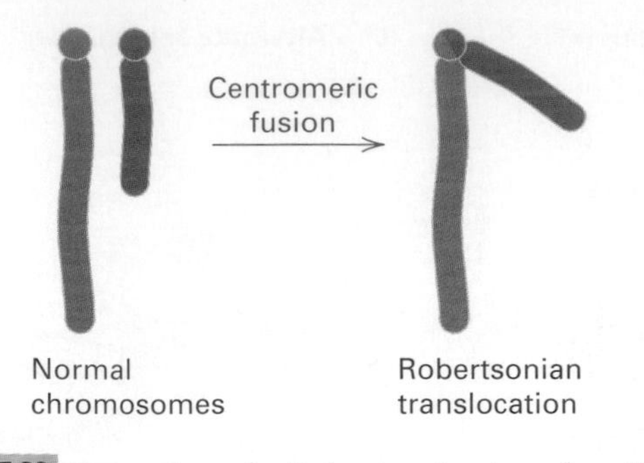

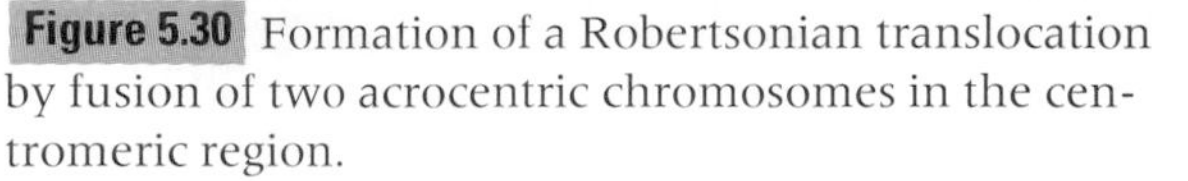

Figure 5.30 Formation of a Robertsonian translocation by fusion of two acrocentric chromosomes in the centromeric region.

acrocentrics in a Robertsonian translocation, the rearrangement leads to a familial type of Down syndrome. An example in which chromosome 21 is joined with chromosome 14 is shown in Figure 5.31 (arrow). The heterozygous carrier is phenotypically normal, but a high risk of Down syndrome results from aberrant segregation in meiosis. Approximately 3 percent of children with Down syndrome are found to have one parent with such a translocation.

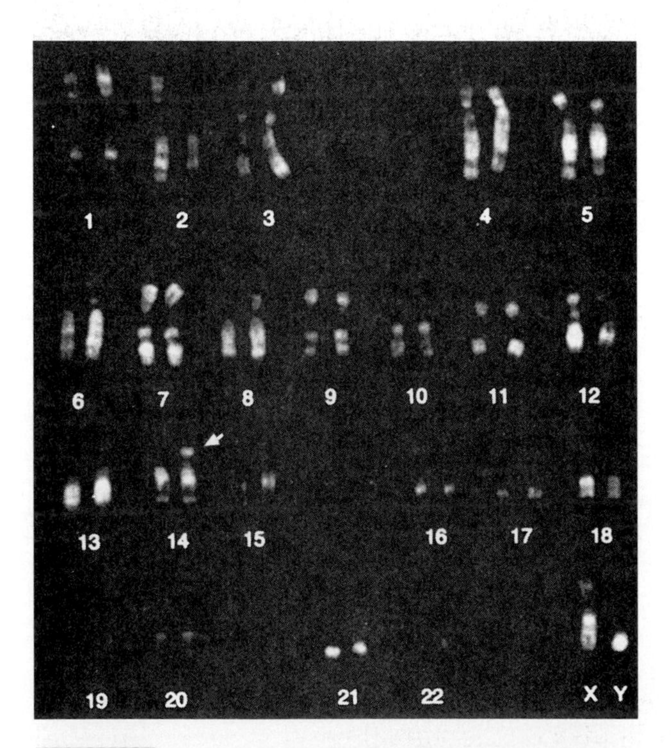

Figure 5.31 A karyotype of a child with Down syndrome, carrying a Robertsonian translocation of chromosomes 14 and 21 (arrow). Chromosomes 19 and 22 are faint in this photo; this has no significance. (Courtesy of Irene Uchida.)

5.7 Cancer is often associated with chromosomal abnormalities.

Cancer refers to an unrestrained proliferation and migration of cells. In all known cases, cancer cells derive from the repeated division of a single mutant cell whose growth has become unregulated, so cancer cells initially constitute a clone. With continued proliferation, many cells within such clones develop chromosomal abnormalities, such as extra chromosomes, missing chromosomes, deletions, duplications, or translocations. The chromosomal abnormalities found in cancer cells are diverse, and they may differ among cancer cells in the same person or among people with the same type of cancer. The accumulation of chromosome abnormalities is evidently one accompaniment of unregulated cell division.

Amid the large number of apparently random chromosome abnormalities found in cancer cells, a small number of aberrations are found consistently in certain types of cancer, particularly in blood diseases such as the leukemias. For example, chronic myelogenous leukemia is frequently associated with a reciprocal translocation between chromosome 22 and either chromosome 8 or chromosome 9. In some cancers with characteristic chromosome abnormalities, the breakpoint in the chromosome is near the chromosomal location of a **cellular oncogene.** An **oncogene** is a gene associated with cancer. Cellular oncogenes, also called *proto-oncogenes,* are the cellular homologs of **viral oncogenes** contained in certain cancer-causing viruses. The distinction is one of location: Cellular oncogenes are part of the normal genome, whereas viral oncogenes are derived from cellular oncogenes through some rare mechanism in which they become incorporated into virus particles. More than 50 different cellular oncogenes are known. They are apparently normal developmental genes that predispose cells to unregulated division when mutated or abnormally expressed. Many of the genes function in normal cells as growth factors that promote and regulate cell division. When a chromosome rearrangement happens near a cellular oncogene (or when the gene is incorporated into a virus), the gene may become expressed abnormally and result in unrestrained proliferation of the cell that contains it. However, abnormal expression by itself is usually not sufficient to produce cancerous growth. One or more additional mutations in the same cell are also required.

Some characteristic chromosome abnormalities found in certain cancers are shown in Figure 5.32, along with the locations of known cellular onco-

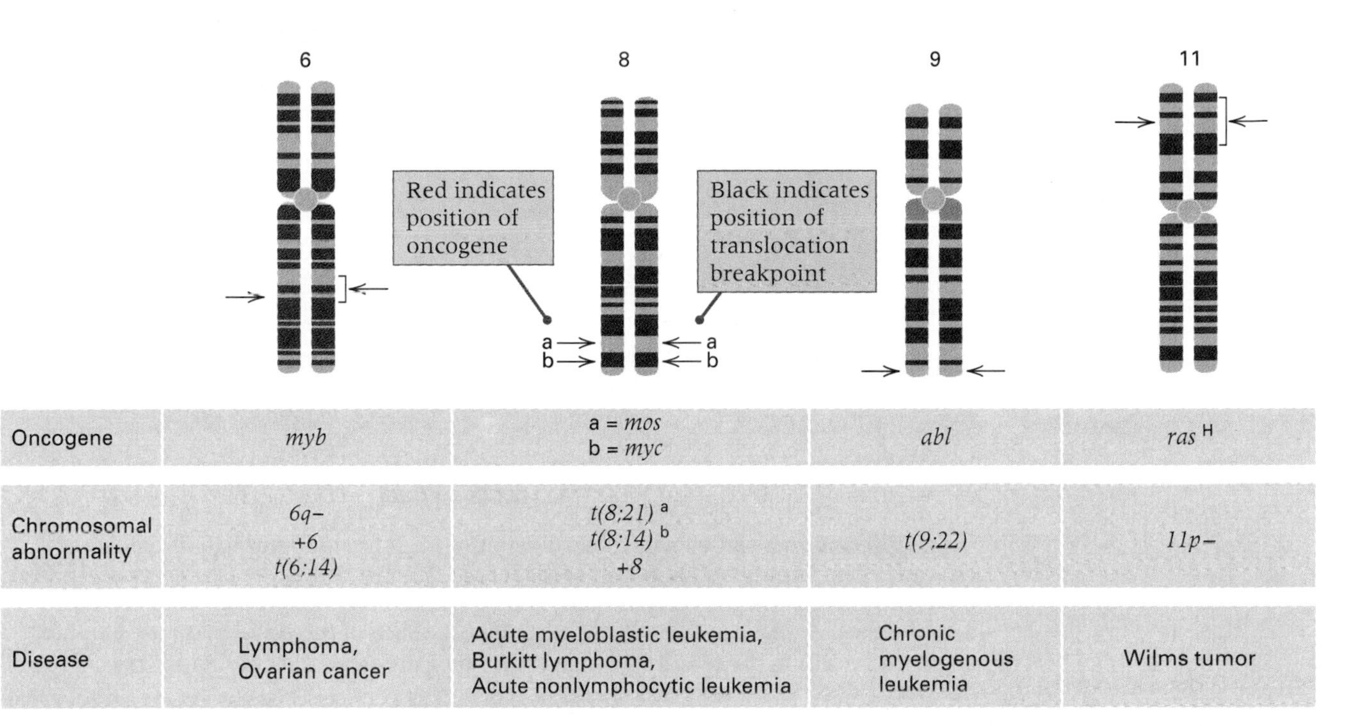

Figure 5.32 Correlation between oncogene positions (red arrows) and chromosome breaks (black arrows) in aberrant human chromosomes frequently found in cancer cells. A plus sign preceding a chromosome number indicates an extra or missing copy of the entire chromosome. A minus sign following a chromosome designation means extra material (or missing material) corresponding to part of the designated chromosome or arm. For example, *11p−* means a deletion of part of the short arm of chromosome 11. The symbol *t* means reciprocal translocation, so *t(9;22)* means a reciprocal translocation between chromosome 9 and chromosome 22.

genes on the same chromosomes. The location of the cellular oncogene is indicated by a red arrow at the left, and the chromosomal breakpoint is indicated by a black arrow at the right. In most cases for which sufficient information is available, the correspondence between the breakpoint and the cellular oncogene location is very close, and in some cases, the breakpoint is within the cellular oncogene itself. The significance of this correspondence is that the chromosomal rearrangement disturbs normal cellular oncogene regulation and ultimately leads to the onset of cancer.

A second class of genes associated with inherited cancers is made up of **tumor-suppressor genes.** These are genes whose presence is necessary to suppress tumor formation. Absence of both normal alleles, through either mutational inactivation or deletion, results in tumor formation. An example of a tumor-suppressor gene is the human gene *Rb-1,* located in chromosome 13. When the normal *Rb-1* gene product is absent, malignant tumors form in the retinas, and surgical removal of the eyes becomes necessary. The disease is known as **retinoblastoma.**

Retinoblastoma is unusual in that the predisposition to retinal tumors is dominant in pedigrees but the *Rb-1* mutation is recessive at the cellular level. The dominance is indicated by the fact that a person who inherits one copy of the *Rb-1* mutation through the germ line is heterozygous, and the penetrance of retinoblastoma is 100 percent. The recessiveness at the cellular level is indicated by the finding that the retinal cells that actually become malignant have the genotype *Rb-1/Rb-1.* The explanation for this apparent paradox is illustrated in Figure 5.33. Part A shows the genotype of an *Rb-1* heterozygote, along with other genes in the same chromosome. Parts B through E show four possible ways in which a second genetic event can result in *Rb-1/Rb-1* cells in the retina. The simplest is a mutation of the wildtype allele in the homolog (part B). The wildtype allele could also be deleted (part C). Part D illustrates a situation in which the normal homolog of the *Rb-1* chromosome is lost and replaced by nondisjunction of the *Rb-1*-bearing chromosome. Recombination in mitosis (part E) is yet another possibility for making *Rb-1* homozygous. Although each of these events takes place at a very low rate per cell division, there are so many cells in the retina (approximately 10^8) that the *Rb-1* allele usually becomes homozygous in at least one cell. (The average number of tumors per retina is three.)

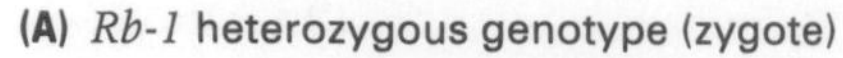

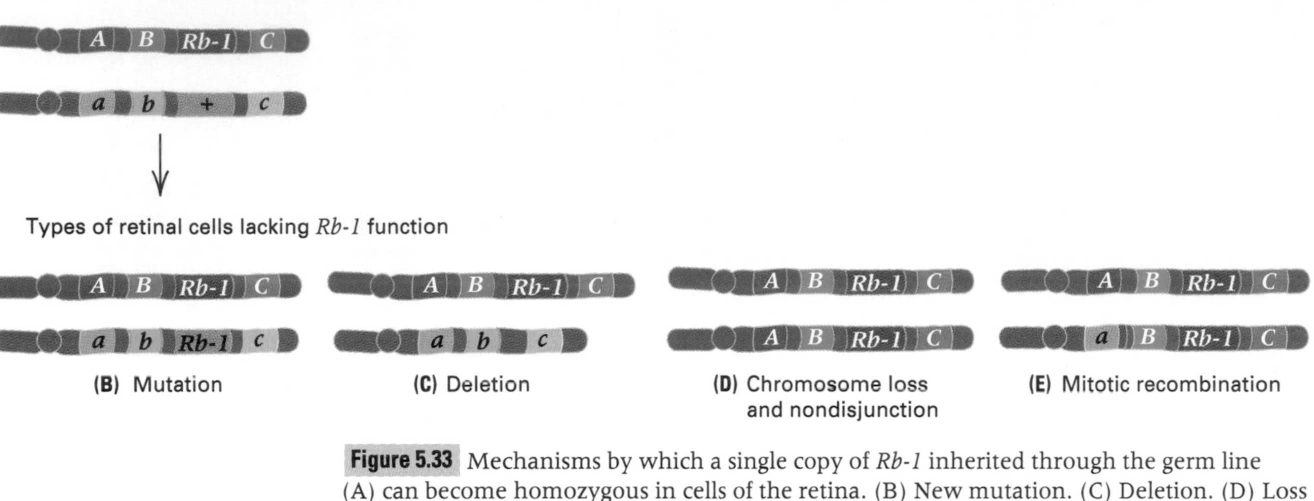

Figure 5.33 Mechanisms by which a single copy of *Rb-1* inherited through the germ line (A) can become homozygous in cells of the retina. (B) New mutation. (C) Deletion. (D) Loss of the normal homologous chromosome and replacement by nondisjunction of the *Rb-1*-bearing chromosome. (E) Recombination in mitosis. Each of these events is rare, but there are so many cells in the retina that on average, there are three such events per eye.

5.8 Transposable elements can move from one chromosome to another.

In the 1940s, in a study of the genetics of kernel mottling in maize (Figure 5.34), Barbara McClintock discovered an element that not only regulated the mottling but also caused breakage of the chromosome carrying the genes for color and consistency of the kernels. The element was called *Dissociation (Ds)*. Mapping data showed that the chromosome breakage always occurs at or very near the location of *Ds*. McClintock's critical observation was that *Ds* does not have a constant location but occasionally moves to a new position (**transposition**), causing chromosome breakage at the new site. Furthermore, *Ds* moves only if a second element, called *Activator (Ac)*, is also present. *Ac* itself moves within the genome and can cause, in the expression of genes at or near its insertion site, alterations similar to the modifications that result from the presence of *Ds*. Other **transposable elements** with characteristics and genetic effects similar to those of *Ac* and *Ds* are known in maize. Much of the color variegation seen in the kernels of varieties used for decorative purposes is attributable to the presence of one or more of these elements.

Since McClintock's discovery, transposable nucleotide sequences have been observed to be widespread in eukaryotes and prokaryotes. In *D. melanogaster*, they constitute from 5 to 10 percent of the genome and comprise about 50 distinct families of sequences. One well-studied family of closely related, but not identical, sequences is called *copia*. This element is present in about 30 copies per genome. The copia element (Figure 5.35) contains about 5000 base pairs with two identical sequences of 267 base pairs located terminally and in the same orientation. Repeated DNA sequences with the same orientation are called **direct repeats** (Figure 5.36A). The ends of each copia direct repeat contain two segments of 17 base pairs, whose sequences are also nearly identical; these shorter segments have opposite orientations. Repeated DNA sequences with opposite orientations are called **inverted repeats** (Figure 5.36B). Other transposable ele-

Figure 5.34 Sectors of purple and yellow tissue in the endosperm of maize kernels resulting from the presence of the transposable elements *Ds* and *Ac*. The heavier sectoring in some ears results from dosage effects of *Ac*. The least-speckled ear has one copy of *Ac*; that in the middle has two (*Ac Ac*); and the most-speckled ear has three (*Ac Ac Ac*). [Courtesy of Jerry L. Kermicle.]

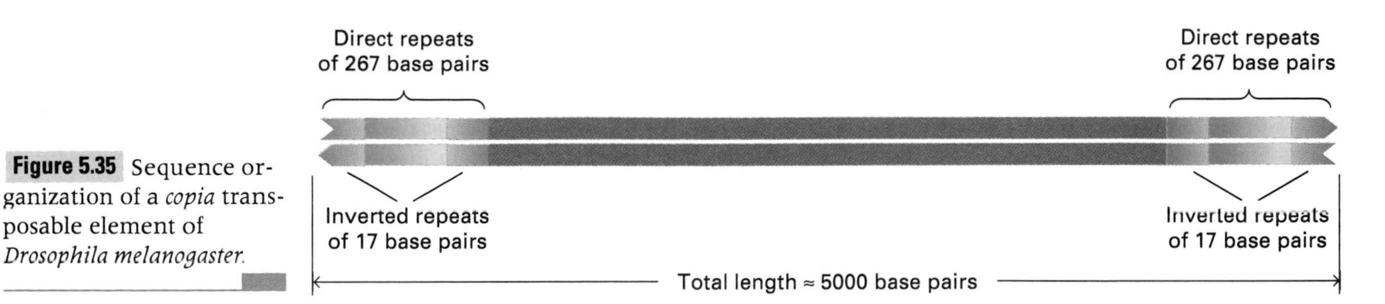

Figure 5.35 Sequence organization of a *copia* transposable element of *Drosophila melanogaster.*

ments have a similar organization with direct or inverted terminal repeats, as do many such elements in other organisms.

In the molecular processes responsible for the movement of transposable elements, transposition of the element is usually accompanied by the duplication of a small number of base pairs originally present at the insertion site, with the result that a copy of this short chromosomal sequence is found immediately adjacent to both ends of the inserted element (Figure 5.37). The length of the duplicated segment ranges from 2 to 12 base pairs, depending on the particular transposable element. Insertion of a transposable element is not a sequence-specific process in that, at each location in the genome, the element is flanked by a different duplicated sequence; however, the *number* of duplicated base pairs is usually the same at each location in the genome and is characteristic of a particular transposable element. Experimental deletion or mutation of part of the terminal base sequences of several different elements has shown that the short, terminal inverted repeats are essential to transposition, probably because they are necessary for binding an enzyme called a **transposase** that is required for transposition. Many transposable elements code for their own transposase by means of a gene located in the central region between the terminal repeats, and hence these elements are able to promote their own transposition. Elements in which the transposase gene has been lost or inactivated by mutation are transposable only if a related element is present in the genome to provide this activity. The inability of the maize *Ds* element to transpose without *Ac* results from the absence of a functional transposase gene in *Ds*. Transposable elements are sometimes referred to as **selfish DNA,** because each type of element maintains itself in the genome through its ability to replicate and transpose.

Transposable elements are responsible for many visible mutations. A transposable element is even responsible for the wrinkled-seed mutation in peas studied by Gregor Mendel. The wildtype allele of the gene codes for starch-branching enzyme I (SBEI), which is used in the synthesis of amylopectin (starch with branched chains). In the wrinkled mutation, a transposable element inserted into the gene renders the enzyme nonfunctional. The pea transposable element has terminal inverted repeats that are very similar to those in the maize *Ac* element, and the insertion site in the *wrinkled* allele is flanked by a duplication of eight base pairs of the SBEI coding sequence. This particular insertion appears to be genetically quite stable, because the transposable element does not seem to have been excised in the long history of wrinkled peas.

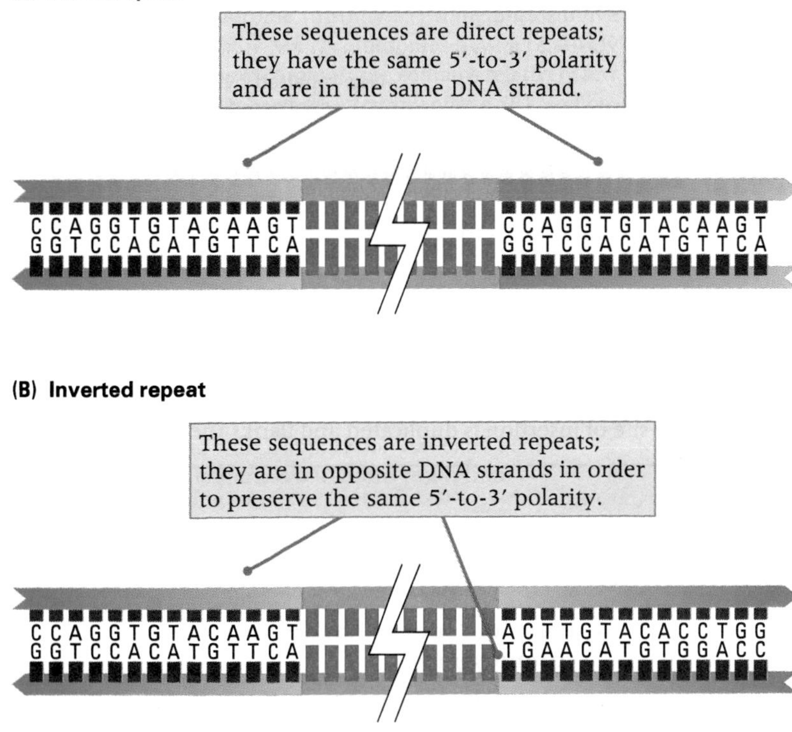

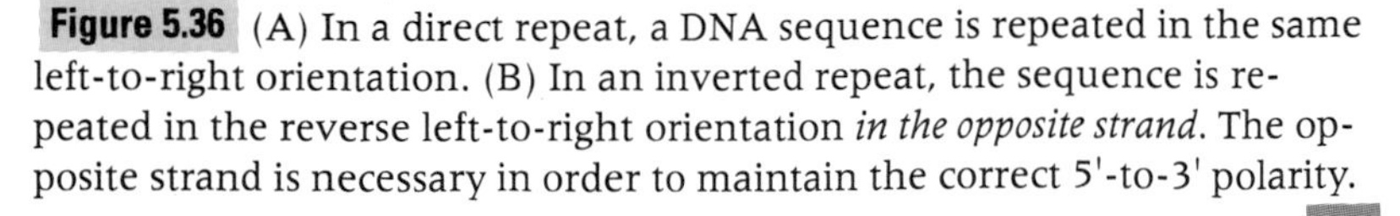
Figure 5.36 (A) In a direct repeat, a DNA sequence is repeated in the same left-to-right orientation. (B) In an inverted repeat, the sequence is repeated in the reverse left-to-right orientation *in the opposite strand*. The opposite strand is necessary in order to maintain the correct 5'-to-3' polarity.

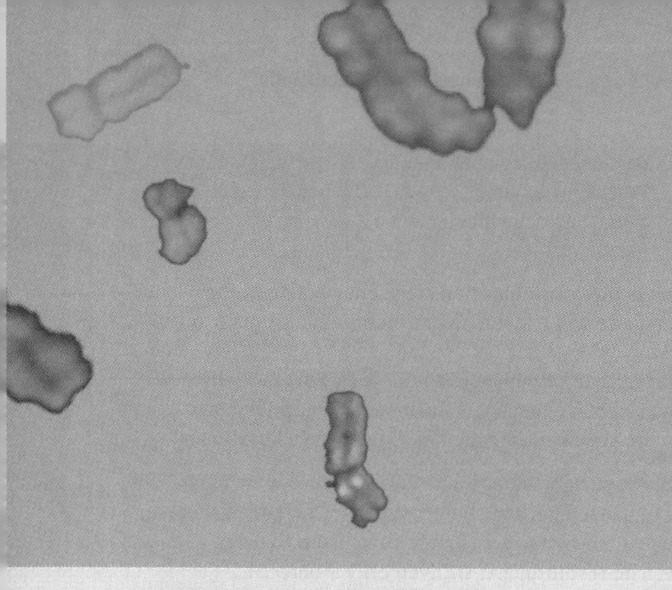

Key Concepts

- Prokaryotes and lower eukaryotes have smaller genomes (less DNA) than higher eukaryotes.
- A DNA strand is a polymer of A, T, G, and C deoxyribonucleotides joined 3' to 5' by phosphodiester bonds.
- The two DNA strands in a duplex are held together by hydrogen bonding between the A–T and the G–C base pairs.
- In DNA replication, each parental strand serves as a template for a daughter strand that is synthesized in the 5' → 3' direction (successive nucleotides are added only at the 3' end).
- Each type of restriction endonuclease enzyme cleaves double-stranded DNA at a particular sequence of bases, usually 4 or 6 nucleotides in length.
- In the polymerase chain reaction, short oligonucleotide primers are used in successive cycles of DNA replication to amplify selectively a particular region of a DNA duplex.
- The DNA fragments produced by a restriction enzyme can be separated by electrophoresis, isolated, sequenced, and manipulated in other ways.

THE TOMB OF THE UNKNOWN soldier. Recent DNA typing has helped to conclusively identify the remains of an "unknown soldier," Lt. Michael Blassie. [© Mark Burnett/Photo Researchers, Inc.]

CHAPTER 6

The Chemical Structure and Replication of DNA

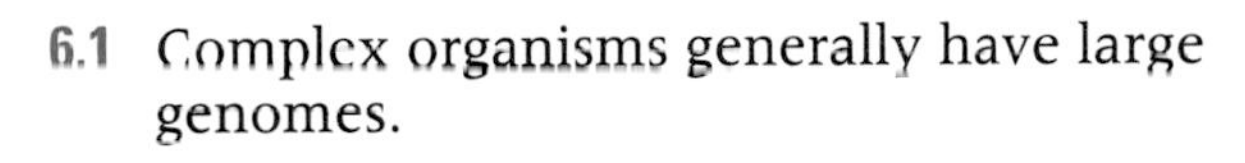

6.1 Complex organisms generally have large genomes.

6.2 DNA is a linear polymer of four deoxyribonucleotides.

6.3 Duplex DNA forms a double helix held together by hydrogen bonds.

6.4 Replication uses each DNA strand as a template for a new one.

6.5 DNA polymerase makes the new DNA strands.

6.6 Knowledge of DNA structure makes possible the manipulation of DNA molecules.

6.7 The polymerase chain reaction makes possible the amplification of a particular DNA fragment.

6.8 Chemical terminators of DNA synthesis are used to determine the base sequence.

The Human Connection

Sickle-Cell Anemia: The First "Molecular Disease"

Chapter Summary

Key Terms

Concepts and Issues: Testing Your Knowledge

Solutions Step by Step

Concepts in Action: Problems for Solution

Further Readings

GeNETics on the Web

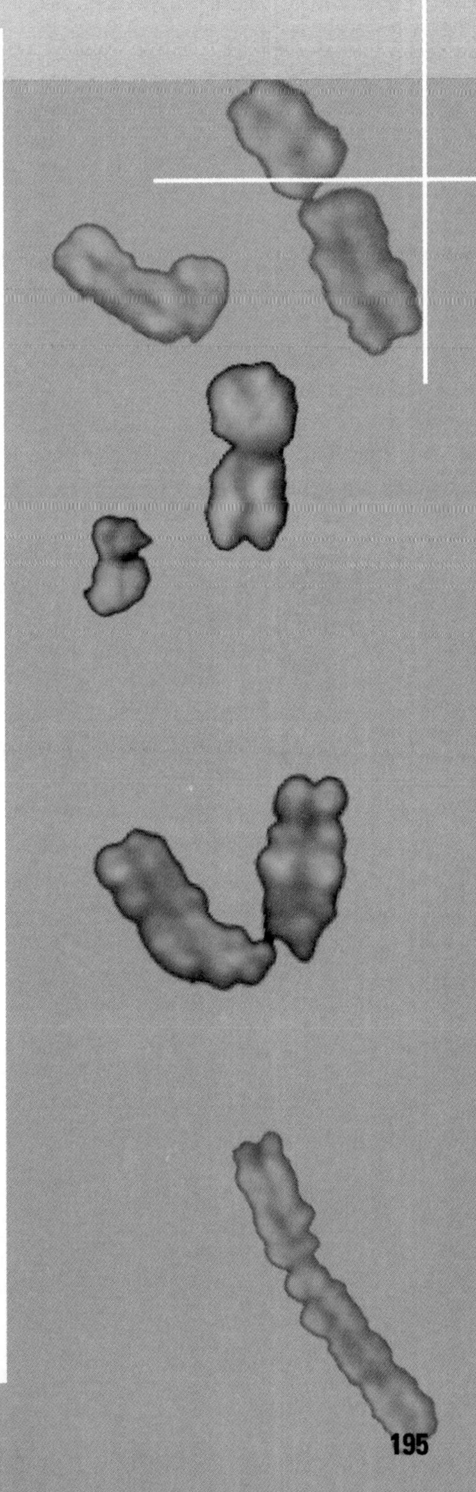

Analysis of the patterns of inheritance and even the phenotypic expression of genes reveals nothing about gene structure at the molecular level, how genes are copied to yield exact replicas of themselves, or how they determine cellular characteristics. Understanding these basic features of heredity requires identification of the chemical nature of the genetic material and the processes through which it is replicated. In Chapter 1, we reviewed the experimental evidence demonstrating that the genetic material is DNA. The structure of DNA was described as a helix of two paired, complementary strands, each composed of an ordered string of nucleotides bearing A (adenine), T (thymine), G (guanine), or cytosine (C). Watson-Crick base pairing between A and T and between G and C in the complementary strands holds the strands together. The complementarity is also the key to replication, because each strand can serve as a template for the synthesis of a new complementary strand. In this chapter, we take a closer look at DNA structure and its replication. We also consider how our knowledge of DNA structure and replication has been used in the development of laboratory techniques for the isolation of fragments containing genes or parts of genes of particular interest and for determining the sequence of bases in DNA fragments.

6.1 Complex organisms generally have large genomes.

The genetic complement of a cell or virus constitutes its **genome.** In eukaryotes, this term is commonly used to refer to one complete haploid set of chromosomes, such as that found in a sperm or egg. Measurement of the nucleic acid content of the genomes of viruses, bacteria, and lower and higher eukaryotes has led to the following generalization:

> Genome size increases roughly with evolutionary complexity.

This generalization is based on three observations: The single nucleic acid molecule of a typical virus is smaller than the DNA molecule in a bacterial chromosome; unicellular eukaryotes, such as the yeasts, contain more DNA than a typical bacterium; and multicellular eukaryotes have the greatest amount of DNA per genome. However, among the multicellular eukaryotes, no correlation exists between evolutionary complexity and the amount of DNA. In higher eukaryotes, DNA content is not directly proportional to the number of genes.

A summary of genome size in a sample of organisms is shown in Table 6.1. Bacteriophage MS2 is one of the smallest viruses; it has only four genes in a single-stranded RNA molecule containing 3569 nucleotides. SV40 virus, which infects monkey and human cells, has a genetic complement of five genes in a circular double-stranded DNA molecule consisting of about 5000 nucleotide pairs. Large DNA molecules are measured in **kilobase pairs (kb),** or thousands of base pairs. The genome of SV40 is about 5 kb. The more complex phages and animal viruses have as many as 250 genes and DNA molecules ranging from 50 to 300 kb. Bacterial genomes are substantially larger. For example, the chromosome of *E. coli* contains about 4300 genes in a DNA molecule composed of about 4600 kb.

Although the genomes of prokaryotes are composed of DNA, their DNA is not packaged into chromosomes. True chromosomes are found only in eukaryotes. The number of chromosomes is characteristic of the particular species, as we saw in Chapter 3. In animals and plants, the DNA content per haploid genome generally tends to increase with the complexity of the organism, but there are many individual exceptions. The number of chromosomes shows no pattern. One of the smallest genomes in a multicellular animal is that of the nematode worm *Caenorhabditis elegans,* with a DNA content about 20 times that of the *E. coli* genome. The *D. melanogaster* and human genomes have about 40 and 700 times as much DNA, respectively, as the *E. coli* genome. The genomes of some amphibia and fish are very large—many times the size of mammalian genomes. Such large genomes are measured in **megabase pairs (Mb),** or millions of base pairs. The human genome is large, but it is by no means the largest among animals or higher plants. At 3000 Mb, the human genome is only 67 percent the size of that in corn and only 4 percent the size of that in the salamander *Amphiuma* (Table 6.1).

Among higher animals and plants, a large genome size does not imply a large number of genes. For example, the 3000-Mb human genome is large enough to contain perhaps 10^6 genes; however, various lines of evidence suggest that the number of genes is no greater than approximately 10^5. Considering the number and size of proteins produced in human cells, it appears that no more than about 4 percent of the human genome actually codes for proteins. Similarly, the genome sizes of closely related species of salamanders can differ by 30-fold in total amount of DNA, yet the genomes are thought to contain about the same number of genes. Therefore, in higher animals and plants, the actual number of genes is much less than the theoretical maximum. The reason for the discrepancy is that in higher organisms, most of the DNA has functions other than coding for the amino acid sequence of proteins. This issue is discussed further in Chapter 9.

Table 6.1
Genome size of some representative viral, bacterial, and eukaryotic genomes

Genome	Approximate length in thousands of nucleotides	Form
Virus		
MS2	4	Single-stranded RNA
SV40	5	Circular double-stranded DNA
ϕX174	5	Circular single-stranded DNA; double-stranded replicative form
M13	6	
λ	50	Linear double-stranded DNA
Herpes simplex	152	
T2,T4,T6	165	
Smallpox	267	
Bacteria		
Mycoplasma hominis	760	Circular double-stranded DNA
Escherichia coli	4,600	
Eukaryotes		Haploid chromosome number
Saccharomyces cerevisiae (yeast)	13,000	16
Caenorhabditis elegans (nematode)	100,000	6
Arabidopsis thaliana (wall cress)	100,000	5
Drosophila melanogaster (fruit fly)	165,000	4
Homo sapiens (human being)	3,000,000	23
Zea mays (maize)	4,500,000	10
Amphiuma sp. (salamander)	76,500,000	14

6.2 DNA is a linear polymer of four deoxyribonucleotides.

DNA is a polymer—that is, a large molecule that contains repeating units and is composed of 2'-deoxyribose (a five-carbon sugar), phosphoric acid, and the four nitrogen-containing bases denoted A, T, G, and C. The chemical structures of the bases are shown in Figure 6.1. Note that two of the bases have a double-ring structure; these are called **purines.** The other two bases have a single-ring structure; these are called **pyrimidines.**

- The purine bases are adenine (A) and guanine (G).
- The pyrimidine bases are thymine (T) and cytosine (C).

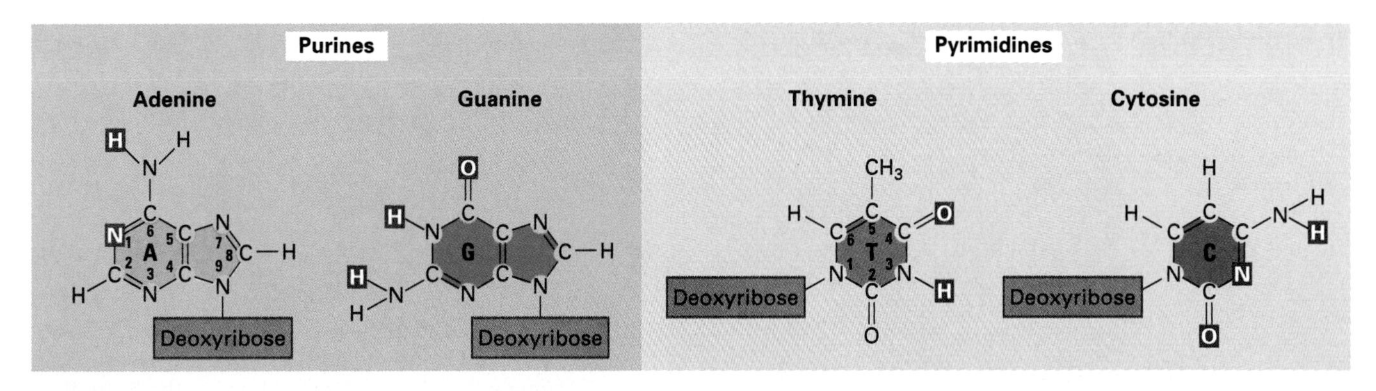

Figure 6.1 Chemical structures of adenine, thymine, guanine, and cytosine, the four nitrogen-containing bases in DNA. In each base, the nitrogen atom linked to the deoxyribose sugar is indicated. The atoms shown in red participate in hydrogen bonding between the DNA base pairs.

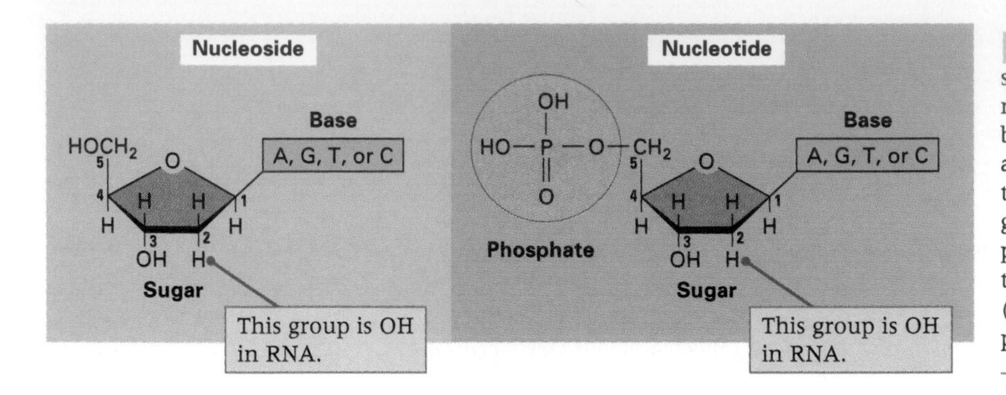

Figure 6.2 A typical nucleotide showing the three major components (phosphate, sugar, and base), the difference between DNA and RNA, and the distinction between a nucleoside (no phosphate group) and a nucleotide (with phosphate). Nucleotides may contain one phosphate unit (monophosphate), two (diphosphate) or three (triphosphate).

In DNA, each base is chemically linked to one molecule of the sugar deoxyribose, forming a compound called a **nucleoside.** When a phosphate group is also attached to the sugar, the nucleoside becomes a **nucleotide** (Figure 6.2). Thus a nucleo*t*ide is a nucleo*s*ide plus a phosphate. In the conventional numbering of the carbon atoms in the sugar in Figure 6.2, the carbon atom to which the base is attached is the 1' carbon. (The atoms in the sugar are given primed numbers to distinguish them from atoms in the bases.)

In nucleic acids, such as DNA and RNA, the nucleotides are joined to form a **polynucleotide chain,** in which the phosphate attached to the 5' carbon of one sugar is linked to the hydroxyl group attached to the 3' carbon of the next sugar in line (Figure 6.3). The chemical bonds by which the sugar components of adjacent nucleotides are

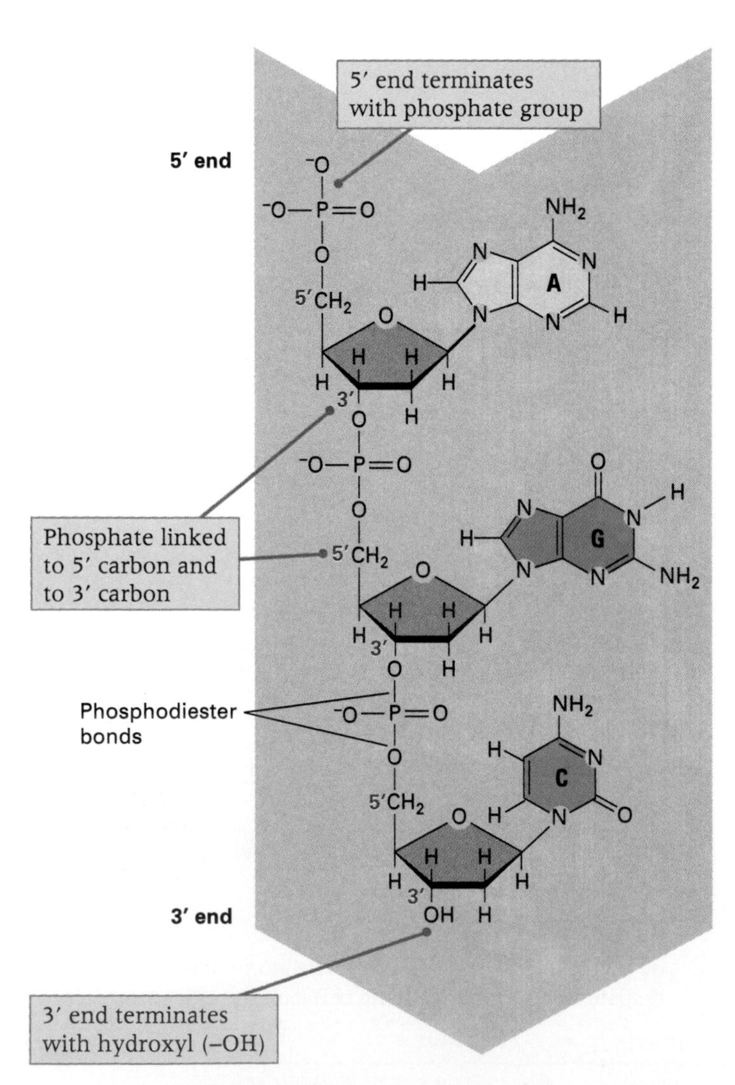

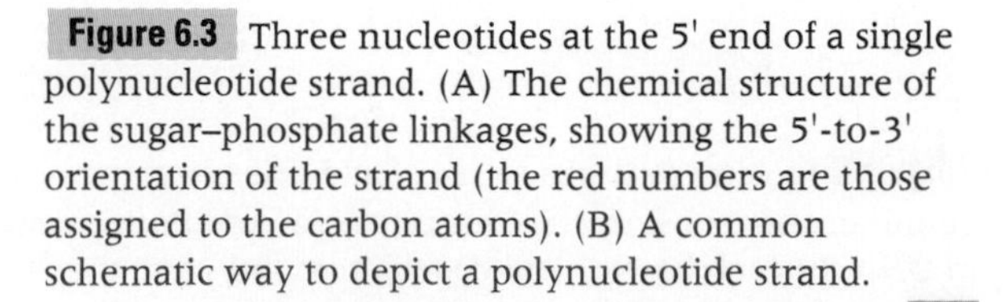

Figure 6.3 Three nucleotides at the 5' end of a single polynucleotide strand. (A) The chemical structure of the sugar–phosphate linkages, showing the 5'-to-3' orientation of the strand (the red numbers are those assigned to the carbon atoms). (B) A common schematic way to depict a polynucleotide strand.

linked through the phosphate groups are called **phosphodiester bonds.** The 5'−3'−5'−3' orientation of these linkages continues throughout the chain, which typically consists of millions of nucleotides. Note that the terminal groups of each polynucleotide chain are a 5' phosphate (**5'-P**) group at one end (depicted as the "tail" of the broad arrow) and a 3'-hydroxyl (**3'-OH**) group at the other (depicted as the "head" of the arrow). The asymmetry of the ends of a DNA strand is the chemical basis of its polarity: One end of the strand is the 5' end (which terminates in a phosphate), whereas the other end is the 3' end (which terminates in a hydroxyl).

6.3 Duplex DNA forms a double helix held together by hydrogen bonds.

Figure 6.4 shows several representations of double-stranded DNA. The duplex molecule of DNA consists of two polynucleotide chains twisted around one another to form a right-handed helix in which adenine and thymine are paired, as are guanine and cytosine (Figure 6.4). Each chain makes one complete turn every 34 Å. The bases are spaced at 3.4 Å, so there are ten bases per helical turn in each strand, or ten base pairs per turn of the double helix. Each base is paired to its partner base in the other strand by hydrogen bonds, which provide the main force holding the strands together. A **hydrogen bond** is a weak bond in which two negatively charged atoms share a hydrogen atom. The paired bases are planar, parallel to one another, and perpendicular to the long axis of the double helix.

The central feature of DNA structure is the A−T and G−C pairing between the bases:

> The purine adenine pairs with the pyrimidine thymine (forming an A−T pair), and the purine guanine pairs with the pyrimidine cytosine (forming a G−C pair).

The principles of A−T and G−C base pairing explain two generalizations about the relative amounts of the bases found in all double-stranded DNA:

- Number of adenine bases [A] equals number of thymine bases [T], so [A] = [T].
- Number of guanine bases [G] equals number of cytosine bases [C], so [G] = [C].

Although [A] = [T] and [G] = [C] in double-stranded DNA, the proportion of bases that are either G or C (called the *percent G + C*) varies among species but is constant in all cells of an organism. For example, human DNA has 39 percent G + C on the average, but there can be large variations in base composition along the chromosomes.

The adenine–thymine base pair and the guanine–cytosine base pair are illustrated in Figure 6.5. Note that an A−T pair has two hydrogen bonds and a G−C pair has three hydrogen bonds. This means that the hydrogen bonding between G and C is stronger in the sense that it requires more energy to break. The specificity of base pairing means that the sequence of bases along one polynucleotide strand of the DNA is matched (complementary) with the base sequence in the other strand. However, the base pairs along a DNA duplex can be arranged in any sequence, and the sequence of bases can vary from one part of the molecule to another and from species to species. Because there is no restriction on the base sequence, DNA has a virtually unlimited capability to code for a variety of different protein molecules.

The backbone of each polynucleotide strand in the double helix in Figure 6.4 consists of deoxyribose sugars alternating with phosphate groups that link the 3' carbon atom of one sugar to the 5' carbon of the next in line. The two polynucleotide strands of the double helix run in opposite directions, as can be seen from the orientation of the deoxyribose sugars in Figure 6.6. The paired strands are said to be **antiparallel.** Figure 6.4 also shows that there are two grooves spiraling along outside of the double helix. These grooves are not symmetrical in size. The large one is called the **major groove,** the smaller one the **minor groove.** Proteins that interact with double-stranded DNA often have regions that make contact with the base pairs by fitting into the major groove, the minor groove, or both.

The diagrams of the DNA duplexes in Figure 6.4A and B are static and so somewhat misleading. DNA is in fact a very dynamic molecule; it is constantly in motion. In some regions, the strands can separate briefly and then come together again in the same conformation or in a different one. Although the right-handed double helix in Figure 6.4 is the standard helix, DNA can form more than 20 slightly different variants of right-handed helices, and some regions can even form helices in which the strands twist to the left. If there are complementary stretches of nucleotides in the same strand, a single strand, separated from its partner, can fold back upon itself like a hairpin. Even triple helices, consisting of three strands, can form in regions of DNA that contain suitable base sequences.

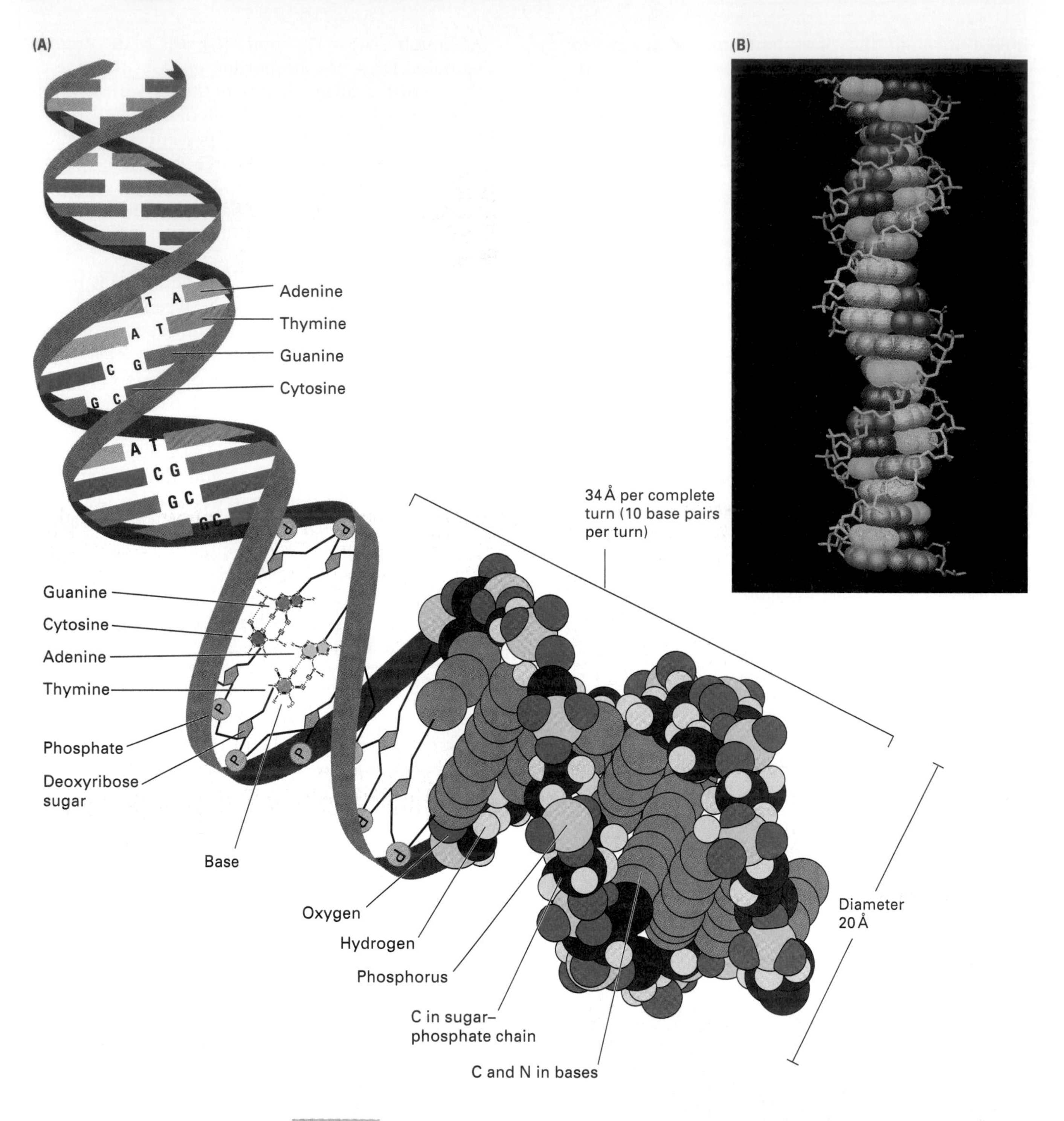

Figure 6.4 Two representations of DNA illustrating the three-dimensional structure of the double helix. (A) In a "ribbon diagram," the sugar–phosphate backbones are depicted as bands, with horizontal lines used to represent the base pairs. (B) A computer model of the standard form of DNA. The stick figures are the sugar–phosphate chains winding around outside the stacked base pairs, forming a major groove and a minor groove. The color coding for the base pairs is A, red or pink; T, dark green or light green; G, dark brown or beige; C, dark blue or light blue. The bases depicted in dark colors are those attached to the blue sugar–phosphate backbone; bases depicted in light colors are attached to the beige backbone. [B, courtesy of Antony M. Dean.]

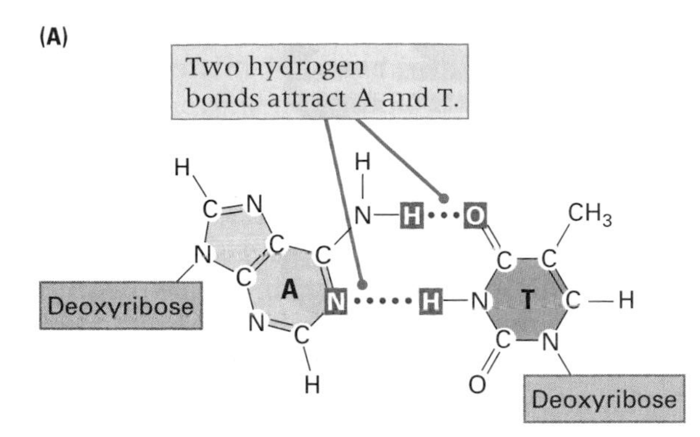

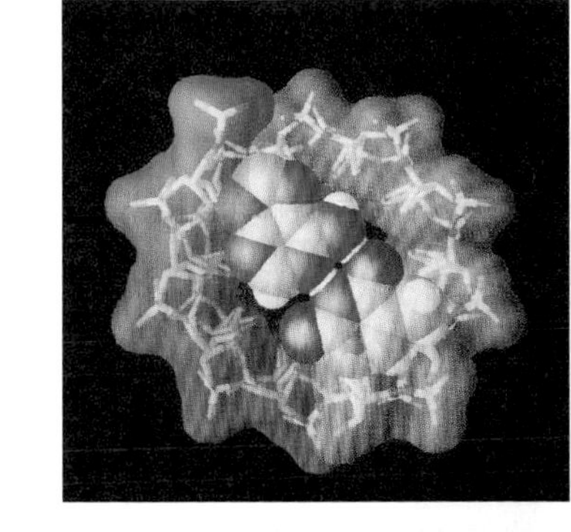

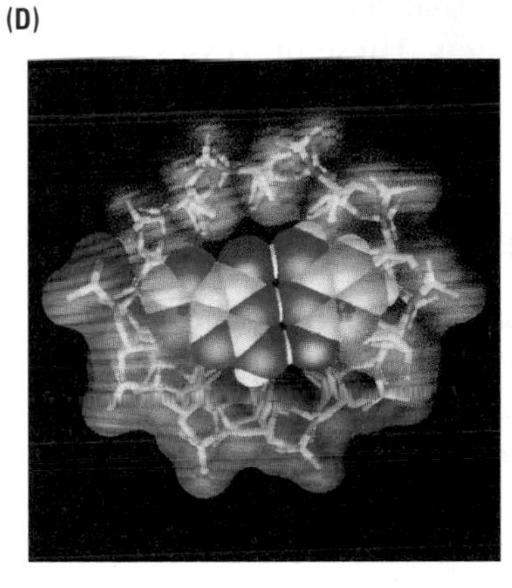

Figure 6.5 Normal base pairs in DNA. On the left, the hydrogen bonds (dotted lines) and the joined atoms are shown in red. (A, B) An A–T base pair. (C, D) A G–C base pair. In the space-filling models (B and D), the colors are C, gray; N, blue; O, red; and H (shown in the bases only), white. Each hydrogen bond is depicted as a white disk squeezed between the atoms sharing the hydrogen. The stick figures on the outside represent the backbones winding around the stacked base pairs. [B and D, courtesy of Antony M. Dean.]

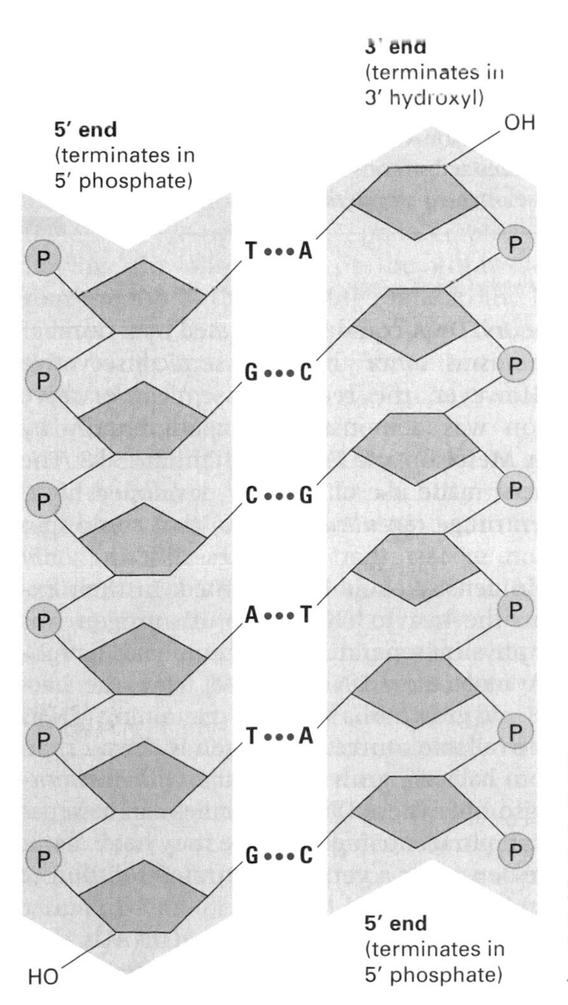

Figure 6.6 A segment of a DNA molecule showing the antiparallel orientation of the complementary strands. The arrows indicate the 5'-to-3' direction of each strand. The phosphate groups (P) join the 3' carbon atom of one deoxyribose to the 5' carbon atom of the adjacent deoxyribose.

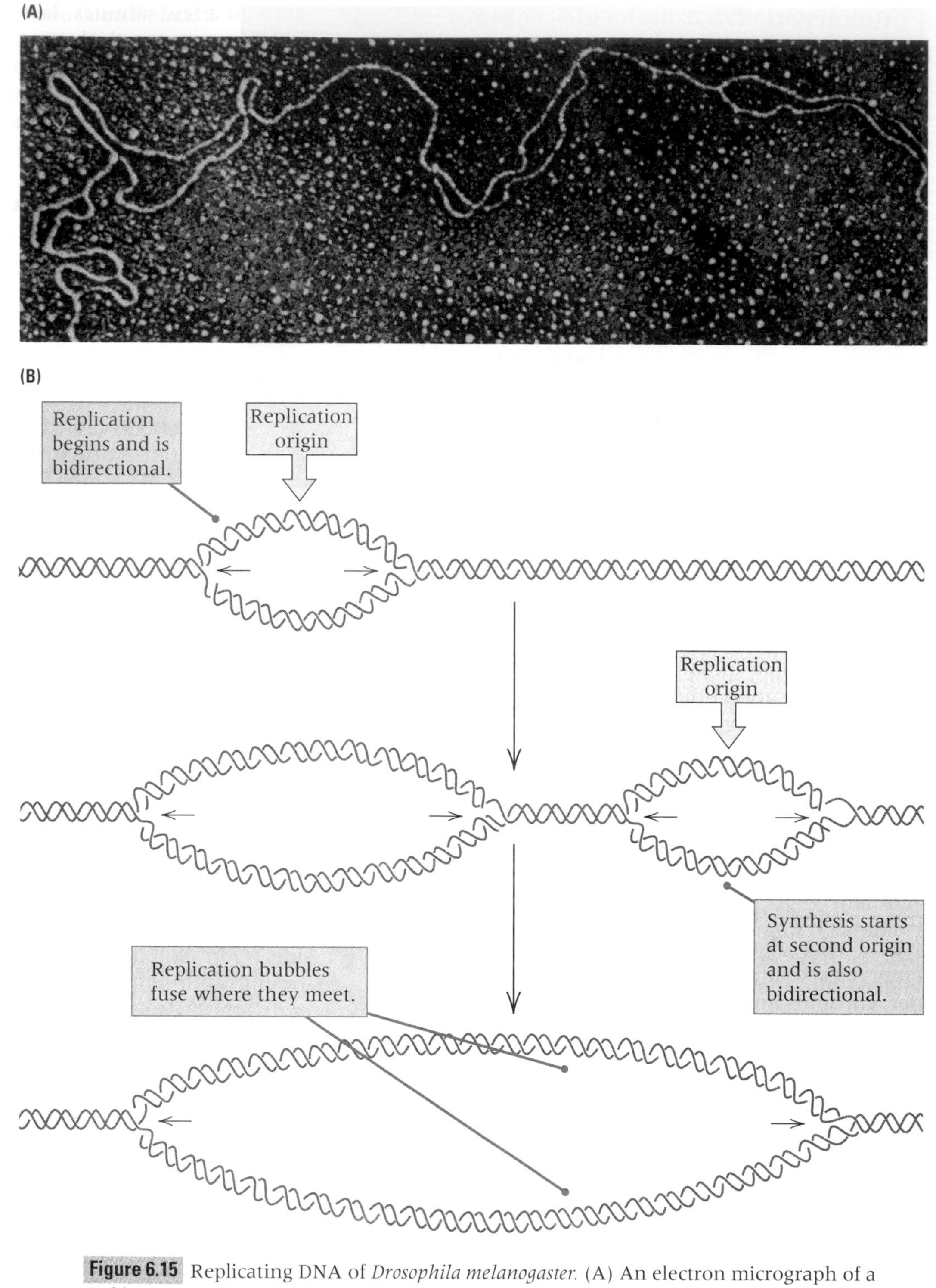

Figure 6.15 Replicating DNA of *Drosophila melanogaster.* (A) An electron micrograph of a 30-kb segment showing five replication loops. (B) An interpretive drawing showing how loops merge. Two replication origins are shown in the drawing. The arrows indicate the direction of movement of the replication forks. [Electron micrograph courtesy of David Hogness.]

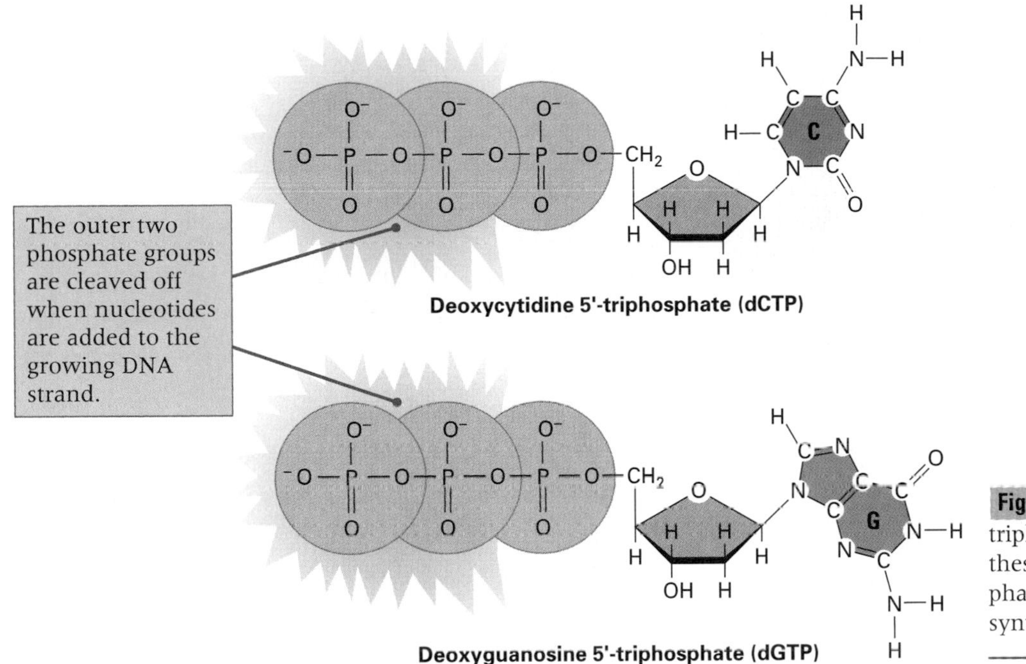

Figure 6.16 Two deoxynucleoside triphosphates used in DNA synthesis. The outermost two phosphate groups are removed during synthesis.

nucleoside 5'-triphosphates and does not take place if any of them are omitted. Details of the structures of dCTP and dGTP are shown in Figure 6.16. The outermost two phosphate groups are cleaved off during DNA synthesis.

- *A preexisting single strand of DNA to be replicated must be present.* Such a strand is called a **template** strand.
- *A nucleic acid segment, which may be very short, must be present and hydrogen-bonded to the template strand.* This segment is called a **primer** (Figure 6.17). *No known DNA polymerase is able to initiate chains,* so the presence of a primer chain with a free 3'-OH group is essential for the initiation of replication. In living cells, the primer is a short segment of RNA; in cell-free replication *in vitro,* the primer may be either RNA or DNA.

The reaction catalyzed by a DNA polymerase is the formation of a phosphodiester bond between the free 3'-OH group of the chain being extended and the innermost phosphorus atom of the nucleoside triphosphate being incorporated at the 3' end (Figure 6.17). What happens is that the 3' hydroxyl group at the 3' terminus of the growing strand attacks the innermost phosphate of the incoming nucleotide and forms a phosphodiester bond, releasing the two outermost phosphates. The result is as follows:

> DNA synthesis proceeds by the elongation of primer chains, *always in the 5' → 3' direction.*

Recognition of the appropriate incoming nucleoside triphosphate in replication depends on base pairing with the opposite nucleotide in the template chain. DNA polymerase usually catalyzes the polymerization reaction that incorporates the new nucleotide at the primer terminus only when the correct base pair is present. The same DNA polymerase is used to add each of the four deoxynucleoside phosphates to the 3'-OH terminus of the growing strand.

Two DNA polymerases are needed for DNA replication in *E. coli*—DNA polymerase I (abbreviated Pol I) and DNA polymerase III (Pol III). Polymerase III is the major replication enzyme. Pol III exists in the cell as a large complex including at least eight other polypeptide chains and is responsible not only for the elongation of DNA molecules but also for the intiation of the replication fork at origins of replication and the addition of deoxynucleotides to the RNA primers. Polymerase I plays an essential, but secondary, role in replication that will be described in a later section. Eukaryotic cells also contain several DNA polymerases. The enzyme responsible for the replication of chromosomal DNA is called polymerase α. Mitochondria have their own DNA polymerase to replicate the mitochondrial DNA.

In addition to their ability to polymerize nucleotides, most DNA polymerases are capable of **nuclease** activities that break phosphodiester bonds in the sugar–phosphate backbones of nucleic acid chains. Many other enzymes have nuclease activity, and they are of two types: (1) **exonucleases,** which can remove a nucleotide only from the end of a chain, and (2) **endonucleases,** which break

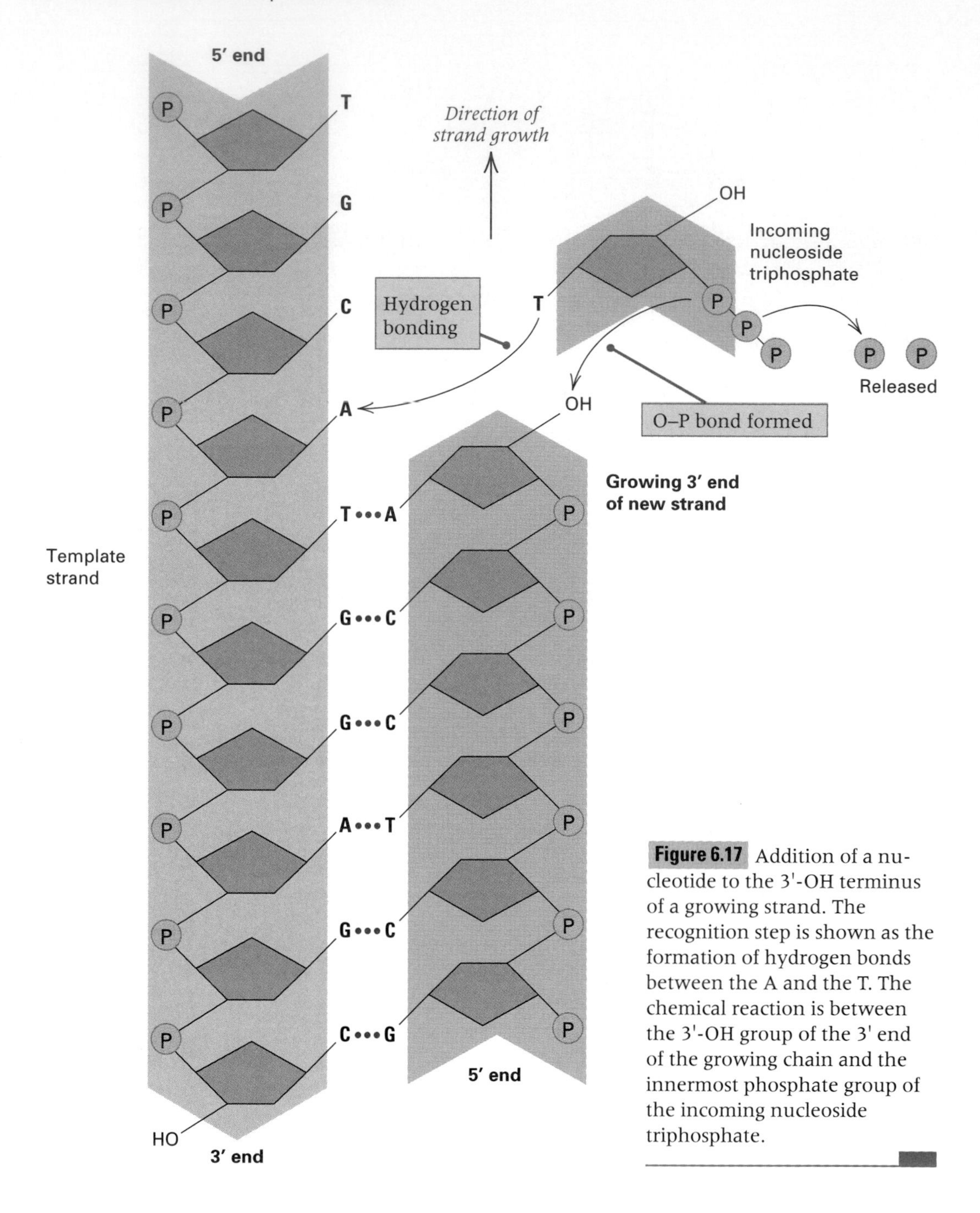

Figure 6.17 Addition of a nucleotide to the 3'-OH terminus of a growing strand. The recognition step is shown as the formation of hydrogen bonds between the A and the T. The chemical reaction is between the 3'-OH group of the 3' end of the growing chain and the innermost phosphate group of the incoming nucleoside triphosphate.

bonds within the chains. DNA polymerases I and III of *E. coli* have an exonuclease activity that acts only at the 3' terminus (a 3' → 5' exonuclease activity). This exonuclease activity provides a built-in mechanism for correcting rare errors in polymerization. Occasionally, a polymerase adds to the growing chain an incorrect nucleotide, which cannot form a proper base pair with the base in the template strand. The presence of an unpaired nucleotide activates the 3' → 5'-exonuclease activity, which cleaves the unpaired nucleotide from the 3'-OH end of the growing chain (Figure 6.18). Because it cleaves off an incorrect nucleotide and gives the polymerase another chance to get it right, the 3' → 5'-exonuclease activity of DNA polymerase is also called the **proofreading** or **editing** function. The proofreading function can "look back" only one base (the one added last). Nevertheless,

> The genetic significance of the proofreading function is that it is an error-correcting mechanism that serves to reduce the frequency of mutation resulting from the incorporation of incorrect nucleotides in DNA replication.

The human connection

Sickle-Cell Anemia: The First "Molecular Disease"

Vernon M. Ingram 1957
Cavendish Laboratory, University of Cambridge, England
Gene Mutations in Human Hemoglobin: The Chemical Difference Between Normal and Sickle-Cell Hemoglobin

The mutation in sickle-cell anemia results in a change in the molecular structure of hemoglobin, but what is the nature of this change? Ingram studied various peptide fragments of hemoglobin and found that the only difference resided in a peptide fragment of eight amino acids ("peptide number 4"). To study this fragment further, he used a method of "fingerprinting," in which digests of peptide 4 containing still smaller fragments were resolved into spots on a sheet of filter paper, first by separating the fragments on the basis of charge along one edge of the paper (electrophoresis) and then by separating on the basis of solubility (chromatography) in the other direction. The complete sequence of peptide 4 was deduced after determining the amino acid sequence of each of the short peptides in the fingerprints. In this case, the normal peptide number 4 has the amino acid sequence

Val–His–Leu–Thr–Pro–Glu–Glu–Lys

(V–H–L–T–P–E–E–K in the single-letter codes), whereas that from sickle-cell hemoglobin has the sequence

Val–His–Leu–Thr–Pro–Val–Glu–Lys

(V–H–L–T–P–V–E–K). The only difference is in the underlined amino acid. This was the first evidence that genes may code for polypeptides in a relatively simple manner, in which successive bits of DNA sequence encode sucessive amino acids in the polypeptide chain. (There were a few minor errors in Ingram's peptide sequences; they have been corrected here.)

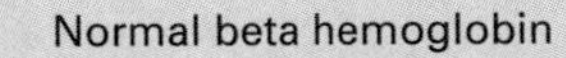

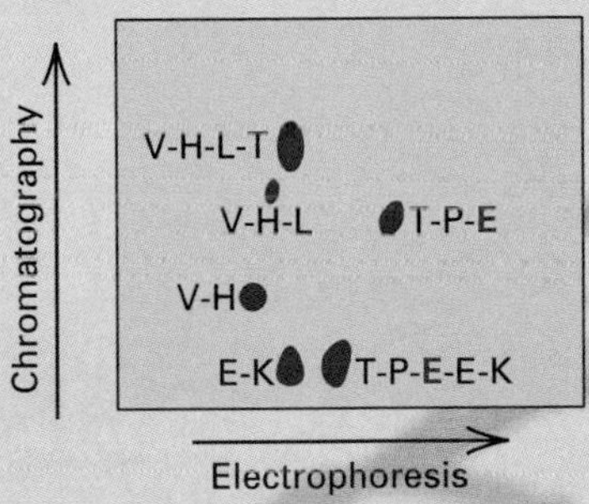

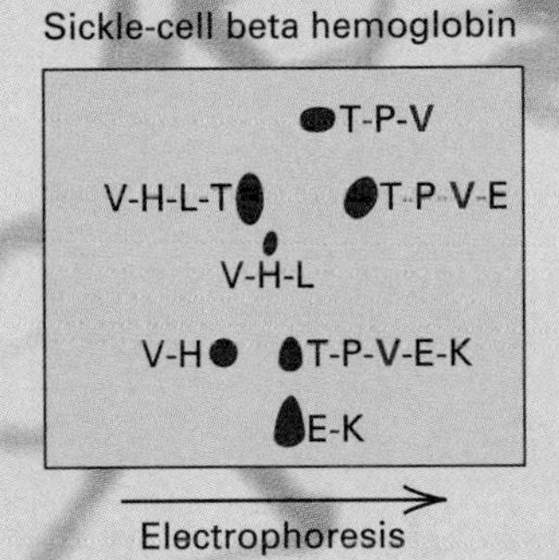

I reported [the previous year] that the globins of normal and sickle-cell anemia differed only in a small portion of their polypeptide chains. I have now found that out of nearly 300 amino acids in the two proteins, only one is different; one of the glumatic acid residues of normal hemoglobin is replaced by a valine residue in sickle-cell anemia hemoglobin. [These results] show, for the first time, that the effect of a single gene mutation is a change in one amino acid of the hemoglobin polypeptide. . . . Tryptic digests of the two proteins . . . were separated on a sheet of paper, using electrophoresis in one direction and chromatography in the other. . . . All peptides had identical chromatographic properties, except for one spot, peptide number 4. . . . Partial hydrolysis of this peptide . . . followed by "fingerprinting" gave the products indicated in the accompanying figure.

The only difference found between the two peptides is that the first glutamic acid residue (E) in normal hemoglobin peptide is replaced by valine (V) in the hemoglobin S peptide. . . . Sickle-cell anemia is an example of a "molecular disease" that is due to an alteration in the structure of a large protein molecule. . . . The difference consists in a replacement of only one of nearly 300 amino acids—a very small change indeed. . . . The results presented are certainly what one would expect on the basis of the widely accepted hypothesis of gene action; the sequence of base pairs along the chain of nucleic acid provides the information which determines the sequence of amino acids in the polypeptide chain for which the particular gene is responsible. A substitution in the nucleic acid leads to a substitution in the polypeptide.

THE DIFFERENCE consists in a replacement of only one of nearly 300 amino acids—a very small change indeed.

Source: Nature 180: 326–328.

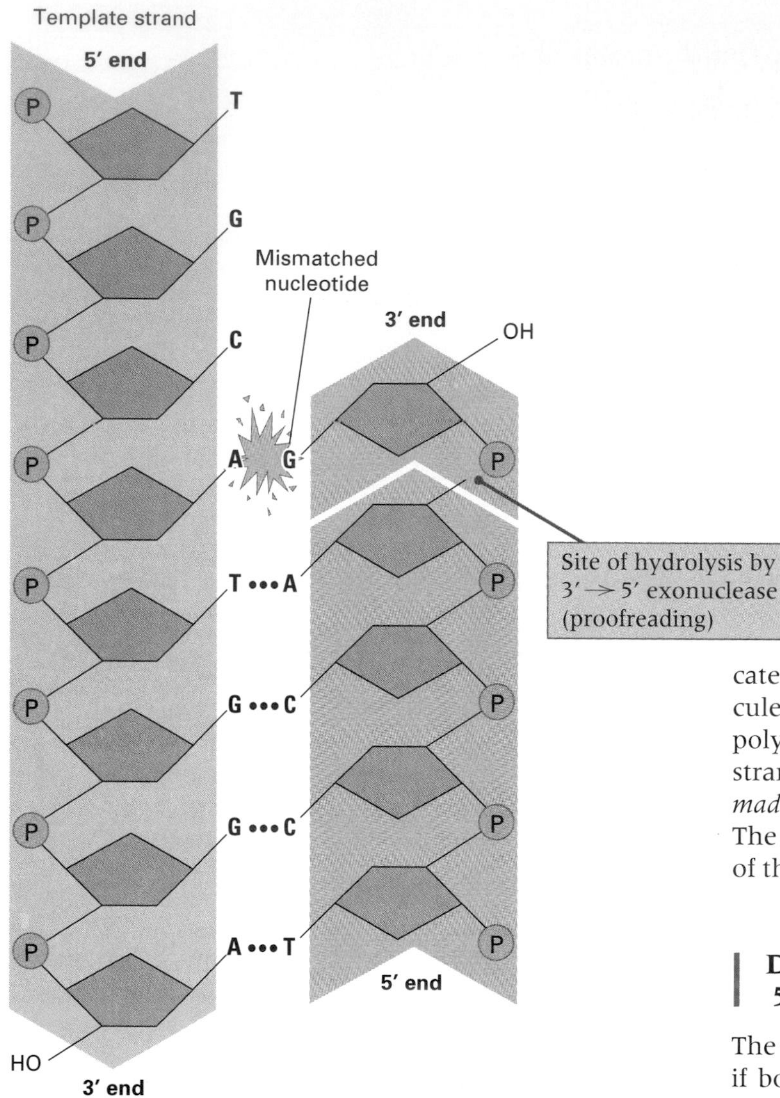

Figure 6.18 The 3'-to-5' exonuclease activity of the proofreading function. The growing strand is cleaved to release a nucleotide containing the base G, which does not pair with the base A in the template strand.

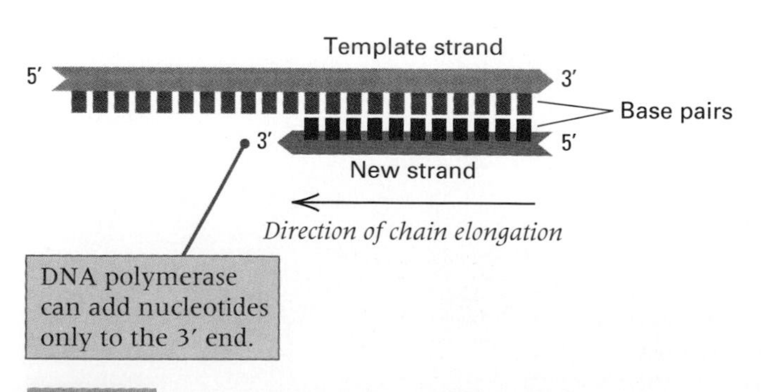

Figure 6.19 The geometry of DNA replication. The new strand (red) is elongated by the addition of successive nucleotides to the 3' end as the polymerase moves along the template strand in the 3' → 5' direction.

Two unexpected features of DNA replication result from functional constraints that are present in all known DNA polymerases. One constraint is that a polymerase can elongate a newly synthesized DNA strand only at its 3' end (Figure 6.19). Hence the polymerase can move along the template strand only in the 3' → 5' direction. The second constraint is that DNA polymerase is unable to initiate new chains but requires a preexisting primer. How the process of DNA replication deals with these constraints is described in the next section.

One strand of replicating DNA is synthesized in pieces.

In the model of replication suggested by Watson and Crick (see Figure 6.7), both daughter strands were supposed to be replicated as continuous units. No known DNA molecule actually replicates in this way. Because DNA polymerase can elongate a newly synthesized DNA strand only at its 3' end, *one of the daughter strands is made in short fragments, which are then joined together.* The reason for this mechanism and the properties of these fragments are described next.

DNA is synthesized only in the 5' → 3' direction.

The model of replication in Figure 6.7 implies that if both daughter strands grew in the same overall direction, then each growing strand would need a 3'-OH terminus because DNA polymerases can add nucleotides only to a 3'-OH group. But the two strands in a DNA duplex are antiparallel, so only one of the growing strands can terminate in a free 3'-OH group; the other must terminate in a free 5' end. The solution to this topological problem is that within a single replication fork, both strands grow in the 5' → 3' orientation, which means that they grow in opposite directions along the parental strands. One strand of the newly made DNA is synthesized continuously (in the lower fork in Figure 6.20). The other strand (in the upper fork in Figure 6.20) is made in small **precursor fragments.** The size of the precursor fragments is from 1000 to 2000 base pairs in prokaryotic cells and from 100 to 200 base pairs in eukaryotic cells. Because synthesis of the discontinuous strand is initiated only at intervals, there is always at least one single-stranded region of the parental strand present on one side of the replication fork. Figure 6.20 also makes it clear that the 3'-OH terminus of the continuously replicated strand is always closer to the replication fork than the 5'-P terminus of the discontinuously repli-

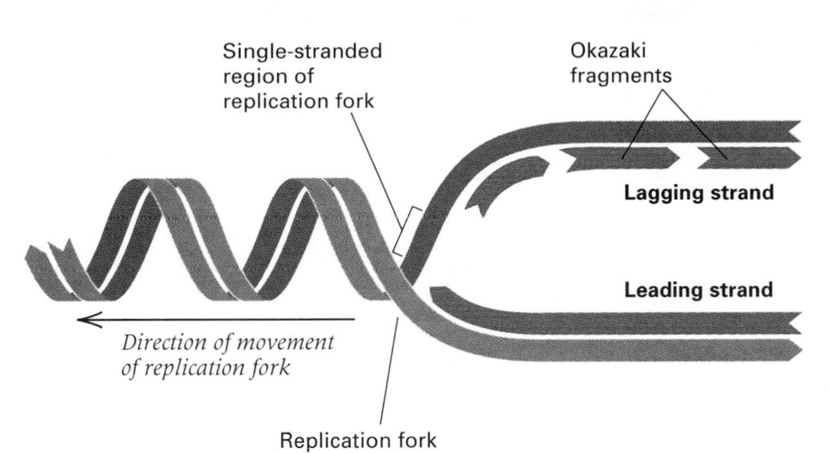

Figure 6.20 Short fragments in the replication fork. For each tract of base pairs, the lagging strand is synthesized later than the leading strand.

cated strand; this is the physical basis of the terms **leading strand** and **lagging strand** that are used for the continuously and discontinuously replicating strands, respectively.

Next we examine how synthesis of a precursor fragment is initiated.

Each new DNA strand or fragment is initiated by a short RNA primer.

As emphasized earlier, DNA polymerases cannot initiate the synthesis of a new strand, so a free 3'-OH is needed. In most organisms, initiation is accomplished by a special type of RNA polymerase. RNA is usually a single-stranded nucleic acid consisting of four types of nucleotides joined together by 3' → 5' phosphodiester bonds (the same chemical bonds as those in DNA). Two chemical differences distinguish RNA from DNA (Figure 6.21). The first difference is in the sugar component. RNA contains **ribose,** which is identical to the deoxyribose of DNA except for the presence of an –OH group on the 2′ carbon atom. The second difference is in one of the four bases: The thymine found in DNA is replaced by the closely related pyrimidine *uracil* (U) in RNA. In RNA synthesis, a DNA strand is used as a template to form a complementary strand in which the bases in the DNA are paired with those in the RNA. Synthesis is catalyzed by an enzyme called an **RNA polymerase.** RNA polymerases differ from DNA polymerases in that they can initiate the synthesis of RNA chains without needing a primer.

DNA synthesis is initiated by using a short stretch of RNA that is base-paired with its DNA template. The size of the primer differs according to the initiation event. In *E. coli,* the length is typically from 2 to 5 nucleotides; in eukaryotic cells, it is usually from 5 to 8 nucleotides. This short stretch of RNA provides a primer onto which a DNA polymerase can add deoxynucleotides (Figure 6.22). The RNA polymerase that produces the primer for DNA synthesis is called *primase,* which is usually found in a multienzyme complex composed of 15 to 20 polypeptide chains called a **primosome.** While it is being synthesized, each precursor fragment in the lagging strand has the structure shown in Figure 6.23.

Precursor fragments are joined together when they meet.

The precursor fragments are ultimately joined to yield a continuous strand of DNA. This strand contains no RNA sequences, so the final stitching together of the lagging strand must require

- Removal of the RNA primer
- Replacement with a DNA sequence
- Joining where adjacent DNA fragments come into contact

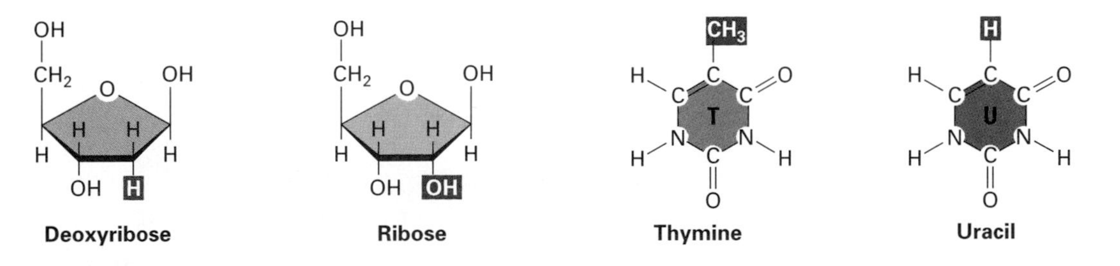

Figure 6.21 Differences between DNA and RNA. The chemical groups in red are the distinguishing features of deoxyribose and ribose and of thymine and uracil.

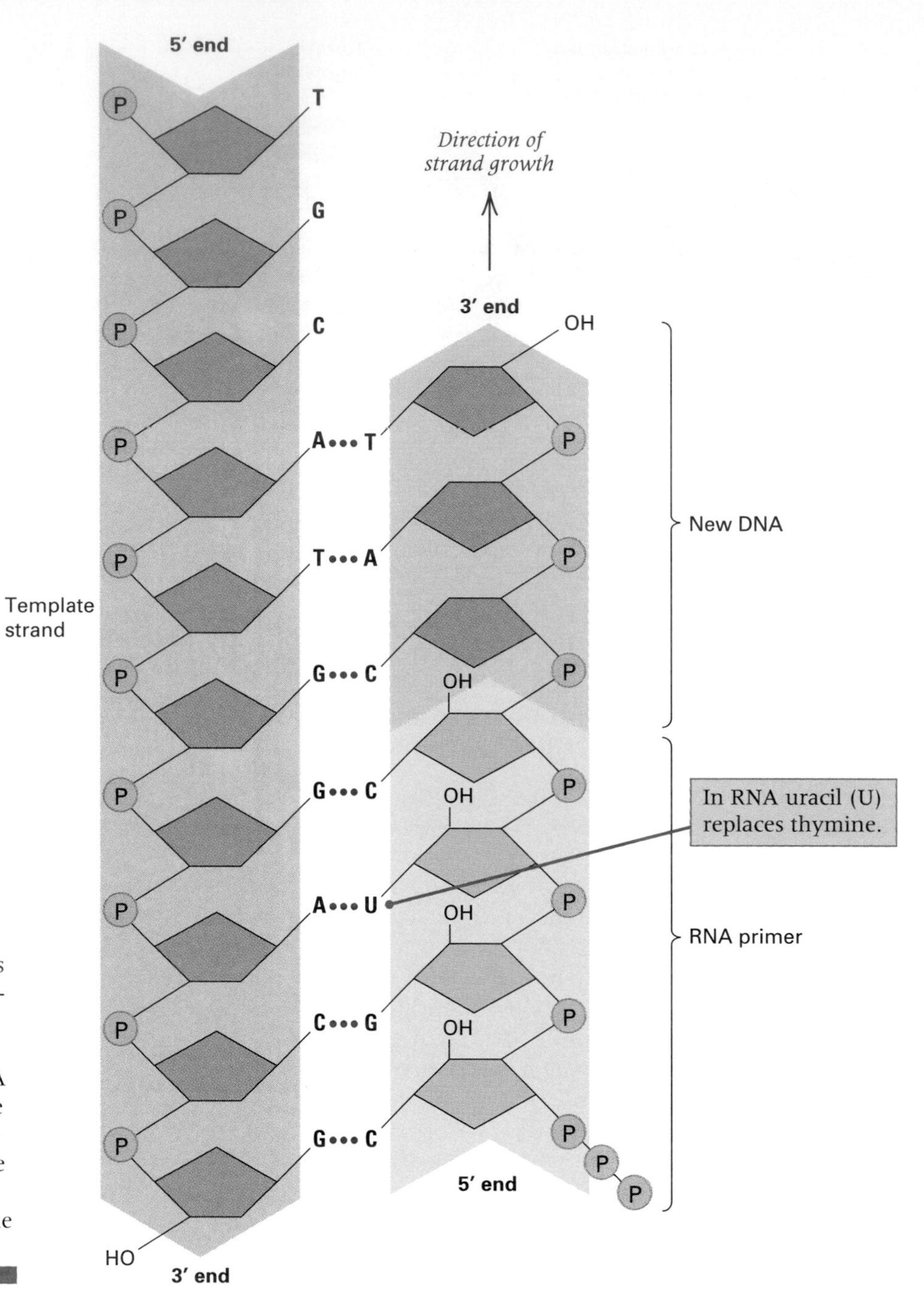

Figure 6.22 Priming of DNA synthesis with an RNA segment. The DNA template strand is shown in blue, newly synthesized DNA in beige. The RNA segment, which was made by an RNA polymerase, is shown in green. In the first step, DNA polymerase adds a deoxyribonucleotide to the 3' end of the primer. Successive deoxyribonucleotides are added to the 3' end of the growing chain.

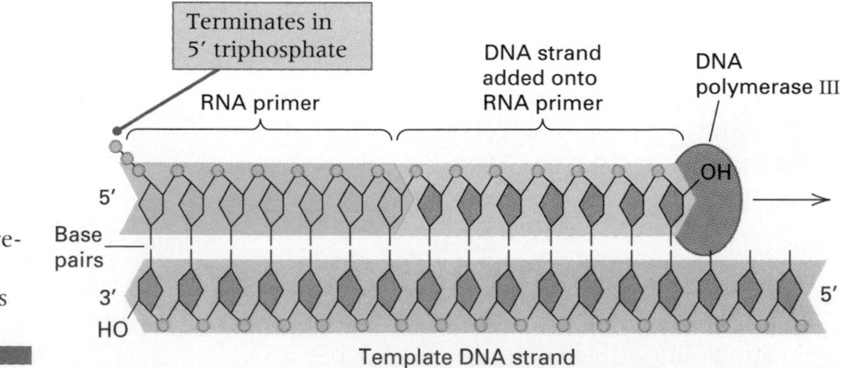

Figure 6.23 Prior to their being joined, each precursor fragment in the lagging strand has the structure shown here. The short RNA primer is shown in green.

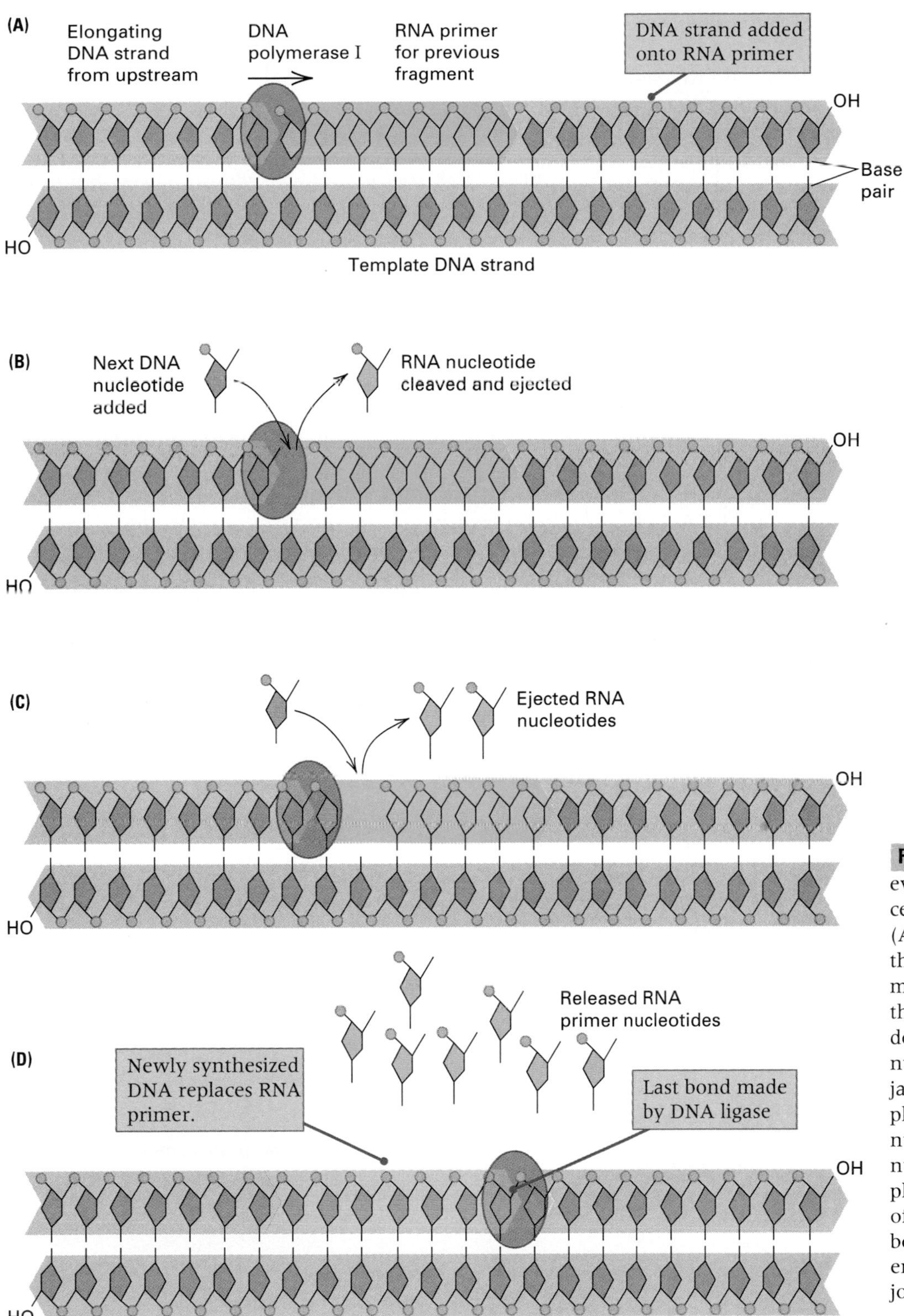

Figure 6.24 Sequence of events in the joining of adjacent precursor fragments. (A) The DNA polymerase of the upstream fragment meets the RNA primer from the next precursor fragment downstream. (B) The RNA nucleotide immediately adjacent is excised and replaced with a DNA nucleotide. (C) Each RNA nucleotide is cleaved and replaced in turn. (D) When all of the RNA nucleotides have been replaced, the adjacent ends of the DNA strand are joined.

In *E. coli,* the first two processes are accomplished by DNA polymerase I, and joining is catalyzed by the enzyme **DNA ligase,** which can link adjacent 3'-OH and 5'-P groups at a nick. How this is done is shown in Figure 6.24. Pol III extends the growing strand until the RNA of the primer of the previously synthesized precursor fragment is reached. Where the DNA and RNA segments meet, there is a single-strand interruption, or **nick.** The *E. coli* DNA ligase cannot seal the nick because a triphosphate is present (it can link only a 3'-OH and a 5'-*mono*phosphate). However, here DNA polymerase I takes over. This enzyme has an exonuclease activity that can remove nucleotides from the 5' end of a base-paired fragment. It is effective with both DNA and RNA. Pol I acts at the nick and displaces it by removing RNA nucleotides one by one as it adds DNA nucleotides to the 3' end of the DNA strand.

When all of the RNA nucleotides have been removed, DNA ligase joins the 3'-OH group to the terminal 5'-P of the precursor fragment. By this sequence of events, the precursor fragment is assimilated into the lagging strand. When the next precursor fragment reaches the RNA primer of the fragment just joined, the sequence begins again. Polymerase I is essential for DNA replication because the exonuclease activity required for removal of the RNA primer and joining of the precursor fragments is not present in Pol III.

Many proteins participate in DNA replication.

The roles of some of the most important components of DNA replication are summarized in Figure 6.25. The replication process includes, in addition to the polymerase III complex and the RNA primase complex, at least one type of topoisomerase, helicase proteins that bind at the replication fork and unwind the double helix, single-strand-binding proteins that bind with and stabilize the single-stranded DNA at the replication fork, the DNA ligase that joins DNA fragments, and the polymerase I complex (not shown in Figure 6.25) that eliminates the RNA primers from precursor fragments before joining can take place.

6.6 Knowledge of DNA structure makes possible the manipulation of DNA molecules.

This and the following sections show how our knowledge of DNA structure and replication has been put to practical use in the development of procedures for the isolation and manipulation of DNA.

Single strands of DNA or RNA with complementary sequences can hybridize.

One of the most important features of DNA is that the two strands of a duplex can be separated by heat without breaking any of the phosphodiester bonds that join successive nucleotides in each strand. If the temperature is maintained sufficiently high, random molecular motion will keep the strands apart. If the temperature is lowered so that hydrogen bonding between complementary base sequences is stable, then the strands will come back together and re-form the DNA duplex. This phenomenon is the basis of one of the most useful methods in molecular genetics, which is called **nucleic acid hybridization.** In this context, molecu-

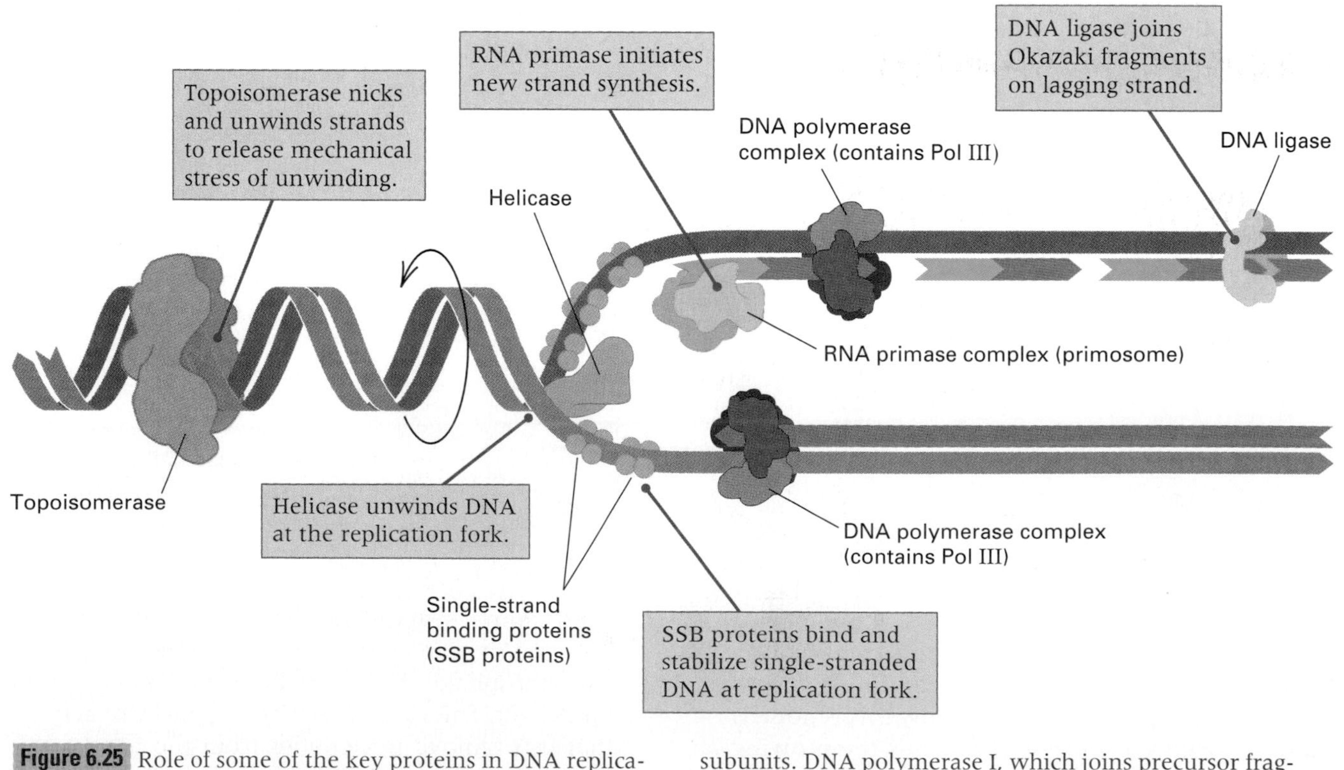

Figure 6.25 Role of some of the key proteins in DNA replication. The DNA polymerase III complex and the primase complex are both composed of multiple different polypeptide subunits. DNA polymerase I, which joins precursor fragments where they meet, is not shown.

lar hybrids are molecules formed when two single nucleic acid strands obtained by denaturing DNA from *different* sources have sufficient sequence complementarity to form a duplex. An example of the use of DNA–DNA hybridization is determination of the fraction of DNA in two different species that have common base sequences. One procedure that uses DNA hybridization is outlined in Figure 6.26. Parts A and B refer to DNA from two bacterial species that have been grown in media that cause the DNA of one species (A) to contain only the nonradioactive isotope of phosphorus, ^{31}P, and cause the DNA of the second species (B) to contain the radioactive isotope ^{32}P. The DNA molecules of the two species are isolated, experimentally broken into many small fragments, and heated or treated with chemicals to separate the strands. The nonradioactive strands are immobilized on a nitrocellulose filter, and the radioactive DNA strands are added. The temperature and salt concentration are adjusted to promote renaturation, and then, after a suitable period of time, the filter is washed. The existence of labeled DNA strands capable of hybridizing with unlabeled DNA strands is detected by the

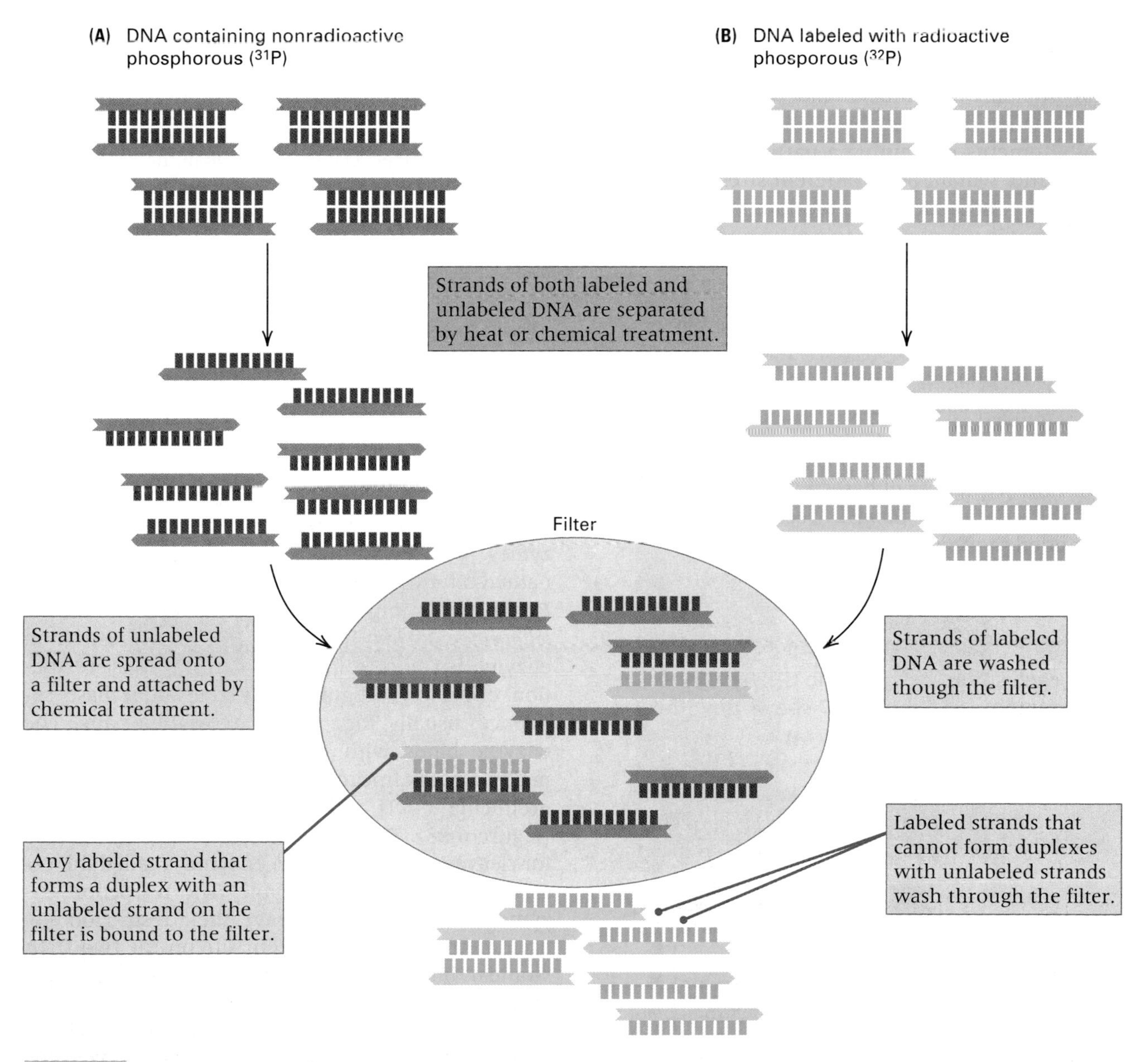

Figure 6.26 Fragments of single-stranded DNA can renature to form a double helix, provided that the sequences of the strands are complementary enough in base sequence. (A) Unlabeled DNA fragments from one source are denatured and the separated strands chemically attached to a special filter. (B) Labeled DNA fragments from a different source are denatured and washed through the filter. Any labeled fragments that cannot find unlabeled pairing partners wash through the filter. Any that do form duplexes with unlabled DNA remain on the filter and can be detected by means of the label.

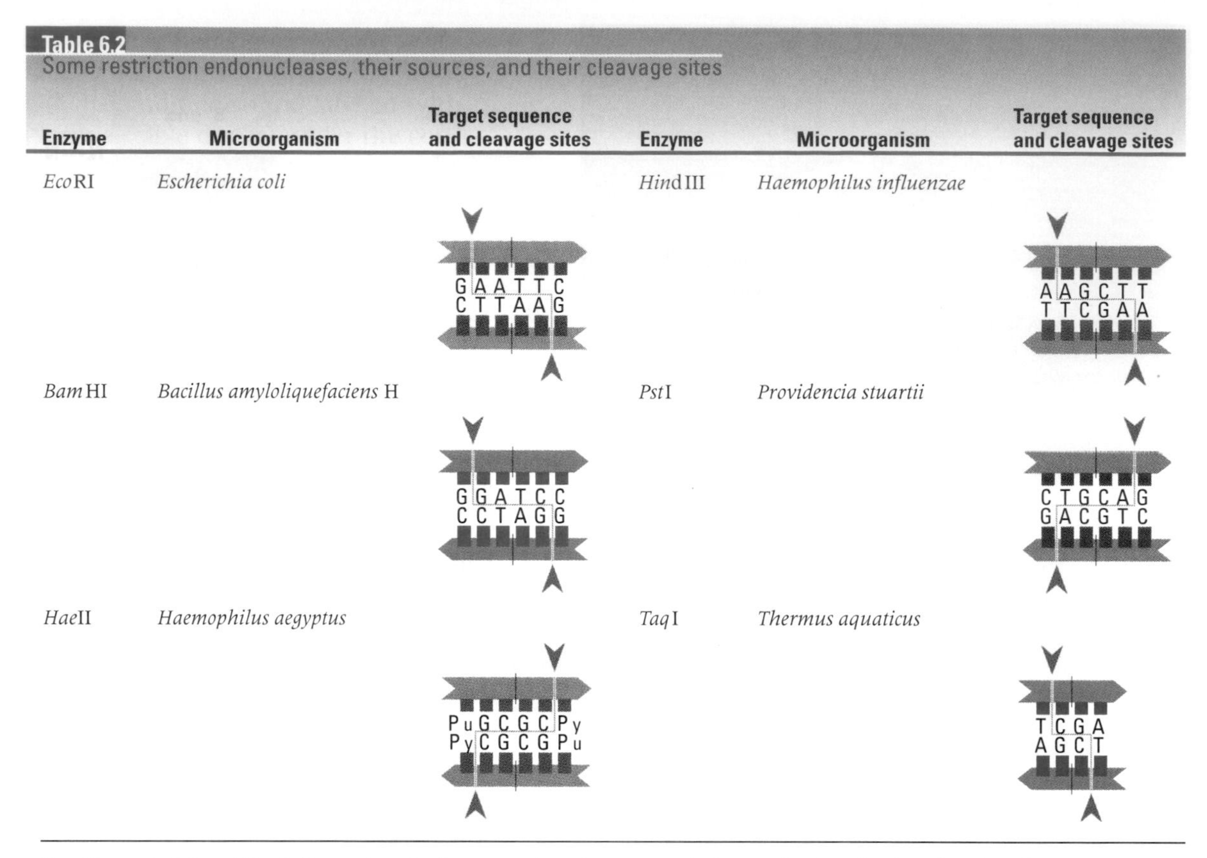

Table 6.2
Some restriction endonucleases, their sources, and their cleavage sites

Enzyme	Microorganism	Target sequence and cleavage sites	Enzyme	Microorganism	Target sequence and cleavage sites
*Eco*RI	*Escherichia coli*	GAATTC CTTAAG	*Hind*III	*Haemophilus influenzae*	AAGCTT TTCGAA
*Bam*HI	*Bacillus amyloliquefaciens* H	GGATCC CCTAGG	*Pst*I	*Providencia stuartii*	CTGCAG GACGTC
*Hae*II	*Haemophilus aegyptus*	PuGCGCPy PyCGCGPu	*Taq*I	*Thermus aquaticus*	TCGA AGCT

Note: The vertical dashed line indicates the axis of symmetry in each sequence. Red arrows indicate the sites of cutting. The enzyme *Taq*I yields cohesive ends consisting of two nucleotides, whereas the cohesive ends produced by the other enzymes contain four nucleotides. Pu and Py refer to any purine and pyrimidine, respectively.

on both strands, provided that the opposite polarity of the strands is taken into account; for example, each strand in the restriction site of *Bam*HI reads 5'-GGATCC-3' (Figure 6.28). A DNA sequence with this type of symmetry is called a *palindrome.* (In ordinary English, a palindrome is a word or phrase that reads the same forward and backward, such as "madam.")

Restriction enzymes have the following important characteristics:

- Most restriction enzymes recognize a single restriction site.
- The restriction site is recognized without regard to the source of the DNA.
- Because most restriction enzymes recognize a unique restriction site sequence, the number of cuts in the DNA from a particular organism is determined by the number of restriction sites that are present.

The DNA fragment produced by a pair of adjacent cuts in a DNA molecule is called a **restriction fragment.** A large DNA molecule will typically be cut into many restriction fragments of different sizes. For example, an *E. coli* DNA molecule, which contains 4.6×10^6 base pairs, is cut into several hundred to several thousand fragments, and mammalian nuclear DNA is cut into more than a million fragments. Although these numbers are large, they are actually quite small relative to the number of sugar–phosphate bonds in the DNA of an organism. Restriction fragments are usually short enough that they can be separated by electrophoresis and manipulated in various ways—for example, using DNA ligase to insert them into self-replicating molecules such as bacteriophage, plasmids, or even small artificial chromosomes. These procedures constitute **DNA cloning** and are the basis of one form of *genetic engineering,* discussed further in Chapter 10.

Because of the sequence specificity, *a particular restriction enzyme produces a unique set of fragments for a particular DNA molecule.* Another enzyme will produce a different set of fragments from the same DNA molecule. Figure 6.29A shows the sites of cutting of *E. coli* phage λ DNA by the enzymes *Eco*RI and *Bam*HI. A map showing the unique sites of cutting of the DNA of a particular organism by a single enzyme is called a **restriction map.** The family of fragments produced by a single enzyme

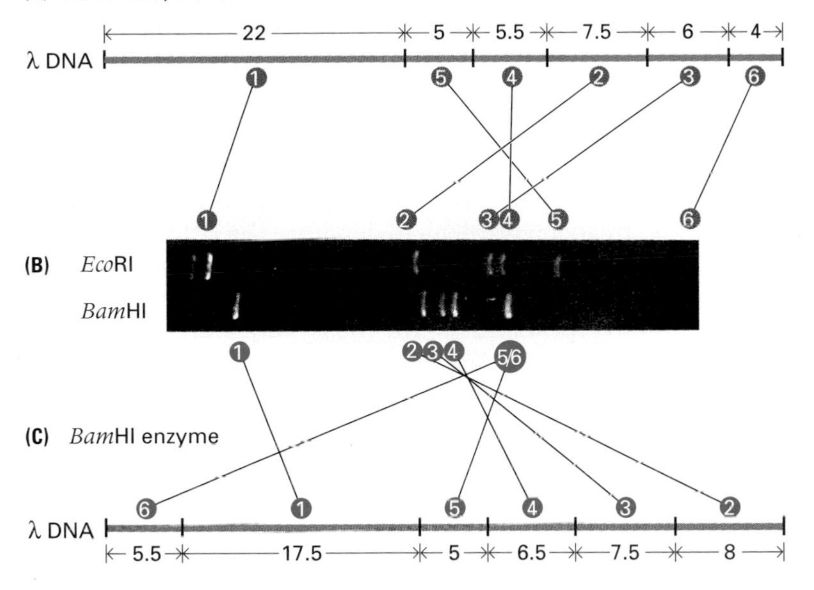

Figure 6.29 (A) Restriction maps of λ DNA for two restriction enzymes, *Eco*RI (A) and *Bam*HI (C). The vertical bars indicate the sites of cutting. The numbers indicate the approximate length of each fragment in kilobase pairs. (B) An electrophoresis gel of *Bam*HI and *Eco*RI enzyme digests of λ DNA. Numbers indicate fragments in order from largest (1) to smallest (6); the circled numbers on the maps correspond to the numbers beside the gel. The DNA has not undergone electrophoresis long enough to separate bands 5 and 6 of the *Bam*HI digest.

can be detected easily by gel electrophoresis of enzyme-treated DNA (Figure 6.29B), and particular DNA fragments can be isolated by cutting out the small region of the gel containing the fragment and removing the DNA from the gel. Gel electrophoresis for the separation of DNA fragments is described next.

Gel electrophoresis separates DNA fragments by size.

As we saw in Chapter 4, the separation of DNA fragments according to size by electrophoresis is one of the principal techniques used in mapping human DNA polymorphisms. The physical basis for separation of the DNA fragments is that DNA molecules are negatively charged and can move in an electric field. If the terminals of an electrical power source are connected to the opposite ends of a horizontal tube containing a DNA solution, then the molecules will move toward the positive end of the tube at a rate that depends on the electric field strength and on the shape and size of the molecules. The movement of charged molecules in an electric field is called *electrophoresis.*

The most common type of electrophoresis used in genetics is **gel electrophoresis.** An experimental arrangement for gel electrophoresis of DNA is shown in Figure 6.30. A thin slab of a gel, usually agarose or acylamide, is prepared containing small slots (called wells) into which samples are placed. An electric field is applied, and the negatively charged DNA molecules penetrate and move through the gel. A gel is a complex molecular network containing narrow, tortuous passages, so smaller DNA molecules pass through more easily, hence the rate of movement increases as the size decreases. For each molecule, the rate of movement depends primarily on size, provided the molecule is linear and not too large. Figure 6.31 shows the result of electrophoresis of a collection of double-stranded DNA molecules in an agarose gel. Each discrete region containing DNA is called a *band.*

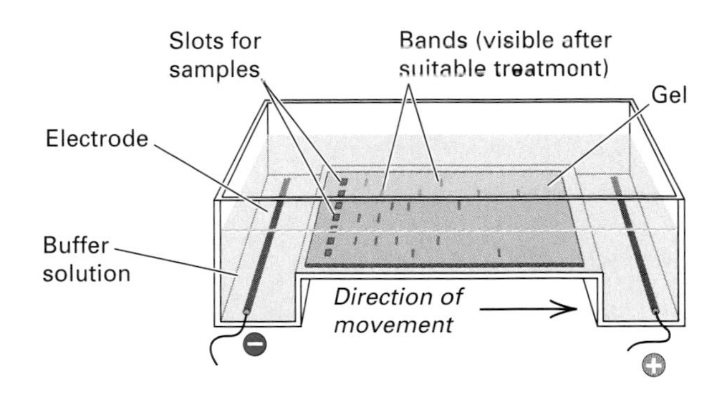

Figure 6.30 Apparatus for gel electrophoresis. Liquid gel is allowed to harden in place, with an appropriately shaped mold placed on top of the gel during hardening in order to make "wells" for the samples. After electrophoresis, the DNA bands, located at various positions in the gel, are made visible by immersing the gel in a solution containing a reagent that binds to or reacts with the separated molecules.

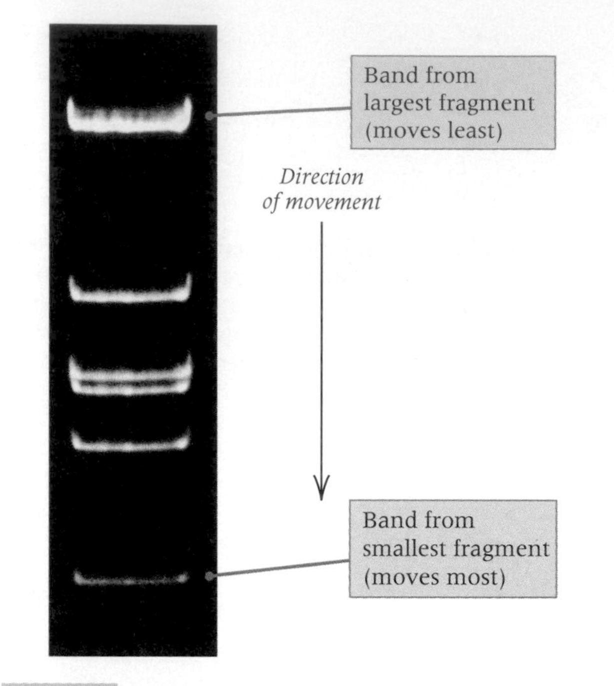

Figure 6.31 Gel electrophoresis of DNA. Molecules of different sizes were mixed and placed in a well. Electrophoresis was in the vertical direction. The DNA has been made visible by the addition of a dye (ethidium bromide) that binds only to DNA and that fluoresces when the gel is illuminated with short-wavelength ultraviolet light.

Specific DNA fragments are identified by hybridization with a probe.

Several techniques enable a researcher to locate a particular DNA fragment in a gel. One of the most generally applicable procedures is the **Southern blot.** In this procedure, a gel in which DNA molecules have been separated by electrophoresis is treated with alkali to denature the DNA and render it single-stranded. Then the DNA is transferred to a sheet of nitrocellulose in such a way that the relative positions of the DNA bands are maintained (Figure 6.32). The nitrocellulose, to which the single-stranded DNA binds tightly, is then exposed to radioactive complementary RNA or DNA (the **probe**) in a way that leads the complementary strands to anneal to form duplex molecules. Radioactivity becomes stably bound (resistant to removal by washing) to the DNA only at positions at which base sequences complementary to the radioactive molecules are present, so that duplex molecules can form. The radioactivity is located by placing the paper in contact with x-ray film; after development of the film, blackened regions indicate positions of radioactivity. For example, a cloned DNA fragment from one species may be used as probe DNA in a Southern blot with DNA from another species; the probe will hybridize only with restriction fragments containing DNA sequences that are sufficiently complementary to allow stable duplexes to form.

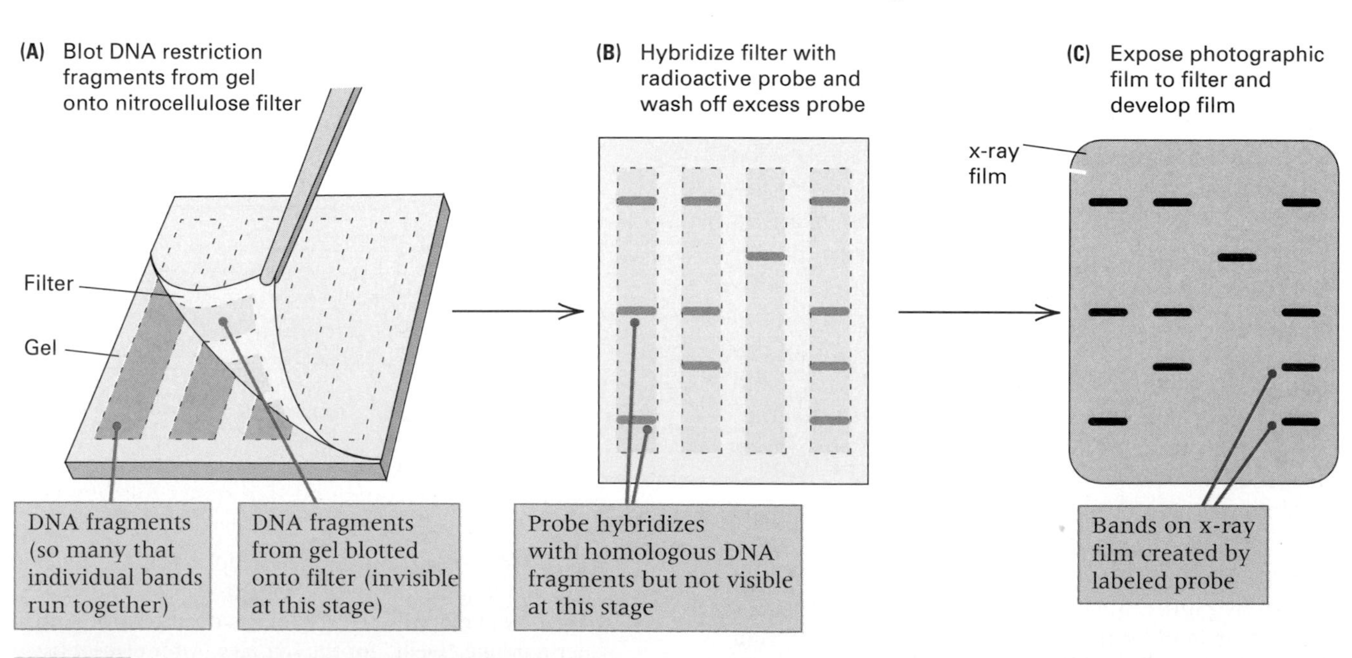

Figure 6.32 Southern blot. (A) DNA restriction fragments are separated by electrophoresis, blotted from the gel onto a nitrocellulose or nylon filter, and chemically attached by the use of ultraviolet light. (B) The strands are denatured and mixed with radioactive probe DNA, which binds with complementary sequences present on the filter. The bound probe remains, whereas unbound probe washes off. (C) Bound probe is revealed by darkening of photographic film placed over the filter. The positions of the bands indicate which restriction fragments contain DNA sequences homologous with those in the probe.

6.7 The polymerase chain reaction makes possible the amplification of a particular DNA fragment.

It is also possible to obtain large quantities of a particular DNA sequence merely by selective replication. The method for selective replication is called the **polymerase chain reaction (PCR),** and it uses DNA polymerase and a pair of short, synthetic oligonucleotides, usually about 20 nucleotides in length, that are complementary in sequence to the ends of the DNA sequence to be amplified and so can serve as primers for strand elongation. Starting with a mixture containing as little as one molecule of the fragment of interest, repeated rounds of DNA replication increase the number of molecules exponentially. For example, starting with a single molecule, 25 rounds of DNA replication will result in $2^{25} = 3.4 \times 10^7$ molecules. This number of molecules of the amplified fragment is so much greater than that of the other unamplified molecules in the original mixture that the amplified DNA can often be used without further purification. For example, a single fragment of 3000 base pairs in *E. coli* accounts for only 0.06 percent of the total DNA in this organism. However, if this single fragment were replicated through 25 rounds of replication, 99.995 percent of the resulting mixture would consist of the amplified sequence.

An outline of the polymerase chain reaction is shown in Figure 6.33. The DNA sequence to be amplified and the oligonucleotide sequences are shown in contrasting colors. The oligonucleotides act as primers for DNA replication because they anneal to the ends of the sequence to be amplified and become the substrates for chain elongation by DNA polymerase. In the first cycle of PCR amplification, the DNA is denatured to separate the strands. The denaturation temperature is usually around 95°C. Then the temperature is decreased to allow annealing in the presence of a vast excess of the primer oligonucleotides. The annealing temperature is typically in the range of 50°C to 60°C, depending largely on the G + C content of the oligonucleotide primers. The temperature is raised slightly, to about 70°C, for the elongation of each primer. The first cycle in PCR produces two copies of each molecule containing sequences complementary to the primers. The second cycle of PCR is similar to the first. The DNA is denatured and then renatured in the presence of an excess of primer oligonucleotides, whereupon the primers are elongated by DNA polymerase; after this cycle there are four copies of each molecule present in the original mixture. The steps of denaturation, renaturation, and replication are repeated from 20 to 30 times, and in each cycle, the number of molecules of the amplified sequence is doubled. The theoretical result of 25 rounds of amplification is 2^{25} copies of each template molecule present in the original mixture.

Hot Springs at Yellowstone National Park in Wyoming, a source of the archaeon *Thermus aquaticus* and the Taq polymerase used in the polymerase chain reaction. [© Robert A. Isaacs/ Photo Researchers, Inc.]

Implementation of PCR with conventional DNA polymerases is not practical, because at the high temperature necessary for denaturation, the polymerase is itself irreversibly unfolded and becomes inactive. However, DNA polymerase isolated from certain archaea is heat-stable because the organisms normally live in hot springs at temperatures well above 90°C, such as are found in Yellowstone National Park. Such organisms are said to be *thermophiles.* The most widely used heat-stable DNA polymerase is called *Taq* polymerase because it was originally isolated from the thermophilic archaea *Thermus aquaticus.*

PCR amplification is very useful for generating large quantities of a specific DNA sequence. The principal limitation of the technique is that the DNA sequences at the ends of the region to be amplified must be known so that primer oligonucleotides can be synthesized. In addition, sequences longer than about 5000 base pairs cannot be replicated efficiently by conventional PCR procedures. On the other hand, there are many applications in which PCR amplification is useful. PCR can be employed to study many different mutant alleles of a gene whose wildtype sequence is known in order to identify the molecular basis of the mutations. Similarly, variation in DNA sequence among alleles present in natural populations can easily be determined using PCR. The PCR procedure has also come into widespread use in clinical laboratories for diagnosis. To take just one very important example, the presence of the human immunodeficiency virus (HIV), which causes acquired immune

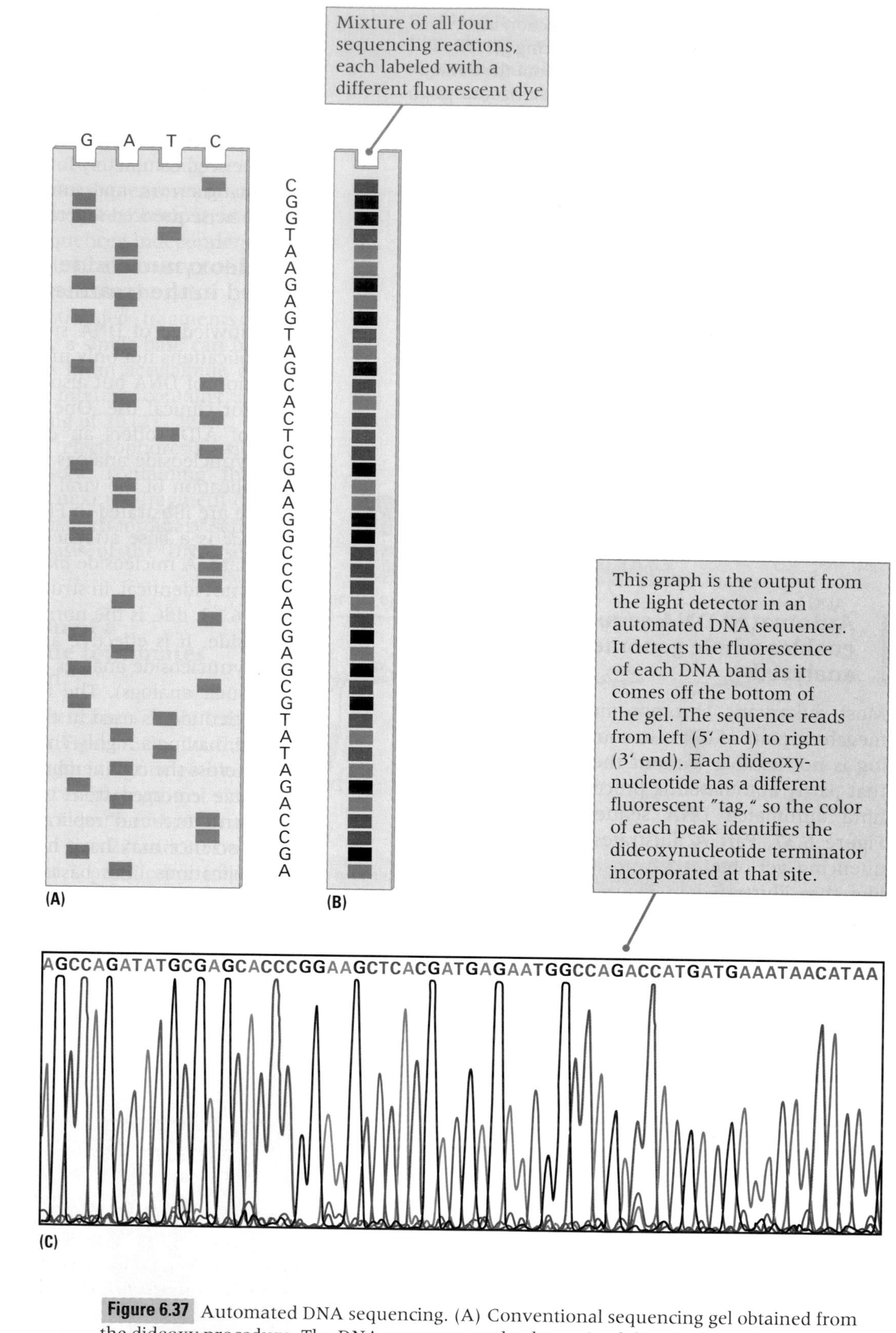

Figure 6.37 Automated DNA sequencing. (A) Conventional sequencing gel obtained from the dideoxy procedure. The DNA sequence can be determined directly from photographic film according to the positions of the bands. (B) Banding pattern obtained when each of the terminating nucleotides is labeled with a different fluorescent dye and the bands are separated in the same lane of the gel. (C) Trace of the fluorescence pattern obtained from the gel in part B by automated detection of the fluorescence of each band as it comes off the bottom of the gel during continued electrophoresis.

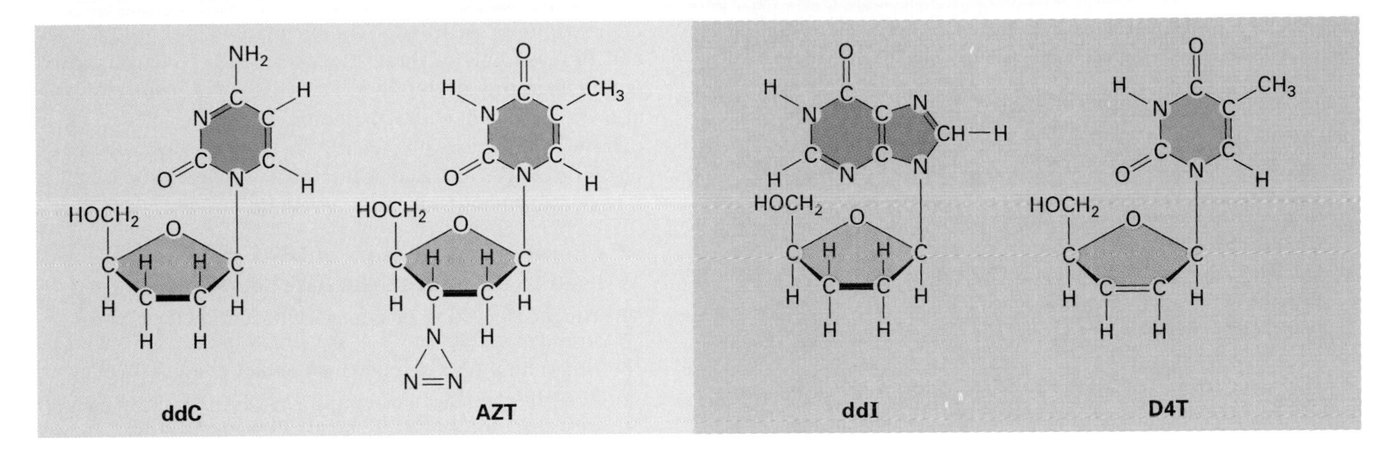

Figure 6.38 A few of the drugs that have been found to be effective in the treatment of AIDS by interfering with the replication of HIV virus. The technical names of the substances are as follows: ddC is 2',3'-dideoxycytidine; AZT is 3'-azido-2',3'-dideoxythymidine; D4T is 2',3'-didehydro-2',3'-dideoxythymidine; and ddI is 2',3'-dideoxyinosine.

Chapter Summary

- **Complex organisms generally have large genomes.**

DNA content varies widely among species of organisms. Small viruses exist whose DNA contains only a few thousand nucleotides, and among higher animals and plants, the DNA content can be as large as 1.5×10^{11} nucleotides. Generally speaking, genome size increases with the complexity of the organism, but there are many exceptions, such as closely related species whose genomes differ as much as tenfold in DNA content.

- **DNA is a linear polymer of four deoxyribonucleotides.**
- **Duplex DNA forms a double helix held together by hydrogen bonds.**

DNA is a double-stranded polymer consisting of deoxyribonucleotides. A nucleotide has three components: a base, a sugar (deoxyribose in DNA, ribose in RNA), and a phosphate. Sugars and phosphates alternate in forming a single polynucleotide chain with one terminal 3'-OH group and one terminal 5'-P group. In double-stranded (duplex) DNA, the two strands are antiparallel: each end of the double helix carries a terminal 3'-OH group in one strand and a terminal 5'-P group in the other strand. Four bases are found in DNA: adenine (A) and guanine (G), which are purines, and cytosine (C) and thymine (T), which are pyrimidines. Equal numbers of purines and pyrimidines are found in double-stranded DNA, because the bases are paired as A—T pairs and G—C pairs. This pairing holds the two polynucleotide strands together in a double helix. The base composition of DNA varies from one organism to the next. The information content of a DNA molecule resides in the sequence of bases along the chain, and each gene consists of a unique sequence.

- **Replication uses each DNA strand as a template for a new one.**

Nucleotides are added one at a time to the growing end of a DNA strand.

DNA replication is semiconservative: The parental strands remain intact.

DNA strands must unwind to be replicated.

Eukaryotic DNA molecules contain multiple origins of replication.

- **DNA polymerase makes the new DNA strands.**

One strand of replicating DNA is synthesized in pieces.

DNA is synthesized only in the 5' → 3' direction.

Each new DNA strand or fragment is initiated by a short RNA primer.

Precursor fragments are joined together when they meet.

Many proteins participate in DNA replication.

The double helix replicates by using enzymes called DNA polymerases, but many other proteins also are needed. Replication is semiconservative in that each parental single strand, called a template strand, is found in one of the double-stranded progeny molecules. Semiconservative replication was first demonstrated in the Meselson-Stahl experiment, which used equilibrium density-gradient centrifugation to separate DNA molecules containing two ^{15}N-labeled strands, two ^{14}N-labeled strands, or one of each. Replication proceeds by a DNA polymerase (1) bringing in a nucleoside triphosphate with a base capable of hydrogen-bonding with the corresponding base in the template strand and (2) joining the 5'-P group of the nucleotide to the free 3'-OH group of the growing strand. (The terminal P—P from the nucleoside triphosphate is cleaved off and released.) Because double-stranded DNA is antiparallel, only one strand (the leading strand) grows in the direction of movement of the replication fork. The other strand (the lagging strand) is synthesized in the opposite direction as short fragments that are subsequently joined together. DNA polymerases cannot initiate synthesis, so a primer is always needed. The primer is an RNA fragment made by an RNA polymerase enzyme; the RNA primer is removed at later stages of replication. DNA molecules of prokaryotes usually have a single replication origin; eukaryotic DNA molecules usually have many origins.

- **Knowledge of DNA structure makes possible the manipulation of DNA molecules.**

Single strands of DNA or RNA with complementary sequences can hybridize.
Restriction enzymes cleave duplex DNA at particular nucleotide sequences.
Gel electrophoresis separates DNA fragments by size.
Specific DNA fragments are identified by hybridization with a probe.

- **The polymerase chain reaction makes possible the amplification of a particular DNA fragment.**

Restriction enzymes cleave DNA molecules at the positions of specific sequences (restriction sites) of usually four or six nucleotides. Each restriction enzyme produces a unique set of fragments for any particular DNA molecule. These fragments can be separated by electrophoresis and used for purposes such as DNA sequencing. The positions of particular restriction fragments in a gel can be visualized by means of a Southern blot, in which radioactive probe DNA is mixed with denatured DNA made up of single-stranded restriction fragments that have been transferred to a filter membrane after electrophoresis. The probe DNA will form stable duplexes (anneal or renature) with whatever fragments contain sufficiently complementary base sequences, and the positions of these duplexes can be determined by autoradiography of the filter. Particular DNA sequences can also be amplified without cloning by means of the polymerase chain reaction (PCR), in which short synthetic oligonucleotides are used as primers to replicate and amplify, repeatedly, the sequence between them.

- **Chemical terminators of DNA synthesis are used to determine the base sequence.**

The incorporation of a dideoxynucleotide terminates strand elongation.
Automated DNA sequencing enables whole genomes to be analyzed.
Dideoxynucleoside analogs are also used in the treatment of diseases.

The base sequence of a DNA molecule can be determined by dideoxynucleotide sequencing. In this method, the DNA is isolated in discrete fragments containing several hundred nucleotide pairs. Complementary strands of each fragment are sequenced, and the sequences of overlapping fragments are combined to yield the complete sequence. The dideoxy sequencing method uses dideoxynucleotides to terminate daughter-strand synthesis and reveal the identity of the base present in the daughter strand at the site of termination. Large-scale genomic sequencing is also underway in a number of organisms through the application of automated DNA sequencing machines.

Key Terms

antiparallel
daughter strand
dideoxyribose
dideoxy sequencing method
DNA cloning
DNA ligase
DNA polymerase
endonuclease
exonuclease
5' end
gel electrophoresis
genome
hydrogen bond
initiation
kilobase pairs (kb)
lagging strand
leading strand
megabase pairs (Mb)
major groove
minor groove
nick
nuclease
nucleic acid hybridization
nucleoside
nucleotide
parental strand
phosphodiester bond
polymerase chain reaction (PCR)
polynucleotide chain
precursor fragment
primosome
primer
probe
proofreading
purine
pyrimidine
replication fork
replication origin
restriction enzyme
restriction fragment
restriction map
restriction site
ribose
RNA polymerase
rolling-circle replication
semiconservative replication
Southern blot
template
θ-replication
3' end
topoisomerase

Concepts and Issues
Testing Your Knowledge

- What are the four bases commonly found in DNA? Which form base pairs?
- What is the relationship between the amount of DNA in a somatic cell and the amount in a gamete?
- What chemical feature at the 3' end of a DNA strand in the process of being synthesized is essential for elongation? Can the strand also be elongated at the 5' end?
- What does it mean to say that the two strands in duplex DNA are antiparallel?
- If the paired strands in duplex DNA were parallel rather than antiparallel, would replication still involve a leading strand and a lagging strand? Explain.
- Why is the polymerase chain reaction so extremely specific in amplifying a single region of DNA? Why is the technique so extremely powerful in multiplying the sequence?
- What feature of DNA replication guarantees that the incorporation of a dideoxynucleotide will terminate strand elongation?

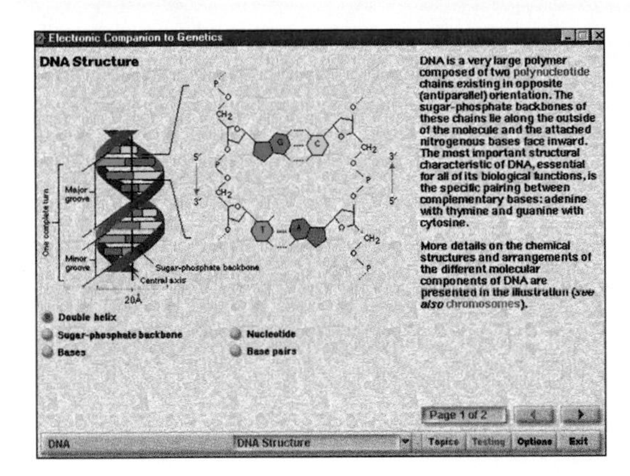

TO ENHANCE YOUR STUDY, turn to *An Electronic Companion to Genetics*™, © 1998, Cogito Learning Media, Inc. This CD-ROM is a multimedia tool that provides interactive explanations of important genetic concepts through animations, diagrams, and videos. It also provides interactive test questions.

Solutions Step by Step

Problem 1

A single-stranded DNA molecule is replicated *in vitro* by extending a primer oligonucleotide in the presence of nucleoside triphosphates whose two outermost phosphate groups (called the γ and β phosphates) are labeled with radioactive ^{32}P but whose innermost phosphate (the α phosphate) is not labeled. Is the resulting double-stranded DNA labeled or unlabeled? Explain.

Solution Recall that the addition of each new nucleotide in strand elongation requires a nucleoside triphosphate, but in the reaction itself the covalent bond connecting the α and β phosphates is cleaved, releasing the still connected β and γ phosphates into the medium. This means that none of the ^{32}P from each labeled triphosphate will be incorporated into the phosphodiester bond of the growing DNA strand, so the daughter duplex will be unlabeled.

Problem 2

The polymerase chain reaction is used to amplify a region of human DNA of length 3 kb from a DNA solution prepared from nuclei of human cells. The human genome has a size of 3 Gb (gigabases), or 3×10^9 base pairs, per haploid genome.

(a) Prior to amplification, what proportion of the DNA in the solution consists of the 3-kb target sequence? Assume that the target sequence is present in one copy per haploid genome.
(b) Each round of amplification doubles the number of target molecules. How many rounds of replication would be required to reach a stage in which the amplified sequence constitutes more than 99.9 percent of all the DNA in the solution?

Solution (a) The original DNA solution contains one 3-kb target sequence per 3-Gb haploid genome. The proportion of DNA consisting of the target sequence is therefore

$$\frac{3 \times 10^3}{3 \times 10^9} = 1 \times 10^{-6} = 0.0001 \text{ percent}$$

(b) Because each round of amplification doubles the number of target molecules, after n rounds of replication there will be 2^n target molecules for each haploid human genome present in the original solution. Each of these has a length of 3000 bp, so the total amount of amplified target DNA will be $2^n \times 3000$ bp. This DNA is newly created and therefore increases the total amount of DNA in the solution. After n rounds of replication, the amount of DNA present per haploid genome is $2^n \times 3000$ bp (the newly created material) + 3×10^9 bp (the original material). The question asks the value of n for which the fraction of newly created DNA constitutes 99.9 percent of the total DNA in solution. The inequality to be solved is

$$\frac{2^n \times 3000 \text{ bp}}{2^n \times 3000 \text{ bp} + 3 \times 10^9 \text{ bp}} \geq 0.999$$

from which we obtain

$$n \geq \frac{1}{\log(2)} \times \log\left[\frac{(3 \times 10^9)(0.999)}{(3000)(1 - 0.999)}\right] = 29.9$$

This means that 30 rounds of amplification increase the percentage of target DNA in the solution by a factor of almost 10^6.

Problem 3

A solution containing single-stranded DNA with the sequence

5'-ATGGTGCACCTGACTCCTGAGGAGAAGTCTNNNNNNNN-3'

undergoes DNA replication *in vitro* in the presence of all four nucleoside triphosphates plus an amount of dideoxyadenosine triphosphate sufficient to compete for incorporation with deoxyadenosine triphosphate. The run of N's represents the nucleotides that bind with the oligonucleotide primer. What DNA fragments are expected?

Solution Replication will proceed normally for all A, G, and C nucleotides in the template strand, but it will terminate at a T wherever a dideoxyadenosine was incorporated instead of deoxyadenosine. The resulting fragments will be as shown, where XXXXXXXX represents the nucleotides in the oligonucleotide primer.

5'-XXXXXXXXA-3'
5'-XXXXXXXXAGA-3'
5'-XXXXXXXXAGACTTCTCCTCA-3'
5'-XXXXXXXXAGACTTCTCCTCAGGA-3'
5'-XXXXXXXXAGACTTCTCCTCAGGAGTCA-3'
5'-XXXXXXXXAGACTTCTCCTCAGGAGTCATTCA-3'
5'-XXXXXXXXAGACTTCTCCTCAGGAGTCATTCACA-3'
5'-XXXXXXXXAGACTTCTCCTCAGGAGTCATTCACACCA-3'
5'-XXXXXXXXAGACTTCTCCTCAGGAGTCATTCACACCAT-3'

geNETics on the web

www.jbpub.com/genetics

These GeNETics on the Web exercises will introduce some of the most useful sites for finding genetic information on the World Wide Web. Genetic sites provide access to a rich storehouse of information on all aspects of genetics. These range from sites written in nontechnical language for the lay person to sites with sophisticated databases designed for the professional geneticist. In carrying out these exercises, you will get a taste of what the Internet can offer a student in genetics.

The keywords shown in color in the following exercises are available on the Jones and Bartlett Publishers' web site as hyperlinks to various genetic sites. To complete the exercises, visit GeNETics on the Web home page at

http://www.jbpub.com/genetics

Select the link to Essential geNETics on the Web. Then choose a chapter. You will find a list of keywords that correspond to the exercises that follow. Select a keyword to link to a web site containing the genetic information necessary to complete the exercise. Each exercise includes a specific assignment that makes use of the information available at the site.

Exercises

- Many people think that birth defects are rare and happen only to other people. But birth defects affect more than 150,000 newborns each year and are the leading cause of infant death and disability. Some 3000 to 5000 different birth defects have been described. The most common are listed at this keyword site. The list includes only those conditions that can be diagnosed immediately at birth; it does not include such conditions as cystic fibrosis, Tay-Sachs disease, and sickle-cell anemia, which become apparent in the first weeks or months after birth. Considering only the structural and metabolic birth defects whose estimated incidence is tabulated, calculate the overall incidence of the conditions taken together. Then rewrite the list of the conditions in order of their incidence, from highest to lowest.
- The concept of the polymerase chain reaction (PCR) occurred to Kary Mullis one night while he was traveling on Route 128 from San Francisco to Mendocino. He immediately realized that this approach would be unique in its ability to amplify, at an exponential rate, a specific nucleotide sequence present in a vanishingly small quantity amid a much larger background of total nucleic acid. Once its feasibility was demonstrated, PCR was quickly recognized as a major technical advance in molecular biology. The new technique earned Mullis the 1993 Nobel Prize in chemistry, and today it is the basis for a large number of experimental and diagnostic procedures. At this keyword site you can learn more about the development of the PCR from Mullis's original conception. Two major innovations were necessary to perfect the process. Write a short report explaining what these innovations were and why they were important.

Pic Site

The Pic Site showcases some of the most visually appealing genetics sites on the World Wide Web. To visit the showcase genetics site, select the Pic Site for Chapter 6.

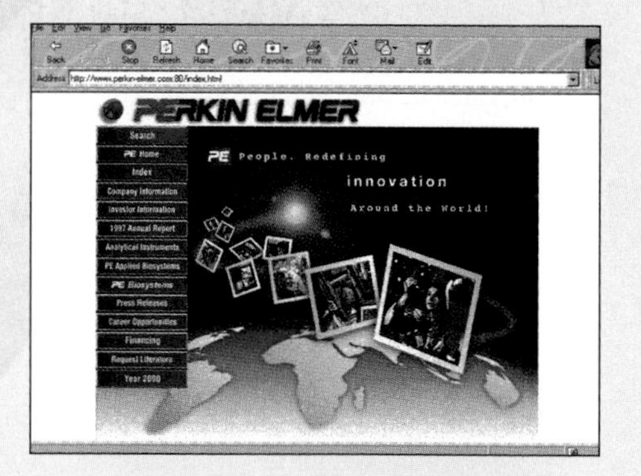

Concepts in Action
Problems for Solution

6.1 State four requirements for the initiation of DNA synthesis. To what chemical group in a DNA chain is an incoming nucleotide added? What chemical group in the incoming nucleotide reacts with the DNA terminus?

6.2 Consider a culture of *E. coli* cells grown for many generations in a ^{15}N-containing medium. The cells are washed and transferred to a ^{14}N-containing medium. After exactly two chromosome replications in the second medium, the DNA is extracted without any breakage whatsoever. What density classes would result and in what proportions?

6.3 What is meant by the statement that the DNA replication fork is asymmetrical?

6.4 The haploid genome of the wall cress, *Arabidopsis thaliana*, contains 100,000 kb and has 10 chromosomes. If a particular chromosome contains 10 percent of the DNA in the haploid genome, what is the approximate length of its DNA molecule in micrometers? (There are 10^4 micrometers per angstrom unit.)

6.5 For the chromosome of *A. thaliana* in the previous problem, estimate the time of replication, assuming that there is only one origin of replication (exactly in the middle), that replication is bidirectional, and that the rate of DNA synthesis is:

(a) 1500 nucleotide pairs per second (typical of bacterial cells).

(b) 50 nucleotide pairs per second (typical of eukaryotic cells).

6.6 The double-stranded DNA molecule of a newly discovered virus was found by electron microscopy to have a length of 68 micrometers (68×10^4 Å).

(a) How many nucleotide pairs are present in one of these molecules?
(b) How many complete turns of the two polynucleotide chains are present in such a double helix?

6.7 Specify the function or enzymatic activity of the following proteins that participate in DNA replication: primase, topoisomerase, DNA ligase, polymerase I (Pol I), and polymerase III (Pol III).

6.8 In early studies of the properties of the precursor fragments in DNA synthesis, it was shown that these fragments could hybridize to both strands of *E. coli* DNA. This was taken as evidence that they are synthesized on both branches of the replication fork. However, one feature of the replication of *E. coli* DNA that was not known at the time invalidates this conclusion. What is this characteristic?

6.9 A friend brings you three samples of nucleic acid and asks you to determine each sample's chemical identity (whether DNA or RNA) and whether the molecules are double-stranded or single-stranded. You use powerful nucleases to completely degrade each sample to its constituent nucleotides and then determine the approximate relative proportions of nucleotides. The results of your assay follow. What can you tell your friend about the nature of these samples?

Sample 1:	dGTP 13%	dCTP 14%	dATP 36%	dTTP 37%
Sample 2:	dGTP 12%	dCTP 36%	dATP 47%	dTTP 5%
Sample 3:	GTP 22%	CTP 47%	ATP 17%	UTP 14%

6.10 A new technique is used for determining the base composition of double-stranded DNA. Rather than giving the relative amounts of each of the four bases, A, T, G, and C, the procedure yields the value of the ratio of A to C. If this ratio is 1/3, what are the relative amounts of the four bases?

6.11 A 3.1-kilobase linear fragment of DNA was digested with *Pst*I and produced a 2.0-kb fragment and a 1.1-kb fragment. When the same 3.1-kb fragment was cut with *Hin*dIII, it yielded a 1.5-kb fragment, a 1.3-kb fragment, and a 0.3-kb fragment. When the 3.1-kb molecule was cut with a mixture of the two enzymes, fragments of 1.5, 0.8, 0.5, and 0.3 kb resulted. Draw a map of the original 3.1-kb fragment, and label the restriction sites and the distances between these sites.

6.12 The *dusky* mutation is an X-linked recessive in *Drosophila* that causes small, dark wings. In a stock of wildtype flies, you find a single male that has the dusky phenotype. In wildtype flies, the *dusky* gene is contained within an 8-kb *Xho*I restriction fragment. When you digest genomic DNA from the mutant male with *Xho*I and probe with the 8-kb restriction fragment on a Southern blot, you find that the size of the labeled fragment is 10 kb. You clone the 10-kb fragment and use it as a probe for a polytene chromosome *in situ* hybridization in a number of different wildtype strains, and you notice that this fragment hybridizes to multiple locations along the polytene chromosomes. Each wildtype strain has a different pattern of hybridization. What do these data suggest about the origin of the *dusky* mutation that you isolated?

Further Readings

Bauer, W. R., F. H. C. Crick, and J. H. White. 1980. Supercoiled DNA. *Scientific American*, July.

Cairns, J. 1966. The bacterial chromosome. *Scientific American*, January.

Danna, K., and D. Nathans. 1971. Specific cleavage of Simian Virus 40 DNA by restriction endonuclease of *Hemophilus influenzae*. *Proceedings of the National Academy of Sciences*, USA 68: 2913.

Davies, J. 1995. Vicious circles: Looking back on resistance plasmids. *Genetics* 139: 1465.

DePamphilis, M. L., ed. 1996. *DNA Replication in Eukaryotic Cells*. Cold Spring Harbor, NY: Cold Spring Harbor Press.

Donovan, S., and J. F. X. Diffley. 1996. Replication origins in eukaroytes. *Current Opinion in Genetics & Development* 6: 203.

Grimaldi, D. A. 1996. Captured in amber. *Scientific American*, April.

Grunstein, M. 1992. Histones as regulators of genes. *Scientific American*, October.

Hubscher, U., and J. M. Sogo. 1997. The eukaryotic DNA replication fork. *News in Physiological Sciences* 12: 125.

Kelley, T., ed. 1988. *Eukaryotic DNA Replication*. Cold Spring Harbor, NY: Cold Spring Harbor Laboratory Press.

Kornberg, A. 1995. *DNA Replication*. 2d ed. New York: Freeman.

Kornberg, R. D., and A. Klug. 1981. The nucleosome. *Scientific American*, February.

Mullis, K. B. 1990. The unusual origin of the polymerase chain reaction. *Scientific American*, April.

Neidhardt, F. C., R. Curtiss III, J. L. Ingraham, E. C. C. Lin, K. B. Low, B. Magasanik, W. S. Reznikoff, M. Riley, M. Schaechter, and H. E. Umbarger, eds. 1996. Escherichia coli *and* Salmonella typhimurium: *Cellular and Molecular Biology* (2 volumes). 2d ed. Washington, DC: American Society for Microbiology.

Singer, M., and P. Berg. 1991. *Genes & Genomes*. Mill Valley, CA: University Science Books.

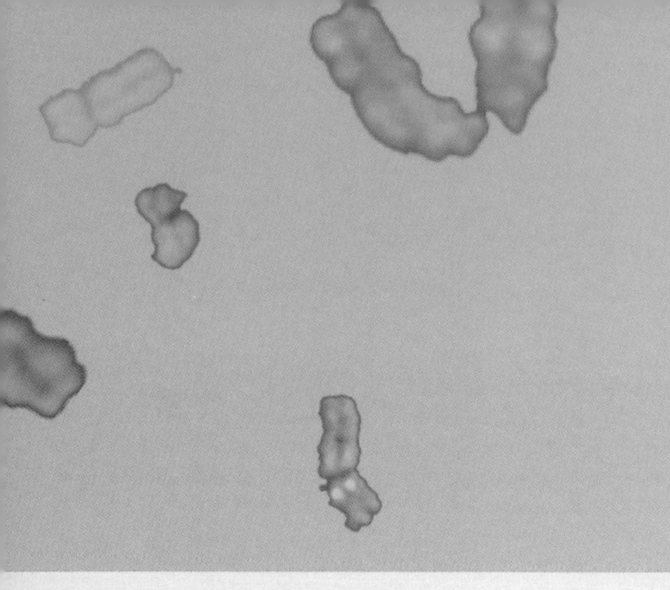

Key Concepts

- Substitution of one base for another is an important mechanism of spontaneous mutation. A single-base substitution in a coding region may result in an amino acid replacement; a single-base deletion or insertion results in a shifted reading frame.
- Transposable element insertion is also an important mechanism of spontaneous mutation.
- Mutations can be induced by various agents, including some classes of chemicals and various types of radiation.
- Most mutagens are also carcinogens.
- Cells contain enzymatic pathways for the repair of different types of damage to DNA. Among the most important repair systems is mismatch repair of duplex DNA, in which a nucleotide containing a mismatched base is excised and replaced with the correct nucleotide.

TEN-MONTH-OLD ONYA-BIRRI, the only albino koala in captivity, with his mother Banjeeri at the San Diego Zoo. In some cases, the albino mutation is a result of the deposition of pigment in the skin or hair; in other cases, the defect is in the ability to synthesize pigments. Only in the latter case are the eyes in pink, which is due to the red blood vessels in the retina at the back of the eyeball. [© AP Photo/San Diego Zoo]

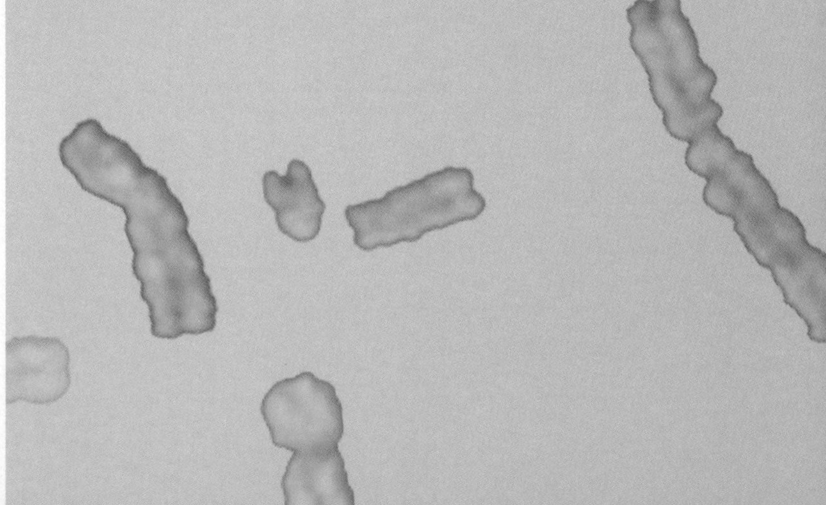

CHAPTER 7

Mutation and DNA Repair

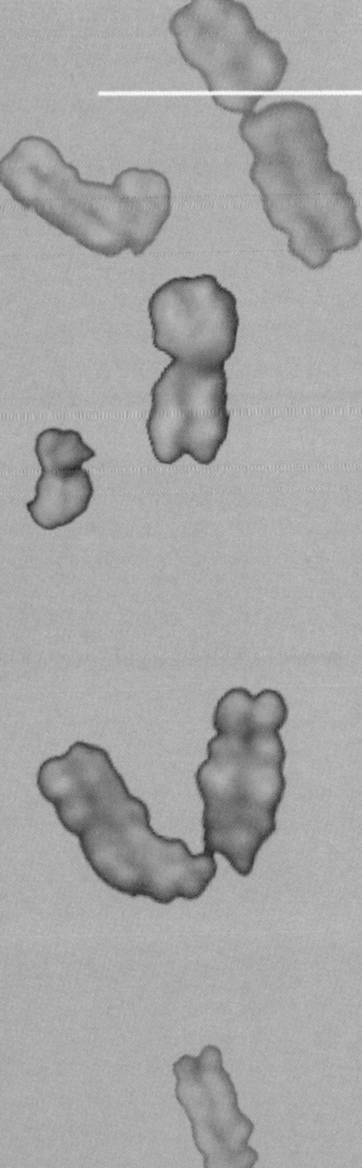

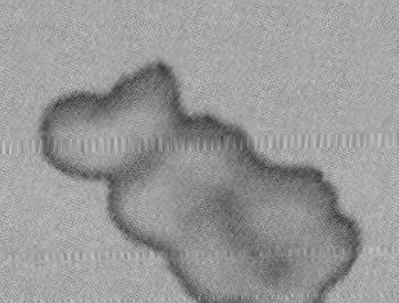

In the preceding chapters, numerous examples were presented in which the information contained in the genetic material had been altered by mutation. A **mutation** is any heritable change in the genetic material. In this chapter, we examine the nature of mutations at the molecular level. You will learn how mutations are created, how they are detected phenotypically, and the means by which many mutations are corrected by special DNA repair enzymes almost immediately after they occur. You will see that mutations can be induced by radiation and a variety of chemical agents that produce strand breakage and other types of damage to DNA.

7.1 Mutations can be classified by their phenotypic effects.

A mutation can happen in any cell at any time. The phenotypic effect may range from a minor alteration that is detectable only by biochemical methods, such as DNA sequencing, to a drastic change in an essential process that causes severe abnormality or death. When genetic phenomena are studied in the laboratory, mutations that produce clearly defined effects are generally used, but most mutations are not of this type. The effect of a mutation is determined by the type of cell containing the mutant allele, by the stage in the life cycle or development of the organism that the mutation affects, and (in diploid organisms) by the dominance or recessiveness of the mutant allele. A recessive mutation is usually not detected until a later generation when two heterozygous genotypes mate. Dominance does not complicate the expression of mutations in bacteria and haploid eukaryotes.

Mutations can be classified in a variety of ways. In multicellular organisms, one distinction is based on the type of cell in which the mutation first occurs: Those that arise in cells that ultimately form gametes are **germ-line mutations,** and all others are **somatic mutations.** A somatic mutation yields an organism that is genotypically, and for many dominant mutations phenotypically, a mixture of normal and mutant tissue. Because reproductive cells are not affected, such a mutant allele is not transmitted to the progeny and may not be detected or be recoverable for genetic analysis. However, in higher plants, somatic mutations can often be propagated by vegetative means (means that do not involve seed production), such as grafting or the rooting of stem cuttings. This process has been the source of valuable new varieties such as the "Delicious" apple and the "Washington" navel orange.

Among the mutations that are most useful for genetic analysis are those whose effects can be turned on or off at will. These are called **conditional mutations** because they produce changes in phenotype in one set of environmental conditions (called the **restrictive conditions**) but not in another (called the **permissive conditions**). A **temperature-sensitive mutation,** for example, is a conditional mutation whose expression depends on temperature. Usually the restrictive temperature is high, and the organism exhibits a mutant phenotype above this critical temperature. The permissive temperature is lower, and under permissive conditions the phenotype is wildtype or nearly wildtype. Temperature-sensitive mutations are frequently used to block particular steps in biochemical pathways in order to test the role of the pathways in various cellular processes, such as DNA replication.

An example of temperature sensitivity is found in the ordinary Siamese cat, with its black-tipped paws, ears, and tail (Figure 7.1). In this breed of cat, the biochemical pathway leading to black pigmentation is temperature-sensitive and is inactivated at normal body temperature. Consequently, the pigment is not present in the hair over most of the body. The tips of the legs, ears, and tail are cooler than the rest of the body, so the pigment *is* deposited in the hair in these areas.

Mutations can also be classified by other criteria, such as the kinds of alterations in the DNA, the kinds of phenotypic effects produced, and whether the mutational events are **spontaneous** in origin or were **induced** by exposure to a known **mutagen** (a mutation-causing agent). Such classifications are often useful in discussing aspects of the

Figure 7.1 A Siamese cat showing the characteristic pattern of pigment deposition.

mutational process. *Spontaneous* usually means that the event that caused a mutation is unknown, and *spontaneous mutations* are those that take place in the absence of any known mutagenic agent. The properties of spontaneous and induced mutations are described in later sections.

7.2 Mutations alter DNA in any of a variety of ways.

All mutations result from changes in the nucleotide sequence of DNA or from deletions, insertions, or rearrangement of DNA sequences in the genome. Some types of major rearrangements in chromosomes were discussed in Chapter 5. In this section, we discuss mutations whose molecular basis can be specified.

A base substitution replaces one base with another.

The simplest type of mutation is a **base substitution,** in which a nucleotide pair in a DNA duplex is replaced with a different nucleotide pair. For example, in an A → G substitution, an A is replaced with a G in one of the DNA strands. This substitution temporarily creates a mismatched G–T base pair; at the very next replication, the mismatch is resolved as a proper G–C base pair in one daughter molecule and as a proper A–T base pair in the other daughter molecule. In this case, the G–C base pair is mutant and the A–T base pair is nonmutant. Similarly, in an A → T substitution, an A is replaced with a T in one strand, creating a temporary T–T mismatch, which is resolved by replication as T–A in one daughter molecule and A–T in the other. In this example, the T–A base pair is mutant and the A–T base pair is nonmutant. The T–A and the A–T are not equivalent, as may be seen by considering the nucleotide context. If the original unmutated DNA strand has the sequence 5'-GAC-3', for example, then the mutant strand has the sequence 5'-GTC-3' (which we have written as T–A), and the nonmutant strand has the sequence 5'-GAC-3' (which we have written as A–T).

Some base substitutions replace one pyrimidine base with the other or one purine base with the other. These are called **transition mutations.** The possible transition mutations are

T → C or C → T
(pyrimidine → pyrimidine)

A → G or G → A
(purine → purine)

Other base substitutions replace a pyrimidine with a purine or the other way around. These are called **transversion mutations.** The possible transversion mutations are

T → A T → G C → A or C → G
(pyrimidine → purine)

A → T A → C G → T or G → C
(purine → pyrimidine)

Altogether, there are four possible transitions and eight possible transversions. Therefore, if base substitutions were strictly random, one would expect a 1 : 2 ratio of transitions to transversions. However,

> Spontaneous base substitutions are biased in favor of transitions. Among spontaneous base substitutions, the ratio of transitions to transversions is approximately 2 : 1.

Examination of the genetic code (Table 1.1 on page 17) shows that the bias toward transitions has an important consequence for base substitutions in the third position of codons. In all codons with a pyrimidine in the third position, the particular pyrimidine present does not matter; likewise, in most codons ending in a purine, either purine will do. This means that most transition mutations in the third codon position do not change the amino acid that is encoded. Such mutations change the nucleotide sequence without changing the amino acid sequence; these are called **silent substitutions** because they are not detectable by changes in phenotype.

Mutational changes in nucleotides that are outside of coding regions can also be silent. In noncoding regions, such as in the DNA between genes, the precise nucleotide sequence is often not critical. These sequences can undergo base substitutions, small deletions or additions, insertions of transposable elements, and other rearrangements, and yet the mutations have no detectable effects on phenotype. On the other hand, some noncoding sequences do have essential functions—for example, the sequences that control the timing and rate of transcription. Mutations in these sequences often do have phenotypic effects.

Most base substitutions in coding regions do result in changed amino acids; these are called **missense mutations.** A change in the amino acid sequence of a protein may alter the biological properties of the protein. The classic example of a phenotypic effect of a single amino acid change is that responsible for the human hereditary disease **sickle-cell anemia.** The molecular basis of sickle-cell anemia is a mutant gene for β-globin, one component of the hemoglobin present in red blood cells. The sickle-cell mutation changes the sixth codon in the coding sequence from the normal

5'-GAG-3', which codes for glutamic acid, into the codon 5'-GUG-3', which codes for valine. In the DNA, the mutant has an A—T base pair (transcribed as the middle A in the codon) replaced with a T—A base pair (transcribed as the middle U in the mutant codon). One consequence of the seemingly simple Glu → Val replacement is that hemoglobin containing the defective β polypeptide chain has a tendency to form long, needle-like crystals. Red blood cells in which crystallization happens become deformed into crescent, sickle-like shapes. Some of the deformed red blood cells are destroyed immediately (reducing the oxygen-carrying capacity of the blood and causing the anemia), whereas others may clump together and clog the blood circulation in the capillaries. The impaired circulation affects the heart, lungs, brain, spleen, kidneys, bone marrow, muscles, and joints. Patients suffer bouts of severe pain. The anemia causes impaired growth, weakness, jaundice, and other symptoms. Affected people are so generally weakened that they are susceptible to bacterial infections, and infections are the most common cause of death in children with the disease.

Sickle-cell anemia is a severe genetic disease that often results in premature death. Yet it is a relatively common disease in areas of Africa and the Middle East in which malaria, caused by the protozoan parasite *Plasmodium falciparum,* is widespread. The association between sickle-cell anemia and malaria is not coincidental: It results from the ability of the mutant hemoglobin to afford some protection against malarial infection. In the life cycle of the parasite, it passes from a mosquito to a human being through the mosquito's bite. The initial stages of infection take place in cells in the liver where specialized forms of the parasite are produced that are able to infect and multiply in red blood cells. Widespread infection of red blood cells impairs the ability of the blood to carry oxygen, causing the weakness, anemia, and jaundice characteristic of malaria. In people with the mutant hemoglobin, however, the infected blood cells undergo sickling and are rapidly removed from circulation. The proliferation of the parasite among the red blood cells is thereby checked, and the severity of the malarial infection is reduced. There is consequently a genetic balancing act between the prevalence of the genetic disease sickle-cell anemia and the parasitic disease malaria. If the mutant hemoglobin becomes too frequent, more lives are lost from sickle-cell anemia than are gained by the protection it affords against malaria; on the other hand, if the mutant hemoglobin becomes too rare, fewer lives are lost from sickle-cell anemia but the gain is offset by more deaths from malaria. The end result of this kind of genetic balancing act is considered in quantitative terms in Chapter 13.

In contrast to the situation with sickle-cell anemia, an amino acid replacement does not always create a mutant phenotype. For instance, replacement of one amino acid for another with the same charge (say, lysine for arginine) may in some cases have no effect on either protein structure or phenotype. Whether the substitution of a similar amino acid for another produces an effect depends on the precise role of that particular amino acid in the structure and function of the protein. Any change in the active site of an enzyme usually decreases enzymatic activity.

Figure 7.2 illustrates the nine possible codons resulting from a single base substitution in the UAU codon for tyrosine. One mutation is silent (it is shown in a box), and six are missense mutations that change the amino acid inserted in the polypeptide at this position. The other two mutations create a stop codon resulting in premature termination of translation and production of a truncated polypeptide. A base substitution that creates a new stop codon is called a **nonsense mutation.** Because nonsense mutations cause premature chain termination, the remaining polypeptide fragment is almost always nonfunctional.

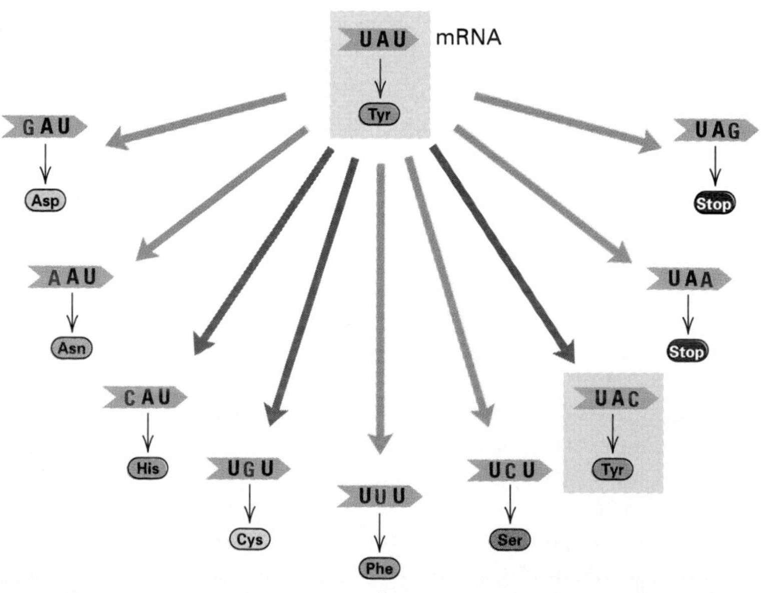

Figure 7.2 Nine codons that can result from a single-base change in the tyrosine codon UAU. Blue arrows indicate transversions, gray arrows transitions. Tyrosine codons are in boxes. Two possible stop ("nonsense") codons are shown in red. Altogether, the codon UAU allows for six possible missense mutations, two possible nonsense mutations, and one silent mutation.

Some mutations add bases to or delete bases from the DNA.

Tandem repeats of short nucleotide sequences are found at a very large number of locations throughout the genome of most higher eukaryotes. In the human genome, a particularly prevalent repeat is that of the dinucleotide TG. A repeat of this type is called a *simple tandem repeat length polymorphism,* or *STRP* (Chapter 4). In most organisms in which STRPs are found, the number of copies of the repeat often differs from one chromosome to the next. Hence populations are usually highly polymorphic for the number of repeating units. The high level of polymorphism makes these repeats very useful in such applications as linkage mapping (Chapter 4), DNA typing (Chapter 13), and family studies to localize genes that influence multifactorial traits (Chapter 14).

STRPs are usually polymorphic because they are susceptible to errors in replication or recombination that change the number of repeats or the length of the run. For example, a run of consecutive TG dinucleotides may have extra copies added or a few deleted. Any short sequence repeated in tandem a number of times may gain or lose a few copies as a result of these types of errors, although the rate of change in the number of repeats depends, in some unknown manner, on the sequence in question as well as on its location in the genome. The umbrella term generally used to describe the processes leading to a change in the number of copies of a short repeating unit is **replication slippage.** One specific process that can result in such additions or deletions is unequal crossing-over, which we discussed in Chapter 5.

An increase in the number of repeating units in an STRP is genetically equivalent to an insertion, and a decrease in the number of copies is equivalent to a deletion. Nonrepeating DNA sequences are also subject to insertion or deletion. The phenotypic consequences of insertion or deletion mutations depend on their location. In nonessential regions, no effects may be seen. STRPs are almost always present in noncoding regions. When insertions of deletions happen in regulatory or coding regions, however, their effects may be significant.

When they take place in coding regions, small insertions or deletions add or delete amino acids to or from the polypeptide, provided that the number of nucleotides added or deleted is an exact multiple of 3 (the length of a codon). Otherwise, the insertion or deletion shifts the phase in which the ribosome reads the triplet codons and, consequently, alters all of the amino acids downstream from the site of the mutation. Mutations that shift the reading frame of the codons in the mRNA are called **frameshift mutations.** A common type of frameshift mutation is a single-base addition or deletion. The consequences of a frameshift can be illustrated by the insertion of an adenine at the position of the arrow in the following mRNA sequence:

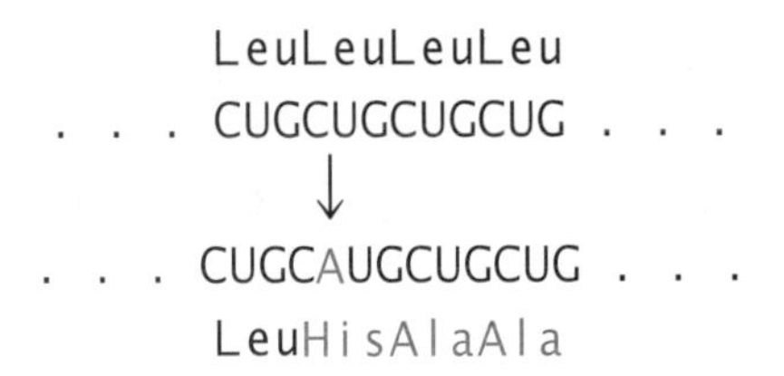

Because of the frameshift, all of the amino acids downstream from the insertion are different from the original. Any addition or deletion that is not a multiple of three nucleotides will produce a frameshift. Unless it is very near the carboxyl terminus of a protein, a frameshift mutation usually results in the synthesis of a nonfunctional protein.

Transposable elements cause insertion mutations.

All organisms contain multiple copies of several different kinds of transposable elements, which are DNA sequences capable of readily changing their positions in the genome. The structure and function of transposable elements were examined in Chapter 5. Transposable elements are important agents of mutation. For example, in some genes in *Drosophila,* approximately half of all spontaneous mutations that have visible phenotypic effects result from insertions of transposable elements.

Among the ways in which transposable elements can cause mutations are the mechanisms illustrated in Figure 7.3. Most transposable elements

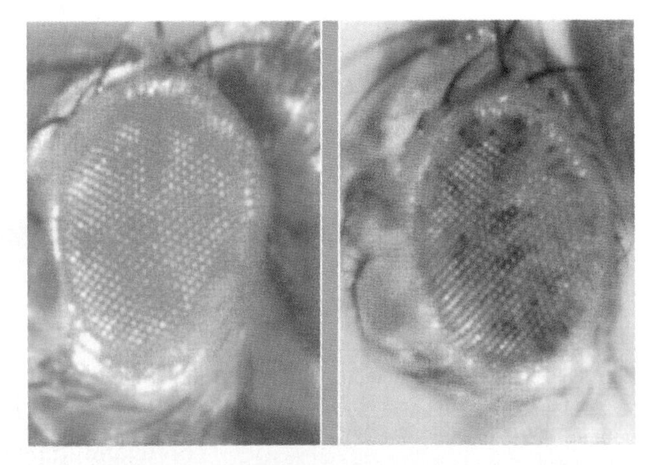

IN THESE *DROSOPHILA* EYES, the patches of red are caused by excision of a transposable element restoring normal eye color to an otherwise peach-colored eye. The eye with the uniform peach color shows the phenotype of the mutant in a genetic background in which transposition cannot occur.

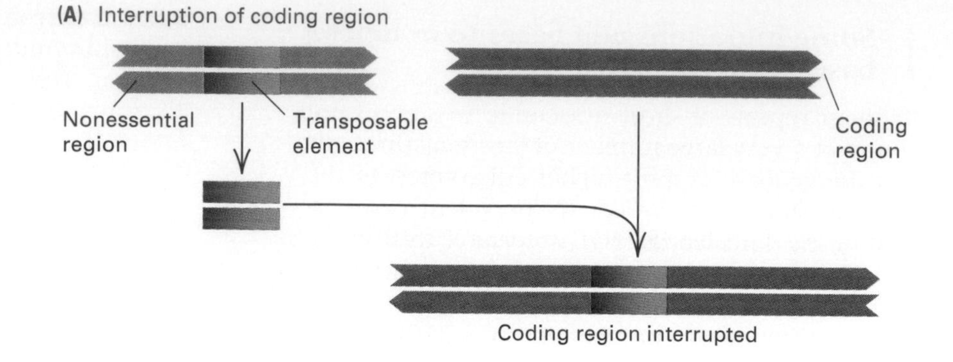

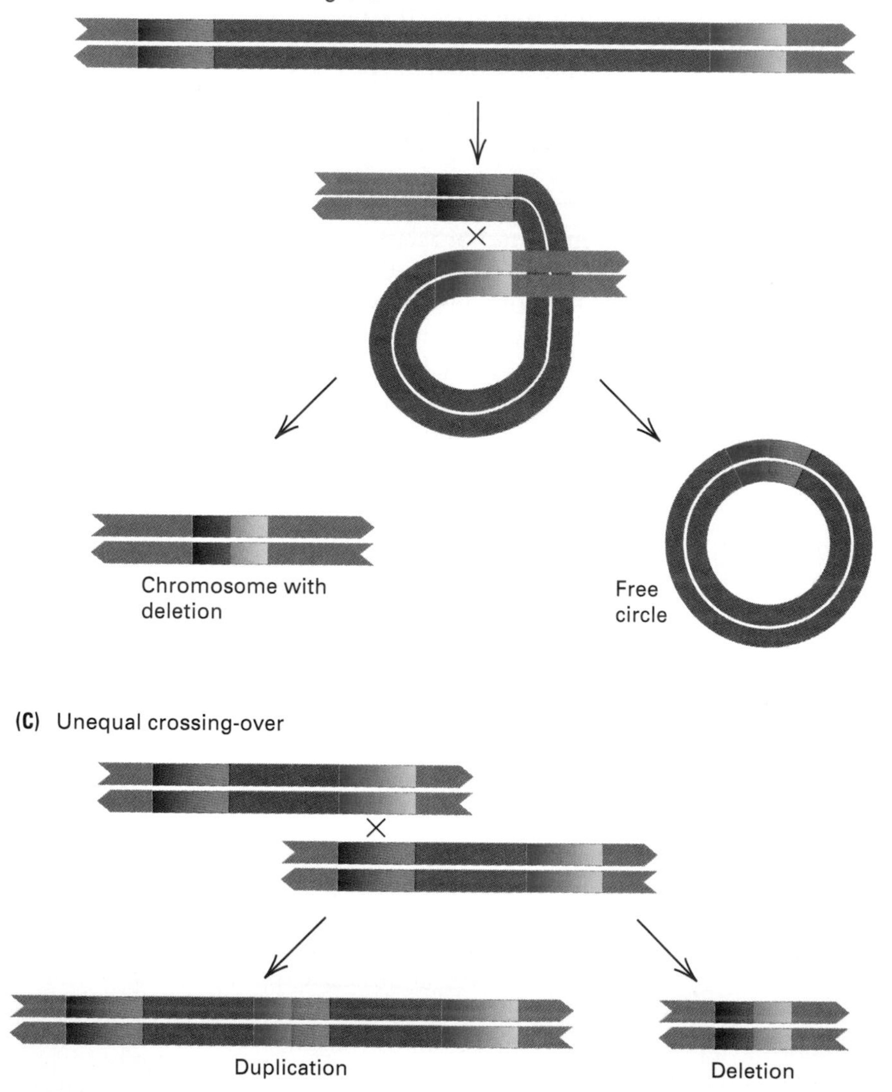

Figure 7.3 Some types of mutations resulting from transposable elements. (A) Insertion into the coding region of a gene eliminates gene function. (B) Crossing-over between two copies of a transposable element present in the same orientation in a single chromatid results in the formation of a single "hybrid" copy and the deletion of the DNA sequence between the copies. (C) Unequal crossing-over between homologous elements present in the same orientation in different chromatids results in products that contain either a duplication or a deletion of the DNA sequences between the elements. Crossing-over takes place at the four-strand stage of meiosis (Chapter 4), but in parts B and C, only the two strands participating in the crossover are shown.

are present in nonessential regions of the genome and usually do little or no harm. When an element transposes, however, it can insert into an essential region and disrupt that region's function. Figure 7.3A shows the result of transposition into a coding region of DNA. The insert interrupts the coding region. Because most transposable elements contain coding regions of their own, either transcription of the transposable element interferes with transcription of the gene into which it is inserted or transcription of the gene terminates within the transposable element. Even if transcription proceeds through the element, the phenotype will be mutant because the coding region then contains incorrect sequences.

Another mechanism of transposable-element mutagenesis results from recombination (Figure 7.3B and C). Transposable elements can be present in multiple copies, often with two or more in the same chromosome. If the copies are near enough together, they can pair during synapsis and undergo crossing-over, with the result that the region of DNA between the elements is deleted. Figure 7.3B shows what happens when elements located within the same chromosome undergo pairing and crossing-over. Alternatively, the downstream element in one chromosome can pair with the upstream element in the homologous chromosome, diagrammed in Figure 7.3C. An exchange within the paired elements results in one chromatid that is missing the region between the elements (bottom right) and one chromatid in which the region is duplicated (bottom left).

7.3 Spontaneous mutations have no assignable cause.

Mutations are statistically random events. There is no way of predicting when, or in which cell, a mutation will take place. However, every gene mutates spontaneously at a characteristic rate, so it is possible to assign probabilities to particular mutational events. In other words, there is a definite probability that a specified gene will mutate in a particular cell, and likewise there is a definite probability that a mutant allele of a specified gene will appear in a population of a designated size. The various kinds of mutational alterations in DNA differ substantially in complexity, so their probabilities of occurrence are quite different. The mutational process is also random in the sense that whether a particular mutation happens is unrelated to any adaptive advantage it may confer on the organism in its environment. A potentially favorable mutation does not arise *because* the organism has a need for it. The experimental basis for this conclusion is presented in the next section.

Mutations arise without reference to the adaptive needs of the organism.

The concept that mutations are spontaneous, statistically random events unrelated to adaptation was not widely accepted until the late 1940s. Before that time, it was believed that mutations occur in bacterial populations *in response to* particular selective conditions. The basis for this belief was the observation that when antibiotic-sensitive bacteria are spread on a solid growth medium containing the antibiotic, some colonies form that consist of cells having an inherited resistance to the drug. The initial interpretation of this observation (and similar ones) was that these adaptive variations were *induced* by the selective agent itself.

Several types of experiments showed that adaptive mutations take place spontaneously and hence were present at low frequency in the bacterial population even *before* it was exposed to the antibiotic. One experiment utilized a technique developed by Joshua and Esther Lederberg called **replica plating** (Figure 7.4). In this procedure, a suspension of bacterial cells is spread on a solid medium. After colonies have formed, a piece of sterile velvet mounted on a solid support is pressed onto the surface of the plate. Some bacteria from each colony stick to the fibers, as shown in Figure 7.4A. Then the velvet is pressed onto the surface of fresh medium, transferring some of the cells from each colony, which give rise to new colonies that have positions identical to those on the first plate. Figure 7.4B shows how this method was used to demonstrate the spontaneous origin of phage T1-r mutants. A master plate containing about 10^7 cells growing on nonselective medium (lacking phage) was replica-plated onto a series of plates that had been spread with about 10^9 T1 phages. After incubation for a time sufficient for colony formation, a few colonies of phage-resistant bacteria appeared in the same positions on each of the selective replica plates. This meant that the T1-r cells that formed the colonies must have been transferred from corresponding positions on the master plate. Because the colonies on the master plate had never been exposed to the phage, the mutations to resistance must have been present, by chance, in a few original cells not exposed to the phage.

The replica-plating experiment illustrates the following principle:

> Selective techniques merely select mutants that preexist in a population.

This principle is the basis for understanding how natural populations of rodents, insects, and disease-causing bacteria become resistant to the chemical substances used to control them. A familiar example is the high level of resistance to insecticides, such as DDT, that now exists in many insect populations, the result of selection for spontaneous mutations affecting behavioral, anatomical, and enzymatic traits that enable the insect to avoid or resist the chemical. Similar problems are encountered in controlling plant pathogens. For example, the introduction of a new variety of a crop plant resistant to a particular strain of disease-causing fungus results in only temporary protection against the disease. The resistance inevitably breaks down because of the occurrence of spontaneous mutations in the fungus that enable it to attack the new plant genotype. Such mutations have a clear selective advantage, and the mutant alleles rapidly become widespread in the fungal population.

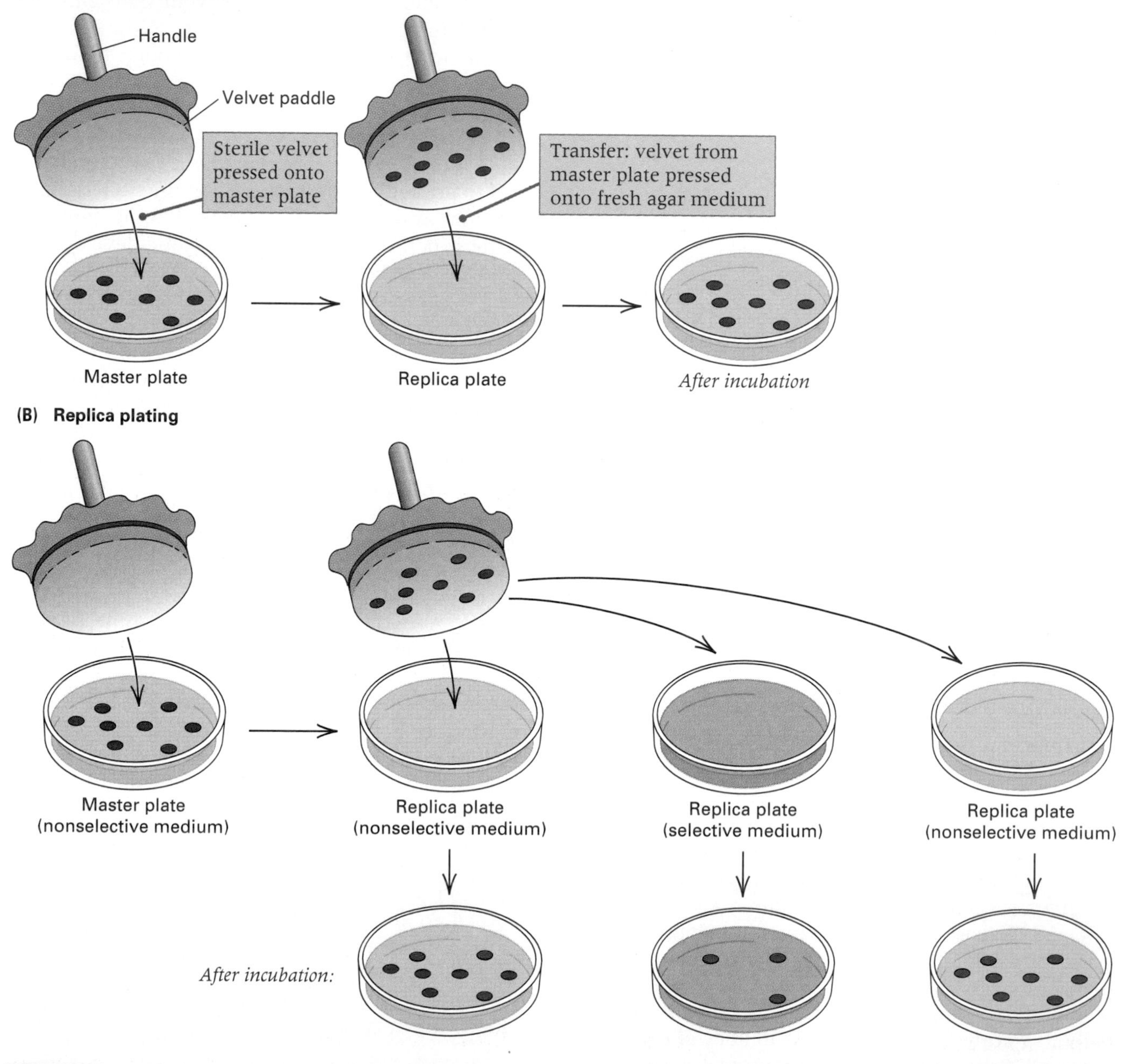

Figure 7.4 Replica plating. (A) In the transfer process, a velvet-covered disk is pressed onto the surface of a master plate in order to transfer cells from colonies on that plate to a second medium. (B) For the detection of mutants, cells are transferred onto two types of plates containing either a nonselective medium (on which all form colonies) or a selective medium (for example, one spread with T1 phages). Colonies form on the nonselective plate in the same pattern as on the master plate. Only mutant cells (for example, T1-r) can grow on the selective plate; the colonies that form correspond to certain positions on the master plate. Colonies consisting of mutant cells are shown in red.

Special methods are necessary to estimate the rate of mutation.

Spontaneous mutations tend to be rare, and the methods used to estimate the frequency with which they arise require large populations and often special techniques. The **mutation rate** is the probability that a gene undergoes mutation in a single generation or in forming a single gamete. Measurement of mutation rates is important in population genetics, in studies of evolution, and in analysis of the effect of environmental mutagens.

One of the earliest techniques for measuring mutation rates was the ***ClB* method.** *ClB* is a special X chromosome of *Drosophila melanogaster;* it has a large inversion (*C*) that prevents the recovery of crossover chromosomes in the progeny from a female heterozygous for the chromosome (for the reasons described in Section 5.6), a recessive lethal (*l*), and the dominant marker *Bar (B),* which reduces the normal round eye to a bar shape. The presence of a recessive lethal in the X chromosome means that males with that chromosome and females homozygous for it cannot survive. The technique is designed to detect mutations that arise in a normal X chromosome.

In the *ClB* procedure, females heterozygous for the *ClB* chromosome are mated with males carrying a normal X chromosome (Figure 7.5). From the F_1 progeny produced, females with the Bar phenotype are selected and then individually mated with normal males. (The presence of the Bar phenotype indicates that the females are heterozygous for the *ClB* chromosome and the normal X chromosome from the male parent.) The critical observation in determining the mutation rate is the fraction of males produced in the F_2 generation. Because the *ClB* males die, all of the males in this generation must contain an X chromosome derived from the X chromosome of the initial normal male (top row of illustration). Furthermore, it must be a nonrecombinant X chromosome because of the inversion (*C*) in the *ClB* chromosome. If the wildtype X chromosome present in the original male sustained a mutation that created a new recessive lethal, then all of the males in the F_2 generation die, as shown in the lower part of the illustration. On the other hand, if the wildtype X chromosome did not undergo mutation to a new recessive lethal, a ratio of two females to one male is expected among the F_2 progeny. Hence the proportion of matings in which no male progeny are detected in the F_2 generation is a measure of the frequency with which the X chromosome present in the original sperm underwent a mutation somewhere along its length to yield a new recessive lethal. This method provides a quantitative estimate of the rate at which mutation to a recessive lethal allele occurs *at any of the large number of genes* in the X chromosome. About 0.15 percent of the X chromosomes acquire new recessive lethals in spermatogenesis. Alternatively, we could say that the mutation rate is 1.5×10^{-3} recessive lethals per X chromosome per generation. Note that the *ClB* method tells us nothing about the mutation rate for a particular gene, because the method does not reveal how many genes on the X chromosome would cause lethality if they were mutant.

Since the time that the *ClB* method was devised, a variety of other methods have been developed for determining mutation rates in *Drosophila* and other organisms. Of significance is the fact that mutation rates vary widely from one gene to another. For example, the yellow-body mutation in *Drosophila* occurs at a frequency of 10^{-4} per gamete per generation, whereas mutations to streptomycin resistance in *E. coli* occur at a frequency of 10^{-9} per cell per generation. Furthermore, the frequency can differ enormously within a single organism, ranging in *E. coli* from 10^{-5} for some genes to 10^{-9} for others.

Mutations are nonrandom with respect to position in a gene or genome.

Certain DNA sequences are called mutational **hotspots** because they are more likely to undergo mutation than others. Mutational hotspots include monotonous runs of a single nucleotide and tandem repeats of short sequences (STRPs), which may gain or lose copies by any of a variety of mechanisms. Hotspots are found at many sites throughout the genome and within genes. For genetic studies of mutation, the existence of hotspots means that a relatively small number of sites account for a disproportionately large fraction of all mutations.

Sites of cytosine methylation are usually highly mutable, and the mutations are usually $G{-}C \rightarrow A{-}T$ transitions. In many organisms, including bacteria, maize, and mammals (but not *Drosophila*), about 1 percent of the cytosine bases are methylated at the carbon-5 position, yielding 5-methylcytosine instead of ordinary cytosine (Figure 7.6). A special enzyme adds the methyl group to the cytosine base in certain target sequences of DNA. In DNA replication, the 5-methylcytosine pairs with guanine and replicates normally. The genetic function of cytosine methylation is not entirely clear, but regions of DNA high in cytosine methylation tend to have reduced gene activity. Examples include the inactive X chromosome in female mammals (Chapter 5) and inactive copies of certain transposable elements in maize.

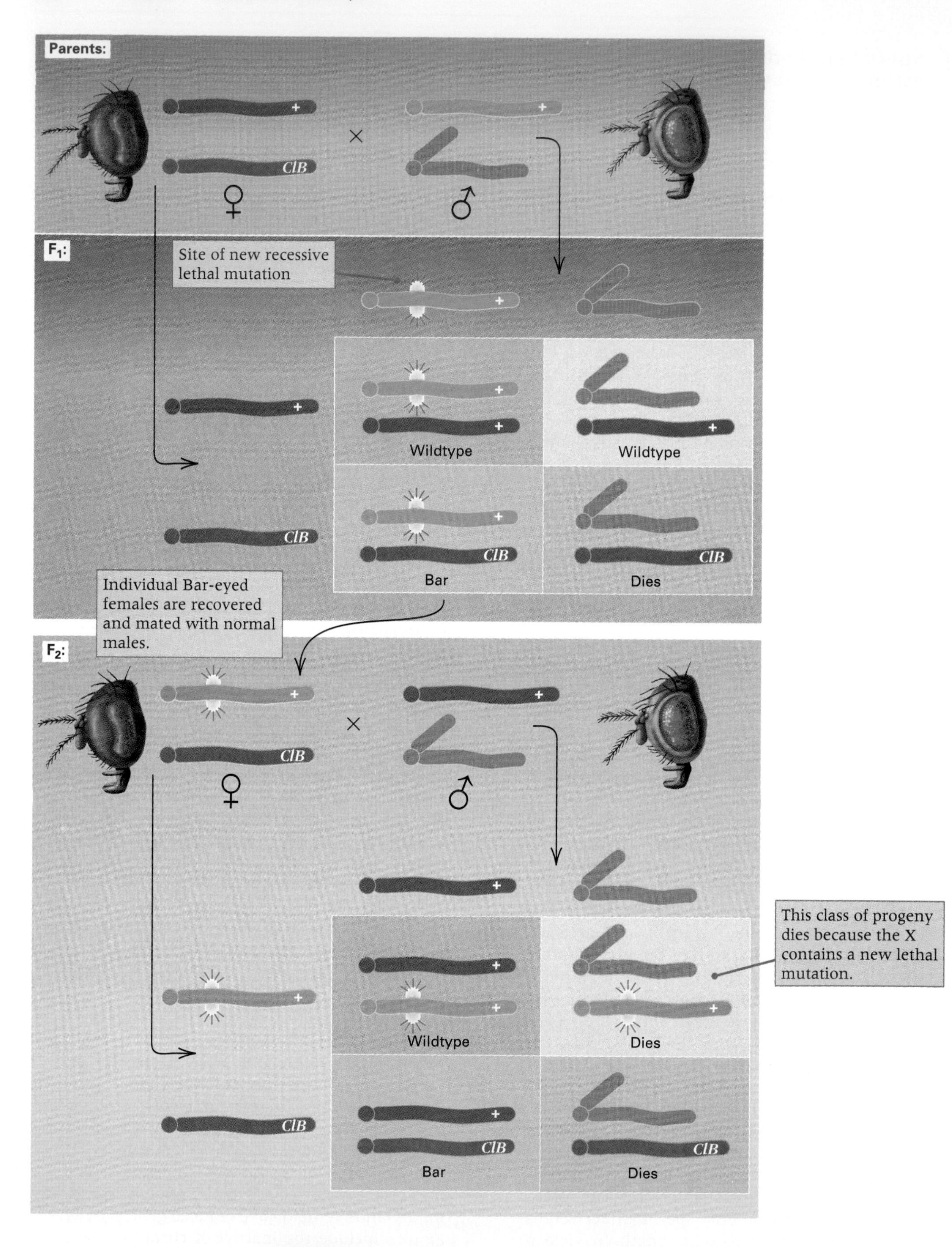

Figure 7.5 The *ClB* method for estimating the rate at which spontaneous recessive lethal mutations arise on the *Drosophila* X chromosome.

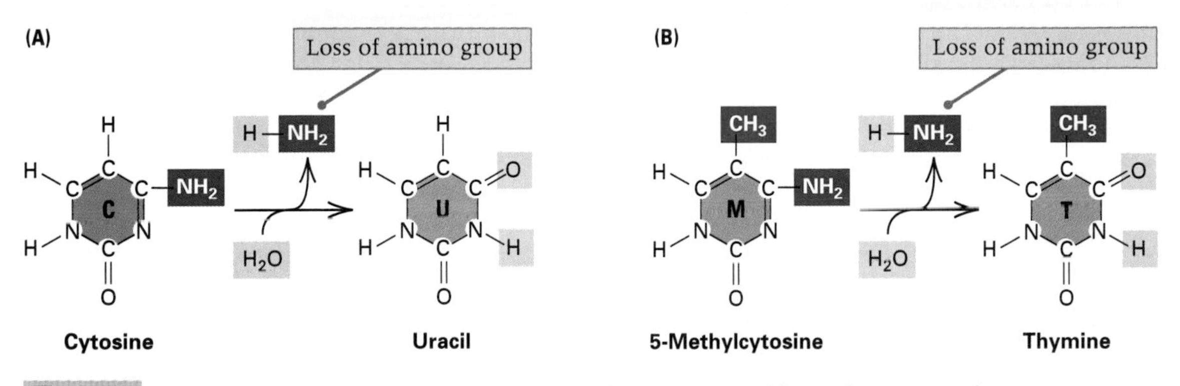

Figure 7.6 (A) Spontaneous loss of the amino group of cytosine to yield uracil. (B) Loss of the amino group from 5-methylcytosine yields thymine.

Cytosine methylation is an important contributor to mutational hotspots, as illustrated in Figure 7.6. Both cytosine and 5-methylcytosine are subject to occasional loss of an amino group. For cytosine, this loss yields uracil. Because uracil pairs with adenine instead of guanine, replication of a molecule containing a G–U base pair would ultimately lead to substitution of an A–T pair for the original G–C pair (by the process G–U → A–U → A–T in successive rounds of replication). However, cells possess a special repair enzyme that specifically removes uracil from DNA. The repair process is shown in Figure 7.7. Part A shows deamination of cytosine leading to the presence of a uracil-contain

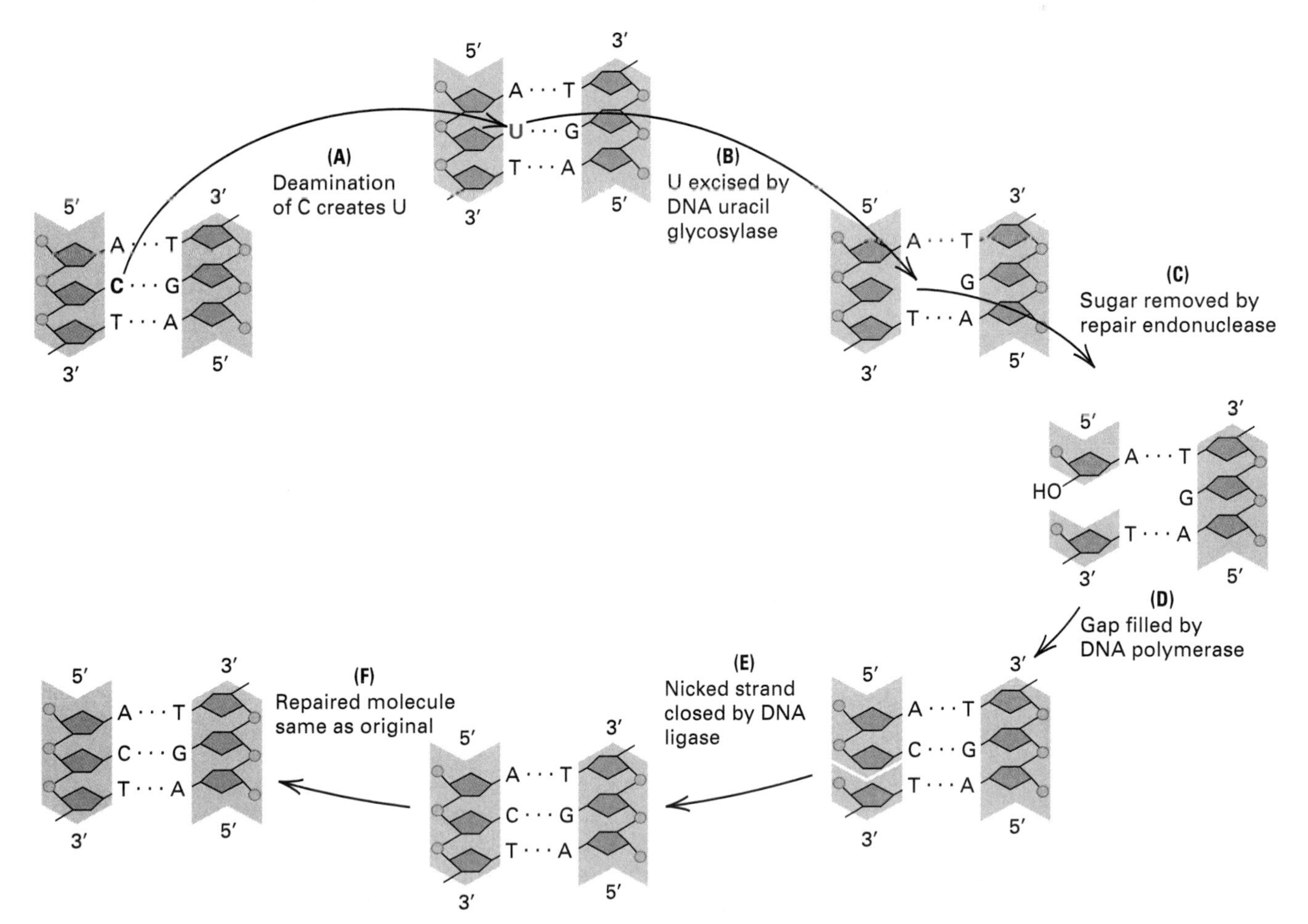

Figure 7.7 Mechanism by which uracil-containing nucleotides are formed in DNA and removed. (A) Deamination of cytosine produces uracil. (B) Uracil is cleaved from the deoxyribose sugar by DNA uracil glycosylase. (C) The deoxyribose sugar that does not contain a base is removed from the DNA backbone by another enzyme. (D) The gap in the DNA strand is filled by DNA polymerase, using the ungapped strand as template. (E) The nick remaining at the end of the gap-filling process is closed by DNA ligase. (F) The repair process restores the original sequence.

ing base. Repair is initiated by an enzyme called **DNA uracil glycosylase,** which recognizes the incorrect G–U base pair and cleaves the offending uracil from the deoxyribose sugar to which it is attached (part B). It is known how this enzyme works. It scans along duplex DNA until a uracil nucleotide is encountered, at which point a specific arginine residue in the enzyme intrudes into the DNA through the minor groove. The intrusion compresses the DNA backbone flanking the uracil and results in the flipping out of the uracil base so that it sticks out of the helix. The flipping out makes it accessible to cleavage by the uracil glycosylase. After the uracil is removed, another enzyme, called the **AP endonuclease,** cleaves the base-less sugar from the DNA (part C), leaving a single-stranded gap that is repaired by one of the DNA polymerases (part D). The gap filling leaves one nick remaining in the repaired strand, which is closed by DNA ligase. The net result of the repair is that the original C → U conversion rarely leads to mutation because the U is removed and replaced with C. Unfortunately, loss of the amino group of 5-methylcytosine yields thymine (Figure 7.6B), which is a normal DNA base and hence is not removed by the DNA uracil glycosylase. Thus the original G–C pair becomes a G–T pair and, in the next round of replication, gives G–C (wildtype) and A–T (mutant) DNA molecules.

7.4 Mutations can be induced by agents that affect DNA.

The first evidence that external agents could increase the mutation rate was presented in 1927 by Hermann Muller, who showed that x rays are mutagenic in *Drosophila.* (He later was awarded the Nobel Prize for this work.) Since then, a large number of physical agents and chemicals have been shown to increase the mutation rate. The use of these mutagens, several of which will be discussed in this section, is a means of greatly increasing the number of mutants that can be isolated. Because some environmental contaminants are mutagenic, as are numerous chemicals found in tobacco products, mutagens are also of great importance in public health.

A base analog masquerades as the real thing.

A **base analog** is a molecule sufficiently similar to one of the four DNA bases that it can be incorporated into a DNA duplex in the course of normal replication. Such a substance must be able to pair with a base in the template strand. Some base analogs are mutagenic because they are more prone to mispairing than are the normal nucleotides. The molecular basis of the mutagenesis can be illustrated with 5-bromouracil (Bu), a commonly used base analog that is efficiently incorporated into the DNA of bacteria and viruses.

The base 5-bromouracil is an analog of thymine, and the bromine atom is about the same size as the methyl group of thymine (Figure 7.8A). When cells are grown in a medium containing 5-bromouracil, a thymine is sometimes replaced by a 5-bromouracil in the replication of the DNA, resulting in the formation of an A–Bu pair at a normal A–T site (Figure 7.8B). The subsequent mutagenic activity of the incorporated 5-bromouracil stems from a rare shift in the configuration of the base. This shift is influenced by the bromine atom and happens in 5-bromouracil more frequently than in thymine. In the mutagenic configuration, 5-bromouracil pairs preferentially with guanine (Figure 7.8C); thus, in the next round of replication, one daughter molecule has a G–C pair at the altered site, yielding an A–T → G–C transition (Figure 7.9).

Other experiments suggest that 5-bromouracil is also mutagenic in another way. The concentration of the nucleoside triphosphates in most cells is regulated by the concentration of thymidine triphosphate (dTTP). This regulation results in appropriate relative amounts of the four triphosphates for DNA synthesis. One part of this complex regulatory process is the inhibition of synthesis of deoxycytidine triphosphate (dCTP) by excess dTTP. The 5-bromouracil nucleoside triphosphate also inhibits production of dCTP. When 5-bromouracil is added to the growth medium, dTTP continues to be synthesized by cells at the normal rate, whereas the synthesis of dCTP is significantly reduced. The ratio of dTTP to dCTP then becomes quite high, and the frequency of misincorporation of T opposite G increases. Although repair systems can remove some incorrectly incorporated thymine, in the presence of 5-bromouracil the rate of misincorporation can exceed the rate of correction. An incorrectly incorporated T pairs with A in the next round of DNA replication, yielding a G–C → A–T transition in one of the daughter molecules. Thus 5-bromouracil usually induces transitions in both directions: A–T → G–C by the route shown in Figure 7.9 and G–C → A–T by the misincorporation route.

Just as a base analog is incorporated into DNA in place of a normal base, a **nucleotide analog** can be incorporated in place of a normal nucleotide. Modern DNA-sequencing methods are based on the use of dideoxynucleotide analogs to terminate strand elongation at predetermined sites (Chapter 6). Dideoxynucleotide analogs are also used in the clinical treatment of various diseases, including cancer and viral infections such as AIDS (Section 6.8).

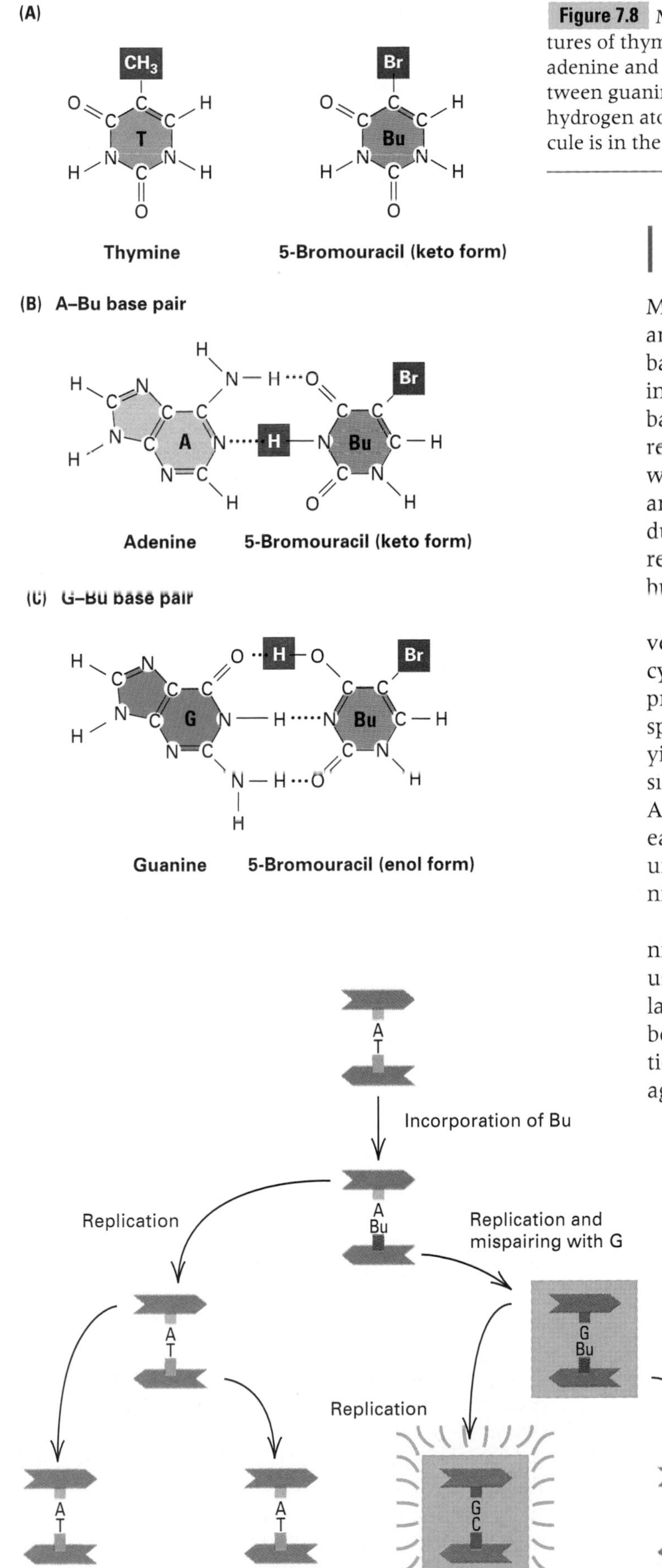

Figure 7.8 Mispairing mutagenesis by 5-bromouracil. (A) Structures of thymine and 5-bromouracil. (B) A base pair between adenine and the keto form of 5-bromouracil. (C) A base pair between guanine and the rare enol form of 5-bromouracil. One of the hydrogen atoms (shown in red) changes position when the molecule is in the keto form.

Highly reactive chemicals damage DNA.

Many mutagens are chemicals that react with DNA and change the hydrogen-bonding properties of the bases. These mutagens are active on both replicating and nonreplicating DNA, in contrast with the base analogs, which are mutagenic only when DNA replicates. Several of these chemical mutagens, of which nitrous acid is a well-understood example, are highly specific in the changes that they produce. Others—for example, the alkylating agents—react with DNA in a variety of ways and produce a broad spectrum of effects.

Nitrous acid (HNO_2) acts as a mutagen by converting amino (NH_2) groups of the bases adenine, cytosine, and guanine into keto (=O) groups. This process of *deamination* alters the hydrogen-bonding specificity of each base. Deamination of adenine yields a base, hypoxanthine, that pairs with cytosine rather than thymine, resulting in an A–T → G–C transition (Figure 7.10). As we noted earlier, deamination of cytosine yields uracil (Figure 7.6A), which pairs with adenine instead of guanine, producing a G–C → A–T transition.

In genetic research, chemical mutagens such as nitrous acid (and many others) are exceedingly useful in prokaryotic systems but are not particularly useful as mutagens in higher eukaryotes because the chemical conditions necessary for reaction are not easily obtained. However, alkylating agents are highly effective in eukaryotes as well as prokaryotes. **Alkylating agents,** such as ethylmethane sulfonate (EMS) and nitrogen mustard (the structures of which are

Figure 7.9 One mechanism for mutagenesis by 5-bromouracil (Bu). An A–T → G–C transition is produced by the incorporation of 5-bromouracil in DNA replication, forming an A–Bu pair. In the mutagenic round of replication, the Bu (in its rare enol form) pairs with G, and in the next round of replication, the G pairs with a C, completing the transition mutation.

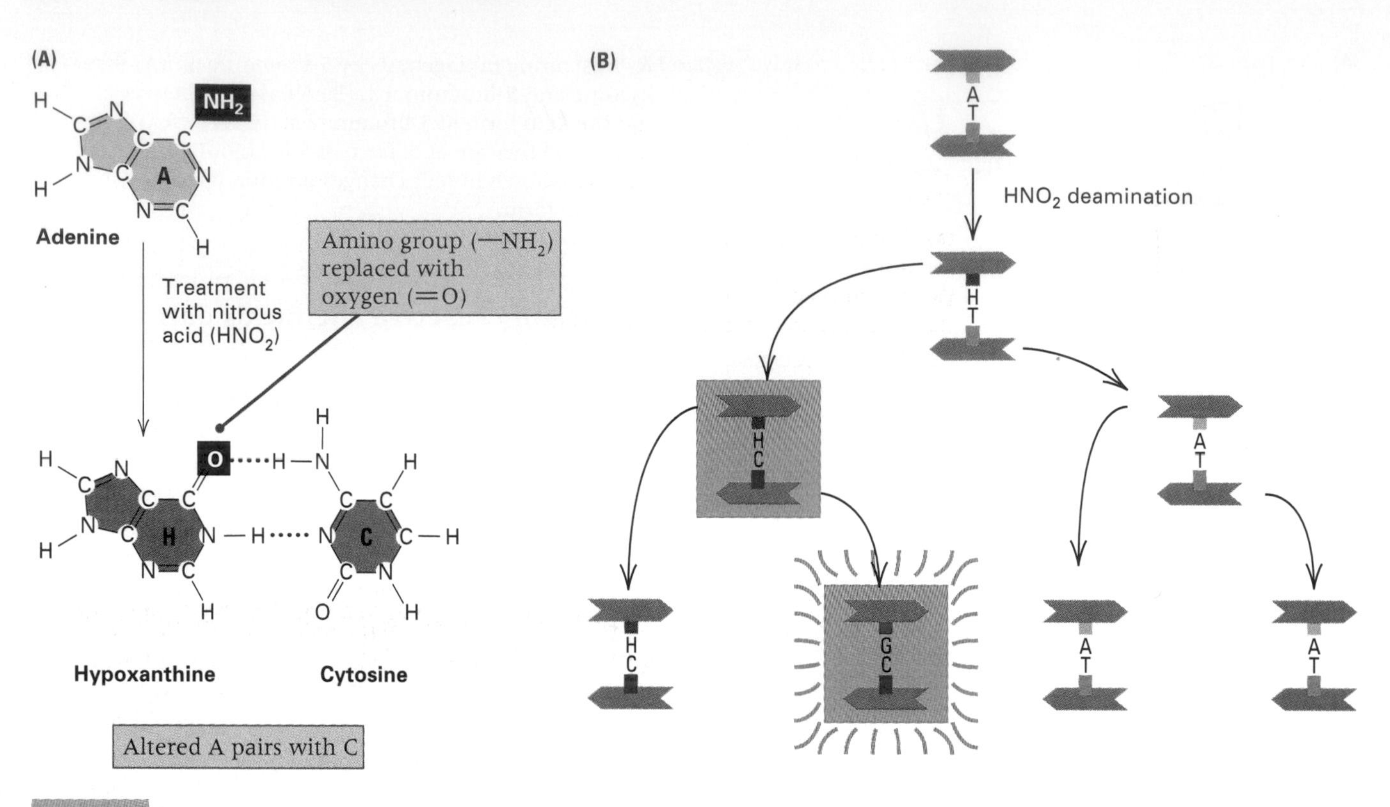

Figure 7.10 Nitrous acid mutagenesis. (A) Conversion of adenine into hypoxanthine, which pairs with cytosine. (The cytosine → uracil conversion is shown in Figure 7.6.) (B) Production of an A—T → G—C transition. In the mutagenic round of replication, the hypoxanthine (H) pairs with C. In the next round of replication, the C pairs with G, completing the transition mutation.

shown in Figure 7.11), are potent mutagens that have been used extensively in genetic research. These agents add various chemical groups to the DNA bases that either alter their base-pairing properties or cause structural distortion of the DNA molecule. Alkylation of either guanine or thymine causes mispairing, leading to the transitions A—T → G—C and G—C → A—T. EMS reacts less readily with adenine and cytosine.

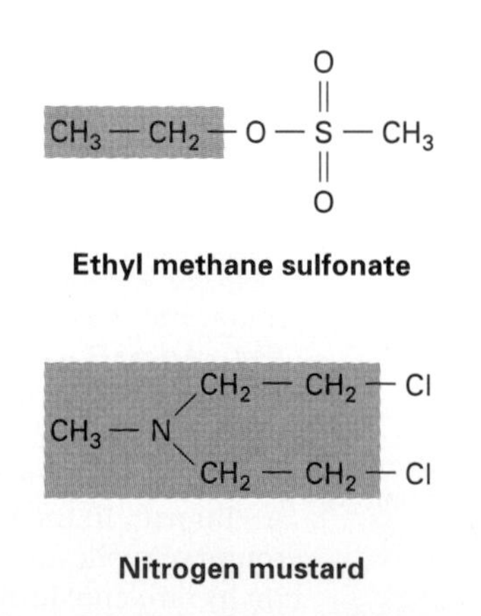

Figure 7.11 The chemical structures of two highly mutagenic alkylating agents; the alkyl groups are shown in shaded boxes.

Another phenomenon that results from the alkylation of guanine is **depurination,** or loss of the alkylated base from the DNA molecule by breakage of the bond joining the purine nitrogen with the deoxyribose. The process is illustrated in Figure 17.12. Depurination is not always mutagenic, because the loss of the purine can be repaired by excision of the deoxyribose (a function of the AP endonuclease discussed in Section 7.3) and gap repair by DNA polymerase. If, however, the replication fork reaches the apurinic site before repair has taken place, then replication almost always inserts an adenine nucleotide in the daughter strand opposite the apurinic site. After another round of replication, the original G—C pair becomes a T—A pair, an example of a transversion mutation.

DNA is susceptible to spontaneous depurination also. In air, the rate of spontaneous depurination is approximately 3×10^{-9} depurinations per purine nucleotide per minute. This rate is at least tenfold greater than any other single source of spontaneous degradation. At this rate, the half-life of a purine nucleotide exposed to air is about 300 years. It is for this reason that reports of PCR amplification of ancient DNA from any material exposed to the elements for more than a few thousand years should be treated with some skepticism.

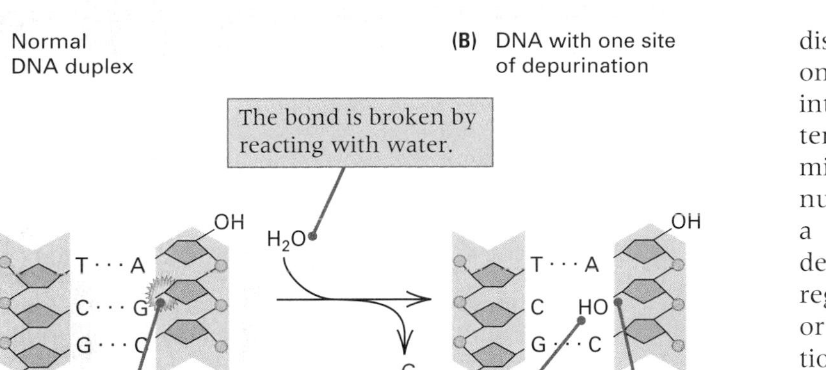

Figure 7.12 Deoxyribose-purine bonds are somewhat unstable and prone to undergo spontaneous reaction with water (hydrolysis), which results in loss of a purine base from the DNA (depurination). (A) Part of a DNA molecule prior to depurination. The bond between the labeled G and the deoxyribose to which it is attached is about to be hydrolyzed. (B) Hydrolysis of the bond releases the G purine, which diffuses away from the molecule and leaves a hydroxyl (–OH) in its place in the depurinated DNA.

Some agents cause base-pair additions or deletions.

The **acridine** molecules, of which proflavin is an example (Figure 7.13A), are planar three-ringed molecules whose dimensions are roughly the same as those of a purine–pyrimidine pair. These substances insert between adjacent base pairs of DNA (Figure 7.13B), a process called **intercalation.** The effect of the intercalation of an acridine molecule is to cause the adjacent base pairs to move apart by a distance roughly equal to the thickness of one base pair. When DNA that contains intercalated acridines is replicated, the template and daughter DNA strands can misalign, particularly in regions where nucleotides are repeated. The result is that a nucleotide can be either added or deleted in the daughter strand. In a coding region, the result of a single-base addition or deletion is a frameshift mutation (Section 7.2).

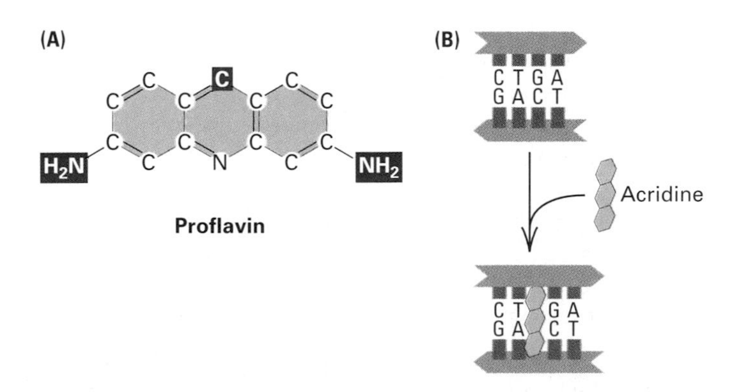

Figure 7.13 (A) The structure of proflavine, an acridine derivative. Other mutagenic acridines have additional atoms on the NH_2 group and on the C of the central ring. Hydrogen atoms are not shown. (B) Separation of two base pairs caused by intercalation of an acridine molecule.

Ultraviolet radiation absorbed by DNA is mutagenic.

Ultraviolet (UV) light produces both mutagenic and lethal effects in all viruses and cells. The effects are caused by chemical changes in the bases resulting from absorption of the energy of the light. The major products formed in DNA after UV irradiation are covalently joined pyrimidines (**pyrimidine dimers**), primarily thymine (Figure 7.14A), that are adjacent in the same polynucleotide strand. This chemical linkage brings the bases closer together, causing a distortion of the helix (Figure 7.14B), which blocks transcription and transiently blocks DNA replication. Pyrimidine dimers can be repaired in ways discussed later in this chapter.

In human beings, the inherited autosomal recessive disease **xeroderma pigmentosum** is the result of a defect in a system that repairs ultraviolet-damaged DNA. Persons with this disease are extremely sensitive to sunlight, which induces excessive skin pigmentation and the development of numerous skin lesions that frequently become cancerous. Removal of pyrimidine dimers does not occur in the DNA of cells cultured from patients with the disease, and the cells are sensitive to much lower doses of ultraviolet light than are cells from normal persons.

Even in normal people, excessive exposure to the UV rays in sunlight increases the risk of skin cancer. Ozone, a form of oxygen (O_3) present at a height of 15 to 30 kilometers in the atmosphere, absorbs UV rays and ameliorates, to some extent, the damaging effects of sunlight. The thinning of the ozone layer (the "ozone hole"), which is especially pronounced at certain latitudes, greatly increases the effective exposure to UV and correspondingly increases the rate of skin cancer. The thinning of the ozone layer can be attributed to the destructive effects of chlorofluorocarbons (CFCs), used as pressurizers in aerosol spray cans and as refrigerants, when these agents ascend to the upper atmosphere.

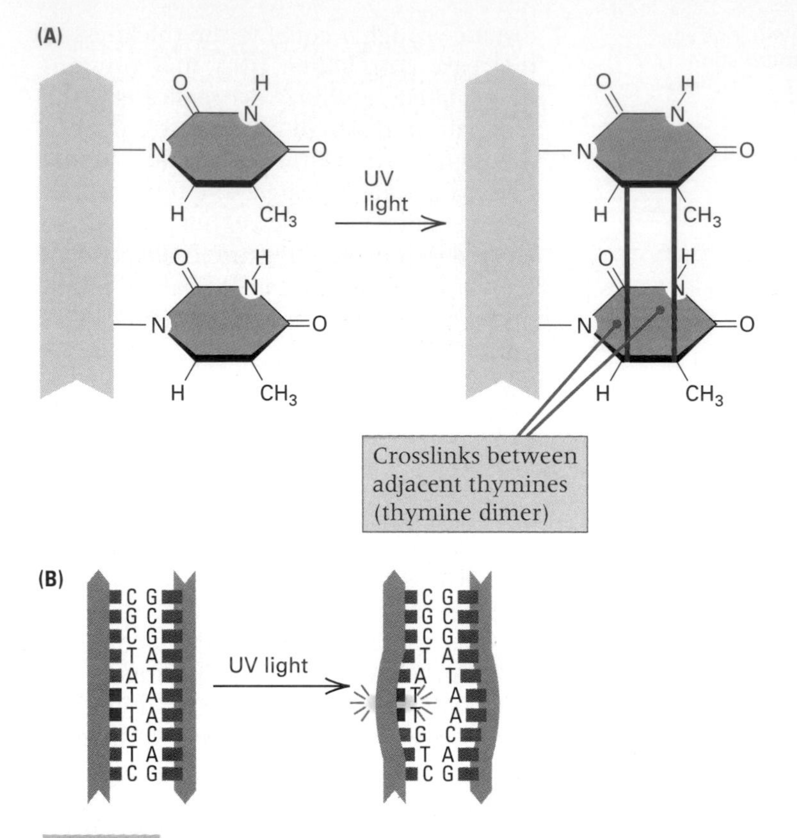

Figure 7.14 (A) Structural view of the formation of a thymine dimer. Adjacent thymines in a DNA strand that have been subjected to ultraviolet (UV) irradiation are joined by formation of the bonds shown in red. Other types of bonds between the thymine rings also are possible. Although they are not drawn to scale, these bonds are considerably shorter than the spacing between the planes of adjacent thymines, so that the double-stranded structure becomes distorted. The shape of each thymine ring also changes. (B) The distortion of the DNA helix caused by two thymines moving closer together when joined in a dimer.

Ionizing radiation is a potent mutagen.

Ionizing radiation includes x rays and the particles and radiation released by radioactive elements (α and β particles and γ rays). When first discovered late in the nineteenth century, the power of x rays to pass through solid materials was regarded as a harmless entertainment and source of great amusement. Witness this account from one history of the period:

> By 1898, personal x rays had become a popular status symbol in New York. The *New York Times* reported that "there is quite as much difference in the appearance of the hand of a washerwoman and the hand of a fine lady in an x-ray picture as in reality." The hit of the exhibition season was Dr. W. J. Morton's full-length portrait of *"the x-ray lady,"* a "fashionable woman who had evidently a scientific desire to see her bones." The portrait was said to be a "fascinating and coquettish" picture, the lady having agreed to be photographed without her stays and corset, the better to satisfy the "longing to have a portrait of well-developed ribs." Dr. Morton said women were not afraid of x rays: "After being assured that there is *no danger* they take the rays without fear."
>
> The titillating possibility of using x rays to see through clothing or to invade the privacy of locked rooms was a familiar theme in popular discussions of x rays and in cartoons and jokes. Newspapers carried advertisements for *"x ray proof underclothing"* for those seeking to protect themselves from x ray inspection.
>
> The luminous properties of radium soon produced a full-fledged radium craze. A famous woman dancer performed *radium dances* using veils dipped in fluorescent salts containing radium. *Radium roulette* was popular at New York casinos, featuring a "roulette wheel washed with a radium solution, such that it glowed brightly in the darkness; an unseen hand cast the ball on the turning wheel and sparks marked its course as it bounded from pocket to glimmery pocket." A patent was issued for a process for making women's gowns luminous with radium, and Broadway producer Florenz Ziegfeld snapped up the rights for his stage extravaganzas.
>
> Even while the unrestrained use of x rays and radium was growing, evidence was accumulating that the new forces might not be so benign after all. Hailed as tools for fighting cancer, they could also cause cancer. Doctors using x rays were the first to learn this bitter lesson. [Quoted from S. Hilgartner, R. C. Bell, and R. O'Connor. 1982. *Nukespeak.* Sierra Club Books, San Francisco, 1982.]

Doctors were indeed the first to learn the lesson. Many suffered severe x-ray burns or required amputation of overexposed hands or arms. Many others died from radiation poisoning or from radiation-induced cancer. By the mid-1930s, the number of x-ray deaths had grown so large that a monument to the "x-ray martyrs" was erected in a hospital courtyard in Germany. Yet the full hazards of x-ray exposure were not widely appreciated until the 1960s.

When ionizing radiation interacts with water or with living tissue, highly reactive ions called **free radicals** are formed. The free radicals react with other molecules, including DNA, which results in the carcinogenic and mutagenic effects. The intensity of a beam of ionizing radiation can be described quantitatively in several ways. There are, in fact, a bewildering variety of units in common use (Table 7.1). Some of the units (becquerel, curie) deal with the number of disintegrations emanating from a material, others (roentgen) with the number of ionizations the radiation produces in air, still others (gray, rad) with the amount of energy imparted to material exposed to the radiation, and some (rem, sievert) with the effects of radiation on living tissue. The types of units have proliferated through the years in attempts to encompass different types of radiation, including nonionizing radiation, in a common frame of reference. The units in Table 7.1 are presented only as an aid in interpreting the multitude of units found in the literature on radiation genetics.

Genetic studies of ionizing radiation support the following general principle:

> Over a wide range of x-ray doses, the frequency of mutations induced by x rays is proportional to the radiation dose.

One type of evidence supporting this principle is the frequency with which X-chromosome recessive lethals are induced in *Drosophila* (Figure 7.15). The mutation rate increases linearly with increasing x-ray dose. For example, an exposure of 10 sieverts increases the frequency from the spontaneous value of 0.15 percent to about 3 percent. The mutagenic and lethal effects of ionizing radiation at low to moderate doses result primarily from damage to DNA. Three types of damage in DNA are produced by ionizing radiation: single-strand breakage (in the sugar–phosphate backbone), double-strand breakage, and alterations in nucleotide bases. The single-stand breaks are usually efficiently repaired, but the other damage is responsible for mutation and lethality. In eukaryotes, ionizing radiation also results in chromosome breaks. Although systems exist for repairing the breaks, the repair often leads to translocations, inversions, duplications, and deletions. In human cells in culture, a dose of 0.2 sievert results in an average of one visible chromosome break per cell.

Although the effects of extremely low levels of radiation are very difficult to measure because of the background of spontaneous mutation, most experiments support the following principle:

Table 7.1
Units of radiation

Unit (abbreviation)	Magnitude
Becquerel (Bq)*	1 disintegration/second = 2.7×10^{-11} Ci
Curie (Ci)	3.7×10^{10} disintegrations/second = 3.7×10^{10} Bq
Gray (Gy)*	1 joule/kilogram = 100 rad
Rad (rad)	100 ergs/gram = 0.01Gy
Rem (rem)	Damage to living tissue done by 1 rad = 0.01 Sv
Roentgen (R)	Produces 1 electrostatic unit of charge per cubic centimeter of dry air under normal conditions of pressure and temperature. (By definition, 1 electrostatic unit repels with a force of 1 dyne at a distance of 1 centimeter.)
Sievert (Sv)	100 rem

*Units officially recognized by the International System of Units as defined by the General Conference on Weights and Measures.

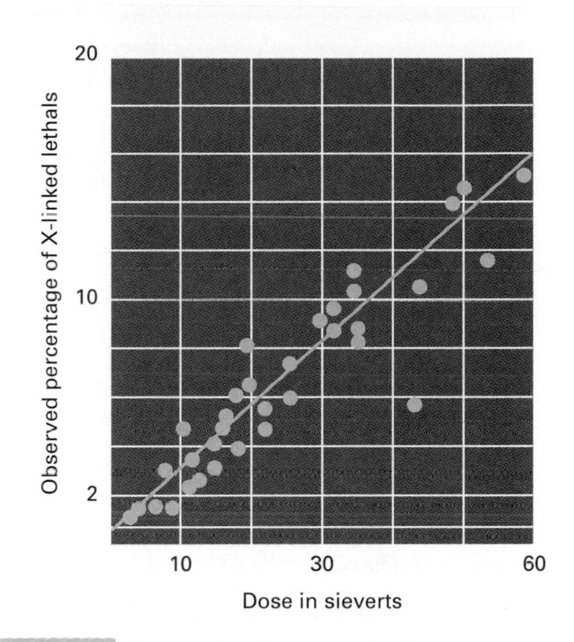

Figure 7.15 The relationship between the percentage of X-linked recessive lethals in *D. melanogaster* and x-ray dose. The frequency of spontaneous X-linked lethal mutations is 0.15 percent per X chromosome per generation.

> There appears to be no threshold exposure below which mutations are not induced. Even very low doses of radiation induce some mutations.

Ionizing radiation is widely used in tumor therapy. The basis for the treatment is the increased frequency of chromosomal breakage (and the consequent lethality) in cells undergoing mitosis compared with cells in interphase. Tumors usually contain many more mitotic cells than do most normal tissues, so more tumor cells than normal cells are destroyed. Because all tumor cells are not in mitosis at the same time, irradiation is carried out at intervals of several days to allow interphase tumor cells to enter mitosis. Over a period of time, most tumor cells are destroyed.

Table 7.2 gives representative values of doses of ionizing radiation received by human beings in the United States in the course of a year. The unit of measure is the millisievert, which equals 0.1 rem. The exposures in Table 7.2 are on a yearly basis, so over the course of a generation, the total exposure is approximately 100 millisieverts. Note that, with the exception of diagnostic x rays, which yield important compensating benefits, most of the total radiation exposure comes from natural sources, particularly radon gas. Less than 20 percent of the average radiation exposure comes from artificial sources. Nevertheless, there are dangers inherent in any exposure to ionizing radiation,

Table 7.2
Annual exposure of human beings in the United States to various forms of ionizing radiation

Source	Dose (in millisieverts*)
Natural radiation	
Radon gas	2.06
Cosmic rays	0.27
Natural radioisotopes in the body	0.39
Natural radioisotopes in the soil	0.28
Total natural radiation	3.00
Other radiation sources	
Diagnostic x rays	0.39
Radiopharmaceuticals	0.14
Consumer products (x rays from TV, in clock dials) and building materials	0.10
Fallout from weapon tests	<0.01
Nuclear power plants	<0.01
Total from non-natural sources	0.63
Total from all sources	3.63

*One millisievert (mSv) equals 1/1000 sievert, or 0.1 rem.

Source: From National Research Council, Committee on the Biological Effects of Ionizing Radiations, Health Effects of Exposure to Low Levels of Ionizing Radiation (*BEIR V*), National Academy Press, 1990.

particularly an increased risk of leukemia and certain other cancers in the exposed persons. In regard to increased genetic diseases in future generations resulting from the mutagenic effects of radiation, the risk of a small amount of additional radiation is low enough that most geneticists are currently more concerned about the effects of the many mutagenic (as well as carcinogenic) chemicals that are introduced into the environment from a variety of sources.

The National Academy of Sciences of the United States regularly updates the estimated risks of radiation exposure. The latest estimates are summarized in Table 7.3. The message is that an additional 10 millisieverts of radiation per generation (about a 10 percent increase in the annual exposure) is expected to cause a relatively modest increase in diseases that are wholly or partly due to genetic factors. The most common conditions in the table are heart disease and cancer. No estimate for the radiation-induced increase is given for either of these traits because the genetic contribution to the total is still uncertain.

The Chornobyl nuclear accident had unexpectedly large genetic effects.

The city of Chornobyl (formerly Chernobyl) in the Ukraine is a symbol for nuclear disaster. On April 26, 1986, a nuclear power plant near the city exploded, heavily contaminating the immediate area and sending clouds of radioactive debris over long distances. It was the largest publicly acknowledged nuclear accident in history. The meltdown is estimated to have released between 50 and 200 million curies of radiation, along with a wide variety of heavy-metal and chemical pollutants. Iodine-131 and cesium-137 were the principal radioactive contaminants. People living in the area were evacuated almost immediately, but many were heavily exposed to radiation. Within a short time there was a notable increase in the frequency of thyroid cancer in children, and other health effects of radiation exposure were also detected.

At first, little attention was given to possible genetic effects of the Chornobyl disas-

Table 7.3
Estimated genetic effects of an additional 10 millisieverts per generation

Type of disorder	Current incidence per million liveborn	Additional cases per million liveborn per 10 mSv per generation	
		First generation	At equilibrium
Autosomal dominant			
Clinically severe	2500	5–20	25
Clinically mild	7500	1–15	75
X-linked	400	<1	<5
Autosomal recessive	2500	<1	Very slow increase
Chromosomal			
Unbalanced translocations	600	<5	Very little increase
Trisomy	3800	<1	<1
Congenital abnormalities			
(multifactorial)	20,000–30,000	10	10–100
Other multifactorial disorders			
Heart disease	600,000	Unknown	Unknown
Cancer	300,000	Unknown	Unknown
Others	300,000	Unknown	Unknown

Source: Health Effects of Exposure to Low Levels of Ionizing Radiation (BEIR V), National Research Council, Washington, D.C.

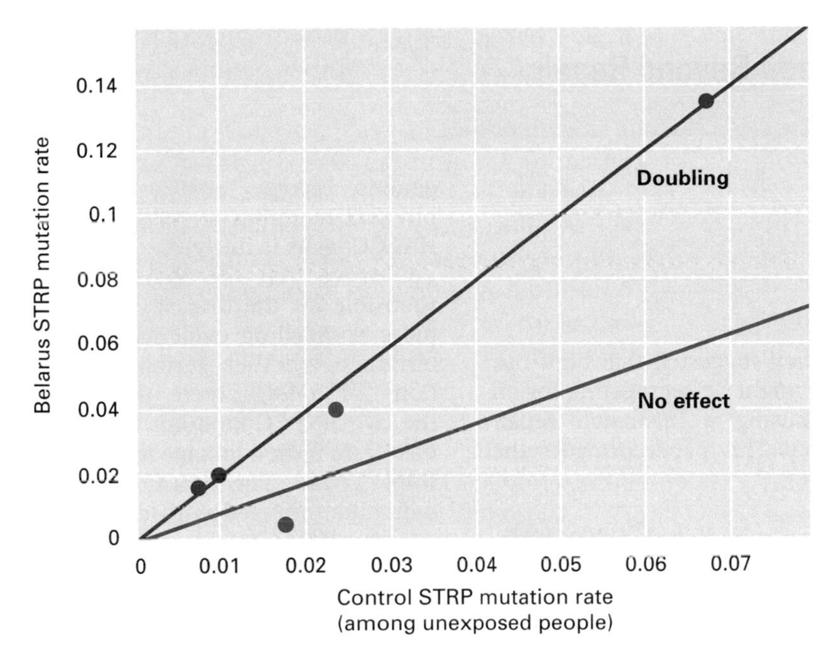

Figure 7.16 Mutation rates of five STRP polymorphisms among people of Belarus who were exposed to radiation from Chornobyl and among unexposed British people. Three of the loci (red dots) show evidence of an approximately twofold increase in the mutation rate. [Data from Y. E. Dubrova, V. N. Nesterov, N. G. Krouchinsky, V. A. Ostapenko, R. Neumann, D. L. Neil, and A. J. Jeffreys. 1996. *Nature* 380: 183.]

ter because relatively few people were acutely exposed and because the radiation dose to people outside the immediate area was considered too small to worry about. In the district of Belarus, some 200 kilometers north of Chornobyl, the average exposure to iodine-131 was estimated at approximately 0.185 sievert per person. This is a fairly high dose, but the exposure is brief because the half-life of iodine-131 is only 8 days. Radiation from the much longer-lived cesium-137, with a half-life of 30 years, was estimated at less than 5 millisieverts per year. On the basis of data from laboratory animals and studies of the survivors of the Hiroshima and Nagasaki atomic bombs, little detectable genetic damage was expected from these exposures.

Nevertheless, 10 years after the meltdown, studies of people living in Belarus did indicate a remarkable increase in the mutation rate. The observations focused on STRP polymorphisms because of their intrinsically high mutation rate due to replication slippage and other factors. Each STRP was examined in parents and their offspring to detect any DNA fragments that increased or decreased in size, which would indicate an increase or decrease in the number of repeating units contained in the DNA fragment. Five STRP polymorphisms were studied, with the results summarized in Figure 7.16. Two of the loci (blue dots) showed no evidence of an increase in the mutation rate. This is the expected result. The unexpected finding was that three of the loci (red dots) did show a significant increase, and the level of increase was consistent with an approximate doubling of the mutation rate. At the present time, it is still not known whether the increase was detected because replication slippage is more sensitive to radiation than other types of mutations or because the effective radiation dose of the Belarus population was much higher than originally thought.

The finding of increased mutation among the people of Belarus is paralleled by observations carried out in two species of voles, small rodents related to hamsters and gerbils, that live in the immediate vicinity of Chornobyl. These animals live in areas still highly contaminated by radiation, heavy metals, and chemical pollutants. The animals eat food that is so heavily contaminated that they themselves become extremely radioactive. Yet they survive and reproduce in this inhospitable environment.

Among the voles of Chornobyl, the rate of mutation in the *cytochrome b* gene present in mitochondrial DNA was estimated at 2×10^{-4} mutations per nucleotide per generation. This rate is at least 200 times higher than that found among animals living in a relatively uncontaminated area just 30 kilometers to the southwest of Chornobyl. It is so high that, on average, each mitochondrial DNA molecule of approximately 17 kb has three new mutations take place in each generation. It is not yet known whether this high rate of mutation is due to radiation exposure, to nonradioactive pollutants, or to the synergistic effects of both.

7.5 Cells have enzymes that repair certain types of DNA damage.

Spontaneous damage to DNA in human cells takes place at a rate of approximately 1 event per billion nucleotide pairs per minute (or, expressed per nucleotide pair, at a rate of 1×10^{-9} per nucleotide pair per minute). This may seem quite a small rate,

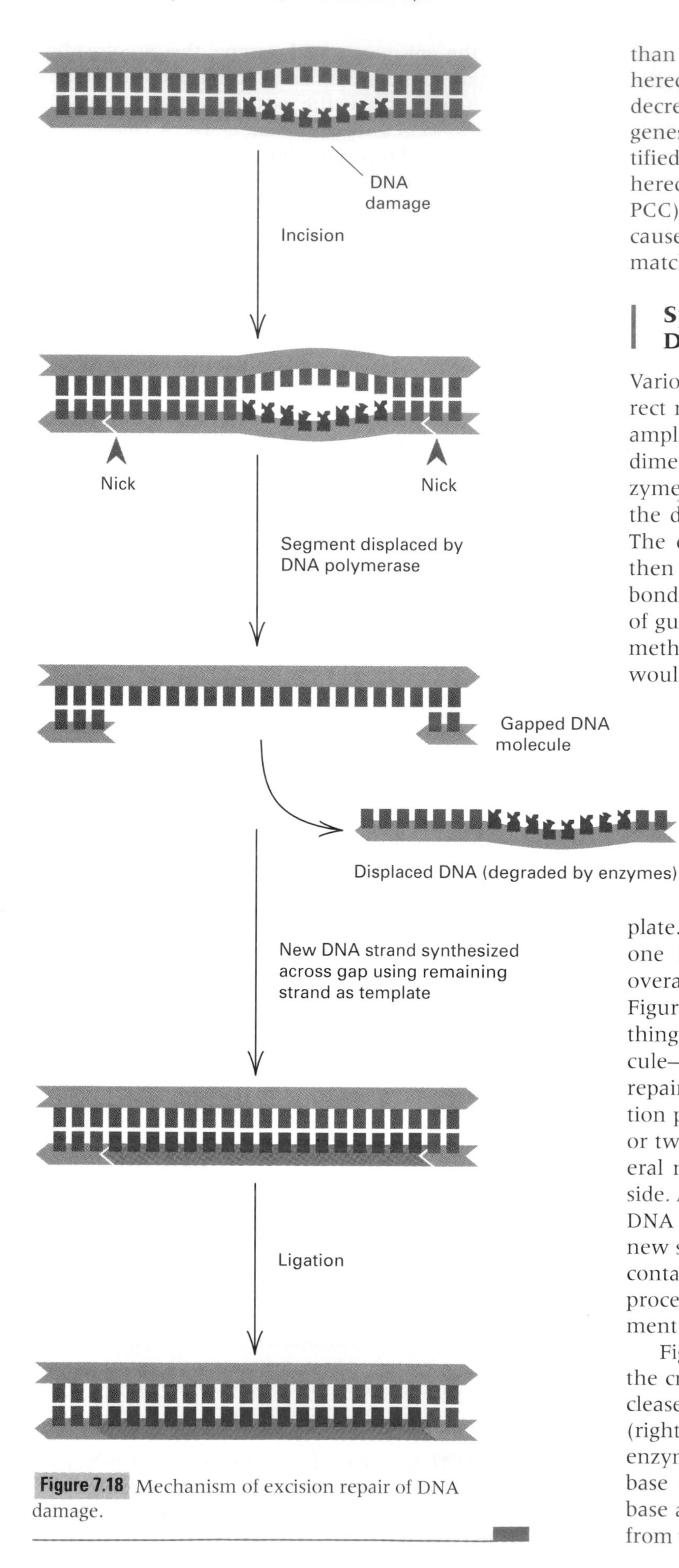

Figure 7.18 Mechanism of excision repair of DNA damage.

than in wildtype yeast. Some forms of human hereditary colorectal cancer are also associated with decreased stability of simple repeats. Four human genes homologous to *mutL* or *mutS* have been identified, any one of which, when mutated, results in hereditary nonpolyposis colorectal cancer (HNPCC). Most cases of this type of cancer may be caused by mutations in one of these four mismatch-repair genes.

Special enzymes repair damage to DNA caused by ultraviolet light.

Various enzymes can recognize and catalyze the direct reversal of specific DNA damage. A classic example is the reversal of UV-induced pyrimidine dimers by **photoreactivation,** in which an enzyme breaks the bonds that join the pyrimidines in the dimer and thereby restores the original bases. The enzyme binds to the dimers in the dark but then utilizes the energy of blue light to cleave the bonds. Another important example is the reversal of guanine methylation in O^6-methyl guanine by a methyl transferase enzyme, which otherwise would pair like adenine in replication.

Excision repair works on a wide variety of damaged DNA.

Excision repair is a ubiquitous multistep enzymatic process by which a stretch of a damaged DNA strand is removed from a duplex molecule and replaced by resynthesis using the undamaged strand as a template. Mismatch repair of single mispaired bases is one important category of excision repair. The overall process of excision repair is illustrated in Figure 7.18. The DNA damage can be due to anything that produces a distortion in the duplex molecule—for example, a pyrimidine dimer. In excision repair, a repair endonuclease recognizes the distortion produced by the DNA damage and makes one or two cuts in the sugar–phosphate backbone, several nucleotides away from the damage on either side. A 3'-OH group is produced at the 5' cut, which DNA polymerase uses as a primer to synthesize a new strand while displacing the DNA segment that contains the damage. The final step of the repair process is the joining of the newly synthesized segment to the contiguous strand by DNA ligase.

Figure 7.19 shows a molecular model based on the crystal structure of the repair enzyme endonuclease III from *E. coli.* The purple region in the DNA (right) is a region of 15 base pairs contacted by the enzyme in carrying out the repair of the damaged base (white). The enzyme removes the damaged base and introduces a single-strand nick at the site from which the damaged base was removed.

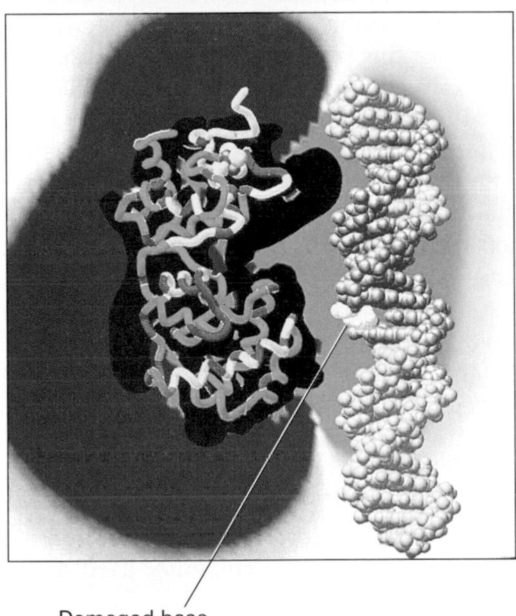

Figure 7.19 Model of endonuclease III from *E. coli* (left) in close proximity with a region of 15 base pairs (purple) of DNA (right) that it contacts in carrying out its function of removing a stretch of DNA containing a damaged base. The damaged base is shown in white. [Courtesy of John A. Tainer. Computer graphics model by Michael Pique, Cindy Fisher, and John A. Tainer. From C.-F. Kuo, D. E. McRee, C. L. Fisher, S. F. O'Handley, R. P. Cunningham and J. A Tainer. 1992. *Science* 258:434.]

Postreplication repair skips over damaged bases.

Sometimes DNA damage persists rather than being reversed or removed, but its harmful effects may be minimized. This often requires replication across damaged areas, so the process is called **postreplication repair.** For example, when DNA polymerase reaches a damaged site (such as a pyrimidine dimer), it stops synthesis of the strand (Figure 7.20A). However, after a brief time, synthesis is reinitiated beyond the damage, and chain growth continues, producing a gapped strand with the damaged spot in the gap (Figure 7.20B). The gap can be filled by strand exchange with the parental strand having the same polarity (part C), and then the secondary gap produced in the undamaged strand is filled by repair synthesis (part D). The products of this exchange and resynthesis are two intact single strands, each of which can then serve in the next round of replication as a template for the synthesis of an undamaged DNA molecule.

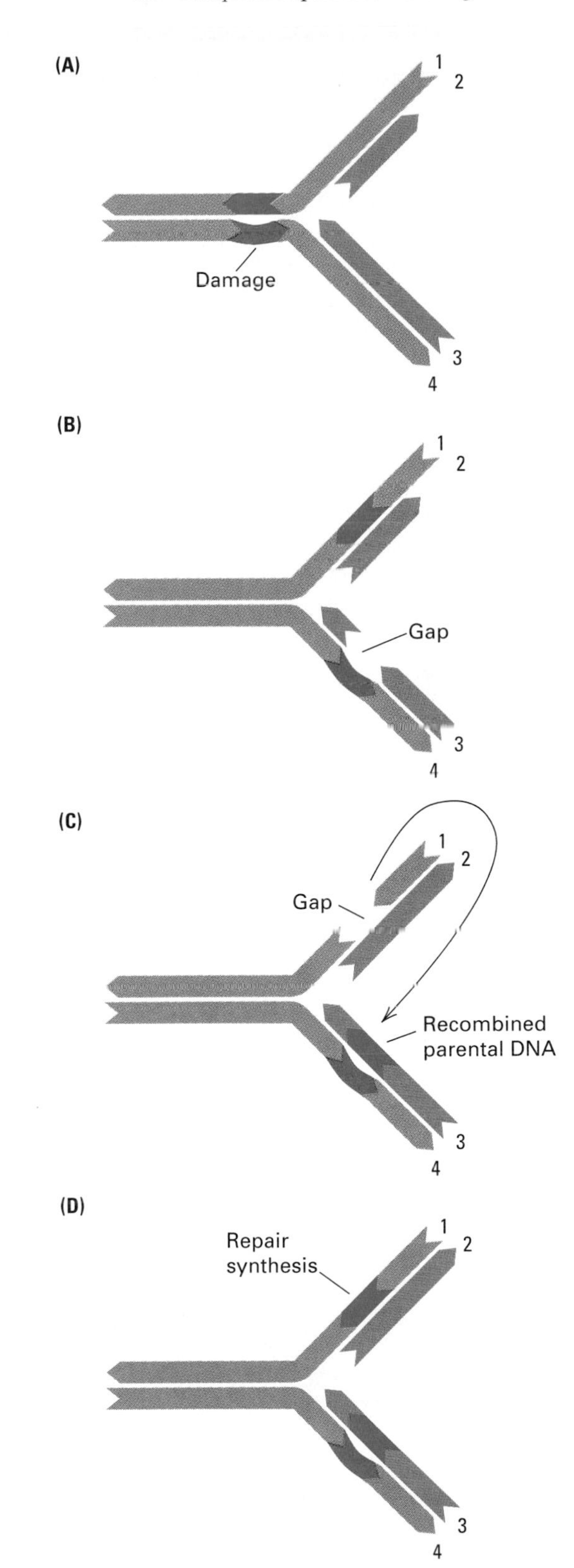

Figure 7.20 Postreplication repair. (A) A molecule with DNA damage in strand 4 is being replicated. (B) By reinitiating synthesis beyond the damage, a gap is formed in strand 3. (C) A segment of parental strand 1 is excised and inserted in strand 3. (D) The gap in strand 1 is next filled in by repair synthesis.

7.6 Reverse mutations are useful in detecting mutagens and carcinogens.

In most of the mutations considered so far, a wildtype gene is changed into a form that results in a mutant phenotype, an event called a **forward mutation.** Mutations are frequently reversible, and an event that restores the wildtype phenotype is called a **reversion.** A reversion may result from a **reverse mutation,** an exact reversal of the alteration in base sequence that occurred in the original forward mutation, restoring the wildtype DNA sequence. A reversion may also result from the occurrence of a second mutation, at some other site in the genome, that in any of several ways compensates for the effect of the original mutation.

Although reversion by the exact reversal mechanism is infrequent, reverse mutation is the basis of a widely used test for the detection of chemical mutagens. In view of the increased number of chemicals used and present as environmental contaminants, tests for the mutagenicity of these substances have become important. Furthermore, most carcinogens are also mutagens, and so mutagenicity provides an initial screening for potential hazardous agents. One simple method for screening large numbers of substances for mutagenicity is a reversion test that uses nutritional mutants of bacteria. In the simplest type of reversion test, a compound that is a potential mutagen is added to solid growth medium, a known number of a mutant bacterium is plated, and the revertant colonies are counted. An increase in the reversion frequency significantly greater than that obtained in the absence of the test compound identifies the substance as a mutagen. However, simple tests of this type fail to demonstrate the mutagenicity of a large number of potent carcinogens. The explanation for this failure is that many substances are not directly mutagenic (or carcinogenic); rather, they require a conversion into mutagens by enzymatic reactions that take place in the livers of animals and that have no counterpart in bacteria. The normal function of these enzymes is to protect the organism from various naturally occurring harmful substances by converting them into soluble nontoxic substances that can be disposed of in the urine. However, when the enzymes encounter certain artificial and natural compounds, they convert these substances, which may not be harmful in themselves, into mutagens or carcinogens. The enzymes of liver cells, when added to the bacterial growth medium, activate the compounds and allow their mutagenicity to be recognized. The addition of liver extract is one step in the currently used Ames test for carcinogens and mutagens.

In two kittens in this litter, the color of the left eye does not match that of the right eye. These are examples of "mosaic" phenotypes, often due to somatic mutation.

In the **Ames test,** histidine-requiring (His^-) mutants of the bacterium *Salmonella typhimurium,* containing either a base substitution or a frameshift mutation, are tested for reversion to His^+. In addition, the bacterial strains have been made more sensitive to mutagenesis by the incorporation of several mutant alleles that inactivate the excision-repair system and that make the cells more permeable to foreign molecules. Because some mutagens act only on replicating DNA, the solid medium used contains enough histidine to support a few rounds of replication but not enough to permit the formation of a visible colony. The medium also contains the potential mutagen to be tested and an extract of rat liver. If the test substance is a mutagen or is converted into a mutagen, some colonies are formed. A quantitative analysis of reversion frequency can also be carried out by incorporating various amounts of the potential mutagen in the medium. The reversion frequency generally depends on the concentration of the substance being tested and, for a known carcinogen or mutagen, correlates roughly with its carcinogenic potency in animals. The Ames test is simple, rapid, inexpensive, and exquisitely sensitive. Some chemicals can be detected to be mutagenic in amounts as small as 10^{-9} g, and a condensate of as little as 1/100 of a cigarette can be shown to be mutagenic. The test is also highly quantitative (Figure 7.21). Chemicals need not be classified simplistically as "mutagenic" or "nonmutagenic." They can be classified according to their potency as mutagens, because more than a millionfold range in potency can be detected in the *Salmonella* test.

An example of the importance of adequate testing involves a chemical known as tris-BP. This substance was used as a flame retardant in children's polyester pajamas from 1972 to 1977. Then it was discovered that tris-BP is a potent mutagen in

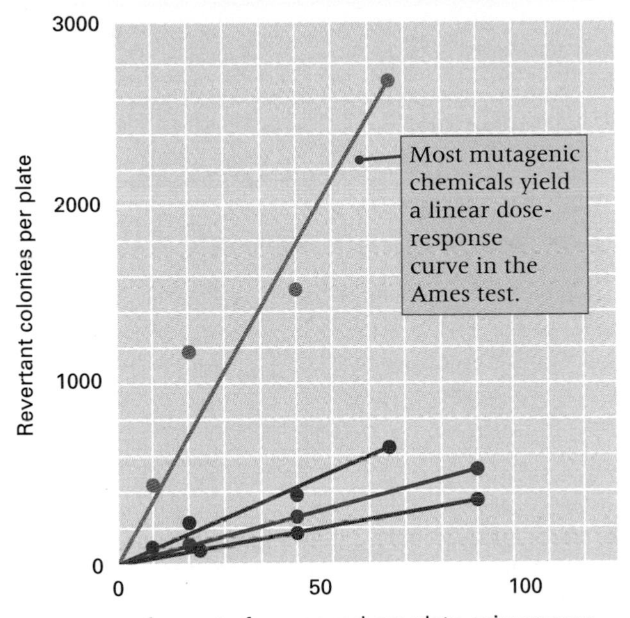

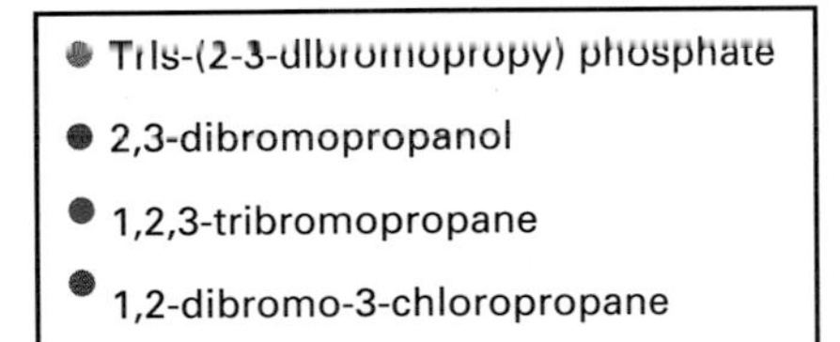

Figure 7.21 Linear dose-response relationships obtained with various chemical mutagens in the Ames test. [Data from B. N. Ames. 1974. *Science* 204: 587–593.]

Salmonella and also in *Drosophila*. Further studies showed that Tris-BP interacts with human DNA and damages mammalian chromosomes. It was also found to be a carcinogen in experimental tests in rats and mice, and it was found to be capable of causing sterility in laboratory animals. Moreover, the substance was shown to be absorbed through the skin, and its breakdown products could be detected in the urine of children who were wearing the treated sleepware. Prior to the time when this information became known and use of the substance in clothing was discontinued, more than 50 million children were exposed to the chemical through contact with their nightclothes.

The Ames test has been used with thousands of substances and mixtures (such as industrial chemicals, food additives, pesticides, hair dyes, and cosmetics), and numerous unsuspected substances have been found to stimulate reversion in this test. A high frequency of reversion does not necessarily indicate that the substance is definitely a carcinogen but only that it has a high probability of being so. As a result of these tests, many industries have reformulated their products: for example, the cosmetics industry has changed the formulation of many hair dyes and cosmetics to render them nonmutagenic. Ultimate proof of carcinogenicity is determined by testing for tumor formation in laboratory animals. However, only a few percent of the substances known from animal experiments to be carcinogens failed to increase the reversion frequency in the Ames test.

Chapter Summary

- **Mutations can be classified by their phenotypic effects.**
- **Mutations alter DNA in any of a variety of ways.**

A base substitution replaces one base with another.
Some mutations add or delete bases from the DNA.
Transposable elements cause insertion mutations.

Mutations can be classified in a variety of ways—in terms of (1) how they come about, (2) the nature of the chemical change, or (3) how they are expressed. Conditional mutations cause a change in phenotype under restrictive but not permissive conditions. For example, temperature-sensitive mutations are expressed only above a particular temperature. Spontaneous mutations are of unknown origin, whereas induced mutations result from exposure to chemical reagents or radiation. All mutations ultimately result from changes in the sequence of nucleotides in DNA, including base substitutions, insertions, deletions, and rearrangements. Base-substitution mutations are single-base changes that may be silent (for example, in a noncoding region or a synonymous codon position), may change an amino acid in a polypeptide chain (a missense mutation), or may cause chain termination by producing a stop codon (a nonsense mutation). A transition is a base-substitution mutation in which a pyrimidine is substituted for another pyrimidine or a purine for another purine. In a transversion, a pyrimidine is substituted for a purine, or the other way around. A mutation may consist of an insertion or deletion of one or more bases; in a coding region, if the number of bases is not a multiple of 3, the mutation is a frameshift.

- **Spontaneous mutations have no assignable cause.**

Mutations arise without reference to the adaptive needs of the organism.
Special methods are necessary to estimate the rate of mutation.
Mutations are nonrandom with respect to position in a gene or genome.

Spontaneous mutations are random. Favorable mutations do not take place in response to harsh environmental conditions. Rather, the environment selects for particular favorable mutations that happen to arise by chance and that have a growth advantage. Spontaneous mutations often arise by errors in DNA replication that fail to be corrected either by the proofreading system or by the mismatch-repair system. The proofreading system removes an incorrectly incorporated base immediately after it is added to the growing end of a DNA strand. The mismatch-repair system removes incorrect bases at a later time. Methylation of parental DNA strands and delayed methylation of daughter strands allows the mismatch-repair system to operate selectively on the daughter DNA strand.

- **Mutations can be induced by agents that affect DNA.**

A base analog masquerades as the real thing.
Highly reactive chemicals damage DNA.
Some agents cause base-pair additions or deletions.
Ultraviolet radiation absorbed by DNA is mutagenic.
Ionizing radiation is a potent mutagen.
The Chornobyl nuclear accident had unexpectedly large genetic effects.

Mutations can be induced chemically by direct alteration of DNA—for example, by nitrous acid, which deaminates bases. Base analogs are incorporated into DNA in replication. They undergo mispairing more often than do the normal bases, which gives rise to transition mutations. The base analog 5-bromouracil is an example of such a mutagen. Alkylating agents often cause depurination; this type of mutation is usually repaired, but often an adenine is inserted opposite a depurination site, which results in a transversion. Acridine molecules intercalate (interleaf) between base pairs of DNA and cause misalignment of parental and daughter strands in DNA replication, giving rise to frameshift mutations, usually of one or two bases. Ionizing radiation results in oxidative free radicals that cause a variety of alterations in DNA, including double-stranded breaks. Although the amount of genetic damage from normal background and other sources of radiation is believed to be low, studies of persons exposed to fallout from the Chornobyl nuclear meltdown indicate an increased mutation rate, at least for simple sequence length polymorphisms (STRPs).

- **Cells have enzymes that repair certain types of DNA damage.**

Mismatch repair fixes incorrectly matched base pairs.
Special enzymes repair damage to DNA caused by ultraviolet light.
Excision repair works on a wide variety of damaged DNA.
Postreplication repair skips over damaged bases.

A variety of systems exist for repairing damage to DNA. In excision repair, a damaged stretch of a DNA strand is excised and replaced with a newly synthesized copy of the undamaged strand. In photoreactivation, there is direct cleavage of the pyrimidine dimers produced by ultraviolet radiation. Damage to DNA resulting in the formation of O^6-methyl guanine is repaired by a special methyl transferase. In all of these systems, the correct template is restored. Postreplication repair is an exchange process in which gaps in one daughter strand produced by failure to replicate across damaged sites are filled by nondefective segments from the parental strand of the other branch of the newly replicated DNA. Thus a new template is produced by this system.

- **Reverse mutations are useful in detecting mutagens and carcinogens.**

Mutant organisms sometimes revert to the wildtype phenotype. Reversion is normally due to an additional mutation at another site. The Ames test measures reversion as an indicator of mutagens and carcinogens; it uses an extract of rat liver, which in mammals occasionally converts intrinsically harmless molecules into mutagens and carcinogens.

Key Terms

acridine
alkylating agent
Ames test
AP endonuclease
base analog
base-substitution mutation
ClB method
conditional mutation
depurination
DNA uracyl glycosidase
excision repair
forward mutation
free radical
germ-line mutation
hotspot
induced mutation
intercalation
ionizing radiation
missense mutation
mutagen
mutation
mutation rate
nitrous acid
nonsense mutation
nucleotide analog
permissive condition
photoreactivation
postreplication repair
pyrimidine dimer
replica plating
replication slippage
restrictive condition
reverse mutation
reversion
sickle-cell anemia
silent mutation
somatic mutation
spontaneous mutation
temperature-sensitive mutation
transition mutation
transversion mutation
xeroderma pigmentosum

Concepts and Issues
Testing Your Knowledge

- If a mutation is a conditional mutation, what determines whether the mutant phenotype will be expressed?
- How can an organism with a temperature-sensitive, recessive-lethal mutation survive as a homozygous genotype?
- What does it mean to say that a particular allele has a mutation rate of 10^{-6} per gene per generation?
- How does replica plating demonstrate that mutations to antibiotic resistance can arise even in cells that have never been exposed to the antibiotic?
- Mutations in genes whose products are involved in DNA repair are often associated with an increased risk of cancer. What does this observation imply about the role of spontaneous mutation in the development of cancer?
- What is "mismatched" in the process of mismatch repair?
- Why were the genetic effects of the Chornobyl nuclear meltdown initially expected to be undetectable?

Solutions
Step by Step

Problem 1

Among genomes as diverse as the bacteriophages λ and T4, the bacteria *Escherichia coli,* and the fungi *Neurospora crassa* and *Saccharomyces cerevisiae,* the spontaneous mutation rate is approximately 0.003 mutation per genome per DNA replication. The genome sizes of λ, *E. coli,* and *S. cerevisiae* are approximately 50 thousand, 4.6 million, and 13.5 million base pairs, respectively.

(a) What proportion of genomes escape spontaneous mutation in each replication?

(b) What is the mutation rate per base pair per DNA replication in each of these genomes?

(c) What might the differences in mutation rate per base pair imply about the evolution of mutation rates?

Solution **(a)** If the mutation rate is 0.003 mutation per genome per DNA replication, then $1 - 0.003 = 99.7$ percent of the genomes contain no new mutations after one round of replication. **(b)** The rate of mutation per base pair per DNA replication for λ is $0.003/50{,}000 = 6 \times 10^{-8}$. For the other genomes, the values are calculated similarly and equal 6.5×10^{-10} for *E. coli* and 2.2×10^{-10} for *S. cerevisiae.* **(c)** The great variation in spontaneous mutation rate per base pair among these genomes suggests that the larger genomes have evolved lower mutation rates per nucleotide pair through either greater fidelity of DNA replication or greater efficiency of repair mechanisms.

Problem 2

The nontemplate DNA strand in an amino-acid–coding portion of a gene is shown here, along with the amino acid sequence of the polypeptide chain encoded in this region. This particular region of the polypeptide is quite tolerant of amino acid replacements, so many missense mutations in the region do not destroy function. Four frameshift mutations are isolated: −A is a deletion of the red A, +A is a duplication of the red A, −G is a deletion of the red G, and +G is a duplication of the red G. As expected, all four mutations lead to a nonfunctional protein. Recombination is used to combine the mutations, in the expectation that a "−" deletion of one nucleotide will be compensated for by a "+" addition slightly farther along, because the second mutation will shift the reading frame back to the normal reading frame, and changes in the amino acids between the mutations should not destroy function. An unexpected result was obtained. Although the combination +A with −G did, indeed, restore protein function, the combination −A with +G was still mutant. How can you account for this result?

GCTACAAAAATGACTGGCACTGCA
AlaThrLysMetThrGlyThrAla

Solution In this kind of problem, the best strategy is first to deduce the amino acid sequence of each mutant polypeptide chain. The answer often becomes clear immediately. In this case, the mutant amino acid sequences for all four mutants and the two types of double mutants are shown. Note that the −A mutation results in a termination codon (UGA), which terminates translation at that codon (indicated by the raised dot). The −A with +G combination also has this termination codon, so it remains mutant; even though the frameshift is corrected by the +G, the correction is ineffectual because translation is already terminated prior to this point. The +A with −G combination is nonmutant, because the frameshifts do compensate, and there is no termination codon in the region between them.

−A	GCUACAAAAUGACUGGCACUGCA AlaThrLys ·
+A	GCUACAAAAAAUGACUGGCACUGCA AlaThrLysAsnAspTrpHisCys
−G	GCUACAAAAAUGACUGCACUGCA AlaThrLysMetThrAlaLeu
+G	GCUACAAAAAUGACUGGGCACUGCA AlaThrLysMetThrGlyHisCys
+A with −G	GCUACAAAAAAUGACUGCACUGCA AlaThrLysAsnAspCysThrAla
−A with +G	GCUACAAAAUGACUGGGCACUGCA AlaThrLys ·

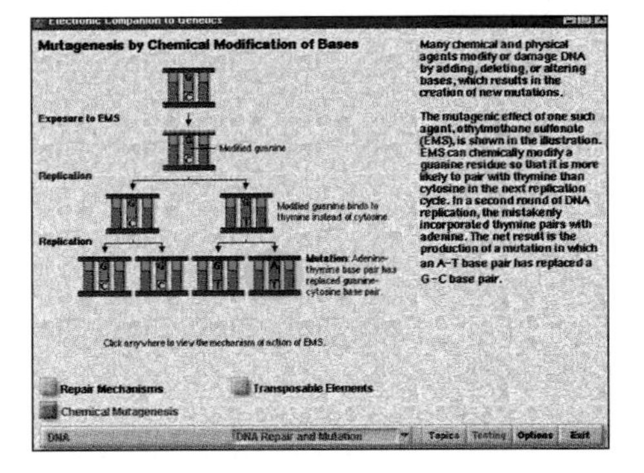

To ENHANCE YOUR STUDY turn to *An Electronic Companion to Genetics*™, © 1998, Cogito Learning Media, Inc. This CD-ROM is a multimedia tool that provides interactive explanations of important genetic concepts through animations, diagrams, and videos. It also provides interactive test questions.

Concepts in Action
Problems for Solution

7.1 Occasionally, a person is found who has one blue eye and one brown eye or who has a sector of one eye a different color from the rest. Can these phenotypes be explained by new mutations? If so, in what types of cells must the mutations occur?

7.2 How many different codons can result from a single-base substitution in DNA coding for the cysteine codon UGC? Classify each as silent, missense, or nonsense.

geNETics on the web

www.jbpub.com/genetics

These GeNETics on the Web exercises will introduce some of the most useful sites for finding genetic information on the World Wide Web. Genetic sites provide access to a rich storehouse of information on all aspects of genetics. These range from sites written in nontechnical language for the lay person to sites with sophisticated databases designed for the professional geneticist. In carrying out these exercises, you will get a taste of what the Internet can offer a student in genetics.

The keywords shown in color in the following exercises are available on the Jones and Bartlett Publishers' web site as hyperlinks to various genetic sites. To complete the exercises, visit GeNETics on the Web home page at

http://www.jbpub.com/genetics

Select the link to Essential geNETics on the web. Then choose a chapter. You find a list of keywords that correspond to the exercises below. Select a keyword to link to a web site containing the genetic information necessary to complete the exercise. Each exercise includes a specific assignment that makes use of the information available at the site.

Exercises

- Several types of hereditary colon cancer are associated with inherited defects in DNA repair, including hereditary nonpolyposis colorectal cancer (HNPCC), which involves homologs of the *E. coli* genes *mutS* and *mutL*. You can learn more about hereditary colon cancer at this keyword site. Prepare a written report answering the following questions. What proportion of all colon cancers (large-bowel cancers) are directly caused by inherited genetic abnormalities? Should people be tested for colon cancer if they have no family history of the disease? How are colon cancers detected? At what age should testing begin for people with a family history of the disease? For people without a family history of the disease? What are some of the issues that must be considered in early testing for the types of hereditary colon cancer that can be detected by blood tests?
- The Chornobyl (also spelled Chernobyl) nuclear accident was by far the most devastating in the history of nuclear power. It released into Earth's atmosphere about 400 times more radioactive material than the atomic bomb dropped on Hiroshima (but less than 1 percent of the amount from atmospheric weapons tests conducted in the 1950s and 1960s). This keyword site summarizes what is known about the nuclear accident and its aftermath. Write a brief report discussing what happened, how it happened, how many people were significantly exposed to radiation, the immediate health effects, and the anticipated long-term health and environmental effects of this environmental disaster.

Pic Site

The Pic Site showcases some of the most visually appealing genetics sites on the World Wide Web. To visit the showcase genetics site, select the Pic Site for Chapter 7.

7.3 Weedy plants that are resistant to the herbicide atrazine have a single amino acid substitution in the gene *psbA* that results in the replacement of a serine with an alanine in the polypeptide. Is the base change in the *psbA* gene that results in this amino acid replacement a transition or a transversion?

7.4 What is the minimum number of single-nucleotide substitutions that would be necessary for each of the following amino acid replacements?

(a) Trp → Lys **(b)** Tyr → Gly
(c) Met → His **(d)** Ala → Asp

7.5 How many amino acids can replace tyrosine by a single-base substitution in the DNA? (Do not assume that you know which tyrosine codon is being used.)

7.6 In the *rIIB* gene of bacteriophage T4, there is a small coding region near the beginning of the gene in which the particular amino acid sequence that is present is not essential for protein function. (Single-base insertions and deletions in this region were used to prove the triplet nature of the genetic code.) Under what condition would a $+1$ frameshift mutation at the beginning of this region not be suppressed by a -1 frameshift mutation at the end of the region?

7.7 What genetic characteristic is often associated with mutations caused by the insertion of a transposable element?

7.8 A population of 1×10^6 bacterial cells undergoes one round of DNA replication and cell division. The forward mutation rate of a gene is 1×10^{-6} per replication.

(a) What is the expected number of mutant cells after cell division?

(b) What is the probability that the population contains no mutant cells?

7.9 Every human gamete contains, very approximately, 80,000 genes. If the forward mutation rate is between 10^{-5} and 10^{-6} new mutations per gene per generation, what percentage of all gametes will carry a gene that has undergone a spontaneous mutation?

7.10 A strain of *E. coli* contains a mutant $tRNA^{Leu}$. Instead of recognizing the codon 5'-CUG-3', as it does in nonmutant cells, the mutant tRNA recognizes the codon 5'-GUG-3'. A missense mutation of another gene, affecting amino acid number 28 along the chain, is suppressed in cells with the mutant $tRNA^{Leu}$.

(a) Assuming normal Watson–Crick base pairs, determine the anticodons of the wildtype and mutant $tRNA^{Leu}$ molecules.
(b) What kind of mutation is present in the mutant $tRNA^{Leu}$ gene?
(c) What amino acid would be inserted at position 28 of the mutant polypeptide chain if the missense mutation were not suppressed?
(d) What amino acid is inserted at position 28 when the missense mutation is suppressed?

7.11 Gene conversion results from mismatch repair in a heteroduplex DNA molecule that gives rise to fungal asci with ratios of alleles such as 3 *A* : 1 *a* and 1 *A* : 3 *a* , instead of 2 *A* : 2 *a* .

(a) Why is a ratio of 2 *A* : 2 *a* expected?
(b) Why, among a large number of gametes chosen at random, is the Mendelian segregation ratio of 1 *A* : 1 *a* still observed in spite of gene conversion?

7.12 You carry out a large-scale cross of genotypes *A m B* × *a + b* of two strains of *Neurospora crassa* and observe a number of aberrant asci, some of which are shown here.

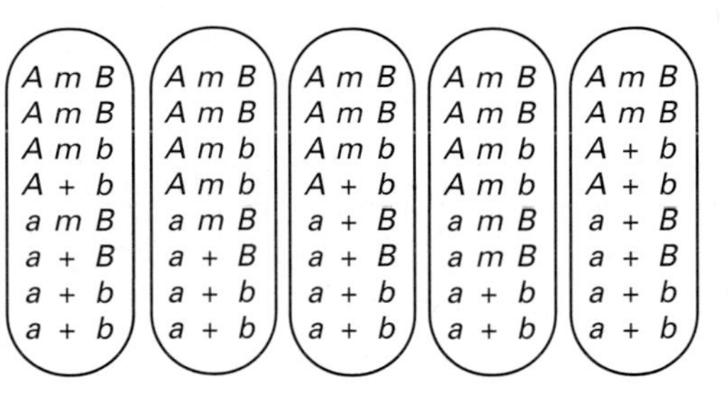

Your colleague who provided the strains insists that they are both deficient in the same gene in the DNA mismatch repair pathway. The results of your cross seem to contradict this assertion.

(a) Which of the asci depicted here offer evidence that your colleague is incorrect?
(b) Which ascus (or asci) would you exhibit as definitive evidence that DNA mismatch repair in these strains is not 100 percent efficient?
(c) Among all asci resulting from meioses in which heteroduplexes form across the *m*/+ region, what proportion of asci will be tetratype for the *A* and *B* markers and will also show 4 : 4 segregation of *m* : +, under the assumption that 30 percent of heteroduplexes in the region are not corrected, 40 percent are corrected to *m*, and 30 percent are corrected to +?

Further Readings

Ames, B. W. 1979. Identifying environmental chemicals causing mutations and cancer. *Science* 204: 587.

Bhatia, P. K., Z. G. Wang, and E. C. Friedberg. 1996. DNA repair and transcription. *Current Opinion in Genetics & Development* 6: 146.

Camerini-Otero, R. D., and P. Hsieh. 1995. Homologous recombination proteins in prokaryotes and eukaryotes. *Annual Review of Genetics* 29: 509.

Chu, G., and L. Mayne. 1996. Xeroderma pigmentosum, Cockayne syndrome and trichothiodystrophy: Do the genes explain the diseases? *Trends in Genetics* 12: 187.

Cox, E. C. 1997. *mutS*, proofreading, and cancer. *Genetics* 146: 443.

Crow, J. F., and C. Denniston. 1985. Mutation in human populations. *Advances in Human Genetics* 14: 59.

De la Chapelle, A., and P. Peltomaki. 1995. Genetics of hereditary colon cancer. *Annual Review of Genetics* 29: 329.

Drake, J. W. 1991. Spontaneous mutation. *Annual Review of Genetics* 25: 125.

Eggleston, A. K., and S. C. West. 1996. Exchanging partners: Recombination in *E. coli*. *Trends in Genetics* 12: 20.

Hoeijmakers, J. H. J. 1993. Nucleotide excision repair I: From *E. coli* to yeast. *Trends in Genetics* 9: 173.

Jackson, S. P. 1996. The recognition of DNA damage. *Current Opinion in Genetics & Development* 6: 19.

Shcherbak, Y. M. 1996. Confronting the nuclear legacy. I. Ten years of the Chernobyl era. *Scientific American*, April.

Singer, B., and J. T. Kusmierek. 1982. Chemical mutagenesis. *Annual Review of Biochemistry* 51: 655.

Sommer, S. S. 1995. Recent human germ-line mutation: Inferences from patients with hemophilia B. *Trends in Genetics* 11: 141.

Weinberg, R. A. 1996. How cancer arises. *Scientific American*, September.

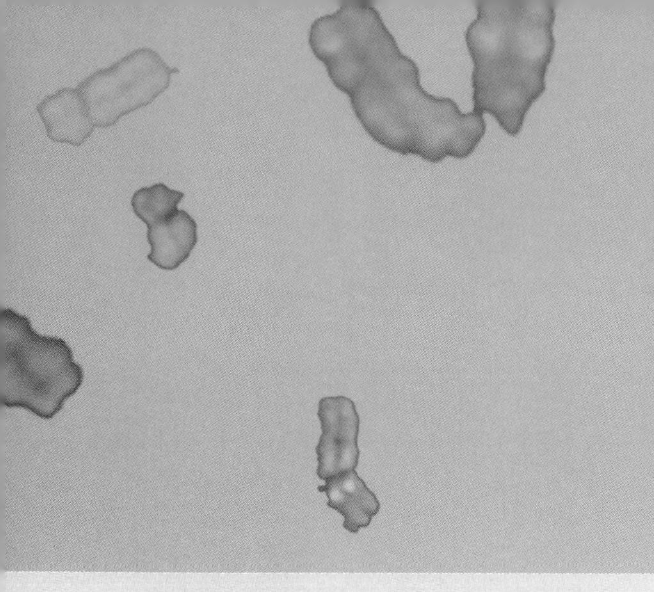

Key Concepts

- Some bacteria are capable of DNA transfer and genetic recombination.
- In *E. coli,* the F (fertility) plasmid can mobilize the chromosome for transfer to another cell in the process of conjugation.
- Some types of bacteriophages can incorporate bacterial genes and transfer them into new host cells in the process of transduction.
- DNA molecules from related bacteriophages that are present in the same host cell can undergo genetic recombination.
- Transposable elements and plasmids are widely used for genetic analysis and DNA manipulation in bacteria.

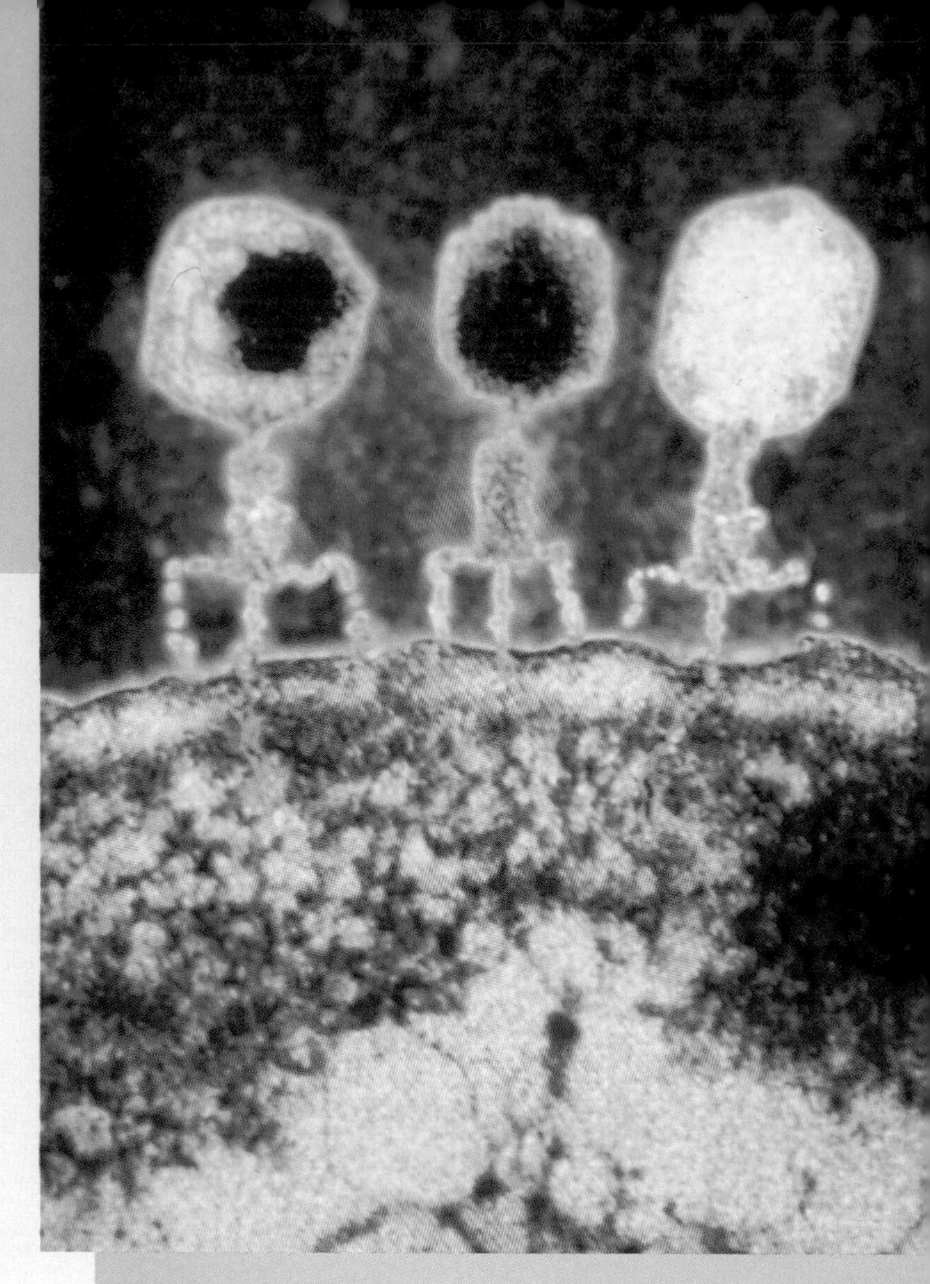

T2 BACTERIOPHAGE attacking an *E. coli* cell. The DNA of the phages can be seen to be entering through the cell wall.

CHAPTER 8

The Genetics of Bacteria and Viruses

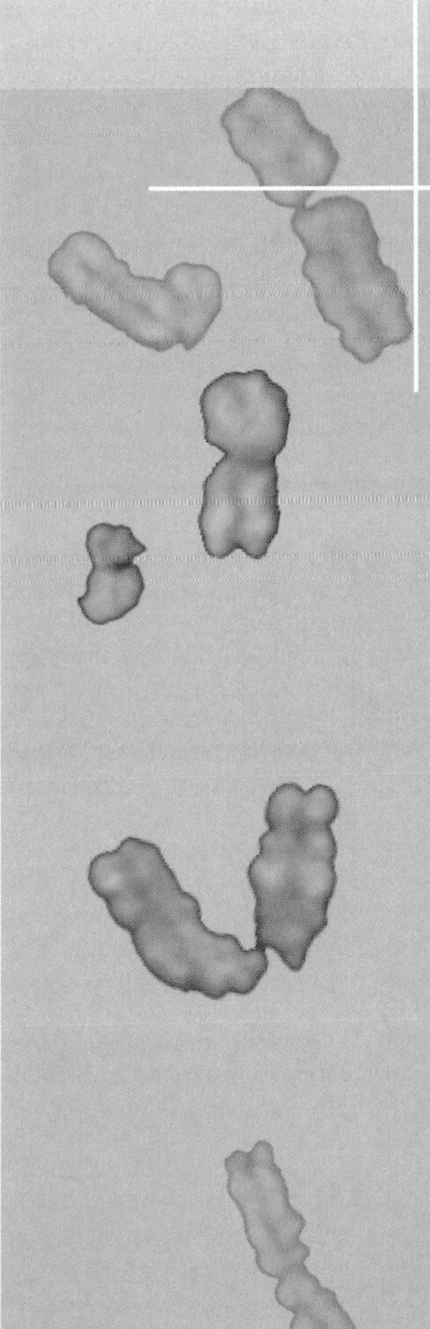

The use of bacteria and the bacteriophages that infect them offers four important advantages in genetic research. First, they are haploid, so dominance or recessiveness of alleles is not a complication in identifying genotype. Second, a new generation is produced in minutes rather than weeks or months, which vastly increases the rate at which data accumulate. Third, they are easy to grow in enormous numbers under controlled laboratory conditions, which facilitates biochemical studies and the analysis of rare genetic events. Fourth, a laboratory population derived from the reproduction of a single cell or phage is a **clone,** in which the individual organisms are genetically identical.

In Chapter 4, we examined the genetic properties of linkage and genetic recombination in eukaryotic organisms. In meiosis and sexual reproduction, genotypic variation among the progeny is achieved both by independent assortment of nonhomologous chromosomes and by crossing-over, resulting in recombination between linked genes. Two important features of crossing-over in eukaryotes are (1) that it results in a reciprocal exchange of material between two homologous chromosomes and (2) that both products of a single exchange can often be recovered.

The characteristics of recombination are quite different in prokaryotes. A bacterial cell contains a major DNA molecule that almost never encounters another complete DNA molecule. Instead, genetic recombination is usually between a chromosomal fragment from one cell and an intact chromosome from another cell. Furthermore, a clear donor-recipient relationship exists: The donor cell is the source of the DNA fragment, which is transferred to the recipient cell by one of several mechanisms, and recombination of genetic material takes place in the recipient. Incorporation of a part of the transferred donor DNA into the chromosome requires at least two (or any even number) of exchange events because the recipient molecule is circular. The usual outcome of these events is the recovery of *only one* of the recombinant products. However, in some situations, the transferred DNA also is circular, and a single exchange results in the incorporation of the complete donor molecule into the chromosome of the recipient.

8.1 Much of our understanding of molecular genetics comes from bacteria and bacteriophages.

Bacteria can be grown both in liquid medium and on the surface of a semisolid growth medium hardened with agar. Bacteria used in genetic analysis are usually grown on an agar surface in plastic petri dishes (called plates). A single bacterial cell placed on a solid medium will grow and divide many times, forming a visible cluster of cells called a *colony* (Figure 8.1). The number of bacterial cells in a suspension can be determined by spreading a known volume of the suspension on an agar surface and counting the colonies that form. Typical *E. coli* cultures contain as many as 10^9 cells/ml. The appearance of colonies, or the ability or inability to form colonies, on a particular medium can sometimes be used to identify the genotype of the cell that produced the colony.

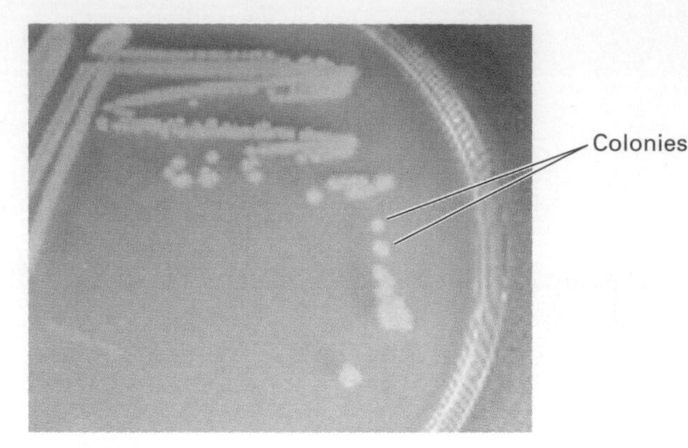

Figure 8.1 A petri dish with bacterial colonies that have formed on a solid medium. The heavy streaks of growth result from colonies so densely packed that there is no space between them.

As in other organisms, genetic analysis in bacteria requires mutants. In bacteria, mutations that affect metabolic pathways or antibiotic resistance are particularly useful. There are three principal types of mutations.

- **Antibiotic-resistant mutants** These mutants are able to grow in the presence of an antibiotic, such as streptomycin (Str) or tetracycline (Tet). For example, streptomycin-sensitive (Str-s) cells have the wildtype phenotype and fail to form colonies on medium containing streptomycin, but streptomycin-resistant (Str-r) mutants can form colonies.

- **Nutritional mutants** Wildtype bacteria can synthesize most of the complex nutrients they need from simple molecules present in the growth medium. The wildtype cells are said to be **prototrophs.** The ability to grow in simple medium can be lost by mutations that disable the enzymes used in synthesizing the complex nutrients. Mutant cells are unable to synthesize an essential nutrient and thus cannot grow unless the required nutrient is supplied in the medium. Such a mutant bacterium is said to be an **auxotroph** for the particular nutrient. For example, a methionine auxotroph cannot grow on a **minimal medium** containing only inorganic salts and a source of energy and carbon atoms (such as glucose), but such Met^- cells can grow if the minimal medium is supplemented with methionine.

- **Carbon-source mutants** These mutants cannot utilize particular substances as sources of energy or carbon atoms. For example, Lac^- mutants cannot utilize the sugar lactose for growth and are unable to form colonies on minimal medium containing lactose as the only carbon source.

A medium on which all wildtype cells form colonies is called a **nonselective medium.** Mutants and wildtype cells may or may not be distinguishable by growth on a nonselective medium. If the medium allows growth of only one type of cell (either wildtype or mutant), it is said to be **selective.** For example, a medium containing streptomycin is selective for the Str-r (resistant) phenotype and selective against the Str-s (sensitive) phenotype, and minimal medium containing lactose as the sole carbon source is selective for Lac^+ cells and against Lac^- cells.

In bacterial genetics, phenotype and genotype are designated in the following way. A phenotype is designated by three letters, the first of which is capitalized; a superscript + or − denotes the presence or absence of the designated character; and s or r denotes sensitivity or resistance, respectively. A genotype is designated by lowercase italicized letters. Thus a cell unable to grow without a supplement of leucine (a leucine auxotroph) has a Leu^- phenotype, and this would usually result from a leu^- mutation in one of the genes required for leucine biosynthesis. Often the − superscript is omitted, but we will use it consistently to avoid ambiguity.

8.2 Transformation results from the uptake of DNA and recombination.

As we saw in Chapter 1, important evidence that DNA is the genetic material came from experiments in which DNA from a heat-killed virulent strain of a pneumonia-causing bacterium was able to convert genetically cells of another strain from nonvirulent into virulent. The process of genetic alteration by pure DNA is *transformation,* and we know much more about it now than was known in 1944 when the experiments were carried out.

In transformation, recipient cells acquire genes from free DNA molecules in the surrounding medium. In laboratory experiments, DNA isolated from donor cells is added to a suspension of recipient cells. In nature, DNA can become available by spontaneous breakage (lysis) of donor cells. Either way, transformation begins with uptake of a DNA fragment from the surrounding medium by a recipient cell and terminates with *one strand* of donor DNA replacing the homologous segment in the recipient DNA. Most bacterial species are probably capable of the recombination step, but the ability of most bacteria to take up DNA efficiently is limited. Even in a species capable of transformation, DNA is able to penetrate only some of the cells in a growing population. However, appropriate chemical treatment of cells of these species yields a population of cells that are competent to take up DNA.

Transformation affords a convenient technique for gene mapping. DNA that is isolated from a donor bacterium is invariably broken into small fragments. With suitable recipient cells and excess DNA, transformation takes place at a frequency of about one transformed cell per 10^3 cells. If two genes, *a* and *b*, are so widely separated in the donor chromosome that they are always contained in two different DNA fragments, then the probability of simultaneous transformation (**cotransformation**) of an $a^- \ b^-$ recipient into wildtype $a^+ \ b^+$ is the product of the probabilities of transformation of each marker separately, or roughly $10^{-3} \times 10^{-3}$, which equals one wildtype transformant per 10^6 recipient cells. However, if the two genes are so near one another that they are often present in a single donor fragment, then the frequency of cotransformation is nearly the same as the frequency of single-gene transformation, or one wildtype transformant per 10^3 recipients. The general principle is as follows:

> Cotransformation of two genes at a frequency substantially greater than the product of the single-gene transformation frequencies implies that the two genes are close together in the bacterial chromosome.

Studies of the ability of various pairs of genes to be cotransformed also yield gene order. For example, if genes *a* and *b* can be cotransformed, and genes *b* and *c* can be cotransformed, but genes *a* and *c* cannot, then the gene order must be *a b c* (Figure 8.2). Note that cotransformation frequencies are not equivalent to the recombination frequencies used in mapping eukaryotes (Chapter 4), because cotransformation frequencies are determined by the size distribution of donor fragments and the likelihood of recombination between bacterial DNA molecules, rather than by the formation of chiasmata between synapsed homologous chromosomes.

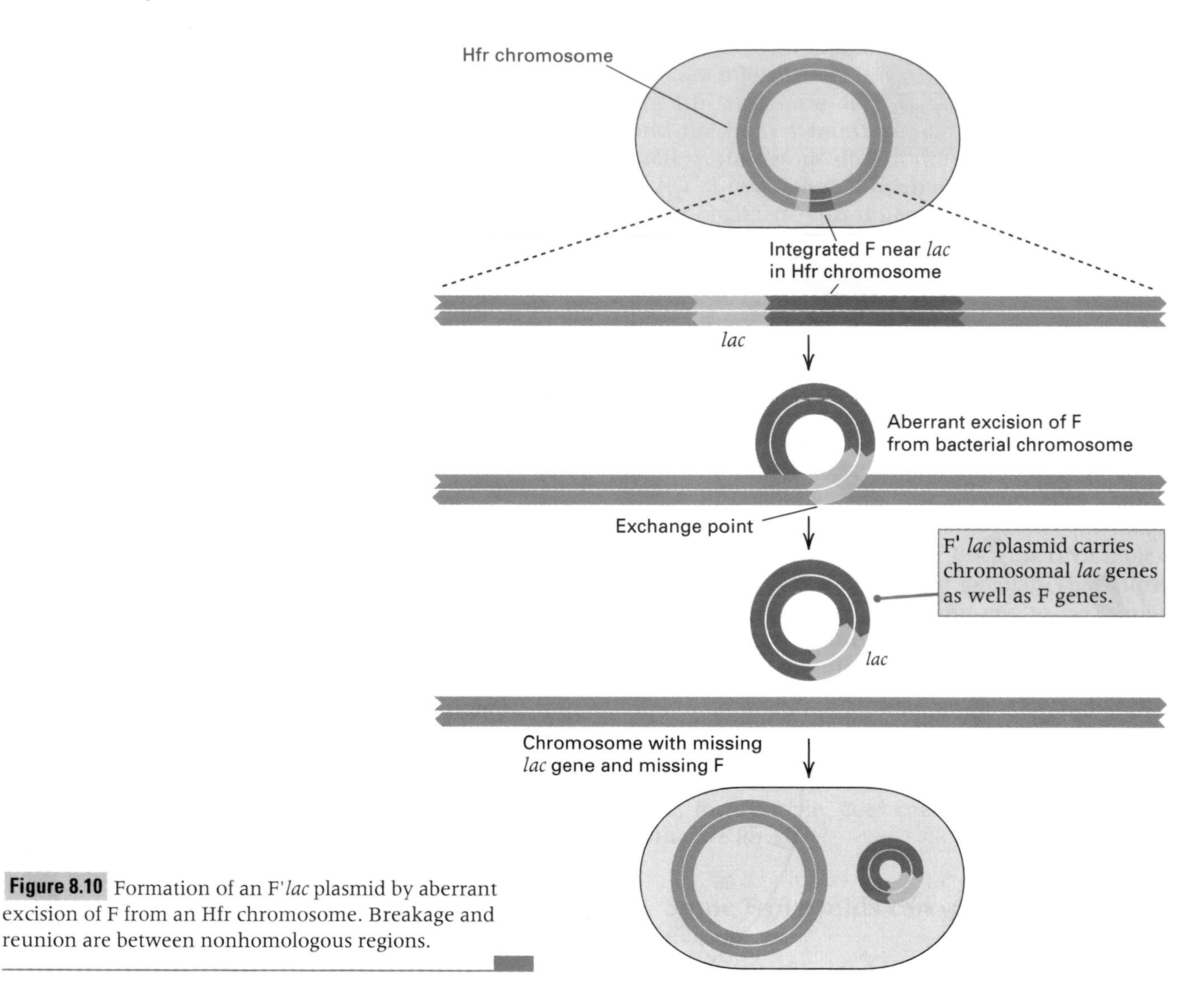

Figure 8.10 Formation of an F'*lac* plasmid by aberrant excision of F from an Hfr chromosome. Breakage and reunion are between nonhomologous regions.

8.4 Some phages can transfer small pieces of bacterial DNA.

In the process of **transduction,** bacterial DNA is transferred from one bacterial cell to another by a phage particle containing the DNA. Such a particle is called a **transducing phage.** Two types of transducing phages are known, generalized and specialized. A **generalized transducing phage** produces some particles that contain only DNA obtained from the host bacterium, rather than phage DNA; the bacterial DNA fragment can be derived from *any* part of the bacterial chromosome. A **specialized transducing phage** produces particles that contain both phage and bacterial genes linked in a single DNA molecule, but the bacterial genes are obtained from a *particular* region of the bacterial chromosome. In this section, we consider *E. coli* phage P1, a well-studied generalized transducing phage. Specialized transducing particles are discussed in Section 8.6.

During infection by P1, the phage makes a nuclease that cuts the bacterial DNA into fragments. Single fragments of bacterial DNA comparable in size to P1 DNA are occasionally packaged into phage particles in place of P1 DNA. The positions of the nuclease cuts in the host chromosome are random, so a transducing particle may contain a fragment derived from any region of the host DNA. A large population of P1 phages will contain a few particles carrying any bacterial gene. On the average, any particular gene is present in roughly one transducing particle per 10^6 phages. When a transducing particle adsorbs to a bacterium, the bacterial DNA contained in the phage head is injected into the cell and becomes available for recombination with the homologous region of the host chromosome. A typical P1 transducing particle contains from 100 to 115 kb of bacterial DNA.

Let us now examine the events that follow infection of a bacterium by a generalized transducing particle obtained, for example, by growth of P1 on wildtype *E. coli* containing a leu^+ gene (Figure 8.11).

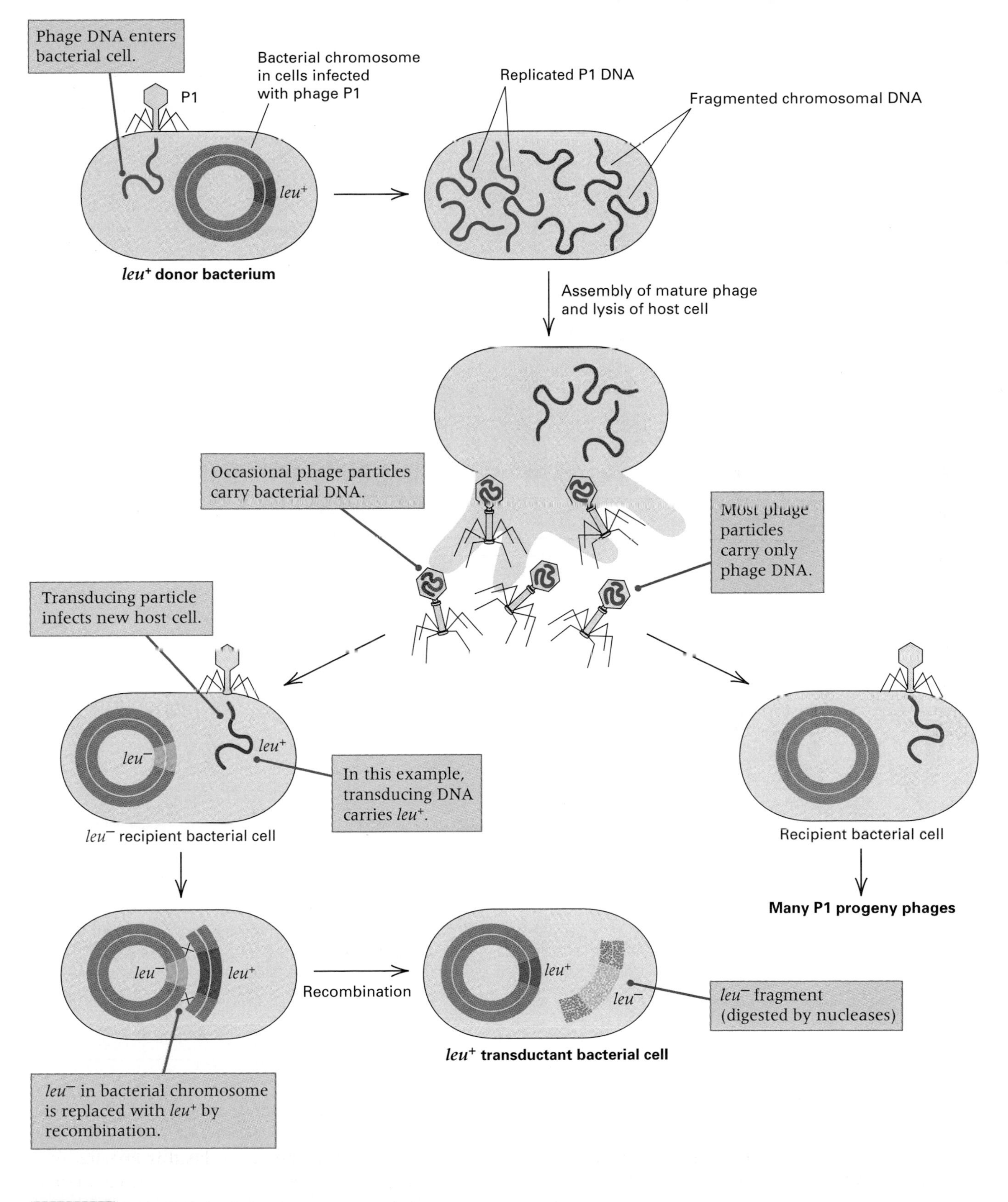

Figure 8.11 Transduction. Phage P1 infects a *leu*$^+$ donor, yielding predominantly normal P1 phages with an occasional one carrying bacterial DNA instead of phage DNA. If the phage population infects a bacterial culture, then the normal phages produce progeny phages, whereas the transducing particle yields a transductant. Note that the recombination step requires two crossovers. For clarity, double-stranded phage DNA is drawn as a single line.

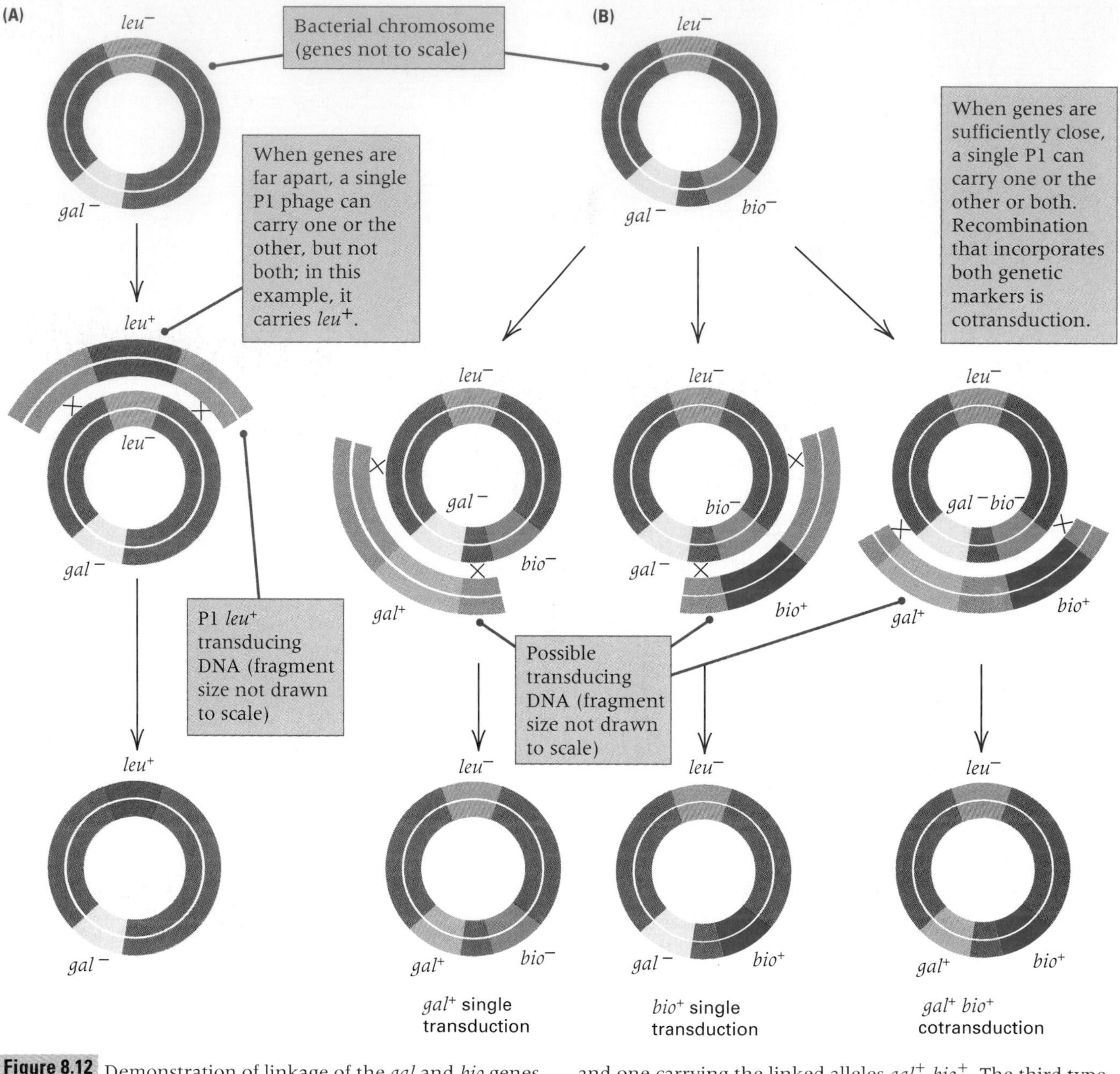

Figure 8.12 Demonstration of linkage of the *gal* and *bio* genes by cotransduction. (A) A P1 transducing particle carrying the leu^+ allele can convert a leu^- gal^- cell into a leu^+ gal^- genotype (but cannot produce a leu^+ gal^+ genotype). (B) The transductants that could be formed by three possible types of transducing particles—one carrying gal^+, one carrying bio^+, and one carrying the linked alleles gal^+ bio^+. The third type results in cotransduction. For clarity, the distance between *gal* and *bio*, relative to that between *leu* and *gal*, is greatly exaggerated, and the size of the DNA fragment in a transducing particle, relative to the size of the bacterial chromosome, is not drawn to scale.

If such a particle adsorbs to a bacterial cell of leu^- genotype and injects the DNA that it contains into the cell, the cell survives because the phage head contained only bacterial genes and no phage genes. A recombination event exchanging the leu^+ allele carried by the phage for the leu^- allele carried by the host converts the genotype of the host cell from leu^- into leu^+. In such an experiment, typically about one leu^- cell in 10^6 becomes leu^+. Such frequencies are easily detected on selective growth medium. For example, if the infected cell is placed on solid medium lacking leucine, it is able to multiply and a leu^+ colony forms. A colony does not form unless recombination inserted the leu^+ allele.

The small fragment of bacterial DNA contained in a transducing particle includes about 50 genes, so transduction provides a valuable tool for genetic linkage studies of short regions of the bacterial genome. Consider a population of P1 prepared from a leu^+ gal^+ bio^+ bacterium. This population contains particles able to transfer any of these alleles to another cell; that is, a leu^+ particle can transduce a leu^- cell to leu^+, or a gal^+ particle can transduce a gal^- cell to gal^+. Furthermore, if a leu^- gal^- culture is infected by phage, both leu^+ gal^- and leu^- gal^+ bacteria are produced. However, leu^+ gal^+ colonies do not arise, because the *leu* and *gal* genes are too far apart to be included in the same DNA fragment (Figure 8.12A).

The situation is quite different for a recipient cell with genotype gal^- bio^-, because the *gal* and *bio* genes are so closely linked that both genes are sometimes present in a single DNA fragment carried in a transducing particle—namely, a *gal-bio* particle (Figure 8.12B). However, not all gal^+ transducing particles also include bio^+, nor do all bio^+ particles include gal^+. The probability of both markers being in a single particle, and hence the probability of simultaneous transduction of both markers (**cotransduction**), depends on how close to each other the genes are. The closer they are, the greater the frequency of cotransduction. Cotransduction of the gal^+-bio^+ pair can be detected by plating infected cells on the appropriate growth medium. If bio^+ transductants are selected by spreading the infected cells on a glucose-containing medium that lacks biotin, both gal^+ bio^+ and gal^- bio^+ colonies grow. If these colonies are tested for the *gal* marker, 12 percent are found to be gal^+ bio^+ and the rest gal^- bio^+; similarly, if gal^+ transductants are selected, 12 percent are found to be gal^+ bio^+. In other words, the **frequency of cotransduction** of *gal* and *bio* is 12 percent, which means that 12 percent of all transducing particles that contain one gene also include the other gene.

Studies of cotransduction can be used to map closely linked genetic markers by means of three-factor crosses analogous to those described in Chapter 4. That is, P1 is grown on wildtype bacteria and used to transduce cells carrying a mutation of each of three closely linked genes. Cotransductants that contain various pairs of wildtype alleles are examined. The gene located in the middle can be identified because its wildtype allele is nearly always cotransduced with the wildtype alleles of the genes that flank it. For example, in Figure 8.12B, a genetic marker located between gal^+ and bio^+ would almost always be present in gal^+ bio^+ transductants.

8.5 Bacteriophage DNA molecules in the same cell can recombine.

The reproductive cycle of a phage is called the **lytic cycle** and is diagrammed in Figure 8.13. In the lytic cycle, phage DNA enters a cell and replicates repeatedly, bacterial ribosomes are used to produce phage protein components, the newly synthesized

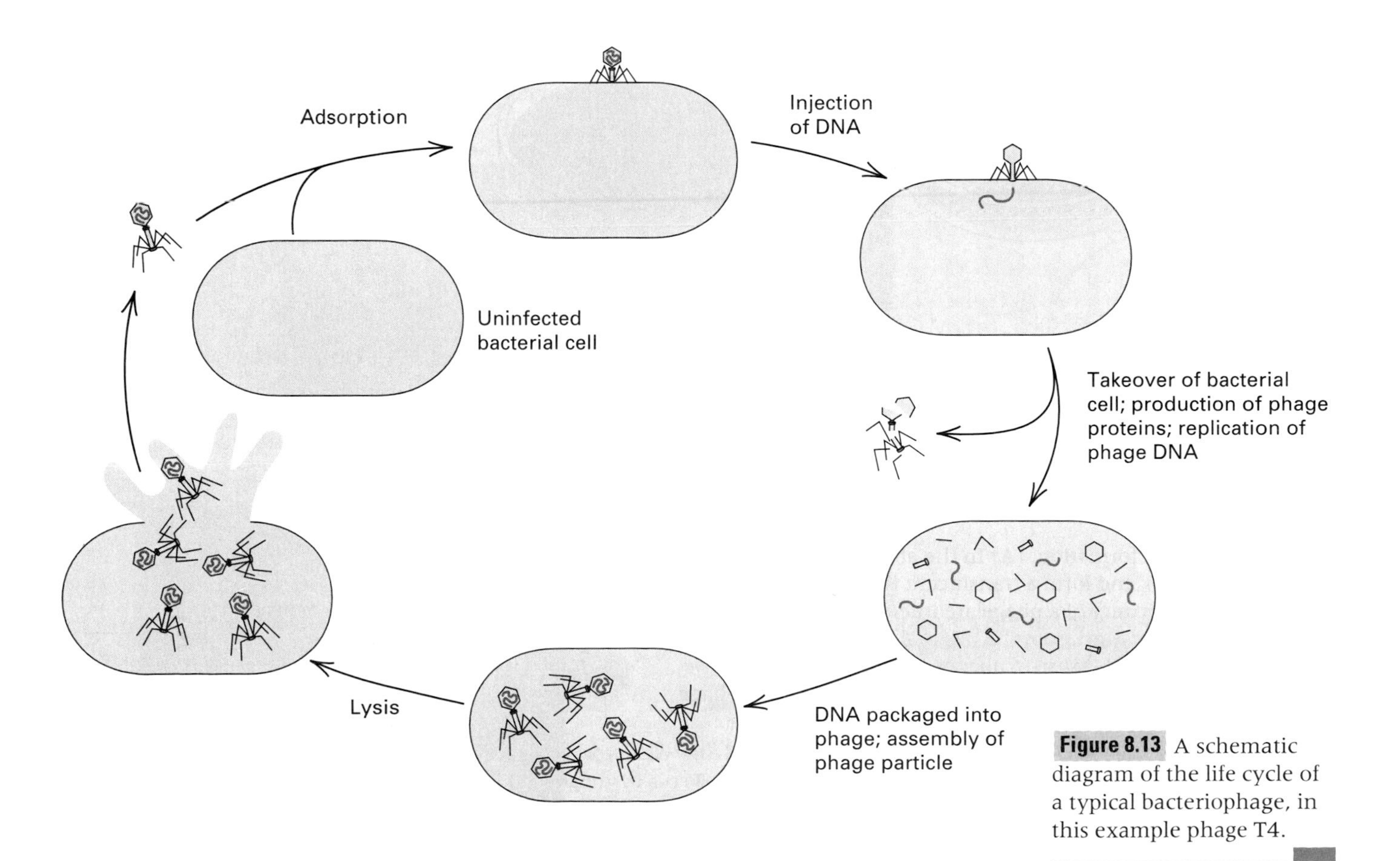

Figure 8.13 A schematic diagram of the life cycle of a typical bacteriophage, in this example phage T4.

such an element, the usual designations for the genes are used. For example, Tn5(*neo-r ble-r str-r*) contains genes for resistance to three different antibiotics: neomycin, bleomycin, and streptomycin. Such genes provide genetic markers, making it easy to detect transposition of the composite element, as is shown in Figure 8.27. An F'*lac* plasmid carrying wildtype alleles of the genes for lactose utilization is transferred by conjugation into a bacterial cell containing a transposable element that carries the *neo-r* (neomycin-resistance) gene. The bacterial cell is allowed to grow, and in the course of multiplication, transposition of the transposon into the F' plasmid occasionally takes place in a progeny cell. Transposition yields an F' plasmid containing both the wildtype *lac* genes and the *neo-r* gene. In a subsequent mating with a Neo-s Lac$^-$ cell, the F'*lac* genes and the *neo-r* marker are transferred together and so are genetically linked.

In nature, sequential insertion of transposons containing different antibiotic-resistance genes into the same plasmid results in the evolution of plasmids that confer resistance to multiple antibiotics. These multiple-resistance plasmids are called **R plasmids.** The evolution of R plasmids is promoted by the use (and regrettable overuse) of antibiotics, which selects for resistant cells because, in the presence of antibiotics, resistant cells have a growth advantage over sensitive cells. The presence of multiple antibiotics in the environment selects for multiple-drug resistance. Serious clinical complications result when plasmids resistant to multiple drugs are transferred to bacterial *pathogens* that cause disease. Infections caused by pathogens containing R factors are difficult to treat because the cells are resistant to the commonly used antibiotics. These resistance plasmids have become so widely distributed among pathogenic bacteria and confer resistance to so many antibiotics that they are causing a worldwide health crisis in

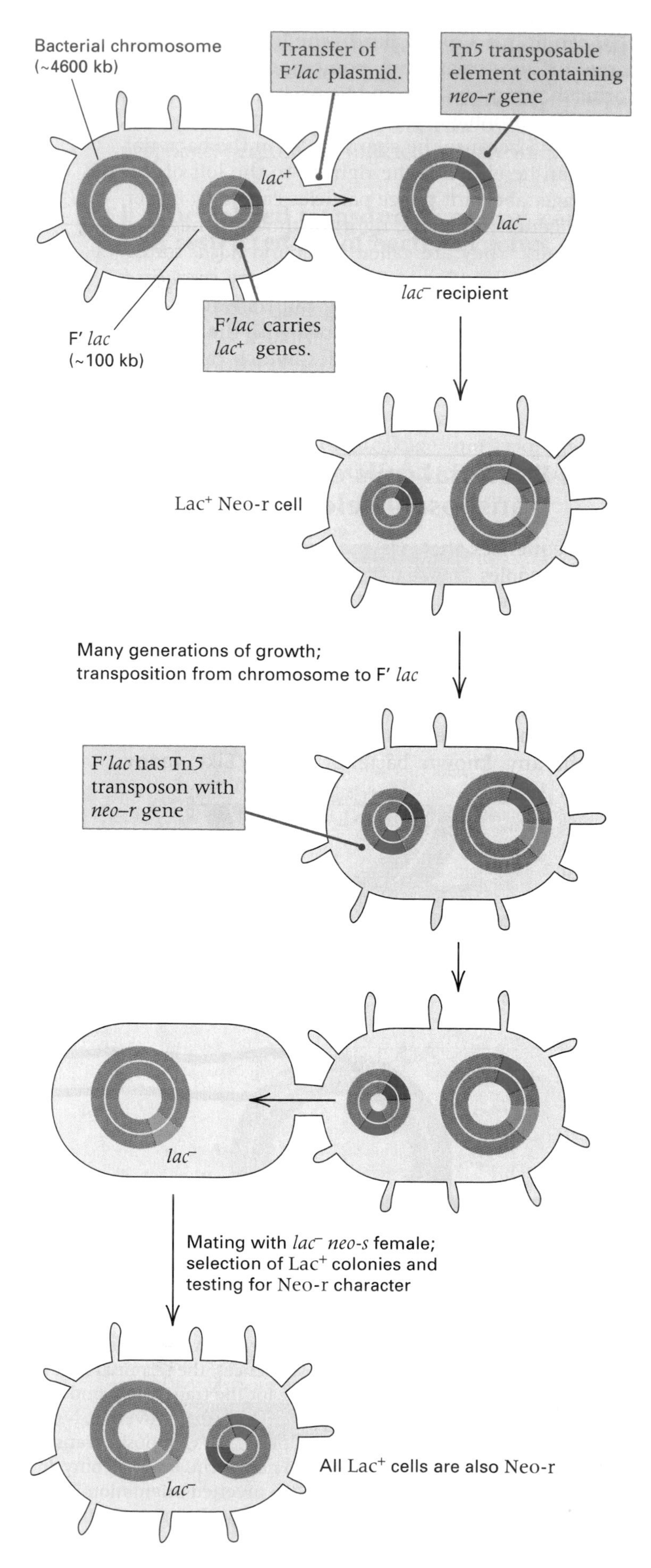

Figure 8.27 An experiment demonstrating the transposition of transposon Tn5, which contains a neomycin-resistance gene, from the chromosome to an F'plasmid containing the bacterial gene for lactose utilization (F'*lac*). The bacterial chromosome is *lac*$^-$ and the cells are unable to grow on lactose unless the F'*lac* is present. After the transposition, the F'*lac* plasmid also contains Tn5, indicated by the linkage of the *neo-r* gene to the F'factor. Note that transposition to the F' plasmid does not eliminate the copy of the transposon in the chromosome.

treatment of tuberculosis. In addition, for some pathogens, there remains not a single antibiotic of any kind to which the bacteria are susceptible.

Bacterial transposons are important tools in genetic analysis.

Transposons can be employed in a variety of ways in bacterial genetic analysis. Three features make them especially useful for this purpose.

1. Transposons can insert at a large number of potential target sites that are essentially random in their distribution throughout the genome.
2. Many transposons code for their own transposases and require only a small number of host genes for mobility.
3. Transposons contain one or more genes for antibiotic resistance that serve as genetic markers for selection.

Genetic analysis using transposons is particularly important in bacterial species that do not have readily exploitable systems for genetic manipulation or large numbers of identified and mapped genes. Many of these species are bacterial pathogens, and transposons can be used to identify and manipulate the disease factors.

The use of a *neo-r* transposable element to identify a particular disease gene in a pathogenic bacterial species is diagrammed in Figure 8.28. In this example, the transposon is introduced into the pathogenic bacterium by a mutant bacteriophage that cannot replicate in the pathogenic host. A variety of other methods of introduction also are possible. After introduction of the transposon and selection on medium containing the antibiotic, the only resistant cells are those in which the transposon is inserted into the bacterial chromosome, because the phage DNA in which it was introduced is incapable of replication. Any of a number of screening methods are then used to identify cells in which a particular disease gene became inactivated (nonfunctional) as a result of the transposon inserting into the gene. This method is known as **transposon tagging** because the gene with the insertion is tagged (marked) with the antibiotic-resistance phenotype of the transposon.

Once the disease gene has been tagged by transposon insertion, it can be used in many ways (for example, in genetic mapping) because the phenotype resulting from the presence of the tagged gene is antibiotic resistance, which is easily identified and selected. The lower part of Figure 8.28 shows how the transposon tag is used to transfer the disease gene into *E. coli*. In the first step, DNA from the

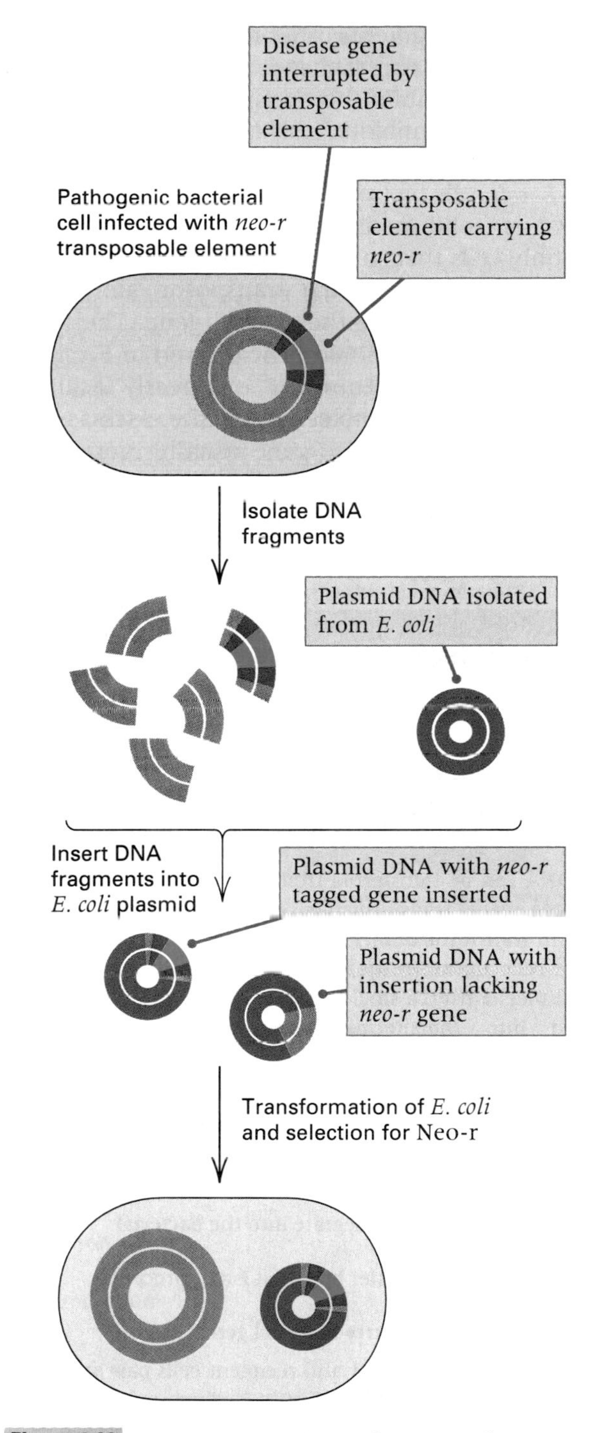

Figure 8.28 Transposon tagging making use of a *neo-r* transposon to mutate a disease gene by inserting into it. The *neo-r* resistance marker is then used to select plasmids containing the disease gene after transfer into *E. coli*, where the gene can be analyzed more safely and easily.

pathogenic strain containing the tagged gene is purified, cut into fragments of suitable size, and inserted into a small plasmid capable of replication in *E. coli*. (Details of these genetic engineering methods are considered in Chapter 10.) The result is a heterogeneous collection of plasmids, most con-

geNETics on the web

www.jbpub.com/genetics

These GeNETics on the Web exercises will introduce some of the most useful sites for finding genetic information on the World Wide Web. Genetic sites provide access to a rich storehouse of information on all aspects of genetics. These range from sites written in nontechnical language for the lay person to sites with sophisticated databases designed for the professional geneticist. In carrying out these exercises, you will get a taste of what the Internet can offer a student in genetics.

The keywords shown in color in the following exercises are available on the Jones and Bartlett Publishers' web site as hyperlinks to various genetic sites. To complete the exercises, visit the GeNETics on the Web home page at

http://www.jbpub.com/genetics

Select the link to Essential geNETics on the web. Then choose a chapter. You find a list of keywords that correspond to the exercises below. Select a keyword to link to a web site containing the genetic information necessary to complete the exercise. Each exercise includes a specific assignment that makes use of the information available at the site.

Exercises

- The Centers for Disease Control has estimated that each year in the United States, approximately two million patients acquire infections while in the hospital, requiring treatment at an estimated cost of $4.5 billion dollars. Among the leading agents for hospital-acquired infections (which are called *nosocomial infections*) is *Staphylococcus aureus*, a bacterium that is particularly troublesome because of its multiple antibiotic resistance. The keyword site staph has information on various types of *Staphylococcus* infections, as well as on the problem of multiple antibiotic resistance. Write a brief report on *S. aureas* infections, including the main types of infections caused by this organism, and on the problem of multiple antibiotic resistance. Why, by the way, is the species called *aureus*?
- Bacteria of the group A *Streptococcus* (GAS) are commonly found in the throat and on the skin. The organism sometimes causes relatively mild infections, including the infections commonly known as *strep throat* and *impetigo*. Rarely the bacteria invade a part of the body that they cannot normally reach, such as the blood, lungs, deep muscle, or fat, and in such cases they may cause a severe—sometimes fatal—infection. The most important of the deep infections are necrotizing fasciitis and one form of toxic shock syndrome. The invasive GAS strains that cause necrotizing fasciitis are sometimes called flesh-eating bacteria. At the keyword site GAS you can get more information about this organism and the infections it causes. On a single sheet of paper, write a report detailing the frequency of invasive GAS infections in the United States, the rate of fatalities among the infections, what types of people are most at risk of invasive GAS disease, and how it can be prevented.

Pic Site

The Pic Site showcases some of the most visually appealing genetics sites on the World Wide Web. To visit the showcase genetics site, select the Pic Site for Chapter 8.

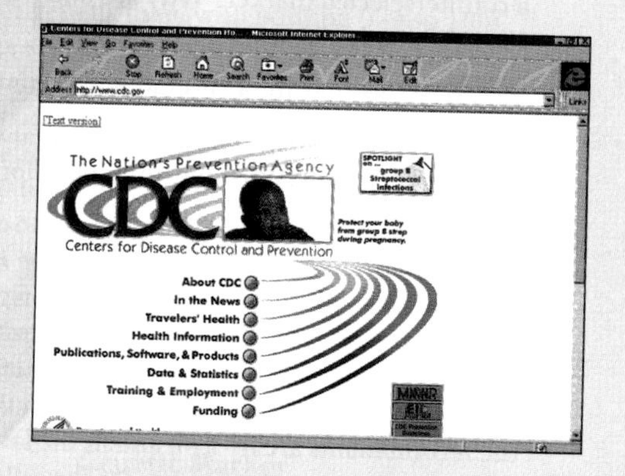

Determine the order in which the substances A, B, C, D, and E are most likely to participate in the biochemical pathway, and indicate the enzymatic steps by arrows. Label each arrow with the name of the gene that codes for the corresponding enzyme.

8.10 The genes *A, B, G, H, I,* and *T* were tested in all possible pairs for cotransduction with bacteriophage P1. Only the following pairs were found able to be cotransduced: *G* and *H*, *G* and *I*, *T* and *A*, *I* and *B*, *A* and *H*. What is the order of the genes along the chromosome? Explain your logic.

8.11 The order of the genes in the λ phage virus is

A B C D E att int xis N CI O P Q S R

(a) Given that the bacterial attachment site, *att*, is between *gal* and *bio* in the bacterial chromosome, what is the prophage gene order?

(b) A mutant phage is discovered that has the reverse gene order in the prophage as in the wildtype prophage. What does this say about the orientation of the *att* site in regard to the termini of the phage chromosome?

(c) A wildtype λ lysogen is infected with another λ phage carrying a genetic marker, *Z*, located between *E* and *att*. The superinfection gives rise to a rare, doubly lysogenic *E. coli* strain that carries both λ and λZ prophage. Assuming that the second phage also entered the chromosome at an *att* site, diagram two possible arrangements of the prophages in the bacterial chromosome and indicate the locations of the bacterial genes *gal* and *bio*.

8.12 An experiment was carried out in *E. coli* to map five genes around the chromosome using each of three different Hfr strains. The genetic markers were *bio, met, phe, his,* and *trp*. The Hfr strains were found to transfer the genetic markers at the times indicated here. Construct a genetic map of the *E. coli* chromosome that includes all five genetic markers, the genetic distances in minutes between adjacent gene pairs, and the origin and direction of transfer of each Hfr. Complete the missing entries in the table, which are indicated by question marks.

Hfr1 markers	*bio*	*met*	*phe*	*his*
Time of entry	26	44	66	?
Hfr2 markers	*phe*	*met*	*bio*	*trp*
Time of entry	?	26	44	75
Hfr3 markers	*phe*	*his*	?	*bio*
Time of entry	6	27	35	?

Further Readings

Adelberg, E. A., ed. 1966. *Papers on Bacterial Genetics.* Boston: Little, Brown.

Campbell, A. 1976. How viruses insert their DNA into the DNA of the host cell. *Scientific American,* December.

Clowes, R. D. 1975. The molecules of infectious drug resistance. *Scientific American,* July.

Davies, J. 1995. Vicious circles: Looking back on resistance plasmids. *Genetics* 139: 1465.

Drlica, K., and M. Riley. 1990. *The Bacterial Chromosome.* Washington, DC: American Society for Microbiology.

Edgar, R. S., and R. H. Epstein. 1965. The genetics of a bacterial virus. *Scientific American,* February.

Hopwood, D. A., and K. E. Chater, eds. 1989. *Genetics of Bacterial Diversity.* New York: Academic Press.

Levy, S. B. 1998. The challenge of antibiotic resistance. *Scientific American,* March.

Losick, R., and D. Kaiser. 1997. Why and how bacteria communicate. *Scientific American,* February.

Low, K. B., and R. Porter. 1978. Modes of genetic transfer and recombination in bacteria. *Annual Review of Genetics* 12: 249.

Miller, R. V. 1998. Bacterial gene swapping in nature. *Scientific American,* January.

Neidhardt, F. C., R. Curtiss III, J. L. Ingraham, E. C. C. Lin, K. B. Low, B. Magasanik, W. S. Reznikoff, M. Riley, M. Schaechter, and H. E. Umbarger, eds. 1996. Escherichia coli *and* Salmonella typhimurium: *Cellular and Molecular Biology* (2 volumes). 2d ed. Washington, DC: American Society for Microbiology.

Novick, R. P. 1980. Plasmids. *Scientific American,* December.

Robertson, B. D., and T. F. Meyer. 1992. Genetic variation in pathogenic bacteria. *Trends in Genetics* 8: 422.

Zinder, N. 1958. Transduction in bacteria. *Scientific American,* November.

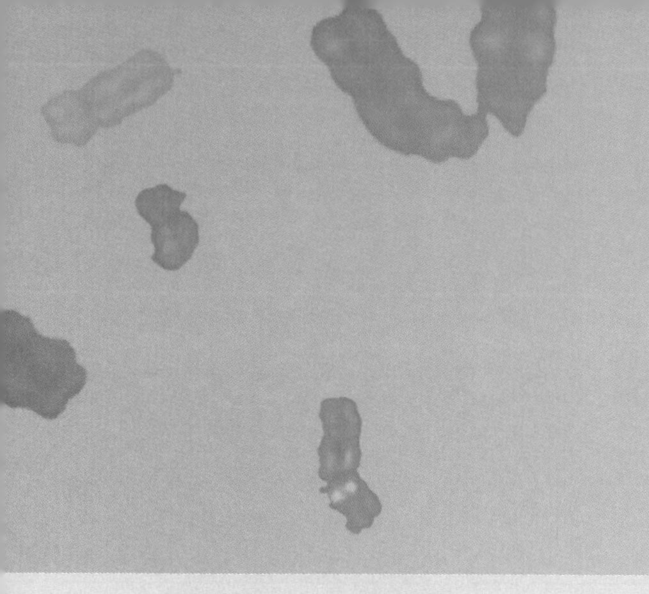

Key Concepts

- In gene expression, information in the base sequence of DNA is used to dictate the linear order of amino acids in a polypeptide by means of an RNA intermediate.
- Transcription of an RNA from one strand of the DNA is the first step in gene expression.
- In eukaryotes, the RNA transcript is modified and may undergo splicing to make the messenger RNA.
- The messenger RNA is translated on ribosomes in groups of three bases (codons), each specifying an amino acid through an interaction with molecules of transfer RNA.

ALL THE CELLS in this plant contain the genetic information for making floral pigments, but the pigment genes are expressed only in the petals.

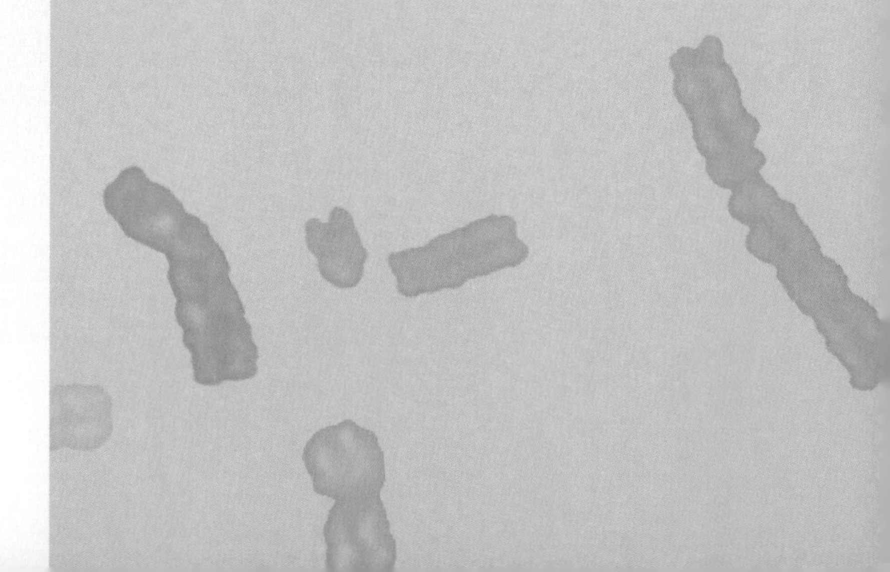

CHAPTER 9

Gene Expression

Figure 9.3 Properties of a polypeptide chain. (A) Formation of a dipeptide by reaction of the carboxyl group of one amino acid (left) with the amino group of a second amino acid (right). A molecule of water (H_2O) is eliminated to form a peptide bond (red line). (B) A tetrapeptide showing the alternation of α-carbon atoms (black) and peptide groups (blue). The four amino acids are numbered below.

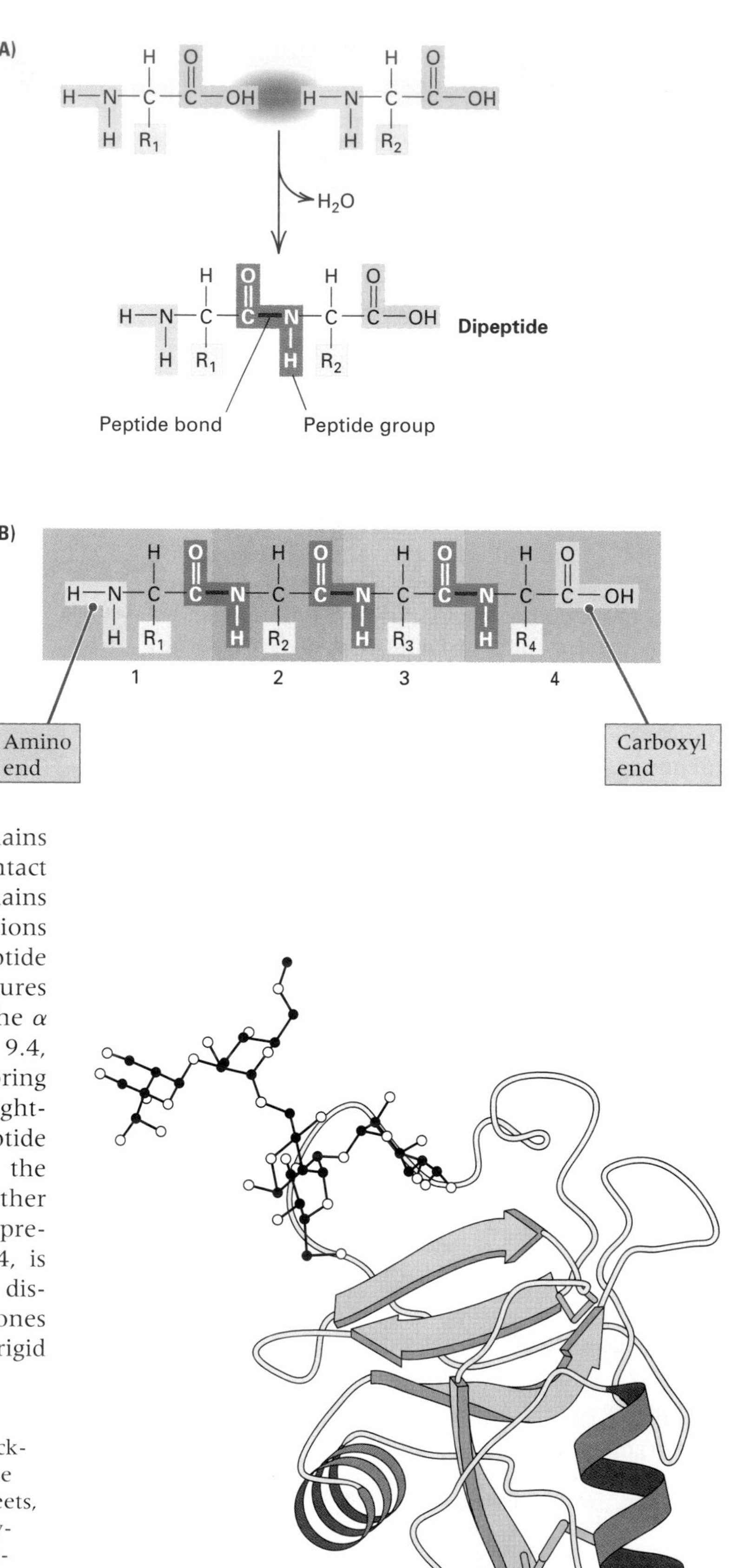

numbered starting at the amino terminus. Therefore, the amino acids are numbered in the order in which they are added to the chain during synthesis.

Most polypeptide chains are highly folded, and a variety of three-dimensional shapes have been observed. The manner of folding is determined primarily by the sequence of amino acids through noncovalent interactions between the side chains, so each polypeptide chain tends to fold into a unique three-dimensional shape as it is being synthesized (in some cases assisted by interactions with other proteins in the cell). On the average, the molecules fold so that amino acids with charged side chains tend to be on the surface of the protein (in contact with water) and those with uncharged side chains tend to be internal. Specific folded configurations also result from hydrogen bonding between peptide groups. Two fundamental polypeptide structures are the α helix and the β sheet (Figure 9.4). The α helix, represented as a coiled ribbon in Figure 9.4, is formed by interactions between neighboring amino acids that twist the backbone into a right-handed helix in which the N–H in each peptide group is hydrogen-bonded with the C=O in the peptide group located four amino acids further along the helix. In contrast, the β sheet, represented as parallel "flat" ribbons in Figure 9.4, is formed by interactions between amino acids in distant parts of the polypeptide chain; the backbones of the polypeptide chains are held flat and rigid

Figure 9.4 A "ribbon" diagram of the path of the backbone of a polypeptide, showing the ways in which the polypeptide is folded. Arrows represent parallel β sheets, each of which is held to its neighboring β sheet by hydrogen bonds. Helical regions are shown as coiled ribbons. The polypeptide chain in this example is a mannose-binding protein. The stick figure at the upper left shows a molecule of mannose bound to the protein. [Adapted from William I. Weis, Kurt Drickamer, and Wayne A. Hendrickson. 1992. *Nature,* 360: 127.]

(forming a "sheet") because alternate N—H groups in one polypeptide backbone are hydrogen-bonded with alternate C=O groups in the polypeptide backbone of the adjacent chain. In each polypeptide backbone, alternate C=O and N—H groups are free to form hydrogen bonds with their counterparts in a different polypeptide backbone on the opposite side, so a β sheet can consist of multiple aligned segments in the same (or different) polypeptide chains. Other types of interactions also are important in protein folding; for example, covalent bonds may form between the sulfur atoms of pairs of cysteines in different parts of the polypeptide. The physical chemistry of protein folding is complex, and the final shape of a protein cannot usually be predicted from the amino acid sequence alone, except in relatively simple cases.

Many protein molecules consist of more than one polypeptide chain. When this is the case, the protein is said to contain *subunits.* The subunits may be identical or different. For example, hemoglobin, the oxygen carrier of blood, consists of four subunits—two of the α polypeptide chain and two of the β polypeptide chain.

9.2 The linear order of amino acids is encoded in a DNA base sequence.

Most genes contain the information for the synthesis of only one polypeptide chain. Furthermore, the linear order of nucleotides in a gene determines the linear order of amino acids in a polypeptide. This point was first proved by studies of the tryptophan synthase gene *trpA* in *E. coli,* a gene in which many mutations had been obtained and accurately mapped genetically. The effects of numerous mutations on the amino acid sequence of the enzyme were determined by directly analyzing the amino acid sequences of the wildtype and mutant enzymes. Each mutation was found to result in a single amino acid substituting for the wildtype amino acid in the enzyme. More important, the order of the mutations in the genetic map was the same as the order of the affected amino acids in the polypeptide chain (Figure 9.5). This attribute of genes and polypeptides is called **colinearity,** which means that the sequence of base pairs in DNA determines the sequence of amino acids in the polypeptide in a colinear, or point-to-point, manner. Colinearity is universally found in prokaryotes. However, we will see later that in eukaryotes, noninformational DNA sequences interrupt the continuity of most genes; in these genes, the order but not the spacing between the mutations correlates with amino acid substitution.

9.3 The base sequence in DNA specifies the base sequence in an RNA transcript.

The first step in gene expression is the synthesis of an RNA molecule copied from the segment of DNA that constitutes the gene. The basic features of the production of RNA are described in this section.

The chemical synthesis of RNA is similar to that of DNA.

Although the essential chemical characteristics of the enzymatic synthesis of RNA are generally similar to those of DNA, described in Chapter 6, there are also some important differences.

- Each RNA molecule produced in transcription derives from a single strand of DNA, because in any particular region of the DNA, usually only one strand serves as a template for RNA synthesis.

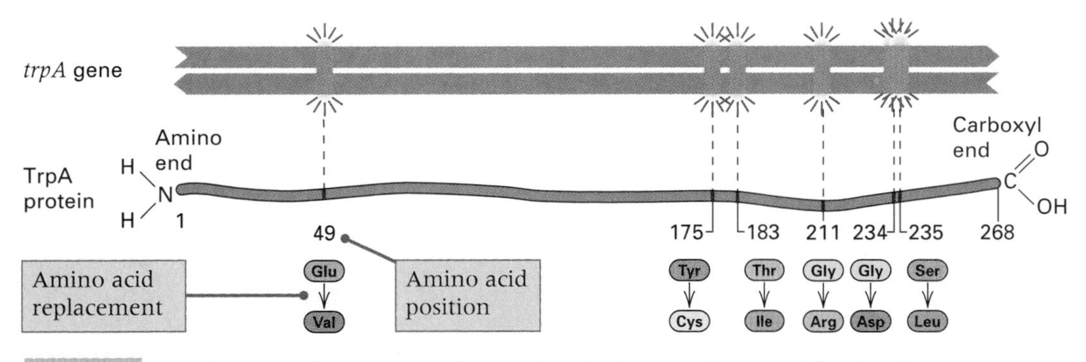

Figure 9.5 Correlation of the positions of mutations in the genetic map of the *E. coli trpA* gene with positions of amino acid replacements in the TrpA protein.

- The precursors in the synthesis of RNA are the four ribonucleoside 5'-triphosphates: adenosine triphosphate (ATP), guanosine triphosphate (GTP), cytidine triphosphate (CTP), and uridine triphosphate (UTP). They differ from the DNA precursors only in that the sugar is ribose rather than deoxyribose and the base uracil (U) replaces thymine (T) (Figure 9.6).
- The sequence of bases in an RNA molecule is determined by the sequence of bases in the DNA template. Each base added to the growing end of the RNA chain is chosen for its ability to base-pair with the DNA template strand. Thus the bases C, T, G, and A in the DNA template cause G, A, C, and U, respectively, to be added to the growing end of the RNA molecule.
- In the synthesis of RNA, a sugar–phosphate bond is formed between the 3'-hydroxyl group of one nucleotide and the 5'-triphosphate of the next nucleotide in line (Figure 9.7A and B). The chemical bond formed is the same as that in the synthesis of DNA, but the enzyme is different. The enzyme used in transcription is RNA polymerase rather than DNA polymerase.
- Nucleotides are added only to the 3'-OH end of the growing chain; as a result, the 5' end of a growing RNA molecule bears a triphosphate group. The 5' → 3' direction of chain growth is the same as that in DNA synthesis (Figure 9.7C).
- RNA polymerase (unlike DNA polymerase) is able to initiate chain growth without a primer (Figure 9.7C).

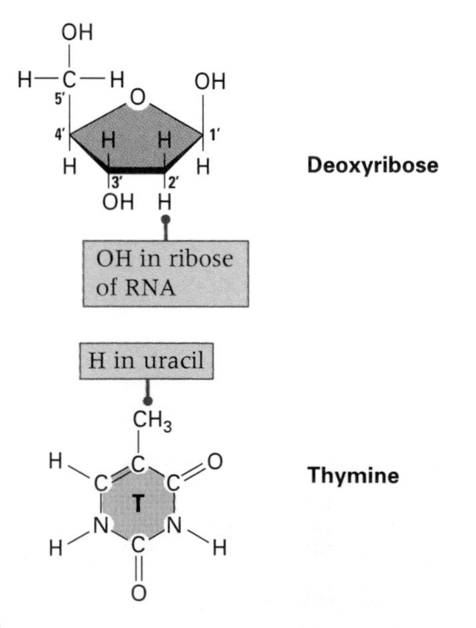

Figure 9.6 Differences between the structures of ribose and deoxyribose and between those of uracil and thymine.

Particular nucleotide sequences define the beginning and end of a gene.

How does RNA polymerase determine which strand of DNA should be transcribed? How does the enzyme recognize where transcription of the template strand should begin? How does the enzyme recognize where transcription should stop? These are critical features in the regulation of RNA synthesis. Overall, transcription can be described as consisting of four discrete stages:

1. Promoter recognition The RNA polymerase binds to DNA wherever the DNA has a particular base sequence called a **promoter.** Many promoter regions have been isolated and their base sequences determined. Typical promoters are from 20 to 200 bases long. There is substantial variation in base sequence among promoters, which results in the promoters having different strengths in binding with the RNA polymerase. However, most promoters have certain sequence motifs in common. Two consensus sequences often found in promoter regions in *E. coli* are illustrated in Figure 9.8. A **consensus sequence** is a sequence of bases determined by majority rule: Each base in the consensus sequence is the base most often observed at that position among a set of observed sequences. Any particular observed sequence may resemble the consensus sequence very well or very poorly.

The consensus promoter regions in *E. coli* are TTGACA, centered approximately 35 base pairs upstream from the transcription start site (conventionally numbered the +1 site), and TATAAT, centered approximately 10 base pairs upstream from the +1 site. The −10 sequence, which is called the **TATA box,** is similar to sequences found at corresponding positions in many eukaryotic promoters. The positions of the promoter sequences determine where, and on which strand, the RNA polymerase begins synthesis, and an A or a G is often the first nucleotide in the transcript.

The strength of the binding of RNA polymerase to different promoters varies greatly, which causes differences in the extent of expression from one gene to another. Most of the differences in promoter strength result from variations in the −35 and −10 promoter elements and in the number of bases between them. Promoter strength among *E. coli* genes differs by a factor of 10^4, and most of the variation can be attributed to the promoter sequences themselves. In general, the more closely the −35 and −10 promoter elements resemble the consensus sequences, the stronger the promoter. The situation is somewhat different in eukaryotes, where other types of DNA sequences (*enhancers*) interact with promoters to determine the level of transcription (Chapter 11).

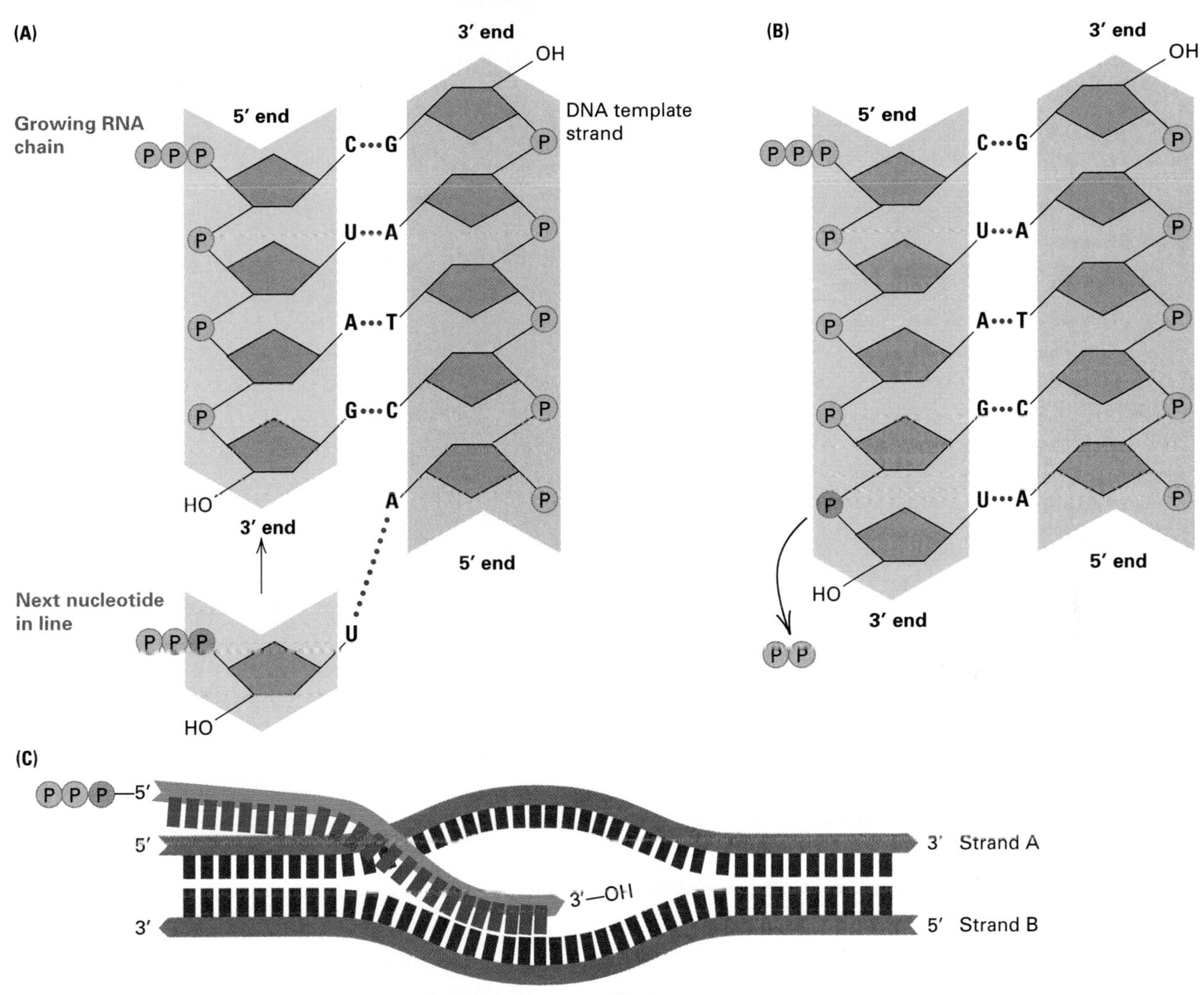

Figure 9.7 RNA synthesis. (A) The polymerization step in RNA synthesis. The incoming nucleotide forms hydrogen bonds (red dots) with a DNA base. The −OH group in the growing RNA chain reacts with the orange P in the next nucleotide in line (B). (C) Geometry of RNA synthesis. RNA is copied from only one strand of a segment of a DNA molecule—in this example, strand B—without the need for a primer. In this region of the DNA, RNA is not copied from strand A. However, in a different region (for example, in a different gene), strand A might be copied rather than strand B. Because RNA elongates in the 5' → 3' direction, its synthesis moves along the DNA template in the 3' → 5' direction; that is, the RNA molecule is antiparallel to the DNA strand being copied.

```
                         -35            Consensus              -10          Transcription start
Gene                     Sequence       sequences              Sequence
                         TTGACA                                TATAAT       +1
lac   TAGGCACCCCAGGC TTTACA CTTTA    TGCTTCCGGCTCG TATGTT GTG TGG A ATTGTGAGC
lacI  GACACCATCGAATG GCGCAA AACTT     TTCGCGGTATGG CATGAT AGCGCCC G GAAGAGAGT
trp   TCTGAAATGAGCTG TTGACA ATTAA     TCATCGAACTAG TTAACT AGTACGC A AGTTCACGT
his   ATATAAAAAAGTTC TTGCTT TCTAA    CGTGAAAGTGGTT TAGGTT AAAAGAC A TCAGTTGAA
leu                G TTGACA TCCGT     TTTTGTATCCAG TAACTC TAAAAGC A TATCGCATT
gal   CTAATTTATTCCAT GTCACA CTTTTCGCATCTTTGTTATGC TATGGT TAT TTC A TACCATAAG
bio   GCCTTCTCCAAAAC GTGTTT TTTGT    TGTTAATTCGGTG TAGACT TGT  AA A CCTAAATCT
recA  TTTCTACAAAACAC TTGATA CTGTA      TGAGCATACAG TATAAT TGC TTC A ACAGAACAT
```

Figure 9.8 Base sequences in promoter regions of several genes in *E. coli*. The consensus sequences located 10 and 35 nucleotides upstream from the transcription start site (+1) are indicated. Promoters vary tremendously in their ability to promote transcription. Much of the variation in promoter strength results from differences between the promoter elements and the consensus sequences at −10 and −35.

2. Chain initiation After the initial binding step, RNA polymerase initiates RNA synthesis at a nearby transcription start site, denoted the +1 site in Figure 9.8. The first nucleoside triphosphate is placed at this site, the next nucleotide in line is attached to the 3' carbon of the ribose, and so forth. Only one of the DNA strands serves as the template for transcription. Because RNA is synthesized in the $5' \rightarrow 3'$ direction, the DNA template is traversed in the $3' \rightarrow 5'$ direction. Therefore, relative to the orientation of the promoter sequences shown in Figure 9.8, the template DNA strand is the *partner* of the strand illustrated. Take the *lac* promoter as an example. Transcription begins on the opposite strand at the nucleotide denoted +1 and proceeds from left to right. Hence the base sequence of the RNA transcript is the same as that of the DNA strand illustrated (except that RNA contains U where T appears in DNA), and so the *lac* RNA sequence begins AAUUGUGAGC . . . .

3. Chain elongation After initiation at the +1 site, RNA polymerase moves along the DNA template strand, adding nucleotides to the growing RNA chain (Figure 9.7C). Each new nucleotide is added to the 3' end of the chain, so RNA chains resemble DNA chains in growing in the $5' \rightarrow 3'$ direction. Figure 9.7C also shows that transcription separates the partner strands of the DNA duplex only in a short region around the point of chain elongation. As the RNA polymerase moves along the template strand, only about 17 base pairs of the DNA duplex (less than two turns of the double helix) are unwound at any time. Once the RNA polymerase has passed, the DNA strands are released and the duplex forms again, with the part of the RNA chain already synthesized trailing off as a separate polynucleotide strand.

4. Chain termination Special sequences also terminate RNA synthesis. When the RNA polymerase reaches a transcription-termination sequence in the DNA, the polymerase enzyme disassociates from the DNA and the newly synthesized RNA molecule is released. Two kinds of termination events are known: (1) those that are self-terminating and depend only on the transcription-termination sequence in the DNA and (2) those that require the presence of a termination protein in addition to the transcription-termination sequence. Self-termination is the usual case and takes place when the polymerase encounters a particular sequence of bases in the template strand that causes the polymerase to stop. An example is shown in Figure 9.9.

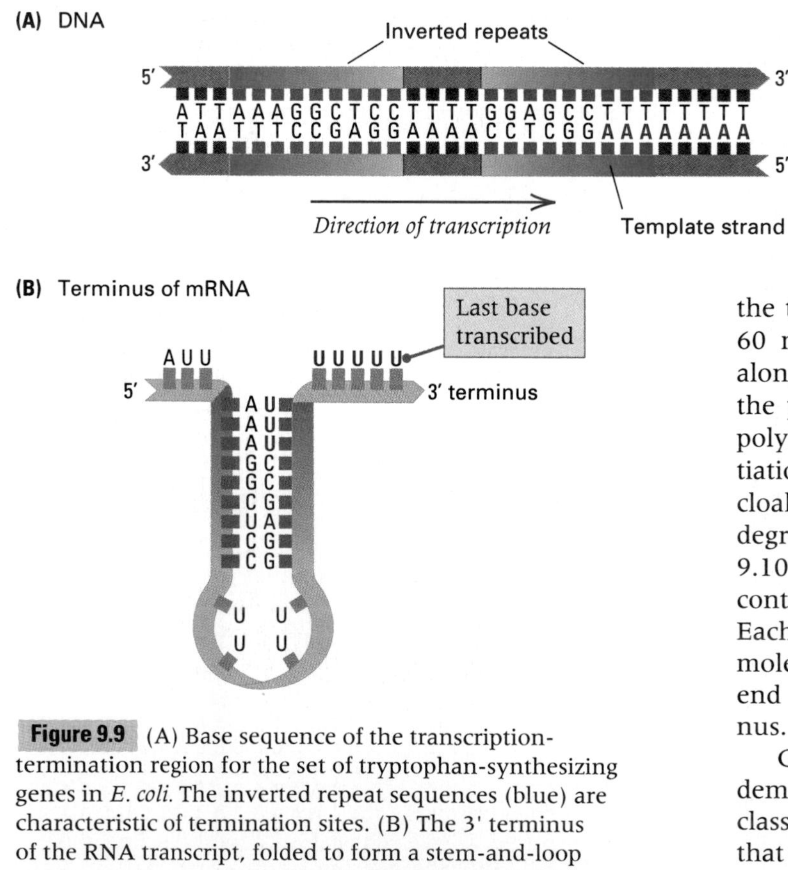

Figure 9.9 (A) Base sequence of the transcription-termination region for the set of tryptophan-synthesizing genes in *E. coli*. The inverted repeat sequences (blue) are characteristic of termination sites. (B) The 3' terminus of the RNA transcript, folded to form a stem-and-loop structure. The sequence of U's found at the end of the transcript in this and many other prokaryotic genes is shown in red. The RNA polymerase, not shown here, terminates transcription when the loop forms in the transcript.

Initiation of a second round of transcription need not await completion of the first. By the time an RNA transcript reaches a size of 50 to 60 nucleotides, the RNA polymerase has moved along the DNA far enough from the promoter that the promoter becomes available for another RNA polymerase to initiate a new transcript. Such reinitiation can take place repeatedly, and a gene can be cloaked with numerous RNA molecules in various degrees of completion. The micrograph in Figure 9.10 shows a region of the DNA of the newt *Triturus* containing tandem repeats of a particular gene. Each gene is associated with many growing RNA molecules. The shortest ones are at the promoter end of the gene, the longest near the gene terminus.

Genetic experiments in *E. coli* yielded the first demonstration of the existence of promoters. A class of Lac^- mutations, denoted p^-, was isolated that was unusual in two respects:

- All p^- mutations were closely linked to the *lacZ* gene.
- Any p^- mutation eliminated activity of a wild-type *lacZ* gene *present in the same DNA molecule.*

Figure 9.10 Electron micrograph of part of the DNA of the newt *Triturus viridescens* containing tandem repeats of genes being transcribed into ribosomal RNA. The thin strands forming each feather-like array are RNA molecules. A gradient of lengths can be seen for each rRNA gene. Regions in the DNA between the thin strands are spacer DNA sequences, which are not transcribed. [Courtesy of Oscar Miller.]

The need for an adjacent genetic configuration to eliminate *lacZ* activity can be seen by examining a cell with two copies of the *lacZ* gene. Such cells can be produced through the use of F'*lacZ* plasmids, which contain a copy of *lacZ* in an F plasmid. Infection with an F'*lacZ* plasmid yields a cell with two copies of *lacZ*—one in the chromosome and another in the F'. Transcription of the *lacZ* gene enables the cell to synthesize the enzyme β-galactosidase. Table 9.1 shows that a wildtype *lacZ* gene ($lacZ^+$) is inactive when a p^- mutation is present in the same DNA molecule (either in the bacterial chromosome or in an F' plasmid). This result can be seen by comparing entries 4 and 5. Analysis of the RNA shows that in a cell with the genotype p^- $lacZ^+$, the $lacZ^+$ gene is not transcribed. On the other hand, cells of genotype p^+lacZ^- produce a mutant RNA. The p^- mutations are called *promoter mutations.*

Table 9.1
Effect of promoter mutations on transcription of the *lacZ* gene

Genotype	Transcription of $lacZ^+$ gene
1. p^+lacZ^+	Yes
2. p^-lacZ^+	No
3. p^+lacZ^+/p^+lacZ^-	Yes
4. p^-lacZ^+/p^+lacZ^-	No
5. p^+lacZ^+/p^-lacZ^-	Yes

Note: $lacZ^+$ is the wildtype gene; the $lacZ^-$ mutant produces a nonfunctional enzyme.

Mutations have also been instrumental in defining the transcription-termination region. For example, mutations have been isolated that create a new termination sequence upstream from the normal one. When such a mutation is present, an RNA molecule is made that is shorter than the wildtype RNA. Other mutations eliminate the terminator, resulting in a longer transcript.

RNA polymerase enzymes are large complexes. The RNA polymerase of the bacterium *E. coli* consists of five protein subunits and is large enough to be seen by electron microscopy (Figure 9.11). In the process of transcription, the enzyme is in contact with a region of about 50 to 60 bases in duplex DNA, but only about 17 bases are unwound at any one time. In *E. coli*, all transcription is catalyzed by this enzyme.

In contrast with bacterial cells, eukaryotic cells have three distinct RNA polymerases. They are denoted RNA polymerase I, II, and III, and each makes a particular class of RNA molecule.

- RNA polymerase I transcribes the two large RNA components of the ribosomes.
- RNA polymerase II is the enzyme used in the transcription of protein-coding genes.
- RNA polymerase III transcribes transfer RNA molecules, the small RNA component of the ribosomes, and various small RNA molecules used in splicing (Section 9.4).

Not only are there multiple types of RNA polymerase in eukaryotes, but each of the RNA polymerases is itself an aggregate of polypeptide subunits that combines with still other proteins

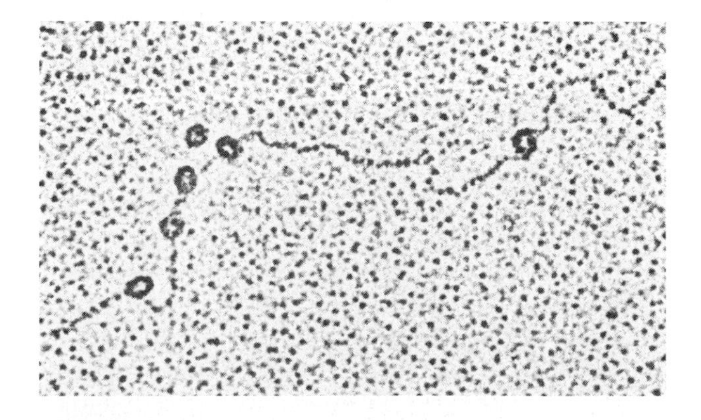

Figure 9.11 *E. coli* RNA polymerase molecules bound to DNA. [Courtesy of Robley Williams.]

to form a **transcription complex** that actually carries out transcription. Regulation of transcription in eukaryotes is discussed in Chapter 11, but the heart of the matter is that different genes that use the same RNA polymerase are transcribed at different times or in different cell types because their transcription requires the formation of different transcription complexes. Regulation by this mechanism is possible because the promoter specificity of a transcription complex can be regulated by the replacement of some of the polypeptide subunits in the transcription complex with alternative polypeptides. Figure 9.12 illustrates the postulated arrangement of polypeptide subunits in a transcription complex of RNA polymerase III in yeast used in the transcription of transfer RNA.

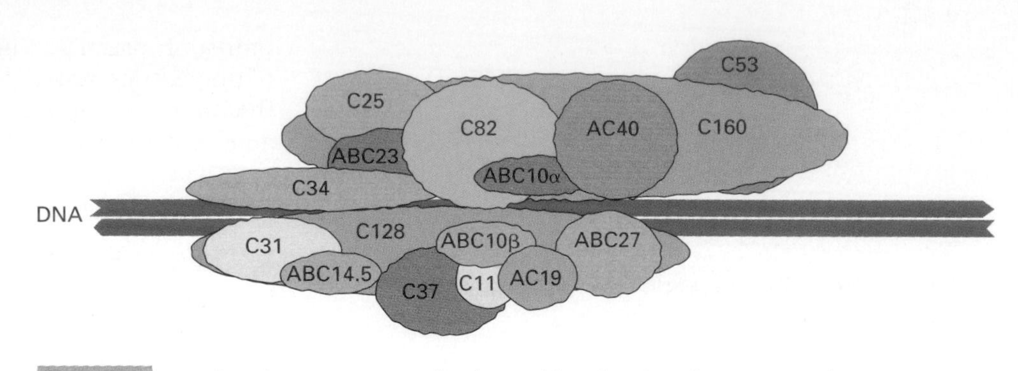

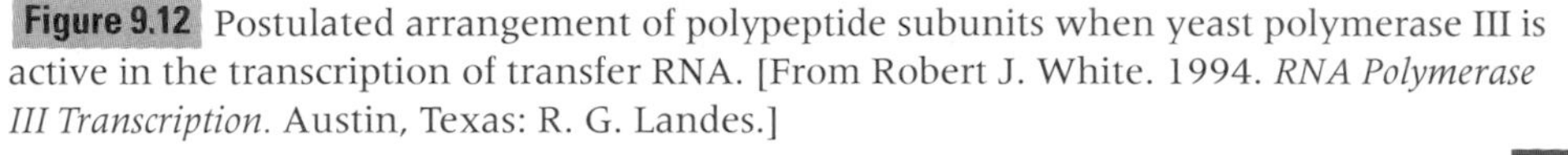
Figure 9.12 Postulated arrangement of polypeptide subunits when yeast polymerase III is active in the transcription of transfer RNA. [From Robert J. White. 1994. *RNA Polymerase III Transcription*. Austin, Texas: R. G. Landes.]

Messenger RNA directs the synthesis of a polypeptide chain.

The process of gene expression begins with transcription of the base sequence of the template strand of DNA into the base sequence of an RNA molecule. In prokaryotes, this RNA molecule is called **messenger RNA,** or **mRNA,** and it is used directly in polypeptide synthesis. In eukaryotes, the RNA molecule is usually processed before it becomes mRNA. The sequence of amino acids in the polypeptide is determined from the sequence of bases in mRNA by the protein-synthesizing machinery of the cell.

Not all base sequences in an mRNA molecule are translated into the amino acid sequences of polypeptides. For example, translation of an mRNA molecule rarely starts exactly at one end and proceeds to the other end; instead, initiation of polypeptide synthesis may begin many nucleotides downstream from the 5' end of the RNA. The untranslated 5' segment of RNA is called a **leader** and in some cases contains regulatory sequences that affect the rate of protein synthesis. The leader is followed by a **coding sequence,** which specifies the polypeptide chain. A typical coding sequence in an mRNA molecule is between 500 and 3000 bases long (depending on the number of amino acids in the protein), but it may be much longer if it encodes a large polypeptide like myosin. The 3' end of an mRNA molecule following the coding sequence also is not translated.

In prokaryotes, most mRNA molecules are degraded within a few minutes after synthesis. In eukaryotes, a typical lifetime is several hours, although some last only minutes whereas others persist for days. In both kinds of organisms, the degradation enables cells to dispose of molecules that are no longer needed and to recycle the nucleotides in synthesizing new RNAs. The short lifetime of prokaryotic mRNA is an important factor in regulating gene activity (Chapter 11).

9.4 RNA processing converts the original RNA transcript into messenger RNA.

Although the process of transcription is very similar in prokaryotes and eukaryotes, there are major differences in the relationship between the transcript and the mRNA used for polypeptide synthesis. In

THIS RACOON IS an example of a eukaryote. Both eukaryotes and prokaryotes use similar processes of transcription, translation, and the standard genetic code, owing to their ancient common ancestry. [© D. P. Burnside/Photo Researchers, Inc.]

prokaryotes, the immediate product of transcription (the **primary transcript**) is mRNA; in contrast, *the primary transcript in eukaryotes must be converted into mRNA.* The conversion of the original transcript into mRNA is called *RNA processing.* It usually consists of three types of events, as illustrated diagrammatically in Figure 9.13:

- The 5' end of the transcript is modified by the addition of a *cap.*
- The 3' end of the transcript is modified by the addition of a *tail.*
- Certain regions internal to the transcript (*introns*) are removed by splicing.

In the first step in RNA processing, the 5' end is altered by the addition of a modified guanosine in an uncommon 5'–5' linkage (instead of the typical 3'–5') linkage; this terminal group is called the **cap.** The 3' terminus of a eukaryotic mRNA molecule also is usually modified by the addition of a polyadenylate sequence called the **poly-A tail,** which can consist of as many as 200 nucleotides. The 5' cap is necessary for the ribosome to bind with the mRNA to begin protein synthesis, and the poly-A tail is thought to help determine mRNA stability.

Splicing removes introns from the RNA transcript.

The important feature of *splicing* of the primary transcript in eukaryotes is also shown in Figure 9.13. The segments that are excised from the primary transcript are called **introns** or **intervening sequences.** Accompanying the excision of introns is a rejoining of the coding segments (**exons**) to form the mRNA molecule. The mechanism of excision and splicing is illustrated schematically in Figure 9.14. Part A shows the consensus sequence found at the **splice donor** (5') end and the **splice acceptor** (3') end of most introns. The symbols are N, any nucleotide; Y, any pyrimidine (C or U); and S, either A or C. After an initial cut in the donor site (part B), the G at the 5' end of the intron forms a

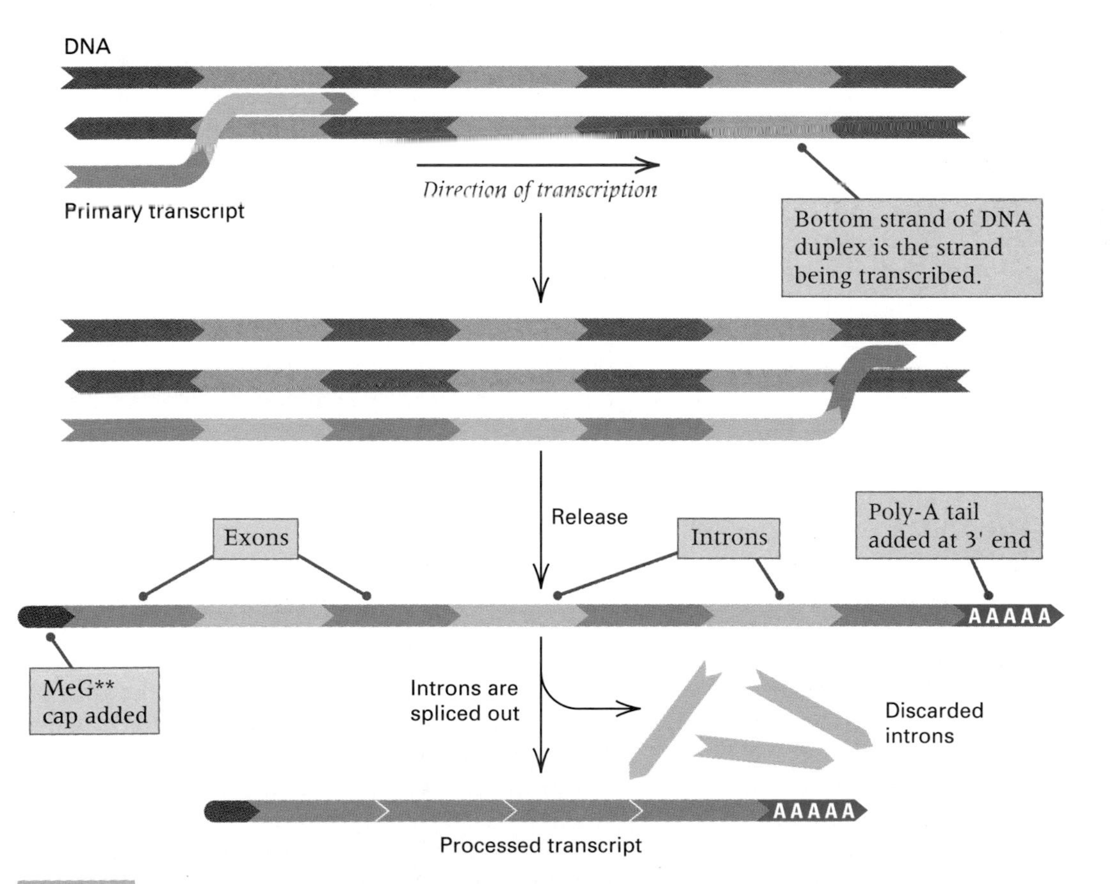

Figure 9.13 A schematic drawing showing the production of eukaryotic mRNA. The primary transcript is capped before it is released from the DNA. MeG denotes 7-methylguanosine (a modified form of guanosine), and the two asterisks indicate two nucleotides whose riboses are methylated. The 3' end is usually modified by the addition of consecutive adenines. Along the way, the introns are excised. These reactions take place within the nucleus.

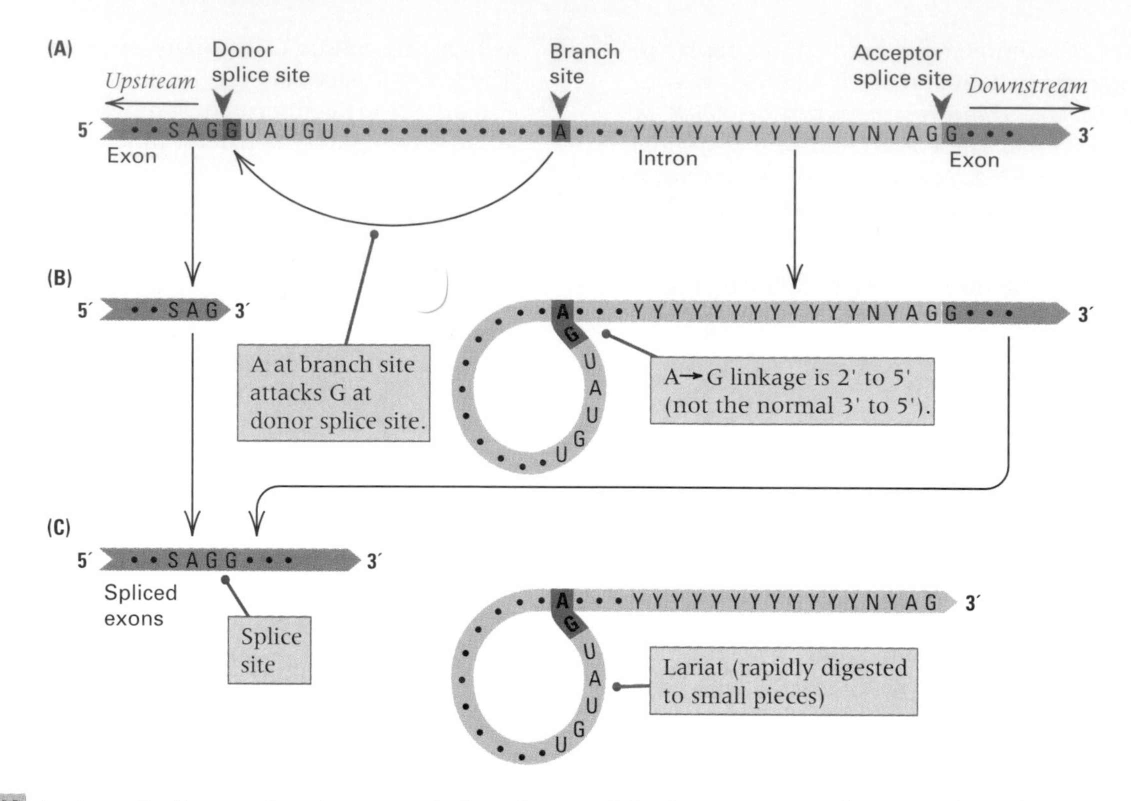

Figure 9.14 A schematic diagram showing removal of one intron from a primary transcript. The A nucleotide at the branch site attacks the terminus of the 5' exon, cleaving the exon-intron junction and forming a loop connected back to the branch site. The 5' exon is later brought to the site of cleavage of the 3' exon, a second cut is made, and the exon termini are joined and sealed. The loop is released as a lariat-shaped structure that is degraded. Because the loop includes most of the intron, the loop of the lariat is usually very much longer than the tail.

loop by becoming attached to an A nucleotide located a short distance upstream from the run of pyrimidines (Y) near the acceptor site. The A–G linkage is unusual in being 2'-to-5' (instead of the usual 3'-to-5'). In the final step (part C), a cut is made in the acceptor site and the intron is freed. The remaining parts (the exons) are joined together. The excision of the introns and the joining of the exons to form the final mRNA molecule is called **RNA splicing.** The freed intron is said to be a **lariat** structure because it has a loop and a tail. The lariat is rapidly degraded into free nucleotides by nucleases.

RNA splicing takes place in nuclear particles known as **spliceosomes.** These abundant particles are composed of protein and several specialized small RNA molecules, which are present in the cell as small nuclear ribonucleoprotein particles; the underlined letters give the acronym for these particles: **snRNPs.** The specificity of splicing comes from the five small snRNP RNAs denoted U1, U2, U4, U5, and U6, which contain sequences complementary to the splice junctions, the branchpoint region of the intron, and/or to one another; as many as 100 spliceosome proteins may also be required for splicing. The ends of the intron are brought together by U1 RNA, which forms base pairs with nucleotides in the intron at both the 5' and the 3' ends. U2 RNA binds to the branchpoint region. U2 RNA interacts with a paired complex of U4/U6 RNAs, resulting in a complex in which U2 RNA ends up paired with U6 RNA and the intron of the transcript (Figure 9.15). All of these dynamic interactions bring the branchpoint region near to the donor splice site and allow the A in the branchpoint to attack the G of the donor splice site, freeing the upstream exon and forming the lariat intermediate (Figure 9.15). U5 RNA helps line up the two exons and somehow facilitates the final step in splicing, which results in scission of the intron from the downstream exon and ligation of the upstream and downstream exons together.

Introns are also present in some genes in organelles, such as mitochondria, but the mechanisms of their excision differ from those of introns in nuclear genes because organelles do not contain spliceosomes. In one class of organelle introns, the intron contains a sequence coding for a protein that participates in removing the intron that codes for it. The situation is even more remarkable in the splic-

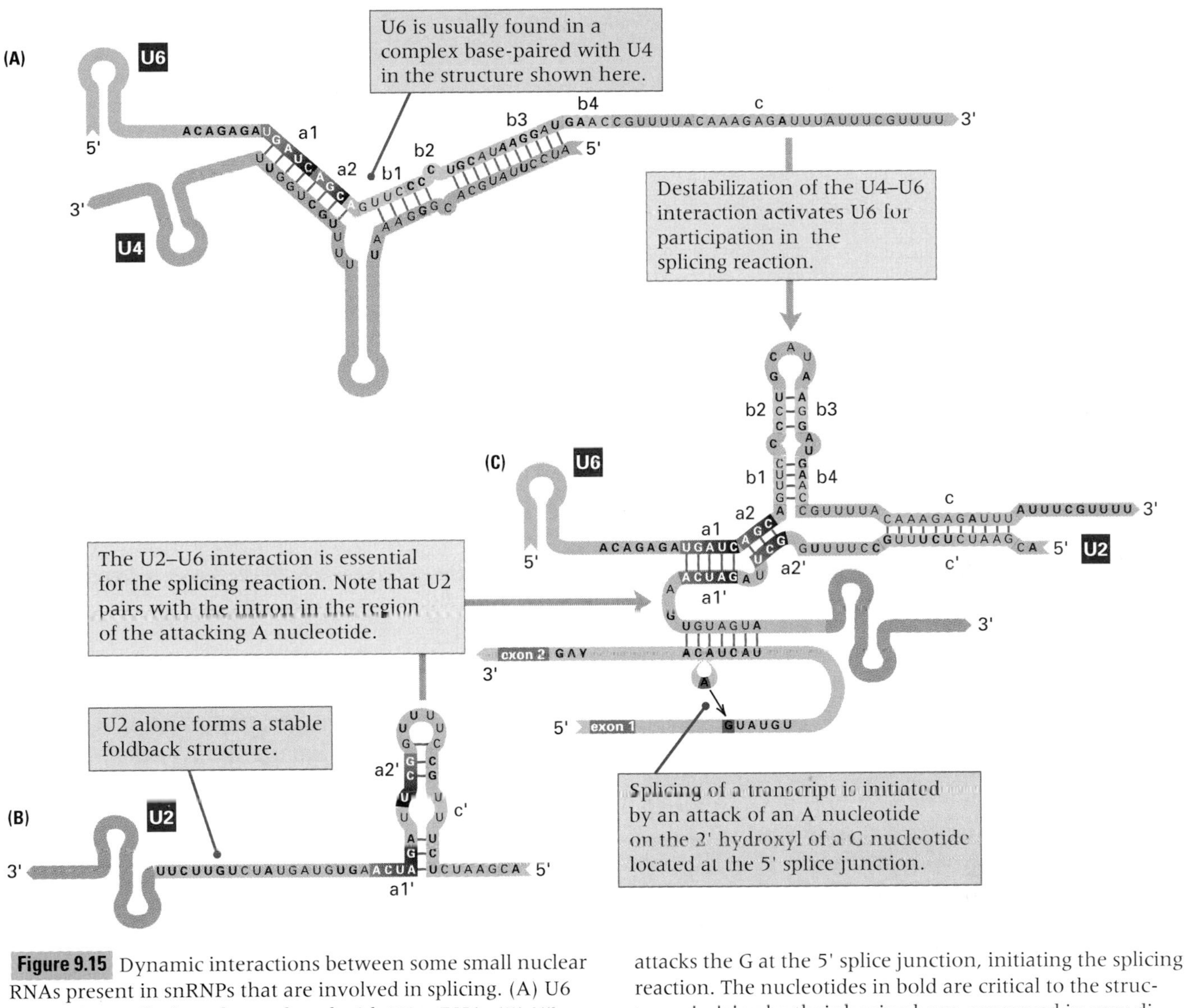

Figure 9.15 Dynamic interactions between some small nuclear RNAs present in snRNPs that are involved in splicing. (A) U6 snRNA is usually found complexed with U4 snRNA. (B) U2 snRNA forms a stable foldback structure on its own. (C) Essential to the splicing reaction is destabilization of the U4–U6 structure and formation of a U2–U6 structure in which U2 is base-paired with part of the intron. An A in the paired region attacks the G at the 5' splice junction, initiating the splicing reaction. The nucleotides in bold are critical to the structures, judging by their having been conserved in very diverse species. Note that G–U base pairs are allowed in double-stranded RNA. [After H. D. Madhani and C. Guthrie. 1994. *Annual Reviews of Genetics* 28: 1.]

ing of a ribosomal RNA precursor in the ciliate *Tetrahymena*. In this case, the splicing reaction is intrinsic to the folding of the precursor; that is, the RNA precursor is *self-splicing* because the folded precursor RNA creates its own RNA-splicing activity. The self-splicing *Tetrahymena* RNA was the first example found of an RNA molecule that could function as an enzyme in catalyzing a chemical reaction; such enzymatic RNA molecules are usually called **ribozymes.**

The existence and the positions of introns in a particular primary transcript are readily demonstrated by renaturing the transcribed DNA with the fully processed mRNA molecule. The DNA–RNA hybrid can then be examined by electron microscopy. An example of adenovirus mRNA (fully processed) and the corresponding DNA are shown in Figure 9.16. The DNA copies of the introns appear as single-stranded loops in the hybrid molecule, because no corresponding RNA sequence is available for hybridization.

The number of introns per RNA molecule varies considerably from one gene to the next. One of the major genes for inherited breast cancer in women (*BRCA1*) contains 21 introns spread across more than 100,000 bases. More than 90 percent of the

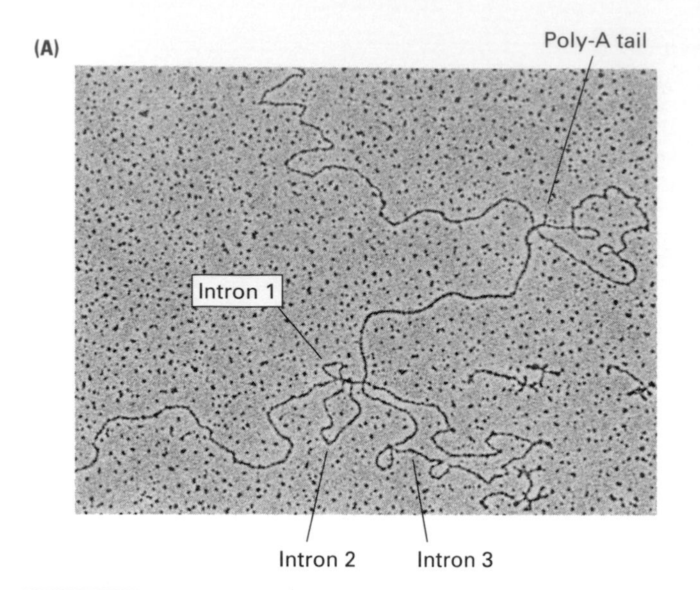

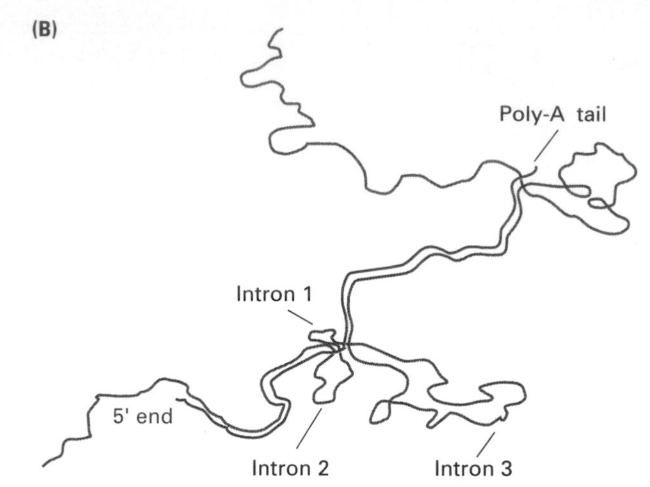

Figure 9.16 (A) An electron micrograph of a DNA–RNA hybrid obtained by annealing a single-stranded segment of adenovirus DNA with one of its mRNA molecules. The loops are single-stranded DNA. (B) An interpretive drawing. RNA and DNA strands are shown in red and black, respectively. Four regions do not anneal, creating three single-stranded DNA segments that correspond to the introns and the poly-A tail of the mRNA molecule. [Electron micrograph courtesy of Tom Broker and Louise Chow.]

primary transcript is excised in processing, yielding a processed mRNA of about 7800 bases, which codes for a polypeptide chain of 1863 amino acids. The polypeptide encoded by *BRCA1* is a tumor suppressor gene (Chapter 5). Among the human genes with a simpler intron-exon structure is that for α-globin, which contains two introns. Introns vary greatly in size as well as in number. In human beings and other mammals, most introns range in size from 100 to 10,000 bases, and in the processing of a typical primary transcript, the amount of discarded RNA ranges from about 50 percent to more than 90 percent of the primary transcript. In lower eukaryotes, such as yeast, nematodes, and fruit flies, genes generally have fewer introns than do genes in mammals, and the introns tend to be much smaller.

Most introns appear to have no function in themselves. A genetically engineered gene that lacks a particular intron usually functions normally. In those cases where an intron seems to be required for function, it is usually not because the interruption of the gene is necessary but rather because the intron happens to include certain nucleotide sequences that regulate the timing or tissue specificity of transcription. The implication is that many mutations in introns, including small deletions and insertions, should have essentially no effect on gene function—and this is indeed the case. Moreover, the nucleotide sequence of a particular intron is found to undergo changes (including small deletions and insertions) extremely rapidly in the course of evolution, and this lack of sequence conservation is another indication that most of the nucleotide sequences present within introns are not critically important.

Mutations that affect any of the critical splicing signals do have important consequences because they interfere with the splicing reaction. Two possible outcomes are illustrated in Figure 9.17. In part A, the intron with the mutated splice site fails to be removed, and it remains in the processed mRNA. The result is the production of a mutant protein with a normal sequence of amino acids up to the splice site and an abnormal sequence afterward. Most introns are long enough so that, by chance, they contain a stop sequence that terminates protein synthesis, and once a stop is encountered, the protein grows no further. A second kind of outcome is shown in part B. In this case, splicing takes place, but at an alternative splice site. (The example shows the alternative site downstream from the mutation, but alternative sites can also be upstream.) The alternative site is called a **cryptic splice site** because it is not normally used. The cryptic splice site is usually a poorer match with the consensus sequence and is ignored when the normal splice site is available. The result of using the alternative splice site is again an incorrectly processed mRNA and a mutant protein. In some splice-site mutations, either outcome (Figure 9.17A or B) is possible: Some transcripts leave the intron unspliced, whereas others are spliced at cryptic splice sites.

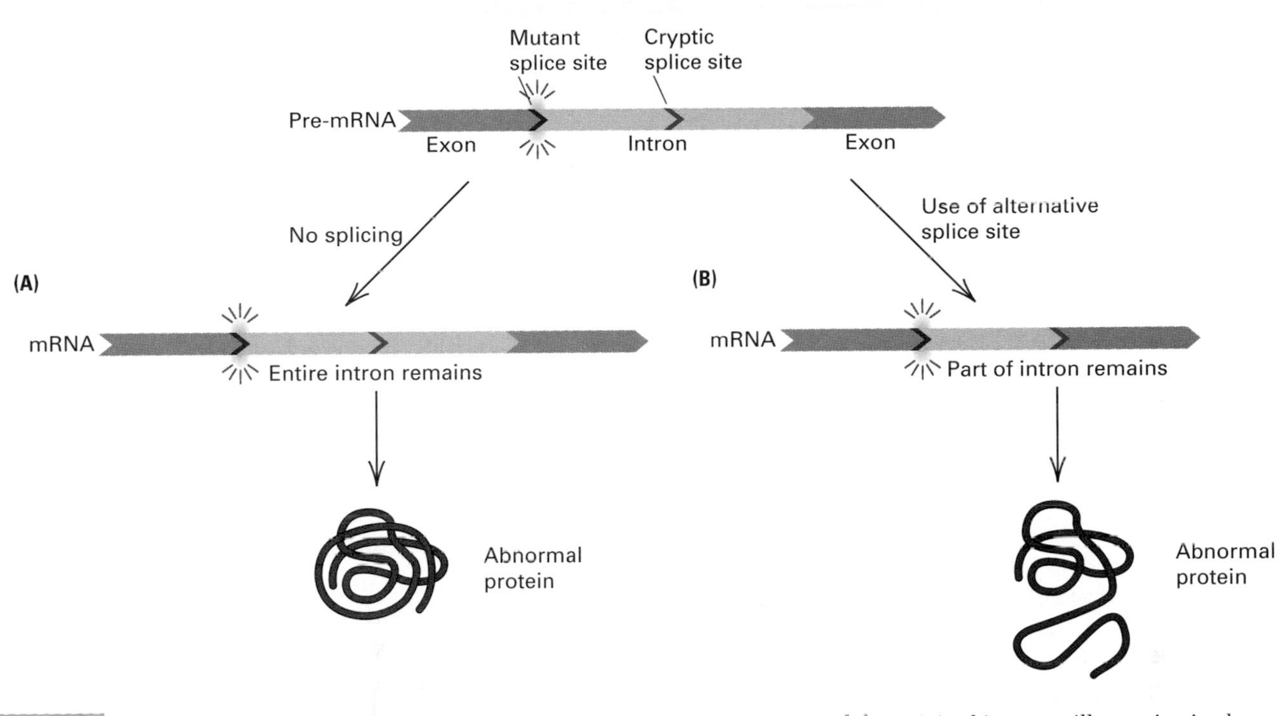

Figure 9.17 Possible consequences of mutation in the donor splice site of an intron. (A) No splicing occurs, and the entire intron remains in the processed transcript. (B) Splicing occurs at a downstream cryptic splice site, and only the upstream part of the original intron still remains in the processed transcript. Neither outcome results in a normal protein product.

Many exons code for distinct protein-folding domains.

Although introns play no essential role in regulating gene expression, they may play a role in gene evolution. In some cases, the exons in a gene code for segments of the completed protein that are more or less independent in their folding characteristics. For example, the central exon of the β-globin gene codes for the segment of the protein that folds around an iron-containing molecule of heme. Relatively autonomous folding units are known as **folding domains,** and the correlation between exons and domains found in some genes suggests that the genes were originally assembled from smaller pieces. In some cases, the ancestry of the exons can be traced. For example, the human gene for the low-density lipoprotein receptor that participates in cholesterol regulation shares exons with certain blood-clotting factors and epidermal growth factors. The model of protein evolution through the combination of different exons is called the **exon-shuffle** model. The mechanism for combining exons from different genes is not known. Although some genes support the model, in other genes the boundaries of the folding domains do not coincide with exons.

The evolutionary origin of introns is unknown. On the one hand, introns may be an ancient feature of gene structure. The existence of self-splicing RNAs means that introns could have existed long before the evolution of the spliceosome mechanism, and therefore some introns may be as old as the genes themselves. Furthermore, the finding that some genes have introns in the same locations in both plants and animals suggests that the introns may have been in place before plants and animals became separate lineages. If introns are ancient, then exon shuffling might have been important early in evolution by creating new genes with novel combinations of exons. It has even been suggested that all forms of early life had introns in their genes and that today's prokaryotes, which lack introns, lost their introns in their evolution. On the other hand, it has also been argued that introns arose relatively late in evolution and became inserted into already-existing genes, particularly in vertebrate genomes.

9.5 Translation into a polypeptide chain takes place on a ribosome.

The synthesis of every protein molecule in a cell is directed by an mRNA originally copied from DNA. Protein production includes two kinds of processes: (1) information-transfer processes, in which the RNA base sequence determines an amino acid sequence, and (2) chemical processes, in which the amino acids are linked together. The complete series of events is called **translation.**

The translation system consists of five major components:

1. **Messenger RNA** Messenger RNA is needed to bring the ribosomal subunits together (described below) and to provide the coding sequence of bases that determines the amino acid sequence in the resulting polypeptide chain.
2. **Ribosomes** These components are particles on which protein synthesis takes place. They move along an mRNA molecule and align successive transfer RNA molecules; the amino acids are attached one by one to the growing polypeptide chain by means of peptide bonds. Ribosomes consist of two separate RNA–protein particles (the small subunit and the large subunit), which come together in polypeptide synthesis to form a mature ribosome. An electron micrograph and a three-dimensional model of a ribosome from *E. coli* are shown in Figure 9.18.
3. **Transfer RNA, or tRNA** The sequence of amino acids in a polypeptide is determined by the base sequence in the mRNA by means of a set of adaptor molecules, the tRNA molecules, each of which is attached to a particular amino acid. Each successive group of three adjacent bases in the mRNA forms a **codon** that binds to a particular group of three adjacent bases in the tRNA (an anticodon), bringing the attached amino acid into line for addition to the growing polypeptide chain.
4. **Aminoacyl-tRNA synthetases** Each enzyme in this set of molecules catalyzes the attachment of a particular amino acid to its corresponding tRNA molecule. A tRNA attached to its amino acid is called an **aminoacylated tRNA** or a **charged tRNA.**
5. **Initiation, elongation, and termination factors** Polypeptide synthesis can be divided into three stages: initiation, elongation, and termination. Each stage requires specialized molecules.

In prokaryotes, all of the components for translation are present throughout the cell; in eukaryotes, they are located in the cytoplasm, as well as in mitochondria and chloroplasts.

The mechanism of protein synthesis is depicted in Figures 9.19 through 9.22. In overview, the process is as follows: An mRNA molecule binds to the surface of a ribosome. The aminoacylated tRNAs are brought along sequentially, one by one, to the ribosome that is translating the mRNA molecule. Peptide bonds are made between successively aligned amino acids, each time joining the amino group of the incoming amino acid to the carboxyl group of the amino acid at the growing end. Finally, the chemical bond between the last tRNA and its attached amino acid is broken and the completed protein is removed.

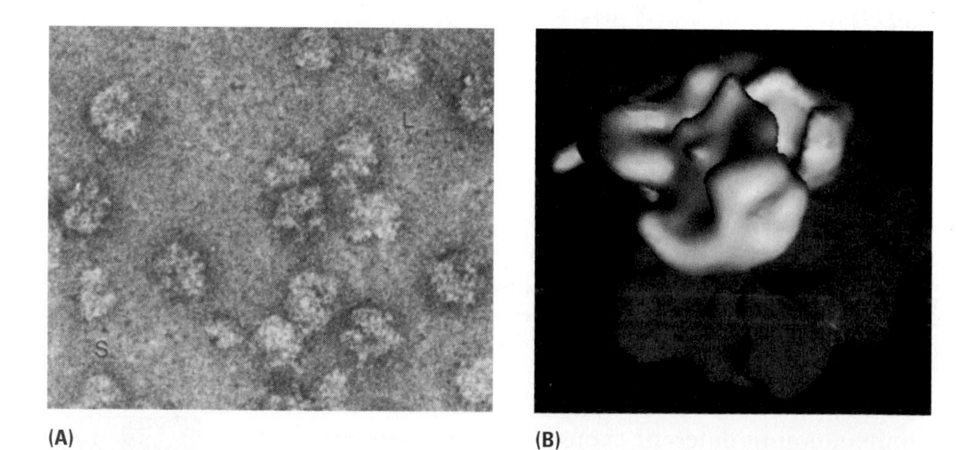

Figure 9.18 Ribosomes. (A) An electron micrograph of 70S ribosomes from *E. coli.* The 70S ribosome consists of one small subunit of size 30S and one large subunit of size 50S. (B) A three-dimensional model of the *E. coli* 70S ribosome based on high-resolution electron microscopy. The 30S subunit is in light green and the 50S subunit is in blue. [A, courtesy of James Lake; B, courtesy of J. Frank, A. Verschoor, Y. Li, J. Zhu, R. K. Lata, M. Radermacher, et al. 1995. *Biochemistry and Cell Biology* 73, 757.]

The human connection

Too Much of a Good Thing

Carl T. Montague and 14 other authors 1997
University of Cambridge, Cambridge, UK, and
5 other research institutions
Congenital Leptin Deficiency Is Associated with Severe Early-Onset Obesity in Humans

Twin, adoption, and population genetic studies indicate that up to 80 percent of the risk for human obesity resides in our genes. This paper shows that one familial form of human obesity has exactly the same genetic basis as a similar condition in the laboratory mouse. The gene is called obese (*ob*), and it was recognized more than 40 years ago. It is one of several genes for obesity in the mouse for which the molecular basis has been identified in recent years. The *ob* gene encodes a small protein hormone called leptin, which is secreted by fat cells. Leptin interacts with receptors in the hypothalamus within the brain to regulate appetite and energy expenditure. In effect, leptin signals the brain that appetite has been satisfied. Leptin is also thought to be necessary for puberty and sexual development in mice, possibly acting to signal that the nutritional state of the organism is sufficient to allow and sustain pubertal growth and sexual development. Homozygous *ob ob* mice are both obese and sterile. Administration of leptin to the affected mice corrects the phenotypic abnormalities, restoring fertility and reducing appetite. Over time, leptin administration restores normal body weight. This, of course, is the hope for affected human children, but the results are not yet in.

We have studied two related children with extreme obesity. . . . The two children are cousins within a highly consanguineous family of Pakistani origin. Although of normal weight at birth, both children suffered from severe, intractable obesity from an early age. . . . Each child has two siblings of normal weight, and none of their parents are morbidly obese. . . . To examine whether a mutation in the gene that encodes leptin might underlie severe obesity, the nucleotide sequence of leptin cDNA was examined. . . . All clones from one child contained a frame-shift mutation in the leptin-coding region consisting of the deletion of a single guanine nucleotide normally present in codon 133. . . . Both children were confirmed to be homozygous for this deletion. . . . As expected, all four parents were heterozygotes. . . . Homozygosity for this frame-shift mutation was associated with . . . serum leptin levels . . . close to the limits of detection. . . . The finding of barely detectable levels of serum leptin in these two extremely obese children was notable. . . . Given the ethical constraints on the study of prepubertal children and our concern to conform to parental wishes, phenotypic characterization was restricted to clinical observations. . . . One child had a normal birthweight of 3.46 kg. . . . At the age of 8 years, she weighs 86 kg (>99.6th centile), and her percentage body fat is 57%. The other child is now aged 2 years. He had a normal weight at birth (3.53 kg). . . . He currently weighs 29 kg (>99.6th centile), with 54% body fat. . . . There is a clear history of hyperphagia [overeating], with both children noted from early infancy to be constantly hungry, demanding food continuously and eating considerably more than their siblings. Thus, in both mice and humans, congenital leptin deficiency is associated with a normal body weight at birth followed by the rapid development of severe obesity associated with hyperphagia and impaired satiety. . . . Now that recombinant human leptin is available for investigation in humans, it should be possible to determine whether the correction of leptin deficiency in the two obese children might have therapeutic benefits.

CONGENITAL LEPTIN deficiency is associated with a normal body weight at birth followed by the rapid development of severe obesity.

Source: Nature 387: 903–908.

The main features of the initiation step in polypeptide synthesis are the binding of mRNA to the small subunit of the ribosome and the binding of a charged tRNA bearing the first amino acid (Figure 9.19A). The ribosome includes three sites for tRNA molecules. They are called the **E (exit)** site, the **P (peptidyl)** site, and the **A (aminoacyl)** site. In the initiation of translation, three initiation factors, the 30S subunit, and a special initiator tRNA charged with methionine come together and combine with an mRNA. (In prokaryotes, the initiating tRNA is formylmethionyl-tRNA, abbreviated thus: $tRNA^{fMet}$.) In prokaryotes, the mRNA binding is facilitated by hydrogen bonding between the 16S RNA present in the 30S subunit and the **ribosome-binding site** of the mRNA; in eukaryotes, the 5' cap on the mRNA is instrumental. Together, the 30S + $tRNA^{Met}$ + mRNA complex recruits a 50S

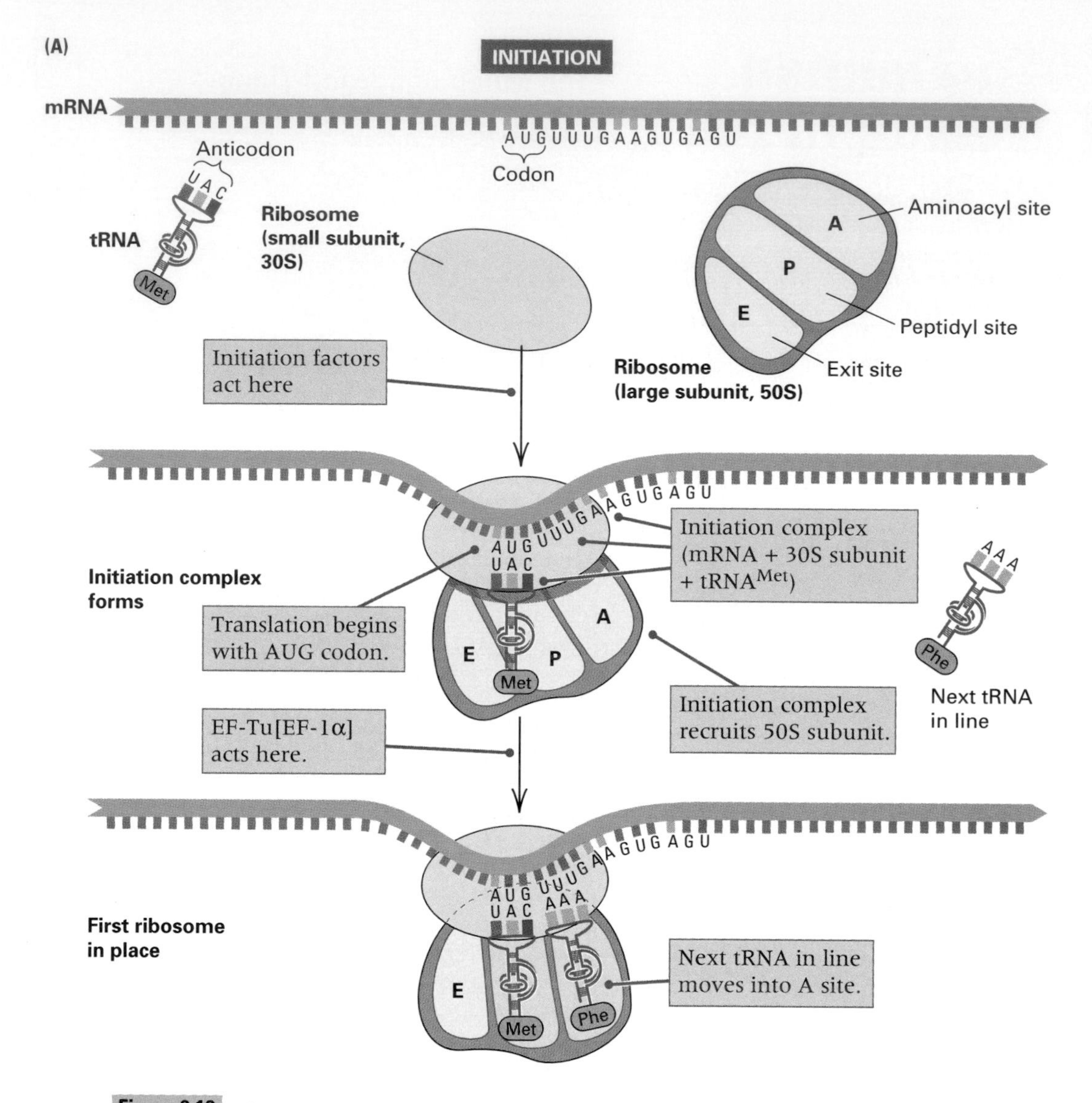

Figure 9.19 (above and facing page) Initiation of protein synthesis. (A) The initiation complex—consisting of the mRNA, one 30S ribosome, and $tRNA^{Met}$—recruits a 50S ribosomal subunit in which the $tRNA^{Met}$ occupies the P (peptidyl) site of the ribosome. A second charged tRNA molecule (in this example $tRNA^{Phe}$) joins the complex in the A (aminoacyl) site. (B) First steps in elongation. The methionine is transferred from the $tRNA^{Met}$ onto the

subunit, in which the $tRNA^{Met}$ is positioned in the P site and aligned with the AUG initiation codon, forming the 70S initiation complex (Figure 9.19A). The tRNA binding is accomplished by hydrogen bonding between the AUG codon in the mRNA and the three-base **anticodon** in the tRNA. In the assembly of the completed ribosome, the initiation factors dissociate from the complex.

The elongation stage of translation consists of three processes: bringing each new aminoacylated tRNA into line, forming the new peptide bond to elongate the polypeptide, and moving the ribosome along the mRNA so that the codons can be translated successively. The first step in elongation is illustrated in Figure 9.19B. A key role is played by the elongation factor EF-Tu, although a second protein, EF-Ts, is also required. (The eukaryotic counterpart of EF-Tu is called EF-1α.) The EF-Tu, bound with guanosine triphosphate (EF-Tu−GTP), brings the next aminoacylated tRNA into the A site on the 50S subunit, which in this example is $tRNA^{Phe}$. This process requires the hydrolysis of GTP to GDP, and once the GDP is formed, the EF-Tu−GDP has low affinity for the ribosome and diffuses away, becoming available for reconversion into EF-Tu−GTP.

Once the A site is filled, a *peptidyl transferase* activity catalyzes a concerted reaction in which the bond connecting the methionine to the $tRNA^{Met}$ is transferred to the amino group of the phenylalanine, forming the first peptide bond. Peptidyl trans-

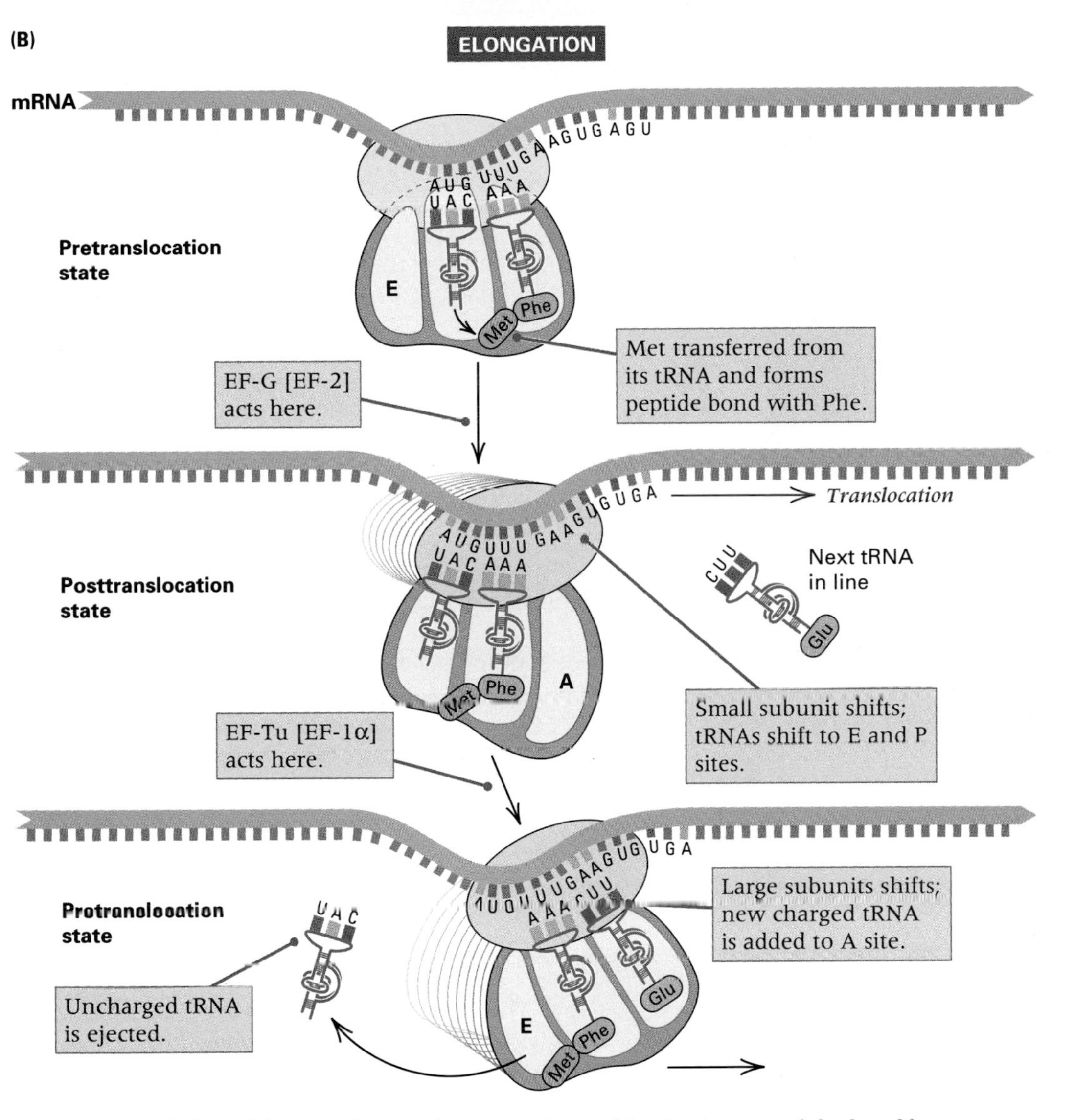

amino group of phenylalanine (the attacking group), resulting in cleavage of the bond between methionine and its tRNA and peptide bond formation in a concerted reaction catalyzed by peptidyl transferase in the 50S subunit. Then the ribosome shifts one codon along the mRNA to the next codon in line.

ferase activity is not due to a single molecule but requires several components of the 50S subunit, including several proteins and an RNA component (called 23S) of the 50S subunit. Some evidence indicates that the actual catalysis is carried out by the 23S RNA, which would suggest that 23S is an example of a ribozyme at work.

Figure 9.20 shows a cutaway view of a 70S ribosome and the bound tRNA molecules in the P site and the A site. The 30S subunit, in light green at the top, binds the mRNA and moves along it in the direction indicated by the arrow. The 50S subunit, in blue at the bottom, contains the tRNA binding sites in the P and A sites. The P site is in red, to the left, and the A site is in dark green, to the right.

Note in Figure 9.19B that the relative positions of the 30S and 50S ribosomal subunits are shifted from one panel to the next. The configuration of the subunits in the top panel is called the *pretranslocation state.* In the middle panel, the 30S subunit shifts one codon to the right. This event is called *translocation.* After translocation, the ribosome is said to be in the *post-translocation state.* In the next step of polypeptide synthesis, shown in the bottom panel in Figure 9.19B, the 50S subunit shifts one step over to the right, which reconfigures the ribosome into the pretranslocation state.

The term **translocation,** as applied to protein synthesis, means the movement of the 30S subunit one codon further along the mRNA. With each

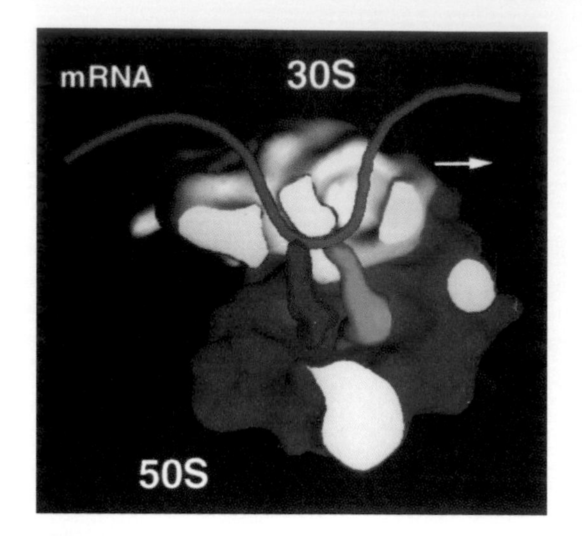

Figure 9.20 A model of a 70S ribosome with some parts cut away to show the orientation of the mRNA relative to the 30S and 50S subunits. The P and A tRNA sites are indicated in red and dark green, respectively. This is the pretranslocation state, in which the E site is unoccupied. [Courtesy of J. Frank, A. Verschoor, Y. Li, J. Zhu, R. K. Lata, M. Radermacher, et al. 1995. *Biochemistry and Cell Biology* 73, 757.]

successive translocation of the 30S subunit, one more amino acid is added to the growing polypeptide chain. The entire cycle of charged tRNA addition, peptide bond formation, and translocation is *elongation.*

The repetitive steps in elongation are outlined in Figure 9.21. Starting with a ribosome in the pretranslocation state (Figure 9.21A), the elongation factor EF-G binds with the ribosome. (The eukaryotic counterpart of EF-G is called EF-2.) Like EF-Tu, the EF-G comes on in the form EF-G−GTP; in fact, it binds to the same ribosomal site as EF-Tu−GTP. Hydrolysis of the GTP to GDP yields the energy to shift the tRNAs in the P and A sites to the E and P sites, respectively, as well as to translocate the 30S subunit one codon along the mRNA (red arrow). The ribosome is thereby converted to the *post-translocation state* (Figure 9.21B), and the EF-G−GDP is released.

At this stage, EF-Tu−GTP comes into play again, and four events happen, as indicated in Figure 9.21C:

- The next aminoacylated tRNA is brought into line (in this case, $tRNA^{Val}$).
- The uncharged tRNA is ejected from the E site.
- In a concerted reaction, the bond connecting the growing polypeptide chain to the tRNA in the P site is transferred to the amino group of the amino acid in the A site, forming the new peptide bond.
- The ribosome transitions to the *pretranslocation* state.

Also, the EF-Tu−GDP is released, making room for the EF-G−GTP, whose function is shown in Figure 9.21D.

- *Translocation* of the 30S ribosome one codon further along the mRNA and return of the ribosome to its *post-translocation* state.

After these steps, the EF-G−GDP is released for regeneration into EF-G−GTP for use in another cycle. The translocation state of the ribosome is not depicted separately in Figure 9.21 because it happens so rapidly. In effect, the ribosome shuttles between the pretranslocation and post-translocation states.

You will note that Figure 9.21D is essentially identical to Figure 9.21B except that the ribosome is one codon farther to the right and the polypeptide is one amino acid greater in length. Hence the ribosome is again available for EF-Tu−GTP to start the next round of elongation. Polypeptide elongation may therefore be considered a cycle of events repeated again and again. The steps B → C → B → C (or, equivalently, C → D → C → D) are carried out repeatedly until a termination codon is encountered.

In Figure 9.21D, for example, the configuration of the ribosome is such that the $tRNA^{Glu}$ and $tRNA^{Val}$ are occupying the E and P sites, respectively. The aminoacylated tRNA that corresponds to the codon AGU is brought into line (it is $tRNA^{Ser}$), and the bond connecting the polypeptide to $tRNA^{Val}$ is transferred to the amino group of Ser, creating a new peptide bond and elongating the polypeptide by one amino acid. At the same time, the ribosome is converted to the pretranslocation state in preparation for translocation. The elongation cycle happens relatively fast. Under optimal conditions, *E. coli* synthesizes a polypeptide at the rate of about 20 amino acids per second; in eukaryotes, the rate of elongation is about 15 amino acids per second.

The elongation steps of protein synthesis are carried out repeatedly until a stop codon for termination is reached. The stop codons are UAA, UAG, and UGA. No tRNA exists that can bind to a stop codon, so the tRNA holding the polypeptide remains in the P site (Figure 9.22). Specific *release factors* act to cleave the polypeptide from the tRNA to which it is attached as well as to disassociate the 70S ribosome from the mRNA, after which the individual 30S and 50S subunits are recycled to initiate translation of another mRNA. The release factor RF-1 recognizes the stop codons UAA and UAG, whereas the release factor RF-2 recognizes UAA and UGA. A third release factor, RF-3, is also required for translational termination.

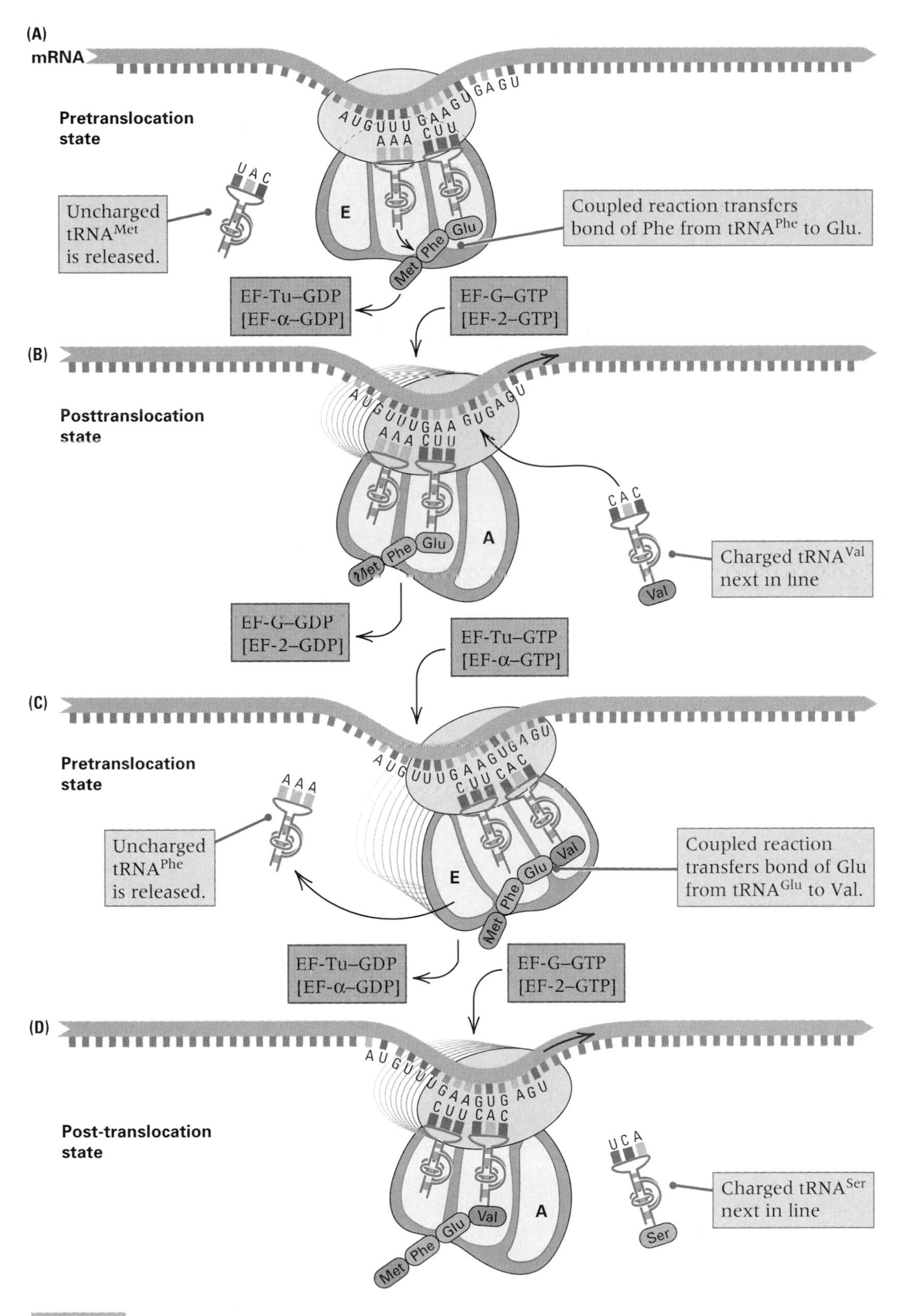

Figure 9.21 Elongation cycle in protein synthesis. (A) Pretranslocation state. (B) Post-translocation state, in which an uncharged tRNA occupies the E site and the polypeptide is attached to the tRNA in the P site. (C) The function of EF-Tu is to release the uncharged tRNA and bring the next charged tRNA into the A site. A peptide bond is formed between the polypeptide and the amino acid held in the A site, in this case Val. Simultaneously, the 50S subunit is shifted relative to the 30S subunit, forming the pretranslocation state. (D) The function of EF-G is to translocate the 30S ribosome to the next codon, once again generating the post-translocation state.

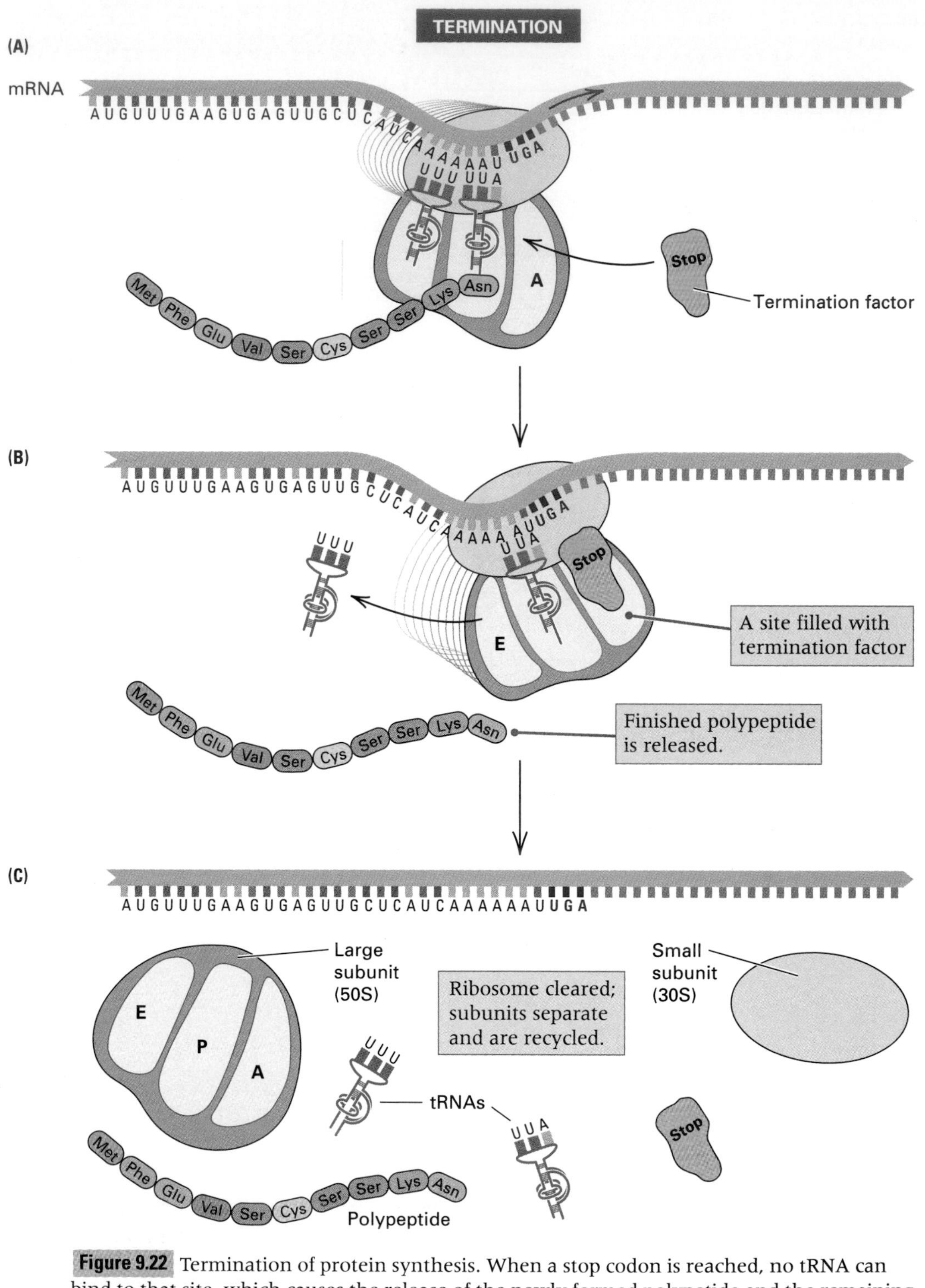

Figure 9.22 Termination of protein synthesis. When a stop codon is reached, no tRNA can bind to that site, which causes the release of the newly formed polypetide and the remaining bound tRNA.

Selection of the initiation codon differs in prokaryotes and eukaryotes.

The process of selecting the correct AUG initiation codon is of some importance in many aspects of gene expression. In prokaryotes, mRNA molecules commonly contain information for the amino acid sequences of several different polypeptide chains; such a molecule is called a **polycistronic mRNA.** (*Cistron* is a term often used to mean a base sequence that encodes a single polypeptide chain.) In a polycistronic mRNA, each polypeptide coding region is preceded by its own ribosome-binding site and AUG initiation codon. After the synthesis of

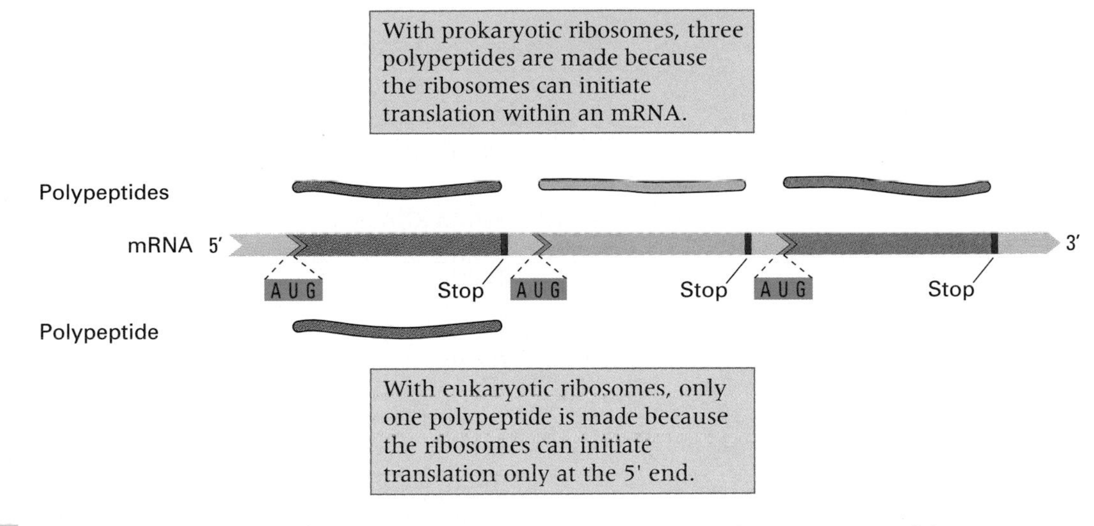

Figure 9.23 Different products are translated from a three-cistron mRNA molecule by the ribosomes of prokaryotes and eukaryotes. The prokaryotic ribosome translates all of the genes, but the eukaryotic ribosome translates only the gene nearest the 5' terminus of the mRNA. Translated sequences are shown in purple, yellow, and orange, stop codons in red, the ribosome binding sites in green, and the spacer sequences in light green.

one polypeptide is finished, the next along the way is translated (Figure 9.23). The genes contained in a polycistronic mRNA molecule often encode the different proteins of a metabolic pathway. For example, in *E. coli*, the ten enzymes needed to synthesize histidine are encoded by one polycistronic mRNA molecule. The use of polycistronic mRNA is an economical way for a cell to regulate the synthesis of related proteins in a coordinated manner. For example, in prokaryotes, the usual way to regulate the synthesis of a particular protein is to control the synthesis of the mRNA molecule that codes for it (Chapter 11). With a polycistronic mRNA molecule, the synthesis of several related proteins can be regulated by a single signal so that appropriate quantities of each protein are made at the same time; this is termed **coordinate regulation.**

In eukaryotes, the 5' terminus of an mRNA molecule binds to the ribosome, after which the mRNA molecule slides along the ribosome until the AUG codon nearest the 5' terminus is in contact with the ribosome. Then protein synthesis begins. There is no mechanism for initiating polypeptide synthesis at any AUG other than the first one encountered. Eukaryotic mRNA is always monocistronic (Figure 9.23).

The definitive feature of translation is that it proceeds in a particular direction along the mRNA and the polypeptide.

> The mRNA is translated from an initiation codon to a stop codon in the 5'-to-3' direction. The polypeptide is synthesized from the amino end toward the carboxyl end by the addition of amino acids, one by one, to the carboxyl end.

For example, a polypeptide with the sequence NH_2–Met–Pro–···–Gly–Ser–COOH would start with methionine, and serine would be the last amino acid added to the chain. The directions of synthesis are illustrated schematically in Figure 9.24.

By convention, in writing nucleotide sequences, we place the 5' end at the left, and in writing amino acid sequences, we place the amino end at the left. Polynucleotides are generally written so that both synthesis and translation proceed from left to right, and polypeptides are written so that synthesis proceeds from left to right. This convention is used in all of our subsequent discussions of the genetic code.

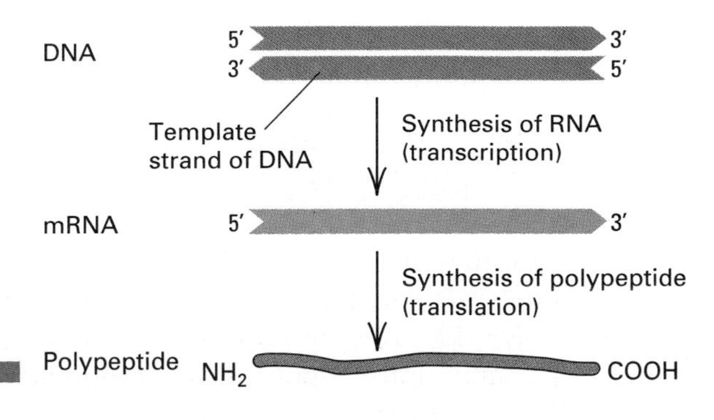

Figure 9.24 Direction of synthesis of RNA with respect to the coding strand of DNA, and of synthesis of protein with respect to mRNA.

9.6 The genetic code for amino acids is a triplet code.

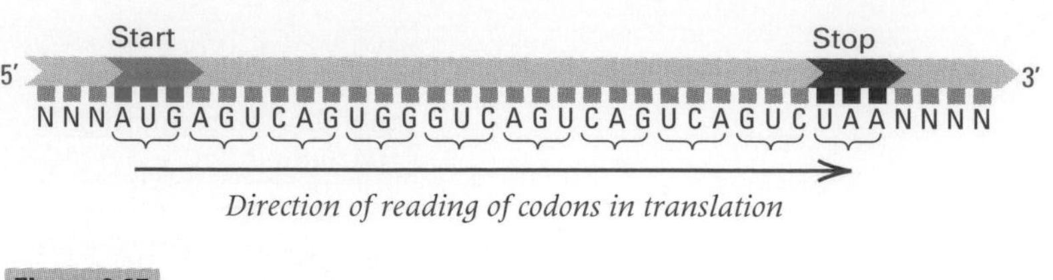

Figure 9.25 Bases in an RNA molecule are read sequentially in the 5' → 3' direction, in groups of three.

Only four bases in DNA are needed to specify the 20 amino acids in proteins because a combination of three adjacent bases is used for each amino acid, as well as for the signals that start and stop protein synthesis. Each sequence of three adjacent bases in mRNA is a codon that specifies a particular amino acid (or chain termination). The **genetic code** is the list of all codons and the amino acid that each one encodes. Before the genetic code was determined experimentally, it was assumed that if all codons had the same number of bases, then each codon would have to contain at least three bases. Codons consisting of pairs of bases would be insufficient, because four bases can form only $4^2 = 16$ pairs; triplets of bases would suffice because four bases can form $4^3 = 64$ triplets. In fact, the genetic code is a triplet code, and all 64 possible codons carry information of some sort. Most amino acids are encoded by more than one codon. Furthermore, in the translation of mRNA molecules, the codons do not overlap but are used sequentially (Figure 9.25).

Genetic evidence for a triplet code came from three-base insertions and deletions.

Although theoretical considerations suggested that each codon must contain at least three letters, codons having more than three letters could not be ruled out. The first widely accepted proof for a triplet code came from genetic experiments using *rII* mutants of bacteriophage T4 that had been induced by replication in the presence of the chemical *proflavin*. These experiments were carried out in 1961 by Francis Crick and collaborators. Proflavin-induced mutations typically resulted in total loss of function. Because proflavin is a large planar molecule, investigators suspected that it caused mutations that inserted or deleted a base pair by interleaving between base pairs in the double helix. Analysis of the properties of these mutations led directly to the deduction that the code is read three nucleotides at a time from a fixed point; in other words, there is a **reading frame** to each mRNA. Mutations that delete or add a base pair shift the reading frame and are called **frameshift mutations.** Figure 9.26 illustrates the profound effect of a frameshift mutation on the amino acid sequence of the polypeptide produced from the mRNA of the mutant gene.

The genetic analysis of the structure of the code began with an *rII* mutation called *FC0,* which was arbitrarily designated (+), as if it had an inserted base pair. (It could also arbitrarily have been designated (−), as if it had a deleted base pair. Calling it (+) was a lucky guess, however, because when *FC0* was sequenced, it did turn out to have a single-base insertion.) If *FC0* has a (+) insertion, then it should be possible to revert the *FC0* allele to "wildtype" by deletion of a nearby base. Selection for r^+ revertants was carried out by isolating plaques formed on a lawn of an *E. coli* strain K12 that was lysogenic for phage λ. The basis of the selection is that *rII* mutants are unable to propagate in K12(λ). Analysis of the revertants revealed that each still carried the original *FC0* mutation, along with a second (suppressor) mutation that reversed the effects of the *FC0* mutation. The suppressor mutations could be separated by recombination from the original mutation by crossing each revertant to wildtype; each suppressor mutation proved to be an *rII* mutation that, by itself, would cause the *r* (rapid lysis) phenotype. If *FC0* had an inserted base, then the suppressors should all result in deletion of a base pair; hence each suppressor of *FC0* was designated (−). The consequences of three such revertants for the translational reading frame are illustrated using ordinary three-letter words in Figure 9.27. The (−) mutations are designated $(-)_1$, $(-)_2$, and $(-)_3$, and

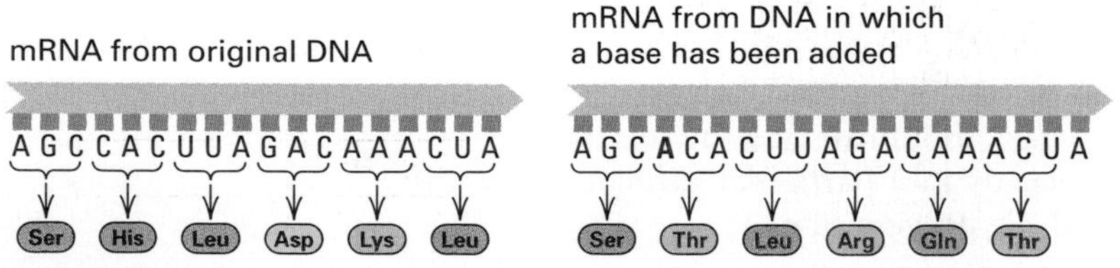

Figure 9.26 The change in the amino acid sequence of a protein caused by the addition of an extra base, which shifts the reading frame. A deleted base also shifts the reading frame.

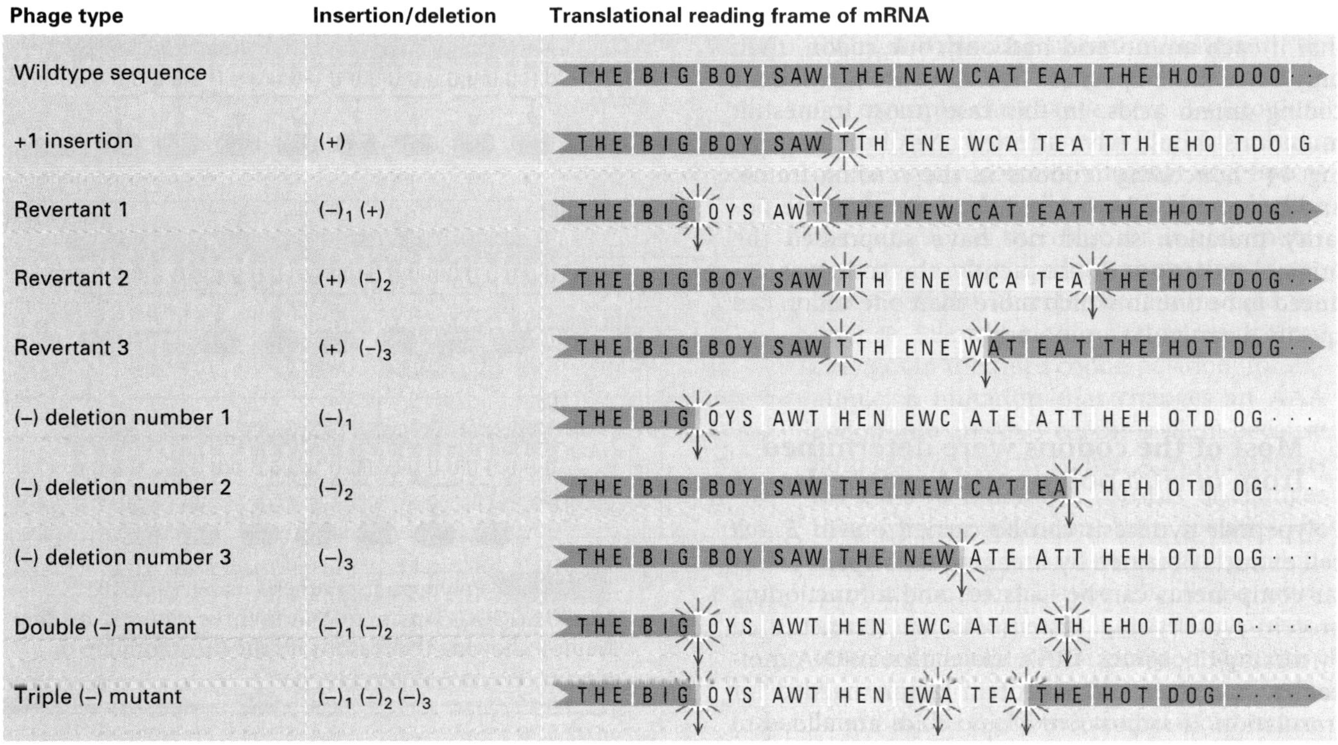

Phage type	Insertion/deletion	Translational reading frame of mRNA
Wildtype sequence		THE BIG BOY SAW THE NEW CAT EAT THE HOT DOG · ·
+1 insertion	(+)	THE BIG BOY SAW TTH ENE WCA TEA TTH EHO TDO G
Revertant 1	$(-)_1$ (+)	THE BIG OYS AWT THE NEW CAT EAT THE HOT DOG · ·
Revertant 2	(+) $(-)_2$	THE BIG BOY SAW TTH ENE WCA TEA THE HOT DOG · ·
Revertant 3	(+) $(-)_3$	THE BIG BOY SAW TTH ENE WAT EAT THE HOT DOG · ·
(−) deletion number 1	$(-)_1$	THE BIG OYS AWT HEN EWC ATE ATT HEH OTD OG · · ·
(−) deletion number 2	$(-)_2$	THE BIG BOY SAW THE NEW CAT EAT HEH OTD OG · · ·
(−) deletion number 3	$(-)_3$	THE BIG BOY SAW THE NEW ATE ATT HEH OTD OG · · ·
Double (−) mutant	$(-)_1$ $(-)_2$	THE BIG OYS AWT HEN EWC ATE ATH EHO TDO G · · ·
Triple (−) mutant	$(-)_1$ $(-)_2$ $(-)_3$	THE BIG OYS AWT HEN EWA TEA THE HOT DOG · · · ·

Figure 9.27 Interpretation of the *rII* frameshift mutations showing that combinations of appropriately positioned single-base insertions (+) and single-base deletions (−) can restore the correct reading frame (green). The key finding was that a combination of three single-base deletions, as shown in the bottom line, also restores the correct reading frame. Two single-base deletions do not restore the reading frame. These classic experiments gave strong genetic evidence that the genetic code is a triplet code.

those parts of the mRNA translated in the correct reading frame are indicated in green.

In the *rII* experiments, all of the individual (−) suppressor mutations were used, in turn, to select other "wildtype" revertants, with the expectation that these revertants would carry new suppressor mutations of the (+) variety, because the (−)(+) combination should yield a phage able to form plaques on K12(λ).

Various double-mutant combinations were made by recombination. Usually any (+) (−) combination, or any (−)(+) combination, resulted in a wildtype phenotype, whereas (+)(+) and (−)(−) double-mutant combinations always resulted in the mutant phenotype. The most revealing result came when triple mutants were made. Usually, the (+)(+)(+) and (−)(−)(−) triple mutants yielded the wildtype phenotype!

The phenotypes of the various (+) and (−) combinations were interpreted in terms of a reading frame. The initial *FC0* mutation, a +1 insertion, shifts the reading frame, resulting in incorrect amino acid sequence from that point on and thus a nonfunctional protein (Figure 9.27). Deletion of a base pair nearby will restore the reading frame, although the amino acid sequence encoded between the two mutations will be different and incorrect. In (+)(+) or (−)(−) double mutants, the reading frame is shifted by two bases; the protein made is still nonfunctional. However, in the (+)(+)(+) and (−)(−)(−) triple mutants, the reading frame is restored, though all amino acids encoded within the region bracketed by the outside mutations are incorrect; the protein made is one amino acid longer for (+)(+)(+) and one amino acid shorter for (−)(−)(−) (Figure 9.27).

The genetic analysis of the (+) and (−) mutations strongly supported the following conclusions:

- Translation of an mRNA starts from a fixed point.
- There is a single reading frame maintained throughout the process of translation.
- Each codon consists of three nucleotides.

Crick and his colleagues also drew other inferences from these experiments. First, in the genetic code, most codons must function in the specification of an amino acid. Second, each amino acid must be

FIVE GENETICALLY-IDENTICAL cloned sheep. Clones can be made by removing an embryo in the early stage of development and splitting. Each separate part is placed in the womb of a sheep, which acts as a surrogate mother. Clones can also be made by injecting the genetic material from a sheep egg or adult cell into a sheep egg from which the genetic material has been removed. This technique allows production of genetically identical animals and is an effective way of propagating any genes introduced into the parents. [© Geoff Tompkinson/Photo Researchers, Inc./Science Photo Library.]

Key Concepts

- In recombinant DNA (gene cloning), DNA fragments are isolated, inserted into suitable vector molecules, and introduced into host cells (usually bacteria or yeast), where they are replicated.
- Recombinant DNA is widely used in research, medical diagnostics, and the manufacture of drugs and other commercial products.
- In reverse genetics, gene function is analyzed by introducing into the germline of an organism an altered gene carrying a designed mutation.
- Specialized methods for manipulating and cloning large fragments of DNA (up to one million base pairs) have made it possible to develop physical maps of the DNA in complex genomes.
- Large-scale automated DNA sequencing has resulted in the complete sequence of the genomes of several species of bacteria, the complete sequence of the genome of the yeast *Saccharomyces cerevisiae* (the first eukaryote completely sequenced), and most of the genome of the nematode *Caenorhabditis elegans*. Large-scale sequencing projects are under way in many other organisms of genetic or commercial interest.

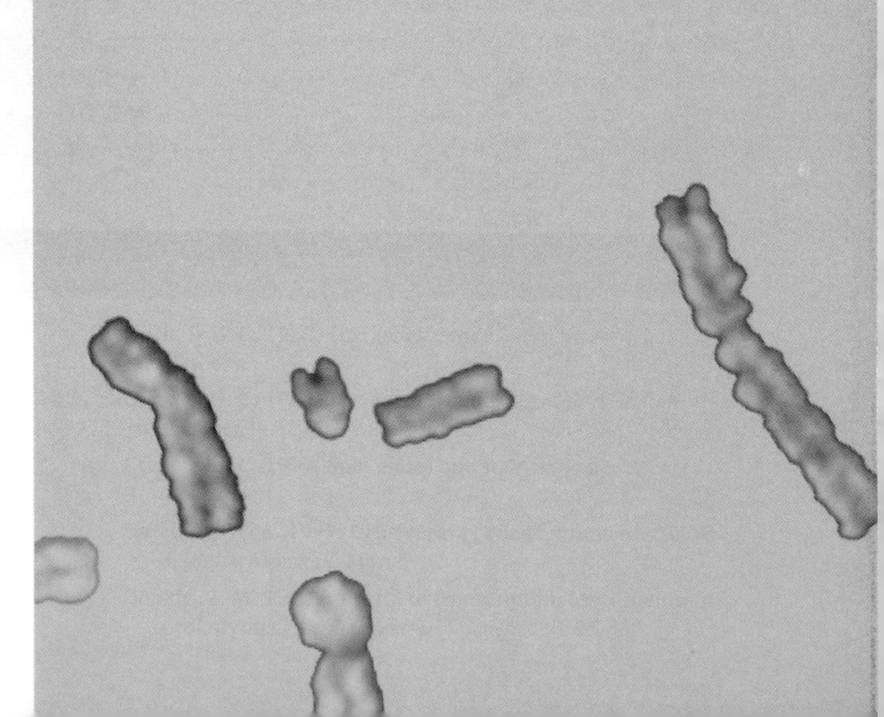

CHAPTER 10

Genetic Engineering and Genome Analysis

Genetic analysis has been instrumental in advancing our understanding of biological processes. Genetics plays an equally important role in the deliberate manipulation of biological systems for scientific or economic reasons in order to create organisms with novel genotypes and phenotypes. The traditional foundations of genetic manipulation have been mutation and recombination. These processes are essentially random and rarely produce organisms with a particular combination of desired characteristics. Selective procedures, often quite complex, are required to identify organisms with the desired characteristics among the many other types of organisms produced. Since the 1970s, techniques have been developed in which the genotype of an organism can instead be modified in a directed and predetermined way. This approach has been called **recombinant DNA, genetic engineering,** and **gene cloning.** It entails isolating DNA fragments, joining them in new combinations, and introducing the recombined molecules back into a living organism. Selection of the desired genotype is still necessary, but the probability of success is usually many orders of magnitude greater than with traditional procedures. The basic technique is quite simple: Two DNA molecules are isolated and cut into fragments by one or more specialized enzymes; then the fragments are joined together *in any desired combination* and introduced back into a cell for replication and reproduction. Recombinant DNA procedures complement the traditional methods of plant and animal breeding in supplying new types of genetic variation for improvement of the organisms.

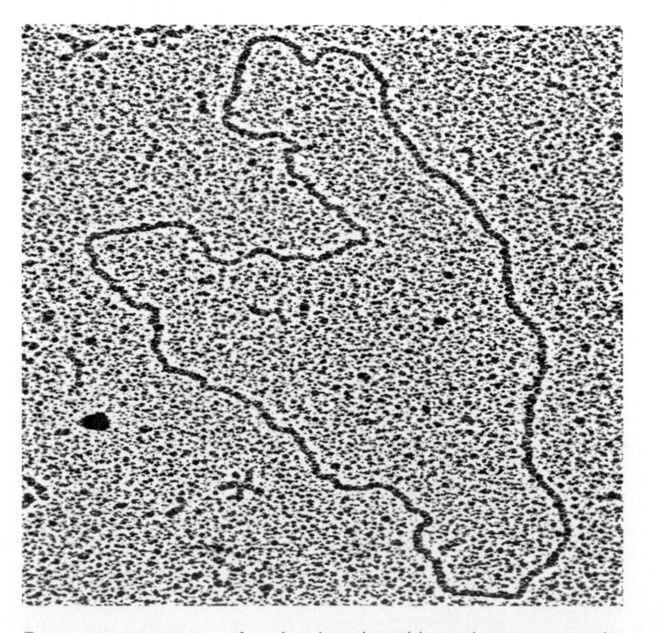

ELECTRON MICROGRAPH of a circular plasmid used as a vector for cloning in *E. coli.*

Current interest in genetic engineering centers on its many practical applications, which include:

1. Isolation of a particular gene, part of a gene, or region of a genome
2. Production of particular RNA and protein molecules in quantities formerly unobtainable
3. Increased efficiency in the production of biochemicals, such as enzymes and drugs
4. Creation of organisms with particular desirable characteristics (for example, plants that require less fertilizer and animals resistant to disease)
5. Diagnosis of allelic status for molecular genetic disease
6. Correction of genetic defects in higher organisms

Some specific examples are considered in this chapter.

10.1 Cloning a DNA molecule takes place in several steps.

In genetic engineering, the immediate goal of an experiment is usually to insert a *particular* fragment of chromosomal DNA into a plasmid or a viral DNA molecule. This is accomplished by techniques for breaking DNA molecules at specific sites and for isolating particular DNA fragments.

Restriction enzymes cleave DNA into fragments with defined ends.

DNA fragments are usually obtained by the treatment of DNA samples with restriction enzymes. *Restriction enzymes* are nucleases that cleave DNA wherever it contains a particular short sequence of nucleotides that matches the *restriction site* of the enzyme (see Section 6.6). Most restriction sites consist of four or six nucleotides, within which the restriction enzyme makes two single-strand breaks, one in each strand, generating 3'-OH and 5'-P groups at each position. Several hundred restriction enzymes, nearly all with different restriction site specificities, have been isolated from microorganisms.

Most restriction sites are symmetrical in the sense that the sequence is identical in both strands of the DNA duplex. For example, the restriction enzyme *Eco*RI, isolated from *E. coli,* has the restriction site 5'-GAATTC-3'; the sequence of the other strand is 3'-CTTAAG-5', which is identical but written with the 3' end at the left. *Eco*RI cuts each strand between the G and the A. The term *palindrome* is used to denote this type of symmetrical sequence.

Soon after restriction enzymes were discovered, observations with the electron microscope indicated that the fragments produced by many restriction enzymes could spontaneously form circles. The circles could be made linear again by heating. On the other hand, if the circles that formed spontaneously were treated with DNA ligase, which joins 3'-OH and 5'-P groups (Section 6.5), then they could no longer be made linear with heat because the ends were covalently linked by the DNA ligase. This observation was the first evidence for three important features of restriction enzymes:

- Restriction enzymes cleave DNA molecules in palindromic sequences.
- The breaks need not be directly opposite one another in the two DNA strands.
- Enzymes that cleave the DNA strands asymmetrically generate DNA fragments with complementary ends.

These properties are illustrated for *Eco*RI in Figure 10.1.

Most restriction enzymes are like *Eco*RI in that they make staggered cuts in the DNA strands, producing single-stranded ends called **sticky ends** that can adhere to each other because they contain complementary nucleotide sequences. Some restriction enzymes (such as *Eco*RI) leave a single-stranded overhang at the 5' end (Figure 10.2A); others leave a 3' overhang. A number of restriction enzymes cleave both DNA strands at the center of symmetry, forming **blunt ends.** Figure 10.2B shows the blunt ends produced by the enzyme *Bal*I. Blunt ends also can be ligated by DNA ligase. However, whereas ligation of sticky ends re-creates the original restriction site, any blunt end can join with any other blunt end and not necessarily create a restriction site.

Most restriction enzymes recognize their restriction sequence without regard to the source of the DNA. Thus,

> Restriction fragments of DNA obtained from one organism have the same sticky ends as restriction fragments from another organism if they were produced by the same restriction enzyme.

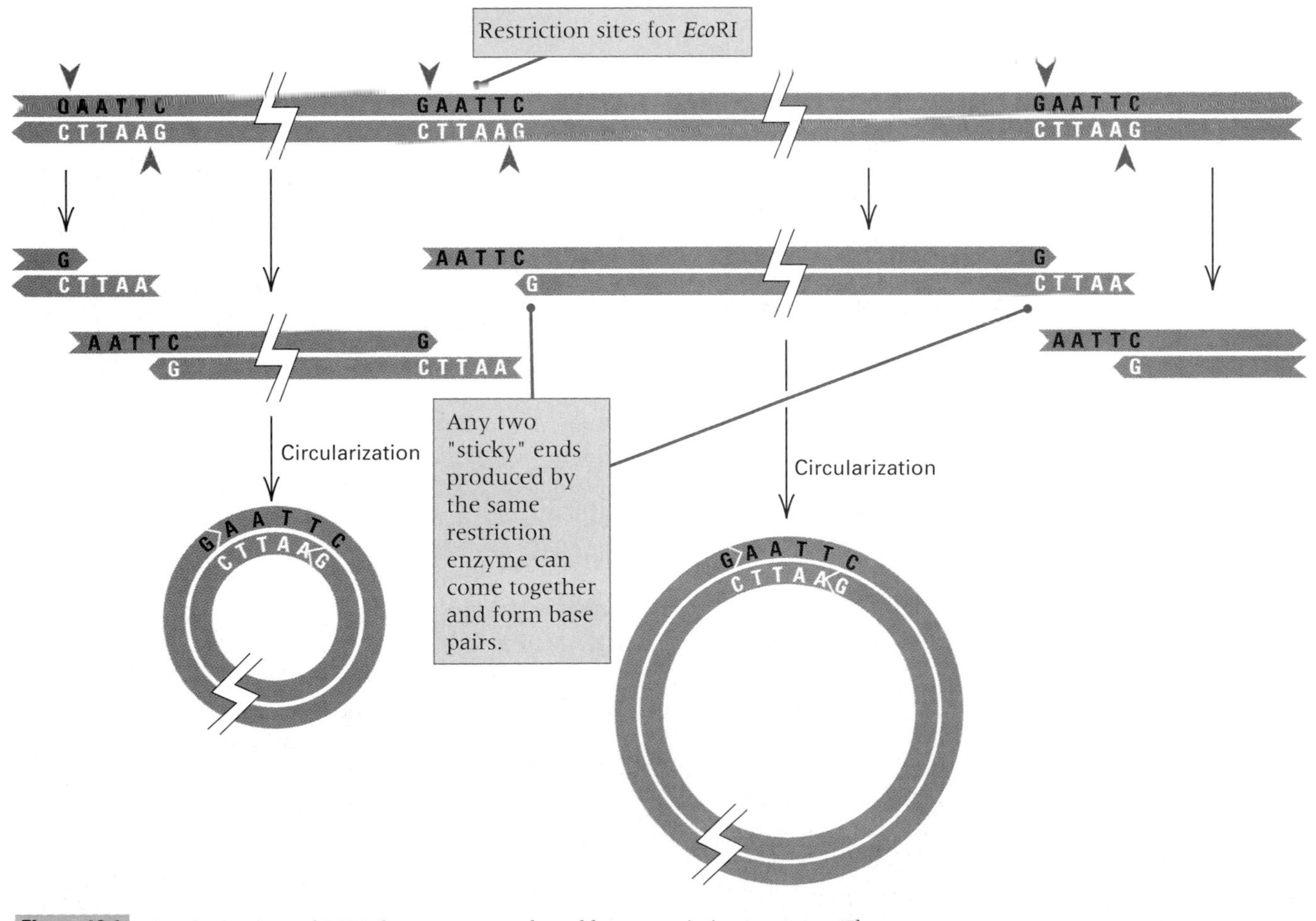

Figure 10.1 Circularization of DNA fragments produced by a restriction enzyme. The arrows indicate the *Eco*RI cleavage sites.

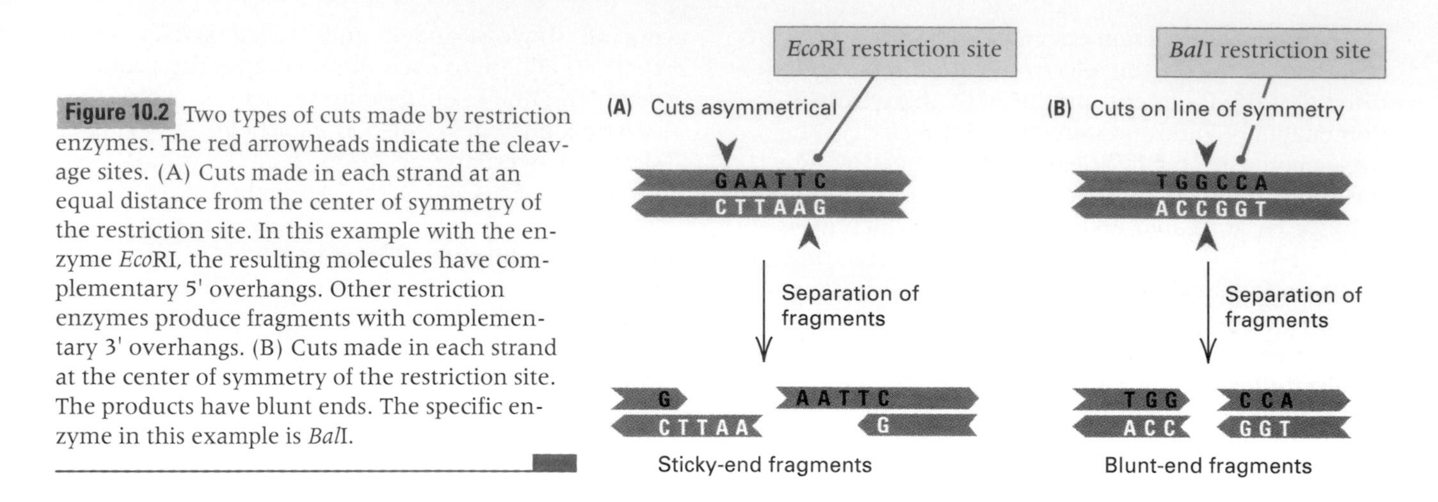

Figure 10.2 Two types of cuts made by restriction enzymes. The red arrowheads indicate the cleavage sites. (A) Cuts made in each strand at an equal distance from the center of symmetry of the restriction site. In this example with the enzyme *Eco*RI, the resulting molecules have complementary 5' overhangs. Other restriction enzymes produce fragments with complementary 3' overhangs. (B) Cuts made in each strand at the center of symmetry of the restriction site. The products have blunt ends. The specific enzyme in this example is *Bal*I.

This principle will be seen to be one of the foundations of recombinant DNA technology.

Because most restriction enzymes recognize a unique sequence, the number of cuts made in the DNA of an organism by a particular enzyme is limited. For example, an *E. coli* DNA molecule contains 4.6×10^6 base pairs, and any enzyme that cleaves a six-base restriction site will cut the molecule into about a thousand fragments. This number of fragments follows from the fact that any particular six-base sequence (including a six-base restriction site) is expected to occur in a random sequence every $4^6 = 4096$ base pairs, on the average, assuming equal frequencies of the four bases. For the same reason, mammalian nuclear DNA would be cut into about a million fragments. These large numbers are still small compared with the number that would be produced if breakage occurred at completely random sequences. Of special interest are the smaller DNA molecules, such as viral or plasmid DNA, which may have from only one to ten sites of cutting (or even none) for particular enzymes. Plasmids that contain a single site for a particular enzyme are especially valuable, as we will see shortly.

Restriction fragments are joined end to end to produce recombinant DNA.

In genetic engineering, a particular DNA segment of interest is joined to a **vector,** a relatively small DNA molecule that is able to replicate inside a cell and that usually contains one or more sequences able to confer antibiotic resistance (or some other detectable phenotype) on the cell. This recombinant molecule is introduced into a cell by means of DNA transformation, where the recombinant molecule is replicated (Figure 10.3). When a stable transformant has been isolated, the genes or DNA sequences linked to the vector are said to be **cloned.** A vector is therefore a DNA molecule into which another DNA fragment can be cloned, or a carrier for recombinant DNA. In the following section, several types of vectors are described.

A vector is a carrier for recombinant DNA.

The most generally useful vectors have three properties:

1. The vector DNA can be introduced into a host cell.
2. The vector contains a replication origin and so can replicate inside the host cell.
3. Cells containing the vector can usually be selected by a straightforward assay, most conveniently by allowing growth of the host cell on a solid selective medium.

At present, the most commonly used vectors are *E. coli* plasmids and derivatives of the bacteriophages λ and M13. Many other plasmids and viruses also have been developed for cloning into cells of animals, plants, and bacteria. Recombinant DNA can be detected in host cells by means of genetic features or particular markers made evident in the formation of colonies or plaques. Plasmid and phage DNA can be introduced into cells by a transformation in which cells gain the ability to take up free DNA by exposure to a $CaCl_2$ solution (Section 8.2). Recombinant DNA can also be introduced into cells by a kind of electrophoretic procedure called *electroporation.* After introduction of the DNA, the cells that contain the recombinant DNA are plated on a solid medium. If the added DNA is a plasmid, colonies consisting of bacterial cells that contain the recombinant plasmid are formed, and the transformants can usually be detected by the phenotype that the plasmid confers on the host cell. For example, plasmid vectors typically include one or more genes for resistance to antibiotics, and plating the

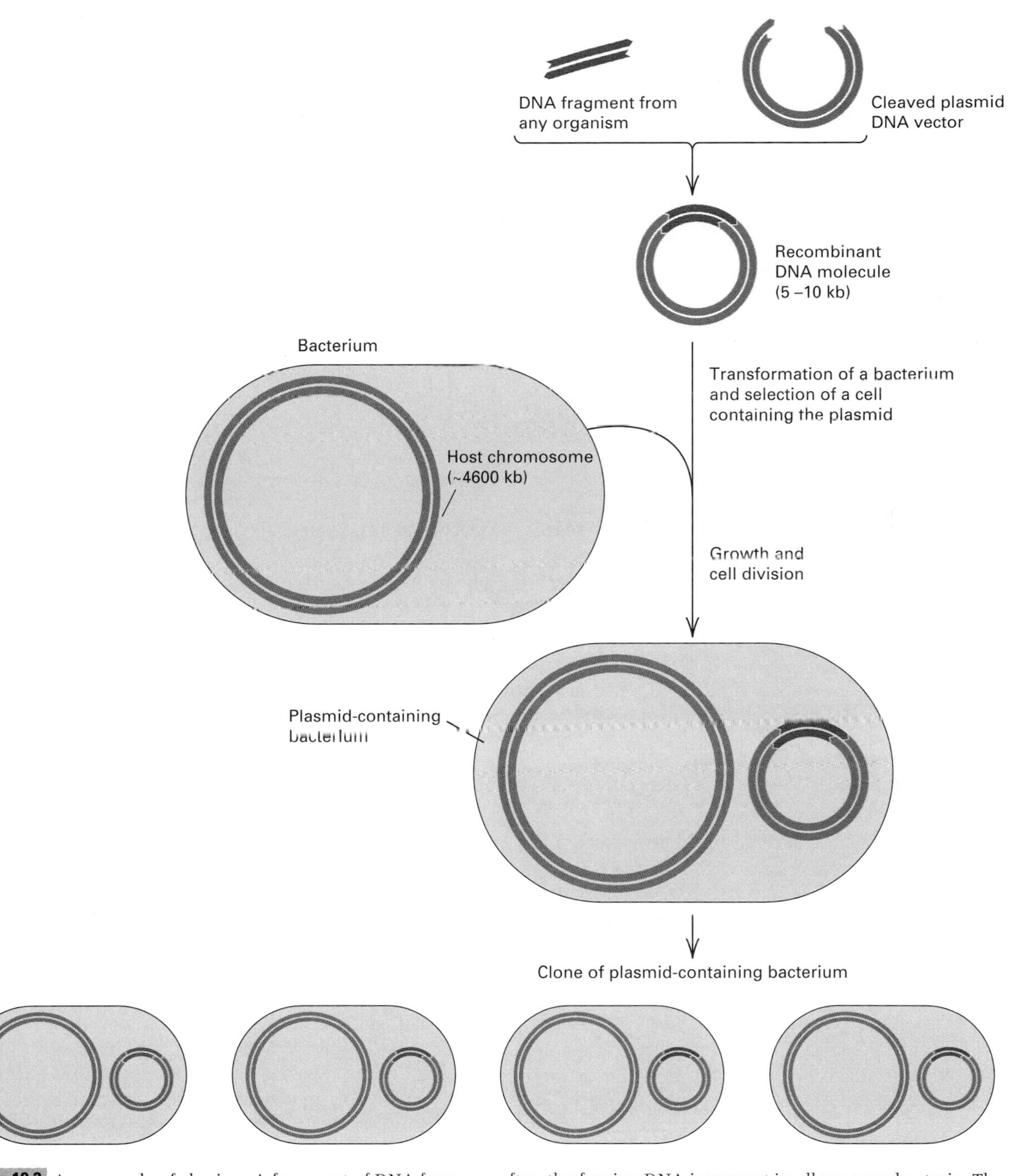

Figure 10.3 An example of cloning. A fragment of DNA from any organism is joined to a cleaved plasmid. The recombinant plasmid is then used to transform a bacterial cell, and thereafter, the foreign DNA is present in all progeny bacteria. The bacterial host chromosome is not drawn to scale. It is typically about 1000 times larger than the plasmid.

transformed cells on a selective medium with antibiotic prevents all but the plasmid-containing cells from growing (Section 8.1). Alternatively, if the vector is phage DNA, the infected cells are plated in the usual way to yield plaques. Variants of these procedures are used to transform animal or plant cells with suitable vectors, but the technical details may differ considerably.

Three types of vectors commonly used for cloning into *E. coli* are illustrated in Figure 10.4. Plasmids (Figure 10.4A) are most convenient for cloning relatively small DNA fragments (5 to10 kb). Somewhat larger fragments can be cloned with bacteriophage λ (Figure 10.4B). The wildtype phage is approximately 50 kb in length, but the central portion of the genome is not essential for

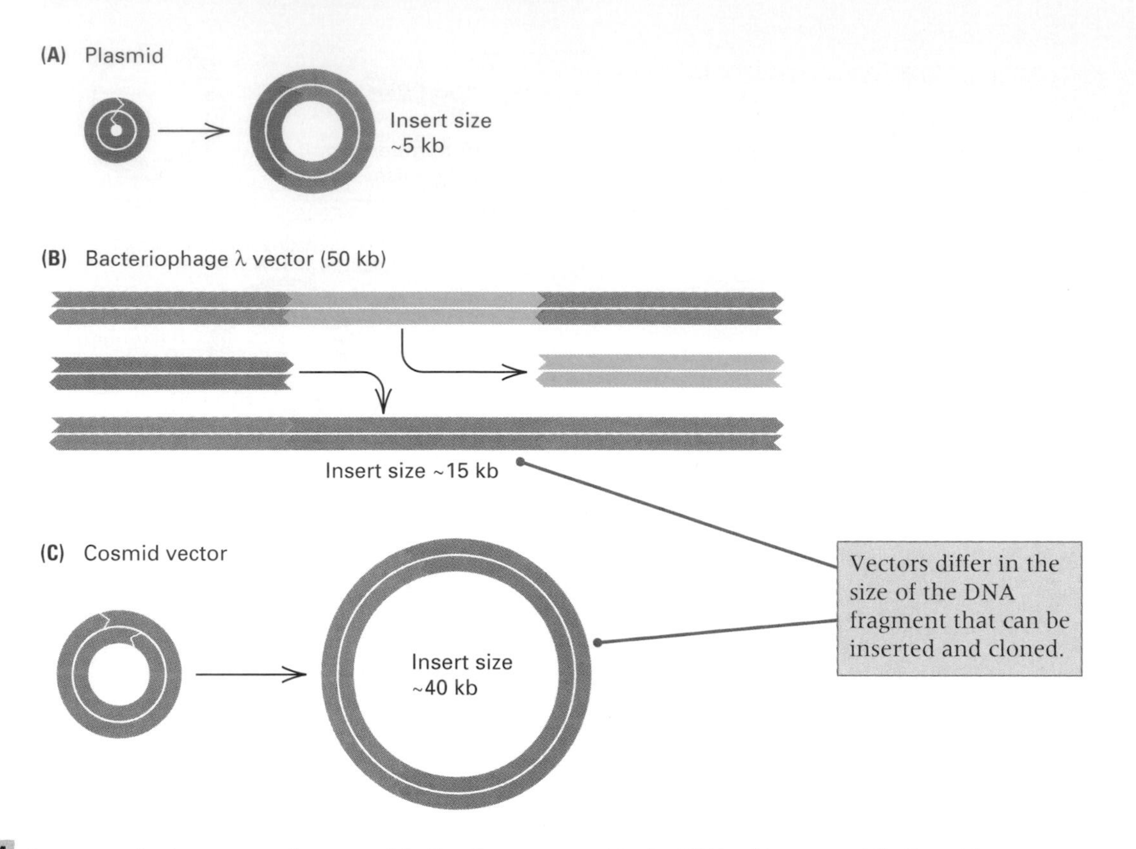

Figure 10.4 Common cloning vectors for use with *E. coli,* drawn approximately to scale. (A) Plasmid vectors are ideal for cloning relatively small fragments of DNA. (B) Bacteriophage λ vectors contain convenient restriction sites for removing the middle section of the phage and replacing it with the DNA of interest. (C) Cosmid vectors are useful for cloning DNA fragments up to about 40 kb; they can replicate as plasmids but contain the cohesive ends of phage λ and so can be packaged in phage particles.

lytic growth and can be removed and replaced with donor DNA. After the donor DNA has been ligated in place, the recombinant DNA is packaged into mature phage *in vitro*, and the phage is used to infect bacterial cells. However, to be packaged into a phage head, the recombinant DNA must be neither too large nor too small, which means that the donor DNA must be roughly the same size as the portion of the λ genome that was removed. Most λ cloning vectors accept inserts ranging in size from 12 to 20 kb. Still larger DNA fragments can be inserted into cosmid vectors (Figure 10.4C). These vectors can exist as plasmids, but they also contain the cohesive ends of phage λ (Section 8.6), which enables them to be packaged into mature phages. The size limitation on cosmid inserts usually ranges from 40 to 45 kb. Vectors for cloning even larger fragments of DNA are referred to as *artificial chromosomes;* these are discussed in Section 10.4.

A variety of strategies can be used to clone a gene.

In genetic engineering, the immediate goal of an experiment is usually to insert a *particular* fragment of chromosomal DNA into a plasmid or a viral DNA molecule. One strategy for isolating DNA from a known gene is *transposon tagging,* which was discussed in Chapter 8. In transposon tagging, the target gene to be cloned is first inactivated by means of a transposon insertion. Because the insertion inactivates the gene, the mutant phenotype and the position of the mutation in the genetic map indicate that the correct gene has been hit. Because the transposon contains an antibiotic-resistance gene, the resistance affords a direct selection for any vector molecules that contain the transposon. The DNA flanking the transposon in the recombinant clone must necessarily derive from the target gene.

Although transposon tagging is an important approach in gene cloning, the method is feasible only in organisms whose genetics is already highly developed, such as maize and *Drosophila.* Transposon tagging is not possible in many organisms of interest, including human beings. Even in organisms in which the technique is available, particular genes may not be clonable with this approach because the gene contains no target sites for the transposon. It was therefore necessary to develop a set of alternative strategies for cloning genes. These are considered next.

DNA fragments are joined with DNA ligase.

The circularization of restriction fragments that have terminal single-stranded regions with complementary bases is illustrated in Figure 10.1. Because a particular restriction enzyme produces fragments with *identical* sticky ends, without regard for the source of the DNA, fragments from DNA molecules isolated from two different organisms can be joined, as shown in Figure 10.5. In this example, the restriction enzyme *Eco*RI is used to digest DNA from any organism of interest and to cleave a bacterial plasmid that contains only one *Eco*RI restriction

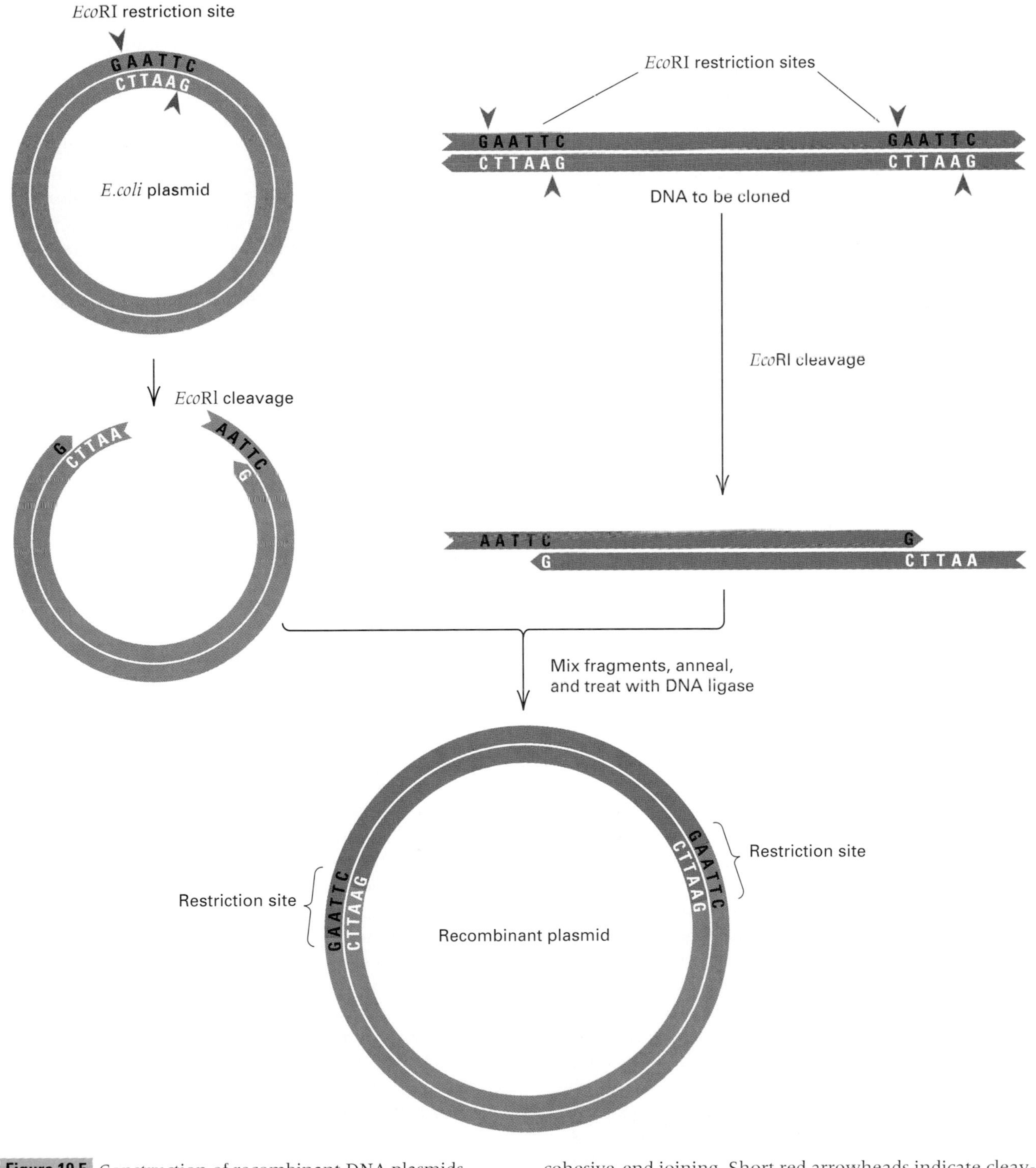

Figure 10.5 Construction of recombinant DNA plasmids containing fragments derived from a donor organism, by the use of a restriction enzyme (in this example *Eco*RI) and cohesive-end joining. Short red arrowheads indicate cleavage sites.

site. The donor DNA is digested into many fragments (one of which is shown) and the plasmid into a single linear fragment. When the donor fragment and the linearized plasmid are mixed, recombinant molecules can form by base pairing between the complementary single-stranded ends. At this point, the DNA is treated with DNA ligase to seal the joints, and the donor fragment becomes permanently joined in a combination that may never have existed before. The ability to join a donor DNA fragment of interest to a vector is the basis of the recombinant DNA technology.

Joining sticky ends does not always produce a DNA sequence that has functional genes. For example, consider a linear DNA molecule that is cleaved into four fragments—A, B, C, and D—whose sequence in the original molecule was ABCD. Reassembly of the fragments can yield the original molecule, but because B and C have the same pair of sticky ends, molecules with the fragment arrangements ACBD and BADC can also form with the same probability as ABCD. Restriction fragments from the vector can also join together in the wrong order, but this potential problem can be eliminated by using a vector that has only one cleavage site for a particular restriction enzyme. Plasmids of this type are available (most have been created by genetic engineering). Many vectors contain unique sites for several different restriction enzymes, but generally only one enzyme is used at a time.

DNA molecules that lack sticky ends also can be joined. A direct method uses the DNA ligase made by *E. coli* phage T4. This enzyme differs from other DNA ligases in that it not only heals single-stranded breaks in double-stranded DNA but also can join molecules with blunt ends (Figure 10.2B).

A recombinant cDNA contains the coding sequence of a eukaryotic gene.

Many genes in higher eukaryotes are very large. They can extend over hundreds of thousands of base pairs. Much of the length is made up of introns, which are excised from the mRNA in processing (Section 9.4). With such large genes, the length of the mRNA is usually much smaller than the length of the gene. Even if the large DNA sequence were cloned, expression of the gene product in bacterial cells would be impossible because bacterial cells are not capable of RNA splicing. Therefore, when a gene is so large that it is difficult to clone and express directly, it would be desirable to clone the coding sequence present in the mRNA to determine the base sequence and study the polypeptide gene product. The method illustrated in Figure 10.6 makes possible the direct cloning of any eukaryotic coding sequence from cells in which the mRNA is present.

Cloning from mRNA molecules depends on an unusual polymerase, **reverse transcriptase,** which can use a single-stranded RNA molecule as a template and synthesize a complementary strand of DNA called **complementary DNA,** or **cDNA.** Like other DNA polymerases, reverse transcriptase requires a primer. The stretch of A nucleotides usually found at the 3' end of eukaryotic mRNA serves as a convenient priming site, because the primer can be an oligonucleotide consisting of poly-T (Figure 10.6). Like any other single-stranded DNA molecule, the single strand of DNA produced from the RNA template can fold back upon itself at the extreme 3' end to form a "hairpin" structure that includes a very short double-stranded region consisting of a few base pairs. The 3' end of the hairpin serves as a primer for second-strand synthesis. The second strand can be synthesized either by DNA polymerase or by reverse transcriptase itself. Reverse transcriptase is the source of the second strand in RNA-based viruses that use reverse transcriptase, such as the human immunodeficieny virus (HIV). Conversion into a conventional double-stranded DNA molecule is achieved by cleavage of the hairpin by a nuclease.

In the reverse transcription of an mRNA molecule, the resulting full-length cDNA contains an uninterrupted coding sequence for the protein of interest. As we saw in Chapter 9, eukaryotic genes often contain DNA sequences, called *introns,* that are initially transcribed into RNA but are removed in the production of the mature mRNA. Because the introns are absent from the mRNA, the cDNA sequence is not identical with that in the genome of the original donor organism. However, if the purpose of forming the recombinant DNA molecule is to identify the coding sequence or to synthesize the gene product in a bacterial cell, then cDNA formed from processed mRNA is the material of choice for cloning. The joining of cDNA to a vector can be accomplished by available procedures for joining blunt-ended molecules (Figure 10.6).

Some specialized animal cells make only one protein, or a very small number of proteins, in large amounts. In these cells, the cytoplasm contains a great abundance of specific mRNA molecules, which constitute a large fraction of the total mRNA synthesized. An example is the mRNA for globin, which is highly abundant in reticulocytes while they are producing hemoglobin. The cDNA produced from purified mRNA from these cells is greatly enriched for the globin cDNA. Genes that are not highly expressed are represented by mRNA molecules whose abundance ranges from low to exceedingly low. The cDNA molecules produced

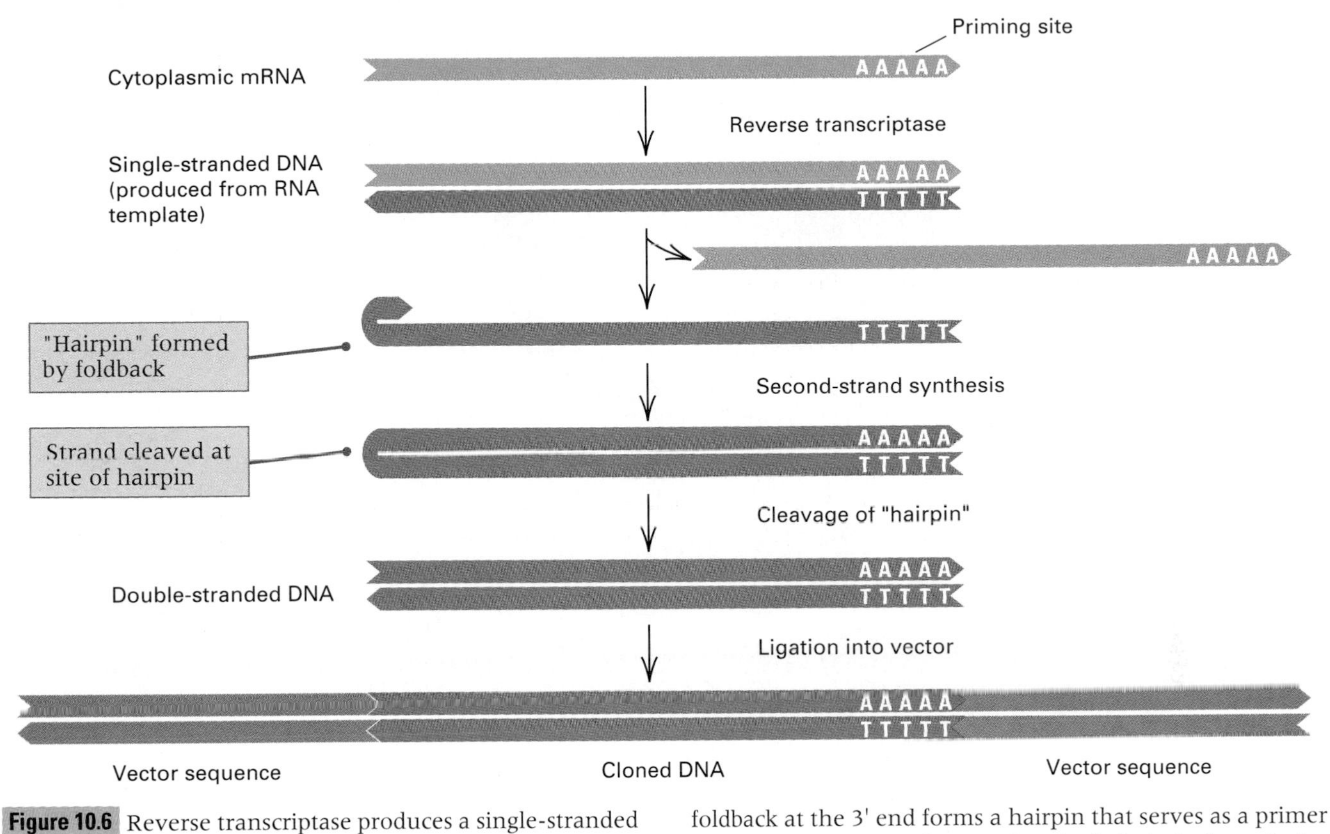

Figure 10.6 Reverse transcriptase produces a single-stranded DNA complementary in sequence to a template RNA. In this example, a cytoplasmic mRNA is copied. As indicated here, most eukaryotic mRNA molecules have a tract of consecutive A nucleotides at the 3' end, which serves as a convenient priming site. After the single-stranded DNA is produced, a foldback at the 3' end forms a hairpin that serves as a primer for second-strand synthesis. After the hairpin is cleaved, the resulting double-stranded DNA can be ligated into an appropriate vector either immediately or after PCR amplification. The resulting clone contains the entire coding region for the protein product of the gene.

from such rare RNAs will also be rare. The efficiency of cloning rare cDNA molecules can be markedly increased by PCR amplification prior to ligation into the vector. The only limitation on the procedure is the requirement that enough DNA sequence be known at both ends of the cDNA for appropriate oligonucleotide primers to be designed. PCR amplification of the cDNA produced by reverse transcriptase is called **reverse transcriptase PCR (RT-PCR).** The resulting amplified molecules contain the coding sequence of the gene of interest with very little contaminating DNA.

Loss of β-galactosidase activity is often used to detect recombinant vectors.

When a vector is cleaved by a restriction enzyme and renatured in the presence of many different restriction fragments from a particular organism, many types of molecules result, including such examples as a self-joined circular vector that has not acquired any fragments, a vector containing one or more fragments, and a molecule consisting only of many joined fragments. To facilitate the isolation of a vector containing a particular gene, some means is needed to ensure (1) that the vector does indeed possess an inserted DNA fragment, and (2) that the fragment is in fact the DNA segment of interest. This section describes several useful procedures for detecting the correct products.

In the use of transformation to introduce recombinant plasmids into bacterial cells, the initial goal is to isolate bacteria that contain the plasmid from a mixture of plasmid-free and plasmid-containing cells. A common procedure is to use a plasmid that possesses an antibiotic-resistance marker and to grow the transformed bacteria on a medium that contains the antibiotic: Only cells that contain plasmid can form a colony. An example of a state-of-the-art cloning vector is the pBluescript plasmid illustrated in Figure 10.7A. The entire plasmid is 2961 base pairs. Different regions contribute to its utility as a cloning vector:

- The plasmid origin of replication is derived from the *E. coli* plasmid ColE1. The ColE1 is a high-copy-number plasmid, and its origin of replication enables pBluescript and its recombinant

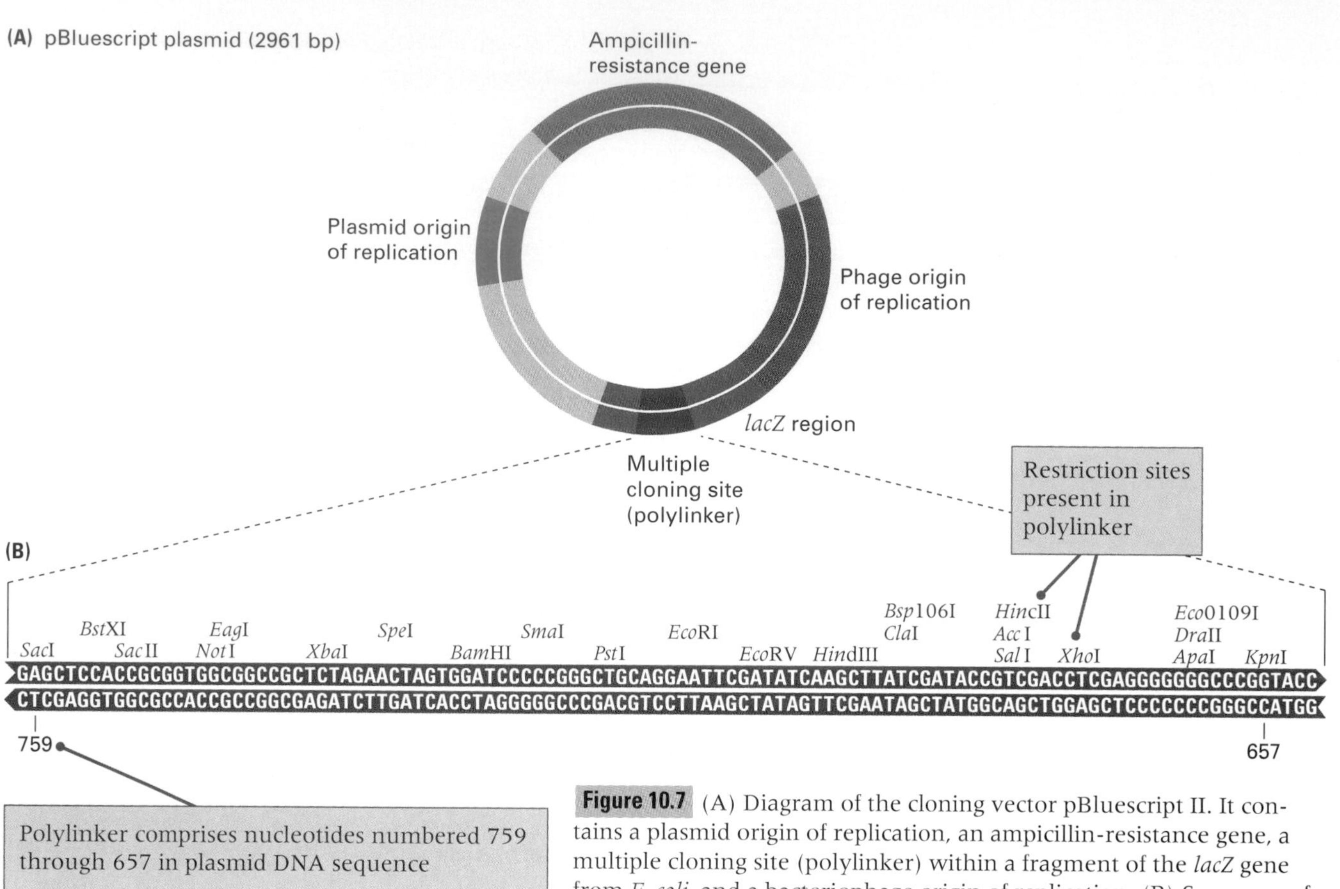

Figure 10.7 (A) Diagram of the cloning vector pBluescript II. It contains a plasmid origin of replication, an ampicillin-resistance gene, a multiple cloning site (polylinker) within a fragment of the *lacZ* gene from *E. coli,* and a bacteriophage origin of replication. (B) Sequence of the multiple cloning site showing the unique restriction sites at which the vector can be opened for the insertion of DNA fragments. The numbers 657 and 759 refer to the position of the base pairs in the complete sequence of pBluescript. [Courtesy of Stratagene Cloning Systems, La Jolla, CA.]

derivatives to exist in approximately 300 copies per cell.

- The ampicillin-resistance gene allows for selection of transformed cells in medium containing ampicillin.

- The cloning site is called a multiple cloning site (MCS), or *polylinker,* because it contains unique cleavage sites for many different restriction enzymes and enables many types of restriction fragments to be inserted. In pBluescript, the MCS is a 108-bp sequence that contains cloning sites for 23 different restriction enzymes (Figure 10.8B).

- The detection of recombinant plasmids is by means of a region containing the *lacZ* gene from *E. coli,* shown in blue in Figure 10.8A. The basis of the selection is illustrated in Figure 10.8. When the *lacZ* region is interrupted by a fragment of DNA inserted into the MCS, the recombinant plasmid yields Lac^- cells because the interruption renders the lacZ region nonfunctional. Nonrecombinant plasmids do not contain a DNA fragment in the MCS and yield Lac^+ colonies. The Lac^+ and Lac^- phenotypes can be distinguished by color when the cells are grown on a special β-galactoside compound called X-gal, which releases a deep blue dye when cleaved. On medium containing X-gal, Lac^+ colonies contain nonrecombinant plasmids and are a deep blue, whereas Lac^- colonies contain recombinant plasmids and are white.

- The bacteriophage origin of replication is from the single-stranded DNA phage f1. When cells that contain a recombinant plasmid are infected with an f1 helper phage, the f1 origin enables a single strand of the inserted fragment, starting with *lacZ,* to be packaged in progeny phage.

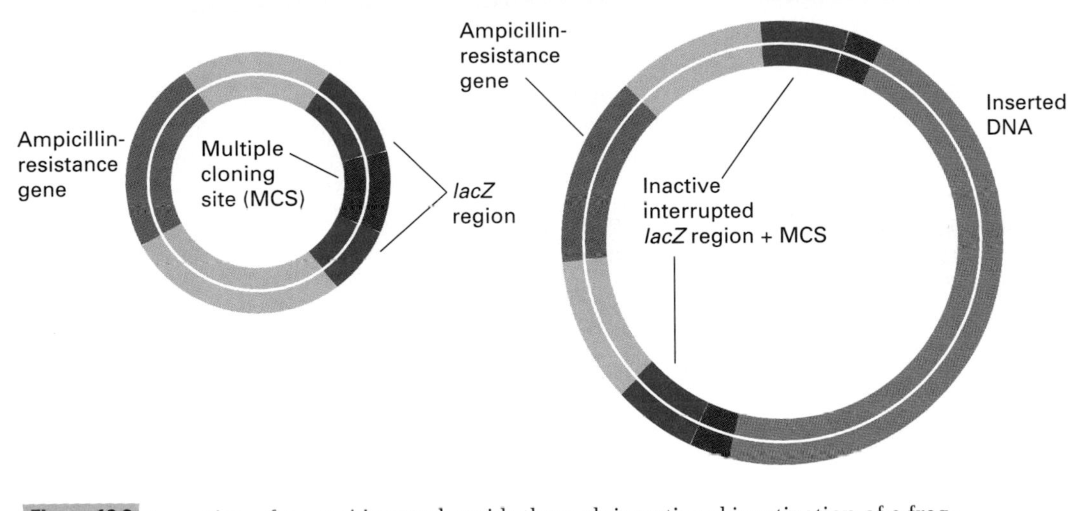

Figure 10.8 Detection of recombinant plasmids through insertional inactivation of a fragment of the *lacZ* gene from *E. coli.* (A) Nonrecombinant plasmid containing an uninterrupted *lacZ* region. The multiple cloning site (MCS) within the region (not drawn to scale) is sufficiently small that the plasmid still confers β-galactosidase activity. (B) Recombinant plasmid with donor DNA inserted into the multiple cloning site. This plasmid confers ampicillin resistance but not β-galactosidase activity, because the donor DNA interrupting the *lacZ* region is large enough to render the region nonfunctional.

This feature is very convenient because it yields single-stranded DNA for sequencing. The plasmid shown in Figure 10.8A is the SK(+) variety. There is also an SK(−) variety in which the f1 origin is in the opposite orientation and packages the complementary DNA strand.

All good cloning vectors have an efficient origin of replication, at least one unique cloning site for the insertion of DNA fragments, and a second gene whose interruption by inserted DNA yields a phenotype indicative of a recombinant plasmid. Once a **library,** or large set of clones, has been obtained in a particular vector, the next problem is how to identify the particular recombinant clones that contain the gene of interest.

Recombinant clones are often identified by hybridization with labeled probe.

Genomic DNA or cDNA clones are often used to isolate additional clones containing DNA fragments with which they have sequences in common. This section discusses procedures that allow detection of the presence of any clone that contains a DNA fragment for which a complementary DNA or RNA sequence, labeled with radioactivity or by some other means, is available. The labeled nucleic acid used to detect the recombinant clones of interest is called the *probe.* A typical example is the use of a probe from a clone of cDNA to identify clones of genomic DNA that include parts of the coding sequence. Another typical example is the use of a probe from a cloned gene from one organism (for example, yeast or *Drosophila*) to identify clones that contain homologous DNA from another organism (for example, mouse or human being).

The procedure of **colony hybridization** makes it possible to detect the presence of any gene for which DNA or RNA labeled with radioactivity or some other means is available (Figure 10.9). Colonies to be tested are transferred (*lifted*) from a solid medium onto a nitrocellulose or nylon filter by gently pressing the filter onto the surface. A part of each colony remains on the agar medium, which constitutes the reference plate. The filter is treated with sodium hydroxide (NaOH), which simultaneously breaks open the cells and separates (*denatures*) the duplex DNA into single strands. The filter is then saturated with labeled probe complementary in base sequence to the gene being sought, and the DNA strands are allowed to form duplex molecules again (*renatured*). After washing to remove unbound probe, the positions of the bound probe identify the desired colonies. For example, with radioactively labeled probe, the desired colonies are located by means of autoradiography. A similar assay is done with phage vectors, but in this case plaques rather than colonies are lifted onto the filters.

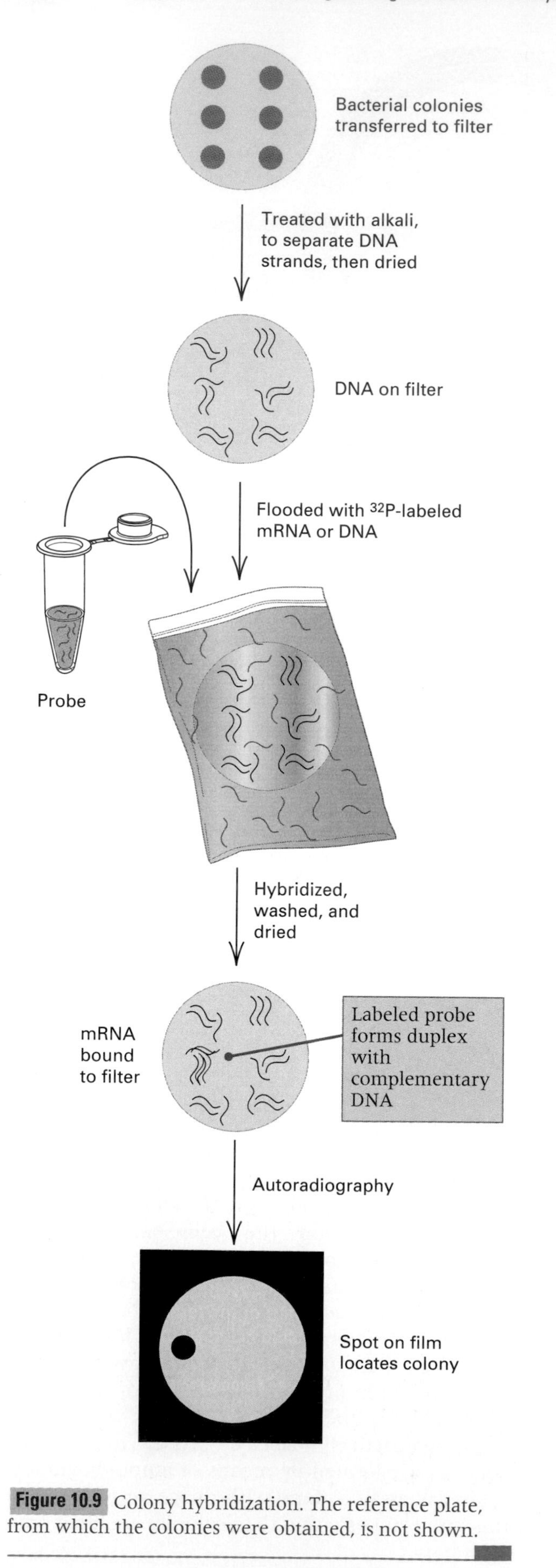

Figure 10.9 Colony hybridization. The reference plate, from which the colonies were obtained, is not shown.

If transformed cells can synthesize the protein product of a cloned gene or cDNA, then immunological techniques may also allow the protein-producing colony to be identified. In one method, the colonies are transferred as in colony hybridization, and the transferred copies are exposed to a labeled antibody directed against the particular protein. Colonies to which the antibody adheres are those that contain the gene of interest.

10.2 Positional cloning is based on the location of a gene in the genetic map.

Because of the large number of polymorphisms in DNA sequence that are found among members of natural populations of virtually every species (Section 4.4), the genetic map of many species is covered almost completely with a dense concentration of molecular markers. The most common type of molecular marker is a *simple tandem repeat polymorphism,* or *STRP,* in which the polymorphism consists of variation in the number of copies of a simple repeating sequence (Section 4.4). STRPs and another type of molecular marker called an *STS (sequence-tagged site),* discussed in Section 10.4, have been used to construct high-density genetic maps of the human and mouse genomes as part of the **Human Genome Project.** In the molecular map of the human genome, there is an average of approximately 100 kb between adjacent genetic markers. This corresponds to about 0.1 cM in genetic distance. In a species in which there is a high density of molecular markers in the genetic map, the genetic map itself often supplies the information needed for cloning a gene. The procedure for cloning a mutation in any gene of interest consists of four steps:

1. Determine the genetic map position of the mutant gene as precisely as possible using molecular markers linked to it.
2. For molecular markers flanking the mutant gene, isolate clones that contain the marker sequences.
3. Verify that the DNA inserts in the isolated clones are mutually overlapping in such a way as to cover the entire region between the flanking markers without any gaps. Such a mutually overlapping set of clones is called a **contig** because adjacent clones are contiguous ("touching"). If the initial set of clones does include one or more gaps, complete the contig by screening the clone library with additional probes derived from the ends of the clones flanking each gap.

4. Identify the clone or clones in the contig that contain the gene of interest. This usually entails determining the nucleotide sequence of all or part of the contig in wildtype and mutant chromosomes to identify the site of the mutation.

Note that the cloning procedure outlined in these steps requires no knowledge of what the gene of interest does. All the method requires is that a mutation in the gene be mapped rather precisely with respect to molecular markers, which, in turn, are used to isolate a contig of clones covering the relevant region. Use of the genetic map to clone genes in this manner is called **positional cloning** or **map-based cloning.** It is one of the most important approaches to cloning genes in all species that have a dense genetic map, and it is particularly important in human molecular genetics. To date, more than 100 different genes that are mutated in human disease have been positionally cloned, and the number is increasing rapidly.

In many well-studied organisms, contigs of clones are already available that cover almost the entire genome. These organisms include bacteria, such as *E. coli*; fungi, including *S. cerevisiae*; the nematode *C. elegans*; the fruit fly *D. melanogaster*; plants, such as *A. thaliana*; the mouse *Mus musculus*; human beings, *Homo sapiens*; and many others. The utility of such contigs is illustrated for a small part of the *Drosophila* genome in Figure 10.10. On the left is a diagram of the banding pattern in the polytene salivary gland chromosome across the region 2E2 through 3B3. (The conventions for naming the bands are discussed in Section 5.6.) These 26 bands comprise approximately 650 kb of DNA. To the right of the bands, the red lines represent the positions of molecular markers in the region, spaced at an average distance of about 30 kb. The blue bars depict clones of *Drosophila* DNA that are known to contain sequences of the molecular markers. Each clone has an insert of approximately 85 kb of *Drosophila* DNA in a bacteriophage P1 vector. The red lines coincide with the positions of the molecular markers, and the points of intersection with the clones show the presence of each sequence within each clone. Seven cloned genes in the region are indicated. In the cloning of any new

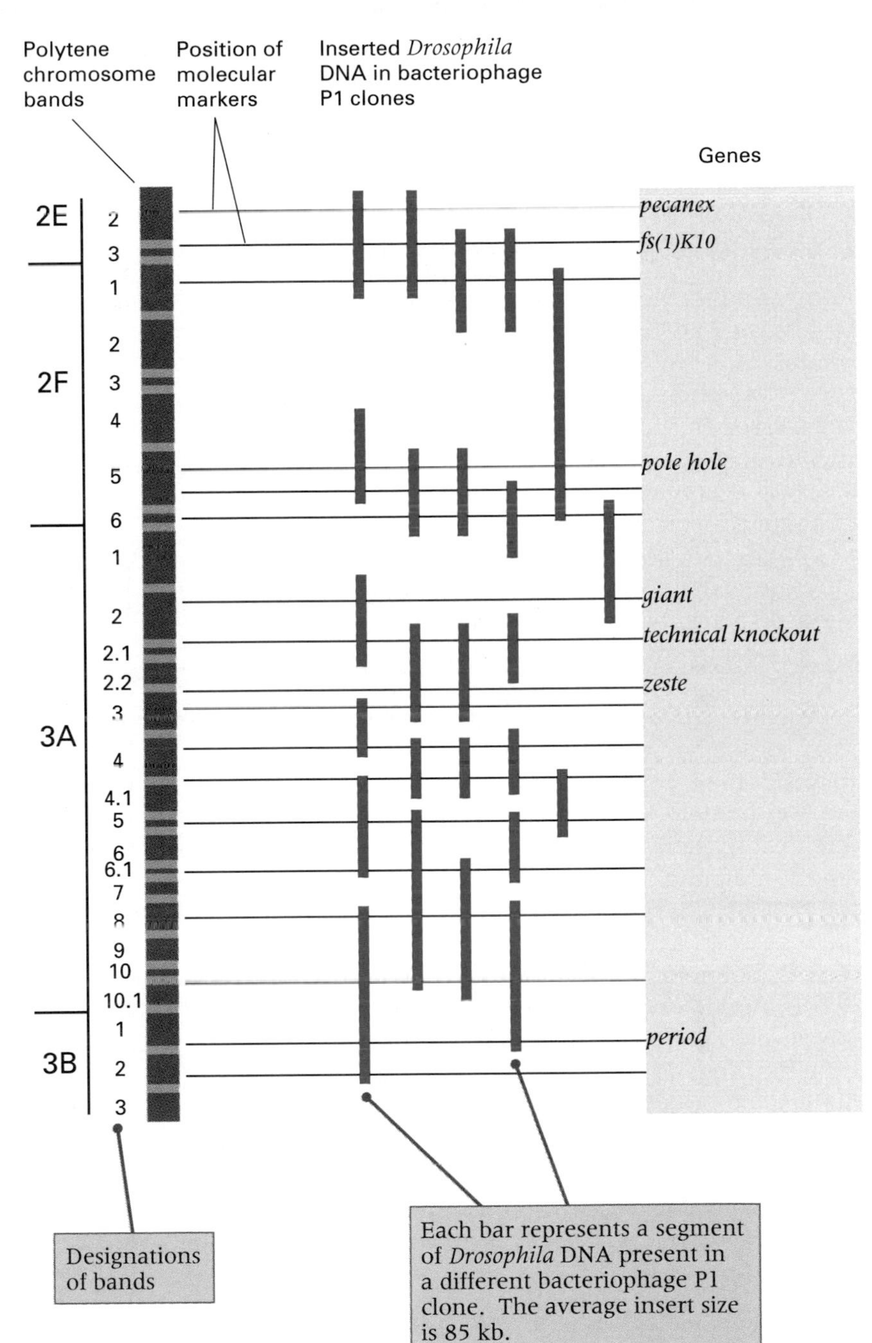

Figure 10.10 Positional cloning in *Drosophila* is made possible by a set of P1 clones that covers virtually the entire genome. A set of P1 clones covering a small region of the X chromosome is represented in blue. The P1 clones contain sequences for a set of molecular markers (red lines) that have known locations on the genetic map as well as in the polytene chromosomes of the salivary glands (left). The locations of several cloned genes in the region are shown. Any new mutation in the region is cloned by first determining its location with respect to flanking molecular markers and then identifying the gene within the P1 clones containing the flanking markers. [After D. L. Hartl, D. I. Nurminsky, R. W. Jones, and E. R. Lozovskaya. 1994. *Proc. Natl. Acad. Sci. USA* 91: 6824.]

mutant gene that maps within the region, the location of the mutation with respect to the markers immediately identifies the clone or clones in which the gene is located.

Close genetic linkages are often conserved among related species.

Comparing the genetic maps of homologous genes in human and mouse chromosomes reveals that groups of genes that are closely linked in one species are often also closely linked in the other. In some cases, the gene order is surprisingly well conserved over quite large regions of a chromosome arm. For example, the genes that make up a region of about 40 percent of human chromosome 1 are present as a single contiguous block in mouse chromosome 1; those that make up a different region of about 40 percent of human chromosome 1 are present as a contiguous block in mouse chromosome 4; and the rest of the genes in human chromosome 1 appear as a single block in mouse chromosome 3. Many of the homologous genes also appear to have retained their primary functions, because mutations in homologous genes in mouse and humans often cause the same phenotypic changes. For example, mutation of the *hairless* gene in the mouse or the homologous gene in the human results in complete absence of body hair; and mutations in the *obese* gene, which encodes the hormone leptin, causes morbid obesity in both mouse and humans. These correspondences between gene position and gene function aid in positional cloning as well as in understanding gene function.

10.3 Reverse genetics creates an organism with a designed mutation.

Mutation has traditionally provided the raw material needed for genetic analysis. The customary procedure has been to use a mutant phenotype to recognize a mutant gene and then to identify the wildtype allele and its normal function. This approach has proved highly successful, as evidenced by numerous examples throughout this book. But the approach also has its limitations. For example, it may prove difficult or impossible to isolate mutations in genes that duplicate the functions of other genes or that are essential for the viability of the organism.

Recombinant DNA technology has made possible another approach in genetic analysis in which wildtype genes are cloned, intentionally mutated in specific ways, and introduced back into the organism to study the phenotypic effects of the mutations. Because the position and molecular nature of each mutation are precisely defined, a very fine level of resolution is possible in determining the functions of particular regions of nucleotide sequence. This type of analysis has been applied to defining the promoter and enhancer sequences that are necessary for transcription, the sequences necessary for normal RNA splicing, particular amino acids that are essential for protein function, and many other problems. The procedure is often called **reverse genetics** because it reverses the usual flow of study: Instead of starting with a mutant phenotype and trying to identify the wildtype gene, reverse genetics starts by making a mutant gene and studies the resulting phenotype. The following sections describe some techniques of reverse genetics.

Recombinant DNA can be introduced into the germ line of animals.

Reverse genetics can be carried out in most organisms that have been extensively studied genetically, including the nematode *Caenorhabditis elegans, Drosophila,* the mouse, and many domesticated animals and plants. In nematodes, the basic procedure is to manipulate the DNA of interest in a plasmid that also contains a selectable genetic marker that will alter the phenotype of the transformed animal. The DNA is injected directly into the reproductive organs and sometimes spontaneously becomes incorporated into the chromosomes in the germ line. The result of transformation is observed and can be selected in the progeny of the injected animals because of the phenotype conferred by the selectable marker.

A somewhat more elaborate procedure is necessary for germ-line transformation in *Drosophila.* The usual method makes use of a 2.9-kb transposable element called the ***P* element,** which consists of a central region coding for transposase flanked by 31-base pair inverted repeats (Figure 10.11A). A genetically engineered derivative of this *P* element, called *wings clipped,* can make functional transposase but cannot itself transpose because of deletions introduced at the ends of the inverted repeats (Figure 10.11B). For germ-line transformation, the vector is a plasmid containing a *P* element that includes, within the inverted repeats, a selectable genetic marker (usually one affecting eye color), as well as a large internal deletion that removes much of the transposase-coding region. By itself, this *P* element cannot transpose because it makes no transposase, but it can be mobilized by the transposase produced by the wings-clipped or other intact *P* elements. In *Drosophila* transformation, any DNA fragment of interest is introduced between the ends of the deleted *P* element. The resulting plasmid and a different plasmid containing the wings-clipped element are injected into the region

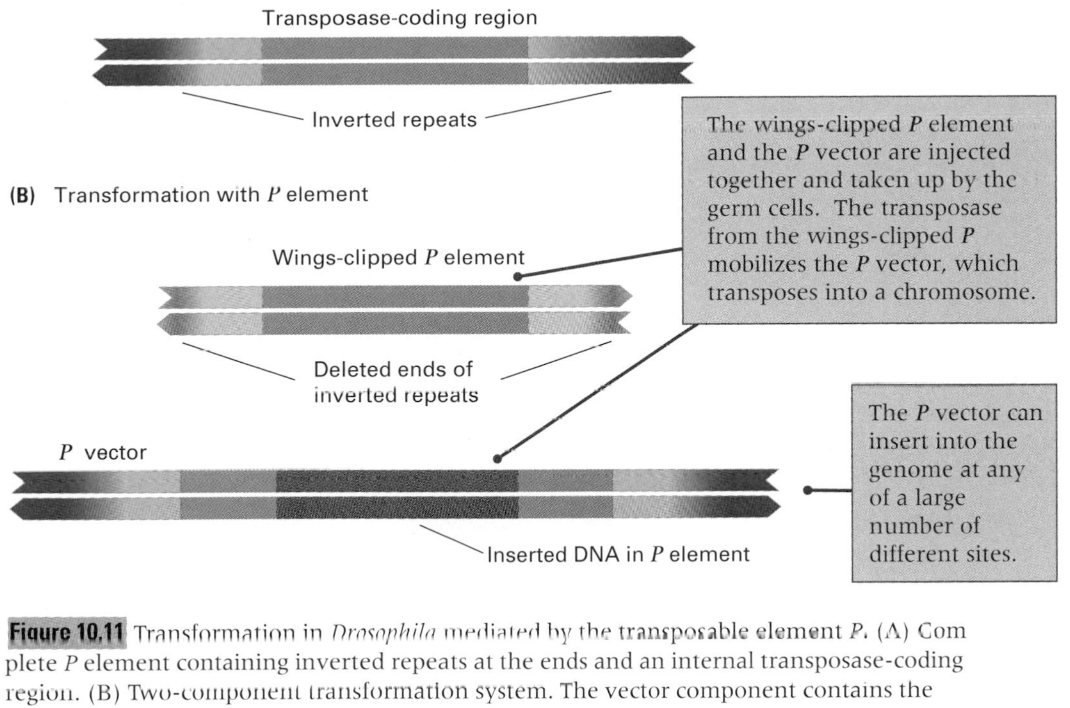

Figure 10.11 Transformation in *Drosophila* mediated by the transposable element *P*. (A) Complete *P* element containing inverted repeats at the ends and an internal transposase-coding region. (B) Two-component transformation system. The vector component contains the DNA of interest flanked by the recognition sequences needed for transposition. The wings-clipped component is a modified *P* element that codes for transposase but cannot transpose itself because critical recognition sequences are deleted.

of the early embryo that contains the germ cells. The DNA is taken up by the germ cells, and the wings-clipped element produces functional transposase (Figure 10.11B). The transposase mobilizes the engineered *P* vector and results in its transposition into an essentially random location in the genome. Transformants are detected among the progeny of the injected flies because of the eye color or other genetic marker included in the *P* vector. Integration into the germ line is typically very efficient: From 10 to 20 percent of the injected embryos that survive and are fertile yield one or more transformed progeny. However, the efficiency decreases with the size of the DNA fragment in the *P* element, and the effective upper limit is approximately 20 kb.

Transformation of the germ line in mammals can be carried out in several ways. The most direct is the injection of vector DNA into the nucleus of fertilized eggs, which are then transferred to the uterus of foster mothers for development. The vector is usually a modified retrovirus. **Retroviruses** have RNA as their genetic material and code for a reverse transcriptase that converts the retrovirus genome into double-stranded DNA that becomes inserted into the genome in infected cells. Genetically engineered retroviruses containing inserted genes undergo the same process. Animals that have had new genes inserted into the germ line in this or any other manner are called **transgenic** animals.

Another method of transforming mammals uses **embryonic stem cells** obtained from embryos a few days after fertilization (Figure 10.12). Although embryonic stem cells are not very hardy, they can be isolated and then grown and manipulated in culture; mutations in the stem cells can be selected or introduced using recombinant DNA vectors. The mutant stem cells are introduced into another developing embryo and transferred into the uterus of a foster mother (Figure 10.12A), where they become incorporated into various tissues of the embryo and often participate in forming the germ line. If the embryonic stem cells carry a genetic marker, such as a gene for black coat color, then mosaic animals can be identified by their spotted coats (Figure 10.12B). Some of these animals, when mated, produce black offspring (Figure 10.12C), which indicates that the embryonic stem cells had become incorporated into the germ line. In this way, mutations introduced into the embryonic stem cells while they were in culture may become incorporated into the germ line of living animals. The method in Figure 10.12 has been used to create strains of mice with mutations in genes associated with such human genetic diseases as cystic fibrosis. These strains serve as mouse models for studying the disease and for testing new drugs and therapeutic methods.

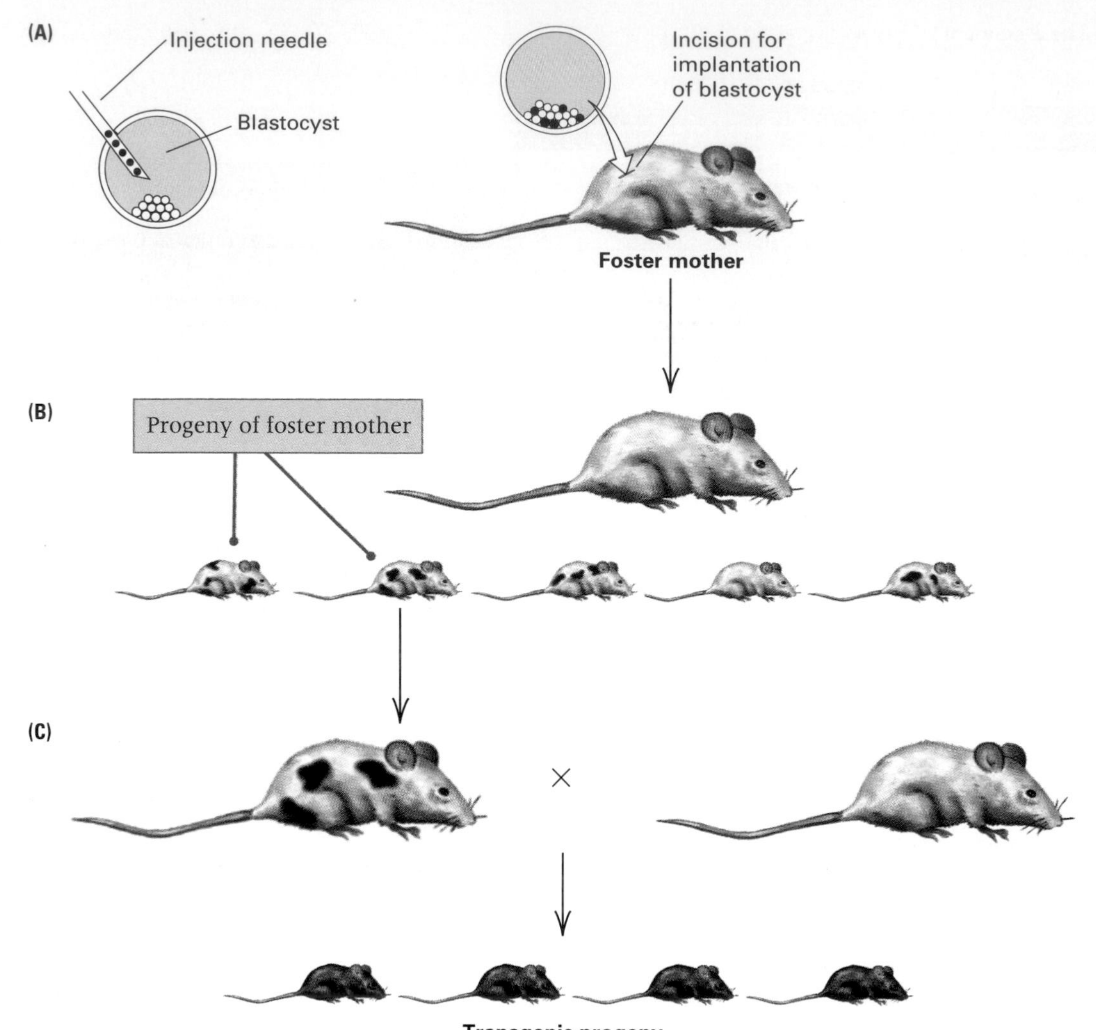

Figure 10.12 Transformation of the germ line in the mouse using embryonic stem cells. (A) Stem cells obtained from an embryo of a black strain are isolated and, after genetic manipulation in culture, mixed with the embryonic stem cells from a white strain and introduced into the uterus of a foster mother. (B) The resulting offspring are often mosaics containing cells from both the black and the white strains. (C) If cells from the black strain colonized the germ line, the offspring of the mosaic animal will be black. [After M. R. Capecchi. 1989. *Trends Genet.* 5: 70.]

The procedure for introducing mutations into specific genes is called **gene targeting.** The specificity of gene targeting comes from the DNA sequence homology needed for homologous recombination. Two examples are illustrated in Figure 10.13, where the DNA sequences present in gene-targeting vectors are shown as looped configurations paired with homologous regions in the chromosome prior to recombination. The targeted gene is shown in pink. In Figure 10.13A, the vector contains the targeted gene interrupted by an insertion of a novel DNA sequence, and homologous recombination results in the novel sequence becoming inserted into the targeted gene in the genome. In Figure 10.13B, the vector contains only flanking sequences, not the targeted gene, so homologous recombination results in replacement of the targeted gene with an unrelated DNA sequence. In both cases, cells with targeted gene mutations can be selected by including an antibiotic-resistance gene, or other selectable genetic marker, in the sequences that are incorporated into the genome through homologous recombination.

Recombinant DNA can also be introduced into plant genomes.

A procedure for the transformation of plant cells makes use of a plasmid found in the soil bacterium *Agrobacterium tumefaciens* and related species. Infection of susceptible plants with this bacterium results in the growth of what are known as crown gall

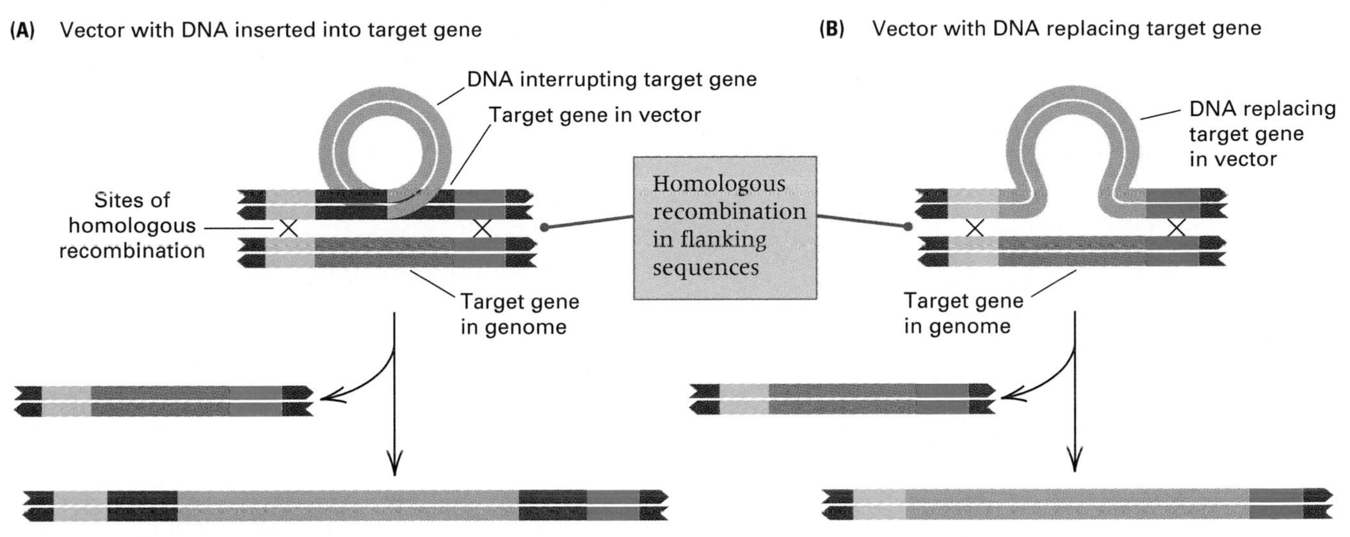

Figure 10.13 Gene targeting in embryonic stem cells. (A) The vector (top) contains the targeted sequence (red) interrupted by an insertion. Homologous recombination introduces the insertion into the genome. (B) The vector contains DNA sequences flanking the targeted gene. Homologous recombination results in replacement of the targeted gene with an unrelated DNA sequence. [After M. R. Capecchi. 1989. *Trends Genet.* 5: 70.]

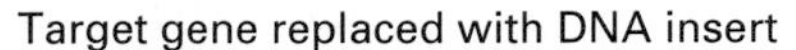

tumors at the entry site, which is usually a wound. Susceptible plants comprise about 160,000 species of flowering plants, known as the dicots, and include the great majority of the most common flowering plants.

The *Agrobacterium* contains a large plasmid of approximately 200 kb called the ***Ti* plasmid,** which includes a smaller (~25 kb) region known as the ***T* DNA** flanked by 25-base pair direct repeats (Figure 10.14A). The *Agrobacterium* causes a profound change in the metabolism of infected cells because of transfer of the *T* DNA into the plant genome. The *T* DNA contains genes coding for proteins that stimulate division of infected cells, hence causing the tumor, and also coding for enzymes that convert the amino acid arginine into an unusual derivative,

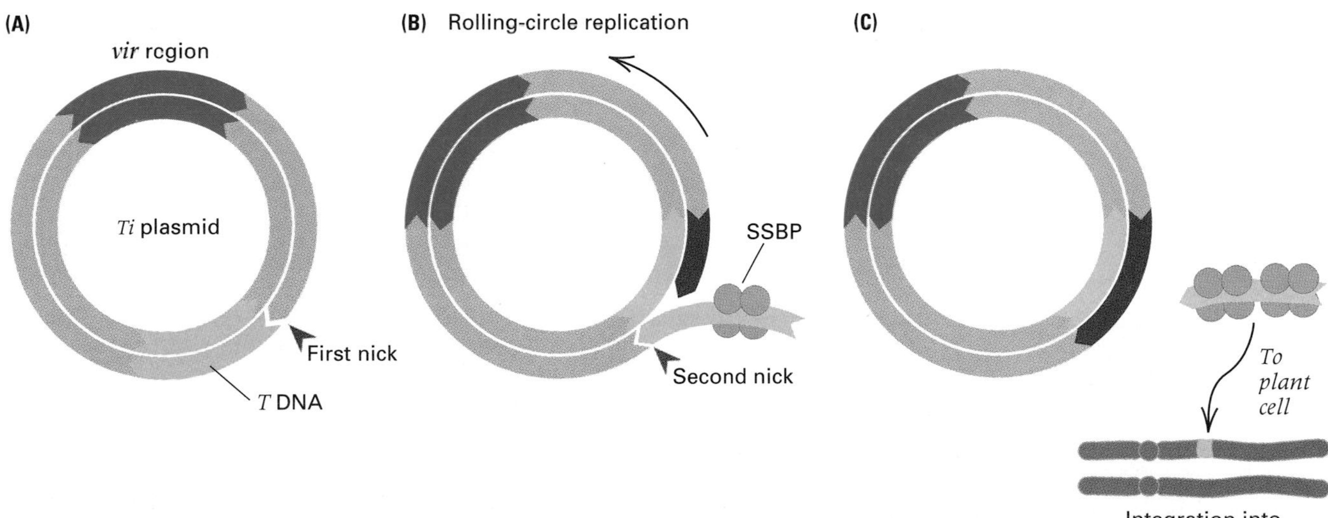

Figure 10.14 Transformation of a plant genome by *T* DNA from the *Ti* plasmid. (A) A nick forms at the 5' end of the *T* DNA. (B) Rolling-circle replication elongates the 3' end and displaces the 5' end, which is stabilized by single-stranded binding protein (SSBP). A second nick terminates replication. (C) The SSBP-bound *T* DNA is transferred to a plant cell and inserts into the genome.

generally *nopaline* or *octopine* (depending on the particular type of *Ti* plasmid), that the bacterium needs in order to grow. The transfer functions are present not in the *T* DNA itself but in another region of the plasmid called the *vir* (stands for *virulence*) region of about 40 kb that includes six genes necessary for transfer.

Transfer of *T* DNA into the host genome is similar in some key respects to bacterial conjugation, which we examined in Chapter 8. In infected cells, transfer begins with the formation of a nick that frees one end of the *T* DNA (Figure 10.14A), which peels off the plasmid and is replaced by rolling-circle replication (Figure 10.14B). The region of the plasmid that is transferred is delimited by a second nick at the other end of the *T* DNA, but the position of this nick is variable. The resulting single-stranded *T* DNA is bound with molecules of a single-stranded binding protein (SSBP) and is transferred into the plant cell and incorporated into the nucleus. There it is integrated into the chromosomal DNA by a mechanism that is still unclear (Figure 10.14C). Although the SSBP has certain similarities in amino acid sequence to the *recA* protein from *E. coli,* which plays a key role in homologous recombination, it is clear that integration of the *T* DNA does not require homology.

Use of *T* DNA in plant transformation is made possible by engineered plasmids in which the sequences normally present in *T* DNA are removed and replaced with those to be incorporated into the plant genome along with a selectable marker. A second plasmid contains the *vir* genes and permits mobilization of the engineered *T* DNA. In infected tissues, the *vir* functions mobilize the *T* DNA for transfer into the host cells and integration into the chromosome. Transformed cells are selected in culture by the use of the selectable marker and then grown into mature plants in accordance with the methods described in Section 5.3.

10.4 Genetic engineering is applied in agriculture, industry, medicine, and research.

At the present time, the main uses of recombinant DNA technology are (1) the efficient production of useful proteins; (2) the creation of cells capable of synthesizing economically important molecules; (3) the generation of DNA and RNA sequences as research tools or in medical diagnosis; (4) the manipulation of the genotype of organisms such as bacteria, plants, and animals; and (5) the potential correction of genetic defects in animals, including human patients (gene therapy). Some examples of these applications follow.

Agricultural crop plants are primary targets of genetic engineering.

Altering the genotypes of agricultural plants is an important application of recombinant DNA technology. By such means, it is possible to transfer genes from one plant species to another. These include genes that affect yield, hardiness, and disease resistance. There are also attempts to alter the surface structure of the roots of grains, such as wheat, by introducing certain genes from legumes (peas, beans), to give the grains the ability that legumes possess to establish root nodules of nitrogen-fixing bacteria. If successful, this would eliminate the need for the addition of nitrogenous fertilizers to soils where grains are grown.

The first engineered recombinant plant of commercial value was developed in 1985. An economically important herbicide is *glyphosate,* a weed killer that inhibits a particular essential enzyme in many plants. However, most herbicides cannot be applied to fields growing crops because both the crop and the weeds would be killed. The target gene of glyphosate is also present in the bacterium *Salmonella typhimurium.* A resistant form of the gene was obtained by mutagenesis and growth of *Salmonella* in the presence of glyphosate. Then the gene was transferred to the *T* DNA of *Agrobacterium.* Transformation of plants with the glyphosate-resistance gene has yielded varieties of maize, cotton, and tobacco that are resistant to the herbicide. Thus fields of these crops can be sprayed with glyphosate at any stage of growth of the crop. The weeds are killed, but the crop remains unharmed.

Genetic engineering can also be used to control insect pests. For example, the black cutworm causes extensive crop damage and is usually combatted with noxious insecticides. The bacterium *Bacillus thuringiensis* produces a protein that is lethal to the black cutworm, but the bacterium does not normally grow in association with the plants that are damaged by the worm. However, the gene coding for the lethal protein has been introduced into the soil bacterium *Pseudomonas fluorescens,* which lives in association with maize and soybean roots. Inoculation of soil with the engineered *P. fluorescens* helps control the black cutworm and reduces crop damage.

Specific plant tissues can be targeted for self-destruction.

One practical application of genetic engineering is in the production of male-sterile plants. Hybrid plants are important in agriculture because hybrid plants are usually superior to their inbred parents in numerous respects, including higher yield and increased disease resistance (Chapter 13). Male sterility is important in hybrid seed production because it promotes efficient hybridization between the inbred lines. Inherited male sterility is widely used in corn breeding, but analogous mutations are not available in many crop plants. To be optimally useful, the male sterility should also be capable of being reversed by other mutations so that the inbred lines can be propagated.

A genetically engineered system of male sterility and fertility restoration has been introduced into a number of plant species, including the oilseed rape, *Brassica napus,* which is a major source of canola vegetable oil. The basis of the sterility is an extracellular RNA nuclease, called *barnase,* produced by the bacterium *Bacillus amyloliquefaciens.* The barnase nuclease is an extremely potent cellular toxin. In the use of barnase to produce male sterility, the coding sequence of the bacterial gene is fused with the promoter of a gene, *TA29,* that has a tissue-specific expression in the tapetal cell layer that surrounds the pollen sacs in the anther (Figure 10.15A). When the artificial *TA29-barnase* gene is transformed into the genome by the use of *Agrobacterium,* the resulting plants are male-sterile because of the destruction of tapetal cell RNA by the barnase enzyme and the resulting lethality of cells in the tapetum (Figure 10.15B).

Fertility restoration makes use of another protein, called *barstar,* also produced by *B. amyloliquefaciens.* Barstar is an intracellular protein that protects against the lethal effects of barnase by forming a stable, enzymatically inactive complex in the cytoplasm. Hence barstar is the bacterial cell's self-defense against its own barnase. Plants transformed with an artificial *TA29-barstar* gene are healthy and fertile; they merely produce barstar in the tapetum because of the tissue specificity of the *TA29* promoter. However, when the male-sterile *TA29-barnase* plants are crossed with those carrying *TA29-barstar,* the resulting genotype that combines *TA29-barnase* with *TA29-barstar* is male-fertile (Figure 10.15C). The reason for the restoration of fertility is that the barstar protein combines with the barnase nuclease and renders it ineffective.

The macroscopic appearance and the microscopic appearance of oilseed rape plants of various genotypes are shown in Figure 10.16. The wildtype phenotype is shown in parts A and B, and numerous pollen grains (PG) are apparent in the cross section of the anther. Parts C and D show the flower and anthers of plants carrying the *TA29-barnase;* the anthers are small, shriveled, and devoid of pollen. Parts E and F show the flower and anthers of plants whose genotypes include *TA29-barnase*

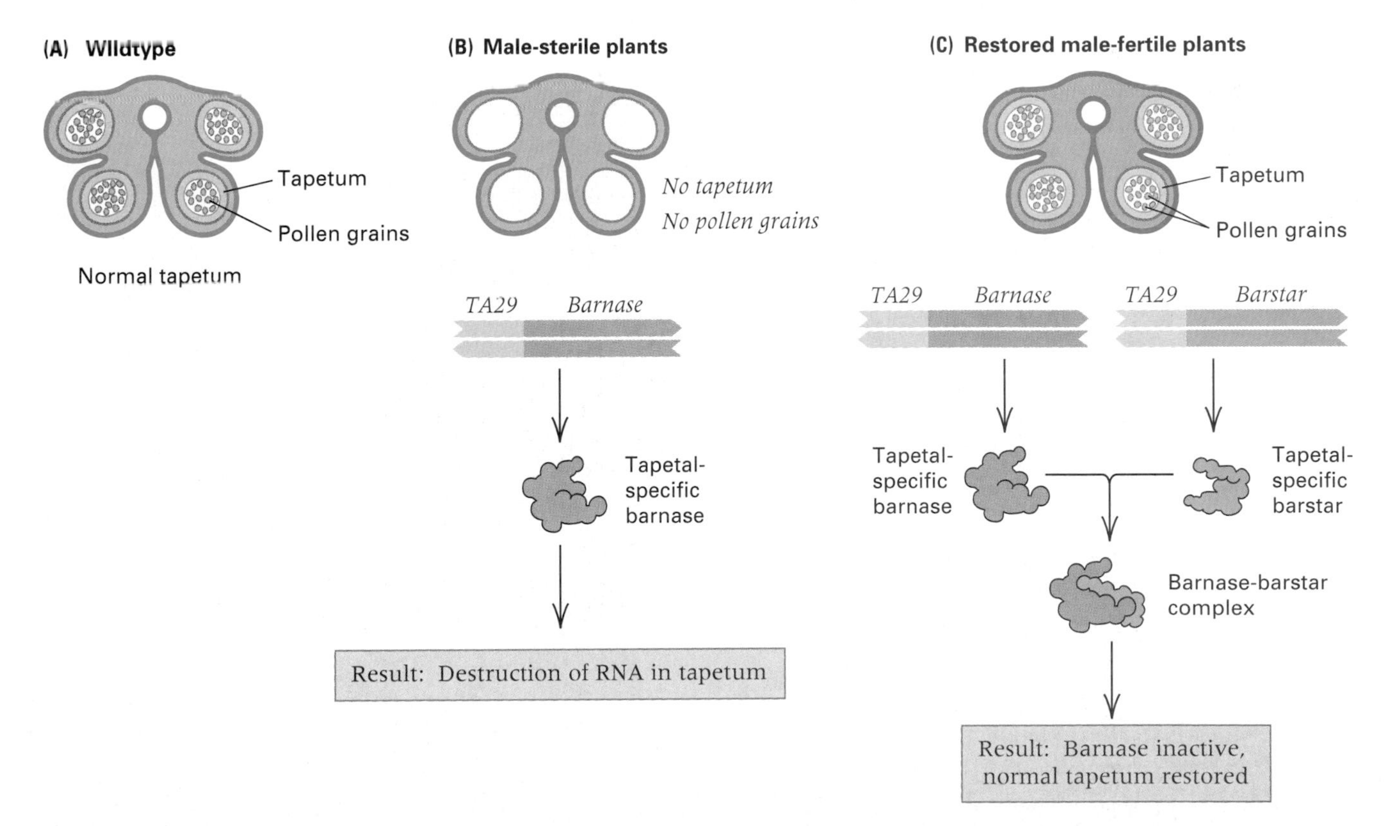

Figure 10.15 Engineered genetic male sterility and fertility restoration. (A) Pollen grains (PG) develop within a thin layer of tapetal cells (T). (B) Expression of the barnase RNA nuclease under the control of the tapetum-specific regulatory region of gene *TA29* destroys the tapetum and renders the plants male-sterile. (C) Expression of barstar (a barnase inhibitor) in tapetal cells inactivates barnase and restores fertility. [Courtesy of Robert B. Goldberg.]

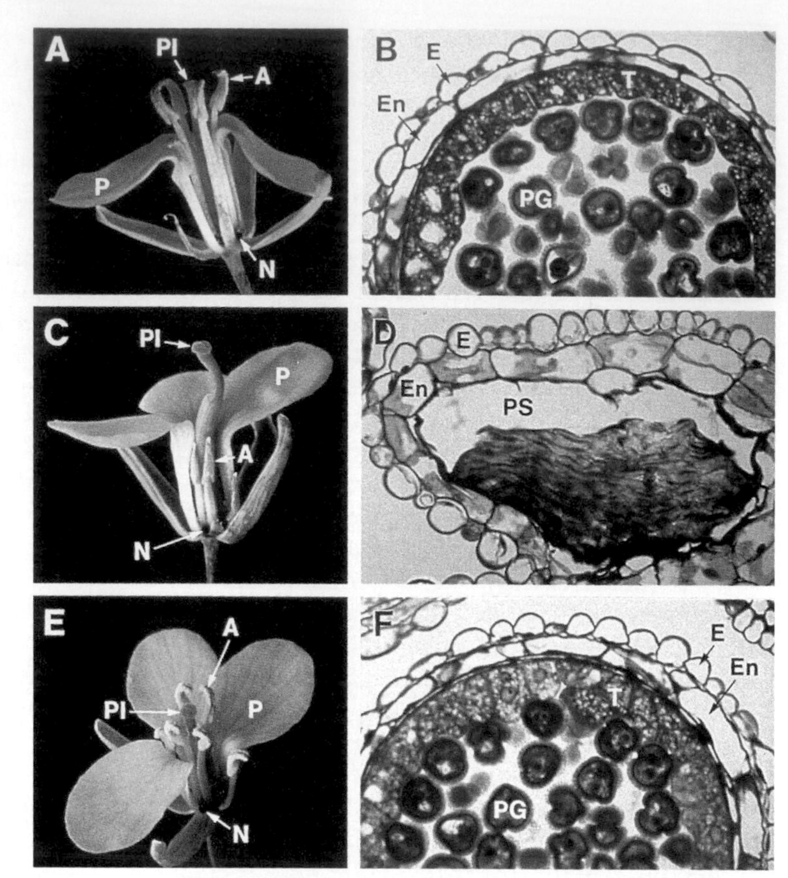

Figure 10.16 Flowers (A, C, and E) and cross sections of anthers (B, D, and F) of oilseed rape plants. The genotypes are as follows: A and B, wildtype; C and D, *TA29-barnase*; E and F, *TA29-barnase* + *TA29-barstar.* Note that the fertility-restored plants (E and F) are virtually indistinguishable from wildtype (A and B). The labeled parts of the flowers are A, anther; P, petal; PI, pistil; and N, nectary. The labeled parts of the cross sections are E, epidermis; En, endothecium; PG, pollen grain; PS, pollen sac; and T, tapetum. [From C. Mariani, V. Gossele, M. De Beuckeleer, M. De Block, R. B. Goldberg, W. De Greef and J. Leemans. 1992. *Nature* 357: 384.]

plus *TA29-barstar;* the phenotype is virtually indistinguishable from wildtype.

Animal growth rate can be genetically engineered.

In many animals, the rate of growth is controlled by the amount of growth hormone produced. Transgenic animals with a growth-hormone gene under the control of a highly active regulatory sequence are often larger than their normal counterparts. An example of a highly active sequence that can drive gene expression is the regulatory region of a gene for *metallothionein.* The metallothioneins are proteins that bind heavy metals. They are ubiquitous in eukaryotic organisms and are often encoded in a family of related genes. Human beings, for example, have more than ten metallothionein genes that can be separated into two major groups according to their sequences. The regulatory region of a metallothionein gene elevates transcription of any gene to which it is attached in response to heavy metals or steroid hormones. When DNA constructs consisting of a rat growth-hormone gene under metallothionein control are used to produce transgenic mice, the resulting animals grow about twice as large as normal mice.

The effect of another growth-hormone construct is shown in Figure 10.17. The fish are coho salmon at 14 months of age. Those on the left are normal, whereas those on the right are transgenic animals that contain a salmon growth-hormone gene driven by a metallothionein regulatory region. Both the growth-hormone gene and the metallothionein gene were cloned from the sockeye salmon. As an index of scale, the largest transgenic fish on the right has a length of about 42 cm. On average, the transgenic fish are 11 times heavier than their normal counterparts; the largest transgenic fish was 37 times the average weight of the nontransgenic

Figure 10.17 Normal coho salmon (left) and genetically engineered coho salmon (right) containing a sockeye salmon growth-hormone gene driven by the regulatory region from a metallothionein gene. The transgenic salmon average 11 times the weight of the nontransgenic fish. The smallest fish on the left is about 4 inches long. [Courtesy R. H. Devlin; see R. H. Devlin, T. Y. Yesaki, C. A. Biagi, E. M. Donaldson, P. Swanson, and W. K. Chan. 1994. *Nature* 371: 209.]

animals. Not only do the transgenic salmon grow faster and become larger than normal salmon, but they also mature faster.

Engineered microbes can help in the degradation of toxic waste.

A species of marine bacteria has been equipped with genes from other bacterial species for the metabolism of some components of petroleum, yielding an organism used in cleaning up oil spills in the oceans. Many biotechnology companies are at work designing bacteria that can synthesize industrial chemicals or degrade industrial wastes. Bacteria have also been created that are able to compost waste more efficiently and to fix nitrogen to improve the fertility of soil. Other bacteria have been engineered to fluoresce when they encounter chemicals found in marijuana or cocaine, and still others have been engineered to aid in recovery of metals from ores. A great deal of effort is currently being expended to create organisms that can convert biological waste into alcohol.

Recombinant DNA permeates modern biomedical research.

Modern molecular genetics could not exist without recombinant DNA technology. It is an essential research tool. Reverse genetics makes it possible to isolate and alter a gene at will and introduce it back into a living cell or even the germ line. Besides saving time and labor, reverse genetics makes it possible to construct mutants that cannot be formed in any other way. An example is the formation of double mutants of animal viruses, which naturally undergo recombination at such a low frequency that mutations can rarely be recombined by genetic crosses. The greatest impact of recombinant DNA on basic research has been in the study of eukaryotic gene regulation and development, and many of the principles of regulation and development summarized in Chapters 11 and 12 were determined by using these methods.

The production of useful proteins is a primary impetus for recombinant DNA.

Among the most important applications of genetic engineering is the production of large quantities of particular proteins that are otherwise difficult to obtain (for example, proteins that are present in only a few molecules per cell or that are produced in only a small number of cells or only in human cells). The method is simple in principle. A DNA sequence coding for the desired protein is cloned in a vector adjacent to an appropriate regulatory sequence. This step is usually done with cDNA because it has all the coding sequences spliced together in the right order. Using a vector with a high copy number ensures that many copies of the coding sequence will be present in each bacterial cell, which can result in synthesis of the gene product at concentrations ranging from 1 to 5 percent of the total cellular protein. In practice, the production of large quantities of a protein in bacterial cells is straightforward, but there are often problems that must be overcome, because in the bacterial cell, which is a prokaryote, the eukaryotic protein may be unstable, may not fold properly, or may fail to undergo necessary chemical modification. Many important proteins are currently produced in bacterial cells, including human growth hormone, blood-clotting factors, and insulin. Patent offices in Europe and the United States have already issued well over 1000 patents for the clinical use of the products of genetically engineered human genes. Figure 10.18 gives the relative numbers of patents issued for various clinical applications.

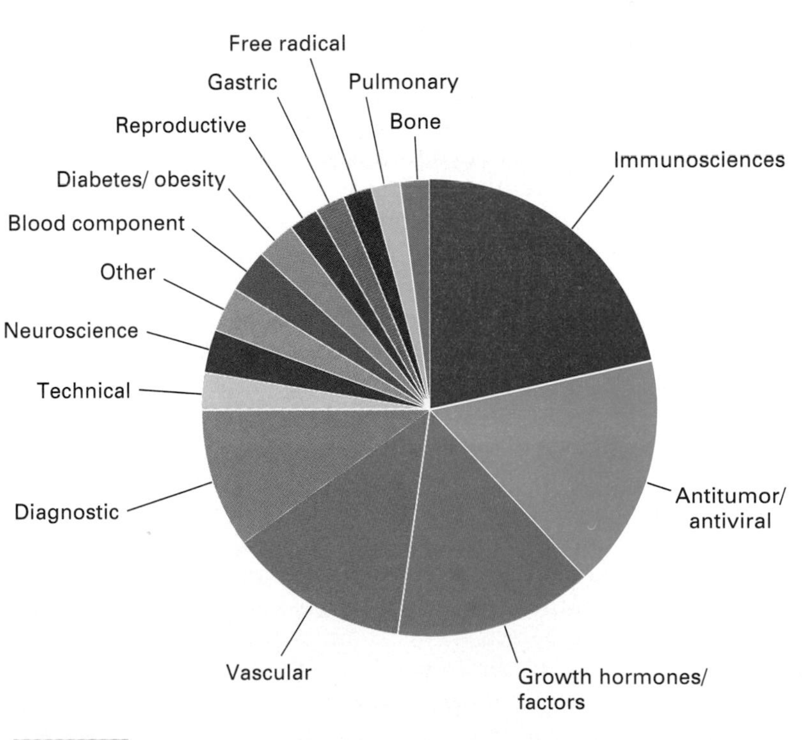

Figure 10.18 Relative numbers of patents issued for various clinical applications of the products of genetically engineered human genes. [Data from *Nature* 1996, 380: 387.]

Animal viruses may prove useful vectors for gene therapy.

The genetic engineering of animal cells often makes use of retroviruses because their reverse transcriptase makes a double-stranded DNA copy of the RNA genome, which then becomes inserted into the chromosomes of the cell. DNA-to-RNA transcription occurs only after the DNA copy is inserted. The infected host cell survives the infection, retaining the retroviral DNA in its genome. These features of retroviruses make them convenient vectors for the genetic manipulation of animal cells, including those of birds, rodents, monkeys, and human beings.

Genetic engineering with retroviruses allows the possibility of altering the genotypes of animal cells. Because a wide variety of retroviruses are known, including many that infect human cells, genetic defects may be corrected by these procedures in the future. However, many retroviruses contain a gene that results in uncontrolled growth of the infected cell, thereby causing a tumor. When retrovirus vectors are used for genetic engineering, the tumor-causing gene is first deleted. This deletion also provides the space needed for the incorporation of foreign DNA sequences. The recombinant DNA procedure employed with retroviruses consists of synthesis in the laboratory of double-stranded DNA from the viral RNA, through the use of reverse transcriptase. The DNA is then cleaved with a restriction enzyme, and foreign DNA is inserted by means of the techniques already described. Transformation yields cells with the recombinant retroviral DNA permanently inserted into the genome. In this way, the genotype of the cells can be altered.

Attempts are currently under way to assess the potential of genetic engineering in the treatment of cystic fibrosis. This disorder results from a mutation in a gene for chloride transport in the lungs, pancreas, and other glands, leading to abnormal secretions and the accumulation of a thick, sticky, honey-like mucus in the lungs. Among its chief symptoms are recurrent respiratory infections. A genetically engineered form of the common cold virus is being used in clinical trials in an attempt to introduce the nonmutant form of the cystic fibrosis gene directly into the lungs of affected patients. This is an example of **gene therapy.** The exciting potential of gene therapy lies in the possibility of correcting genetic defects—for example, restoring the ability to make insulin to persons with diabetes and correcting immunological deficiencies in patients with defective immunity. However, a number of major problems stand in the way of gene therapy becoming widely used. At this time, there is no completely reliable way to ensure that a gene will be inserted only into the appropriate target cell or target tissue, and there is no completely reliable means of regulating the expression of the inserted genes.

A major breakthrough in disease prevention has been the development of synthetic vaccines. Production of certain vaccines is difficult because of the extreme hazards of working with large quantities of the active virus; an example is the human immunodeficiency virus (HIV1) that causes acquired immune deficiency syndrome (AIDS). The danger is minimized by cloning and producing viral antigens in a nonpathogenic organism. Vaccinia virus, the agent used in smallpox vaccination, has been very useful for this purpose. Viral antigens are often on the surface of virus particles, and some of these antigens can be engineered into the coat of vaccinia. For example, engineered vaccinia with surface antigens of hepatitis B, influenza virus, and vesicular stomatitis virus (which kills cattle, horses, and pigs) have produced useful vaccines in animal tests. A surface antigen of *Plasmodium falciparum,* the parasite that causes malaria, has also been placed in the vaccinia coat, a development that may ultimately lead to an antimalaria vaccine.

Recombinant DNA yields probes for the detection of mutant genes in hereditary disease.

Probes derived by recombinant DNA methods are widely used in prenatal detection of disease, including cystic fibrosis, Huntington disease, sickle-cell anemia, and hundreds of other genetic disorders. In some cases, probes derived from the gene itself are used; in other cases, molecular markers, such as STRPs, that are genetically linked to the disease gene are employed (Section 4.4). If the disease gene itself, or a region close to it in the chromosome, differs from the normal chromosome in the positions of one or more cleavage sites for restriction enzymes, then these differences can be detected with Southern blots using cloned DNA from the region as the probe (Section 6.6) or in DNA amplified by the polymerase chain reaction (Section 6.7). The genotype of the fetus can be determined directly. These techniques are very sensitive and can be carried out as soon as tissue from the fetus (or even from the embryonic membranes) can be obtained.

10.5 An entire genome can be physically mapped and its DNA sequence determined.

Gene structure and function are understood in considerable detail because of the manipulative capability of recombinant DNA technology combined with genetic analysis. Most recombinant DNA techniques are limited to the manipulation of DNA fragments smaller than about 50 kb. Although this range includes most cDNAs and many genes, some genes are much larger than 50 kb and so must be analyzed in smaller pieces. However, recent advances in recombinant DNA technology have made it possible to clone and analyze very large DNA molecules. These methods are suitable not only for the study of very large genes but also for the analysis of entire genomes. This section discusses complex genomes and the recombinant DNA technology that has made their analysis possible.

The smallest complex genomes are about 100 million base pairs.

The simplest genomes are those of small bacterial viruses whose DNA is smaller than 10 kb—for example, the *E. coli* phage ϕX174 whose genome contains 5386 base pairs. Toward the larger end of the spectrum is the human genome, consisting of 3×10^9 base pairs. The range is so large that comparisons are often made in terms of information content, with the base pairs of DNA analogous to the letters in a book. The analogy is apt because the range of sizes is from a few pages of text to an encyclopedia.

The book analogy is used to compare genome sizes in Figure 10.19. The volumes are about the size of a big-city telephone book: 1500 pages per volume with 25,000 characters (nucleotides) per page. In these terms, the 50-kb genome of bacteriophage λ fills two pages, and the 4.6-Mb genome of *E. coli* requires about 200 pages. (The abbreviation Mb stands for megabase pairs, or one million base pairs; 4.6 Mb equals 4600 kb.) Among eukaryotes, the size range for organisms of genetic interest varies considerably. Yeast would take up about 500 pages (genome size 12.5 Mb), the nematode

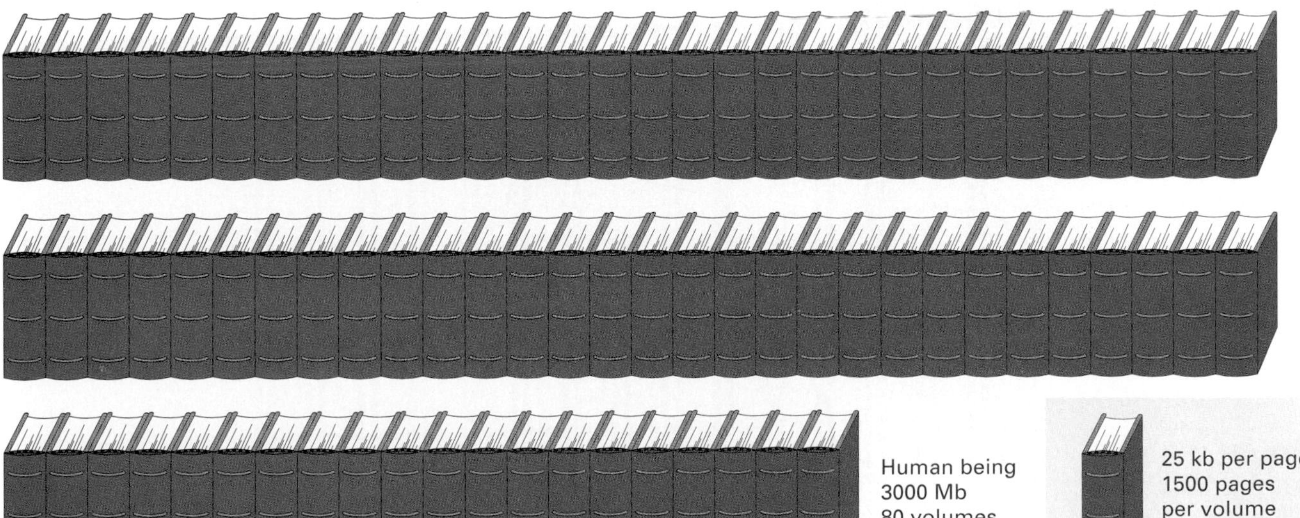

Figure 10.19 Relative sizes of genomes were they printed at 25,000 characters per page and bound in 1500-page volumes. One volume would contain about as many characters as a telephone book 2.5 inches thick. The *E. coli* genome would require about 200 pages, yeast 500 pages, and so forth.

Caenorhabditis elegans and the flowering plant *Arabidopsis thaliana* three volumes (100 Mb) each, *Drosophila melanogaster* five volumes (165 Mb), and human beings 80 volumes (3000 Mb). To put the matter in practical terms, in order to have a 95 percent chance of including any given single-copy sequence in a set of clones, a geneticist would need 1000 cosmid clones of yeast, 7500 clones of *A. thaliana* or *C. elegans*, 12,000 clones of *Drosophila*, and 225,000 clones of human DNA. The larger genomes require a formidable number of clones. An alternative to larger numbers of clones is larger fragments within the clones. Clones that contain large fragments of DNA can be made and analyzed by the procedures described in the following sections.

Special conditions allow production and isolation of large DNA fragments.

As we saw in Chapter 3, molecules up to about 2 Mb can be separated by special types of electrophoresis, and Figure 3.14 shows complete separation of all 16 chromosomes from yeast. They range in size from 230 kb to 1.5 Mb (Figure 10.20). However, even with large-fragment electrophoresis, it is difficult to separate DNA fragments greater than about 2 Mb, and in only a few organisms are any chromosomes so small. For example, in *Drosophila*, the smallest wildtype chromosome is about 6 Mb, and even the smallest rearranged and deleted chromosome is 1 Mb. Most chromosomes

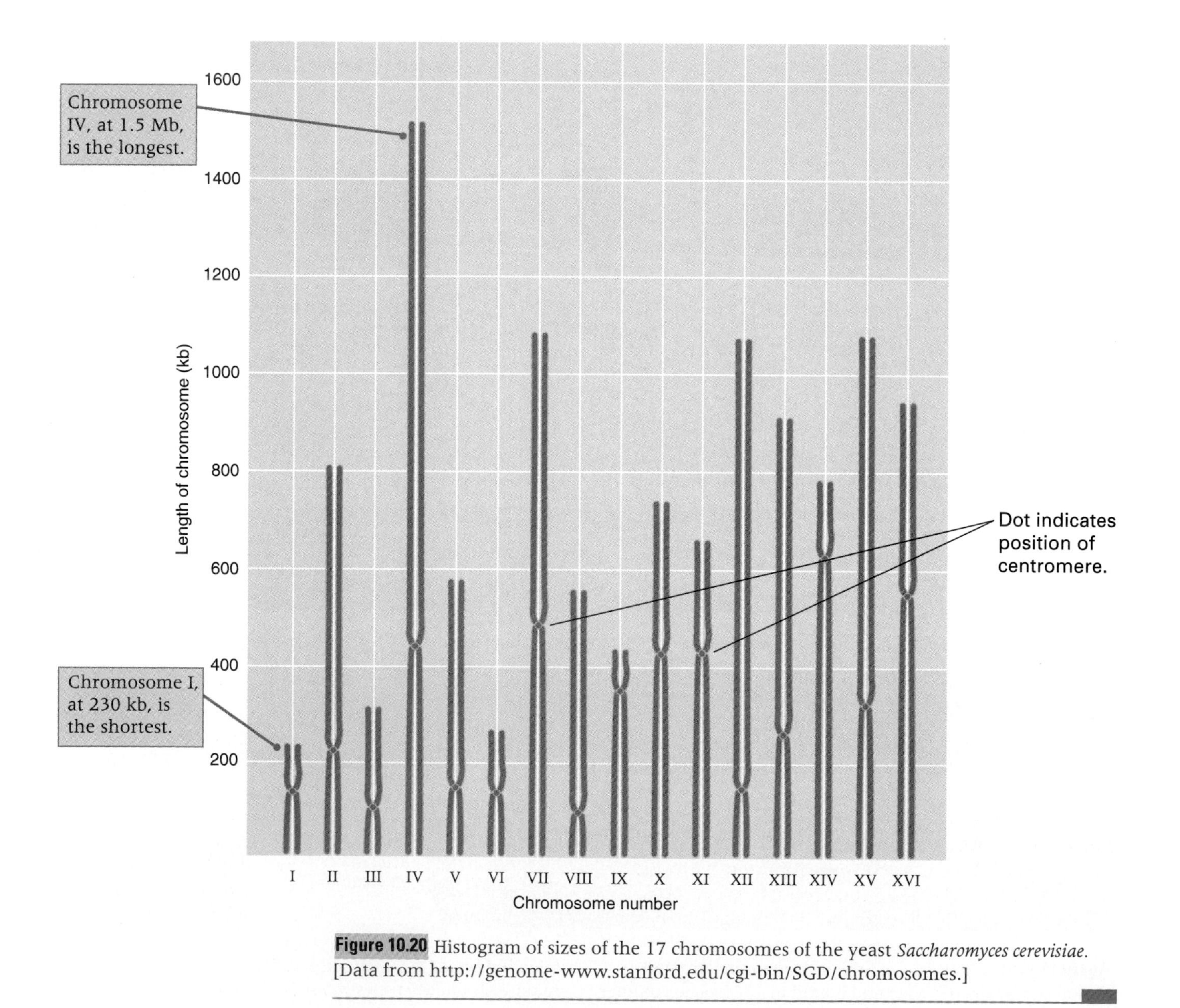

Figure 10.20 Histogram of sizes of the 17 chromosomes of the yeast *Saccharomyces cerevisiae*. [Data from http://genome-www.stanford.edu/cgi-bin/SGD/chromosomes.]

in higher eukaryotes are very much larger. The long arm of the smallest human chromosome (chromosome 21) is approximately 42 Mb. Therefore, even with the ability to separate large DNA fragments by electrophoresis, it is necessary to be able to cut the DNA in the genome into fragments of manageable size. This can be done with a restriction enzyme whose restriction site consists of eight bases rather than the usual six or four. For example, the restriction sites of the eight-cutter enzymes *Not*I and *Sfi*I are

```
                ↓
NotI    5'-GCGGCCGC-3'
        3'-CGCCGGCG-5'
                  ↑

                    ↓
SfiI    5'-GGCCNNNNNGGCC-3'
        3'-CCGGNNNNNCCGG-5'
                ↑
```

Both enzymes cleave double-stranded DNA at the positions of the restriction sites, and the arrows denote the position at which the backbone is cut in each DNA strand. (The N's in the *Sfi*I restriction site mean that any nucleotide can be present at this site.) In a genome with equal proportions of the four nucleotides and random nucleotide sequences, the average size of both *Not*I and *Sfi*I fragments is $4^8 = 65{,}536$ nucleotide pairs, or about 66 kb. Many genomes are relatively A + T-rich, and the average *Not*I and *Sfi*I fragment size is larger than 66 kb. In vertebrate genomes, there is a bias against long runs of G's and C's, so many *Not*I and *Sfi*I fragments are considerably larger than 66 kb. The use of eight-cutter restriction enzymes therefore allows complex DNA molecules to be cleaved into a relatively small number of large fragments that can be separated, cloned, and analyzed individually.

Artificial chromosomes are vectors for large DNA fragments.

Large DNA molecules can be cloned intact in bacterial cells with the use of specialized vectors that can accept large inserts. The vectors that can accept large DNA fragments are called artificial chromosomes. They fall into three classes: **P1 artificial chromosomes (PACs), bacterial artificial chromosomes (BACs),** and **yeast artificial chromosomes (YACs)** (Figure 10.21). PACs and BACs are basically large circular plasmids, whereas the parent YAC replicates as a circular plasmid in bacteria but as a linear chromosome in yeast after receiving inserted DNA. All three types of vectors are being used extensively in the Human Genome Project.

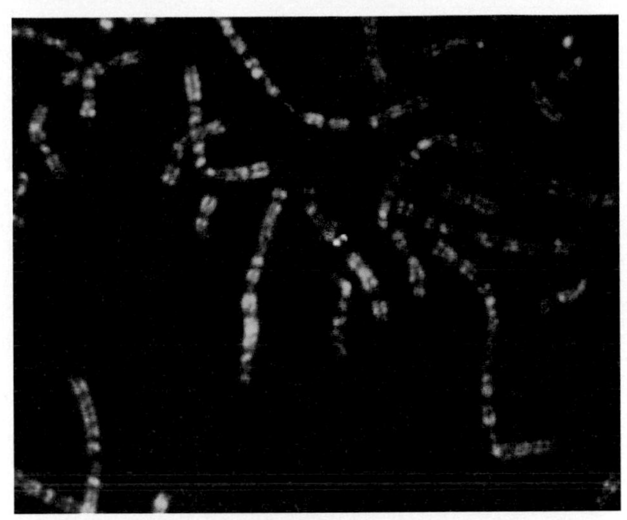

Two YAC CLONES, labeled with red or yellow fluorescent tags, hybridize to nearby sites on the same chromosome, but they clearly carry DNA from different parts of the chromosome. [Courtesy of Terry Featherstone.]

P1 is a bacteriophage that infects cells of *E. coli.* The genome of the wildtype P1 phage is approximately 100 kb, but a smaller region is sufficient to support replication and persistence as a plasmid in *E. coli.* One type of PAC vector is illustrated in Figure 10.21A. The essential elements of the vector are contained in 17 kb of DNA, which is sufficient for replication of the vector as a plasmid, amplification of the vector using the lytic replicon, and selection of the plasmid using kanamycin resistance. Selection for DNA inserts occurs through interruption of the *sacB* gene. When the *sacB* gene is expressed, the cells die if grown in medium containing sucrose (ordinary table sugar). Insertion of DNA at the cloning site separates the *sacB* gene from its promoter, the gene cannot be expressed, and so vectors with inserts in the cloning site allow growth in sucrose. PAC libraries often have an average insert size of 145 kb, and the largest inserts are about 300 kb.

The BAC vector (Figure 10.21B) is based on the F factor of *E. coli,* which we discussed in Chapter 8 in the context of its role in conjugation. The essential functions included in the 6.8-kb vector are genes for replication (*repE* and *oriS*), for regulating copy number (*parA* and *parB*), and for chloramphenicol resistance. BAC vectors with inserts greater than 300 kb can be maintained.

The YAC vector (Figure 10.21C) incorporates essential features of linear chromosomes and includes four types of genetic elements: (1) a cloning site; (2) a yeast centromere, replication origins, and genetic markers that are selectable in yeast; (3) an *E. coli* origin of replication and genetic markers that

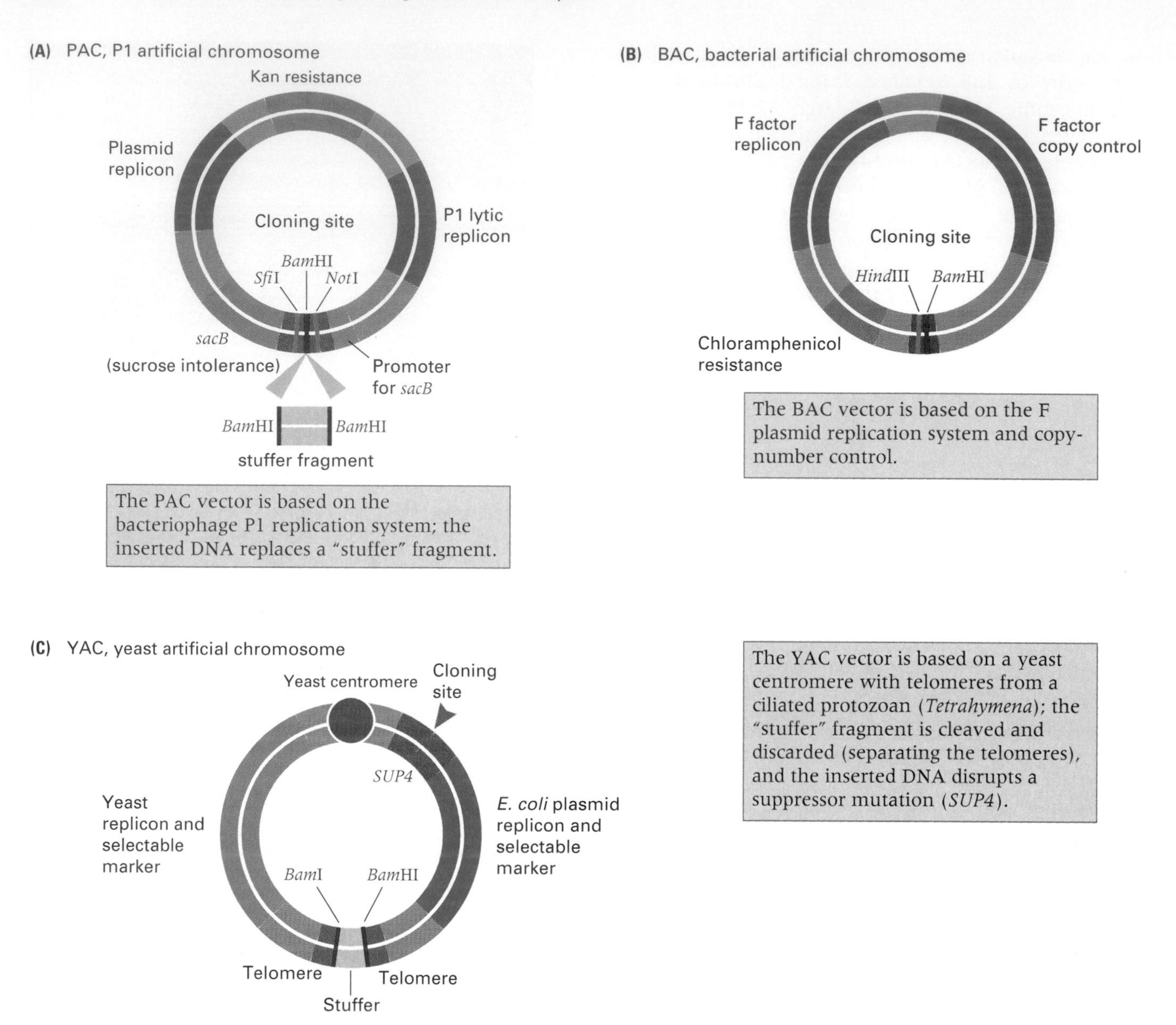

Figure 10.21 Artificial chromosome vectors. (A) P1 artificial chromosomes contain a plasmid replicon that results in 1 to 2 copies per cell, but there is also an inducible lytic replicon by which the number of copies can be amplified. Donor DNA is selected by insertional inactivation of the *sacB* gene, which allows survival in medium containing sucrose. (B) Bacterial artificial chromosomes are based on a vector that contains the F factor replicon as well as genes for regulating copy number. (C) Yeast artificial chromosomes are based on a vector that can be replicated in *E. coli* but that also contains a yeast centromere, replication origins, and selectable marker, as well as telomeres from *Tetrahymena.*

are selectable in *E. coli*; and (4) a pair of telomere sequences from the ciliated protozoan *Tetrahymena*. A YAC vector is therefore a *shuttle vector* that can replicate and be selected both in yeast and in *E. coli*. YAC chromosomes with inserts as large as 1 Mb can be recovered intact in yeast cells.

DNA fragments in the appropriate size range can be produced by breaking larger molecules into fragments of the desired size either by physical means, by treatment with restriction enzymes that have infrequent cleavage sites (for example enzymes such as *Not*I and *Sfi*I, that require eight-base restriction sites), or by treatment with ordinary restriction enzymes under conditions in which only a fraction of the restriction sites are cleaved (*partial digestion*). Cloning the large molecules consists of mixing the large fragments of source DNA with the vector; ligation with DNA ligase; introduction of the recombinant molecules into bacterial or yeast cells, depending on the vector; and selection for the clones of interest. These methods are generally similar to those described in Section 10.1 for the production of recombinant molecules containing small inserts of cloned DNA.

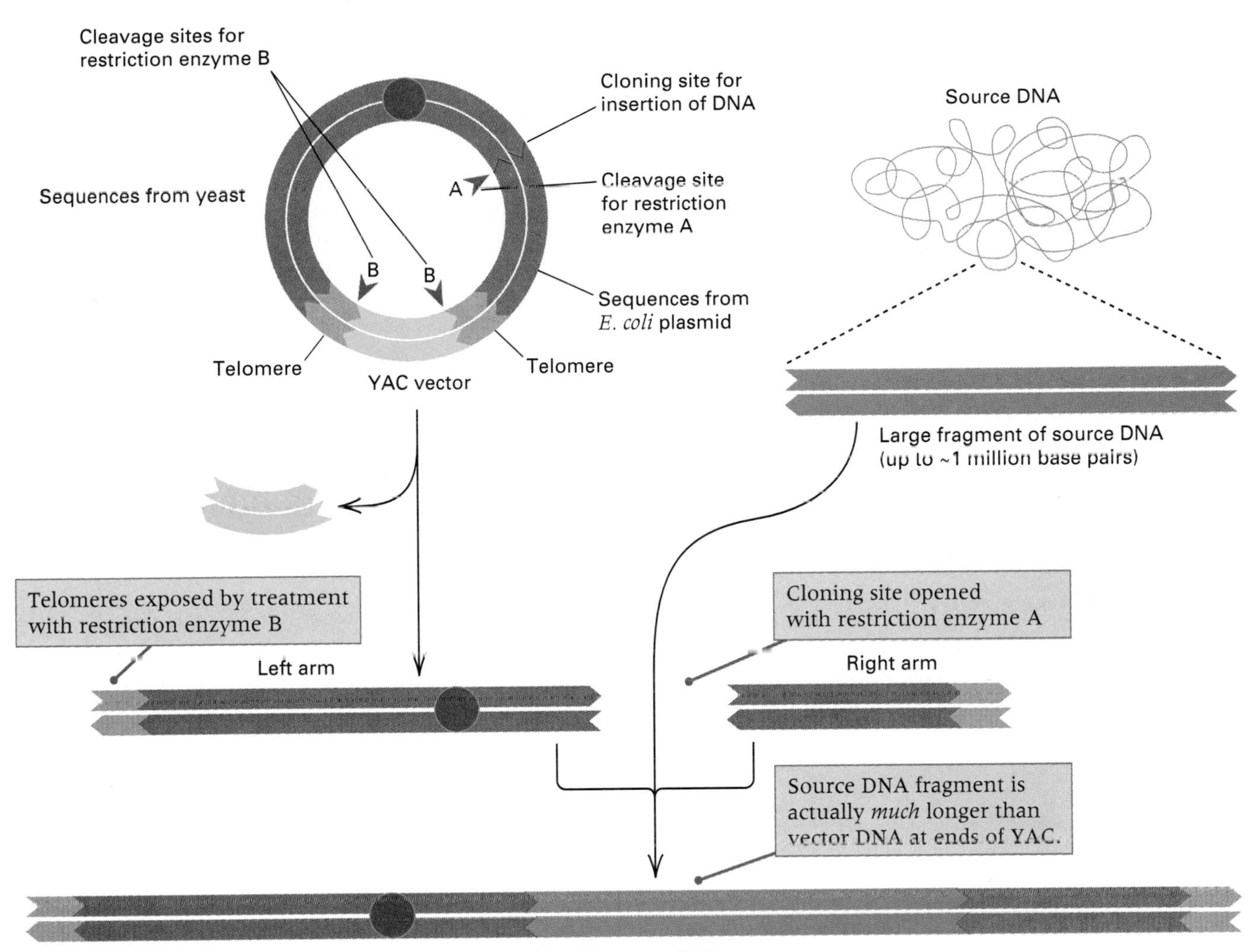

Figure 10.22 Cloning large DNA fragments in yeast artificial chromosomes (YACs). The vector (upper left) contains sequences that allow replication and selection in both *E. coli* and yeast, a yeast centromere, and telomeres from *Tetrahymena.* In producing the YAC clones, the vector is cut by two restriction enzymes (A and B) to free the chromosome arms. These arms are ligated to the ends of large fragments of source DNA, and yeast cells are transformed. Many ligation products are possible, but only those that consist of source DNA flanked by the left and right vector arms form stable artificial chromosomes in yeast.

Use of the YAC vector in cloning is illustrated in Figure 10.22. The circular vector is isolated after growth in *E. coli* and cleaved with two different restriction enzymes—one that cuts only at the cloning site (denoted A) and one that cuts near the tip of each of the telomeres (denoted B). Discarding the segment between the telomeres results in the two telomere-bearing fragments shown, which form the arms of the yeast artificial chromosomes. Ligation of a mixture containing source DNA and YAC vector arms results in a number of possible products. However, transformation of yeast cells and selection for the genetic markers in the YAC arms yields only the ligation product shown in Figure 10.22, in which a fragment of source DNA is sandwiched between the left and right arms of the YAC vector. This product is recovered because it is the only true chromosome that possesses a single centromere and two telomeres. Products with two YAC left arms are dicentric, those with two YAC right arms are acentric, and both of these types of products are genetically unstable. YACs that have donor DNA inserted at the cloning site can be identified because the inserted DNA interrupts a yeast gene present at the cloning site and renders it nonfunctional. Yeast cells containing YACs with inserts of particular sequences of donor DNA can be identified in a number of ways, including colony hybridization (Figure 10.9), and these cells can be grown and manipulated in order to isolate and

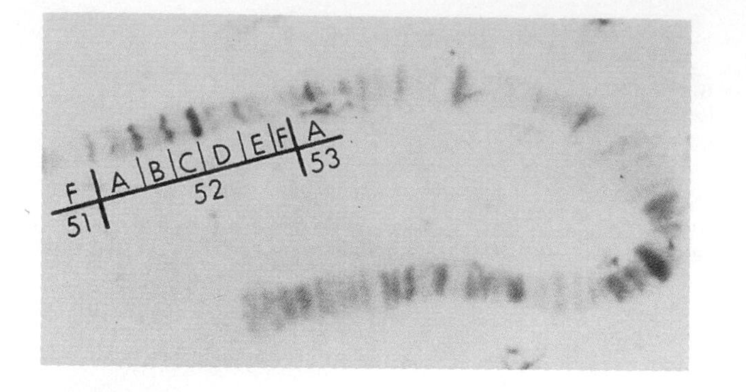

Figure 10.23 Hybridization *in situ* between *Drosophila* DNA cloned into a yeast artificial chromosome and the giant salivary gland chromosomes. The yeast artificial chromosome contains DNA sequences derived from numerous adjacent salivary bands—in this example, the bands in regions 52B through 52E in the right arm of chromosome 2. On average, each salivary band contains about 20 kb of DNA, but the bands vary widely in DNA content.

study the YAC insert. Figure 10.23 shows a region of the *Drosophila* salivary gland chromosomes that hybridizes with a YAC containing an insert of 300 kb. The entire genome of *Drosophila* could be contained in only 550 YAC clones of this size.

The landmarks in physical maps range from chromosome bands to DNA sequences.

The development of methods for isolating and cloning large DNA fragments has stimulated major efforts to map and sequence the human genome, which is the principal goal of the Human Genome Project. This project also aims to map and sequence the genomes of a number of model genetic organisms, including *E. coli,* yeast, *Arabidopsis thaliana,* the nematode *C. elegans, Drosophila,* and the laboratory mouse. The first stage in the analysis of complex genomes is usually the production of a **physical map,** which is a diagram of the genome depicting the physical locations of various landmarks along the DNA. The landmarks in a physical map usually consist of the locations of particular DNA sequences, such as coding regions or sequences present in particular cloned DNA fragments. If the landmarks are the locations of the cleavage sites for restriction enzymes, then the physical map is also a restriction map (Section 6.6). More useful landmarks are the positions of molecular markers, such as STRPs (simple tandem repeat polymorphisms), that have also been located in the genetic map (Section 4.4). Molecular markers serve to unify the genetic map and the physical map of an organism. The utility of a physical map is that it affords a single framework for organizing and integrating diverse types of genetic information, including the positions of chromosome bands, chromosome breakpoints, mutant genes, transcribed regions, and DNA sequences.

A more complex type of physical map is illustrated in Figure 10.24. The map covers a small part of human chromosome 16, and it illustrates how the physical map is used to organize and integrate several different levels of genetic information. A map of the metaphase banding pattern of chromosome 16 is shown across the top. (The entire chromosome contains about 95 Mb of DNA.) Beneath the cytogenetic map, the somatic cell hybrid map shows the locations of chromosome breakpoints observed in cultured hybrid cells that contain only a part of chromosome 16. The genetic linkage map depicts the locations of various genetic markers studied in pedigrees; most of the genetic markers are molecular markers, such as STRPs. One region of the genetic linkage map shows the location of a YAC clone, which has been assigned a physical location in the cytogenetic map by *in situ* hybridization. The large DNA insert in the YAC clone is also represented in a set of overlapping PAC or BAC clones; these clones define a contig covering a contiguous region of the genome without any gaps.

The various levels of the physical map in Figure 10.24 are connected by a special type of genetic marker shown at the bottom: a **sequence-tagged site (STS).** An STS marker is a DNA sequence, present once per haploid genome, that can be amplified with a suitable pair of oligonucleotide primers by means of the polymerase chain reaction (PCR), described in Section 6.7. Hence an STS marker defines a unique site in the genome whose presence in a cloned DNA fragment can be detected by PCR amplification. In Figure 10.24, for example, the STS marker located within the region of interest is present in two clones of the PAC/BAC contig, as well as in the YAC clone. Furthermore, the STS can be positioned on the somatic cell hybrid map by carrying out the PCR reaction with DNA from hybrid cells that contain rearranged or deleted chromosomes, and the amplified PCR product (or a clone containing the STS) can be used with *in situ* chromosome hybridization to localize the STS on the cytogenetic map. In this manner, STS markers can be used to integrate different types of information in a physical map. At present, more than 95 percent of the human genome is covered by a set of YAC clones interconnected by 30,000 STS markers. There is an average spacing between adjacent STS markers of 100 kb in the 3000-Mb human genome. Nearly 24,000 STS markers have been assigned to particular chromosomes, and a large fraction of these have been assigned to particular contigs within the chromosome.

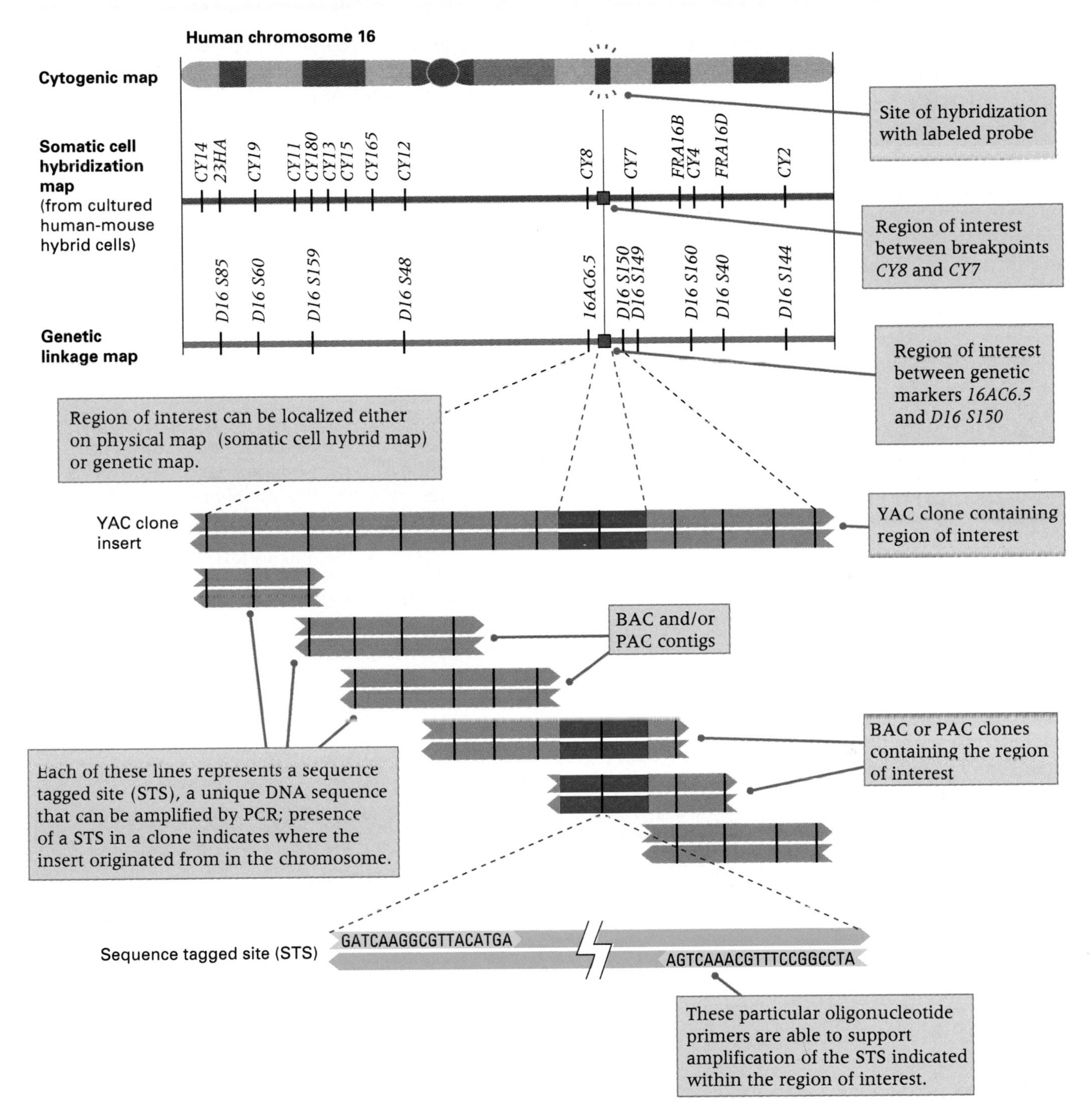

Figure 10.24 Integrated physical map of a small part of human chromosome 16. The map contains information about (1) the banding pattern of the metaphase chromosome (cytogenetic map), (2) the position of a particular sequence in the chromosome derived from *in situ* hybridization, (3) the locations of chromosome breakpoints in cultured cells (somatic cell hybrid map), (4) the positions of genetic markers analyzed by means of recombination in pedigrees (genetic linkage map), (5) the position of a YAC clone, and (6) a set of overlapping PAC or BAC clones forming a contig (coverage of a contiguous region of the genome without any gaps). The various levels of the map are integrated by sequence-tagged sites (STSs), sequences present once per haploid genome that can be amplified with the polymerase chain reaction. A YAC, PAC, or BAC clone will typically contain multiple, different STSA. [Marker data from *Human Genome: 1991–92 Program Report,* United States Department of Energy.]

10.6 Many large-scale DNA sequencing projects are under way.

Large-scale DNA sequencing is well under way in many organisms and has already been completed in many others, including more than 20 prokaryotes and numerous viruses. A selected list of these organisms is given in Table 10.1, which includes three of the genetic model organisms in the Human Genome Project (*E. coli, S. cerevisiae,* and *C. elegans*). Species for which sequencing is soon to be completed include the agents that cause bubonic plague, malaria, gonorrhea, leprosy, and necrotizing fasciitis (caused by the Group A Streptococcus, "flesh-eating" bacterium). Other organisms whose sequencing is under way include *Drosophila,* rice, two fish species, the mouse, and the human. There are clear biases in the genomes chosen for sequencing; they tend to be disease-causing organisms, model organisms, and several archaea. In terms of practical usefulness, a knowledge of the genomes of pathogens may help us devise better ways to make vaccines and limit the death and destruction that these organisms cause; a knowledge of the genomes of model organisms and that of human beings may reveal important new information about disease, development, aging, and many other biological processes. In a wider view, genomic sequencing will inevitably lead to a greater understanding of ourselves, our world, and the evolutionary history of organisms, and we have already learned a great deal.

The complete sequence of the *E. coli* genome is known.

A diagram of the genome of *E. coli,* based on the complete DNA sequence, is illustrated in Figure 10.25. The coordinates of the circle are given in minutes on the genetic map (0–100) as well as in base pairs. The "replichores" are the two halves of the circle replicated bidirectionally starting from the origin. The gold bars on the outside denote genes whose transcription is from left to right; the yellow bars on the inside denote genes transcribed from right to left. The green arrows show the positions of tRNA genes, red arrows rRNA genes. The circle just inside the red arrows shows the positions of a 40-base-pair repetitive sequence of unknown function that is present 581 times. The rays of the yellow "sunburst" in the middle show the usage of codons among all of the coding sequences. The length of each ray is proportional to the degree to which codons are used randomly. Short rays indicate genes with a highly biased usage of codons, which is usually associated with a high level of gene expression. This strain has a genome of 4.6 megabases.

The archaea, like bacteria, are prokaryotes—they lack internal membranes. Comparison of the genome of *Methanococcus jannaschii,* an archaeon, to bacterial and eukaryotic genome and gene sequences indicates that the machinery for DNA replication and transcription resembles that of eukaryotes, whereas the metabolism and the genes for metabolism strongly resemble those of bacteria. About half of the genes are unique to archaea, however.

Neither the archaeal nor the bacterial genomes contain genes that encode proteins related to the actin-based cytoskeleton. This finding, together with evolutionary trees based on rDNA sequences, suggest that there may be a still-unidentified group of organisms that participated in the origin of complex eukaryotes.

Table 10.1
Some Examples of Sequenced Genomes

Domain*	Organism	Size (Mb)	Year	Comments
B	*Hemophilus influenzae*	1.83	1995	One cause of meningitis
B	*Mycoplasma genitalium*	0.58	1995	Smallest cellular genome
A	*Methanococcus jannaschii*	1.66	1996	Deep sea vent organism; optimum growth temperature 85°C; produces methane
E	*Saccharomyces cerevisiae*	13.0	1997	Baker's and brewer's yeast
B	*Helicobacter pylori*	1.66	1997	Causes duodenal ulcers
B	*Escherichia coli*	4.60	1997	Resides in human intestinal tract
B	*Borrelia burgdorferi*	1.44	1997	Causes Lyme disease; carries 11 plasmids; spirochaete
A	*Archaeoglobus fulgidus*	2.18	1997	Deep sea vent organism; optimum growth temperature 83°C
B	*Treponema pallidum*	1.14	1998	Causes syphilis; spirochaete
B	*Mycobacterium tuberculosis*	4.40	1998	Causes tuberculosis
B	*Vibrio cholera*	2.50	1998	Causes cholera
E	*Caenorhabditis elegans*	100.0	1999	Soil nematode

*A, Archaea; B, Bacteria; E, Eukarya

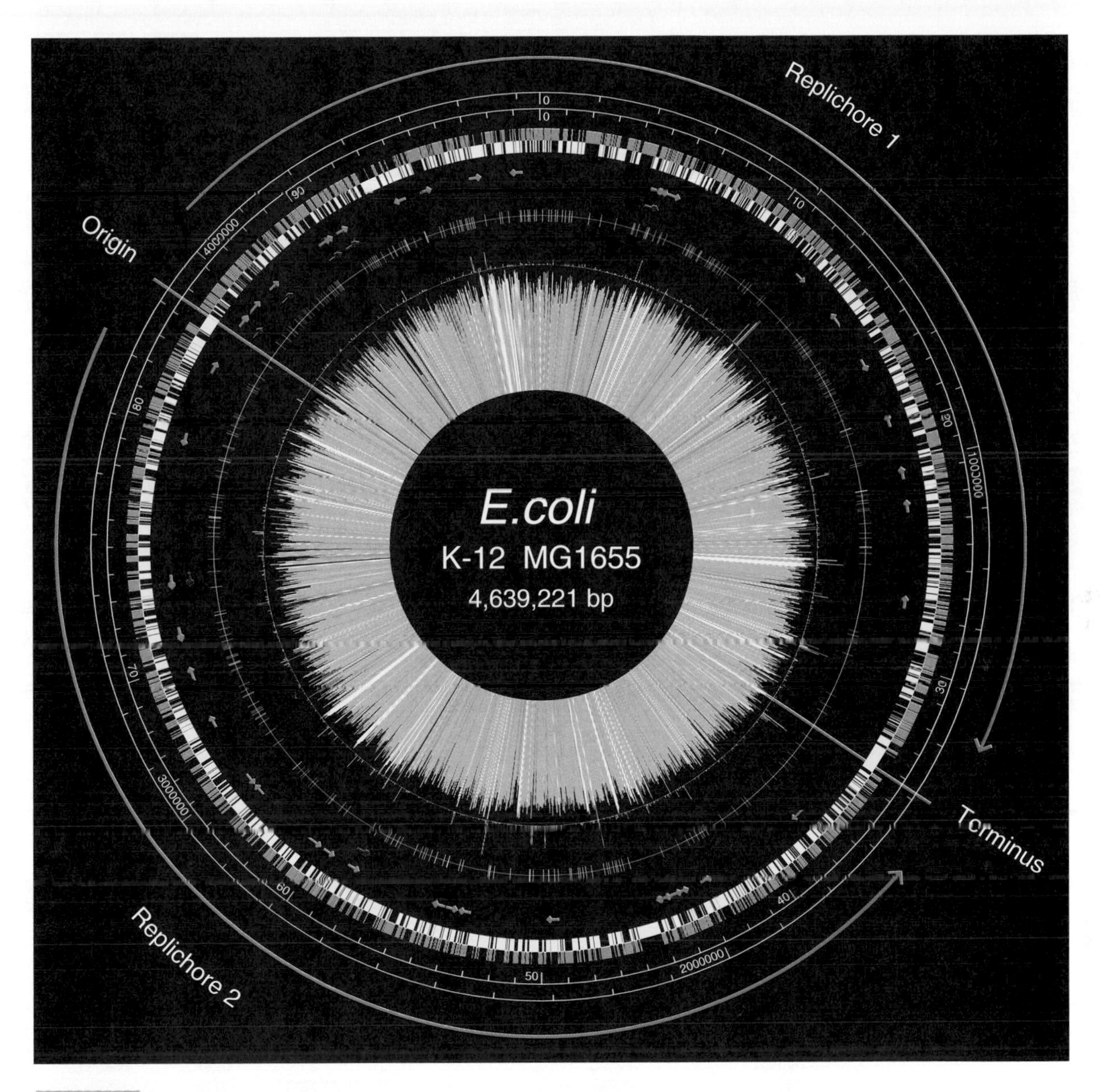

Figure 10.25 Diagram of the DNA sequence organization of *Escherichia coli* strain K-12. The coordinates are given in base pairs as well as in minutes on the genetic map. The coding sequences are shown as gold and yellow bars, which are transcribed in a clockwise (gold) or counterclockwise (yellow) direction. Green and red arrows denote genes for transfer RNAs and ribosomal RNAs, respectively. The gold rays of the "sunburst" are proportional to the degree of randomness of codon usage in the coding sequences. Genes with the longest rays use the codons in the genetic code almost randomly. The origin and terminus of DNA replication are indicated. Bidirectional replication creates two "replichores." The peaks on the circle immediately outside the sunburst indicate coding sequences with high similarity to previously described bacteriophage proteins. [Courtesy of Frederick R. Blattner and Guy Plunkett III. From F. R. Blattner, et al. 1997. *Science* 277: 1453.]

The yeast genome was the first eukaryotic genome sequenced.

A summary of the analysis of the 666,448 nucleotide pairs in the sequence of yeast chromosome XI is illustrated in Figure 10.26. Like other yeast chromosomes, chromosome XI has a high density of coding regions. A coding region includes an **open reading frame (ORF),** which is a region of sequence containing an uninterrupted run of amino-acid-coding triplets (codons) with no "stop" codons that would terminate translation. An ORF of greater than random length is likely to code for a protein of some kind. The coding region associated with an ORF also includes the flanking regulatory sequences. In chromosome XI, approximately 72

Figure 10.26 Genetic organization of yeast chromosome XI. [Courtesy of Bernard Dujon. From B. Dujon, et al. 1994. *Nature* 369: 371.]

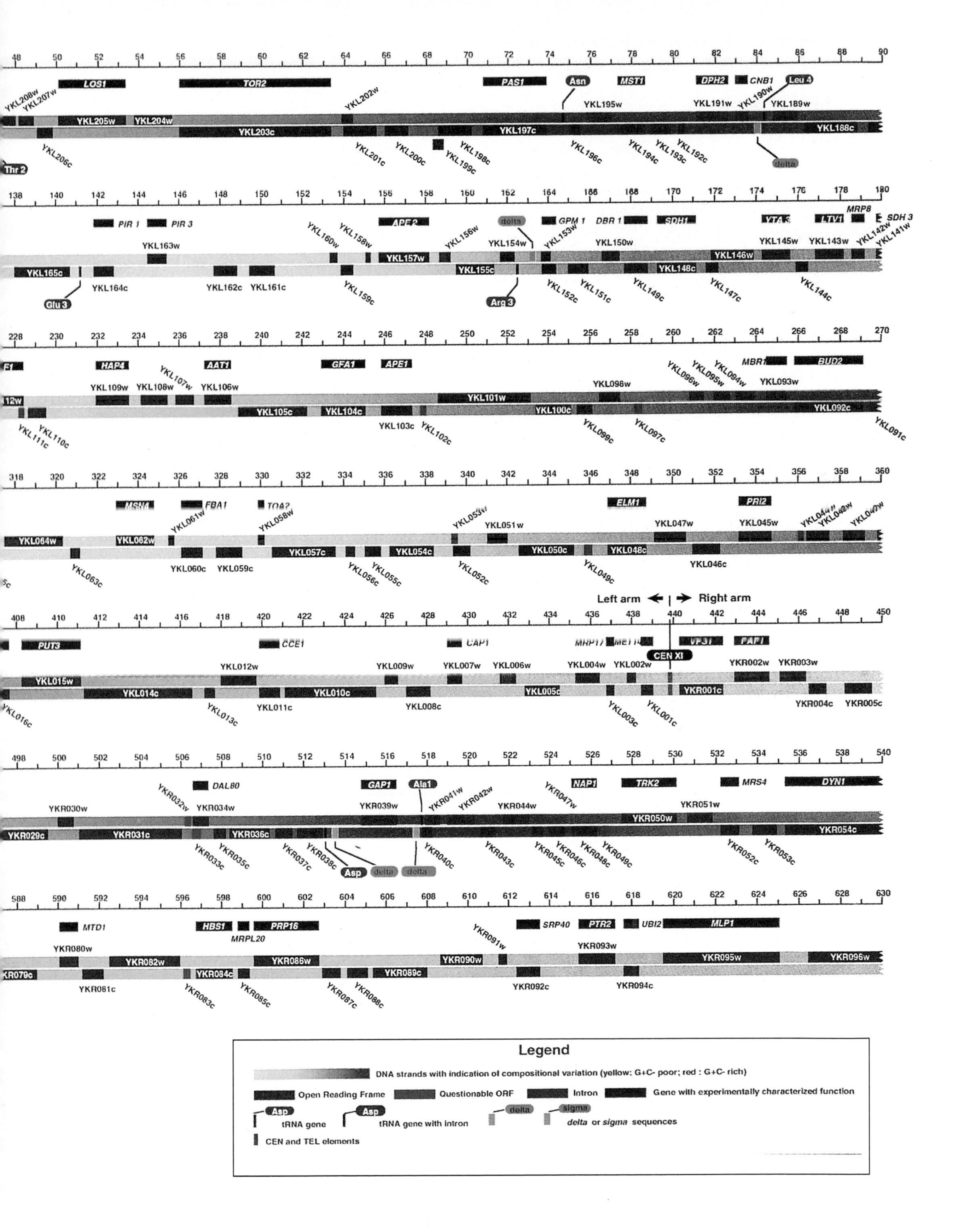
48 50 52 54 56 58 60 62 64 66 68 70 72 74 76 78 80 82 84 86 88 90
LOS1
TOR2
PAS1
Asn
MST1
DPH2
CNB1
Leu 4
YKL208w
YKL207w
YKL202w
YKL195w
YKL191w
YKL190w
YKL189w
YKL205w
YKL204w
YKL203c
YKL197c
YKL188c
YKL206c
YKL201c
YKL200c
YKL198c
YKL199c
YKL196c
YKL194c
YKL193c
YKL192c
delta
Thr 2
138 140 142 144 146 148 150 152 154 156 158 160 162 164 166 168 170 172 174 176 178 180
PIR 1
PIR 3
APE 2
delta
GPM 1
DBR 1
SDH1
VTA 3
LTV1
MRP8
SDH 3
YKL163w
YKL160w
YKL158w
YKL156w
YKL154w
YKL153w
YKL150w
YKL145w
YKL143w
YKL142w
YKL141w
YKL157w
YKL146w
YKL165c
YKL155c
YKL148c
YKL164c
YKL162c
YKL161c
YKL159c
YKL152c
YKL151c
YKL149c
YKL147c
YKL144c
Glu 3
Arg 3
228 230 232 234 236 238 240 242 244 246 248 250 252 254 256 258 260 262 264 266 268 270
HAP4
AAT1
GFA1
APE1
MBR1
BUD2
YKL107w
YKL109w
YKL108w
YKL106w
YKL098w
YKL096w
YKL095w
YKL094w
YKL093w
YKL101w
YKL105c
YKL104c
YKL100c
YKL092c
YKL111c
YKL110c
YKL103c
YKL102c
YKL099c
YKL097c
YKL091c
318 320 322 324 326 328 330 332 334 336 338 340 342 344 346 348 350 352 354 356 358 360
MSN4
FBA1
TOA2
ELM1
PRI2
YKL061w
YKL058w
YKL053w
YKL051w
YKL047w
YKL045w
YKL044w
YKL042w
YKL040w
YKL064w
YKL062w
YKL057c
YKL054c
YKL050c
YKL048c
YKL060c
YKL059c
YKL046c
YKL063c
YKL056c
YKL055c
YKL052c
YKL049c
Left arm
Right arm
408 410 412 414 416 418 420 422 424 426 428 430 432 434 436 438 440 442 444 446 448 450
PUT3
CCE1
CAP1
MRP17
MET14
VPS21
FAF1
CEN XI
YKL012w
YKL009w
YKL007w
YKL006w
YKL004w
YKL002w
YKR002w
YKR003w
YKL015w
YKL014c
YKL010c
YKL005c
YKR001c
YKL011c
YKL008c
YKR004c
YKR005c
YKL016c
YKL013c
YKL003c
YKL001c
498 500 502 504 506 508 510 512 514 516 518 520 522 524 526 528 530 532 534 536 538 540
DAL80
GAP1
Ala1
NAP1
TRK2
MRS4
DYN1
YKR032w
YKR034w
YKR030w
YKR039w
YKR041w
YKR042w
YKR044w
YKR047w
YKR051w
YKR050w
YKR029c
YKR031c
YKR036c
YKR054c
YKR033c
YKR035c
YKR037c
YKR038c
Asp
delta
delta
YKR040c
YKR043c
YKR045c
YKR046c
YKR048c
YKR049c
YKR052c
YKR053c
588 590 592 594 596 598 600 602 604 606 608 610 612 614 616 618 620 622 624 626 628 630
MTD1
HBS1
PRP16
MRPL20
SRP40
PTR2
UBI2
MLP1
YKR080w
YKR091w
YKR093w
YKR082w
YKR086w
YKR090w
YKR095w
YKR096w
YKR079c
YKR084c
YKR089c
YKR081c
YKR083c
YKR085c
YKR087c
YKR088c
YKR092c
YKR094c
Legend
DNA strands with indication of compositional variation (yellow: G+C- poor; red : G+C- rich)
Open Reading Frame
Questionable ORF
Intron
Gene with experimentally characterized function
Asp
tRNA gene
Asp
tRNA gene with intron
delta
sigma
delta or sigma sequences
CEN and TEL elements

The human connection

Back to the Future

José M. Fernández-Cañón[1], Begoña Granadino[1], Daniel Beltrán-Valero de Bernabé[1], Mónica Renedo[2], Elena Fernández-Ruiz[2], Miguel A. Peñalva[1], and Santiago Rodríguez de Córdoba[1] 1996
[1]Consejo Superior de Investigaciones Cientificas, Madrid, Spain
[2]Universidad Autónoma de Madrid, Madrid, Spain
The Molecular Basis of Alkaptonuria

Alkaptonuria (AKU) occupies an honored place in the annals of human genetics. It was first recognized as an inherited metabolic disease by Archibald Garrod in 1902 (see Chapter 1). Affected individuals accumulate and excrete large quantities of homogentisic acid, an intermediate in the degradation of tyrosine, which results in black urine and a debilitating arthritis. Garrod later postulated that an inherited disease in which cellular metabolism is abnormal results from an inherited defect in an enzyme. The authors of this excerpt recognized that the pathway for catabolism of tyrosine is the same in the fungus *Aspergillus nidulans* and in humans. They cloned and sequenced the *Aspergillus* gene for homogentisate dioxygenase (HGO), which converts homogentisic acid to maleylacetoacetate, and used the predicted amino acid sequence of HGO to identify a human cDNA in the dbEST (database of Expressed Sequence Tags) encoding a putative human homolog of the *Aspergillus* enzyme. This study brings the story of alkaptonuria full circle. It is an excellent example of current approaches in human genetics research, combining research on the model organism *Aspergillus,* use of evolutionary relationships as manifested in the similarities in protein sequence, the dbEST of collected human cDNA sequences, and the full armamentarium of modern molecular genetics.

The human EST clone . . . [encodes] a 445-amino-acid deduced polypeptide that shows 52% identity to the *Aspergillus nidulans* hmgA enzyme. This notable conservation in amino acid sequence indicated that the polypeptide is almost certainly a human protein with HGO [homogentisate dioxygenase] activity. To test this possibility we expressed the polypeptide encoded by the EST . . . as a fusion protein [for ease of purification] and found that, in fact, it had HGO activity. . . . The coding region of the human HGO gene is encoded by 14 exons spanning 60 kb of DNA. . . . The human gene responsible for AKU [alkaptonuria] has been mapped to chromosome 3q2. . . . The human HGO gene and the AKU gene map to the same chromosomal location. . . . To establish whether HGO was mutated in AKU, we used genomic DNA obtained from all members of [a Spanish AKU pedigree]. . . . Both parents were heterozygous at the HGO locus, . . . [with one] normal allele . . . and a second allele that we denoted HGO^{P230S}. The HGO^{P230S} allele contained a . . . C → T transition in exon 10, changing a proline residue at position 230 to a serine residue. Pro230 is a conserved residue between the human and fungal proteins. . . . The HGO^{P230S} allele cosegregated with the disease. . . . The two alkaptonuric children, but not the healthy child, are homozygous for the HGO^{P230S} allele. . . . To establish whether the Pro230Ser missense mutation . . . results in loss of enzyme activity, we constructed another fusion protein. The HGO^{P230S} construct showed no HGO activity. . . . Our results from the analysis of the fusion proteins indicate that the single amino acid substitution Pro230Ser in human HGO is sufficient to determine loss of enzyme activity. . . . The molecular basis of alkaptonuria is established here.

THE MOLECULAR BASIS of alkaptonuria is established here.

Source:Nature Genetics 14: 19–24.

percent of the sequence is present within 331 coding regions averaging 2 kb. The average length of ORF codes for a sequence of 488 amino acids, but the longest ORF codes for the protein dynein with 4092 amino acids. Worthy of note in the chromosome XI sequence is the low number of introns (sequences that are transcribed but removed from the RNA in producing the mRNA); only about 2 percent of the genes have introns. The chromosome also includes sequences for 16 transfer-RNA genes used in protein translation as well as representatives of the transposable elements δ and σ.

The yeast sequence contains evidence of an ancient duplication of the entire genome, although only a small fraction of the genes are retained in duplicate. Protein pairs derived from the ancient duplication make up 13 percent of all yeast proteins. These include pairs of cytoskeletal proteins, ribosomal proteins, transcription factors, glycolytic enzymes, cyclins, proteins of the secretory pathway, and protein kinases. When yeast protein sequences are compared with mammalian protein sequences in GenBank (the international repository of protein sequences of all organisms), a mammalian homolog is found for about 31 percent of yeast proteins. This is a minimum estimate of homology, because the mammalian sequences available for comparison are only a fraction of those present in mammalian genomes. The homologous proteins are of many types. Examples include proteins that catalyze metabolic reactions, subunits of [illegible] transcription factors, translation initiation and elongation factors, enzymes of DNA synthesis and repair, nuclear-pore proteins, and structural proteins and enzymes of mitochondria and peroxisomes. In some instances the similarities are known to reflect conservation of function, because in more than 70 cases, a human amino-acid coding sequence will substitute for a yeast sequence. These include coding sequences for cyclins, DNA ligase, the *RAS* proto-oncogenes, translation initiation factors, and proteins involved in signal transduction.

The most common method of cloning human disease genes is positional cloning, which means cloning by map position (Figure 10.24). Usually nothing is known about the gene except that, when defective, it results in disease. The first clue to function often comes from recognizing homology to a yeast gene. Striking examples are the human genes that cause hereditary nonpolyposis colon cancer and Werner's syndrome, a disease associated with premature aging. In cells of patients with hereditary nonpolyposis colon cancer, short repeated DNA sequences are unstable. These findings stimulated studies of stability of repeated DNA sequences in yeast mutants, which revealed that repeated DNA sequences are unstable in yeast cells that are deficient in the repair of mismatched nucleotides in DNA, including *msh2* and *mlh1* mutants (Chapter 7). The prediction that the cancer genes might also encode proteins for mismatch repair was later verified when the cancer genes were cloned. Cells of patients with Werner's syndrome of premature aging show a limited life span in culture. The human gene encodes a protein highly similar to a DNA helicase encoded by *SGS1* in yeast. The *sgs1* mutant yeast cells show accelerated aging and a reduced life span, as well as other cellular phenotypes, including relocation of proteins from telomeres to the nucleolus and nucleolar fragmentation. Examples like these demonstrate how research on model organisms can have direct applicability to human health and disease.

The target date for completion of the human genome sequence is 2005.

A substantial fraction of the human genome is thought to be inessential. For this reason, the initial efforts in sequencing the human genome have concentrated on the cDNAs, which encode proteins. Sequences of cDNA clones have been ob[illegible]ed from libraries representing 37 distinct organs and tissues. Computer matching to DNA sequences already known from humans and other organisms made it possible to assign functions to many of these cDNAs. Figure 10.27 shows a breakdown of cDNAs by function. Approximately 40 percent of human genes are implicated in basic energy metabolism, cell structure, homeostasis, or cell division; a further 22 percent are concerned with RNA and protein synthesis and processing; and 12 percent are associated with signaling and communication between cells. Figure 10.28 summarizes the results of examining the tissue-specific cDNA libraries. For each organ or tissue type, the first number is the total number of cDNA clones sequenced, and the number in parentheses is the number of distinct cDNA sequences found among the total from that organ or tissue type.

Already more than one million sequences have been deposited in a database of human cDNA sequences, and more than 15,000 chromosome assignments of these sequences have been made. The database of human cDNAs is called dbEST, where db stands for database and EST for **Expressed Sequence Tag,** because each cDNA "tags" a gene that is expressed in a particular tissue. It is estimated that the collection of ESTs represents more than 60,000 distinct human genes—well on the way toward all of the 80,000 to 100,000 genes thought to be present in the human genome.

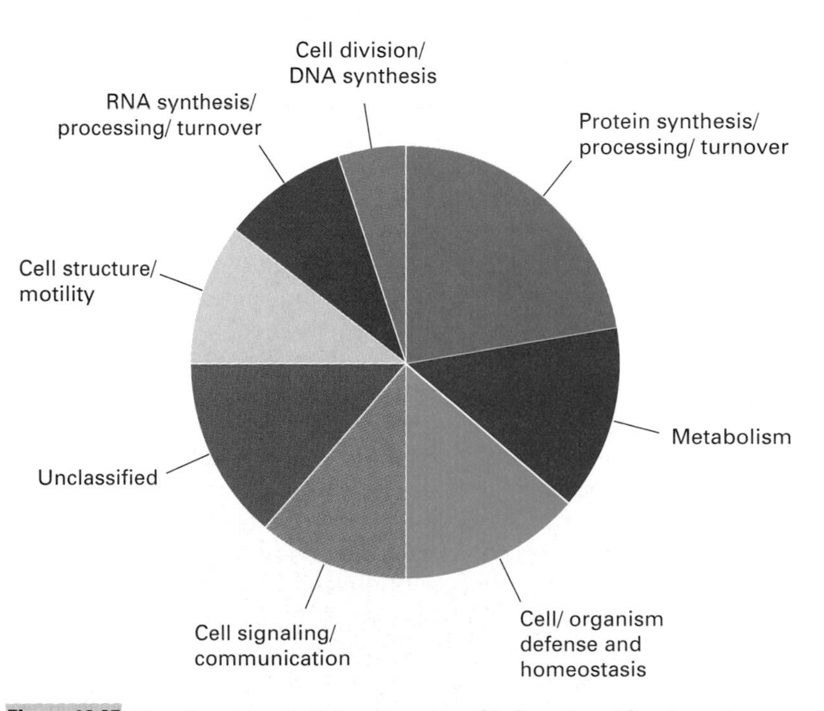

Figure 10.27 Classification of cDNA sequences by function. The chart is based on over 13,000 distinct, randomly selected human cDNA sequences. [Data courtesy of Craig Venter and the Institute for Genomic Research.]

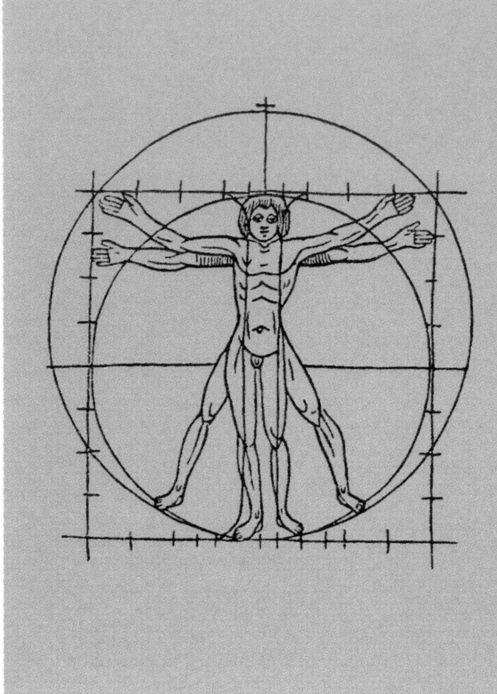

Source of RNA	Sequences	Source of RNA	Sequences
Skin	3,043 (629)	Brain	67,679 (3,195)
Blood cells	23,505 (2,194)	Eye	1,932 (574)
Bone	5,736 (904)	Salivary gland	186 (17)
Thyroid	2,381 (584)	Esophagus	194 (76)
Parathyroid	197 (46)	Adipose tissue	2,412 (581)
Endothelial cells	5,736 (1,031)	Heart	9,400 (1,195)
Liver	37,807 (2,091)	Peritoneum	884 (163)
Gall bladder	3,754 (768)	Spleen	7,924 (924)
Colon	4,832 (879)	Adrenal gland	3,427 (658)
Smooth muscle	297 (127)	Pancreas	5,534 (1,094)
Small intestine	1,009 (297)	Kidney	3,213 (712)
Prostate	7,971 (1,203)	Epididymus	1,716 (370)
Testis	7,117 (1,232)	Skeletal muscle	4,693 (735)
Ovary	3,848 (504)	Synovial membrane	3,889 (813)
Uterus	6,392 (1,059)	Embryo	19,291 (1,989)
Placenta	12,148 (1,290)	Thymus	2,412 (261)

Figure 10.28 Classification of cDNA sequences by organ or tissue type. In each category, the initial number is the total number of cDNA clones examined. The number in parentheses is the number of distinct sequences found per organ or tissue type. [Data from M. D. Adams, et al. 1995. *Nature* 377 (Suppl.): 3.]

10.7 DNA sequencing is highly automated.

Large-scale sequencing efforts are not feasible without the use of instruments that partly automate the process. When a conventional sequencing gel is obtained from the dideoxy procedure described in Section 6, each lane contains the products of DNA synthesis carried out in the presence of a small amount of a dideoxy nucleotide (dideoxy-G, -A, -T or -C), which, when incorporated into a growing DNA strand, terminates further elongation. The products of each reaction are separated by electrophoresis in individual lanes, and the gel is placed in contact with photographic film so that radioactive atoms present in one of the normal nucleotides will darken the film and reveal a band at the position to which each incomplete DNA strand migrated in electrophoresis. After the film is developed, the DNA sequence is read from the pattern of bands, as shown by the sequence at the right of the gel.

In automated DNA sequencing, illustrated in Figure 6.37 on page 228, the nucleotides that terminate synthesis are labeled with different fluorescent dyes (G, black; A, green; T, red; C, purple). Because the colors distinguish the products of DNA synthesis that terminate with each nucleotide, all the synthesis reactions can be carried out simultaneously in the same tube and the reaction products separated by electrophoresis in a single lane. In principle, the sequence could again be read directly from the gel. However, a substantial improvement in efficiency is accomplished by continuing the electrophoresis until each band, in turn, comes off the bottom of the gel. As each band comes off the bottom of the gel, the fluorescent dye that it contains is excited by laser light, and the color of the fluorescence is read automatically by a photocell and recorded in a computer. A trace of the fluorescence pattern that emerges at the bottom of the gel after continued electrophoresis is examined, and the nucleotide sequence is read directly from the colors of the alternating peaks along the trace. When used to maximal capacity, an automated sequencing instrument can generate as much as 10 Mb of nucleotide sequence per year. The amount of finished sequence is considerably smaller, because in a sequencing project, each DNA strand needs to be sequenced completely for the sake of minimizing sequencing errors, and some troublesome regions need to be sequenced several times.

TECHNICIANS PICK OUT BACTERIA colonies from petri dishes in order to isolate colonies which contain human cDNA inserted into bacterial plasmids. This is the first step in purifying and cloning large amounts of human cDNA for the Human Genome Project. Mapping human genes will enable the development of genetic therapies for inherited diseases. [© Hank Morgan/ Photo Researchers, Inc.]

Chapter Summary

- **Cloning a DNA molecule takes place in several steps.**

  Restriction enzymes cleave DNA into fragments with defined ends.
  Restriction fragments are joined end to end to produce recombinant DNA.
  A vector is a carrier for recombinant DNA.
  A variety of strategies can be used to clone a gene.
  DNA fragments are joined with DNA ligase.
  A recombinant cDNA contains the coding sequence of a eukaryotic gene.
  Loss of β-galactosidase activity is often used to detect recombinant vectors.
  Recombinant clones are often identified by hybridization with labeled probe.

- **Positional cloning is based on the location of a gene in the genetic map.**

  Close genetic linkages are often conserved among related species.

Recombinant DNA technology allows the genotype of an organism to be modified in a directed, predetermined way by enabling different DNA molecules to be joined into novel genetic units, altered as desired, and reintroduced into the organism. Restriction enzymes play a key role in the technique because they can cleave DNA molecules within particular base sequences. Many restriction enzymes generate DNA fragments with complementary single-stranded ends (sticky ends), which can anneal and be ligated together with similar fragments from other DNA molecules. The carrier DNA molecule used to propagate a desired DNA fragment is called a vector. The most common vectors are plasmids, phages, viruses, and yeast artificial chromosomes (YACs). Transformation is an essential step in the propagation of recombinant molecules because it enables the recombinant DNA molecules to enter and replicate in host cells, such as those of bacteria, yeast, or mammals. Plasmid vectors become permanently established in the host cell; phages can multiply and produce a stable population of phages carrying source DNA; retroviruses can be used to establish a gene in an animal cell; and YACs include source DNA within an artificial chromosome containing a functional centromere and telomeres. Particular DNA sequences can also be amplified without cloning by means of the polymerase chain reaction (PCR), in which short synthetic oligonucleotides are used as primers to replicate and amplify the sequence between them repeatedly.

- **Reverse genetics creates an organism with a designed mutation.**

  Recombinant DNA can be introduced into the germ line of animals.
  Recombinant DNA can also be introduced into plant genomes.

Recombinant DNA can also be used to transform the germ line of animals and genetically engineer plants. These techniques form the basis of reverse genetics, in which genes are deliberately mutated in specified ways and introduced back into the organism to determine the effects on phenotype. Reverse genetics is commonplace in genetic analysis in bacteria, yeast, nematodes, *Drosophila,* the mouse, and other organisms. In *Drosophila,* transformation employs a system of two vectors based on the transposable *P* element. One vector contains sequences that produce the *P* transposase; the other contains the DNA of interest between the inverted repeats of *P* and other sequences needed for mobilization by transposase and insertion into the genome. Germ-line transformation in the mouse makes use of retrovirus vectors or embryonic stem cells. Dicot plants are transformed with transposable *T* DNA derived from a plasmid found in species of *Agrobacterium,* using a two-component system consisting of transposase producer and transforming vectors analogous to *P* element transformation in *Drosophila.*

- **Genetic engineering is applied in agriculture, industry, medicine, and research.**

  Agricultural crop plants are primary targets of genetic engineering.
  Specific tissues can be targeted for destruction.
  Animal growth rate can be genetically engineered.
  Engineered microbes can help in the degradation of toxic waste.
  Recombinant DNA permeates modern biomedical research.
  The production of useful proteins is a primary impetus for recombinant DNA.

Animal viruses may prove useful vectors for gene therapy.

Recombinant DNA yields probes for the detection of mutant genes in hereditary disease.

Practical applications of recombinant DNA technology include the efficient production of useful proteins, the creation of novel genotypes for the synthesis of economically important molecules, the generation of DNA and RNA sequences for use in medical diagnosis, the manipulation of the genotype of domesticated animals and plants, the development of new types of vaccines, and the potential correction of genetic defects (gene therapy). Production of eukaryotic proteins in bacterial cells is sometimes hampered by protein instability, inability to fold properly, or failure to undergo necessary chemical modification. These problems are often eliminated by production of the protein in yeast or mammalian cells.

- **An entire genome can be physically mapped and its DNA sequence determined.**

The smallest complex genomes are about 100 million base pairs.

Special conditions allow production and isolation of large DNA fragments.

Artificial chromosomes are vectors for large DNA fragments.

The landmarks in physical maps range from chromosome bands to DNA sequences.

- **Many large-scale sequencing projects are under way.**

The complete sequence of the *E. coli* genome is known.

The yeast genome was the first eukaryotic genome sequenced.

The target date for completion of the human genome sequence is 2005.

- **DNA sequencing is highly automated.**

Cloning in bacteriophage P1 (PACs), derivatives of the bacterial F plasmid (BACs), or yeast artificial chromosomes (YACs) allows very large DNA molecules to be isolated and manipulated and has stimulated a major effort to analyze complex genomes, such as those in nematodes, *Drosophila*, the mouse, and human beings. These efforts include the development of detailed physical maps that integrate many levels of genetic information, such as the positions of sequence-tagged sites or the positions and lengths of contigs. DNA sequencing of yeast chromosomes has revealed an unexpectedly high density of genetic information. The development of instruments for partly automated DNA sequencing has made it feasible completely to sequence larger chromosomes from other organisms and even entire genomes. More than a million partial cDNAs from human tissues and organs have been sequenced and identify about 60,000 different genes.

Key Terms

bacterial artificial chromosome (BAC)
blunt ends
cDNA
cloned DNA sequence
colony hybridization
complementary DNA
contig
embryonic stem cell
expressed sequence tag (EST)
gene cloning
gene targeting
gene therapy
genetic engineering
Human Genome Project
library
map-based cloning
open reading frame (ORF)
P1 artificial chromosome (PAC)
physical map
positional cloning
P transposable element
recombinant DNA
retrovirus
reverse genetics
reverse transcriptase
reverse transcriptase PCR (RT-PCR)
sequence-tagged site (STS)
sticky end
T DNA
Ti plasmid
transgenic organism
vector
yeast artificial chromosome (YAC)

Concepts and Issues
Testing Your Knowledge

- What does the term *recombinant DNA* mean? What are some of the practical uses of recombinant DNA?
- What features are essential in a bacterial cloning vector? How can a vector have more than one cloning site?
- What is the reaction catalyzed by the enzyme reverse transcriptase? How is this enzyme used in recombinant DNA technology?
- What is a transgenic organism? What are some of the practical uses of transgenic organisms?
- Distinguish between a physical map and a genetic map. If an organism had no recombination, how would the physical map of its genome differ from the genetic map of its genome?
- In the context of genome analysis, what are PACs, BACs, and YACs?

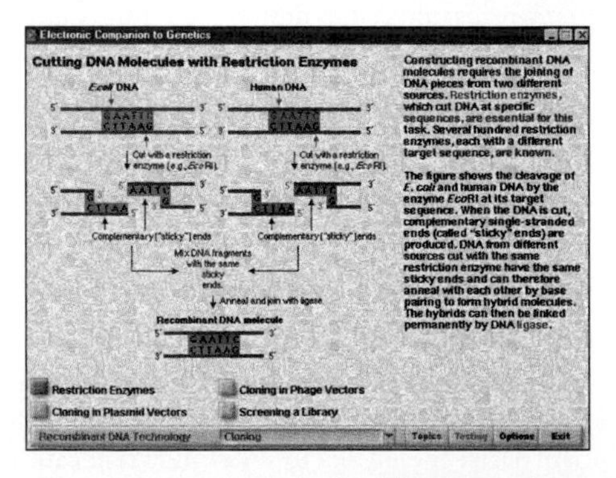

To ENHANCE YOUR STUDY turn to *An Electronic Companion to Genetics*™, © 1998, Cogito Learning Media, Inc. This CD-ROM is a multimedia tool that provides interactive explanations of important genetic concepts through animations, diagrams, and videos. It also provides interactive test questions.

Solutions
Step by Step

Problem 1

Many engineered vectors have a polylinker, or multiple cloning site (MCS), instead of a single restriction site available for insertion of DNA. A polylinker is a short DNA sequence containing multiple different restriction sites, each unique to the vector, any combination of which can be cleaved with the appropriate restriction enzymes to yield a linear vector molecule with defined restriction sites at the ends. The accompanying diagram illustrates five restriction sites in the polylinker of a vector; the vector sequences adjacent to the polylinker are denoted X and Y. Also shown is the restriction map of a genomic target sequence whose ends are denoted A and B. The restriction enzymes *Sac*I, *Bam*HI, *Bgl*II, and *Eco*RI all produce sticky ends; *Sma*I produces blunt ends.

(a) If the vector and target were both digested with *Sma*I and the resulting fragments ligated in such a way that each vector molecule was ligated with one and only one target fragment, in what orientation would the A–B fragment be ligated into the polylinker?
(b) What restriction enzyme (or enzymes) would you use to digest the vector and target so that after mixing of the fragments and ligation, the cloned DNA would have the sequence X–B–A–Y?
(c) What restriction enzyme (or enzymes) would you use to digest the vector and target so that after mixing of the fragments and ligation, the cloned DNA would have the sequence X–A–B–Y?

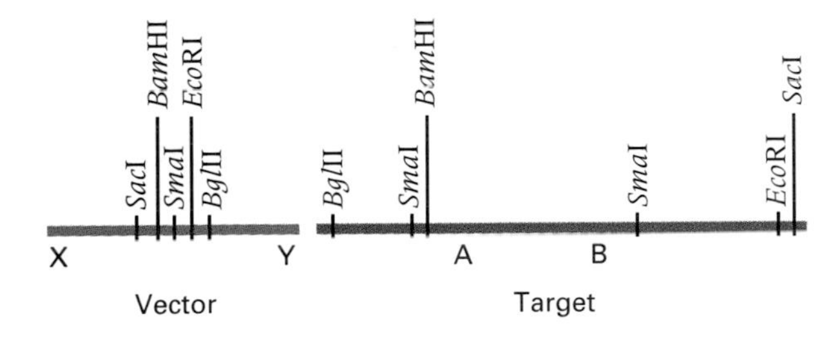

Solution (a) *Sma*I produces blunt ends, and any blunt end can be ligated onto any other. Hence the *Sma*I fragment A–B can be ligated into the polylinker in either of two orientations. The resulting clones are expected to be X–A–B–Y and X–B–A–Y in equal frequency. **(b)** A restriction enzyme produces fragments whose ends are identical, so either end can be ligated onto a complementary sticky end. To force the orientation X–A–B–Y, one needs to cleave the vector and the target with two restriction enzymes that produce different sticky ends. The site nearest X in the vector must match the site nearest A in the target, and the site nearest Y in the vector must match the site nearest B in the vector. In this case, if the vector and target are both cleaved with *Bam*HI and *Eco*RI, the resulting clone is expected to have the orientation X–A–B–Y. **(c)** Following the logic of part b, digestion of the vector and target with *Bgl*II and *Sac*I will yield clones with the orientation X–B–A–Y.

Problem 2

The accompanying table shows the presence or absence of seven sequence-tagged sites (STSs), numbered 1 through 7, in each of six BAC clones, designated A through F. A + in the table means that PCR primers specific to the STS are able to amplify the sequence from the clone, – means that the STS cannot be amplified from the clone. Use these data to make a map of the order of the BAC clones and of the STS markers located within the clones. In orienting the map, put BAC clone A at the top.

	STS marker						
BAC clone	1	2	3	4	5	6	7
A			[illegible]	[illegible]			
B	–	+	–	–	–	–	+
C	+	–	–	–	+	–	–
D	+	–	–	–	–	+	–
E	–	–	–	+	–	+	–
F	–	+	–	–	+	–	–

Solution The principle in ordering the BAC clones is that the DNA fragments present in any two (or more) clones that contain the same STS marker must overlap. The table contains no information bearing on the relative sizes of the BAC clones or on the distance between the STS markers. Thus each clone is drawn as though it had the same size, and the STS markers are spaced equally. Considering the STS markers shared between clones, and starting with clone A, we obtain the map shown here. The mirror-image map with STS 3 on the right and STS 7 on the left is also a valid solution.

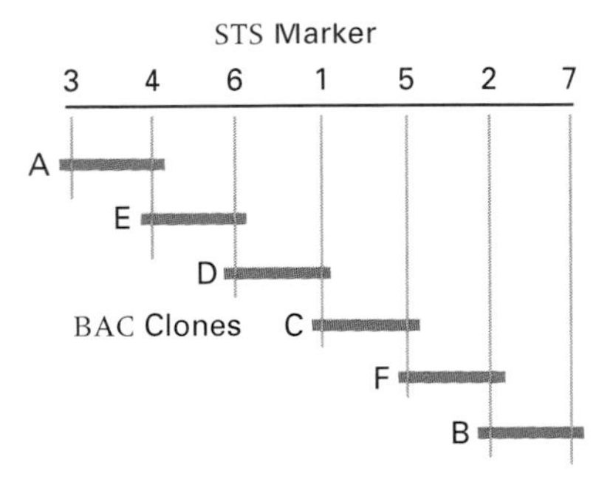

Concepts in Action
Problems for Solution

10.1 Reverse transcriptase, like most enzymes that make DNA, requires a primer. Explain why, when cDNA is to be made for the purpose of cloning a eukaryotic gene, a convenient primer is a short sequence of poly(dT)? Why does this method not work with a prokaryotic messenger RNA?

10.2 A DNA molecule has 23 occurrences of the sequence 5'-AATT-3' along one strand. How many times does the same sequence occur along the other strand?

geNETics on the web

www.jbpub.com/genetics

These GeNETics on the Web exercises will introduce some of the most useful sites for finding genetic information on the World Wide Web. Genetic sites provide access to a rich storehouse of information on all aspects of genetics. These range from sites written in nontechnical language for the lay person to sites with sophisticated databases designed for the professional geneticist. In carrying out these exercises, you will get a taste of what the Internet can offer a student in genetics.

The keywords shown in color in the following exercises are available on the Jones and Bartlett Publishers' web site as hyperlinks to various genetic sites. To complete the exercises, visit the GeNETics on the Web home page at

http://www.jbpub.com/genetics

Select the link to Essential geNETics on the web. Then choose a chapter. You find a list of keywords that correspond to the exercises below. Select a keyword to link to a web site containing the genetic information necessary to complete the exercise. Each exercise includes a specific assignment that makes use of the information available at the site.

Exercises

- Methods for cloning and analyzing large DNA fragments have been essential in studies of certain genes in human beings that are extraordinarily large. Among these is the X-linked gene coding for the muscle protein dystrophin, defects in which cause Duchenne muscular dystrophy. This X-linked recessive disorder is among the most severe common developmental abnormalities. The incidence is about 1 per 5000 male live births. About one-third of the cases are familial, the rest due to new mutations. Use the search engine at this site with the keyword DMD, and write a description of what is known about the genetic structure of the dystrophic gene. What other type of dystrophy is associated with this locus?
- The Human Genome Project is an effort to determine the sequence of the 3 billion nucleotide bases in human DNA, to identify all the estimated 80,000 genes in the genome, and to develop methods for storage, distribution, and analysis of the data. The vastly increased availability of genetic information has important implications for individuals, their families, and society as a whole. The ethical, legal, and social implications of the project are referred to by the acronym ELSI. Use this keyword site to find more information about ELSI, particularly about the high-priority areas of focus for research and policy activities. Write a short paper identifying these high-priority areas, and give an example of an ELSI issue for each.

Pic Site

The Pic Site showcases some of the most visually appealing genetics sites on the World Wide Web. To visit the showcase genetics site, select the Pic Site for Chapter 10.

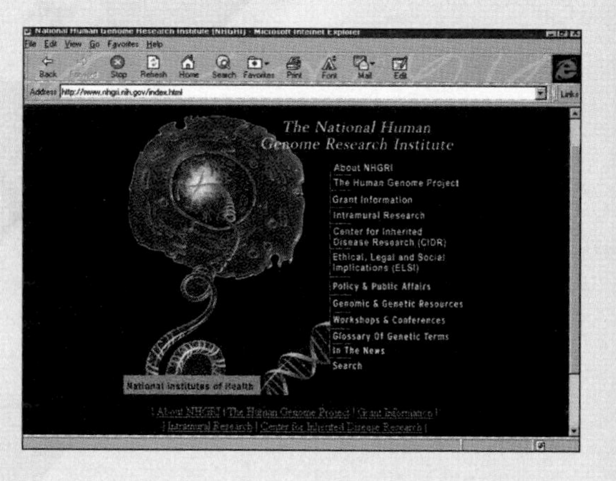

10.3 A *kan-r tet-r* plasmid is treated with the restriction enzyme *Bgl*I, which cleaves the *kan* (kanamycin) gene. The DNA is annealed with and ligated to a *Bgl*I digest of *Neurospora* DNA and then used to transform *E. coli*:

(a) What antibiotic would you put in the growth medium to ensure that each colony has the plasmid?
(b) What antibiotic-resistance phenotypes would be found among the resulting colonies?
(c) Which phenotype is expected to contain *Neurospora* DNA inserts?

10.4 You want to introduce the human insulin gene into a bacterial host in hopes of producing a large amount of human insulin. Should you use the genomic DNA or the cDNA? Explain your reasoning.

10.5 After doing a restriction digest with the enzyme *Sse*I, which has the recognition site 5'-CCTGCA ↓ GG-3' (the arrow indicates the position of the cleavage), you wish to separate the fragments in an agarose gel. In order to choose the proper concentration of agarose, you need to know the expected size of the fragments. Assuming equivalent amounts of each of the four nucleotides in the target DNA, what average size fragment would you expect?

10.6 You are given a plasmid containing part of a gene of *D. melanogaster*. The gene fragment is 303 base pairs long. You would like to amplify it using the polymerase chain reaction (PCR). You design oligonucleotide primers 19 nucleotides in length that are complementary to the plasmid sequences immediately adjacent to both ends of the cloning site. What would be the exact size of the resulting PCR product?

10.7 *Arabidopsis thaliana* has among the smallest genomes in higher plants, with a haploid genome size of about 100 Mb. If this genome is digested with *Not*I (an eight-base cutter) approximately how many DNA fragments would be produced? Assume equal and random frequencies of the four nucleotides.

10.8 How many DNA fragments would you expect if you digested the *E. coli* genome (containing 4.6 Mb) with the enzyme *Bam*HI, which has a six-base recognition sequence?

10.9 How frequently would the restriction enzymes *Taq*I (restriction site TCGA) and *Mae*III (restriction site GTNAC, in which N is any nucleotide) cleave double-stranded DNA molecules containing each of the following random sequences?

(a) 1/6 A, 1/6 T, 1/3 G, and 1/3 C
(b) 1/3 A, 1/3 T, 1/6 G, and 1/6 C

10.10 How many clones are needed to establish a library of DNA from a species of lemur with a diploid genome size of 6×10^9 base pairs if (1) fragments of average size 2×10^4 base pairs are used, and (2) one wants 99 percent of the genomic sequences to be in the library? (*Hint:* If the genome is cloned at random with *x*-fold coverage, the probability that a particular sequence will be missing is e^{-x}.)

10.11 Suppose that you digest the genomic DNA of a particular organism with *Sau*3A (↓ GATC), where the arrow represents the cleavage site. Then you ligate the resulting fragments into a unique *Bam*HI (G ↓ GATCC) cloning site of a plasmid vector. Would it be possible to isolate the cloned fragments from the vector using *Bam*HI? From what proportion of clones would it be possible?

10.12 Digestion of a DNA molecule with *Hin*dIII yields two fragments of 2.2 kb and 2.8 kb. *Eco*RI cuts the molecule, creating 1.8-kb and 3.2-kb fragments. When treated with both enzymes, the same DNA molecule produces four fragments of 0.8 kb, 1.0 kb, 1.2 kb and 2.0 kb. Draw a restriction map of this molecule.

Further Readings

Azpirozleehan, R., and K. A. Feldmann. 1997. T-DNA insertion mutagenesis in *Arabidopsis*: Going back and forth. *Trends in Genetics* 13: 152.

Bishop, J. E., and M. Waldholz. 1990. *Genome.* New York: Simon and Schuster.

Blaese, R. M. 1997. Gene therapy for cancer. *Scientific American,* June.

Botstein, D., A. Chervitz, and J. M. Cherry. 1997. Yeast as a model organism. *Science* 277: 1259.

Capecchi, M. R. 1994. Targeted gene replacement. *Scientific American,* March.

Chilton, M.-D. 1983. A vector for introducing new genes into plants. *Scientific American,* June.

Cohen, S. N. 1975. The manipulation of genes. *Scientific American,* July.

Cooke, H. 1987. Cloning in yeast: An appropriate scale for mammalian genomes. *Trends in Genetics* 3: 173.

Curtiss, R. 1976. Genetic manipulation of microorganisms: Potential benefits and hazards. *Annual Review of Microbiology* 30: 507.

Dujon, B. 1996. The yeast genome project: What did we learn? *Trends in Genetics* 12: 263.

Felgner, P. L. 1997. Nonviral strategies for gene therapy. *Scientific American,* June.

Friedmann, T. 1997. Overcoming the obstacles to gene therapy. *Scientific American,* June.

Gasser, C. S., and R. T. Fraley. 1992. Transgenic crops. *Scientific American,* June.

Gossen, J., and J. Vigg. 1993. Transgenic mice as model systems for studying gene mutations *in vivo. Trends in Genetics* 9: 27.

Havukkala, I. J. 1996. Cereal genome analysis using rice as a model. *Current Opinion in Genetics & Development* 6: 711.

Houdebine, L. M., ed. 1997. *Transgenic Animals: Generation and Use.* New York: Gordon and Breach.

Mariani, C., V. Gossele, M. De Beuckeleer, M. De Block, R. B. Goldberg, W. De Greef, and J. Leemans. 1992. A chimaeric ribonuclease-inhibitor gene restores fertility to male sterile plants. *Nature* 357: 384.

Meisler, M. H. 1992. Insertional mutation of "classical" and novel genes in transgenic mice. *Trends in Genetics* 8: 341.

Rennie, J. 1994. Grading the gene tests. *Scientific American,* June.

Sambrook, J., E. F. Fritsch, and T. Maniatis. 1989. *Molecular Cloning: A Laboratory Manual.* 2d ed. Cold Spring Harbor, NY: Cold Spring Harbor Laboratory.

Smith, D. H. 1979. Nucleotide sequence specificity of restriction enzymes. *Science* 205: 455.

Sternberg, N. L. 1992. Cloning high molecular weight DNA fragments by the bacteriophage P1 system. *Trends in Genetics* 8: 11.

Tanksley, S. D., and S. R. McCouch. 1977. Seed banks and molecular maps: Unlocking genetic potential from the wild. *Science* 277: 1063.

Watson, J. D. 1995. *Recombinant DNA.* 2d ed. New York: Freeman.

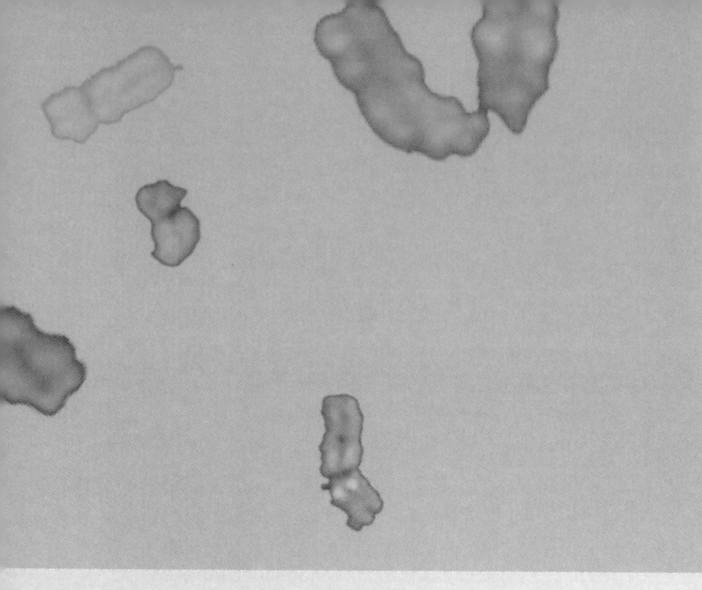

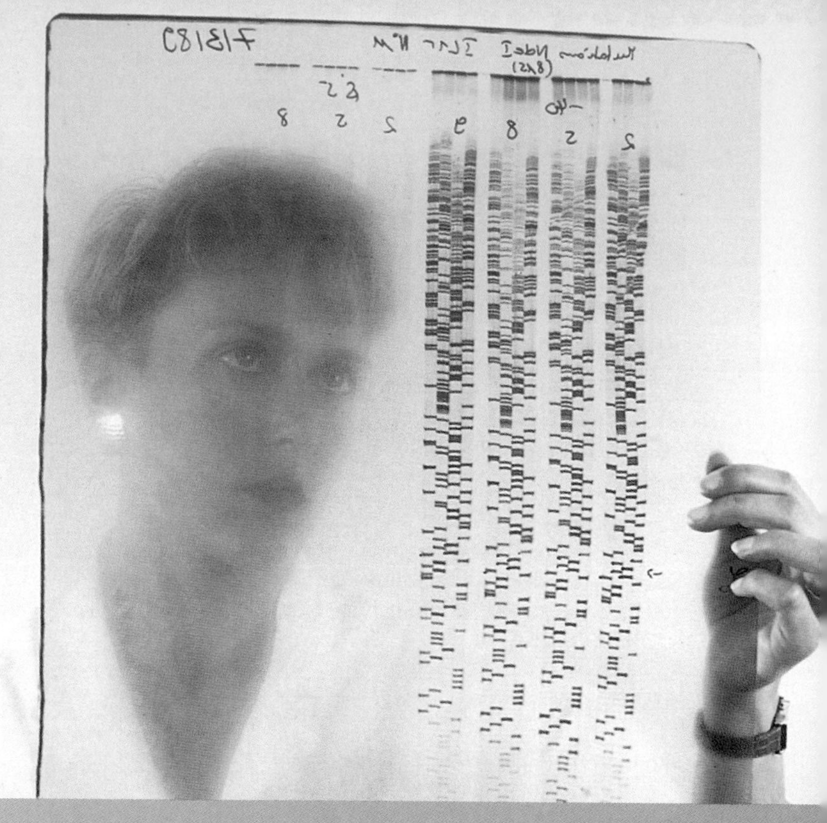

Most of the key DNA sequences involved in gene regulation have been identified. [© Jean Claude Revy/Phototake.]

Key Concepts

- Genes can be regulated at any level, including transcription, RNA processing, translation, and post-translation.
- Control of transcription is an important mechanism of gene regulation.
- Transcriptional control can be negative ("on unless turned off") or positive ("off unless turned on"); many genes include regulatory regions for both types of regulation.
- Most genes have multiple, overlapping regulatory mechanisms that operate at more than one level, from transcription through post-translation.
- In prokaryotes, the genes coding for related functions are often clustered in the genome and controlled jointly by a regulatory protein that binds with an "operator" region at one end of the cluster. This type of gene organization is known as an operon.
- In eukaryotes, genes are not organized into operons. Genes at dispersed locations in the genome are coordinately controlled by one or more "enhancer" DNA sequences located near each gene that interact with transcriptional activator proteins to allow gene expression.

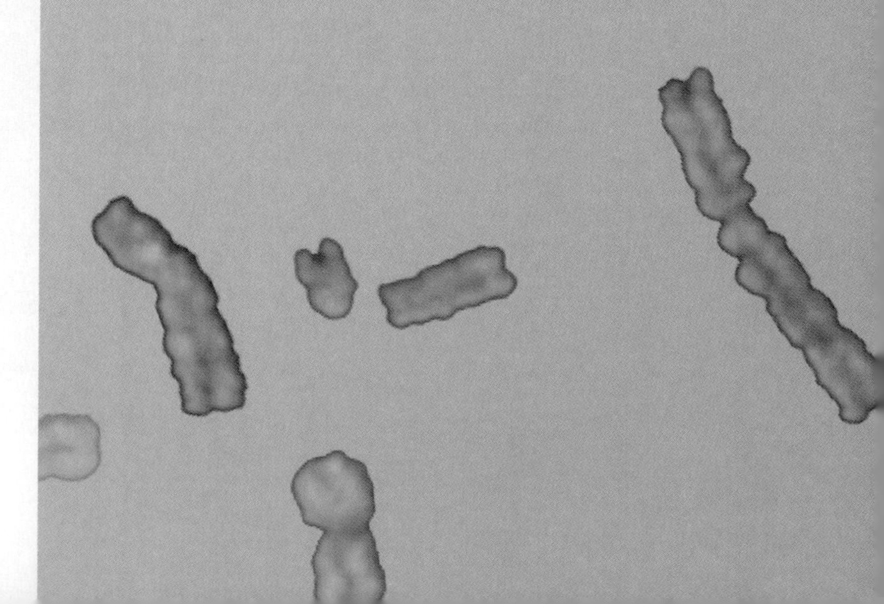

CHAPTER

11

The Regulation of Gene Activity

Not all genes are expressed continuously. The level of gene expression may differ from one cell type to the next or according to stage in the cell cycle. For example, the genes for hemoglobin are expressed at high levels only in precursors of the red blood cells. The activity of a gene also varies according to the function of the cell. A complex vertebrate animal contains approximately 200 different types of cells with specialized functions. With minor exceptions, all cell types contain the same genetic complement. The cell types differ only in which genes are active. In general, the synthesis of particular gene products is controlled by mechanisms collectively called **gene regulation.**

11.1 There are many possible control points for the regulation of gene expression.

In many cases, gene activity is regulated at the level of transcription, either through signals originating within the cell itself or in response to external conditions. For example, many gene products are needed only on occasion, and transcription can be regulated in an on-off manner that enables the gene products to be present only when required by external conditions. The flow of genetic information is regulated in other ways also. Important control points for gene expression include the following:

1. *DNA rearrangements,* in which gene expression changes depending on the position of DNA sequences in the genome
2. *Transcriptional regulation* of the synthesis of RNA transcripts by controlling initiation or termination
3. *RNA processing,* or regulation through RNA splicing or alternative patterns of splicing
4. *Translational control* of polypeptide synthesis
5. *Stability of mRNA,* because mRNAs that persist in the cell have longer-lasting effects than those that are degraded rapidly

The regulatory systems of prokaryotes and eukaryotes are somewhat different from each other. Prokaryotes are generally free-living unicellular organisms that grow and divide indefinitely, as long as environmental conditions are suitable and the supply of nutrients is adequate. Their regulatory systems are geared to provide the maximum growth rate in a particular environment, except when such growth would be detrimental. Prokaryotes can also use the coupling between transcription and translation (Chapter 9) for regulation, but the absence of introns eliminates RNA splicing as a possible control point.

The regulatory requirements of multicellular eukaryotes are different from those of prokaryotes. In a developing organism, not only must a cell grow and divide, but the progeny cells must also undergo considerable changes in morphology and biochemistry, and each progeny cell must maintain its altered state. Furthermore, during embryonic development, most eukaryotic cells are challenged less by the environment than are bacteria, because the composition and concentration of the "growth medium" inside an embryo does not change drastically with time. Finally, in an adult organism, growth and cell division in most cell types have stopped, and each cell needs only to maintain itself and its specialized characteristics.

11.2 Regulation of transcription is a common mechanism in prokaryotes.

In bacteria and phages, on-off gene activity is often controlled through transcriptional regulation; that is, synthesis of a particular mRNA takes place when the gene product is needed and inhibited when the product is not needed. In discussing transcription, we use the term *off* only for convenience: It should be kept in mind that *off* usually means "very low." In bacteria, few examples are known of a system being completely switched off. When transcription is in the "off" state, a basal level of gene expression almost always remains, often averaging one transcriptional event or fewer per cell generation. The "off" state therefore usually allows some synthesis of the gene product, but in very limited amounts. Extremely low levels of expression are also found in certain classes of genes in eukaryotes, including many genes that participate in embryonic development. Regulatory mechanisms other than the on-off type also are known in both prokaryotes and eukaryotes. For example, the activity of a system may be modulated from fully on to partly on, rather than to off.

In bacterial systems, when several enzymes act in sequence in a single metabolic pathway, usually either all or none of these enzymes are produced. This phenomenon is called **coordinate regulation,** and it usually results from control of the synthesis of a single mRNA molecule that includes the coding sequences of all of the enzymes participating in the metabolic pathway. A single mRNA that codes for multiple polypeptides is called a **polycistronic mRNA** because the term *cistron* was once widely used to describe the coding sequence for a single polypeptide. Coordinate regulation of this type is not found in eukaryotes because eukaryotic mRNA typically code for only one polypeptide product, as discussed in Chapter 9.

Several mechanisms of regulation of transcription are common. The particular one used often depends on whether the enzymes being regulated act in degradative or biosynthetic metabolic pathways. For example, in a multistep degradative (catabolic) system, the availability of the molecule to be degraded helps determine whether the enzymes in the pathway will be synthesized. In the presence of the molecule, the enzymes of the degradative (catabolic) pathway are synthesized; in its absence, they are not. Such a system, in which the presence of a small molecule results in enzyme synthesis, is said to be **inducible.** The small molecule is called the **inducer.** The opposite situation is often found in the control of the synthesis of enzymes that participate in biosynthetic (anabolic) pathways; in these cases, the final product of the pathway is frequently the regulatory molecule. In the presence of the final product, the enzymes of the biosynthetic pathway are not synthesized; in its absence, they are synthesized. Such a system, in which the presence of a small molecule results in failure to synthesize enzymes, is said to be **repressible.** The small molecule that participates in the regulation is called the **co-repressor.**

The molecular mechanisms for each of the regulatory patterns vary quite widely, but they usually fall into one of two major categories called **negative regulation** and **positive regulation.** In a negatively regulated system (Figure 11.1A), a **repressor** protein present in the cell prevents transcription. In an inducible system that is negatively regulated, the repressor protein acts by itself to prevent transcription. The inducer antagonizes the repressor, allowing the initiation of transcription. In a repressible system, an **aporepressor** protein combines with the co-repressor molecule to form the functional repressor, which prevents transcription. In the absence of the co-repressor, the aporepressor is unable to prevent transcription. On the other hand, in a positively regulated system (Figure 11.1B), mRNA synthesis takes place only if a regulatory protein binds to a region of the gene that activates transcription. Such a protein is usually referred to as a **transcriptional activator.** Negative and positive regulation are not mutually exclusive, and some systems are both positively and negatively regulated, utilizing two regulators to respond to different conditions in the cell. Negative regulation is more common in prokaryotes, positive regulation in eukaryotes.

A degradative system may be regulated either positively or negatively. In a biosynthetic pathway, the final product usually negatively regulates its own synthesis; in the simplest type of negative regulation, absence of the product increases its synthesis (through production of the necessary enzymes), and presence of the product decreases its synthesis

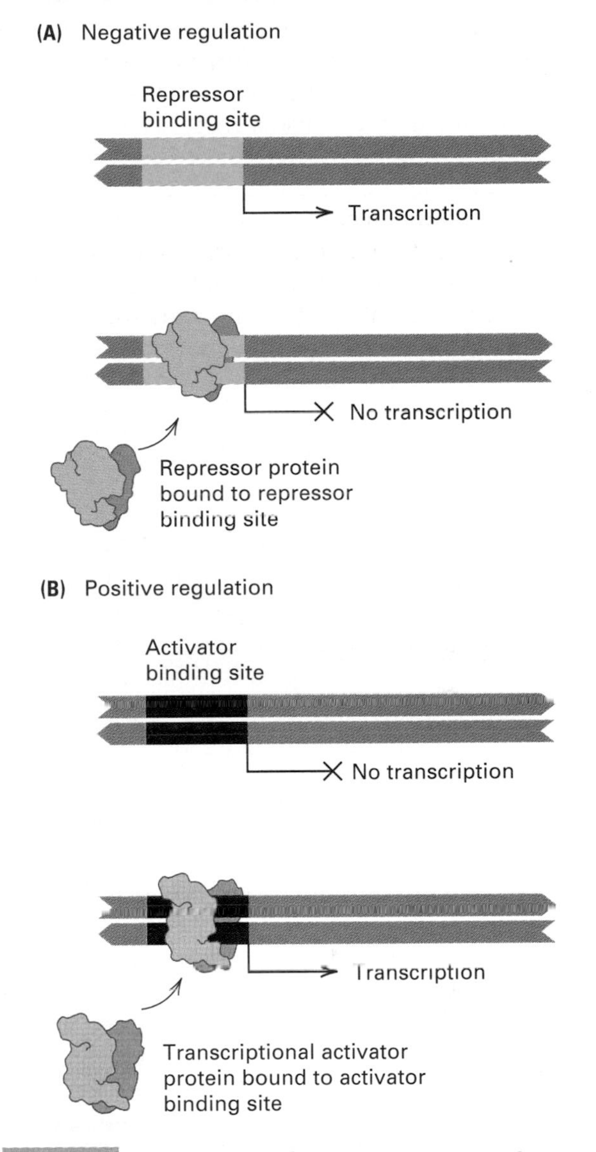

Figure 11.1 The distinction between negative and positive regulation. (A) In negative regulation, the "default" state of the gene is one in which transcription takes place. The binding of a repressor protein to the DNA molecule prevents transcription. (B) In positive regulation, the default state is one in which transcription does not take place. The binding of a transcriptional activator protein stimulates transcription. A single genetic element may be regulated both positively and negatively; in such a case, transcription requires the binding of the transcriptional activator and the absence of repressor binding.

(through repression of enzyme synthesis). Even in a system in which a single protein molecule (not necessarily an enzyme), is translated from a monocistronic mRNA molecule, the protein may be **autoregulated,** which means that the protein regulates its own transcription. In negative autoregulation, the protein inhibits transcription, and high concentrations of the protein result in less transcription of the mRNA that codes for the protein. In positive autoregulation, the protein stimulates transcription: As more protein is made, transcription increases to the maximum rate. Positive autoregulation is a common way for weak induction to be amplified. Only a weak signal is necessary to get production of the protein started, but then the positive autoregulation stimulates the production to the maximum level.

The next three sections are concerned with several systems of regulation in prokaryotes. They serve as an introduction to the remainder of the chapter, which deals with regulation in eukaryotes.

11.3 Lactose degradation is regulated by the lactose operon.

Metabolic regulation was first studied in detail in the system in *E. coli* that is responsible for degradation of the sugar lactose, and most of the terminology used to describe regulation has come from genetic analysis of this system.

The first regulatory mutations that were discovered affected lactose metabolism.

In *E. coli,* two proteins are necessary for the metabolism of lactose: the enzyme **β-galactosidase,** which cleaves lactose (a β-galactoside sugar) to yield galactose and glucose, and a transporter molecule, **lactose permease,** which is required for the entry of lactose into the cell. The existence of two different proteins in the lactose-utilization system was first shown by a combination of genetic experiments and biochemical analysis.

First, hundreds of mutants unable to use lactose as a carbon source, designated Lac$^-$ mutants, were isolated. Some of the mutations were in the *E. coli* chromosome, and others were in an F' *lac,* a plasmid carrying the genes for lactose utilization. By performing F' × F$^-$ matings, investigators constructed partial diploids with the genotypes F' *lac*$^-$ / *lac*$^+$ or F' *lac*$^+$ / *lac*$^-$. (The genotype of the plasmid is given to the left of the slash and that of the chromosome to the right.) It was observed that all of these partial diploids always had a Lac$^+$ phenotype (that is, they made both β-galactosidase and permease). In these experiments, none of the mutants produced an inhibitor that prevented functioning of the *lac*$^+$ genes in either the F' *lac*$^+$ or the chromosomal *lac*$^+$. Other partial diploids were then constructed in which both the F' *lac* plasmid and the chromosome carried a *lac*$^-$ allele. When these were tested for the Lac$^+$ phenotype, it was found that all of the mutants initially isolated could be placed into two complementation groups, called *lacZ* and *lacY,* a result that implies that the *lac* system consists of at least two genes. Complementation is indicated by the observation that the partial diploids F' *lacY*$^-$ *lacZ*$^+$ / *lacY*$^+$ *lacZ*$^-$ and F' *lacY*$^+$ *lacZ*$^-$ / *lacY*$^-$ *lacZ*$^+$ had a Lac$^+$ phenotype, producing both β-galactosidase and permease. However, the genotypes F' *lacY*$^-$ *lacZ*$^+$ / *lacY*$^-$ *lacZ*$^+$ and F' *lacY*$^+$ *lacZ*$^-$ / *lacY*$^+$ *lacZ*$^-$ had the Lac$^-$ phenotype; they were unable to synthesize permease and β-galactosidase, respectively. Hence the *lacZ* gene codes for β-galactosidase and the *lacY* gene for permease. (A third gene that participates in lactose metabolism was later discovered; it was not included among the early mutants because it is not essential for growth on lactose.) A final important result—that the *lacY* and *lacZ* genes are adjacent—was deduced from a high frequency of cotransduction observed in genetic mapping experiments.

Lactose-utilizing enzymes can be inducible (regulated) or constitutive.

The on-off nature of the lactose-utilization system is evident in the following observations:

- If a culture of Lac$^+$ *E. coli* is grown in a medium lacking lactose or any other β-galactoside, the intracellular concentrations of β-galactosidase and permease are exceedingly low—roughly one or two molecules per bacterial cell. However, if lactose is present in the growth medium, the number of each of these molecules is about 10^3-fold higher.
- If lactose is added to a Lac$^+$ culture growing in a lactose-free medium (also lacking glucose, a point we will discuss shortly), both β-galactosidase and permease are synthesized nearly simultaneously, as shown in Figure 11.2. Analysis of the total mRNA present in the cells before and after the addition of lactose shows that almost no *lac* mRNA (the polycistronic mRNA that codes for β-galactosidase and permease) is present before lactose is added and that the addition of lactose triggers synthesis of the *lac* mRNA.

These two observations led to the view that transcription of the lactose genes is **inducible transcription** and that lactose is an *inducer* of transcription. Some analogs of lactose are also inducers, such as a sulfur-containing analog denoted IPTG (isopropylthiogalactoside), which is convenient for experiments because it induces, but is not cleaved by, β-galactosidase. In the presence of IPTG in the medium, the inducer is taken up by the cells and maintained at a constant level, whether or not the β-galactosidase enzyme is present.

Mutants were also isolated in which *lac* mRNA was synthesized (and hence β-galactosidase and

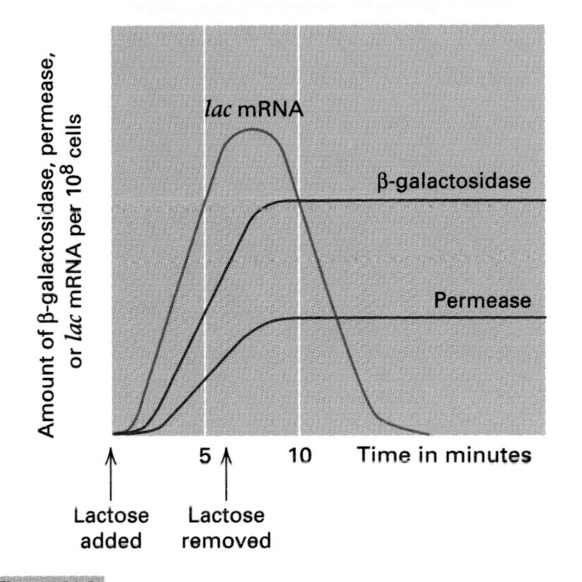

Figure 11.2 The "on-off" nature of the *lac* system. The *lac* mRNA appears soon after lactose or another inducer is added; *β*-galactosidase and permease appear at nearly the same time but are delayed with respect to mRNA synthesis because of the time required for translation. When lactose is removed, no more *lac* mRNA is made, and the amount of *lac* mRNA decreases because of the degradation of mRNA already present. Both *β*-galactosidase and permease are stable proteins: their amounts remain constant even when synthesis ceases. However, their concentration per cell gradually decreases as a result of repeated cell divisions.

permease produced) in both the presence and the *absence* of an inducer. The mutants that eliminated regulation provided the key to understanding induction; because of their constant synthesis, the mutants were termed **constitutive.** Mutants were also obtained that failed to produce *lac* mRNA (and hence *β*-galactosidase and permease) even when the inducer was present. These uninducible mutants fell into two classes, $lacI^s$ and $lacP^-$. The characteristics of the mutants are shown in Table 11.1 and discussed in the following sections.

Repressor shuts off messenger RNA synthesis.

In Table 11.1, genotypes 3 and 4 show that $lacI^-$ mutations are recessive. In the absence of inducer, a $lacI^+$ cell does not make *lac* mRNA, whereas the mRNA is made in a $lacI^-$ mutant. These results suggest that

> The *lacI* gene is a regulatory gene whose product is the repressor protein that keeps the system turned off. Because the repressor is necessary to shut off mRNA synthesis, regulation by the repressor is negative regulation.

A $lacI^-$ mutant lacks the repressor and hence is constitutive. Wildtype copies of the repressor are present in a $lacI^+$ / $lacI^-$ partial diploid, so transcription is repressed. It is important to note that the single $lacI^+$ gene prevents synthesis of *lac* mRNA from both the F' plasmid and the chromosome. Therefore, the repressor protein must be diffusible within the cell to shut off mRNA synthesis from both DNA molecules present in a partial diploid.

On the other hand, genotypes 7 and 8 indicate that the $lacI^s$ mutations are dominant and act to shut off mRNA synthesis from both the F' plasmid and the chromosome, whether or not the inducer is present (the superscript in $lacI^s$ signifies *super-repressor.*) The $lacI^s$ mutations result in repressor molecules that fail to recognize and bind the inducer and thus permanently shut off *lac* mRNA synthesis. Genetic mapping experiments placed the *lacI* gene adjacent to the *lacZ* gene and established the gene order *lacI lacZ lacY.* How the *lacI* repressor prevents synthesis of *lac* mRNA will be explained shortly.

The lactose operator is an essential site for repression.

Entries 1 and 2 in Table 11.1 show that $lacO^c$ mutants are dominant. However, the dominance is evident only in certain combinations of *lac* mutations, as can be seen by examining the partial diploids shown in entries 5 and 6. Both combinations are Lac^+ because a functional *lacZ* gene is present. However, in the combination shown in entry 5, synthesis of *β*-galactosidase is inducible even though a $lacO^c$ mutation is present. The difference between the two combinations in entries 5 and 6 is

Table 11.1
Characteristics of partial diploids containing several combinations of *lacI, lacO,* and *lacP* alleles

Genotype	Synthesis of *lac* mRNA	Lac phenotype
1. F' $lacO^c\ lacZ^+/lacO^+\ lacZ^+$	Constitutive	+
2. F' $lacO^+\ lacZ^+/lacO^c\ lacZ^+$	Constitutive	+
3. F' $lacI^-lacZ^+/lacI^+\ lacZ^+$	Inducible	+
4. F' $lacI^+\ lacZ^+/lacI^-\ lacZ^+$	Inducible	+
5. F' $lacO^c\ lacZ^-/lacO^+\ lacZ^+$	Inducible	+
6. F' $lacO^c\ lacZ^+/lacO^+\ lacZ^-$	Constitutive	+
7. F' $lacI^s\ lacZ^+/lacI^+\ lacZ^+$	Uninducible	−
8. F' $lacI^+\ lacZ^+/lacI^s\ lacZ^+$	Uninducible	−
9. F' $lacP^-\ lacZ^+/lacP^+\ lacZ^+$	Inducible	+
10. F' $lacP^+\ lacZ^+/lacP^-\ lacZ^+$	Inducible	+
11. F' $lacP^+\ lacZ^-/lacP^-\ lacZ^+$	Uninducible	−
12. F' $lacP^+\ lacZ^+/lacP^-\ lacZ^-$	Inducible	+

that in entry 5, the *lacO*^{c} mutation is present in the same DNA molecule as the *lacZ*$^-$ mutation, whereas in entry 6, *lacO*^{c} is contained in the same DNA molecule as *lacZ*$^+$. The key feature of these results is that

> A *lacO*^{c} mutation causes constitutive synthesis of β-galactosidase only when the *lacO*^{c} and *lacZ*$^+$ alleles are contained in the same DNA molecule.

The *lacO*^{c} mutation is said to be ***cis*-dominant,** because only genes in the *cis* configuration (in the same DNA molecule as that containing the mutation) are expressed in dominant fashion. Confirmation of this conclusion comes from an important biochemical observation: The mutant enzyme coded by the *lacZ*$^-$ sequence is synthesized constitutively in a *lacO*^{c} *lacZ*$^-$ / *lacO*$^+$ *lacZ*$^+$ partial diploid (entry 5), whereas the wildtype enzyme (coded by the *lacZ*$^+$ sequence) is synthesized only if an inducer is added. All *lacO*^{c} mutations are located between the *lacI* and *lacZ* genes; hence the gene order of the four genetic elements of the *lac* system is

lacI *lacO* *lacZ* *lacY*

An important feature of all *lacO*^{c} mutations is that they cannot be complemented (a characteristic feature of all *cis*-dominant mutations); that is, a *lacO*$^+$ allele cannot alter the constitutive activity of a *lacO*^{c} mutation. This observation implies that the *lacO* region does not encode a diffusible product and must instead define a site in the DNA that determines whether synthesis of the product of the adjacent *lacZ* gene is inducible or constitutive. The *lacO* region is called the **operator.** In a subsequent section, we will see that the operator is in fact a *binding site* in the DNA for the repressor protein.

The lactose promoter is an essential site for transcription.

Entries 11 and 12 in Table 11.1 show that *lacP*$^-$ mutations, like *lacO*^{c} mutations, are *cis*-dominant. The *cis*-dominance can be seen in the partial diploid in entry 11. The genotype in entry 11 is uninducible, in contrast to the partial diploid of entry 12, which is inducible. The difference between the two genotypes is that in entry 11, the *lacP*$^-$ mutation is in the same DNA molecule with *lacZ*$^+$, whereas in entry 12, the *lacP*$^-$ mutation is combined with *lacZ*$^-$. This observation means that a wildtype *lacZ*$^+$ remains inexpressible in the presence of *lacP*$^-$; no *lac* mRNA is transcribed from that DNA molecule. The *lacP*$^-$ mutations map between *lacI* and *lacO*, and the order of the five genetic elements of the *lac* system is

lacI *lacP* *lacO* *lacZ* *lacY*

As expected because of the *cis*-dominance of *lacP*$^-$ mutations, they cannot be complemented; that is, a *lacP*$^+$ allele on another DNA molecule cannot supply the missing function to a DNA molecule carrying a *lacP*$^-$ mutation. Thus *lacP,* like *lacO,* must define a site that determines whether synthesis of *lac* mRNA will take place. Because synthesis does not occur if the site is defective or missing, *lacP* defines an essential site for mRNA synthesis. The *lacP* region is called the **promoter.** It is a site at which RNA polymerase binding takes place to allow initiation of transcription.

The lactose operon contains linked structural genes and regulatory sequences.

The genetic regulatory mechanism of the *lac* system was first explained by the *operon model* of François Jacob and Jacques Monod, which is illustrated in Figure 11.3. (The figure uses the alternative abbreviations *i, o, p, z, y,* and *a* for *lacI, lacO, lacP, lacZ, lacY,* and *lacA.*) The operon model has the following features:

1. The lactose-utilization system consists of two kinds of components: *structural genes* (*lacZ* and *lacY*), which encode proteins needed for the transport and metabolism of lactose, and *regulatory elements* (the repressor gene *lacI,* the promoter *lacP,* and the operator *lacO*).
2. The products of the *lacZ* and *lacY* genes are coded by a single polycistronic mRNA molecule. The linked structural genes, together with *lacP* and *lacO,* constitute the *lac* **operon.** (The third protein, encoded by *lacA,* is also translated from the polycistronic mRNA. This protein is the enzyme β-galactoside transacetylase; it is used in the metabolism of certain β-galactosides other than lactose and will not be of further concern here.)
3. The promoter mutations (*lacP*$^-$) eliminate the ability to synthesize *lac* mRNA.
4. The product of the *lacI* gene is a repressor, which binds to a unique sequence of DNA bases constituting the operator.
5. When the repressor is bound to the operator, initiation of transcription of *lac* mRNA by RNA polymerase is prevented.
6. Inducers stimulate mRNA synthesis by binding to and inactivating the repressor. In the presence of an inducer, the operator is not bound

with the repressor, and the promoter is available for the initiation of mRNA synthesis.

Note that regulation of the operon requires that the *lacO* operator either overlap or be adjacent to the promoter of the structural genes, because binding with the repressor prevents transcription. Proximity of *lacI* to *lacO* is not strictly necessary, because the *lacI* repressor is a soluble protein and is therefore diffusible throughout the cell. The presence of inducer has a profound effect on the DNA binding properties of the repressor; the inducer-repressor complex has an affinity for the operator that is approximately 10^3 smaller than that of the repressor alone.

The ratio of the numbers of copies of β-galactosidase, permease, and transacetylase is 1.0 : 0.5 : 0.2 when the operon is induced. These differences are partly due to the order of the genes in the mRNA. Downstream cistrons are less likely to be translated because of failure of reinitiation when an upstream cistron has finished translation.

The operon model is supported by a wealth of experimental data and explains many of the features of the *lac* system, as well as numerous other negatively regulated genetic systems in prokaryotes. One aspect of the regulation of the *lac* operon—the effect of glucose—has not yet been discussed. Examination of this feature indicates that the *lac* operon is also subject to positive regulation, as we will see in the next section.

The lactose operon is also subject to positive regulation.

The function of β-galactosidase in lactose metabolism is to form glucose by cleaving lactose. (The other cleavage product, galactose, also is ultimately converted into a glucose derivative by the enzymes of the galactose operon.) If both glucose and lactose are present in the growth medium, activity of the *lac* operon is not needed. In fact, in the presence of glucose, no β-galactosidase is formed until virtually all of the glucose in the medium has been consumed. The lack of synthesis of β-galactosidase is a result of the lack of synthesis of *lac* mRNA. No *lac* mRNA is made in the presence of glucose, because in addition to an inducer to inactivate the *lacI* repressor, another element is needed for initiating *lac* mRNA synthesis; the activity of this element is regulated by the concentration of glucose.

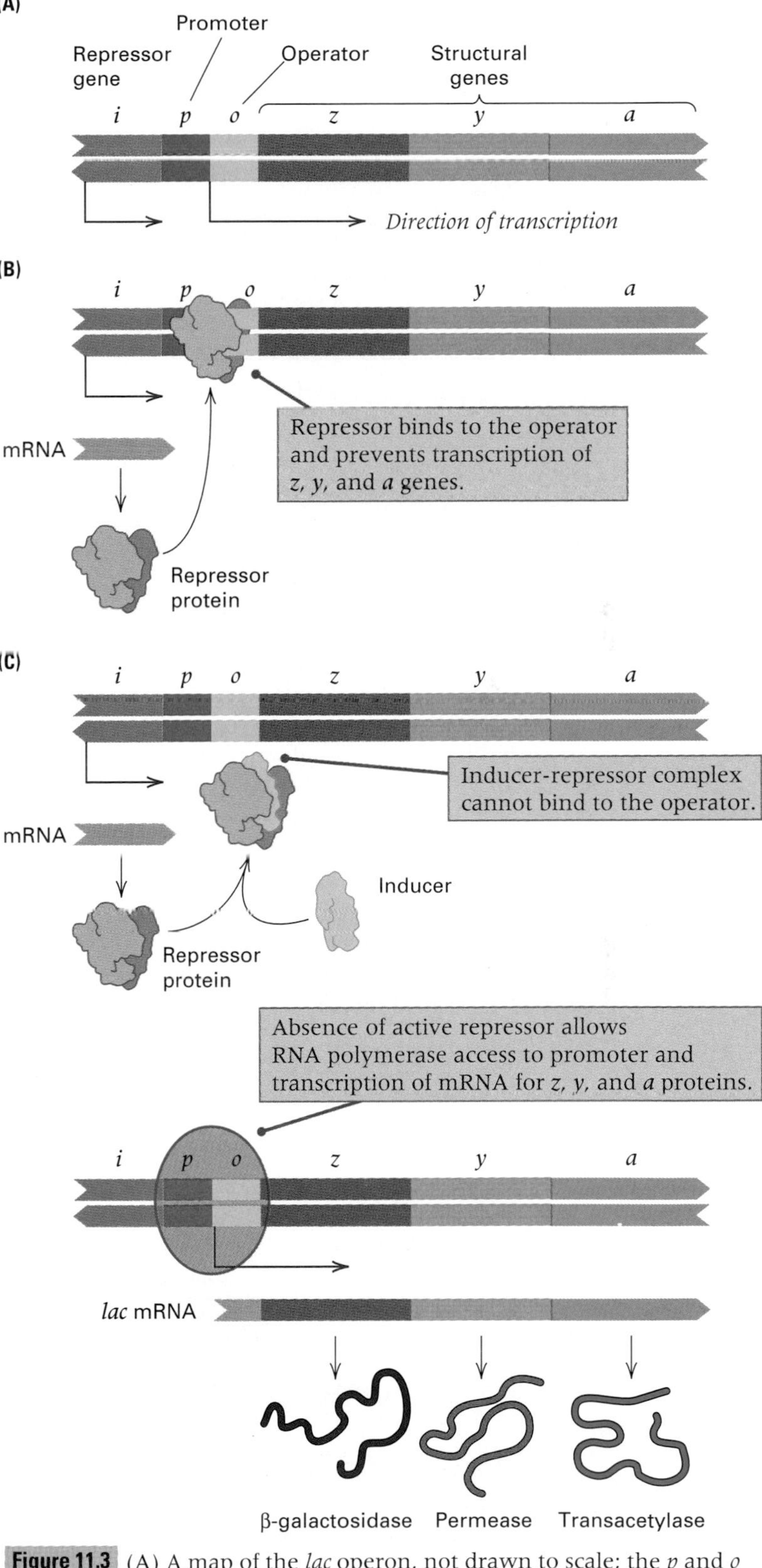

Figure 11.3 (A) A map of the *lac* operon, not drawn to scale; the *p* and *o* sites are actually much smaller than the other genes and together comprise only 83 base pairs. (B) A diagram of the *lac* operon in the repressed state. (C) A diagram of the *lac* operon in the induced state. The inducer alters the shape of the repressor so that the repressor can no longer bind to the operator. The common abbreviations *i, p, o, z, y,* and *a* are used instead of *lacI, lacO,* and so forth. The *lacA* gene is not essential for lactose utilization.

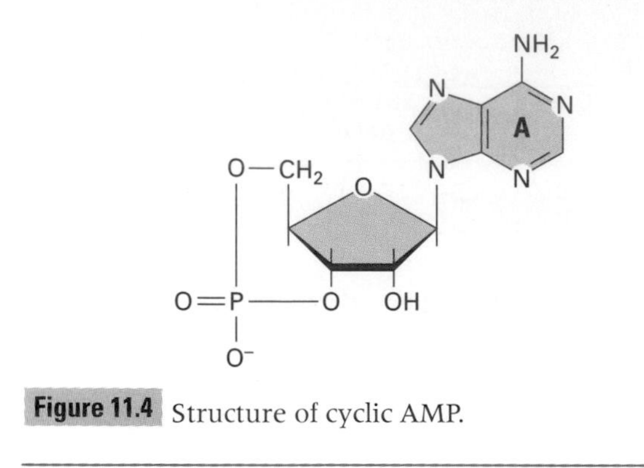

Figure 11.4 Structure of cyclic AMP.

The inhibitory effect of glucose on expression of the *lac* operon is indirect. The small molecule *cyclic adenosine monophosphate* (cAMP), shown in Figure 11.4, is widely distributed in animal tissues and in multicellular eukaryotic organisms, in which it is important in mediating the action of many hormones. It is also present in *E. coli* and many other bacteria, where it has a different function. Cyclic AMP is synthesized by the enzyme *adenyl cyclase,* and the concentration of cAMP is regulated indirectly by glucose metabolism. When bacteria are growing in a medium that contains glucose, the cAMP concentration in the cells is quite low. In a medium containing glycerol or any carbon source that requires aerobic metabolism for degradation, or when the bacteria are otherwise starved of an energy source, the cAMP concentration is high (Table 11.2). Glucose levels help regulate the cAMP concentration in the cell, and *cAMP regulates the activity of the lac operon* (as well as that of several other operons that control degradative metabolic pathways).

Table 11.2 Concentration of cyclic AMP in cells growing in media with the indicated carbon sources

Carbon source	cAMP concentration
Glucose	Low
Glycerol	High
Lactose	High
Lactose + glucose	Low
Lactose + glycerol	High

E. coli and many other bacterial species contain a protein called the *cyclic AMP receptor protein* (CRP), which is encoded by a gene called *crp*. Mutations of either the *crp* or the adenyl cyclase gene prevent synthesis of *lac* mRNA, which indicates that both CRP function and cAMP are required for *lac* mRNA synthesis. CRP and cAMP bind to one another, forming a complex denoted **cAMP-CRP,** which is an active regulatory element in the *lac* system. The requirement for cAMP-CRP is independent of the *lacI* repression system, because *crp* and adenyl cyclase mutants are unable to make *lac* mRNA even if a $lacI^-$ or a $lacO^c$ mutation is present. The reason is that for transcription to occur, the cAMP-CRP complex must be bound to a base sequence in the DNA in the promoter region (Figure 11.5). Unlike the repressor, which is a *negative* regulator, the cAMP-CRP complex is a *positive* regulator. The positive and negative regulatory systems of the *lac* operon are independent of each other.

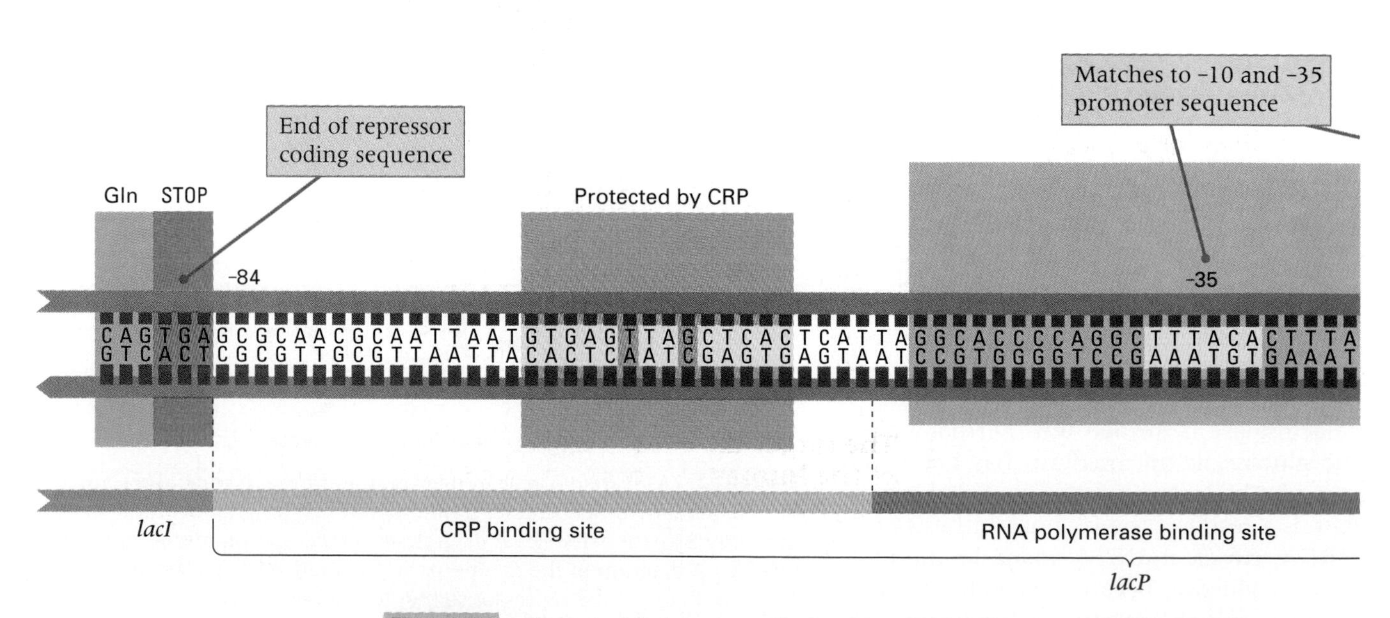

Figure 11.6 (above and facing page) The base sequence of the control region of the *lac* operon. Sequences protected from DNase digestion by binding of the stipulated proteins are indicated in the upper part. The end of the *lacI* gene is shown at the extreme left; the

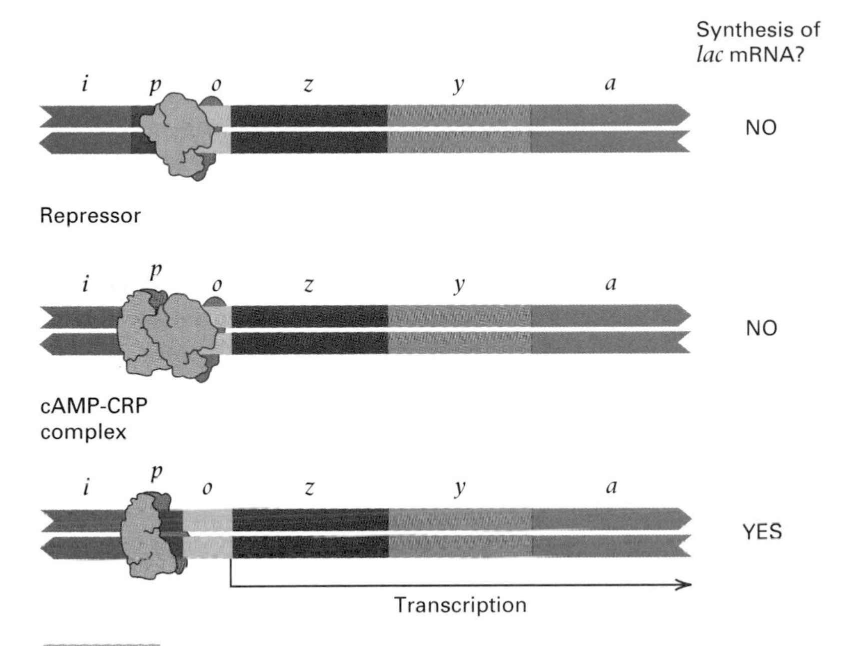

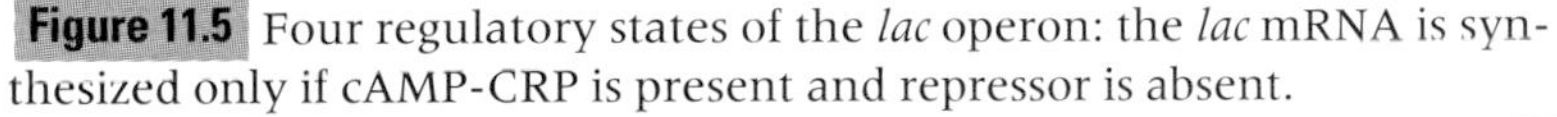
Figure 11.5 Four regulatory states of the *lac* operon: the *lac* mRNA is synthesized only if cAMP-CRP is present and repressor is absent.

Experiments carried out *in vitro* with purified *lac* DNA, *lac* repressor, cAMP-CRP, and RNA polymerase have established two further points:

1. In the absence of the cAMP-CRP complex, RNA polymerase binds only weakly to the promoter, but its binding is stimulated when cAMP-CRP is also bound to the DNA. The weak binding rarely leads to initiation of transcription, because the correct interaction between RNA polymerase and the promoter does not occur.
2. If the repressor is bound to the operator, then RNA polymerase cannot stably bind to the promoter.

These results explain how lactose and glucose function together to regulate transcription of the *lac* operon. The relationship of these elements to one another, to the start of transcription, and to the base sequence in the region is depicted in Figure 11.6.

A great deal is also known about the three-dimensional structure of the regulatory states of the *lac* operon. Figure 11.7 shows that there is actually a 93-base-pair loop of DNA that forms in the operator region when it is in contact with the repressor. This loop corresponds to the *lac* operon region −82 to +11 (numbered as in Figure 11.6). The DNA region in red corresponds, on the right-hand side, to the operator region centered at +11 and, on the left-hand side, to a second repressor-binding site immediately upstream and adjacent to the CRP binding site. The *lac* repressor tetramer (violet) is shown bound to these sites. The DNA loop is formed by the region between the repressor binding sites and includes, in medium blue, the CRP binding site, to which the CRP protein (dark blue) is shown bound. The DNA regions in green are the −10 and −35 sites in the *lacP* promoter indicated in Figure 11.6. In this configuration, the *lac* operon is not transcribed. Removal of the repressor opens up the loop and allows transcription to occur.

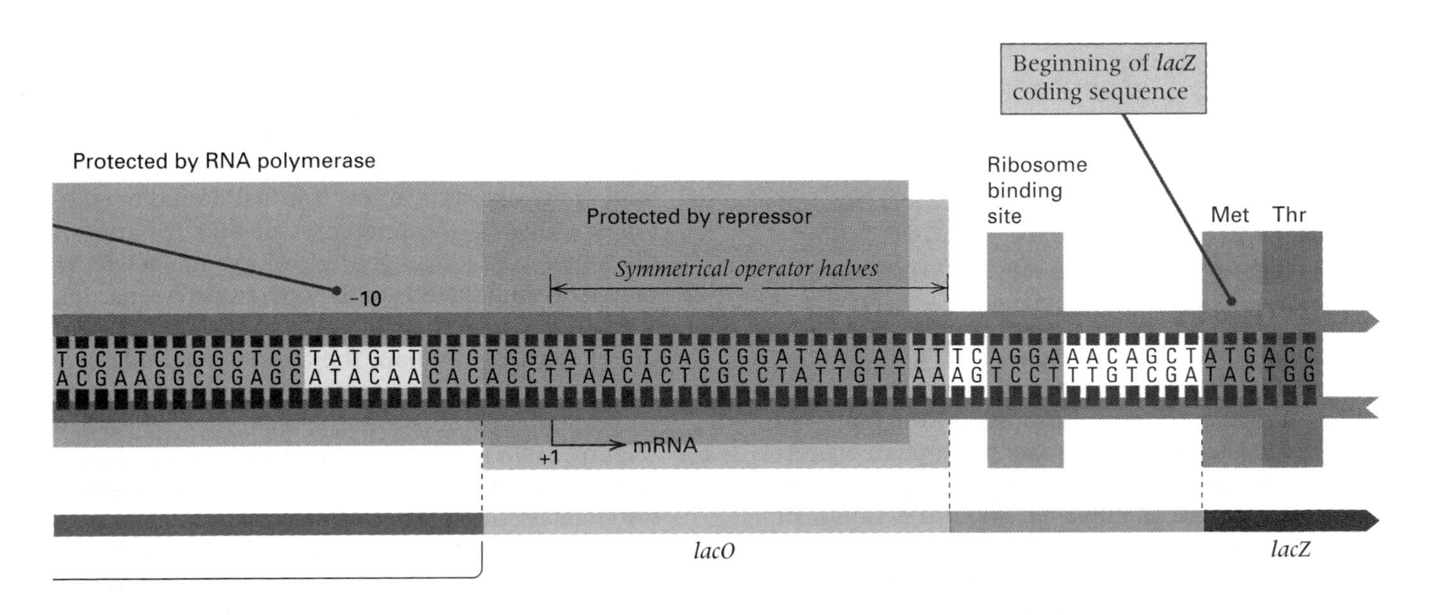

ribosome binding site is the site at which the ribosome binds to the *lac* mRNA. The consensus sites for CRP binding and for RNA polymerase promoter binding are indicated along the bottom.

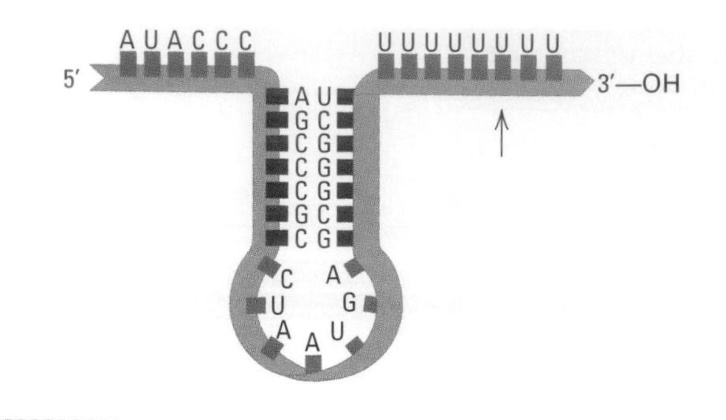

Figure 11.10 The terminal region of the *trp* attenuator sequence. The arrow indicates the final uridine in attenuated RNA. Nonattenuated RNA continues past that base. The bases with red tabs form the hypothetical stem sequence that is shown.

of termination is an RNA molecule containing only 140 nucleotides that stops short of the genes coding for the *trp* enzymes. The 28-base region in which termination takes place is called the **attenuator.** The base sequence of this region (Figure 11.10) contains the usual features of a termination site, including a potential stem-and-loop configuration in the mRNA followed by a sequence of eight uridines.

The leader region whose translation terminates transcription contains several notable features (Figure 11.11):

1. An AUG codon and a downstream UGA stop codon in the same reading frame defining a region coding for a polypeptide consisting of 14 amino acids; this is called the **leader polypeptide.**
2. Two adjacent tryptophan codons that are located in the leader polypeptide at positions 10 and 11. We will see the significance of these repeated codons shortly.
3. Four segments of the leader RNA (denoted in Figure 11.12 as regions 1, 2, 3, and 4) that are capable of base-pairing with each other. In one configuration, region 1 pairs with region 2, and region 3 with region 4. The details of this configuration are shown in Figure 11.12. When pairing takes place in this configuration, transcription is terminated at the run of uridines preceding nucleotide 140. This type of pairing occurs in purified *trp* leader mRNA.
4. An alternative type of pairing can also take place, in which region 2 pairs with region 3. It is this alternative mode of base pairing that allows transcription to proceed through the rest of the operon.

Whether transcription is terminated depends on whether the leader peptide is translated. In other words, *translation* controls *transcription.* The specific mechanism is shown in Figure 11.13. As the leader region is transcribed, translation of the leader polypeptide is initiated. Because there are two tryptophan codons in the coding sequence, translation of the sequence is sensitive to the concentration of charged $tRNA^{Trp}$. If the supply of tryptophan is adequate for translation, the ribosome passes through the Trp codons and into region 2 (Figure 11.13B). Because the presence of a ribosome eliminates the possibility of base pairing in a region of about 10 bases on each side of the codons being translated, the presence of a ribosome in region 2 prevents its becoming paired with region 3. In this case, region 3 pairs with region 4 and forms the terminator shown in Figure 11.13B (and in detail in Figure 11.12B), and transcription is terminated at the run of uridines that follows region 4.

On the other hand, when the level of charged $tRNA^{Trp}$ is insufficient to support translation, the translation of the leader peptide is stalled at the tryptophan codons (Figure 11.13C). The stalling prevents the ribosome from proceeding into region 2, which is then free to pair with region 3. Pairing of regions 2 and 3 prevents formation of the terminator structure, so the complete *trp* mRNA molecule is made, including the coding sequences for the structural genes.

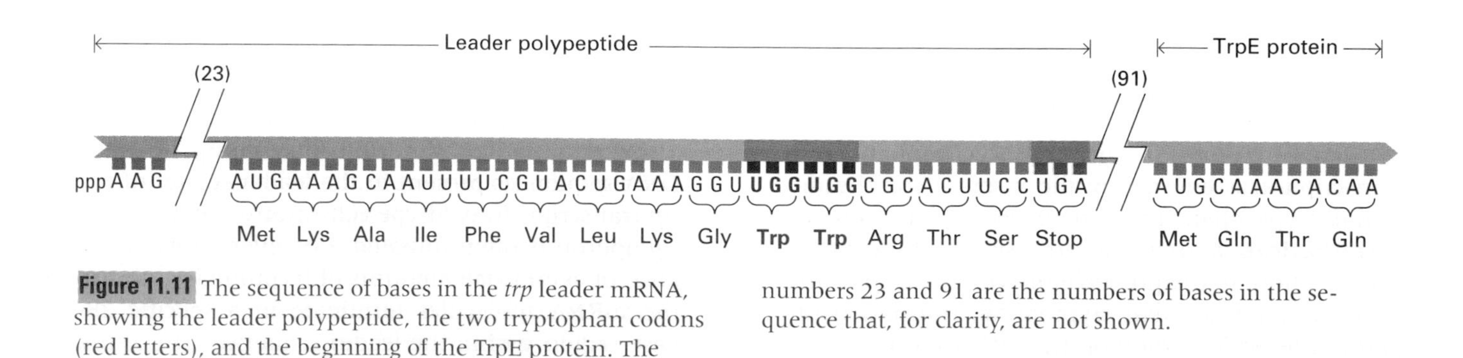

Figure 11.11 The sequence of bases in the *trp* leader mRNA, showing the leader polypeptide, the two tryptophan codons (red letters), and the beginning of the TrpE protein. The numbers 23 and 91 are the numbers of bases in the sequence that, for clarity, are not shown.

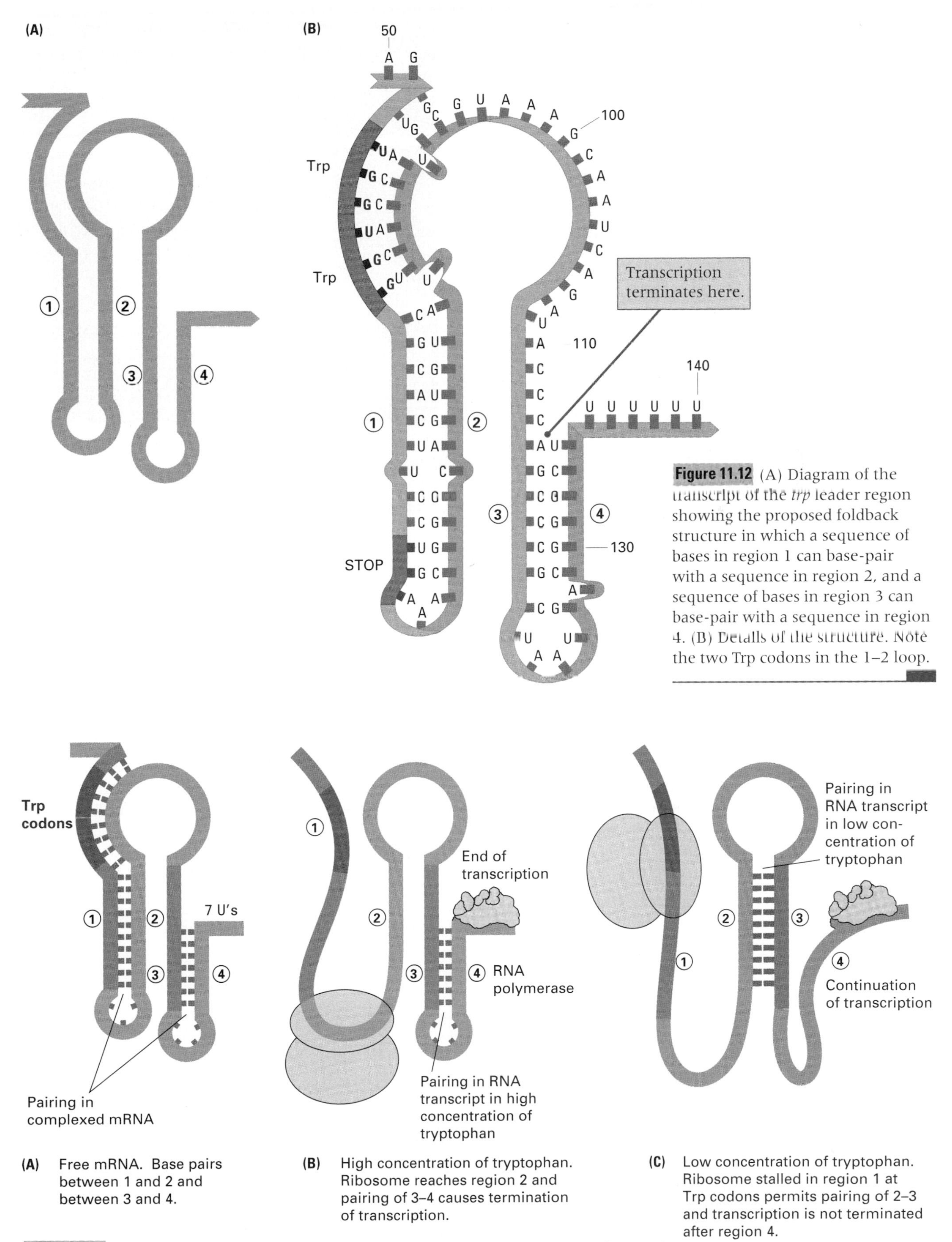

Figure 11.12 (A) Diagram of the transcript of the *trp* leader region showing the proposed foldback structure in which a sequence of bases in region 1 can base-pair with a sequence in region 2, and a sequence of bases in region 3 can base-pair with a sequence in region 4. (B) Details of the structure. Note the two Trp codons in the 1–2 loop.

(A) Free mRNA. Base pairs between 1 and 2 and between 3 and 4.

(B) High concentration of tryptophan. Ribosome reaches region 2 and pairing of 3–4 causes termination of transcription.

(C) Low concentration of tryptophan. Ribosome stalled in region 1 at Trp codons permits pairing of 2–3 and transcription is not terminated after region 4.

Figure 11.13 The explanation for attenuation in the *E. coli trp* operon. The tryptophan codons in part A are those highlighted in red letters in Figure 11.11.

In summary, attenuation is a fine-tuning mechanism of regulation superimposed on the basic negative control of the *trp* operon:

> When charged tryptophan tRNA is present in amounts that support translation of the leader polypeptide, transcription is terminated, and the *trp* enzymes are not synthesized. When the level of charged tryptophan tRNA is too low, transcription is not terminated, and the *trp* enzymes are made. At intermediate concentrations, the fraction of transcription initiation events that result in completion of *trp* mRNA depends on how frequently translation is stalled, which in turn depends on the intracellular concentration of charged tryptophan tRNA.

Many operons responsible for amino acid biosynthesis (for example, the leucine, isoleucine, phenylalanine, and histidine operons) are regulated by attenuators that function by forming alternative paired regions in the transcript. In the histidine operon, the coding region for the leader polypeptide contains seven adjacent histidine codons (Figure 11.14A). In the phenylalanine operon, the coding region for the leader polypeptide contains seven phenylalanine codons divided into three groups (Figure 11.14B). This pattern, in which codons for the amino acid produced by enzymes of the operon are present at high density in the leader peptide mRNA, is characteristic of operons in which attenuation is operative. Through these codons, the cell monitors the level of aminoacylated tRNA charged with the amino acid that is the end product of each amino acid biosynthetic pathway. Note that

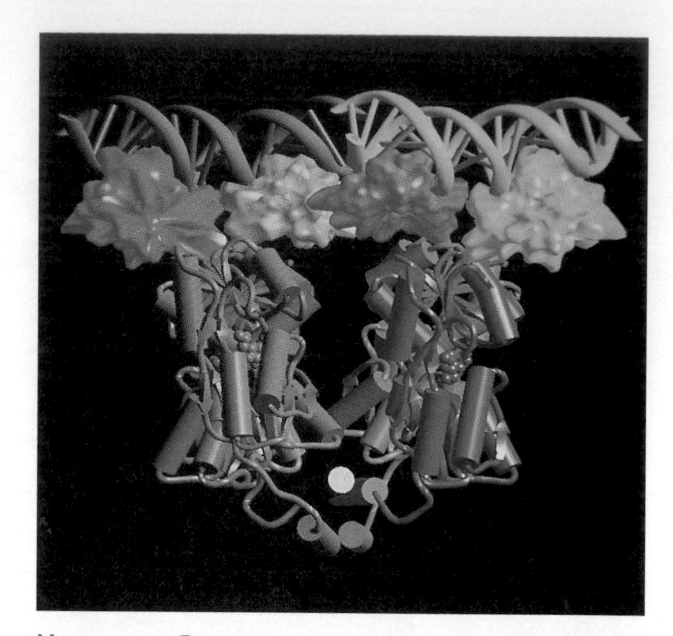

Model of the *E. coli lac* operon showing how the repressor tetramer subunits contact the operator sequences. The operator sequences are represented by the red and blue helices across the top. [Courtesy of Thomas A. Steitz.]

> Attenuation cannot take place in eukaryotes because transcription and translation are uncoupled; transcription takes place in the nucleus and translation in the cytoplasm.

Regulation of the *lac* and *trp* operons exemplifies some of the important mechanisms that control transcription of genes in prokaryotes. In the following section, we will see that similar mechanisms are used in the control of genes in bacteriophages.

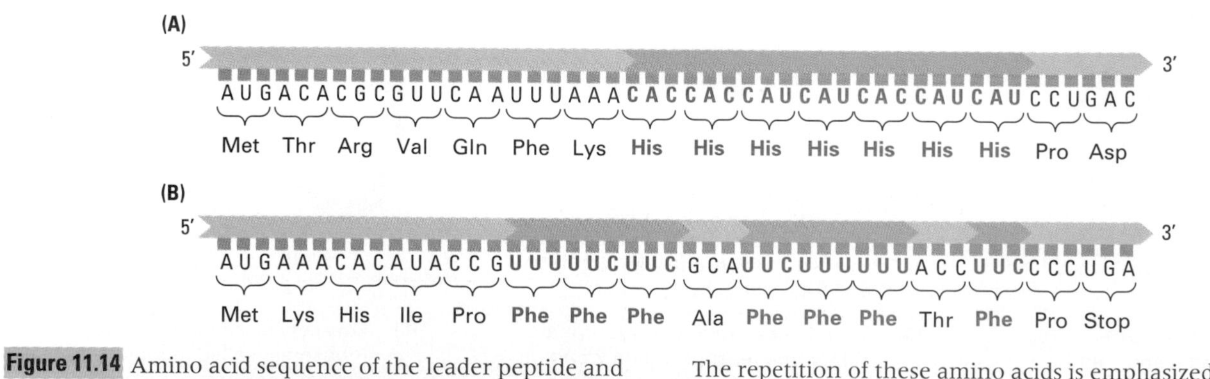

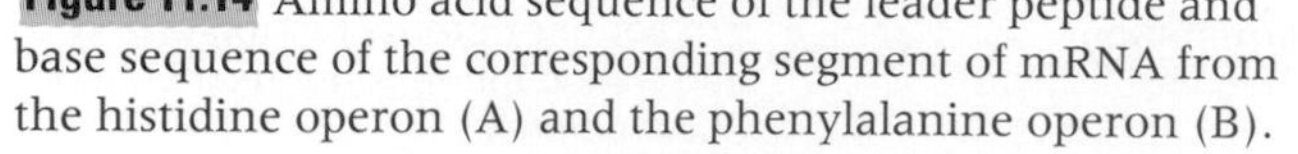

Figure 11.14 Amino acid sequence of the leader peptide and base sequence of the corresponding segment of mRNA from the histidine operon (A) and the phenylalanine operon (B). The repetition of these amino acids is emphasized in red letters.

11.5 Operons regulate the lysogenic and lytic cycles of bacteriophage λ.

When Jacob and Monod first proposed the operon model with negative regulation by repression, they suggested that the model could account not only for regulation in inducible and repressible operons for metabolic enzymes, but also for the lysogenic cycle of temperate bacteriophages. They proposed that λ bacteriophage was kept quiescent and prevented from replicating within bacterial lysogens by a repressor. This explanation ultimately proved to be correct, although the biochemical mechanism of attaining the repressed, lysogenic state is more complicated than was initially thought.

When λ bacteriophage infects *E. coli*, each infected cell can undergo one of two possible outcomes: (1) a lytic infection, resulting in lysis and production of phage particles, or (2) a lysogenic infection, resulting in integration of the λ molecule into the *E. coli* chromosome and formation of a lysogen. Because of this dichotomy, λ normally produces turbid (not completely clear) plaques on a lawn of *E. coli*. The initial infection and lysis produce a cleared region in the bacterial lawn, but a few lysogens grow within the cleared region, partially repopulating the cleared zone and producing a turbid plaque.

Mutations in regulatory genes in λ were first identified in phage mutants that give clear rather than turbid plaques. The mutants proved to fall into four classes: λ*vir*, cI^-, cII^-, and $cIII^-$. The characteristics of these mutants are shown in Table 11.3. The genetic positions of the *cI*, *cII*, and *cIII* regions are shown in the simplified genetic map of λ bacteriophage in Figure 11.15, in which the genes are grouped by functional categories. Recall from Chapter 8 that upon infection, the λ DNA molecule circularizes, bringing the *R* and *A* genes adjacent to one another.

Among the mutants in Table 11.3, the "clear" mutants proved to be analogous to *lacI* and *lacO* mutants in *E. coli*. The λ*vir* mutant is dominant to the wildtype λ^+ in mixed infection, as indicated by the combination of infecting phage designated 1 in Table 11.3. This combination of phage carries out a productive infection and prevents lysogeny by the wildtype λ^+. The λ*vir* mutant is therefore analogous to the $lacO^c$ mutation. However, λ*vir* proves to be a *double* mutant, bearing mutations in two different operators, O_L and O_R, as depicted in Figure 11.16. The cI^- mutations are recessive, as can be seen in entry 2 in Table 11.3. These cI^- mutations are analogous to $lacI^-$ mutations in that the cI^+ gene encodes the λ repressor, which is diffusible. The cII^+ and $cIII^+$ genes encode not for repressor but rather for proteins needed in establishing lysogeny. The cII^- and $cIII^-$ mutations are also recessive in mixed infections (entries 3 and 4 in Table 11.3).

Table 11.3 Characteristics of mixed infections containing several combinations of λ*vir*, cI^- cII^-, and $cIII^-$ mutants

Infecting phages	Clear or turbid plaques
1. $\lambda vir + \lambda^+$	Clear
2. $cI^- + cI^+$	Turbid
3. $cII^- + cII^+$	Turbid
4. $cIII^- + cIII^+$	Turbid

The molecular basis on which the decision between the lysogenic and the lytic cycle is determined can be explained with reference to Figure 11.16. Upon infection of *E. coli* by λ, the λ molecule circularizes, and RNA polymerase binds at P_L and P_R and initiates transcription of the *N* and *cro* genes. N protein acts to prevent termination of the transcripts from P_L and P_R, allowing production of cII protein; cII protein activates transcription at P_E and P_I, thus allowing production of the cI and int proteins. The cI protein shuts down further transcription from P_L and P_R and stimulates transcription at P_M, increasing its own synthesis. Lysogeny is achieved if the concentration of cI protein reaches

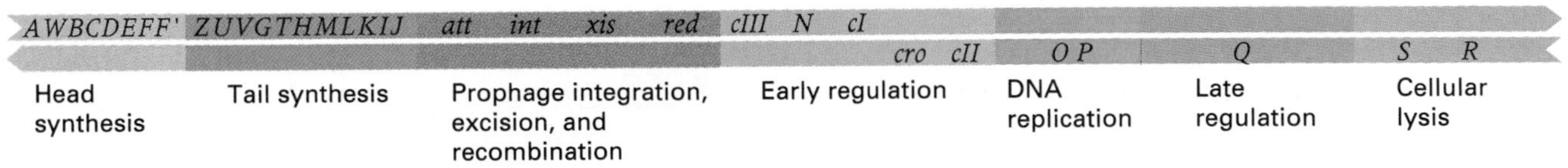

Figure 11.15 Genetic map of λ bacteriophage. The map is drawn to emphasize the functional organization of genes within the phage genome and to call attention to the regulatory features. For a more detailed map, see Figure 8.21.

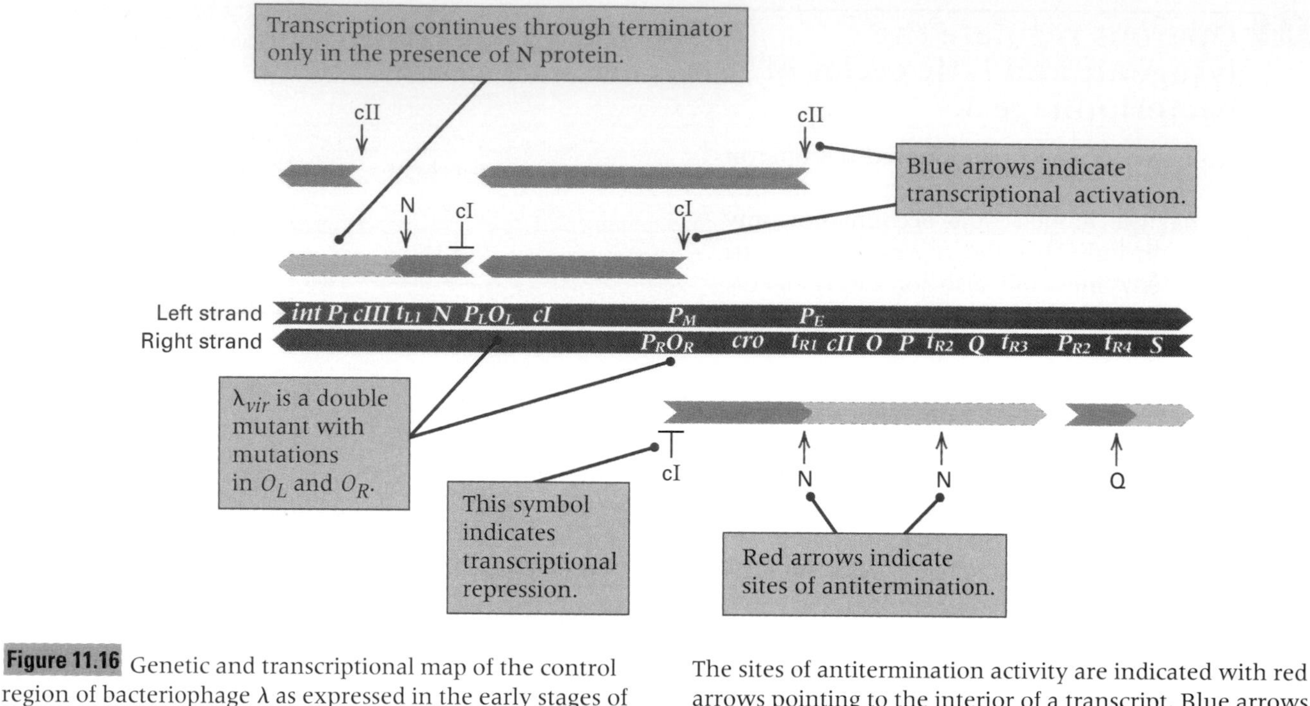

Figure 11.16 Genetic and transcriptional map of the control region of bacteriophage λ as expressed in the early stages of a lysogenic infection. The green arrows show the origin, direction, and extent of transcription. Light green arrows indicate portions of transcripts that are synthesized as a result of antitermination activity of the N or Q proteins. The sites of antitermination activity are indicated with red arrows pointing to the interior of a transcript. Blue arrows pointing to the origin of a transcript indicate transcriptional activation by cI or cII proteins. The sites of transcriptional repression by cI protein are indicated.

levels high enough to prevent transcription from P_L and P_R and to allow int protein to catalyze site-specific recombination between the circular λ molecule and the *E. coli* chromosome at their respective attachment (*att*) sites.

The alternative pathway to lysogeny, which leads to lytic development, takes place when the cro protein dominates. The cro protein also can bind to O_R and, in doing so, blocks transcription from P_M. If this occurs, then the concentration of repressor cannot rise to the levels required to block transcription from P_L and P_R. Transcription will continue from P_R and P_{R2}, and N and Q proteins will prevent termination in the rightward (and also leftward) transcripts. Because the λ DNA molecule is in a circular configuration, rightward transcription moves through genes *S* and *R* and thence through the head and tail genes (*A* through *J* in Figure 11.15). The production of proteins needed for cellular lysis and formation of phage particles ensues, followed by phage assembly and cellular lysis to release phage.

Thus the regulation of lysogeny in λ involves a sort of contest between the cro and cI proteins to determine which protein first achieves a concentration high enough to dominate the outcome of infection. The mechanism is that the cro and cI proteins compete for binding to O_L and O_R (each operator has three protein-binding subsites that participate in the competition). If cro wins the contest, the lytic cycle results in that particular cell; if cI wins, a lysogen is formed. Hence

> The cI and cro proteins function as a genetic switch; cI turns on lysogeny, and cro turns on the lytic cycle.

Additional factors that determine whether cI or cro controls the outcome of a particular infection include the nutritional state of the infected cell. This in turn influences the stabilities and levels of the regulatory proteins. Thus the regulation of even apparently "simple" systems can be quite complex.

11.6 Eukaryotes make use of a great variety of genetic regulatory mechanisms.

The great complexity of multicellular eukaryotes requires a great diversity of genetic regulatory mechanisms. Complex eukaryotes have elaborate developmental programs and numerous specialized cell types and so have different requirements for regulation than do prokaryotic cells. Many different types of regulatory mechanisms have been studied in eukaryotes. The examples we will examine reveal the general features of eukaryotic gene regulation.

Genetic organization differs in eukaryotes and prokaryotes.

Numerous differences exist between prokaryotes and eukaryotes with regard to transcription and translation. Some of the differences most relevant to regulation follow.

1. In a eukaryote, usually only a single type of polypeptide chain can be translated from a completed mRNA molecule. Polycistronic mRNA of the type seen in prokaryotes is not found in eukaryotes.

2. The DNA of eukaryotes is bound to histones, forming chromatin, and to numerous nonhistone proteins. Only a small fraction of the DNA is bare. In bacteria, some proteins are bound with the DNA molecule, but much of the DNA is free.

3. A significant fraction of the DNA of eukaryotes consists of moderately or highly repetitive nucleotide sequences. Some of the repetitive sequences are repeated in tandem copies, but others are not. Other than duplicated rRNA and tRNA genes and a few transposable elements, such as insertion sequences, bacteria contain little repetitive DNA.

4. A large fraction of eukaryotic DNA is untranslated. Most of the nucleotide sequences do not code for proteins. In the human genome, the proportion of amino-acid–coding DNA is only about 4 percent.

5. Some eukaryotic genes are expressed and regulated by the use of mechanisms for rearranging certain DNA segments in a controlled way and for increasing the number of specific genes when needed.

6. Genes in eukaryotes are split into exons and introns, and the introns must be removed in the processing of the RNA transcript before translation begins.

7. In eukaryotes, transcription and translation cannot take place simultaneously. They are uncoupled. The mRNA is synthesized in the nucleus and must be transported through the nuclear envelope to the cytoplasm to be translated. Bacterial cells do not have a nucleus separated from the cytoplasm, so transcription and translation are coupled.

We shall see in the following sections how some of these features are incorporated into particular modes of regulation.

11.7 Some genes are regulated by alteration of the DNA.

Some genes in eukaryotes are regulated by alteration of the DNA. For example, certain sequences may be amplified or rearranged in the genome or the bases may be chemically modified. Some of the alterations are reversible, but others permanently change the genome of the cells. However, the permanent changes take place only in somatic cells, and so they are not genetically transmitted to the offspring through the germ line.

Gene amplification increases dosage and the amount of gene product.

Some gene products are required in much larger quantities than others. One means of maintaining particular ratios of certain gene products (other than by differences in transcription and translation efficiency) is by **gene dosage**. For example, if two genes, *A* and *B*, are transcribed at the same rate and the translation efficiencies are the same, 20 times as much of product A as of product B can be made if there are 20 copies of gene *A* per copy of gene *B*. The histone genes exemplify a gene-dosage effect: to synthesize the huge amount of histone required to form chromatin, most cells contain hundreds of times as many copies of histone genes as of genes required for DNA replication. In this case, the high expression is automatic because the repeated genes are part of the normal chromosome complement.

In some cases, gene dosage is increased temporarily by a process called **gene amplification,** in which the number of genes increases in response to some signal. An example of gene amplification is found in the development of the oocytes of the toad *Xenopus laevis*. The formation of an egg from its precursor, the oocyte, is a complex process that requires a huge amount of protein synthesis. To achieve the necessary rate, a great many ribosomes are needed. Ribosomes contain molecules of rRNA, and the number of rRNA genes in the genome is insufficient to produce the required number of ribosomes for the oocyte in a reasonable period of time. In the development of the oocyte, the number of rRNA genes increases by about 4000 times. The precursor to the oocyte, like all somatic cells of the toad, contains about 600 rRNA-gene (rDNA) units; after amplification, about 2×10^6 copies of each unit are present. This large amount enables the oocyte to synthesize 10^{12} ribosomes, which are required for protein synthesis in the early development of the embryo, at a time when no ribosomes are being formed. The amplification proceeds through the formation of extrachromosomal circles of rDNA that multiply by rolling-circle replication. Following fertilization, the extrachromosomal rDNA is gradually degraded. Amplification of rRNA genes during oogenesis is found in many organisms, including insects, amphibians, and fish.

Yeast cells change mating type by programmed DNA rearrangements.

Rearrangement of DNA sequences in the genome is an unusual but important mechanism by which some genes are regulated. An example is the phenomenon known as **mating-type interconversion** in the yeast *Saccharomyces cerevisiae.* Yeast has two mating types, denoted **a** and α. Mating between haploid **a** and haploid α cells produces the **a** / α diploid, which can undergo meiosis and sporulation to produce four-spored asci that contain haploid **a** and α spores in the ratio 2 : 2. If a single yeast spore of either the **a** or the α genotype is cultured in isolation from other spores, mating between progeny cells would not be expected because the progeny cells would have the mating type of the original parent. However, *S. cerevisiae* has a mating system called **homothallism,** in which some cells undergo a conversion into the opposite mating type to allow matings between cells in what would otherwise be a pure culture of one mating type or the other.

The outlines of mating-type interconversion are shown in Figure 11.17. An original haploid spore (in this example, α) undergoes germination to produce two progeny cells. Both the mother cell (the original parent) and the daughter cell have mating type α, as expected from a normal mitotic division. However, in the next cell division, a switching (interconversion) of mating type takes place in the mother cell and its *new* progeny cell, in which the original α mating type becomes replaced with the **a** mating type. After the second cell division is complete, the α and **a** cells are able to undergo mating because they now are opposite in mating type. Fusion of the nuclei produces the **a** / α diploid, which undergoes mitotic divisions and later sporulation to produce **a** and α haploid spores again.

The genetic basis of mating-type interconversion is outlined in Figure 11.18. The gene that controls mating type is the *MAT gene* in chromosome III, which can have either of two allelic forms, **a** or α. If the allele in a haploid cell is *MAT***a,** the cell has mating type **a,** and if the allele is *MAT*α, the cell has mating type α. However, both genotypes normally contain both **a** and α genetic information in the form of unexpressed **cassettes** present in the same chromosome. The *HML*α cassette contains the α DNA sequence about 200 kb away from the *MAT* gene, and the *HMR***a** cassette contains the **a** DNA sequence about 150 kb away from *MAT* on the other side. When mating-type interconversion takes place, a different yeast gene, *HO,* produces a specific endonuclease that cuts both strands of the DNA in the *MAT* region. The double-stranded break initiates a process in which genetic information in the unexpressed cassette containing the opposite mating type becomes inserted into *MAT.* In this process, the DNA sequence in the donor cassette is

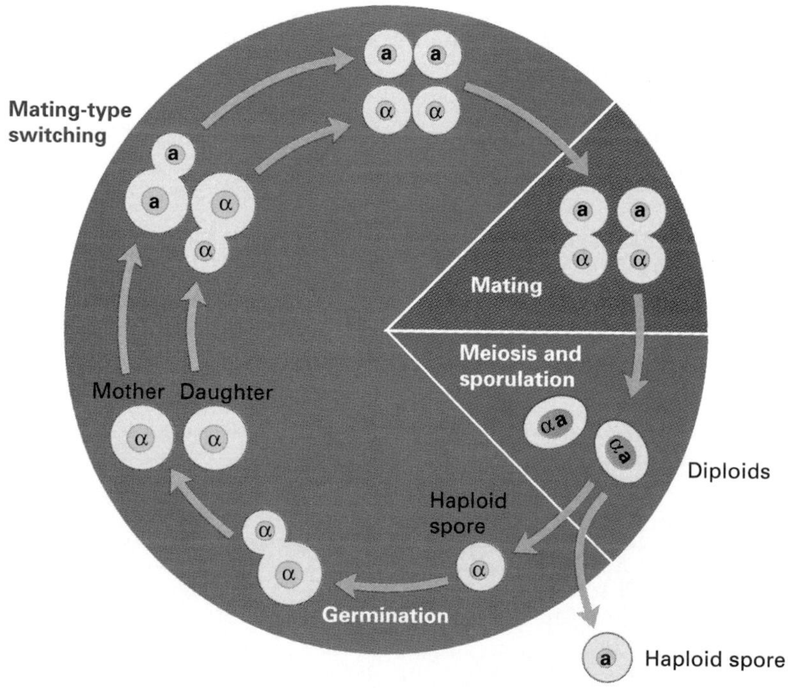

Figure 11.17 Mating-type switching in the yeast *Saccharomyces cerevisiae.* Germination of a spore (in this example, one of mating type α) forms a mother cell and a bud that grows into a daughter cell. In the next division, the mother cell and its new daughter cell switch to the opposite mating type (in this case **a**). The result is two α and two **a** cells. Cells of opposite mating type can fuse to form **a** / α diploid zygotes. In a similar fashion, germination of an **a** spore is accompanied by switching to the α mating type.

duplicated, so the mating type becomes converted but the same genetic information is retained in unexpressed form in the cassette. The terminal regions of *HML, MAT,* and *HMR* are identical (illustrated in blue in Figure 11.18), and these regions are critical in making possible recognition of the regions for interconversion. The unique part of the α region is 747 base pairs in length; that of the **a** region is 642 base pairs. The molecular details of the conversion process are similar to those of meiotic recombination, which is discussed in Chapter 4.

Figure 11.18 illustrates two sequential mating-type interconversions. In the first, an α cell (containing the *MAT*α allele) undergoes conversion into **a,** using the DNA sequence contained in the *HMR***a** cassette. The converted cell has the genotype *MAT***a.** In a later generation, a descendant **a** cell may become converted into mating type α, using the unexpressed DNA sequence contained in *HML*α. This cell has the genotype *MAT*α. Mating-type switches can take place repeatedly in the lineage of any particular cell.

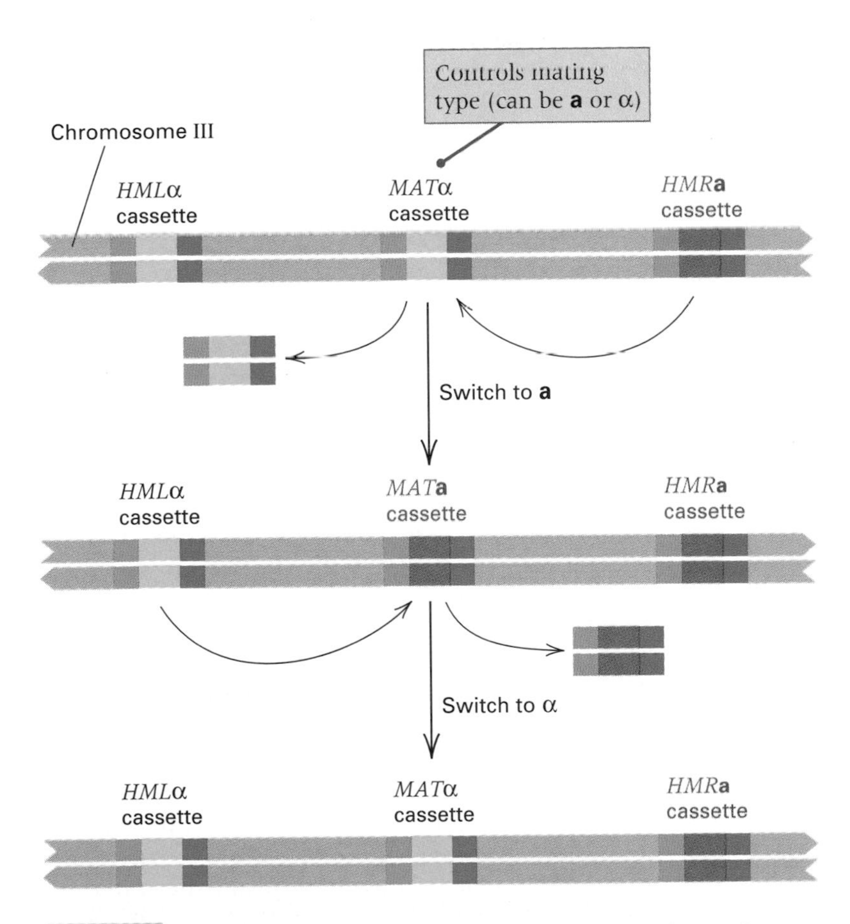

Figure 11.18 Genetic basis of mating-type interconversion. The mating type is determined by the DNA sequence present at the *MAT* locus. The *HML* and *HMR* loci are cassettes containing unexpressed mating-type genes, either α or **a.** In the interconversion from α to **a,** the α genetic information present at *MAT* is replaced with the **a** genetic information from *HMR***a.** In the switch from **a** to α, the **a** genetic information at *MAT* is replaced with the α genetic information from *HML*α.

DNA splicing takes place in the formation of antibodies.

Another important example of programmed DNA rearrangement occurs in vertebrates in cells that form the immune system. In this case, the precursor cells contain numerous DNA sequences that can serve as alternatives for various regions in the final gene. In the maturation of each cell, a combination of the alternatives is created by DNA cutting and rejoining, producing a great variety of possible genes that enable the immune system to recognize and attack most bacteria and viruses.

It has been estimated that a normal mammal is capable of producing more than 10^8 different antibodies, each of which can combine specifically with a particular antigen. Antibodies are proteins, and each unique antibody has a different amino acid sequence. If antibody genes were conventional in the sense that each gene coded for a single polypeptide, then mammals would need more than 10^8 genes for the production of antibodies. This is considerably more genes than are present in the entire genome. In fact, mammals use only a few hundred genes for antibody production, and the huge number of different antibodies derives from remarkable events that take place in the DNA of certain somatic cells. These events are discussed in this section.

Antibodies are produced by a type of white blood cell called a B cell. Each B cell can produce a single type of antibody, but the antibody is not secreted until the cell has been stimulated by the appropriate antigen. The antibody produced by a particular stimulated B cell may be any one of five distinct classes of antibodies designated IgG, IgM, IgA, IgD, and IgE (Ig stands for *immunoglobulin*). Each antibody class serves specialized functions in the immune response and exhibits certain structural differences. All antibodies are composed of two types of polypeptide chains that differ in size: a large one called the *heavy (H) chain* and a small one called the *light (L) chain.*

Immunoglobulin G (IgG) is the most abundant class of antibodies and has the simplest molecular structure. Its molecular organization is illustrated in Figure 11.19. An IgG molecule consists of two heavy and two light chains held together by disulfide bridges (two joined sulfur atoms) and has the overall shape of the letter Y. The sites on the antibody that carry its specificity and combine with the antigen are located in the upper half of the arms above the fork of the Y. Each IgG molecule with a different antigen specificity has a different amino acid sequence for the heavy and light chains in this part of the molecule. These specificity regions are called the **variable regions** of the heavy and light chains (blue pointers in Figure 11.19). The remaining regions of the polypeptide are the **constant regions,** which are called constant because they have virtually the same amino acid sequence in all IgG molecules.

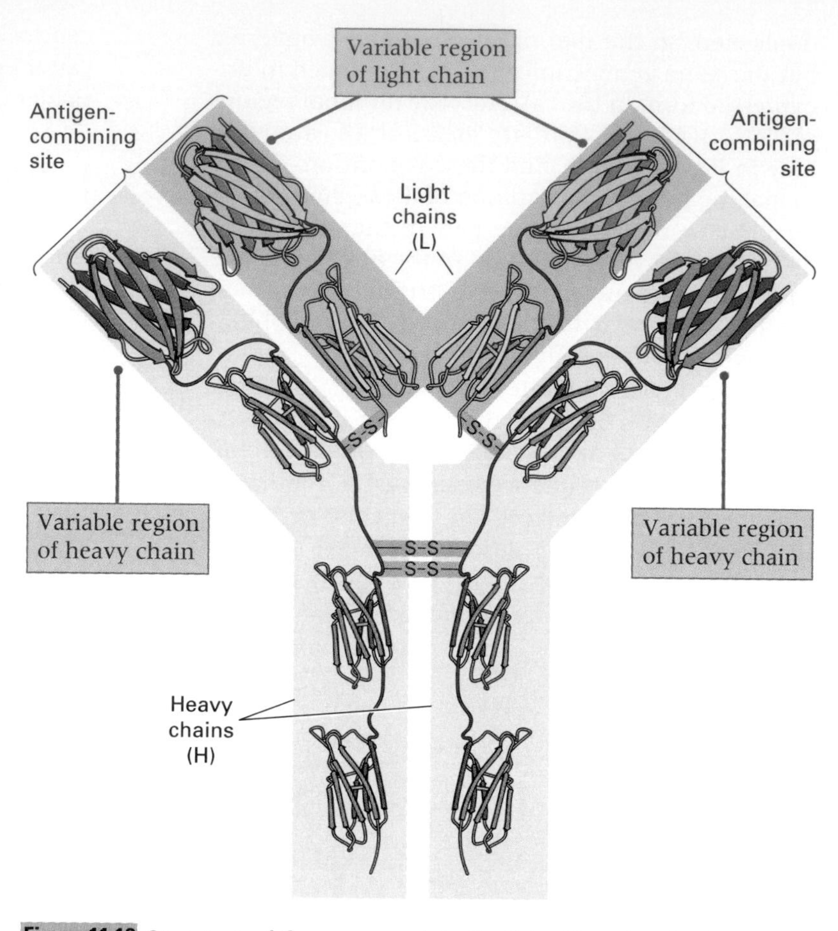

Figure 11.19 Structure of the immunoglobulin G (IgG) molecule showing the light chains (L) and heavy chains (H). Variable and constant regions are indicated.

Initial understanding of the genetic mechanisms responsible for variability in the amino acid sequences of antibody polypeptide chains came from cloning a gene for the light chain of IgG. The critical observation was made by comparing the nucleotide sequence of the gene in embryonic cells or germ cells with that in mature antibody-producing cells. In the genome of a B cell that was actively producing the antibody, the DNA segments corresponding to the constant and variable regions of the light chain were found to be very close together, as expected of DNA that codes for different parts of the same polypeptide. However, in embryonic cells, these same DNA sequences were located far apart. Similar results were obtained for the variable and constant regions of the heavy chains: Segments encoding these regions were close together in B cells but widely separated in embryonic cells.

Extensive DNA sequencing of the genomic region that codes for antibody proteins revealed not only the reason for the different gene locations in B cells and germ cells but also the mechanism for the origin of antibody variability. Cells in the germ line contain a small number of genes corresponding to the constant region of the light chain, which are close together along the DNA. Separated from them, but on the same chromosome, is another cluster consisting of a much larger number of genes that correspond to the variable region of the light chains. In the differentiation of a B cell, one gene for the constant region is spliced (cut and joined) to one gene for the variable region, and this splicing produces a complete light-chain antibody gene. A similar splicing mechanism yields the constant and variable regions of the heavy chains.

The formation of a finished antibody gene is slightly more complicated than this description implies, because light-chain genes consist of three parts and heavy-chain genes consist of four parts. Gene splicing in the origin of a light chain is illustrated in Figure 11.20. For each of two parts of the variable region, the germ line contains multiple coding sequences called the **V (variable)** and **J (joining)** regions. In the differentiation of a B cell, a deletion makes possible the joining of one of the V regions with one of the J regions. The DNA joining process is called **combinatorial joining** because it can create many combinations of the V and J regions. When transcribed, this joined V-J sequence forms the 5' end of the light-chain RNA transcript. Transcription continues on through the DNA region coding for the constant (C) part of the gene. RNA splicing subsequently attaches the C region, creating the light-chain mRNA.

Combinatorial joining also takes place in the genes for the antibody heavy chains. In this case, the DNA splicing joins the heavy-chain counterparts of V and J with a third set of sequences, called D (for diversity), located between the V and J clusters.

The amount of antibody variability that can be created by combinatorial joining is calculated as fol-

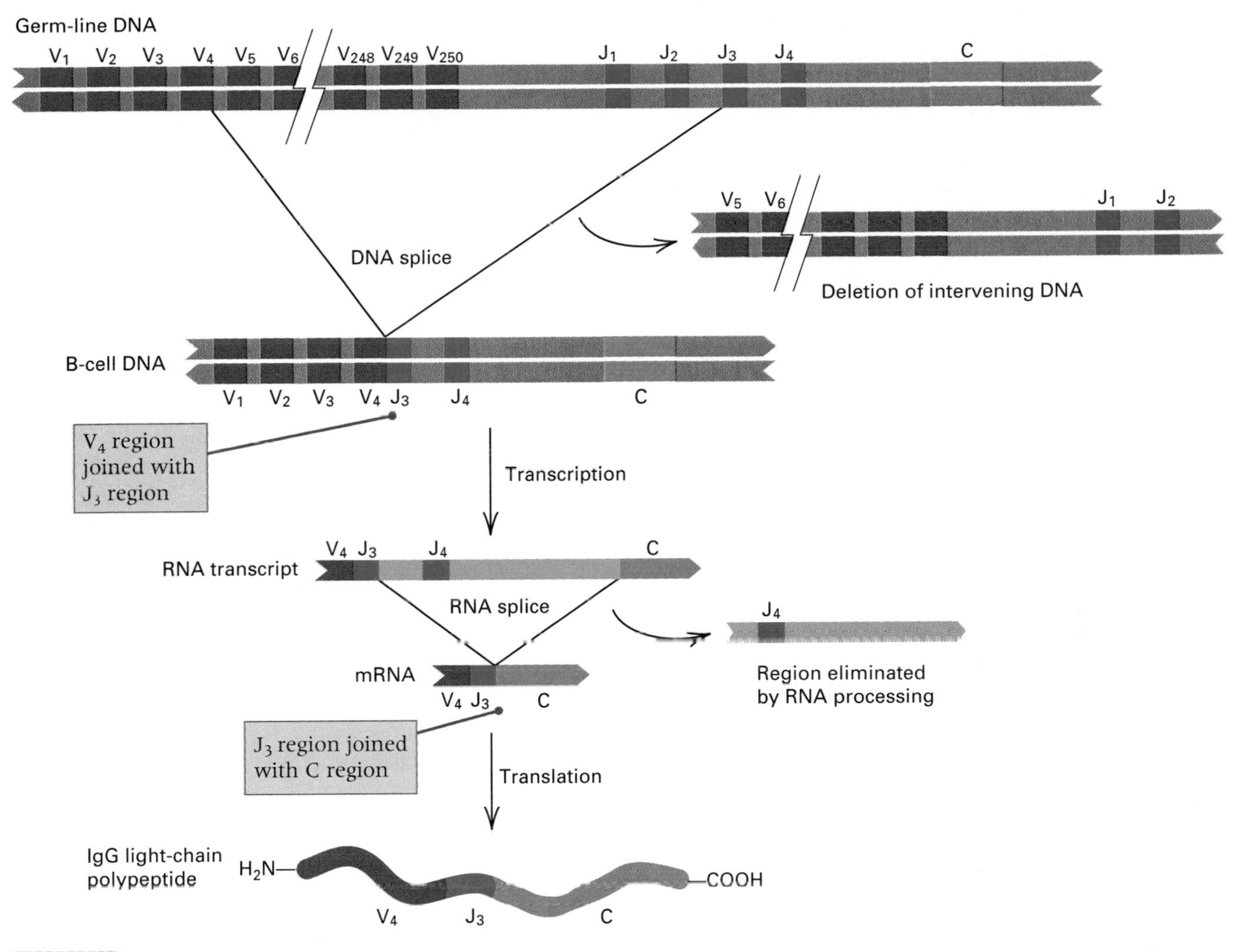

Figure 11.20 Formation of a gene for the light chain of an antibody molecule. One variable (V) region is joined with one randomly chosen J region by deletion of the intervening DNA. The remaining J regions are eliminated from the RNA transcript during RNA processing.

lows: In mice, the light chains are formed from combinations of about 250 V regions and 4 J regions, giving $250 \times 4 = 1000$ different chains. For the heavy chains, there are approximately 250 V, 10 D, and 4 J regions, producing $250 \times 10 \times 4 = 10{,}000$ combinations. Because any light chain can combine with any heavy chain, there are at least $1000 \times 10{,}000 = 10^7$ possible types of antibodies. The number of DNA sequences used for antibody production is quite small (about 500), but the number of possible antibodies is very large.

The value of 10^7 different antibody types is an underestimate, because there are three additional sources of antibody variability:

1. The junction for V-J (or V-D-J) splicing in combinatorial joining can be between different nucleotides of a particular V-J combination in light chains (or a particular V-D or D-J combination in heavy chains). The different splice junctions can generate different codons in the spliced gene. In this manner, variability in the junction of V-J joining can result in polypeptides that differ in amino acid sequence.

2. As part of the recombination reaction in V-J (or V-D-J) joining that fuses a V segment to a J segment (or a particular V-D or D-J combination in heavy chains), a hairpin loop that is later cleaved is formed at the right end of the V (or D) segment. Cleavage of the loop may be asymmetrical, which can result in the insertion of a few base pairs, thus changing the coding in that region.

3. The V regions are susceptible to a high rate of *somatic mutation,* which occurs in B-cell development. These mutations allow different B-cell clones to produce different polypeptide sequences, even if they have undergone exactly the same V-J joining. The mechanism for this high mutation rate is unknown.

Gene splicing occurs during formation of T-cell receptors.

Immunity is also mediated by a different type of white blood cell called a T cell. A T cell carries an antigen receptor on its surface that combines with an antigen, stimulating the T cell to respond. Like the antibody molecules produced in B cells, the T-cell receptors are highly variable in amino acid sequence, and this enables the T cells to respond to many antigens. Although the polypeptide chains in T-cell receptors are different from those in antibody molecules, they have a similar organization in that they are formed from the aggregation of two pairs of polypeptide chains. A particular T cell may carry either of two types of receptors. The majority carry the $\alpha\beta$ receptor, composed of polypeptide chains designated α and β, and the rest carry the $\gamma\delta$ receptor, composed of chains designated γ and δ. Each receptor polypeptide includes a variable region and a constant region. As their variability and similarity in organization suggest, the T-cell receptor genes are formed by somatic rearrangement of components analogous to those of the V, D, J, and C regions in the B cells. For example, in the mouse, the β chain of the T-cell receptor is spliced together from one each of approximately 20 V regions, 2 D regions, 12 J regions, and 2 C regions. Note that there are far fewer V regions for T-cell receptor genes than there are for antibody genes, yet T-cell receptors seem able to recognize just as many foreign antigens as B cells. The extra variation results from a higher rate of somatic mutation in the T-cell receptor genes.

DNA and its expression can be modified by methylation.

In most eukaryotes, a small proportion of the cytosine bases are modified by the addition of a methyl (CH_3) group to the number-5 carbon atom (Figure 11.21). The cytosines are incorporated in their normal, unmodified form in DNA replication, but they are modified later by an enzyme called a **DNA methylase.** Cytosines in 5'-CG-3' dinucleotides are modified preferentially. When a CG dinucleotide that is methylated in both strands undergoes DNA replication, the result is two daughter molecules, each of which contains one parental strand with a methylated CG and one daughter strand with an unmethylated CG. The DNA methylase recognizes the half-methylation in these molecules and methylates the cytosines in the daughter strands. Methylation of CG dinucleotides in the sequence CCGG can easily be detected by the use of the restriction enzymes *Msp*I and *Hpa*II. Both enzymes cleave the sequence CCGG. However, *Msp*I cleaves irrespective of whether the interior C is methylated, whereas *Hpa*II cleaves only unmethylated DNA. Therefore, *Msp*I restriction sites that are not cleaved by *Hpa*II are sites at which the interior C is methylated (Figure 11.22).

Many eukaryotic genes have CG-rich regions upstream of the coding region, providing potential sites for methylation that may affect transcription. A number of observations suggest that high levels of methylation are associated with genes for which the rate of transcription is low. One example is the inactive X chromosome in mammalian cells, which is extensively methylated. Another example is the *Ac* transposable element in maize. Certain *Ac* elements lose activity of the transposase gene without any change in DNA sequence. These elements prove to have heavy methylation in a region particularly rich in the CG dinucleotides. Return to normal activity of the methylated *Ac* elements coincides with loss of methylation through the action of demethylating enzymes in the nucleus. Although there is a correlation between methylation and gene inactivity, it is possible that heavy methylation is a result of gene inactivity rather than a cause of it. However, treatment of cells with the

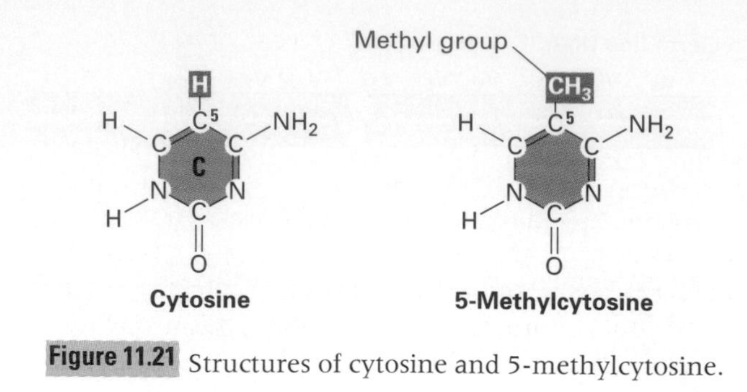

Figure 11.21 Structures of cytosine and 5-methylcytosine.

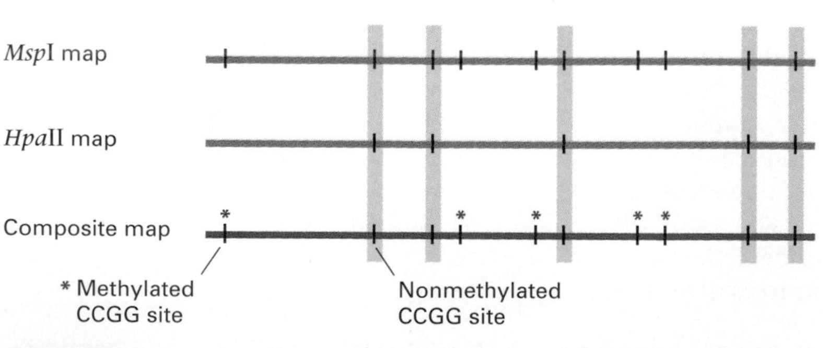

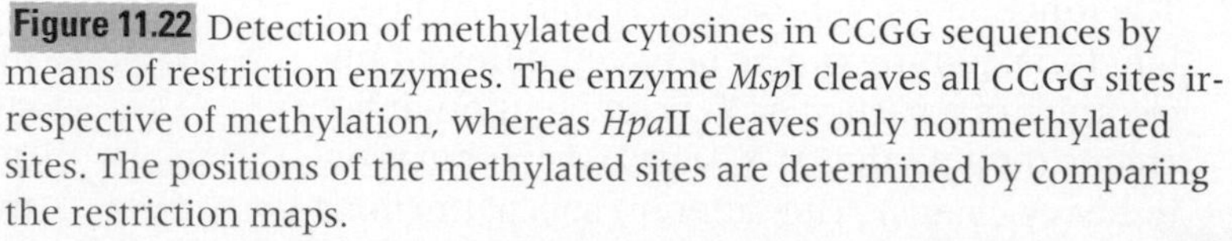

Figure 11.22 Detection of methylated cytosines in CCGG sequences by means of restriction enzymes. The enzyme *Msp*I cleaves all CCGG sites irrespective of methylation, whereas *Hpa*II cleaves only nonmethylated sites. The positions of the methylated sites are determined by comparing the restriction maps.

cytosine analog *azacytidine* reverses methylation and can restore gene activity. For example, some clones of rat pituitary tumor cells express the gene for prolactin, whereas other related clones do not. The gene is methylated in the nonproducing cells but is not methylated in the producers. Reversal of methylation in the nonproducing cells with azacytidine results in prolactin expression. On the other hand, not all organisms have methylation. For example, *Drosophila* DNA is not methylated. In organisms in which the DNA is methylated, methylation increases susceptibility to certain kinds of mutations, which are discussed in Chapter 7.

11.8 Transcriptional control is a frequent mode of regulation in eukaryotes.

Many eukaryotic genes code for essential metabolic enzymes or cellular components and are expressed constitutively at relatively low levels in all cells; these genes are called **housekeeping genes.** The expression of other genes differs from one cell type to the next or among different stages of the cell cycle; these genes are often regulated at the level of transcription. In prokaryotes, the levels of expression in induced and uninduced cells may differ by a thousandfold or more. Such extreme levels of induction are uncommon in eukaryotes, except for some genes in lower eukaryotes such as yeast. Most eukaryotic genes are induced by factors ranging from 2 to 10. In this section, we consider some components of transcriptional regulation in eukaryotes.

Genes controlling yeast mating type regulate transcription.

As we noted earlier, the mating type of a yeast cell is controlled by the allele of the *MAT* gene that is present (refer to Figure 11.18 for the genetic basis of mating-type interconversion in yeast). Both **a** (mating type **a**) and α (mating type α) express a set of haploid-specific genes. They differ in that **a** express a set of **a**-specific genes and α express a set of α-specific genes. The haploid-specific genes that cells of both mating types express include *HO,* which encodes the HO endonuclease used in mating-type interconversion, and *RME1,* which encodes a repressor of meiosis-specific genes. The functions of the mating-type-specific genes include (1) secretion of a mating peptide that arrests cells of the opposite mating type before DNA synthesis and prepares them for cell fusion, and (2) production of a receptor for the mating peptide secreted by the opposite mating type. Therefore, when **a** and α cells are in proximity, they prepare each other for mating and undergo fusion.

Regulation of mating type is at the level of transcription according to the regulatory interactions diagrammed in Figure 11.23. These regulatory interactions were originally proposed on the basis of

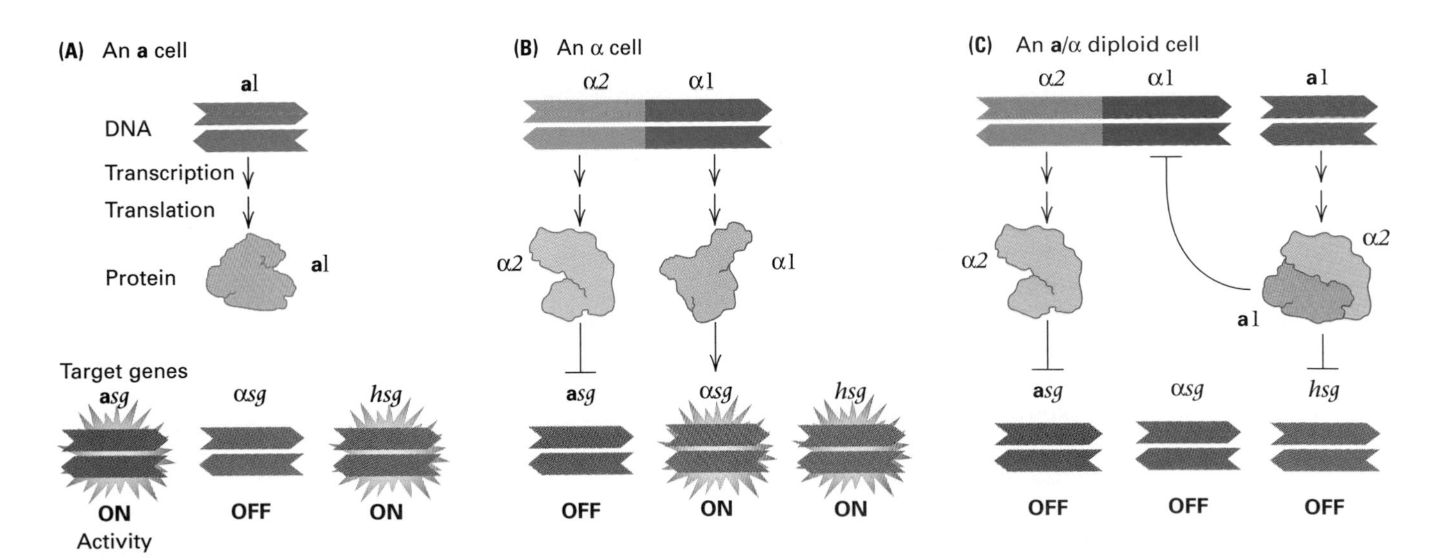

Figure 11.23 Regulation of mating type in yeast. The symbols *asg*, *αsg*, and *hsg* denote sets of **a**-specific genes, α-specific genes, and haploid-specific genes, respectively; sets of genes represented with a "sunburst" are *on*, and those unmarked are *off.* (A) In an **a** cell, the **a**1 peptide is inactive, and the sets of genes manifest their basal states of activity (*asg* and *hsg on* and *αsg off*), so the cell is an **a** haploid. (B) In an α cell, the α2 peptide turns the *asg off* and the α1 peptide turns the *αsg on,* so the cell is an α haploid. (C) In an **a**/α diploid, the **a**1 and α2 peptides form a complex that turns the *hsg off,* the α2 peptide turns the *asg off,* and the *αsg* manifest their basal activity of *off,* so physiologically the cell is non-**a,** non-α, and nonhaploid (that is, it is a normal diploid).

the phenotypes of various types of mutants, and most of the details have been confirmed by direct molecular studies. The symbols *asg*, *αsg*, and *hsg* stand for the **a**-specific genes, the α-specific genes, and the haploid-specific genes, respectively; each set of genes is represented as a single segment (lack of a "sunburst" indicates that transcription does not take place). In a cell of mating-type **a** (Figure 11.23A), the *MAT***a** region is transcribed and produces a polypeptide called **a**1. By itself, **a**1 has no regulatory activity, and in the absence of any regulatory signal, *asg* and *hsg* are transcribed, but not *αsg*. In a cell of mating-type α (Figure 11.23B), the *MATα* region is transcribed, and two regulatory proteins denoted α1 and α2 are produced: α1 is a *positive regulator* of the α-specific genes, and α2 is a *negative regulator* of the **a**-specific genes. The result is that *αsg* and *hsg* are transcribed, but transcription of *asg* is turned off. Both α1 and α2 bind with particular DNA sequences upstream from the genes that they control.

In the **a**/α diploid (Figure 11.23C), both *MAT***a** and *MATα* are transcribed, but the only polypeptides produced are **a**1 and α2. The reason is that the **a**1 and α2 polypeptides combine to form a negative regulatory protein that represses transcription of the α1 gene in *MATα* and of the haploid-specific genes. The α2 polypeptide acting alone is a negative regulatory protein that turns off *asg*. Because α1 is not produced, transcription of *αsg* is not turned on. In sum, then, the *αsg* are not turned on because α1 is absent, the *asg* are turned off because α2 is present, and the *hsg* are turned off by the α2/**a**1 complex. The **a**/α diploid does not transcribe either the mating-type-specific set of genes or the haploid-specific genes. This ensures that mating-type switching ceases (because the HO endonuclease is absent) and that meiosis can occur (because expression of *RME1* is turned off). Thus the homothallic **a**α diploid is stable and able to undergo meiosis.

The repression of transcription of the haploid-specific genes mediated by the **a**1/α2 protein is an example of negative control of the type already familiar from the *lac* and *trp* systems in *E. coli*. The interesting twist in the yeast example is that the α2 protein has a regulatory role of its own in repressing transcription of the **a**-specific genes. Why does the α2 protein, on its own, not repress the haploid-specific genes as well? The answer lies in the specificity of its DNA binding. By itself, the α2 protein has low affinity for the target sequences in the haploid-specific genes. However, the **a**1/α2 heterodimer has both high affinity and high specificity for the target DNA sequences in the haploid-specific genes. The three-dimensional structure of the **a**1/α2 protein in complex with target DNA is shown in Figure 11.24. Upon binding, the **a**1/α2 complex produces a pronounced 60° bend in the DNA molecule, which may play a role in transcriptional repression.

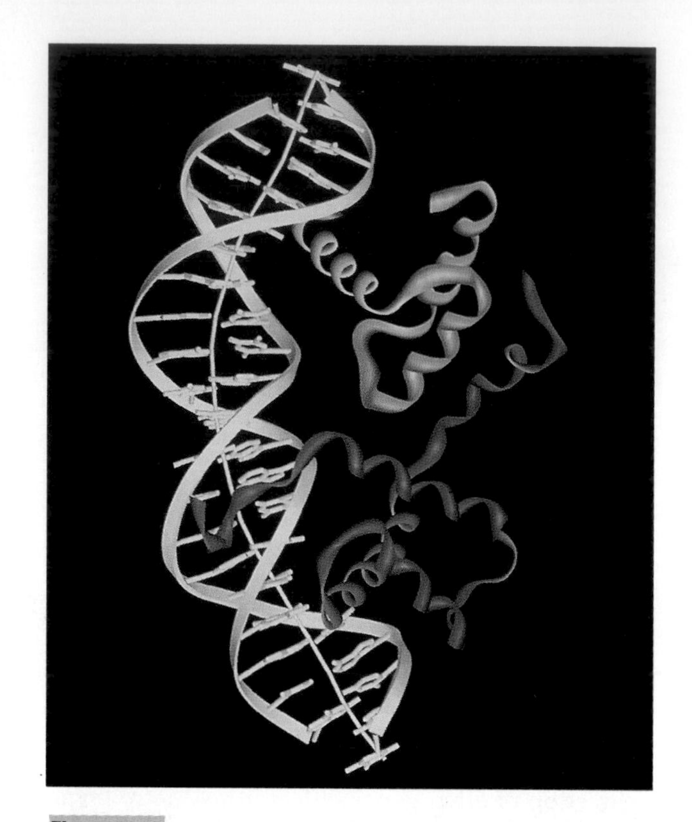

Figure 11.24 Structure of the **a**1/α2 protein bound with DNA. The **a**1 protein is shown in blue and the α2 protein in red. Contact with the DNA target results in a sharp bend in the DNA. [Courtesy of Cynthia Wolberger. From T. Li, M. R. Stark, A. D. Johnson, and C. Wolberger. 1995. *Science* 270:262.]

Transcriptional regulation of the mating-type genes includes negative control (**a**-specific genes and haploid-specific genes) and positive control (α-specific genes). Although the regulation of transcription in eukaryotes is both positive and negative, positive regulation is more common. The regulatory proteins required are the subject of the next section.

Positive transcriptional regulation makes use of activator proteins.

The α1 protein is an example of a *transcriptional activator protein*, which must bind with an upstream DNA sequence to prepare a gene for transcription. Transcriptional activator proteins may participate in the assembly of a transcriptional complex including RNA polymerase, or they may initiate transcription by an already assembled complex. In either case, the activator proteins are essential for the transcription of genes that are positively regulated.

Many transcriptional activator proteins can be grouped into categories on the basis of shared characteristics of their amino acid sequences. For example, one category has a *helix-loop-helix* motif, which consists of a sequence of amino acids forming a pair of α-helices separated by a bend; the helices are so situated that they can fit neatly into the grooves of a double-stranded DNA molecule. The helix-loop-helix motif is the basis of the DNA-binding ability, although the sequence specificity of the binding results from other parts of the protein. The MyoD protein that activates transcription of muscle-specific genes in mammals has a helix-loop-helix motif, as do many other transcriptional activator proteins in both prokaryotes and eukaryotes.

A second large category of transcriptional activator proteins includes a DNA-binding motif called a *zinc finger* because the folded structure incorporates a zinc ion. The human breast cancer gene *BRCA1* codes for a zinc-finger protein. Another example is the product of the yeast gene *GAL4*, which codes for a transcriptional activator of the genes for the utilization of galactose. The protein functions as a dimer composed of two identical GAL4 polypeptides oriented with their zinc-binding domains at the extreme ends (Figure 11.25 shows the zinc ions in yellow). The DNA sequence recognized by the protein is a symmetrical sequence 17 base pairs in length, which includes at each end a CGG triplet that makes direct contact with the zinc-containing domains.

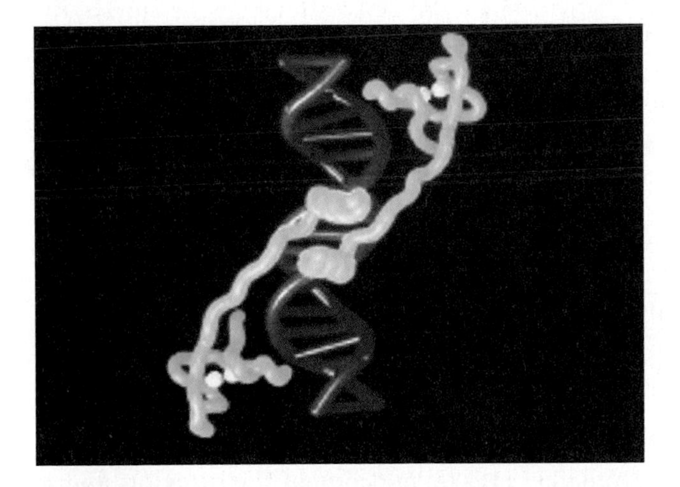

Figure 11.25 Three-dimensional structure of the GAL4 protein (blue) bound to DNA (red). The protein is composed of two polypeptide subunits held together by the coiled regions in the middle. The DNA-binding domains are at the extreme ends, and each physically contacts three base pairs in the major groove of the DNA. The zinc ions in the DNA-binding domains are shown in yellow. [Courtesy of Dr. Stephen C. Harrison. See also R. Marmorstein, M. Carey, M. Ptashne, and S. C. Harrison. 1992. *Nature* 356: 408–414.]

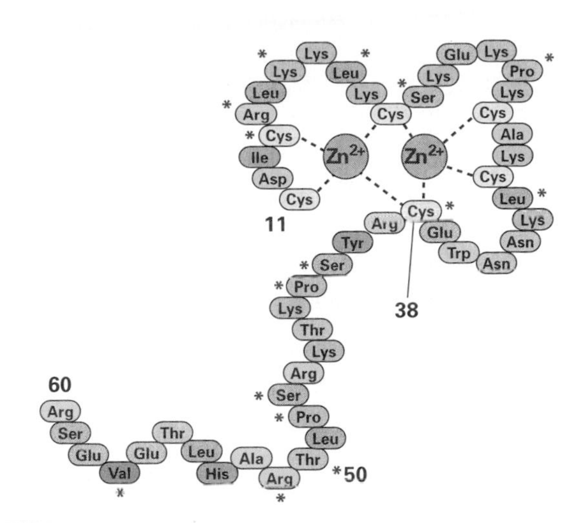

Figure 11.26 DNA-binding domain present in the GAL4 transcriptional activator protein in yeast. The four cysteine residues bind a zinc ion and form a peptide loop called a zinc finger. The zinc finger is a common motif found in DNA-binding proteins. The amino acids marked by a red asterisk have been identified by mutations as sites at which amino acid replacements can abolish the DNA-binding activity of the protein. The result is that the target genes cannot be activated.

A more detailed illustration of the DNA-binding domain of the GAL4 protein is shown in Figure 11.26. Each of two zinc ions (Zn^{2+}) is chelated by bonds with four cysteine residues in characteristic positions at the base of a loop that extends for an additional 841 amino acids beyond those shown. The amino acids marked by red asterisks are the sites of mutations that result in mutant proteins unable to activate transcription. Replacements at amino acid positions 15 (Arg → Gln), 26 (Pro → Ser), and 57 (Val → Met) are particularly interesting because they provide genetic evidence that zinc is necessary for DNA binding. In particular, the mutant phenotypes can be rescued by extra zinc in the growth medium, because the molecular defect reduces the ability of the zinc-finger part of the molecule to chelate zinc; extra zinc in the medium overcomes the defect and restores the ability of the mutant activator protein to attach to its particular binding sites in the DNA.

Some transcriptional activator proteins interact with hormones.

Among the known regulators of transcription in higher eukaryotes, the **hormones**—small molecules or polypeptides that are carried from hormone-producing cells to target cells—have perhaps been studied in most detail. One class of hormones consists of small molecules synthesized from cholesterol; these steroid hormones include the

The human connection

X-ing Out Gene Activity

Mary F. Lyon 1961
Medical Research Council,
Harwell, England
Gene Action in the X Chromosome of the Mouse (Mus musculus L.)

How do organisms solve the problem that females have two X chromosomes whereas males have only one? Unless there were some type of correction (called dosage compensation), the unequal number would mean that for all the genes in the X chromosome, cells in females would have twice as much gene product as cells in males. It would be difficult for the developing organism to cope with such a large difference in dosage for so many genes. The problem of dosage compensation has been solved by different organisms in different ways. The hypothesis put forward in this paper is that in the mouse (and by inference in other mammals), the mechanism is very simple: One of the X chromosomes, chosen at random in each cell lineage early in development, becomes inactivated and remains inactivated in all descendant cells in the lineage. In certain cells the inactive X chromosome becomes visible in interphase as a deeply staining "sex-chromatin body." We now know that there are a few genes in the short arm of the X chromosome that are not inactivated. There is also good evidence that the inactivation of the rest of the X chromosome takes place sequentially from an "X-inactivation center." We also know that in marsupial mammals, such as the kangaroo, it is always the paternal X chromosome that is inactivated.

It has been suggested that the so-called sex chromatin body is composed of one heteropyknotic [that is, deeply staining during interphase] X chromosome. . . . The present communication suggests that evidence of mouse genetics indicates: (1) that the heteropyknotic X chromosome can be either paternal or maternal in origin, in different cells of the same animal; (2) that it is genetically inactivated. The evidence has two main parts. First, the normal phenotype of XO females in the mouse shows that only one active X chromosome is necessary for normal development, including sexual development. The second piece of evidence concerns the mosaic phenotype of female mice heterozygous for some sex-linked [X-linked] mutants. All sex-linked mutants so far known affecting coat colour cause a "mottled" or "dappled" phenotype, with patches of normal and mutant colour. . . . It is here suggested that this mosaic phenotype is due to the inactivation of one or other X chromosome early in embryonic development. If this is true, pigment cells descended from the cells in which the chromosome carrying the mutant gene was inactivated will give rise to a normal-coloured patch and those in which the chromosome carrying the normal gene was inactivated will give rise to a mutant-coloured patch. . . . Thus this hypothesis predicts that for all sex-linked genes of the mouse in which the phenotype is due to localized gene action the heterozygote will have a mosaic appearance. . . . The genetic evidence does not indicate at what early stage of embryonic development the inactivation of the one X chromosome occurs. . . . The sex-chromatin body

IT IS HERE SUGGESTED that this [X-linked] mosaic phenotype is due to the inactivation of one or other X chromosome early in embryonic development.

is thought to be formed from one X chromosome in the rat and in the opossum. If this should prove to be the case in all mammals, then all female mammals heterozygous for sex-linked mutant genes would be expected to show the same phenomena as those in the mouse. The coat of the tortoiseshell cat, being a mosaic of the black and yellow colours of the two homozygous genotypes, fulfills this expectation.

Source: Nature 190: 372

principal sex hormones. Many of the steroid hormones act by turning on the transcription of specific sets of genes. If a hormone regulates transcription, then it must somehow signal the DNA. The signaling mechanism for the steroid hormone cortisol is outlined in Figure 11.27. The hormone penetrates a target cell through diffusion, because steroids are hydrophobic (nonpolar) molecules that pass freely through the cell membrane into the cytoplasm. There it encounters a receptor molecule that is complexed with another protein called Hsp82, which functions to mask the receptor. Once cortisol binds to the receptor, it liberates the receptor from Hsp82, and the hormone-receptor complex migrates to the nucleus where it binds to its DNA target sequences to activate transcription. Nontarget cells do not contain the receptors and so are unaffected by the hormone.

A well-studied example of induction of transcription by a hormone is stimulation of the synthesis of ovalbumin in the chicken oviduct by the steroid sex hormone *estrogen.* When hens are injected with estrogen, oviduct tissue responds by synthesizing ovalbumin mRNA. This synthesis continues as long as estrogen is administered. When the hormone is withdrawn, the rate of synthesis decreases. Before injection of the hormone, and 60 hours after the injections have stopped, no ovalbu-

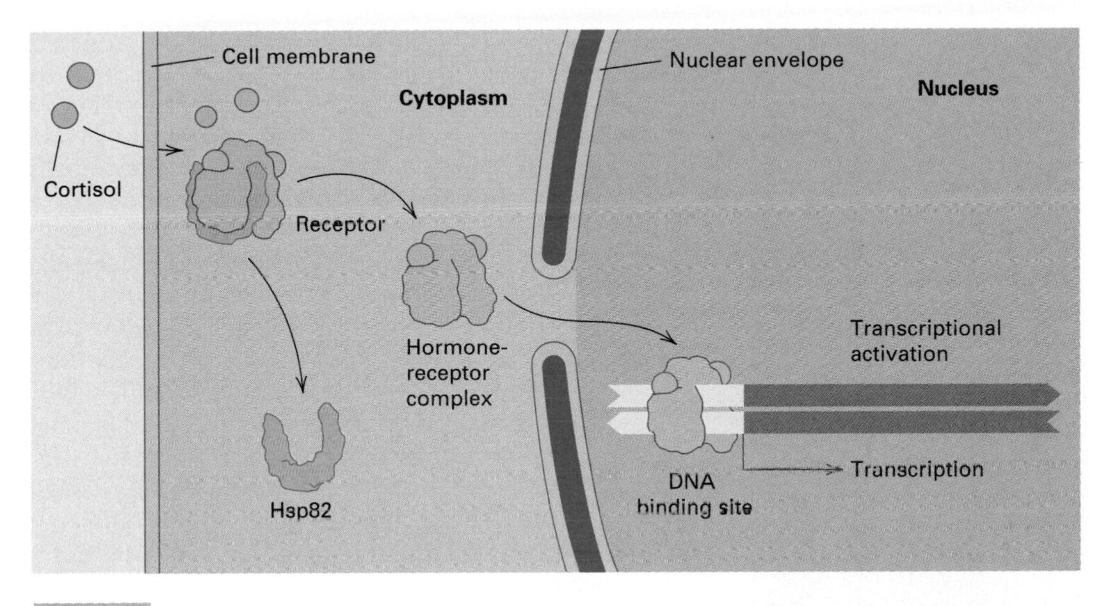

Figure 11.27 A schematic diagram showing how a steroid hormone reaches a DNA molecule and triggers transcription by binding with a receptor to form a transcriptional activator. Entry into the cytoplasm is by passive diffusion.

min mRNA is detectable. When estrogen is given to hens, only the oviduct synthesizes mRNA because other tissues lack the hormone receptor.

A transcriptional activator protein often binds with an enhancer.

Hormone receptors and other transcriptional activator proteins bind with particular DNA sequences known as **enhancers.** Enhancer sequences are typically rather short (usually fewer than 20 base pairs) and are found at a variety of locations around the gene affected. Most enhancers are upstream of the transcriptional start site (sometimes many kilobases away), others are in introns within the coding region, and a few are even located at the 3' end of genes. One of the most thoroughly studied enhancers is in the mouse mammary tumor virus, which determines transcriptional activation by the glucocorticoid steroid hormone. The consensus sequence of the enhancer is AGAQCAGQ, in which Q stands for either A or T. The virus contains five copies of the enhancer positioned throughout its genome, providing five binding sites for the hormone-receptor complex that activates transcription.

Enhancers are essential components of gene organization in eukaryotes because they enable genes to be transcribed only when proper transcriptional activators are present. Some enhancers respond to molecules from outside the cell (for example, steroid hormones that form receptor-hormone complexes). Other enhancers respond to molecules that are produced inside the cell—for example, during development—and these enhancers enable the genes under their control to participate in cellular differentiation or to be expressed in a tissue-specific manner. Many genes are under the control of several different enhancers, so they can respond to a variety of different molecular signals, both external and internal.

Figure 11.28 illustrates several of the genetic elements found in a typical eukaryotic gene. The

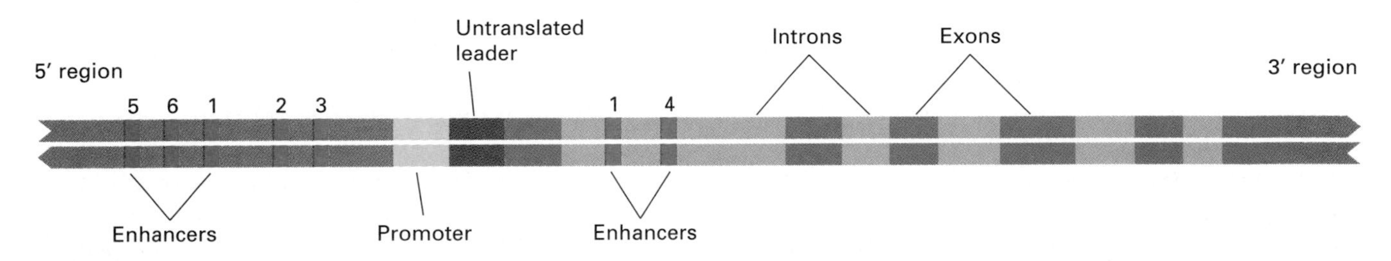

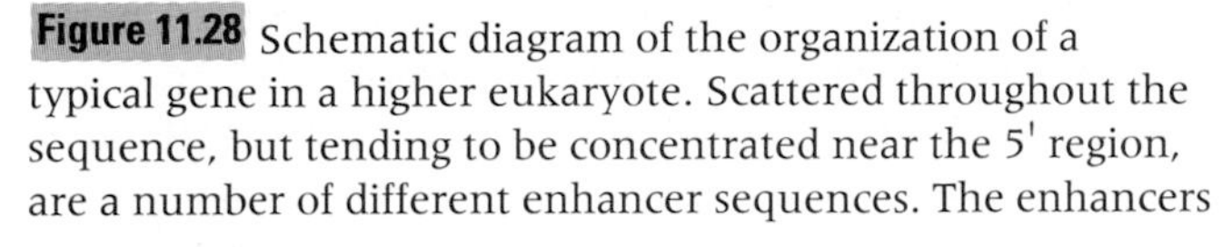

Figure 11.28 Schematic diagram of the organization of a typical gene in a higher eukaryote. Scattered throughout the sequence, but tending to be concentrated near the 5' region, are a number of different enhancer sequences. The enhancers are binding sites for transcriptional activator proteins that make possible the temporal and tissue-specific regulation of the gene.

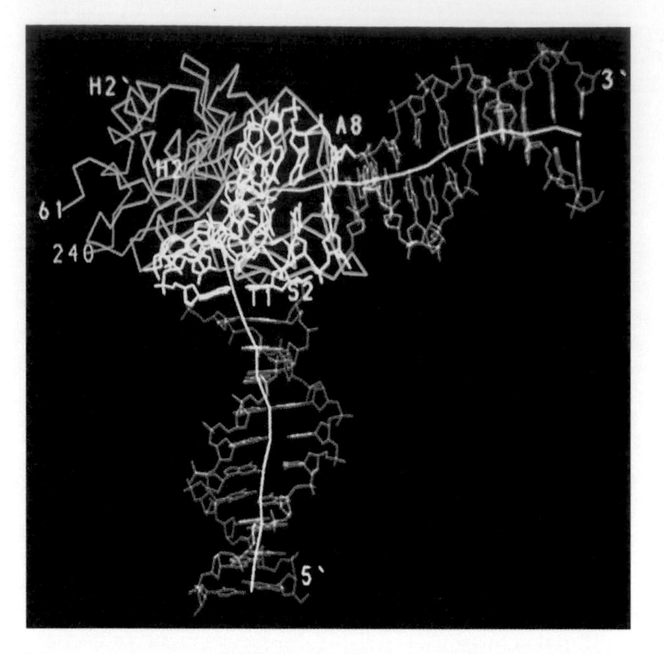

MODEL SHOWING HOW TATA-box-binding protein binds to its target in duplex DNA. Note the pronounced kink in the DNA duplex resulting from interacting with the protein. [Courtesy of Joseph L. Kim and Stephen K. Burley.]

transcriptional complex binds to the promoter (*P*) to initiate RNA synthesis. The coding regions of the gene (exons) are interrupted by one or more intervening sequences (introns) that are eliminated in RNA processing. Transcription is regulated by means of enhancer elements (numbered 1 through 6) that respond to different molecules that serve as induction signals. The enhancers are located both upstream and downstream of the promoter, and some (in this example, enhancer 1) are present in multiple copies.

11.9 The eukaryotic transcription complex includes many different proteins.

Many enhancers stimulate transcription by means of **DNA looping,** which refers to physical interactions between relatively distant regions along the DNA. The mechanism is illustrated in Figure 11.29. Part A shows the physical relationship between an enhancer and a promoter. In this example, the enhancer is upstream (5') of the promoter. Transcription requires that a group of protein factors forming the **transcription complex** bind with the promoter. Among the proteins in the transcription complex are **general transcription factors** that are common to the promoters of many different genes. The general transcription factors in eukaryotes have been highly conserved in evolution. One of the factors is TFIID, which includes a *TATA-box-binding protein (TBP)* that binds with the promoter in the region of the TATA box. In addition to TBP, the TFIID complex may also include a number of other proteins, called *TBP-associated factors (TAFs),* that act as intermediaries through which the effects of the transcriptional activator are transmitted. (Not all of the TBP is found in association with TAFs.) Transcription also requires an *RNA polymerase holoenzyme,* which consists of PolII (itself composed of 12 protein subunits) combined with at least 9 other protein subunits. In yeast these subunits include the transcription factors TFIIB, TFIIF, and TFIIH, as well as other proteins. Other general transcription factors have also been identified (for example, TFIIA and TFIIE), but it is not known whether these are components of larger complexes or whether they join the transcription complex as it is being assembled at the promoter.

Illustrated in Figure 11.29 is a mechanism of transcriptional activation called activation by **recruitment.** The key players, shown in Figure 11.29A, are the transcriptional activator protein and the TFIID and RNA polymerase holoenzyme complexes. The actual structures of TFIID and the holoenzyme complexes are not known, but for concreteness they are shown as multisubunit aggregates. To activate transcription (Figure 11.29B), the transcriptional activator protein binds to an enhancer in the DNA and to one of the TAF subunits in the TFIID complex. This interaction attracts ("recruits") the TFIID complex to the region of the promoter (Figure 11.29C). Attraction of the TFIID to the promoter also recruits the holoenzyme (Figure 11.29D) as well as any remaining general transcription factors, and in this manner the transcriptional complex is assembled for transcription to begin.

Experimental evidence for transcriptional activation by recruitment has come from studies of a number of artificial proteins created by fusing a DNA-binding domain with one of the protein subunits in TFIID. Such fusion proteins act as transcriptional activators wherever they bind to DNA (provided that a promoter is nearby), because the TFIID is "tethered" to the DNA-binding domain, and so the "recruitment" of TFIID is automatic. Similarly, fusion proteins that are tethered to a subunit of the holoenzyme can recruit the holoenzyme to the promoter. In this case, TFIID and the remaining general transcription factors are also attracted to the promoter, and the transcriptional complex is assembled. These experiments suggest that a transcriptional activator protein can activate transcription by interacting with subunits of either the TFIIA complex or the holoenzyme.

As Figure 11.29 suggests, the fully assembled transcription complex in eukaryotes is a very large structure. A real example, taken from early

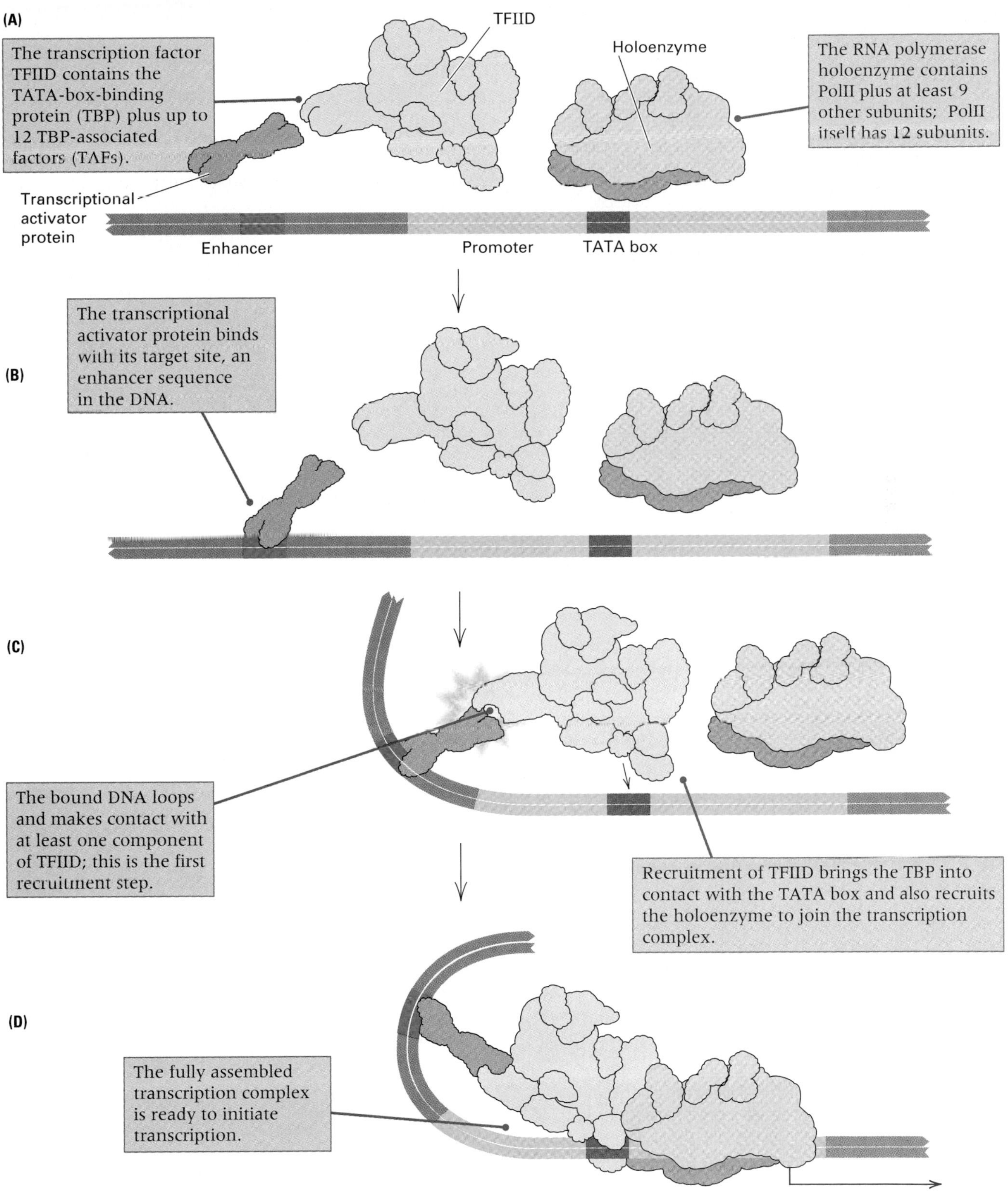

Figure 11.29 Transcriptional activation by recruitment. (A) Relationship between enhancer and promoter and the protein factors that bind to them. (B) Binding of the transcriptional activator protein to the enhancer. (C) Bound transcriptional activator protein makes physical contact with a subunit in the TFIID complex, which contains the TATA-box-binding protein, and attracts ("recruits") the complex to the promoter region. (D) The PolII holoenzyme and any remaining general transcription factors are recruited by TFIID, and the transcription complex is fully assembled and ready for transcription. In the cell, not all of the PolII is found in the holoenzyme, and not all of the TBP is found in TFIID. In this illustration, transcription factors other than those associated with TFIID and the holoenzyme are not shown.

development in *Drosophila,* is shown in Figure 11.30. In this case, the enhancers, located a considerable distance upstream from the gene to be activated, are bound by the transcriptional activator proteins BCD and HB, which are products of the genes *bicoid (bcd)* and *hunchback (hb),* respectively; these transcriptional activators function in establishing the anterior-posterior axis in the embryo. (Early *Drosophila* development is discussed in Chapter 12.) Note the position of the TATA box in the promoter of the gene. The TATA box binding is the function of the TBP. The functions of a number of other components of the transcription complex have also been identified. For example, the TFIIH contains both helicase and kinase activity to melt the DNA and to phosphorylate RNA polymerase II. Phosphorylation allows the polymerase to leave the promoter and elongate mRNA. The looping of the DNA effected by the transcriptional activators is an essential feature of the activation process. Transcriptional activation in eukaryotes is a complex process, especially when compared to the prokaryotic RNA polymerase, which consists of a core $\alpha_2\beta\beta'$ tetramer and a σ factor.

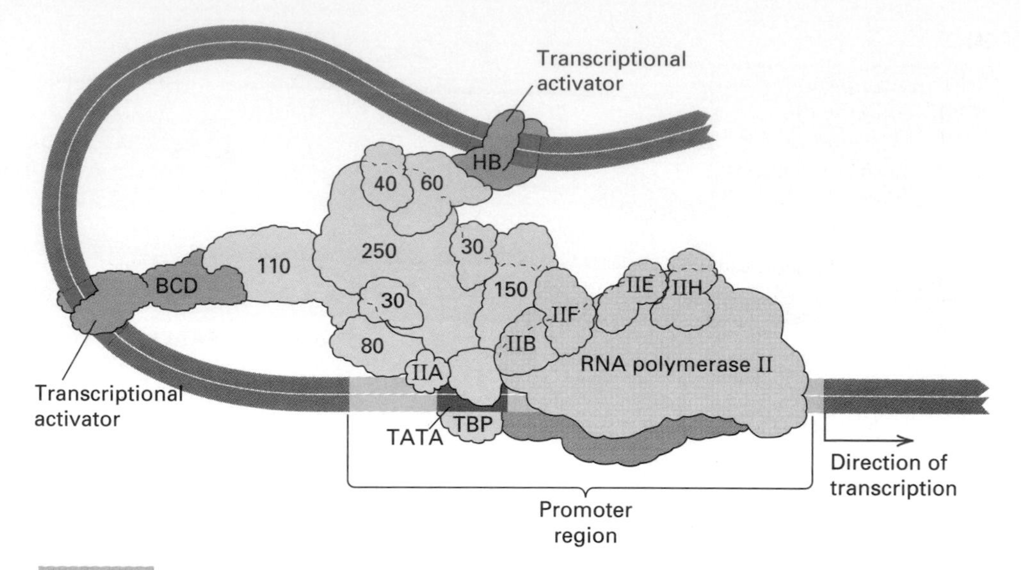

Figure 11.30 An example of transcriptional activation during *Drosophila* development. The transcriptional activators in this example are bicoid protein (BCD) and hunchback protein (HB). The numbered subunits are TAFs (TBP-associated factors) that, together with TBP (TATA-box-binding protein), correspond to TFIID. BCD acts through a 110-kilodalton TAF, and HB through a 60-kilodalton TAF. The transcriptional activators act via enhancers to cause recruitment of the transcriptional apparatus. The fully assembled transcription complex includes TBP and TAFs, RNA polymerase II, and general transcription factors TFIIA, TFIIB, TFIIE, TFIIF, and TFIIH.

A strategically placed enhancer can act as a genetic switch.

The versatility of some enhancers results from their ability to interact with two different promoters in a competitive fashion; that is, at any one time, the enhancer can stimulate one promoter or the other, but not both. An example of this mechanism is illustrated in Figure 11.31, in which *P1* and *P2* are alternative promoters that compete for an enhancer located between them. When complexed with a transcriptional activator specific for promoter *P1,* the enhancer binds preferentially with promoter *P1* and stimulates transcription (Figure 11.31A). When complexed with a different transcriptional activator specific for promoter *P2,* the enhancer binds preferentially with promoter *P2* and stimulates transcription from it (Figure 11.31B). In this way, competition for the enhancer serves as a sort of switch mechanism for the expression of the *P1* or *P2* promoter. This regulatory mechanism is present in chickens and results in a change from the production of embryonic β-globin to that of adult β-globin in development. In this case, the embryonic globin gene and the adult gene compete for a single enhancer, which in the course of development changes its preferred binding from the embryonic promoter to the adult promoter. In human beings, enhancer competition appears to control the developmental switch from the fetal γ-globin to that of the adult β-globin polypeptide chains. In persons in whom the β-globin promoter is deleted or altered in sequence and unable to bind with the enhancer, there is no competition for the enhancer molecules,

Figure 11.31 (facing page) Genetic switching regulated by competition for an enhancer. Promoters *P1* and *P2* compete for a single enhancer located between them. When complexed with an appropriate transcriptional activator protein, the enhancer binds preferentially with promoter *P1* (A) or promoter *P2* (B). Binding to the promoter recruits the transcription complex. If either promoter is mutated or deleted, then the enhancer binds with the alternative promoter. The location of the enhancer relative to the promoters is not critical.

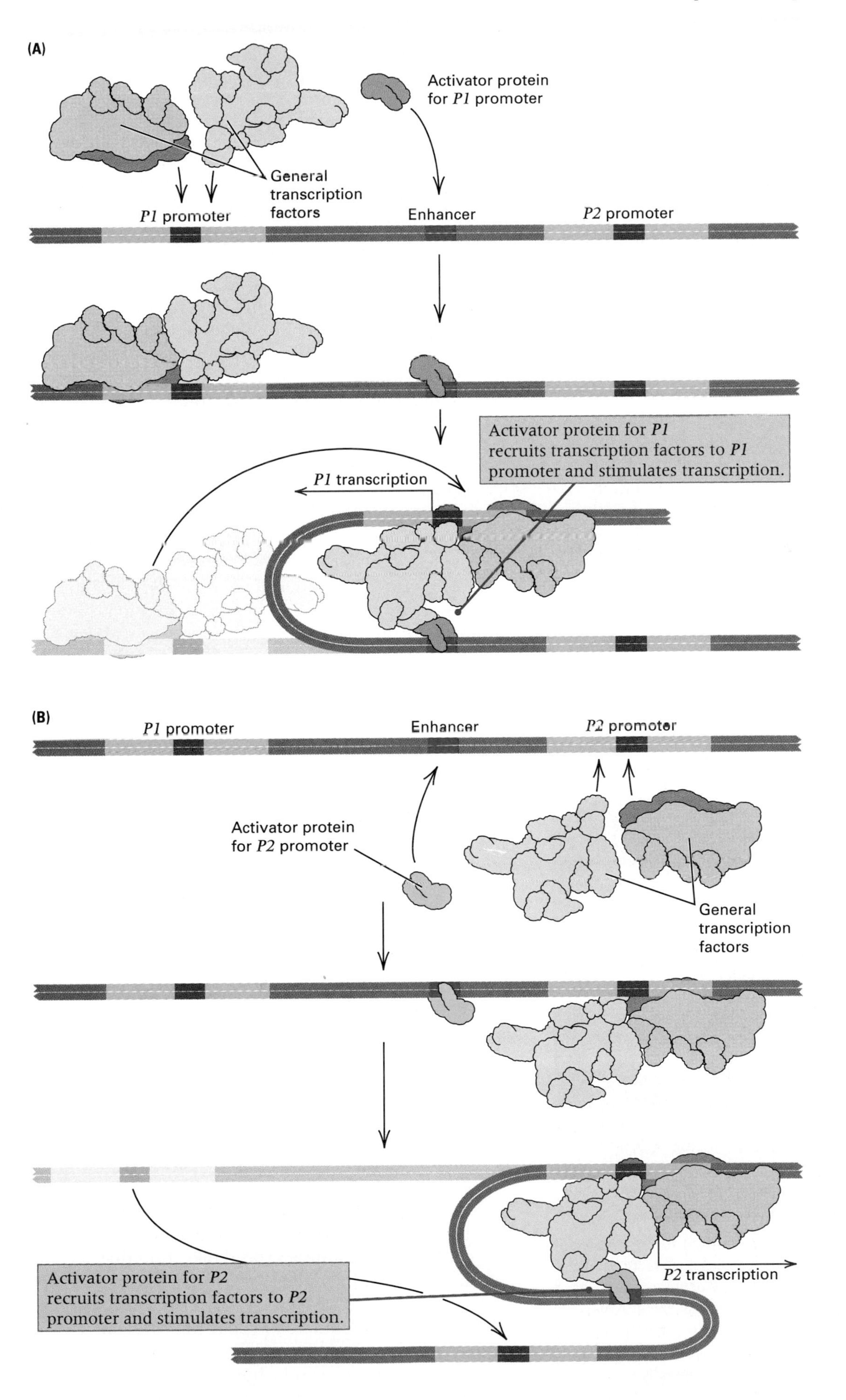
(A)
Activator protein for *P1* promoter
General transcription factors
P1 promoter
Enhancer
P2 promoter
Activator protein for *P1* recruits transcription factors to *P1* promoter and stimulates transcription.
P1 transcription
(B)
P1 promoter
Enhancer
P2 promoter
Activator protein for *P2* promoter
General transcription factors
Activator protein for *P2* recruits transcription factors to *P2* promoter and stimulates transcription.
P2 transcription

and the γ-globin genes continue to be expressed in adult life when normally they would not be transcribed. Adults with these types of mutations have fetal hemoglobin instead of the adult forms. The condition is called hereditary persistence of fetal hemoglobin (or high-F disease), but the clinical manifestations are very mild.

Figure 11.32 (facing page) Use of alternative promoters in the gene for alcohol dehydrogenase in *Drosophila.* (A) The overall gene organization includes two introns within the amino acid coding region. (B) Transcription in larvae uses the promoter nearest the 5' end of the coding region. (C) Transcription in adults uses a promoter farther upstream, and much of the larval leader sequence is removed by splicing.

Combinatorial control means that a few genes can control many others.

Because the genome of a complex eukaryote contains many possible enhancers that respond to different signals or cellular conditions, in principle each gene could be regulated by its own distinct combination of enhancers. This **combinatorial control** is a powerful means of increasing the complexity of possible regulatory states by employing several simple regulatory states in combination.

To consider a simple example, suppose that a gene has a single binding site for only one type of regulatory molecule. Then there are only two regulatory states (call them + and −) according to whether the binding site is occupied. If the gene has single binding sites for each of two different regulatory molecules, then there are four possible regulatory states (+ +, + −, − +, and − −). Single binding sites for three different regulatory molecules yield eight combinatorial states, and in general n different types of regulatory molecules yield 2^n states. If transcription is regulated according to the particular pattern of which binding sites are occupied, then a small number of types of regulatory molecules can result in a large number of different patterns of regulation. For example, because mammals contain approximately 200 different cell types, each gene would theoretically need as few as 8 +/− types of binding sites to specify whether it should be on or off in each cell type, because $2^8 = 256$.

A CORN PLANT requires exposure to sunlight to create chlorophyll, the light-trapping pigment molecules necessary for photosynthesis. A blue-green pigment, phytochrome, acts as a signaling molecule, activating genes that specify enzymes and other proteins with roles in germination, stem elongation, branching, leaf expansion, and formation of flowers, fruits, and seeds. [© Richard H. Gross/Biological Photography.]

The actual regulatory situation is certainly more complex than the simple calculation based on +/− switches would imply, because genes are not merely on or off: their level of activity is adjusted. For example, a gene may have multiple binding sites for an activator protein, and this allows the level of gene expression to be modulated according to the number of binding sites that are occupied. Furthermore, each cell type is programmed to respond to a variety of conditions both external and internal, so the total number of regulatory states must be considerably greater than the number of cell types. On the other hand, the +/− example demonstrates that combinatorial control is extremely powerful in multiplying the number of regulatory possibilities, and therefore a large number of regulatory states does not necessarily imply a hopelessly complex regulatory apparatus.

A gene can have two or more promoters that are regulated differently.

Some eukaryotic genes have two or more promoters that are active in different cell types. The different promoters result in different primary transcripts that contain the same protein-coding regions. An example from *Drosophila* is shown in Figure 11.32. The gene codes for alcohol dehydrogenase, and its organization in the genome is shown in Figure 11.32A; there are three protein-coding regions interrupted by two introns. Transcription in larvae (part B) uses a different promoter from that used in transcription in adults (part C). The adult transcript has a longer 5' leader sequence, but most of this sequence is eliminated in splicing. Alternative promoters allow independent regulation of transcription in larvae and adults.

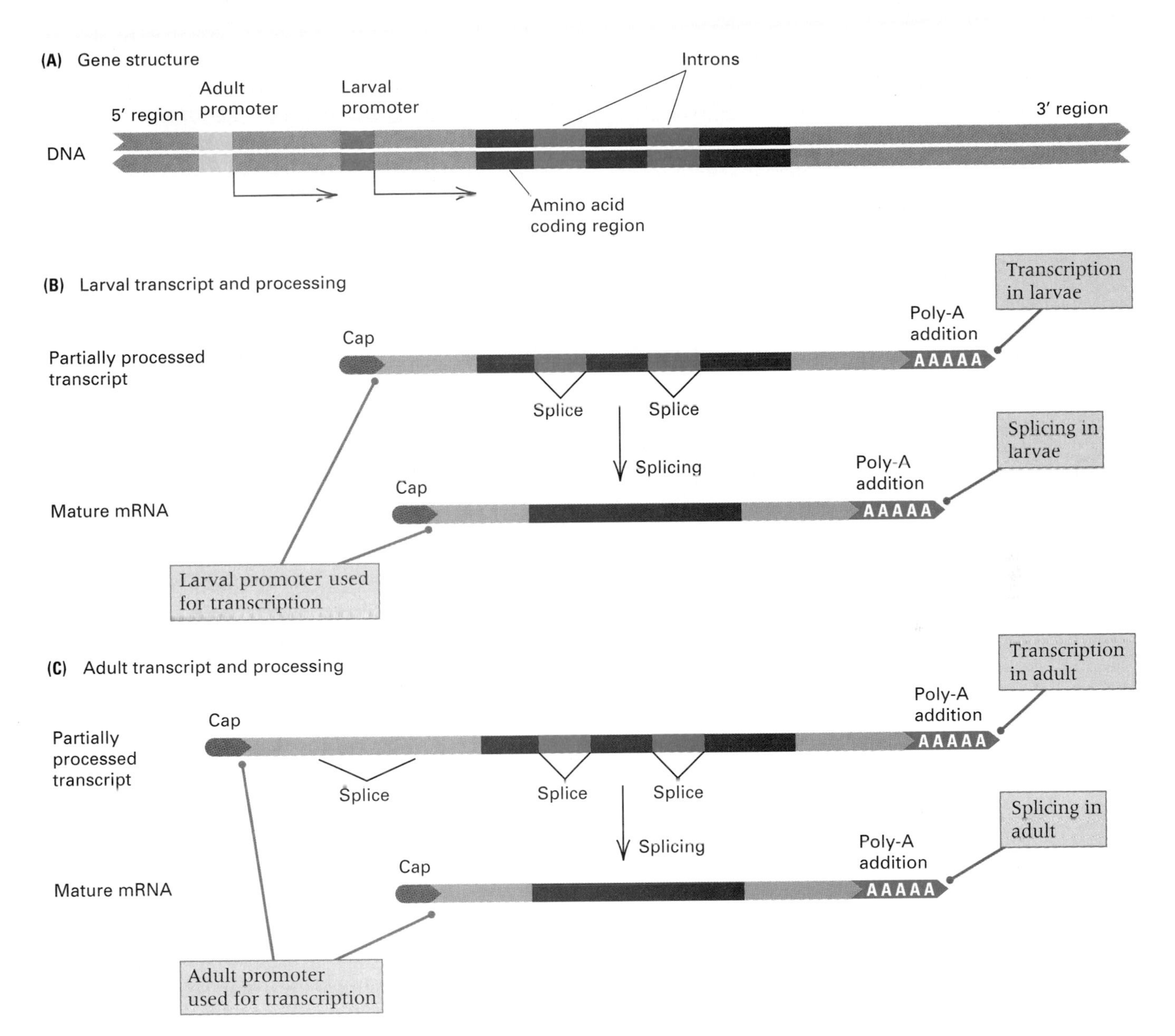

11.10 Some transcripts are spliced in any of a variety of ways.

Even when the same promoter is used to transcribe a gene, different cell types can produce different quantities of the protein, or even different proteins, because of differences in the mRNA produced in processing. The reason is that the same transcript can be spliced differently from one cell type to the next. The different splicing patterns may include exactly the same protein-coding exons, in which case the protein is identical but the rate of synthesis differs because the mRNA molecules are not translated with the same efficiency. In other cases, the protein-coding part of the transcript has a different splicing pattern in each cell type, and the resulting mRNA molecules code for proteins that are not identical even though they share certain exons.

In the synthesis of α-amylase in the mouse, different mRNA molecules are produced from the same gene because of different patterns of intron removal in RNA processing. The mouse salivary gland produces more of the enzyme than the liver, although the same coding sequence is transcribed. In each cell type, the same primary transcript is synthesized, but two different splicing mechanisms are used. The initial part of the primary transcript is shown in Figure 11.33. The coding sequence begins 50 base pairs inside exon 2 and is formed by joining exon 3 and subsequent exons. In the salivary gland, the primary transcript is processed such that exon S is joined with exon 2 (that is, exon L is removed as part of introns 1 and 2). In the liver, exon L is joined with exon 2, and exon S is removed along with intron 1 and the leader L. The exons S and L become alternative 5' leader sequences of the amylase mRNA, and the alternative mRNAs are translated at different rates.

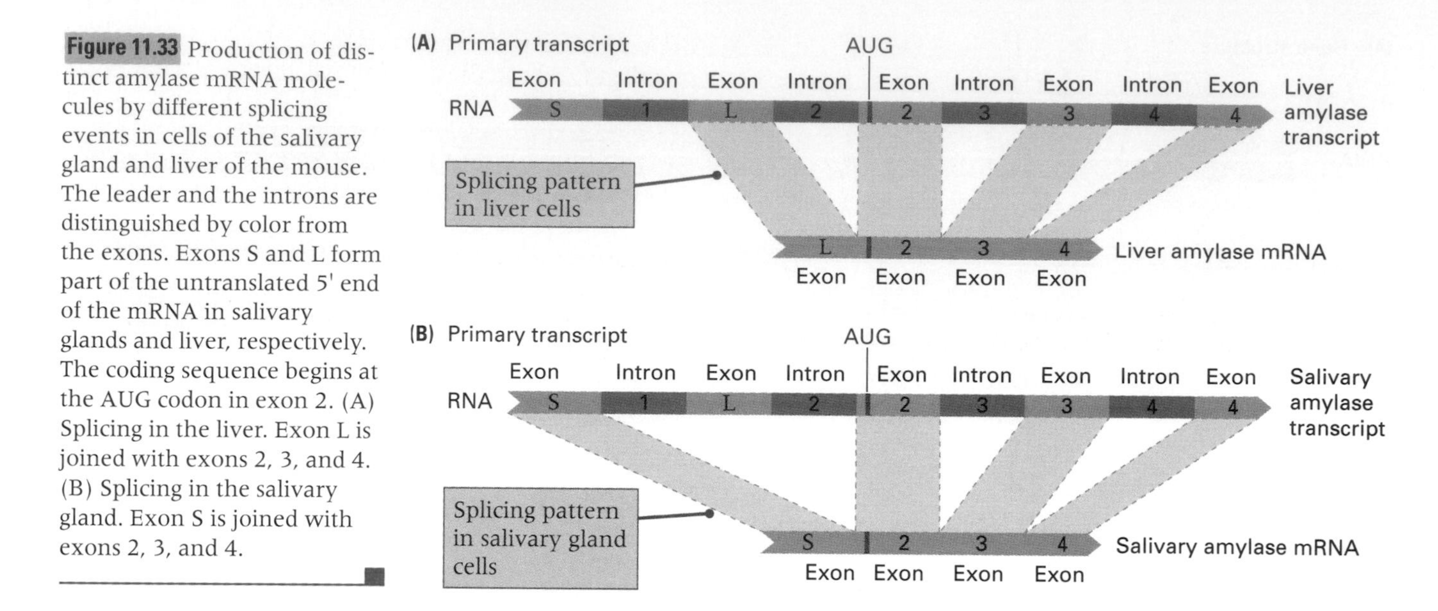

Figure 11.33 Production of distinct amylase mRNA molecules by different splicing events in cells of the salivary gland and liver of the mouse. The leader and the introns are distinguished by color from the exons. Exons S and L form part of the untranslated 5' end of the mRNA in salivary glands and liver, respectively. The coding sequence begins at the AUG codon in exon 2. (A) Splicing in the liver. Exon L is joined with exons 2, 3, and 4. (B) Splicing in the salivary gland. Exon S is joined with exons 2, 3, and 4.

11.11 Gene regulation can also be exerted at the level of translation.

In bacteria, most mRNA molecules are translated about the same number of times, with little variation from gene to gene. In eukaryotes, translation is sometimes regulated. The principal types of translational control are

1. Inability of an mRNA molecule to be translated unless a molecular signal is present
2. Regulation of the lifetime of a particular mRNA molecule
3. Regulation of the overall rate of protein synthesis
4. Aborted translation of the principal open reading frame because of the presence of smaller open reading frames upstream in the mRNA

In this section, we will consider examples of the first three modes of regulation.

An important example of translational regulation is that of **masked mRNA.** Unfertilized eggs are biologically static, but shortly after fertilization, many new proteins must be synthesized—among them, for example, the proteins of the mitotic apparatus and the cell membranes. Unfertilized sea urchin eggs can store large quantities of mRNA for many months in the form of mRNA-protein particles made during formation of the egg. This mRNA is translationally inactive, but within minutes after fertilization, translation of these molecules begins. Here, the timing of translation is regulated; the mechanisms for stabilizing the mRNA, for protecting it against RNases, and for activatiing it are unknown.

Translational regulation of another type occurs in mature unfertilized eggs. These cells need to maintain themselves but do not have to grow or undergo a change of state. Thus the rate of protein synthesis in eggs is generally low. This is not a consequence of an inadequate supply of mRNA but apparently results from failure to form the ribosome-mRNA complex.

The synthesis of some proteins is regulated by direct action of the protein on the mRNA. For instance, the concentration of one type of antibody molecule is kept constant by self-inhibition of translation; that is, the antibody molecule itself binds specifically to the mRNA that encodes it and thereby inhibits initiation of translation.

A dramatic example of translational control is the extension of the lifetime of silk fibroin mRNA in the silkworm. During cocoon formation, the silk gland of the silkworm predominantly synthesizes a single type of protein, silk fibroin. Because the worm takes several days to construct its cocoon, it is the total amount (not the rate) of fibroin synthesis that must be great. The silkworm achieves this in two ways. First, the silk gland cells become highly polyploid, accumulating thousands of copies of each chromosome. Second, each fibroin gene synthesizes thousands of mRNA molecules, each with a lifetime of several days. (A typical eukaryotic mRNA molecule has a lifetime of about 3 hours before it is degraded.) During this time each mRNA molecule is translated repeatedly. Altogether, the silk gland makes about 10^{15} molecules of fibroin over a period of 4 days. If the lifetime of the mRNA were not extended, synthesis of the required fibroin would take about 100 days.

Another example of an mRNA molecule with an extended lifetime is the mRNA encoding casein, the major protein of milk, in mammary glands.

When the hormone prolactin is received by the gland, the lifetime of casein mRNA increases. Synthesis of the mRNA also continues, so the overall rate of production of casein is markedly increased by the hormone. When stimulation by prolactin ceases, the concentration of casein mRNA decreases because the RNA is degraded more rapidly, and lactation terminates.

Is there a general principle of regulation?

With microorganisms, whose environment frequently changes drastically and rapidly, a general principle of regulation is that bacteria make what they need when it is required and in appropriate amounts. Although such extraordinary efficiency is common in prokaryotes, it is rare in eukaryotes. For example, efficiency appears to be violated by the large amount of DNA in the genome that has no protein-coding function and by the large amount of intron RNA that is discarded.

What is abundantly clear is that there is no universal regulatory mechanism. Many control points are possible, and different genes are regulated in different ways. Furthermore, evolution has not always selected for simplicity in regulatory mechanisms but merely for something that works. If a cumbersome regulatory mechanism were to arise, it would in time evolve, be refined, and become more effective, but it would not necessarily become simpler. On the whole, regulatory mechanisms include a variety of seemingly *ad hoc* processes, each of which has stood the test of evolutionary time, primarily because it works.

Chapter Summary

- **There are many possible control points for the regulation of gene expression.**

Most cells do not synthesize molecules that are not needed. Levels of gene regulation include transcription, RNA processing, and translation. The processes of gene regulation generally differ between prokaryotes and eukaryotes.

- **Regulation of transcription is a common mechanism in prokaryotes.**

In prokaryotes, the synthesis of degradative enzymes that are needed only on occasion, such as the enzymes required to metabolize lactose, is typically regulated by an off-on mechanism. When lactose is present, the genes coding for the enzymes required to metabolize lactose are transcribed; when lactose is absent, transcription is prevented. Lactose metabolism is negatively regulated. Two enzymes required to degrade lactose—permease, necessary for the entry of lactose into bacteria, and β-galactosidase, the enzyme that does the degrading—are coded by a single polycistronic mRNA molecule, *lac* mRNA. Immediately downstream from the promoter for *lac* mRNA is a regulatory sequence of bases called an operator. In the absence of lactose, a repressor protein binds tightly to the operator region and prevents RNA polymerase from initiating transcription at the promoter. Lactose is an inducer of transcription because it can bind with the repressor, changing its shape and preventing the repressor from binding with the operator. Therefore, in the presence of lactose, there is no bound repressor, and the *lac* promoter is always accessible to RNA polymerase. The promoter region, the operator region, and the structural genes are adjacent to one another and together constitute the *lac* operon. Repressor mutations have been isolated that inactivate the repressor protein, and operator mutations are known that prevent recognition of the operator by an active repressor; such mutations cause continuous production of *lac* mRNA and are said to be constitutive. Promoter mutations can prevent recognition of the promoter by RNA polymerase and thereby block transcription of *lac* mRNA.

- **Lactose degradation is regulated by the lactose operon.**

The first regulatory mutations that were discovered affected lactose metabolism.
Lactose-utilizing enzymes can be inducible (regulated) or constitutive.
Repressor shuts off messenger RNA synthesis.
The lactose operator is an essential site for repression.
The lactose promoter is an essential site for transcription.
The lactose operon contains linked structural genes and regulatory sequences.
The lactose operon is also subject to positive regulation.

The *lac* operon, like many other operons coding for sugar-degrading enzymes, is also positively regulated by another mechanism. Initiation of transcription requires the binding of a particular protein molecule, called CRP, to a specific region of the promoter. This binding takes place only after CRP has first bound cyclic AMP (cAMP) and formed a cAMP-CRP complex. In the presence of glucose, the cellular concentration of cAMP is too low to form enough cAMP-CRP to make transcription possible. In the absence of glucose, enough cAMP-CRP is formed for induction.

- **Tryptophan biosynthesis is regulated by the tryptophan operon.**

Attenuation allows for fine-tuning of transcriptional regulation.

Biosynthetic enzyme systems exemplify another type of transcriptional regulation in prokaryotes. In the synthesis of tryptophan, transcription of the genes encoding the *trp* enzymes is controlled by the concentration of tryptophan in the growth medium. When excess tryptophan is present, it binds with the *trp* aporepressor to form the active repressor that prevents transcription. The *trp* operon is also regulated by attenuation, in which transcription is initiated continually but the transcript forms a hairpin structure resulting in premature termination. The frequency of termination of transcription is determined by the availability of tryptophan:

Decreasing concentrations of tryptophan lower the chance of termination, which increases the rate of tryptophan synthesis. Analogous attenuators also regulate operons for the synthesis of other amino acids.

- **Operons also regulate the lysogenic and lytic cycles of bacteriophage λ.**

Bacteriophage λ adopts either the quiescent lysogenic state or the lytic cycle as a result of competition between two repressors, cI and cro. If cI repressor dominates, the λ genome becomes integrated into the bacterial chromosome. A lysogen is formed, in which cI continues to be synthesized and prevents transcription of all other λ genes. If cro repressor dominates, cI repressor is no longer synthesized, and the lytic cycle ensues. Although there are elements of positive transcriptional regulation in the regulatory circuitry, the predominant mode of regulation is negative.

- **Eukaryotes make use of a great variety of genetic regulatory mechanisms.**

Genetic organization differs in eukaryotes and prokaryotes.

- **Some genes are regulated by alteration of the DNA.**

Gene amplification increases dosage and the amount of gene product.
Yeast cells change mating type by programmed DNA rearrangements.
DNA splicing takes place in the formation of antibodies.
Gene splicing occurs during formation of T-cell receptors.
DNA and its expression can be modified by methylation.

Eukaryotes employ a variety of genetic regulatory mechanisms, occasionally including changes in DNA. The number of copies of a gene may be increased by DNA amplification, there may be programmed rearrangements of DNA sequences, or regulation may be mediated by the methylation of cytosine bases. Reversible DNA rearrangements take place in mating-type switching in yeast, and irreversible DNA rearrangements take place in the formation of antibody-coding genes in certain cells in the immune system.

- **Transcriptional control is a frequent mode of regulation in eukaryotes.**

Genes controlling yeast mating type regulate transcription.
Positive transcriptional regulation makes use of activator proteins.
Some transcriptional activator proteins interact with hormones.
A transcriptional activator protein often binds with an enhancer.

- **The eukaryotic transcription complex includes many different proteins.**

A strategically placed enhancer can act as a genetic switch.
Combinatorial control means that a few genes can control many others.
A gene can have two or more promoters that are regulated differently.

Many genes in eukaryotes are regulated at the level of transcription. Although regulation may be either negative or positive, positive regulation is the usual situation. In the control of yeast mating type by the *MAT* locus, the **a**-specific genes and the haploid-specific genes are negatively regulated and the α-specific genes are positively regulated. In general, positive regulation is through transcriptional activator proteins that contain particular structural motifs that bind to DNA—for example, the helix-loop-helix motif or the zinc-finger motif. Hormones also can regulate transcription. Steroid hormones bind with receptor proteins in the nucleus and form transcriptional activators. The DNA-binding sites of transcriptional activators are sequences known as enhancers, which are usually short sequences that may be present at a variety of positions around the genes that they regulate. There are many types of enhancers responsive to different transcriptional activators. Combining different types of enhancers provides a large number of possible types of regulation.

- **Some transcripts are spliced in any of a variety of ways.**

- **Gene regulation can also be exerted at the level of translation.**

Is there a general principle of regulation?

Some genes contain alternative promoters used in different tissues; other genes use a single promoter, but the transcripts are spliced in different ways. Alternative splicing can result in mRNA molecules that are translated with different efficiencies, or even in different proteins if there is alternative splicing of the protein-coding exons. Regulation can also be at the level of translation—for example, through masked mRNAs or through factors that affect mRNA stability.

Key Terms

aporepressor
attenuation
attenuator
autoregulation
β-galactosidase
cAMP-CRP complex
cassette
cis-dominant
combinatorial control
constant antibody regions (C)
constitutive mutant
coordinate regulation
co-repressor
DNA looping
DNA methylase
enhancer
gene amplification
gene dosage
general transcription factor
gene regulation
homothallism
hormone
housekeeping gene
inducer
inducible transcription
lactose permease
leader polypeptide
masked mRNA
mating-type interconversion
negative regulation
operator
operon
polycistronic mRNA
positive regulation
promoter
recruitment
repressible transcription
repressor
transcriptional activator protein
transcription complex
variable antibody regions (V)

Concepts and Issues
Testing Your Knowledge

- What is *positive regulation* of transcription? What is *negative regulation* of transcription? Give an example of each type of regulation.
- What class of *lac* mutants demonstrated that the presence of lactose in the growth medium was not necessary for expression of the genes for lactose utilization?
- How does an operon result in coordinate control of the genes included? Are operons found in eukaryotic organisms?
- In what sense does attenuation provide a "fine-tuning" mechanism for operons that control amino acid biosynthesis?
- What is a *transcriptional enhancer?* What role do enhancers play in gene regulation?
- How does the possibility of alternative splicing affect the generality of the statement that one gene encodes one polypeptide chain?
- Would you expect DNA splicing of the type observed in antibody formation to be a reversible process or an irreversible process? Explain.

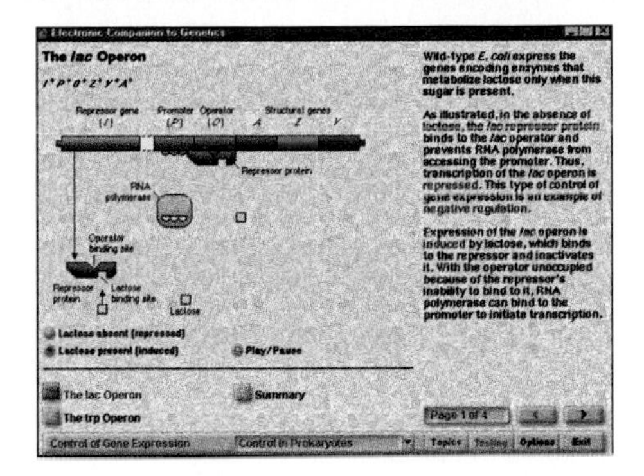

TO ENHANCE YOUR STUDY turn to *An Electronic Companion to Genetics*™, © 1998, Cogito Learning Media, Inc. This CD-ROM is a multimedia tool that provides interactive explanations of important genetic concepts through animations, diagrams, and videos. It also provides interactive test questions.

Solutions
Step by Step

Problem 1

The LacY permease is able to transport lactobionic acid as well as lactose, but LacZ cleaves lactobionic acid very inefficiently. Certain mutant forms of LacZ have an altered substrate specificity that allows the mutant enzyme to cleave lactobionic acid. These mutants are able to grow in medium containing lactobionic acid as the sole source of carbon and energy. However, they cannot grow unless they also have a *lacO*^{c} mutation or, when the operator is wildtype, unless IPTG (isopropylthiogalactoside) is present in the medium. How can you explain these results?

Solution In order to grow on lactobionic acid, even the mutant LacZ strain must have both *lacZ* and *lacY* expressed. Evidently lactobionic acid is not an inducer of the *lac* operon, and therefore a *lacO*$^+$ mutant cannot grow in lactobionic acid unless IPTG, an inducer of the *lac* operon, is also present to induce transcription of the operon. The finding with *lacO*^{c} mutants is consistent with this explanation, because when LacY and the mutant LacZ are produced constitutively, the strains can grow in lactobionic acid.

Problem 2

The *arabinose* operon encodes the enzymes necessary for degradation of the 5-carbon sugar arabinose. It is one of the few degradative operons in *E. coli* that are controlled primarily by positive regulation of transcription. The protein AraC, encoded in the gene *araC*, is an arabinose-responsive transcriptional activator protein. In the presence of arabinose, AraC has a high affinity for the arabinose operator region, where it binds and induces transcription of the operon in a polycistronic mRNA that includes coding regions for the degradative enzymes AraB, AraA, and AraD (collectively called AraBAD).The operon is also positively regulated by the cyclic AMP receptor protein (CRP), in much the way the *lac* operon is regulated by CRP. Fill each box in the table shown here with a + or a − sign. A + (or −) in the column for AraC or CRP means that under the specified growth condition, the AraC or CRP protein is (or is not) bound with the arabinose operator; a + (or −) in the column for AraBAD means that under the specified growth condition, the AraBAD proteins are (or are not) induced.

	Operator binding		Enzyme induction
Growth medium	AraC	CRP	AraBAD
Lactose	❑	❑	❑
Arabinose	❑	❑	❑
Arabinose + glucose	❑	❑	❑

Solution According to the description of how the operon is regulated, AraC binds the operator only when arabinose is present in the medium. The binding of AraC will induce transcription of the araBAD polycistronic mRNA, but AraC alone is not sufficient. As in the *lac* operon, transcription can occur only if the cAMP-CRP complex is also bound with the operator. The cAMP-CRP complex will form when cells have a high level of cyclic AMP, such as when growing in lactose or arabinose medium, but not when the level of cyclic AMP is low, such as under growth in glucose medium. The transcription of the araBAD mRNA results in the synthesis of the AraBAD enzymes. These considerations indicate how the table should be completed.

	Operator binding		Enzyme induction
Growth medium	AraC	CRP	AraBAD
Lactose	−	+	−
Arabinose	+	+	+
Arabinose + glucose	+	−	−

geNETics on the web

www.jbpub.com/genetics

These GeNETics on the Web exercises will introduce some of the most useful sites for finding genetic information on the World Wide Web. Genetic sites provide access to a rich storehouse of information on all aspects of genetics. These range from sites written in nontechnical language for the lay person to sites with sophisticated databases designed for the professional geneticist. In carrying out these exercises, you will get a taste of what the Internet can offer a student in genetics.

The keywords shown in color in the following exercises are available on the Jones and Bartlett Publishers' web site as hyperlinks to various genetic sites. To complete the exercises, visit the GeNETics on the Web home page at

http://www.jbpub.com/genetics

Select the link to Essential geNETics on the web. Then choose a chapter. You find a list of keywords that correspond to the exercises below. Select a keyword to link to a web site containing the genetic information necessary to complete the exercise. Each exercise includes a specific assignment that makes use of the information available at the site.

Exercises

- The "Boy in the Bubble," David Vetter, lived in a plastic bubble isolation unit in Houston, Texas, for a prolonged period before his death. He suffered from a form of severe combined immunodeficiency disease (SCID) due to an X-linked mutation in the gene *IL-2r* encoding a component of the receptor for the lymphokine IL-2, a cell-growth factor that affects the growth and differentiation of T cells, B cells, natural killer cells, glioma cells, and cells of the monocyte lineage. This keyword site tells the story of David Vetter 13 years after his death. Write a brief essay focusing on some of the controversial issues that this case has raised.

- Severe combined immunodeficiency disease (SCID) is a rare disease affecting about 40 children a year worldwide. Because they lack a functional immune system, these children usually die from infections such as chicken pox or pneumonia before 2 years of age. Approximately 40 percent of affected children have a defect in the enzyme adenosine deaminase (ADA), an "inborn error of metabolism." Use the keyword **ADA** at the search engine at this site to learn more about this form of SCID. On the basis of the information given, write a short report discussing the options for treatment of ADA-SCID and some of the problems, or write a short summary of some of the different types of ADA mutants that have been found in patients with ADA-SCID.

Pic Site

The Pic Site showcases some of the most visually appealing genetics sites on the World Wide Web. To visit the showcase genetics site, select the Pic Site for Chapter 11.

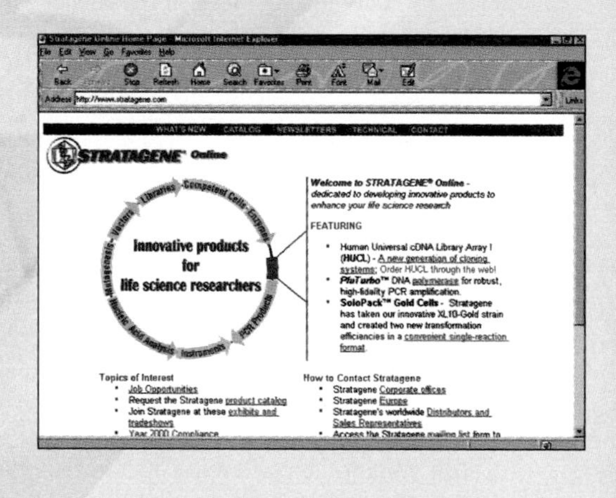

Concepts in Action
Problems for Solution

11.1 Is it inevitable that once mRNA for a protein is made, the protein will be synthesized? What sort of mechanisms might control the translational process?

11.2 Is it necessary for the gene that codes for the repressor of a bacterial operon to be near the structural genes? Why or why not?

11.3 Why are mutations of the *lac* operator often called *cis*-dominant? Why are some constitutive mutations of the *lac* repressor (*lacI*) called *trans*-recessive? Can you think of a way in which a noninducible mutation in the *lacI* gene might be *trans*-dominant?

11.4 Why is the *lac* operon of *E. coli* not inducible in the presence of glucose?

11.5 A mutation imparting constitutive synthesis of an enzyme of arginine biosynthesis in *Citrobacteria* was found. The enzyme is normally repressible by arginine.

(a) What two kinds of regulatory mutations might cause this phenotype?
(b) What kind of mutation in an enzyme of arginine biosynthesis might cause this phenotype?

11.6 If a wildtype *E. coli* strain is grown in a medium without lactose or glucose, how many proteins are bound to the *lac* operon? How many are bound if the cells are grown in glucose?

11.7 In regard to mating-type switching in yeast, what phenotype would you expect from a type α cell that has a deletion of the *HMR***a** cassette?

11.8 Cells of genotype *lacI*$^-$ *lacO*$^+$ *lacZ*$^+$ *lacY*$^+$ Hfr are mated with cells of genotype *lacI*$^+$ *lacO*$^+$ *lacZ*$^+$ *lacY*$^+$ F$^-$. In the absence of any inducer in the medium, no β-galactosidase is made. However, when the *lacI*$^+$ *lacO*$^+$ *lacZ*$^+$ *lacY*$^+$ Hfr strain is mated with a strain of genotype *lacI*$^-$ *lacO*$^+$ *lacZ*$^-$ *lacY*$^-$ F$^-$ under the same conditions, β-galactosidase is synthesized for a short time after the Hfr and F$^-$ cells have been mixed. Explain this observation.

11.9 When glucose is present in an *E. coli* cell, is the concentration of cyclic AMP high or low? Can a mutant with either an inactive *adenyl cyclase* gene or an inactive *crp* gene synthesize β-galactosidase? Does the binding of cAMP-CRP to DNA affect the binding of a repressor?

11.10 Two genotypes of *E. coli* are grown and assayed for levels of the enzymes of the *lac* operon. Using the information provided in the accompanying table for these two genotypes, predict the enzyme levels for the other genotypes listed in parts (a) through (d). The levels of activity are expressed in arbitrary units relative to those observed under the induced conditions.

	Uninduced level		Induced level	
Genotype	Z	Y	Z	Y
$I^+\ O^+\ Z^+\ Y^+$	0.1	0.1	100	100
$I^+\ O^c\ Z^+\ Y^+$	25	25	100	100

(a) $I^-\ O^+\ Z^+\ Y^+$
(b) F' $I^+\ O^+\ Z^-\ Y^-$ / $I^-\ O^+\ Z^+\ Y^+$
(c) F' $I^+\ O^+\ Z^-\ Y^-$ / $I^+\ O^c\ Z^+\ Y^+$
(d) F' $I^+\ O^+\ Z^-\ Y^-$ / $I^-\ O^c\ Z^+\ Y^+$

11.11 Imagine a bacterial species in which the methionine operon is regulated only by an attenuator and there is no repressor. In its mode of operation, the methionine attenuator is exactly analogous to the *trp* attenuator of *E. coli*. The relevant portion of the attenuator sequence in the RNA is

5'-AAAAUGAUGAUGAUGAUGAUGAUGAUGGACUAA-3'

The translation start site is located upstream from this sequence, and the region shown is in the correct reading frame. What phenotype (constitutive, wildtype, or Met$^-$) would you expect for each of the types of mutant RNA below? Explain your reasoning.

(a) The red A is deleted.
(b) Both the red A and the underlined A are deleted.
(c) The first three A's in the sequence are deleted.

11.12 Temperature-sensitive mutations in the *lacI* gene of *E. coli* render the repressor nonfunctional (unable to bind the operator) at 42°C but leave it fully functional at 30°C. Would β-galactosidase be expected to be produced:

(a) In the presence of lactose at 30°C?
(b) In the presence of lactose at 42°C?
(c) In the absence of lactose at 30°C?
(d) In the absence of lactose at 42°C?

Further Readings

Cohen, S., and G. Jürgens. 1991. *Drosophila* headlines. *Trends in Genetics* 7: 267.

Gellert, M. 1992. V(D)J recombination gets a break. *Trends in Genetics* 8: 408.

Gralla, J. D. 1996. Activation and repression of *E. coli* promoters. *Current Opinion in Genetics & Development* 6: 526.

Guarente, L. 1984. Yeast promoters: Positive and negative elements. *Cell* 36: 799.

Holliday, R. 1989. A different kind of inheritance. *Scientific American*, June.

Khoury, G., and P. Gruss. 1983. Enhancer elements. *Cell* 33: 83.

Klar, A. J. S. 1992. Developmental choices in mating-type interconversion in fission yeast. *Trends in Genetics* 8: 208.

Laird, P. W., and R. Jaenisch. 1996. The role of DNA methylation in cancer genetics and epigenetics. *Annual Review of Genetics* 30: 441.

Maniatis, T., and M. Ptashne. 1976. A DNA operator-repressor system. *Scientific American*, January.

Marmorstein, R., M. Carey, M. Ptashne, and S. C. Harrison. 1992. DNA recognition by GAL4: Structure of a protein-DNA complex. *Nature* 356: 408.

Miller, J., and W. Reznikoff, eds. 1978. *The Operon*. Cold Spring Harbor, NY: Cold Spring Harbor Laboratory.

Neidhardt, F. C., R. Curtiss III, J. L. Ingraham, E. C. C. Lin, K. B. Low, B. Magasanik, W. S. Reznikoff, M. Riley, M. Schaechter, and H. E. Umbarger, eds. 1996. *Escherichia coli and Salmonella typhimurium: Cellular and Molecular Biology* (2 volumes). 2d ed. Washington, DC: American Society for Microbiology.

Novina, C. D., and A. L. Roy. 1996. Core promoters and transcriptional control. *Trends in Genetics* 12: 351.

Ptashne, M. 1992. *A Genetic Switch*. 2d ed. Cambridge, MA: Cell Press.

Ptashne, M., and A. Gann. 1997. Transcriptional activation by recruitment. *Nature* 386: 569.

Struhl, K. 1995. Yeast transcriptional regulatory mechanisms. *Annual Review of Genetics* 29: 651.

Tijan, R. 1995. Molecular machines that control genes. *Scientific American*, February.

Ullman, A., and A. Danchin. 1980. Role of cyclic AMP in regulatory mechanisms of bacteria. *Trends in Biochemical Sciences* 5: 95.

Yanofsky, C. 1981. Attenuation in the control of expression of bacterial operons. *Nature* 289: 751.

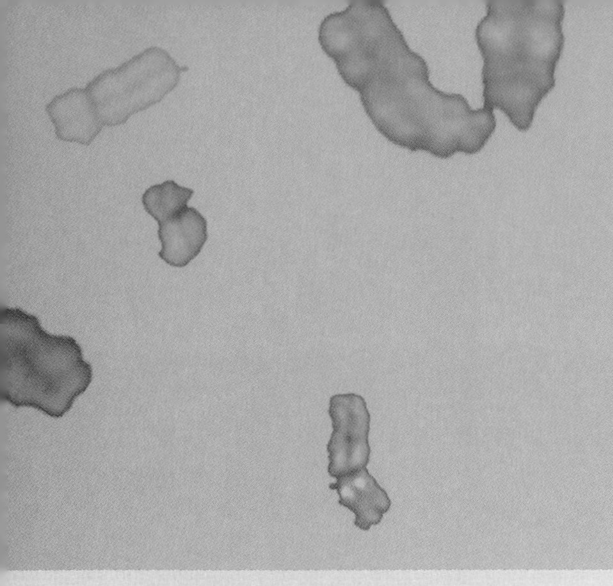

Key Concepts

- In animal cells, maternal gene products in the oocyte control the earliest stages of development, including the establishment of the main body axes.
- Developmental genes are often controlled by gradients of gene products, either within cells or across parts of the embryo.
- Regulation of developmental genes is hierarchical: Genes expressed early in development regulate the activities of genes expressed later.
- Regulation of developmental genes is combinatorial: Each gene is controlled by a combination of other genes.
- Many of the fundamental processes of pattern formation appear to be similar in animals and plants.

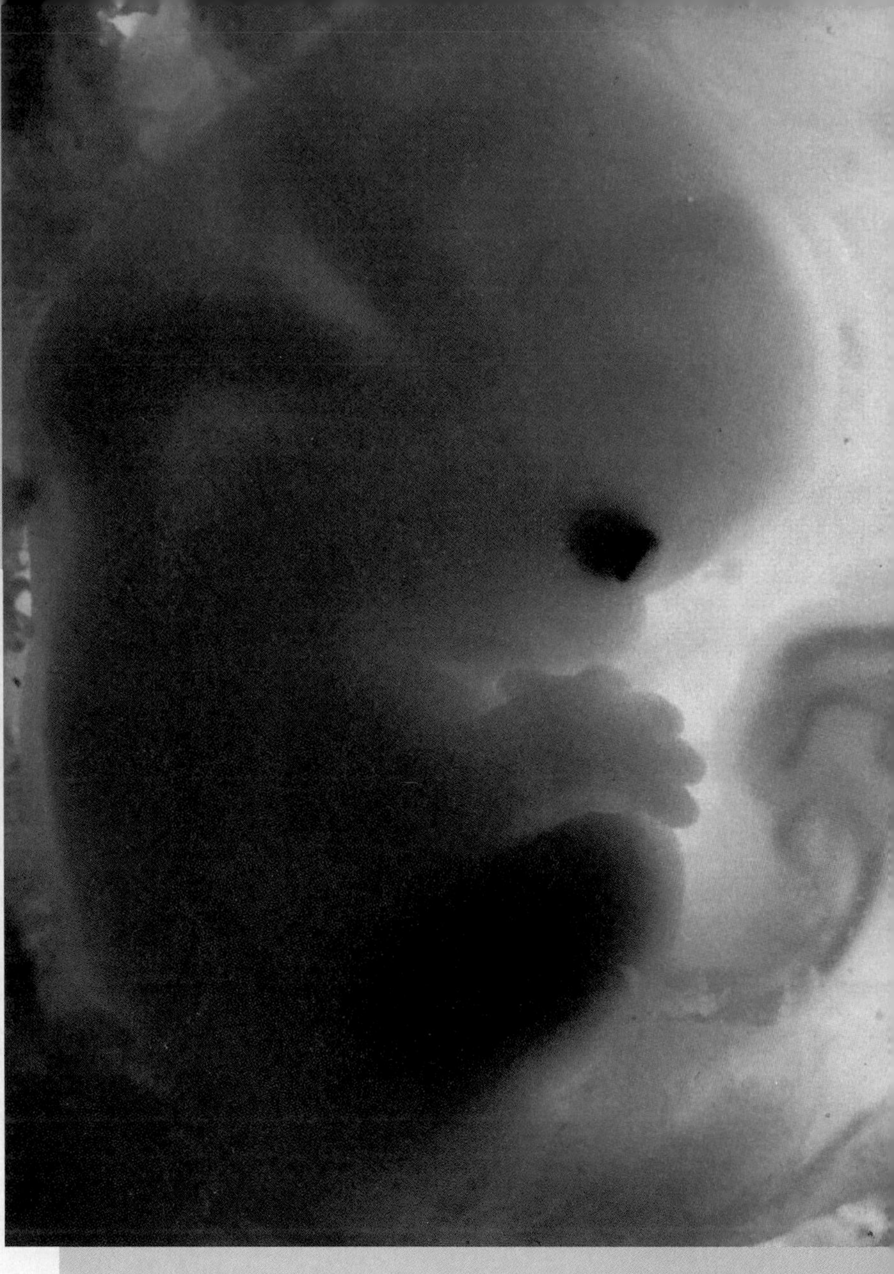

A HUMAN FETUS after 7 weeks of development [© Edelmann/La Villete/Photo Researchers, Inc.]

CHAPTER

12

The Genetic Control of Development

Understanding gene regulation is central to understanding how an organism as complex as a human being develops from a fertilized egg. Genes are expressed according to a prescribed program in development to ensure that the fertilized egg divides repeatedly and that the resulting cells become specialized in an orderly way to give rise to the fully differentiated organism. Within what is usually a wide range of environments, the genotype determines not only the events that take place in development but also the temporal order in which the events unfold.

Genetic approaches to the study of development often make use of mutations that alter developmental patterns. These mutations interrupt developmental processes and make it possible to identify factors that control development and to study the interactions among them. This chapter demonstrates how genetics is used in the study of development. Many of the examples come from the nematode *Caenorhabditis elegans* and the fruit fly *Drosophila melanogaster* because these organisms have been studied intensively from the standpoint of developmental genetics.

The key process in development is **pattern formation,** which means the emergence of the spatially organized and specialized cells in the embryo. One might draw an analogy between pattern formation in development and a pattern formed by fitting the pieces of a jigsaw puzzle together, but the analogy does not quite work. In a jigsaw puzzle, the pattern exists in its own right, independent of the shape and position of the pieces; the cuts in the pattern are superimposed on a preexisting picture that is merely reassembled. In biological development, the emergence of the image on each piece is *caused by* the shape and position of the piece. Also unlike a jigsaw puzzle, the picture changes throughout life as the organism undergoes growth and aging.

12.1 The spatial organization of gene products can determine the fate of daughter cells.

Conceptually, the relationship between genes and development is straightforward:

> The genome contains a developmental program that unfolds and results in the expression of different sets of genes in different types of cells.

In other words, as development unfolds, cell types progressively appear that differ qualitatively in the genes that are expressed. Whether a particular gene is expressed may depend on the presence or absence of a particular transcription factor, a change in chromatin structure, or the synthesis of a particular receptor molecule. These and other molecular mechanisms define the pattern of interactions by which one gene controls another. Collectively, these regulatory interactions ultimately determine the growth, identity, and patterning of the structures they affect. Many interactions that are important in development make use of more general regulatory elements. For example, developmental events often depend on regional differences in the concentration of molecules within a cell or within an embryo, and the activity of enhancers may be modulated by the local concentration of these substances.

The formation of an embryo is also affected by the environment in which development takes place. Although genotype and environment work together in development, the genotype determines the developmental potential of the organism. Genes determine whether a developing embryo will become a nematode, a fruit fly, a chicken, or a mouse. However, the expression of the genetically determined developmental potential is also influenced by the environment—in some cases very dramatically. Fetal alcohol syndrome is one example in which an environmental agent, in this case chronic alcohol poisoning, affects various aspects of fetal growth and development.

Development includes many examples in which genes are selectively turned on or off by the action of regulatory proteins that respond to environmental signals. The identification of genetic regulatory interactions that operate during development is an important theme in developmental genetics. The implementation of different regulatory interactions means that genetically identical cells can become qualitatively different in the genes that are expressed.

Several common mechanisms by which genes become activated or repressed at particular stages in development or in particular tissues are illustrated in the hypothetical example in Figure 12.1. In this example, the developmentally regulated genes are denoted *r* (red), *g* (green), *b* (blue), and *p* (purple). The initial event is transcription of the *r* gene in the oocyte (part A). Unequal partitioning of the *r* gene product into one region of the egg establishes a polarity or regionalization of the egg with respect to the presence of the *r* gene product (part B). When cleavage takes place (part C), the polarized cell produces daughter cells either with (L, left) or without (R, right) the *r* gene product. The *r* gene product is a transcriptional activator of the *g* gene, so the *g* gene product is expressed in cell L but not in cell R;

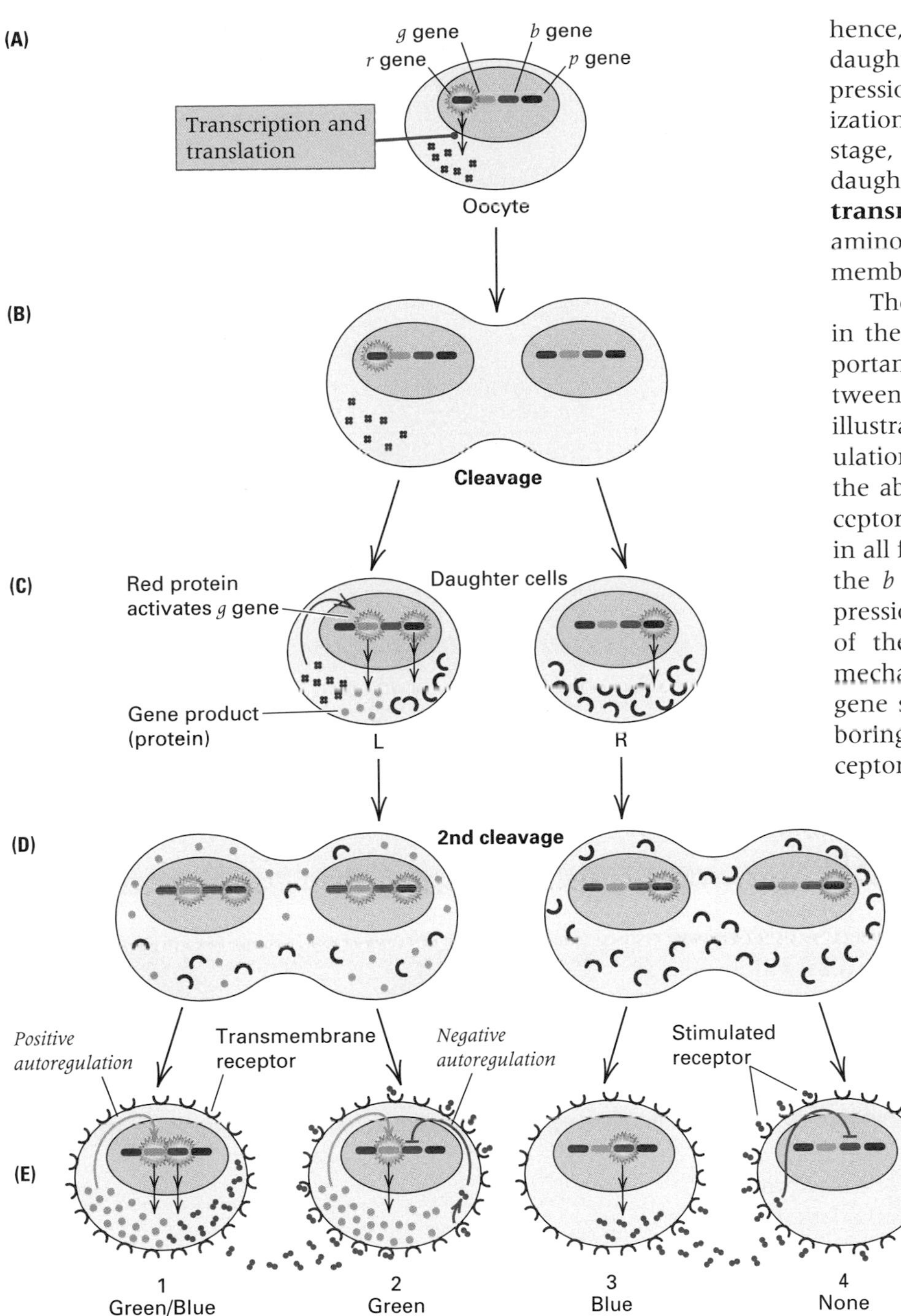

Figure 12.1 Hypothetical example illustrating some regulatory mechanisms that result in differences in gene expression in development. (A) The *r* gene is expressed in the oocyte. (B) Polarized presence of the *r* gene product in the egg. (C) Presence of the *r* gene product in cell L stimulates transcription of the *g* gene. The *p* gene codes for a transmembrane receptor expressed in both L and R cells. (D and E) The *g* gene has positive autoregulation and so continues to be expressed in cells 1 and 2. Expression of the *b* gene in one cell represses its expression in the neighboring cell through stimulation of the *p* gene transmembrane receptor. The result of these mechanisms is that cells 1 through 4 differ in their combinations of *b* gene and *g* gene activity.

hence, even at the two-cell stage, the daughter cells may differ in gene expression as a result of the initial polarization of the egg. Also in the two-cell stage, the *p* gene is expressed in both daughter cells. This gene codes for a **transmembrane receptor** containing amino acid sequences that span the cell membrane.

The presence of receptor molecules in the cell membrane provides an important mechanism for signaling between cells in development, as illustrated in this example with the regulation of the *b* gene in Figure 12.1E. In the absence of the transmembrane receptor, the *b* gene would be expressed in all four cells. However, expression of the *b* gene in one cell inhibits its expression in the neighboring cell because of the transmembrane receptor. The mechanism is that the product of the *b* gene stimulates the receptor of neighboring cells, and stimulation of the receptor represses transcription of *b*. The process in which gene expression in one cell inhibits expression of the same or differing genes in neighboring cells is known as *lateral inhibition.*

At the same time that expression of the *b* gene is regulated by lateral inhibition, the initial activation of the *g* gene in cell L is retained in daughter cells 1 and 2 because the product of the *g* gene has a positive autoregulatory activity and stimulates its own transcription. Positive autoregulation is one way in which cells can amplify weak or transient regulatory signals into permanent changes in gene expression. The result of lateral inhibition of *b* expression and *g* autoregulation is that cells 1 through 4 in Figure 12.1E, though genetically identical, are developmentally different because they express different combinations of the *g* and *b* genes. Specifically, cell 1 has *g* and *b* activity, cell 2 has *g* activity only, cell 3 has *b* activity only, and cell 4 has neither.

Among the most intensively studied developmental genes are those that act early in development, before tissue differentiation, because these genes establish the overall pattern of development.

12.2 Cell fate is progressively restricted in animal development.

The early development of the animal embryo establishes the basic developmental plan for the whole organism. Fertilization initiates a series of mitotic **cleavage divisions** in which the embryo becomes multicellular. There is little or no increase in overall size compared with the egg, because the cleavage divisions are accompanied by little growth and merely partition the fertilized egg into progressively smaller cells. The cleavage divisions form the **blastula,** which is essentially a ball of about 10^4 cells containing a cavity (Figure 12.2). Completion of the cleavage divisions is followed by the formation of the **gastrula** through an infolding of part of the blastula wall and extensive cellular migration. In the reorganization of cells in the gastrula, the cells become arranged in several distinct layers. These layers establish the basic body plan of an animal. In higher plants, as we shall see later, the developmental processes differ substantially from those in animals.

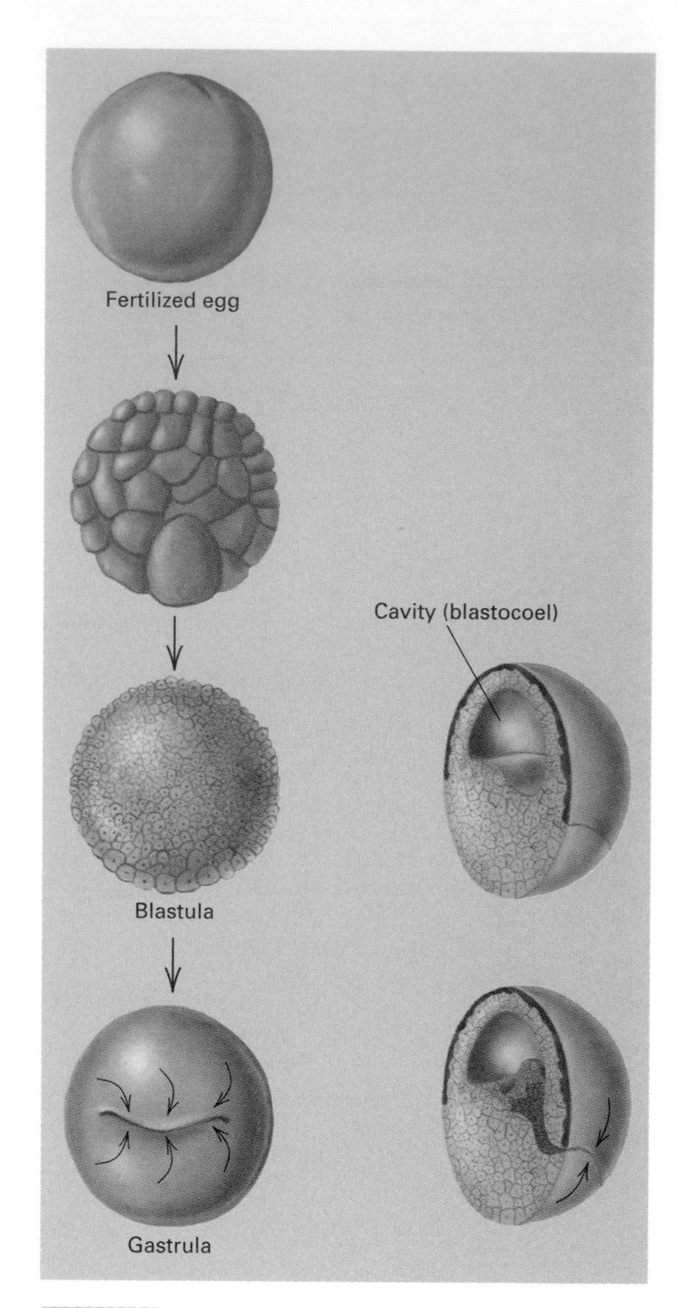

Figure 12.2 Early development of the animal embryo. The cleavage divisions of the fertilized egg result first in a clump of cells and then in a hollow ball of cells (the blastula). Extensive cell migrations form the gastrula and establish the basic body plan of the embryo.

Cell fate is determined by autonomous development and/or intercellular signaling.

Two principal mechanisms progressively restrict the **fate,** or developmental outcome, of cells within a lineage.

- Developmental restriction may be **autonomous,** which means that it is determined by genetically programmed changes in the cells themselves.
- Cells may respond to **positional information,** which means that developmental restrictions are imposed by the position of cells within the embryo. Positional information may be mediated by signaling interactions between neighboring cells or by gradients in concentration of morphogens. A **morphogen** is a molecule that participates directly in the control of growth and form during development.

Restriction of developmental fate is usually studied by transplanting cells of the embryo to new locations to determine whether they can substitute for the cells that they displace. Alternatively, individual cells are isolated from early embryos and cultured in laboratory dishes to study their developmental potential. In some eukaryotes, such as the soil nematode *Caenorhabditis elegans,* many lineages develop autonomously according to genetic programs that are induced by interactions with neighboring cells very early in embryonic development. Subsequent cell interactions also are important, and each stage in development is set in motion by the successful completion of the preceding stage. Figure 12.3 illustrates the first three cell divisions in the development of *C. elegans,* which result in eight embryonic cells that differ in genetic activity and developmental fate. The determination of cell fate in these early divisions is in part autonomous and in part a result of interactions between cells. Part B shows the lineage relationships between the cells. Cell-autonomous mechanisms are illustrated by the transmission of cytoplasmic particles called polar granules from the cells P0 to P1 to P2 to P3. Normal segregation of the polar granules is a function of microfilaments in the cytoskeleton. Cell-signaling mechanisms are illustrated by the effects of P2 on EMS and on ABp. The EMS fate is determined by the activity of the *mom-2* gene in P2. The P2 cell also produces a signaling molecule, APX-1, which determines the fate of ABp through the cell-surface receptor GLP-1. (The specific mechanisms that determine early cell fate in *C. elegans* strongly resemble some of the general mechanisms outlined in Figure 12.1.) In contrast to *C. elegans*, in which many developmental decisions are cell-autonomous, in *Drosophila* and *Mus* (the mouse), regulation by cell-to-cell signaling is more the rule than the exception. The use of cell signaling to regulate development provides a sort of insurance that helps to overcome the death of individual cells in development that might happen by accident.

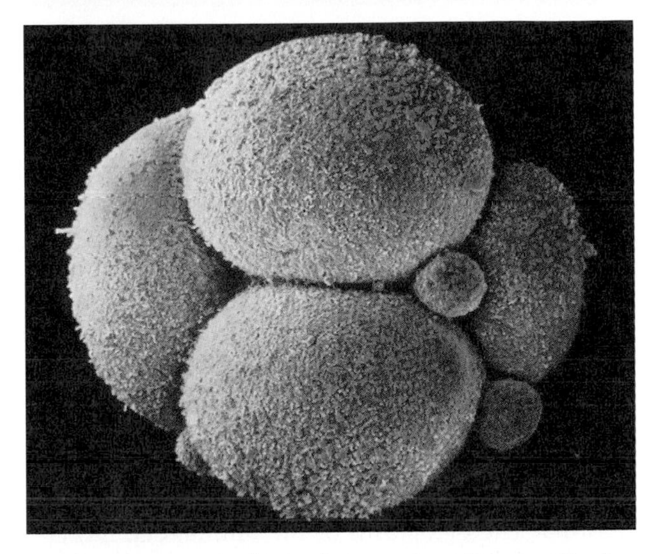

A HUMAN EMBRYO at the four-cell stage, about 40 hours after fertilization. This embryo has only undergone two cell divisions and will continue to divide to eventually transform into a human fetus composed of millions of cells. At this four-cell stage, the embryo has not yet implanted in the womb. [© Dr. Yorgos Nikas/Photo Researchers, Inc./ Science Photo Library.]

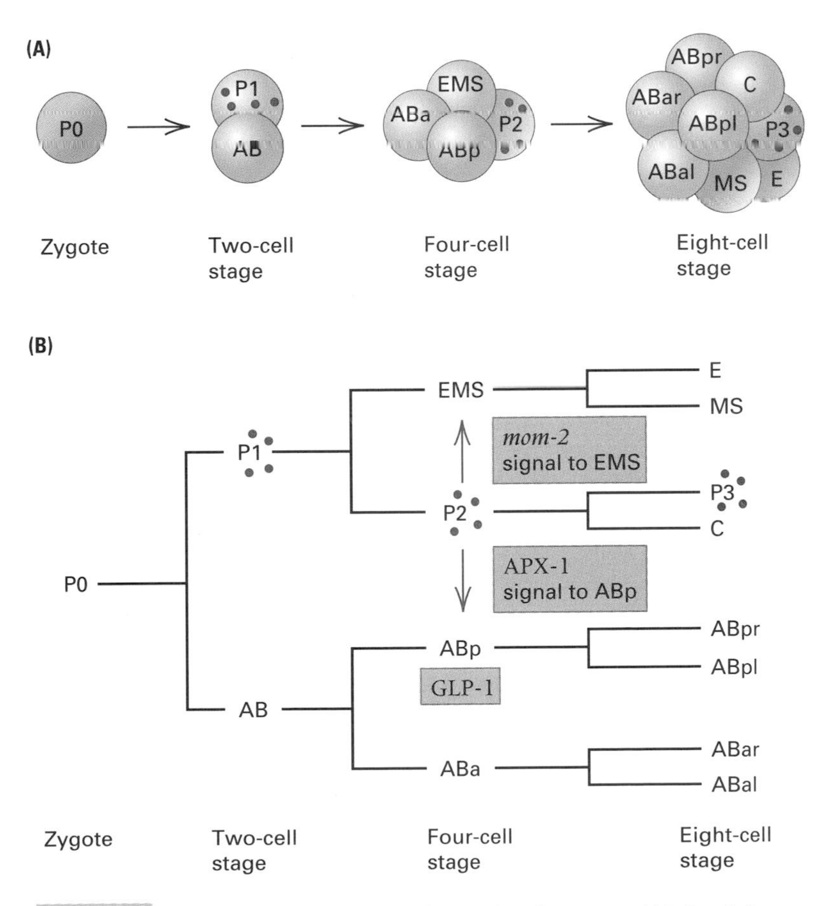

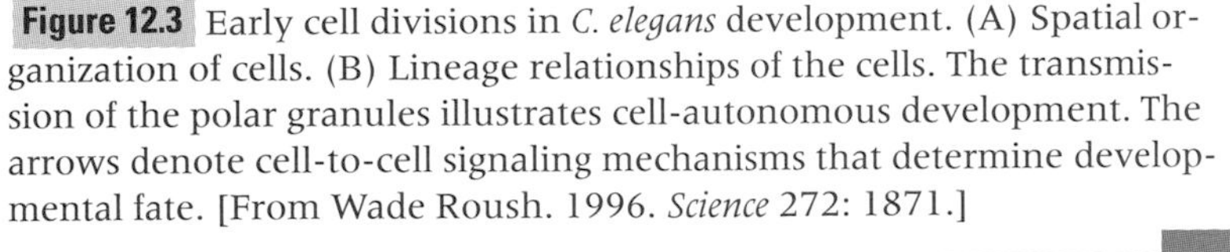

Figure 12.3 Early cell divisions in *C. elegans* development. (A) Spatial organization of cells. (B) Lineage relationships of the cells. The transmission of the polar granules illustrates cell-autonomous development. The arrows denote cell-to-cell signaling mechanisms that determine developmental fate. [From Wade Roush. 1996. *Science* 272: 1871.]

The zygotic genome is activated at different times in different animals.

In most animals, the earliest events in embryonic development do not depend on genetic information in the cell nucleus of the zygote. For example, fertilized frog eggs with the nucleus removed are still able to carry out the cleavage divisions and form rudimentary blastulas. Similarly, when gene transcription in sea urchin or amphibian embryos is blocked shortly after fertilization, there is no effect on the cleavage divisions or on blastula formation, but gastrulation does not take place. The reason why the genes in the zygote are not needed in the early stages of development is that the oocyte cytoplasm produced by the mother contains all the necessary macromolecules.

Following the period early in development in which the genes in the zygote nucleus are not needed, the embryo becomes dependent on the activity of its own genes. In mice—and probably in all mammals—the zygotic genes are needed much earlier than in lower vertebrates. The shift from control by the maternal genome to control by the zygote genome begins in the two-cell stage of development, when proteins coded by the zygote nucleus are first detectable. Inhibitors of transcription stop development of the mouse embryo when applied at any time after the first cleavage division. However, even in mammals, the earliest stages of development are greatly influenced by the cytoplasm of the oocyte, which determines the planes of the initial cleavage divisions and other events that ultimately affect cell fate.

The initial events in development are controlled by gene products in the oocyte.

The oocyte is a diploid cell during most of oogenesis. The cytoplasm of the oocyte is extensively organized and regionally differentiated (Figure 12.4). This spatial differentiation ultimately determines the different developmental fates of cells in the blastula. The cytoplasm of the animal egg has two essential functions:

1. Storage of the molecules needed to support the cleavage divisions and the rapid RNA and protein synthesis that take place in early embryogenesis

2. Organization of the molecules in the cytoplasm to provide the positional information that results in differences between cells in the early embryo

Figure 12.4 The animal oocyte is highly organized internally, which is revealed by the visible differences between the dorsal (dark) and ventral (light) parts of these *Xenopus* eggs. The regional organization of the oocyte determines many of the critical events in early development. [Courtesy of Michael W. Klymkowsky.]

To establish the proper composition and organization of the oocyte requires the participation of many genes within the oocyte itself and gene products supplied by adjacent helper cells of various types (Figure 12.5). Numerous maternal genes are transcribed in oocyte development, and the mature oocyte typically contains an abundance of transcripts that code for proteins needed in the early embryo. Some of the maternal mRNA transcripts are stored complexed with proteins in special ribonucleoprotein particles that cannot be translated, and release of the mRNA, enabling it to be translated, does not happen until after fertilization.

Developmental instructions in oocytes are determined in part by the presence of distinct types of molecules at different positions within the cell and in part by gradients of morphogens that differ in concentration from one position to the next. Although the oocyte contains the products of many genes, only a small number of genes are expressed exclusively in the formation of the oocyte. Most genes expressed in oocyte formation are also important in the development of other tissues or at later times in development. Therefore, it is not only gene products but also their spa-

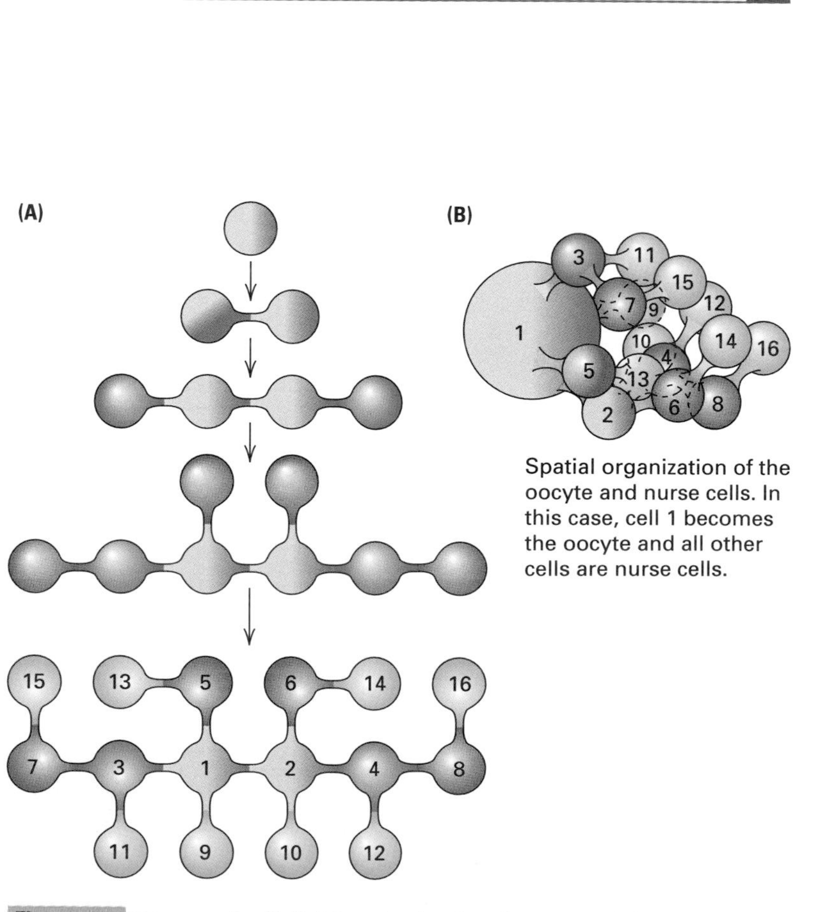

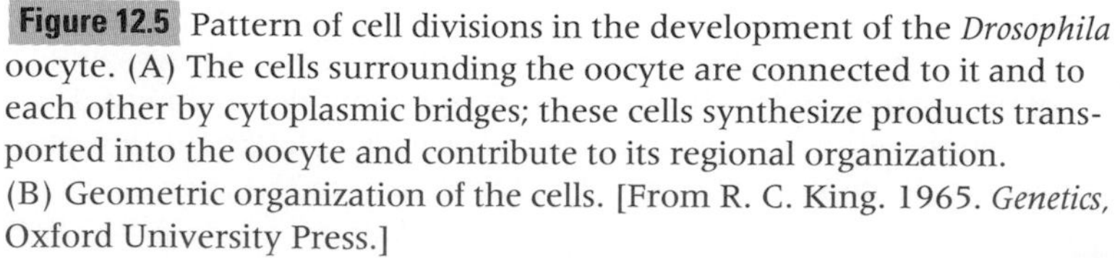

Figure 12.5 Pattern of cell divisions in the development of the *Drosophila* oocyte. (A) The cells surrounding the oocyte are connected to it and to each other by cytoplasmic bridges; these cells synthesize products transported into the oocyte and contribute to its regional organization. (B) Geometric organization of the cells. [From R. C. King. 1965. *Genetics*, Oxford University Press.]

tial organization of the gene products within the cell that give the oocyte its unique developmental potential.

12.3 The developmental fate of cells is often under genetic control.

The mechanisms that control early development can be studied genetically by isolating mutants with early developmental abnormalities and altered cell fates. In most organisms, it is difficult to trace the lineage of individual cells in development because the embryo is not transparent, the cells are small and numerous, and cell migrations are extensive. The **lineage** of a cell refers to the ancestor-descendant relationships among a group of cells. A cell lineage can be illustrated with a **lineage diagram,** a sort of cell pedigree that shows each cell division and that indicates the terminal differentiated state of each cell (Figure 12.6). This is a lineage diagram of a hypothetical cell A in which the cell fate is either programmed cell death or one of the terminally differentiated cell types designated W, X, and Y. The letter symbols are the kind normally used for cells in nematodes, in which the name denotes the cell lineage according to ancestry and position in the embryo. For example, the cells A.a and A.p are the anterior and posterior daughters of cell A, respectively, and A.aa and A.ap are the anterior and posterior daughters of cell A.a.

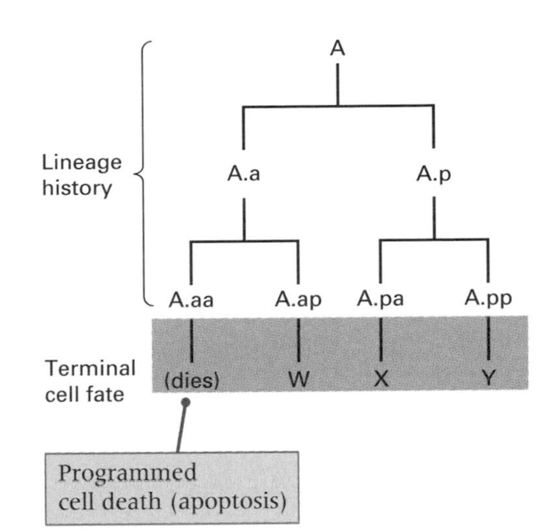

Figure 12.6 Hypothetical cell-lineage diagram. Different terminally differentiated cell fates are denoted W, X, and Y. One cell in the lineage (cell A.aa) undergoes programmed cell death. The lowercase letters *a* and *p* denote anterior and posterior daughter cells, respectively. For example, cell A.ap is the posterior daughter of the anterior daughter of cell A.

Development in the nematode follows a fixed program of lineage diversification.

The soil nematode *Caenorhabditis elegans* (Figure 12.7) is popular for genetic studies because it is small, easy to culture, and has a short generation time with a large number of offspring. The worms are grown on agar surfaces in petri dishes and feed on bacterial cells such as *E. coli.* Because they are microscopic in size, as many as 10^5 animals can be contained in a single petri dish. Sexually mature adults of *C. elegans* are capable of laying more than 300 eggs within a few days. At 20°C, it requires about 60 hours for the eggs to hatch, undergo four larval molts, and become sexually mature adults.

Nematodes are diploid organisms with two sexes. In *C. elegans,* the two sexes are the hermaphrodite and the male. The hermaphrodite contains two X chromosomes (XX), produces both functional eggs and functional sperm, and is capable of self-fertilization. The male produces only sperm and

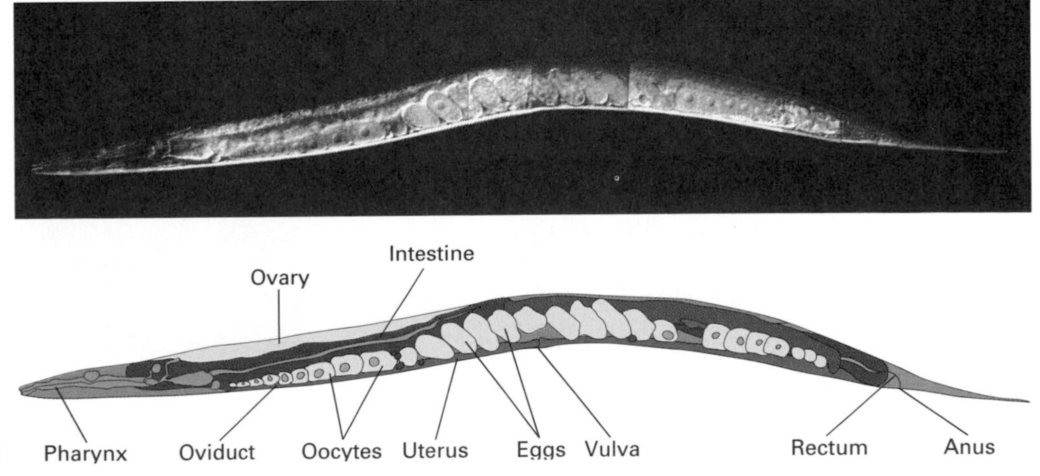

Figure 12.7 The soil nematode *Caenorhabditis elegans.* This organism offers several advantages for the genetic analysis of development, including the fact that each individual of each sex exhibits an identical pattern of cell lineages in the development of the somatic cells. DNA sequencing of the 100-megabase genome is nearing completion. [Photograph courtesy of Tim Schedl.]

fertilizes the hermaphrodites. The sex-chromosome constitution of *C. elegans* consists of a single X chromosome; there is no Y chromosome, and the male karyotype is XO.

The transparent body wall of the worm has made it possible to study the division, migration, and death or differentiation of all cells present in the course of development. Nematode development is unusual in that the pattern of cell division and differentiation is virtually identical from one individual to the next; that is, cell division and differentiation are highly stereotyped. The result is that each sex shows the same geometry in the number and arrangement of somatic cells. The hermaphrodite contains exactly 959 somatic cells, and the male contains exactly 1031 somatic cells. The complete developmental history of each somatic cell is known.

Nematode development is largely autonomous, which means that in most cells, the developmental program unfolds automatically without the need for interactions with other cells. However, in the early embryo, some of the developmental fates are established by interactions among the cells. In later stages of development of these cells, the fates established early are reinforced by still other interactions between cells. Worm development also provides important examples of the effects of intercellular signaling on determination (for example, those in Figure 12.3).

Mutations can affect cell lineages in development.

Many mutations that affect cell lineages have been studied in nematodes, and they reveal several general features by which genes control development.

- The division pattern and fate of a cell are generally affected by more than one gene and can be disrupted by mutations in any of them.
- Most genes that affect development are active in more than one type of cell.
- Complex cell lineages often include simpler, genetically determined lineages within them; these components are called *sublineages* because they are expressed as an integrated pattern of cell division and terminal differentiation.
- The lineage of a cell may be triggered autonomously within the cell itself or by signaling interactions with other cells.
- Regulation of development is controlled by genes that determine the different sublineages that cells can undergo and the individual steps within each sublineage.

The next section deals with some of the types of mutations that affect cell lineages and development.

Developmental mutations can be classified by their effect on lineages.

Mutations can affect cell lineages in two major ways. One is through nonspecific metabolic blocks—for example, in DNA replication—that prevent the cells from undergoing division or differentiation. The other is through specific molecular defects that result in patterns of division or differentiation characteristic of cells normally found elsewhere in the embryo or at a different time in development. From the standpoint of genetic analysis of development, the latter class of mutations is the more informative, because the mutant genes must be involved in developmental processes rather than in general "housekeeping" functions found in all cells.

One of the most interesting findings of the cell-lineage studies in *C. elegans* development was the prevalence of cells that, at a predetermined time, seemed to commit suicide. During embryogenesis, 671 cells are generated, of which 113 undergo programmed cell death, also known as **apoptosis.** Programmed cell death is an important feature of normal development in many organisms. In many cases the signaling molecules that determine apoptosis have been identified (a number are known to be transcription factors). Failure of programmed cell death often results in specific developmental

FEATHER COLOR IN birds is under genetic control. Feather color is critically important in mate recognition and mating success. [© Jeanne White/Photo Researchers, Inc.]

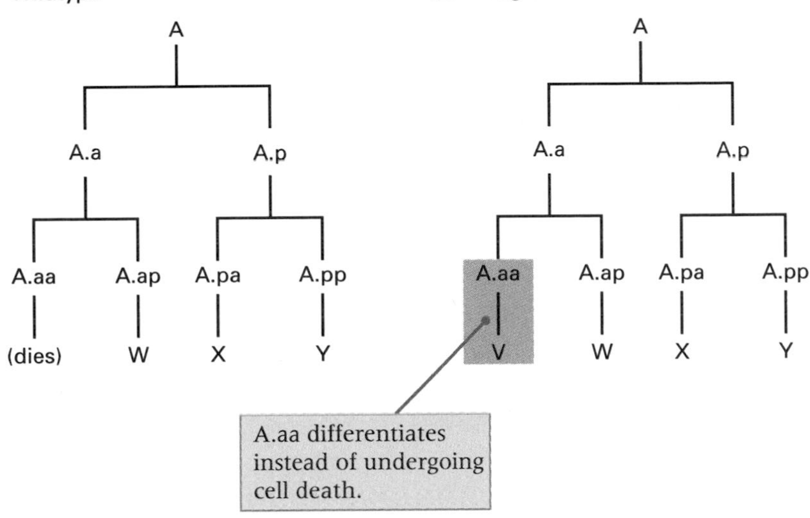

Figure 12.8 Apoptosis (programmed cell death) and heterochrony. (A) A wildtype lineage. (B) Failure of apoptosis, in which cell A.aa does not die but instead differentiates into a V-type cell. In some cases in which programmed cell death fails, the surviving cells differentiate into identifiable types.

abnormalities. For example, compared with the wildtype lineage in Figure 12.8A, the lineage in part B is abnormal in that cell A.aa fails to undergo apoptosis and, instead, differentiates into the cell type V. Phenotypically, when apoptosis fails and the surviving cells differentiate into recognizable cell types, the result is the presence of supernumerary cells of that type. For example, in the *ced-3* gene mutants (*ced* = *cell death abnormal*) in *C. elegans,* a particular cell that normally undergoes programmed cell death survives and often differentiates into a supernumerary neuron. Programmed cell death is usually not essential; mutants that cannot execute programmed cell death are viable and fertile but slightly impaired in development and some sensory capabilities. On the other hand, mutants in *Drosophila* that fail to execute apoptosis are lethal, and in mammals, including human beings, failure of programmed cell death results in severe developmental abnormalities or, in some instances, leukemia or other forms of cancer.

Transmembrane receptors often mediate signal transduction.

Control genes that cause cells to diverge in developmental fate are not always easy to recognize. For example, a mutant allele may identify a gene that is necessary for the expression of a particular developmental fate, but the gene may not actually control or determine the developmental fate of the cells in which it is expressed. This possibility complicates the search for genes that control major developmental decisions.

Genes that control decisions about cell fate can sometimes be identified by the unusual characteristic that dominant or recessive mutations have opposite effects; that is, if alternative alleles of a gene result in opposite cell fates, then the product of the gene must be both necessary and sufficient for expression of the fate. Identification of possible regulatory genes in this way excludes the large number of genes whose functions are merely necessary, but not sufficient, for the expression of cell fate. Recessive mutations in genes that control development often result from **loss of function** in that the mRNA or the protein is not produced; loss-of-function mutations are exemplified by nonsense mutations that cause polypeptide chain termination in translation (Chapter 7). Dominant mutations in developmental-control genes often result from **gain of function** in that the gene is overexpressed or is expressed at the wrong time.

In *C. elegans,* only a small number of genes have dominant and recessive alleles that affect the same cells in opposite ways. Among them is the *lin-12* gene, which controls developmental decisions in a number of cells. One example involves the cells denoted Z1.ppp and Z4.aaa in Figure 12.9A. These cells lie side by side in the embryo, but they have quite different lineages. Normally, one of the cells differentiates into an *anchor cell* (AC), which participates in development of the vulva, and the other one differentiates into a *ventral uterine precursor cell* (VU). Either Z1.ppp or Z4.aaa may become the anchor cell with equal likelihood.

Direct cell-cell interaction between Z1.ppp and Z4.aaa controls the AC-VU decision. If either cell is burned away (ablated) by a laser microbeam, the remaining cell differentiates into an anchor cell (Figure 12.9B). This result implies that the preprogrammed fate of both Z1.ppp and Z4.aaa is that of an anchor cell. When either cell becomes committed to the anchor-cell fate, its contact with the other cell elicits the ventral-uterine-precursor-cell fate. As noted, recessive and dominant mutations

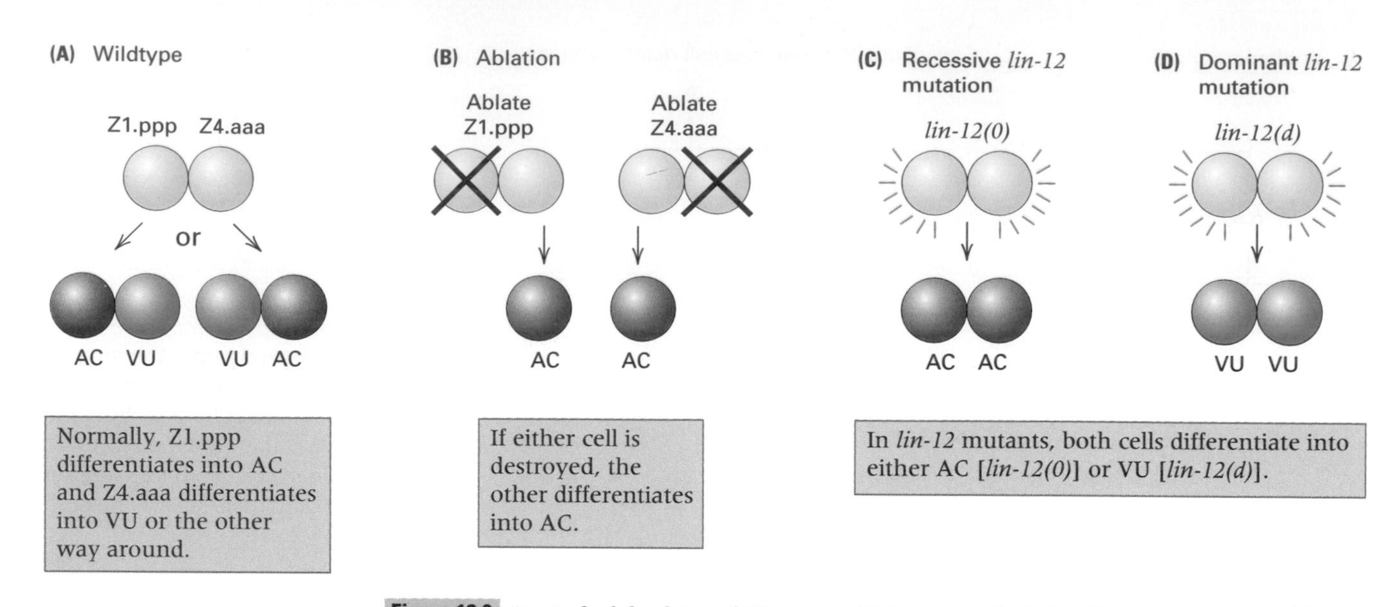

Figure 12.9 Control of the fates of Z1.ppp and Z4.aaa in vulval development. (A) In wildtype cells, there is an equal chance that either cell will become the anchor cell (AC); the other becomes a ventral uterine precursor cell (VU). (B) If either cell is destroyed (ablated) by a laser beam, the other differentiates into the anchor cell. (C) Genetic control of cell fate by the *lin-12* gene. In recessive loss-of-function mutations [*lin-12(0)*], both cells become anchor cells. (D) In dominant gain-of-function mutations [*lin-12(d)*], both cells become ventral uterine precursor cells.

of *lin-12* have opposite effects. Mutations in which *lin-12* activity is lacking or greatly reduced are denoted *lin-12(0)*. These mutations are recessive, and in the mutants both Z1.aaa and Z4.aaa become anchor cells (Figure 12.9C). In contrast, *lin-12(d)* mutations are those that cause *lin-12* activity to be overexpressed. These mutations are dominant or partly dominant, and in the mutants both Z1.aaa and Z4.ppp become ventral uterine precursor cells (Figure 12.9D).

The effects of *lin-12* mutations suggest that the wildtype gene product is a receptor of a developmental signal. The molecular structure of the *lin-12* gene product is typical of a transmembrane receptor protein, and it shares domains with other proteins important in developmental control (Figure 12.10). The transmembrane region separates the LIN-12 protein into an extracellular part (the amino end) and an intracellular part (the carboxyl end). The extracellular part contains 13 repeats of a domain found in a mammalian peptide hormone, epidermal growth factor (EGF), as well as in the product of the *Notch* gene in *Drosophila,* which controls the decision between epidermal-cell and neural-cell fates. Nearer the transmembrane region, the amino end contains three repeats of a cysteine-rich

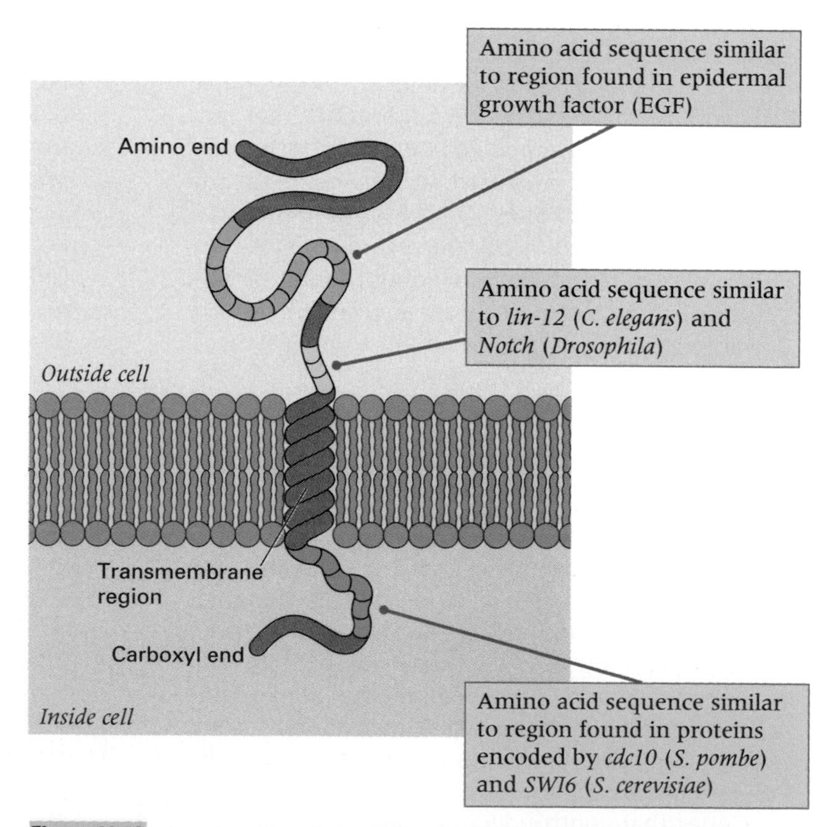

Figure 12.10 The structure of the LIN-12 protein is that of a receptor protein containing a transmembrane region and various types of repeated units that resemble those in epidermal growth factor (EGF) and other developmental control genes.

domain also found in the *Notch* gene product. Inside the cell, the carboxyl part of the LIN-12 protein contains six repeats of a domain also found in the SWI6 proteins, which control cell division in yeast.

The anchor cell expresses a signaling gene, called *lin-3,* that illustrates another case in which loss-of-function and gain-of-function alleles have opposite effects on phenotype. In the anchor cell, the gene *lin-3* controls the fate of certain cells in the development of the vulva. Figure 12.11 illustrates five precursor cells, P4.p through P8.p, that participate in the development of the vulva. Each precursor cell has the capability of differentiating into one of three fates, called the 1°, 2°, and 3° lineages, which differ according to whether descendant cells remain in a syncytium (S) or divide longitudinally (L), transversely (T), or not at all (N). The precursor cells normally differentiate as shown in Figure 12.11, giving five lineages in the order 3°-2°-1°-2°-3°. The vulva itself is formed from the 1° and 2° cell lineages. The spatial arrangement of some of the key cells is shown in Figure 12.12. The black arrow indicates the anchor cell, and the white lines show the pedigrees of 12 cells. The four cells in the middle derive from P6.p, and the four on each side derive from P5.p and P7.p.

The important role of the *lin-3* gene product (LIN-3) is suggested by the opposite phenotypes of loss-of-function and gain-of-function alleles. Loss of LIN-3 results in the complete absence of vulval development, whereas overexpression of LIN-3 results in excess vulval induction. LIN-3 is a typical example of an interacting molecule, or **ligand,** that interacts with an EGF-type transmembrane receptor. In this case the receptor is located in cell P6.p and is the product of the gene *let-23.* The LET-23 protein is a tyrosine-kinase receptor that, when stimulated by the LIN-3 ligand, stimulates a series of intracellular signaling events that ultimately results in the synthesis of transcription factors that determine the 1° fate. Among the genes that are induced is a gene for yet another ligand, which stimulates receptors on the cells P5.p and P7.p, causing these cells to adopt the 2° fate (horizontal arrows in Figure 12.11).

In vulval development, the adoption of the 3° lineages by the P4.p and P8.p cells is determined not by a positive signal but by

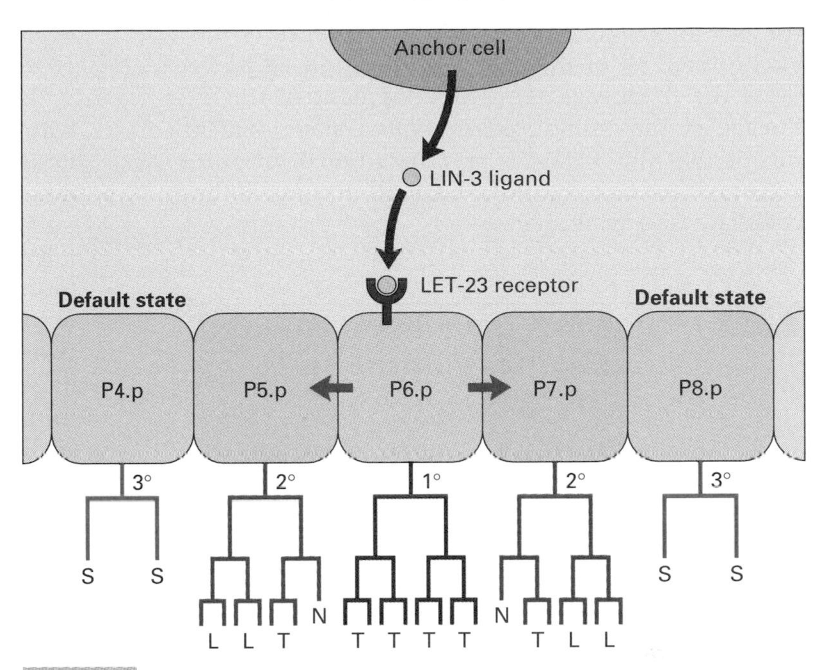

Figure 12.11 Determination of vulval differentiation by means of intercellular signaling. Cells P4.p through P8.p in the hermaphrodite give rise to lineages in the development of the vulva. The three types of lineages are designated 1°, 2°, and 3°. The 1° lineage is induced in P6.p by the ligand LIN-3 produced in the anchor cell (AC), which stimulates the LET-23 receptor tyrosine kinase in P6.p. The P6.p cell, in turn, produces a ligand that stimulates receptors in P5.p and P7.p to induce the 2° fate. On the other hand, the 3° fate is the default or baseline condition, which P4.p and P8.p adopt normally and all cells adopt in the absence of AC.

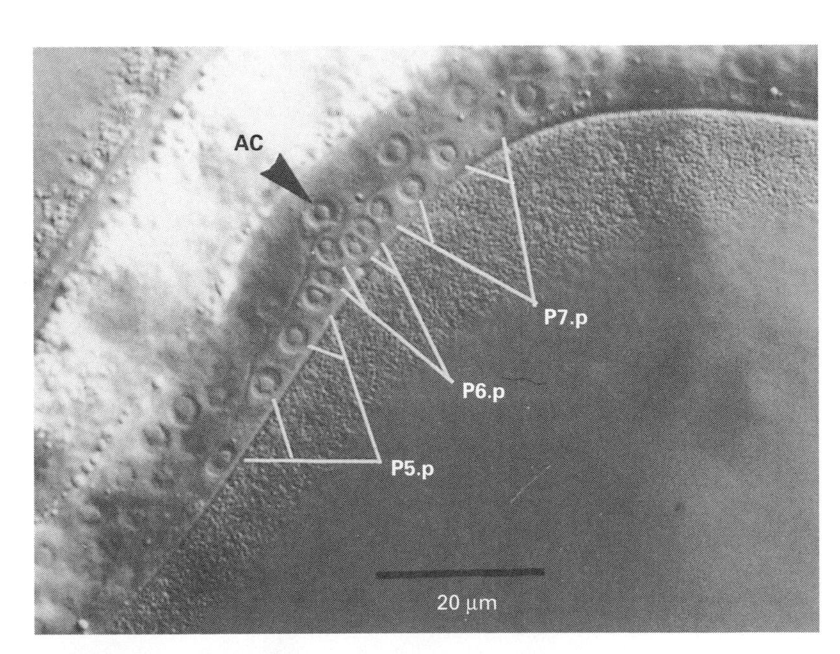

Figure 12.12 Spatial organization of cells in the vulva, including the anchor cell (black arrowhead) and the daughter cells produced by the first two divisions of P5.p through P7.p (white tree diagrams). The length of the scale bar equals 20 micrometers. [Courtesy of G. D. Jongeward, T. R. Clandinin, and P. W. Sternberg. 1995. *Genetics* 139: 1553.]

the lack of a signal, because in the absence of the anchor cell, all of the cells P4.p through P8.p express the 3° lineage. Thus development of the 3° lineage is the uninduced or *default* state, which means that the 3° fate is preprogrammed into the cell and must be overridden by another signal if the cell's fate is to be altered.

12.4 Mutations have been isolated for many developmental genes in *Drosophila*.

Many important insights into developmental processes have been gained from genetic analysis in *Drosophila*. The developmental cycle of *D. melanogaster*, summarized in Figure 12.13, includes egg, larval, pupal, and adult stages. Early development includes a series of cell divisions, migrations, and infoldings that result in the gastrula. About 24 hours after fertilization, the first-stage larva, composed of about 10^4 cells, emerges from the egg. Each larval stage is called an *instar*. Two successive larval molts that give rise to the second- and third-instar larvae are followed by pupation and a complex metamorphosis that gives rise to the adult fly composed of more than 10^6 cells. In wildtype strains reared at 25°C, development requires from 10 to 12 days.

Early development in *Drosophila* takes place within the egg case (Figure 12.14A). The first nine mitotic divisions occur in rapid succession without division of the cytoplasm and produce a cluster of nuclei within the egg (Figure 12.14B). The nuclei migrate to the periphery, and the germ line is formed from about 10 **pole cells** set off at the posterior end (Figure 12.14C); the pole cells undergo two additional divisions and are reincorporated into the embryo by invagination. The nuclei within the embryo undergo four more mitotic divisions without division of the cytoplasm, forming the **syncytial blastoderm,** which contains about 6000 nuclei (Figure 12.14D). Cellularization of the blastoderm takes place from about 150 to 180 minutes after fertilization by the synthesis of membranes that separate the nuclei. The **blastoderm** formed by cellularization (Figure 12.14E) is a flattened hollow ball of cells that corresponds to the blastula in other animals.

The experimental destruction of patches of cells within a *Drosophila* blastoderm results in localized defects in the larva and adult. The spatial correlation of position of the cells destroyed with type of defect results in a **fate map** of the blastoderm,

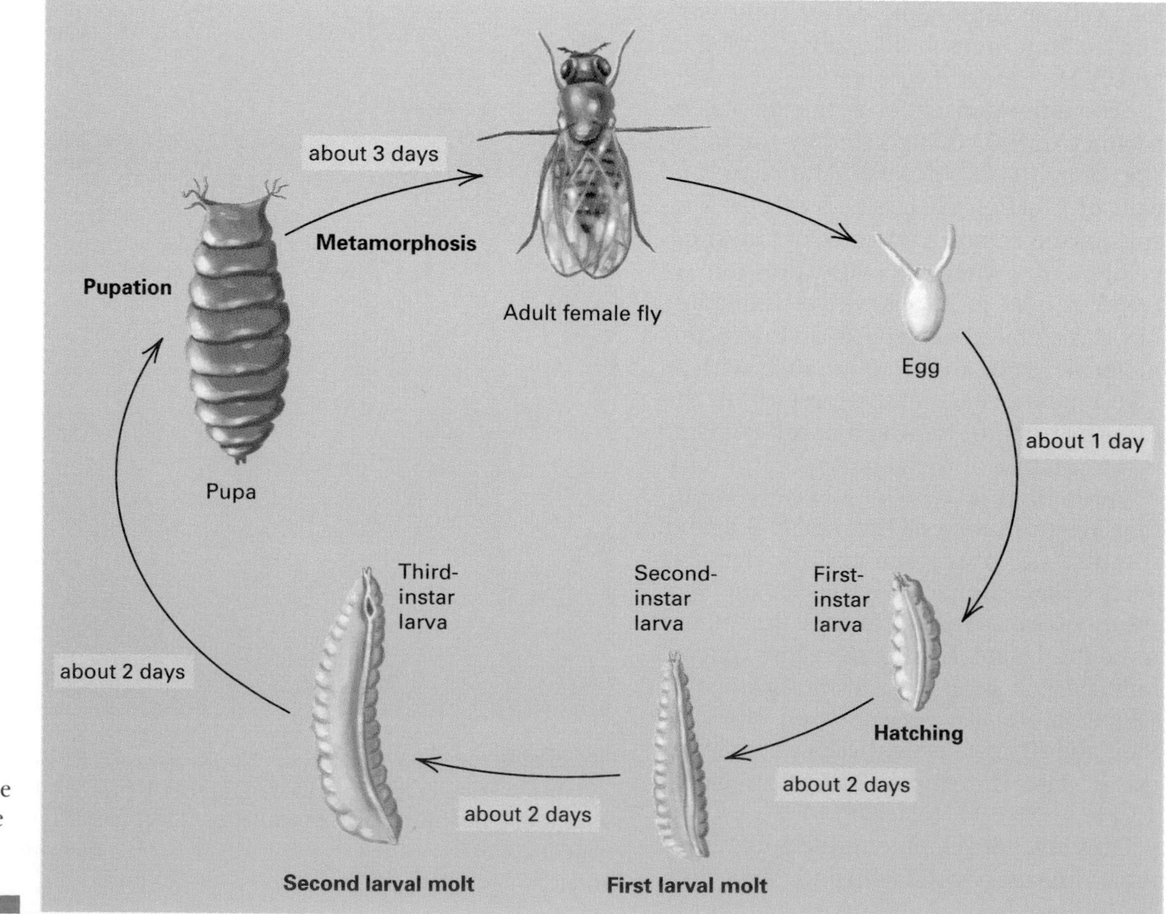

Figure 12.13 Developmental program of *Drosophila melanogaster*. The durations of the stages are at 25°C.

Figure 12.14 Early development in *Drosophila.* (A) The nucleus in the fertilized egg. (B) Mitotic divisions take place synchronously within a syncytium. (C) Some nuclei migrate to the periphery of the embryo, and at the posterior end, the pole cells (which form the germ line) become cellularized. (D) Additional mitotic divisions occur within the syncytial blastoderm. (E) Membranes are formed around the nuclei, giving rise to the cellular blastoderm.

which specifies the cells in the blastoderm that give rise to the various larval and adult structures (Figure 12.15). Use of genetic markers in the blastoderm has allowed further refinement of the fate map. Much like cells in the early blastula of *Caenorhabditis,* cells in the blastoderm of *Drosophila* have predetermined developmental fates, with little ability to substitute in development for other, sometimes even adjacent, cells. Evidence that blastoderm cells in *Drosophila* have predetermined fates comes from experiments in which cells from a genetically marked blastoderm are implanted into host blastoderms. Blastoderm cells implanted into the equivalent regions of the host become part of the normal adult structures. However, blastoderm cells implanted into different regions develop autonomously and are not integrated into host structures.

Because of the relatively high degree of determination in the blastoderm, genetic analysis of *Drosophila* development has tended to focus on the early stages of development, when the basic body plan of the embryo is established and key regulatory processes become activated. The following sections summarize the genetic control of these early events.

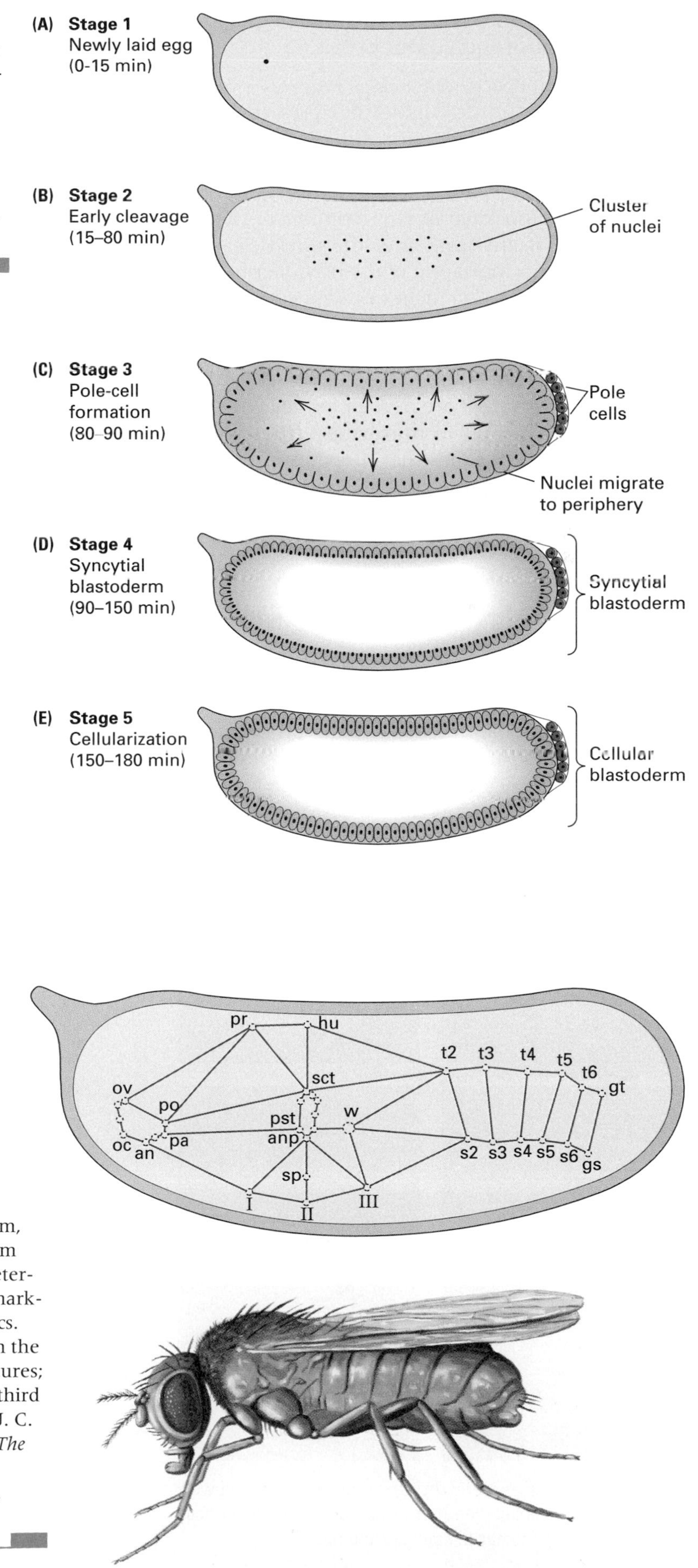

Figure 12.15 Fate map of the *Drosophila* blastoderm, which shows the adult structures that derive from various parts of the blastoderm. The map was determined by correlating the expression of genetic markers in different adult structures in genetic mosaics. The abbreviations stand for various body parts in the adult fly. For example, ov and oc are head structures; w is the wing; I, II, and III the first, second, and third legs; and gs and gt are genital structures. [After J. C. Hall, W. M. Gelbart, and D. R. Kankel. 1976. In *The Genetics and Biology of Drosophila,* vol. 1a, M. Ashburner and E. Novitski, eds. Academic Press, pp. 265–314.]

Mutations in a maternal-effect gene result in defective oocytes.

Early development in *Drosophila* requires translation of maternal mRNA molecules present in the oocyte. Blockage of protein synthesis during this period arrests the early cleavage divisions. Expression of the zygote genome is also required, but the timing varies. Blockage of transcription of the zygote genome at any time after the ninth cleavage division prevents formation of the blastoderm.

Because the earliest stages of *Drosophila* development are programmed in the oocyte, mutations that affect oocyte composition or structure can upset development of the embryo. Genes that function in the mother that are needed for development of the embryo are called **maternal-effect genes,** and developmental genes that function in the embryo are called **zygotic genes.** The interplay between the two types of genes is as follows:

> The zygotic genes interpret and respond to the positional information laid out in the egg by the maternal-effect genes.

Mutations in maternal-effect genes are easy to identify because homozygous females produce eggs unable to support normal embryonic development, whereas homozygous males produce normal sperm. Therefore, reciprocal crosses give dramatically different results. For example, a recessive maternal-effect mutation, *m*, will yield the following results in crosses:

$$m/m \text{ ♀} \times +/+ \text{ ♂} \rightarrow +/m \text{ progeny}$$
(abnormal development)

$$+/+ \text{ ♀} \times m/m \text{ ♂} \rightarrow +/m \text{ progeny}$$
(normal development)

The $+/m$ progeny of the reciprocal crosses are genetically identical, but development is upset when the mother is homozygous m/m.

The reason why maternal-effect genes are needed in the mother is that the maternal-effect genes establish the polarity of the *Drosophila* oocyte even before fertilization takes place. They are active during the earliest stages of embryonic development, and they determine the basic body plan of the embryo. Maternal-effect mutations provide a valuable tool for investigating the genetic control of pattern formation and for identifying the molecules important in morphogenesis.

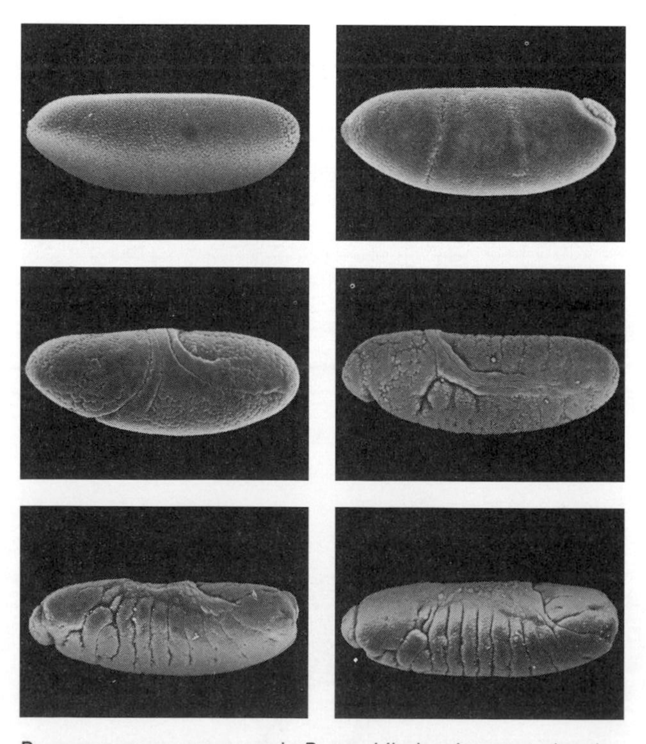

Representative stages in early *Drosophila* development, showing the pattern of segmentation that gives rise to the larval body plan. [Courtesy of Thomas C. Kaufman and Rudi Turner.]

Embryonic pattern formation is under genetic control.

The *Drosophila* embryo features 14 superficially similar repeating units visible as a pattern of stripes along the main trunk (Figure 12.16). The stripes can be recognized externally by the bands of *denticles,* which are tiny, pigmented, tooth-like projections from the surface of the larva. The 14 stripes in the embryo correspond to the segments in the larva that form from the embryo. Each **segment** is defined morphologically as the region between successive indentations formed by the sites of muscle attachment in the larval cuticle. The designations of the segments are indicated in Figure 12.16. There are three head segments (C1–C3), three thoracic segments (T1–T3), and eight abdominal segments (A1–A8.). In addition to the segments, another type of repeating unit is also important in development. These repeating units are called **parasegments**; each parasegment consists of the posterior region of one segment and the anterior region of the adjacent segment. Parasegments have a transient existence in embryonic development. Although they are not visible morphologically, they are important in gene expression because the patterns of expression of many genes coincide with the boundaries of the parasegments rather than with the boundaries of the segments.

The early stages of pattern formation are determined by genes that are often called **segmentation genes** because they determine the origin and fate of the segments and parasegments. There are four classes of segmentation genes that differ in their times and patterns of expression in the embryo.

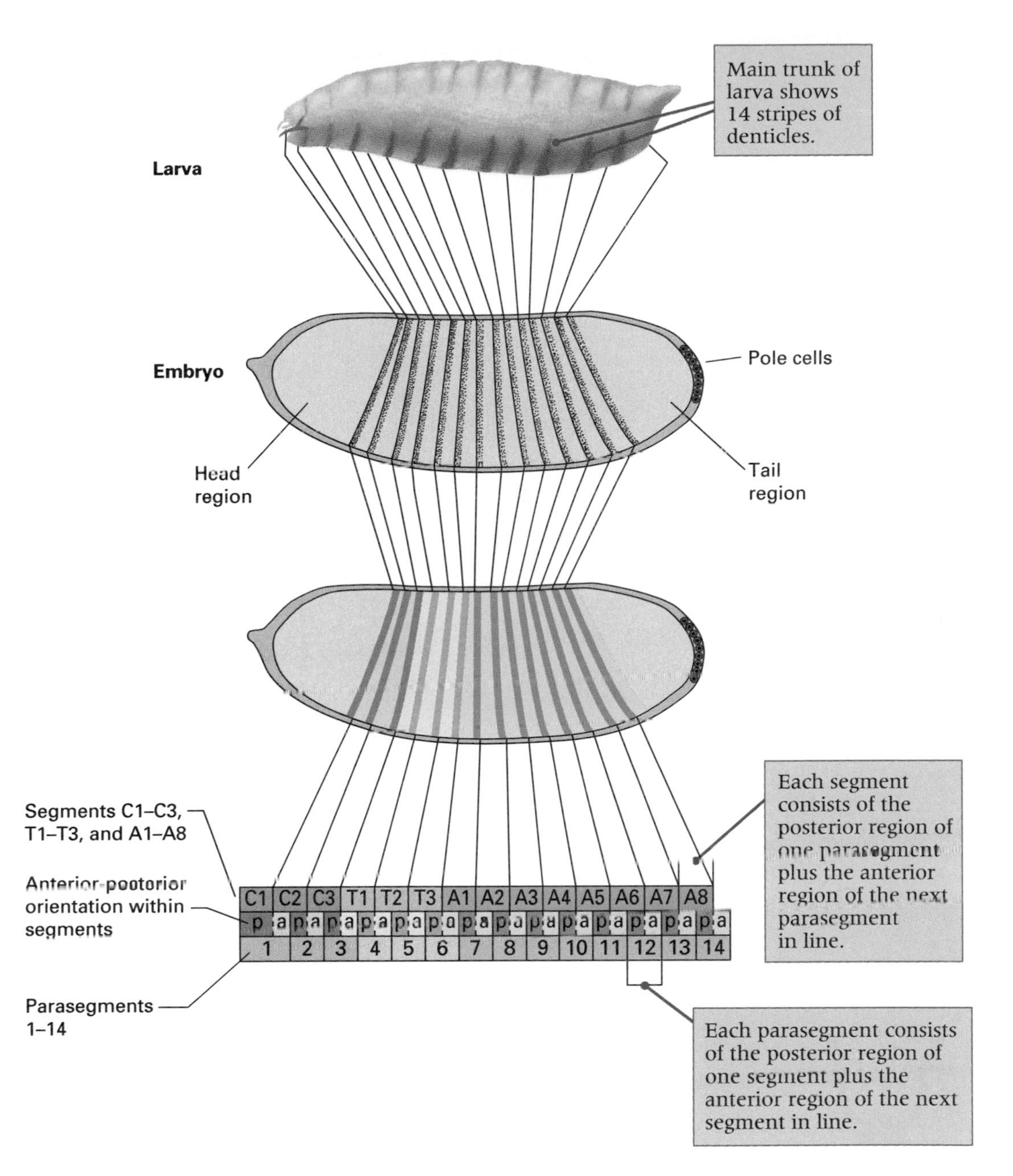

Figure 12.16 Segmental organization of the *Drosophila* embryo and larva. The segments are defined by successive indentations formed by the sites of muscle attachment in the larval cuticle. The parasegments are not apparent morphologically but include the anterior and posterior regions of adjacent segments. The distinction is important, because the patterns of expression of segmentation genes are more often correlated with the parasegment boundaries than with the segment boundaries.

1. The *coordinate genes* determine the principal coordinate axes of the embryo: the anterior-posterior axis, which defines the front and rear, and the dorsal-ventral axis, which defines the top and bottom.
2. The *gap genes* are expressed in contiguous groups of segments along the embryo (Figure 12.17A), and they establish the next level of spatial organization. Mutations in gap genes result in the absence of contiguous body segments, so gaps appear in the normal pattern of structures in the embryo.
3. The *pair-rule genes* determine the separation of the embryo into discrete segments (Figure 12.17B). Mutations in pair-rule genes result in missing pattern elements in alternate segments. The reason for the two-segment periodicity of pair-rule genes is that the genes are expressed in a zebra-stripe pattern along the embryo.

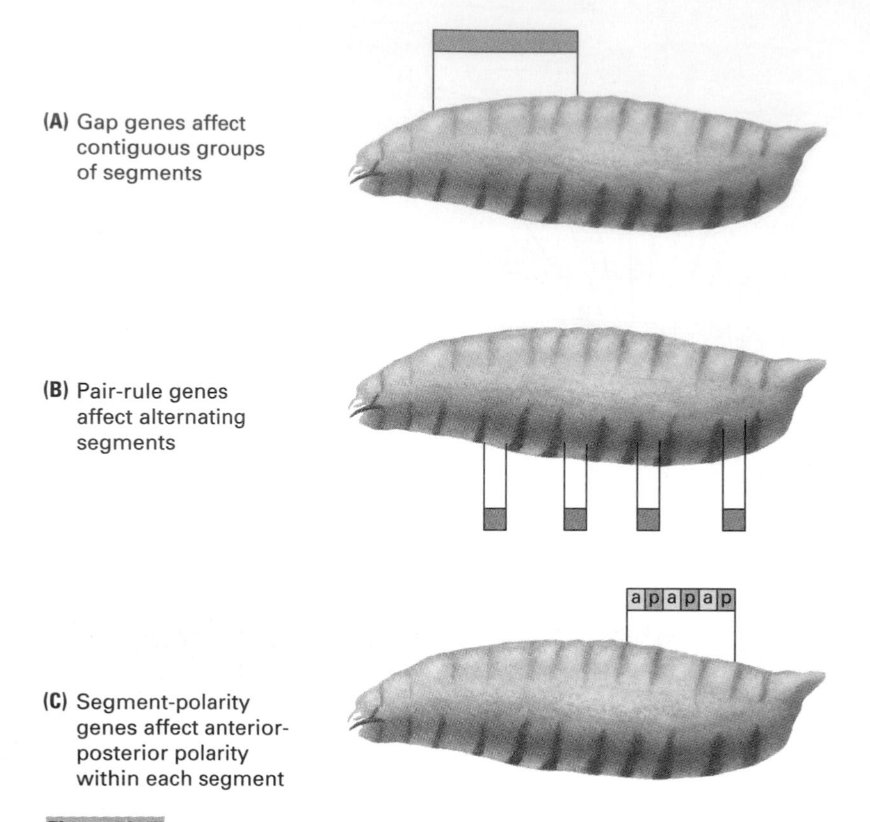

Figure 12.17 Patterns of expression of different types of segmentation genes. (A) The gap genes are expressed in a set of contiguous segments. (B) The pair-rule genes are expressed in alternating segments. (C) The segment-polarity genes are expressed in each segment and determine the anterior-posterior pattern of differentiation within each parasegment.

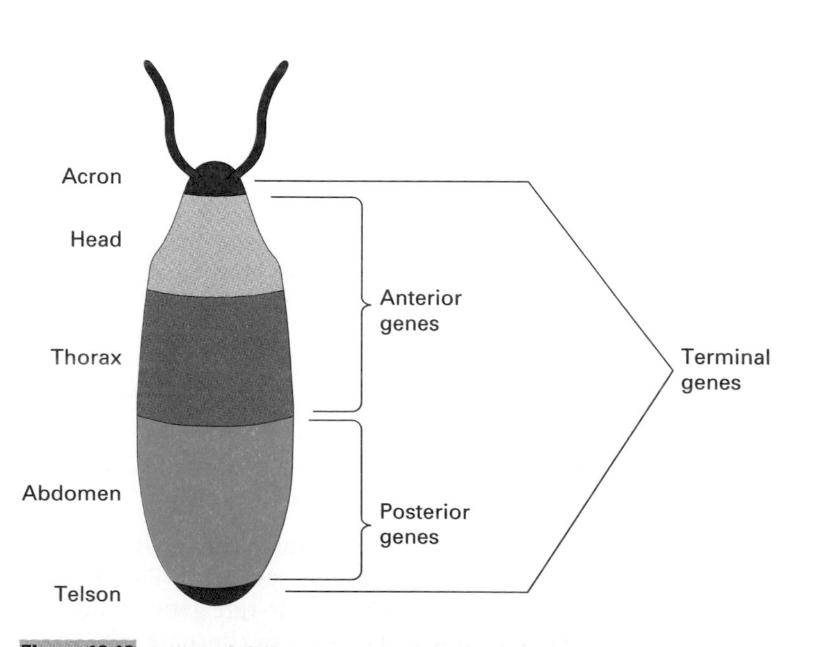

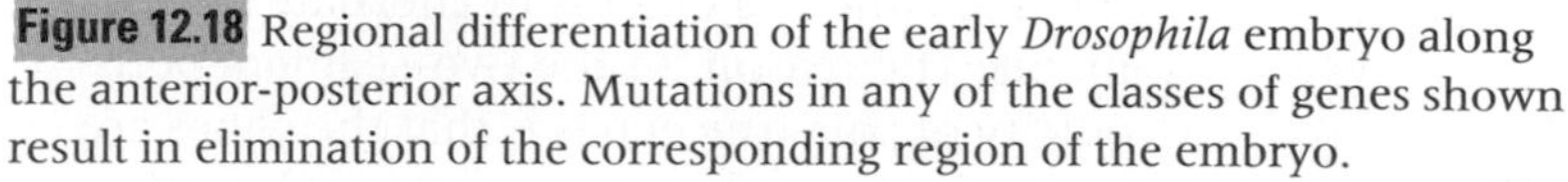

Figure 12.18 Regional differentiation of the early *Drosophila* embryo along the anterior-posterior axis. Mutations in any of the classes of genes shown result in elimination of the corresponding region of the embryo.

4. The *segment-polarity genes* determine the pattern of anterior-posterior development within each segment of the embryo (Figure 12.17C). Mutations in segment-polarity genes affect all segments or parasegments in which the normal gene is active. Many segment-polarity mutants have the normal number of segments, but part of each segment is deleted and the remainder is duplicated in mirror-image symmetry.

Evidence for the existence of the four classes of segmentation genes—coordinate genes, gap genes, pair-rule genes, and segment-polarity genes—is presented in the following sections.

Coordinate genes establish the main body axes.

The **coordinate genes** are maternal-effect genes that establish early polarity through the presence of their products at defined positions within the oocyte or through gradients of concentration of their products. The genes that determine the anterior-posterior axis can be classified into three groups according to the effects of mutations in them, as illustrated in Figure 12.18.

1. The first group of coordinate genes includes the *anterior genes,* which affect the head and thorax. The key gene in this class is *bicoid.* Mutations in *bicoid* produce embryos lacking the head and thorax that occasionally have abdominal segments in reverse polarity duplicated at the anterior end. The *bicoid* gene product is a transcription factor for genes determining anterior structures. Because the *bicoid* mRNA is localized in the anterior part of the early-cleavage embryo, these genes are activated primarily in the anterior region. The *bicoid* mRNA is produced in nurse cells (the cells surrounding the oocyte in Figure 12.5) and exported to a localized region at the anterior pole of the oocyte. The protein product is less localized and, during the syncytial cleavages, forms an anterior-posterior concentration gradient with the maximum at the anterior tip of the embryo. The Bicoid protein is a principal morphogen in determining

the blastoderm fate map. The protein is a transcriptional activator containing a helix-turn-helix motif for DNA binding (Chapter 11). Genes affected by the Bicoid protein contain multiple upstream binding domains that consist of nine nucleotides resembling the consensus sequence 5'-TCTAATCCC-3'. Binding sites that differ by as many as two base pairs from the consensus sequence bind the Bicoid protein with high affinity, and sites that contain four mismatches bind with low affinity. The combination of high- and low-affinity binding sites determines the concentration of Bicoid protein needed for gene activation; genes with many high-affinity binding sites can be activated at low concentrations, but those with many low-affinity binding sites need higher concentrations. Such differences in binding affinity mean that the level of gene expression can differ from one regulated gene to the next along the Bicoid concentration gradient. It is the local concentration of the Bicoid protein that regulates the expression of critical gap genes along the embryo—for example, *hunchback*.

2. The second group of coordinate genes includes the *posterior genes*, which affect the abdominal segments (Figure 12.18). Some of the mutants also lack pole cells. One of the posterior mutations, *nanos*, yields embryos with defective abdominal segmentation but normal pole cells. In this case, the *nanos* mRNA is localized tightly to the posterior pole of the oocyte, and the gene product is a repressor of translation. Among the genes whose mRNA is not translated in the presence of nanos protein is the gene *hunchback*. Hence *hunchback* expression is controlled jointly by the Bicoid and Nanos proteins, Bicoid protein activating transcription in an anterior-posterior gradient and Nanos protein repressing translation in the posterior region.

3. The third group of coordinate genes includes the *terminal genes*, which simultaneously affect the most anterior structure (the acron) and the most posterior structure (the telson) (Figure 12.18). The key gene in this class is *torso*, which codes for a transmembrane receptor that is uniformly distributed throughout the embryo in the early developmental stages. The torso receptor is activated by a signal released only at the poles of the egg by the nurse cells in that location.

Apart from the three sets of genes that determine the anterior-posterior axis of the embryo, a fourth set of genes determines the dorsal-ventral axis. The morphogen for dorsal-ventral determination is the product of the gene *dorsal*, which is present in a pronounced ventral-to-dorsal gradient in the late syncytial blastoderm.

Gap genes regulate other genes in broad anterior–posterior regions.

The main role of the coordinate genes is to regulate the expression of a small group of approximately six gap genes along the anterior-posterior axis. The genes are called **gap genes** because mutations in them result in the absence of pattern elements derived from a group of contiguous segments (Figure 12.17). Gap genes are zygotic genes. The gene *hunchback* serves as an example of the class because *hunchback* expression is controlled by offsetting effects of Bicoid and Nanos. Transcription of *hunchback* is stimulated in an anterior-to-posterior gradient by the Bicoid transcription factor, but posterior *hunchback* expression is prevented by translational repression owing to the posteriorly localized Nanos protein. In the early *Drosophila* embryo in Figure 12.19, the gradient of *hunchback* expression is indicated by the green fluorescence of an antibody specific to the hunchback gene product. The superimposed red fluorescence results from antibody specific to the product of *Krüppel*, another gap gene. The region of overlapping gene expression appears in yellow. The products of both *hunchback* and *Krüppel* are transcription factors of the zinc-finger type (Chapter 11). Other gap genes also are transcription factors. Together, the gap genes have a pattern of regional specificity and partly overlapping domains of expression that enable them to act in combinatorial fashion to control the next set of genes in the segmentation hierarchy, the pair-rule genes.

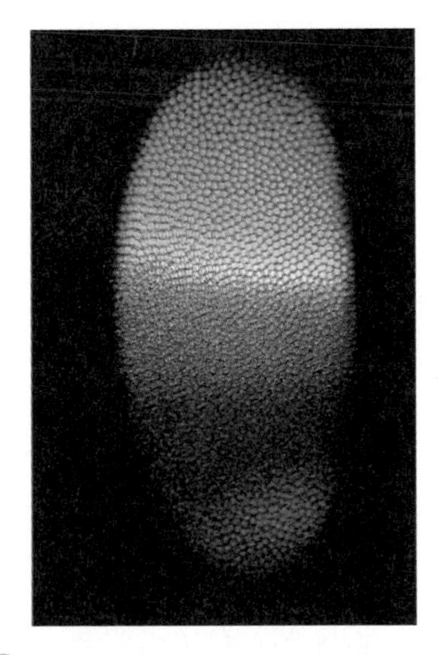

Figure 12.19 An embryo of *Drosophila*, approximately 2.5 hours after fertilization, showing the regional localization of the *hunchback* gene product (green), the *Krüppel* gene product (red), and their overlap (yellow). [Courtesy of James Langeland, Stephen Paddock, and Sean Carroll.]

The human connection — Maternal Impressions

David H. Skuse[1], Rowena S. James[2], Dorothy V. M. Bishop[3], Brian Coppin[4], Paola Dalton[2], Gina Aamodt-Leeper[1], Monique Barcarese-Hamilton[1], Catharine Creswell[1], Rhona McGurk[1], and Patricia A. Jacobs[2] 1997
[1]Institute of Child Health, London, England
[2]Salisbury Distirct Hospital, Salisbury, Wiltshire, UK
3Medical Research Council Applied Psychology Unit, Cambridge, UK
4 Princess Anne Hospital, Southampton, UK

Evidence from Turner's Syndrome of an Imprinted X-linked Locus Affecting Cognitive Function

Genetic imprinting is a difference in gene expression due to parental origin. The mechanism of imprinting is thought to involve DNA methylation, in part because DNA methylation patterns can be inherited. Even though only a handful of loci are known to be imprinted in mammals, the existence of imprinting means that the maternal and paternal gametes do not contribute to embryonic development in completely equivalent ways. These authors realized that phenotypic differences due to X-chromosome imprinting might be revealed in Turner syndrome (45,X). In this chromosomal disorder, one X chromosome is missing as a result of nondisjunction. In 70 percent of Turner-synrdome females, the single X chromosome is maternal in origin (45,X^m); in the remaining 30 percent it is paternal (45,X^p). Although the 45,X chromosomal types are the same, the X^m and X^p chromosomes differ in parental origin. This controversial paper finds evidence that there may be significant behavioral differences (social interaction skills) in Turner syndrome patients, depending on the paternal origin of the chromosome. As the authors point out, this finding may have important implications for social development in chromosomally normal boys and girls.

Here we report a study of 80 females with Turner's syndrome, in 55 of which the X was maternally derived (45,X^m) and in 25 of which it was paternally derived (45,X^p). Members of the 45,X^p group were significantly better adjusted, with superior verbal and higher-order executive function skills, which mediate social interaction. Our observations suggest that there is a genetic locus for social cognition, which is imprinted and is not expressed from the maternally derived X chromosome. . . . If expressed only from the X chromosome of paternal origin, the existence of this locus could explain why 46,XY males (whose single X chromosome is maternal) are more vulnerable to developmental disorders of language and social cognition, such as autism, than are 46,XX females. . . . From a first-stage screening survey of parents and teachers using standardized tests, we discovered that 40% of 45,X^m girls of school age had received a statement of special educational needs, indicating academic failure, compared with 16% of 45,X^p subjects ($P < 0.05$); the figure in the general population is just 2%. We also found that clinically significant social difficulties affected 72.4% of the 45,X^m subjects over 11 years of age (21 of 29), compared with 28.6% of 45,X^p females (4 of 14) ($P < 0.02$). . . . Pilot interviews and observations showed that 45,X^m females in particular lacked flexibility and responsiveness in social interactions. We therefore devised a questionnaire relevant to social cognition. . . . The results for subjects from 6 to 18 years of age confirm there are significant differences between 45,X^m and 45,X^p in the predicted direction. . . . Normal boys also obtained significantly higher scores on the questionnaire than did normal girls, indicating significantly poorer social cognition. The magnitude and direction of this difference are compatible with the hypothesis that there is an imprinted locus on the X chromosome that influences the development of social cognitive skills (although the finding is of course also compatible with other explanations of gender differences in behavior). . . . Males are substantially more vulnerable to a variety of developmental disorders of speech, language impairment and reading disability, as well as more severe conditions such as autism. Our findings are consistent with the hypothesis that the locus described, which we propose to be silent both in males and in 45,X^m females, acts synergistically with susceptibility loci elsewhere in the genome to increase the male-to-female prevalence ratio of such disorders.

THE EXISTENCE OF this locus could explain why 46,XY males . . . are more vulnerable to developmental disorders of language and social cognition than are 46,XX females.

Source: Nature 387: 705–708

Pair-rule genes are expressed in alternating segments or parasegments.

The coordinate and gap genes determine the polarity of the embryo and establish broad regions within which subsequent development takes place. As development proceeds, the progressively more refined organization of the embryo is correlated with the patterns of expression of the segmentation genes. Among these are the **pair-rule genes,** in which the mutant phenotype has alternating segments absent or malformed (Figure 12.17). Approximately eight pair-rule genes have been identified. For example, mutations of the pair-rule gene *even-skipped* affect even-numbered segments, and those of another pair-rule gene, *odd-skipped*, affect odd-numbered segments. The function of the pair-rule genes is to give the early *Drosophila* larva a segmented body pattern with both repetitiveness and individuality of segments. For example, there are eight abdominal segments that are repetitive in that they are regularly spaced and share several common features, but they differ in the details of their differentiation.

One of the earliest pair-rule genes expressed is *hairy*, whose pattern of expression is under both positive and negative regulation by the products of *hunchback*, *Krüppel*, and other gap genes. Expression of *hairy* yields seven stripes (Figure 12.20). The striped pattern of pair-rule gene expression is typical, but the stripes of expression of one gene are usually slightly out of register with those of another. Together with the continued regional expression of the gap genes, the combinatorial patterns of gene expression in the embryo are already complex and linearly differentiated. Figure 12.21 shows an embryo stained for the products of three genes—*hairy* (green), *Krüppel* (red), and *giant* (blue). The regions of overlapping expression appear as color mixtures—orange, yellow, light green, or purple. Even at the early stage in Figure 12.21, there is a unique combinatorial pattern of gene expression in every segment and parasegment. The complexity of combinatorial control can be appreciated by considering that the expression of the *hairy* gene in stripe 7 depends on a promoter element smaller than 1.5 kb that contains a series of binding sites for the protein products of the genes *caudal*, *hunchback*, *knirps*, *Krüppel*, *tailless*, *huckbein*, *bicoid*, and perhaps still other proteins yet to be identified. The combinatorial patterns of gene expression of the pair-rule genes define the boundaries of expression of the segment-polarity genes, which function next in the hierarchy.

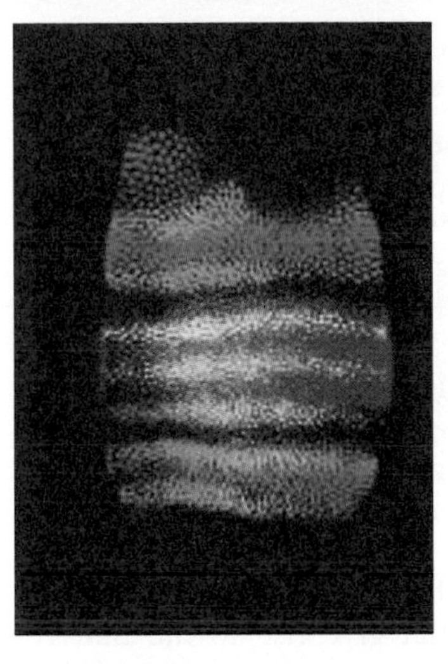

Figure 12.21 Combined patterns of expression of *hairy* (green), *Krüppel* (red), and *giant* (blue) in a *Drosophila* embryo approximately 3 hours after fertilization. Already there is considerable linear differentiation apparent in the patterns of gene expression. [Courtesy of James Langeland, Stephen Paddock, and Sean Carroll.]

Segment-polarity genes govern differentiation within segments.

Whereas the pair-rule genes determine the body plan at the level of segments and parasegments, the **segment-polarity genes** create a spatial differentiation within each segment. Approximately 14 segment-polarity genes have been identified. The mutant phenotype has repetitive deletions of pattern along the embryo (Figure 12.17) and usually a mirror-image duplication of the part that remains. Among the earliest segment-polarity genes expressed is *engrailed*, whose stripes of expression approximately coincide with the boundaries of the parasegments and so divide each segment into anterior and posterior domains (Figure 12.22).

Expression of the segment-polarity genes finally establishes the early polarity and linear differentiation of the embryo into segments and parasegments. These interactions govern the activities of the second set of developmental genes, the *homeotic genes*, which control the pathways of differentiation in each segment or parasegment.

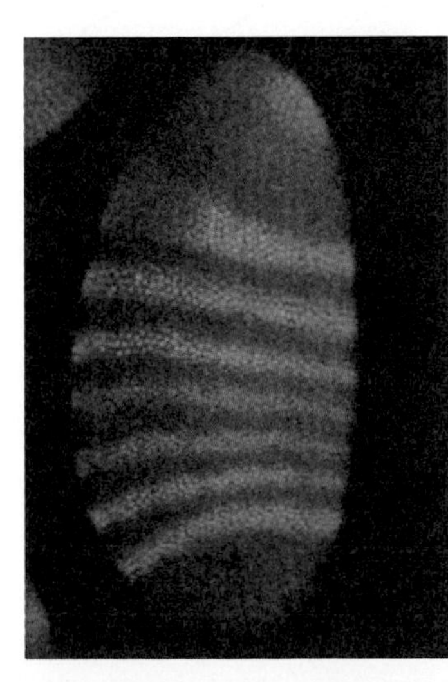

Figure 12.20 Characteristic seven stripes of expression of the gene *hairy* in a *Drosophila* embryo approximately 3 hours after fertilization. (Courtesy of James Langeland, Stephen Paddock, and Sean Carroll.)

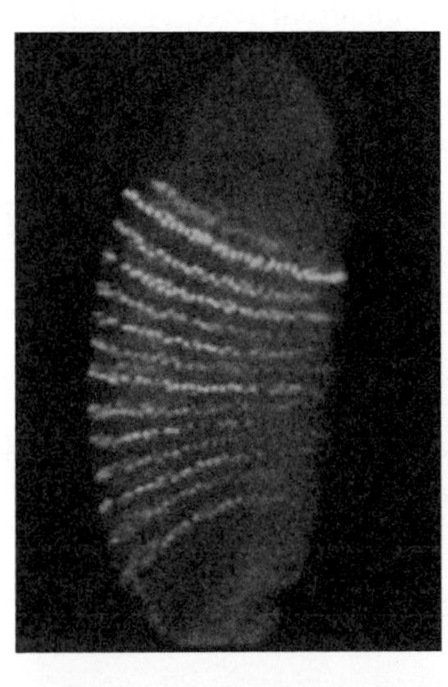

Figure 12.22 Expression of the segment-polarity gene *engrailed* partitions the early *Drosophila* embryo into 14 regions. These eventually differentiate into three head segments, three thoracic segments, and eight abdominal segments. [Courtesy of James Langeland, Stephen Paddock, and Sean Carroll.]

Homeotic genes function in the specification of segment identity.

As with most other insects, the larvae and adults of *Drosophila* have a segmented body plan consisting of a head, three thoracic segments, and eight abdominal segments (Figure 12.23). Metamorphosis makes use of about 20 structures called **imaginal disks** present inside the larvae (Figure 12.24). Formed early in development, the imaginal disks ultimately give rise to the principal structures and tissues in the adult organism. Examples of imaginal disks include the pair of wing disks (one on each side of the body) that give rise to the wings and their attachments on thoracic segment T2; and the pair of haltere disks that give rise to the halteres (flight balancers) on thoracic segment T3 (Figure 12.23). Developmentally, the halteres are highly modified wings. During the pupal stage, when many larval tissues and organs break down, the imaginal disks progressively unfold and differentiate into adult structures. The morphogenic events that take place in the pupa are initiated by the hormone ecdysone, secreted by the larval brain.

As in the early embryo, overlapping patterns of gene expression and combinatorial control guide later events in *Drosophila* development. The pattern of expression of a key gene in wing development, *vestigial,* in a wing disk is shown in Figure 12.25. This pattern is the summation of *vestigial* response to two separate signaling pathways summarized in Figure 12.26. Pathway A consists of the products of the genes *Apterous, fringe, Serrate,* and so forth; pathway B consists of the products of the genes *engrailed, hedgehog,* and so forth. The Suppressor-of-Hairy protein binds to a so-called *boundary enhancer* in the vestigial gene and induces gene expression in the cross-shaped pattern indicated. The rest of the vestigial pattern (part B) is due to activation of a separate *quadrant enhancer* by the protein Mad. Such overlapping patterns of gene expression of *vestigial* and other genes in wing development ultimately yield the exquisitely fine level of cellular and morphological differentiation observed in the adult animal.

Among the genes that transform the periodicity of the *Drosophila* embryo into a body plan with

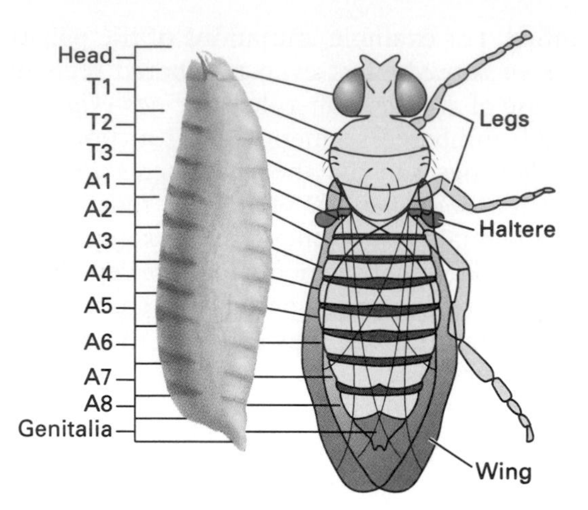

Figure 12.23 Relationship between larval and adult segmentation in *Drosophila.* Each of the three thoracic segments in the adult carries a pair of legs. The wings develop on the second thoracic segment (T2) and the halteres (flight balancers) on the third thoracic segment (T3).

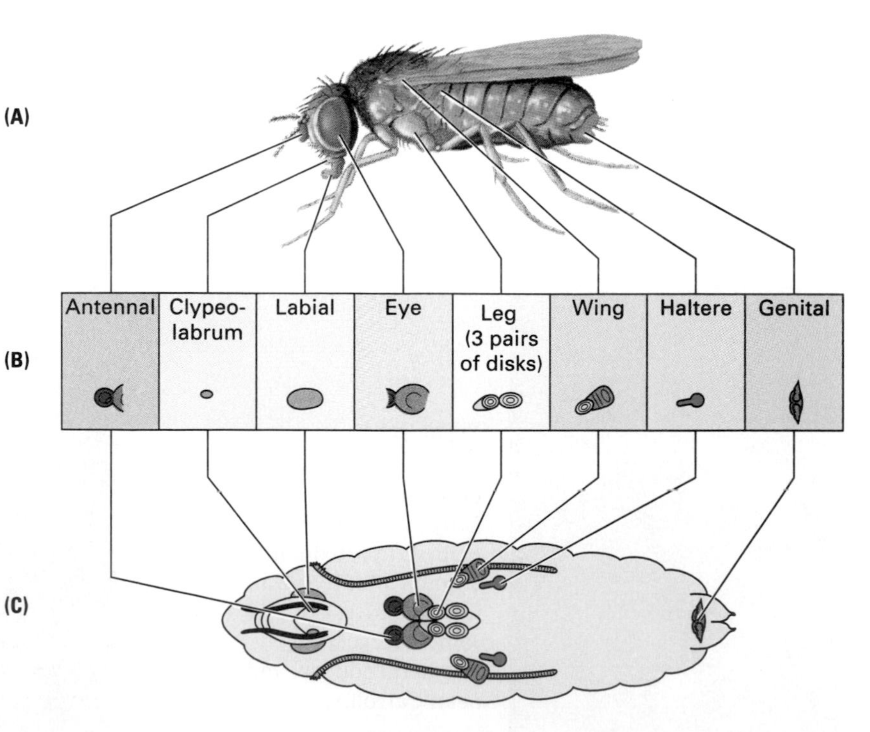

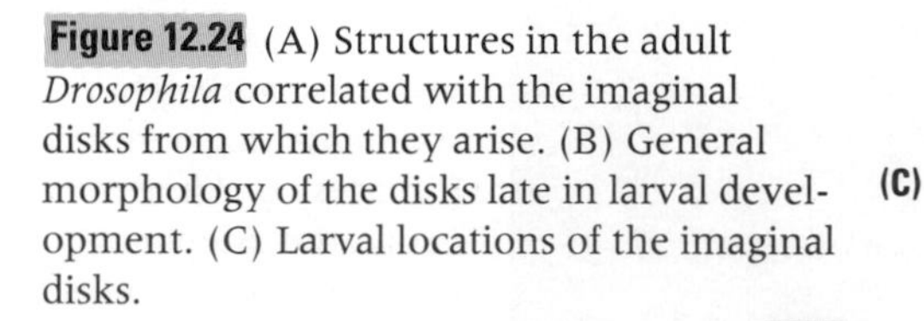

Figure 12.24 (A) Structures in the adult *Drosophila* correlated with the imaginal disks from which they arise. (B) General morphology of the disks late in larval development. (C) Larval locations of the imaginal disks.

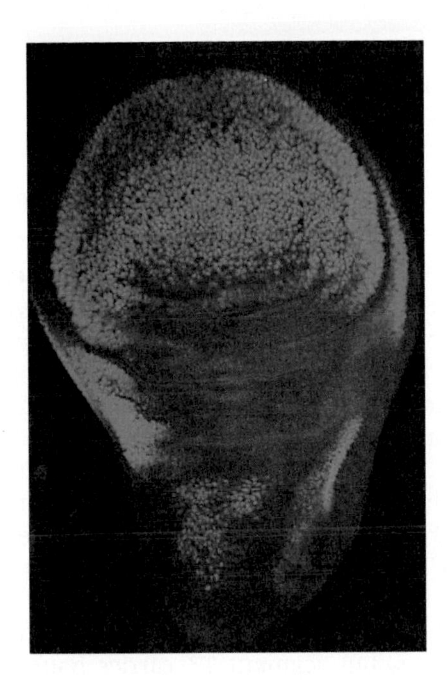

Figure 12.25 Expression of the vestigial gene (green) in the developing wing imaginal disk. The circular area of expression in the upper center will give rise to the wing proper. [Courtesy of Stephen Paddock and Sean Carroll.]

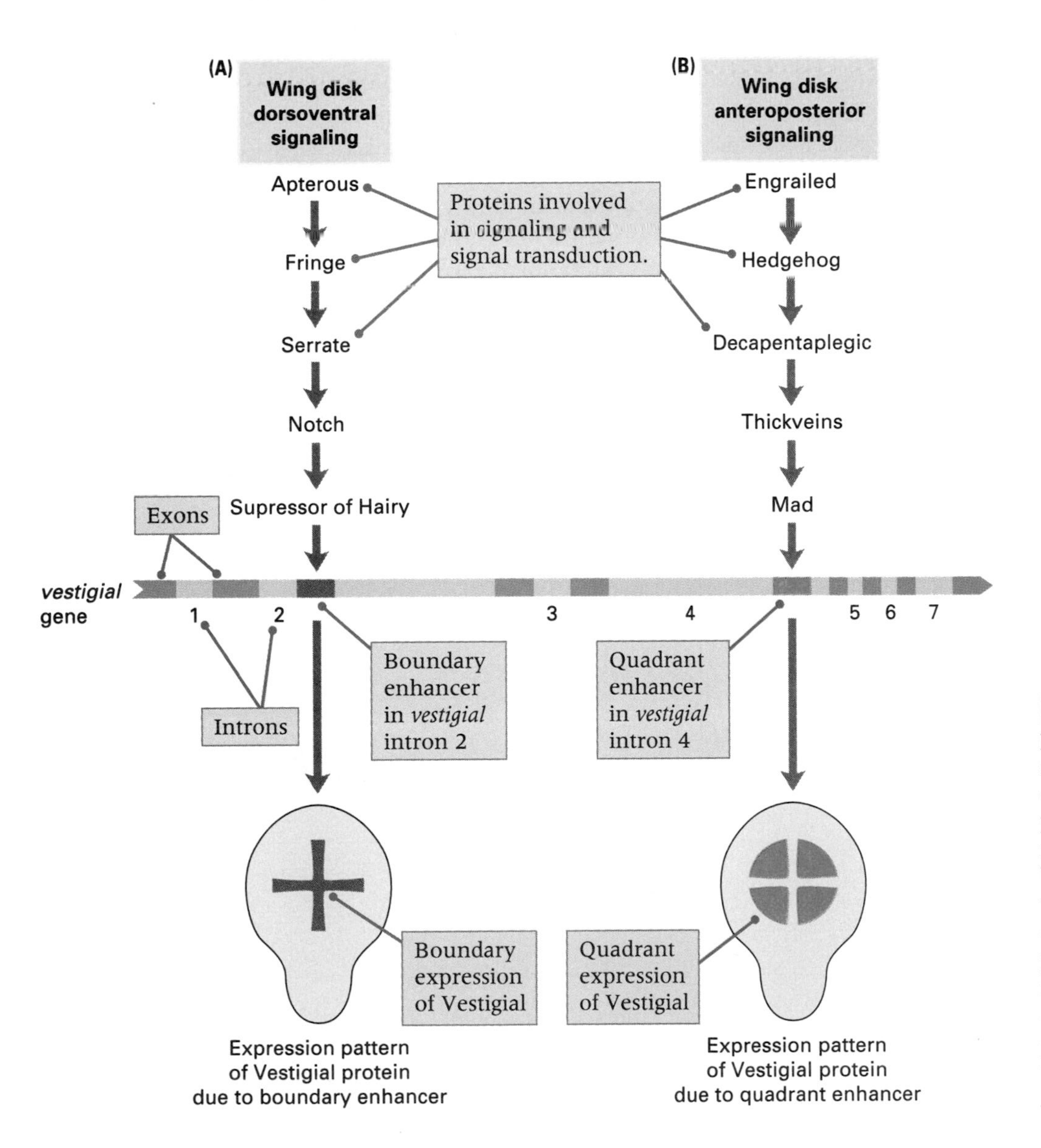

Figure 12.26 The uniform pattern of *vestigial* expression in the wing imaginal disk results from the superposition of two separate patterns. (A) The boundary expression pattern is determined by a dorsoventral signaling pathway. (B) The quadrant expression pattern is determined by the anteroposterior signaling pathway.

linear differentiation that includes the development of wings from segment T2 and halteres from segment T3 are two small sets of **homeotic,** or ***HOX,* genes.** Homeotic mutations result in the transformation of one body segment into another, which is recognized by the misplaced development of structures that are normally present elsewhere in the embryo. One class of homeotic mutation is illustrated by *bithorax,* which causes transformation of the anterior part of the third thoracic segment into the anterior part of the second thoracic segment, with the result that the halteres are transformed into an extra pair of wings (Figure 12.27). The other class of homeotic mutation is illustrated by *Antennapedia,* which results in transformation of the antennae into legs. The *HOX* genes represented by *bithorax* and *Antennapedia* are in fact gene clusters. The cluster containing *bithorax* is designated BX-C (stands for *bithorax*-complex), and that containing *Antennapedia* is called ANT-C (stands for *Antennapedia*-complex). Both gene clusters were initially discovered through their homeotic effects in adults. Later they were shown to affect the identity of larval segments. The BX-C is primarily concerned with the development of larval segments T3 through A8 (Figure 12.23), with principal effects in T3 and A1. The ANT-C is primarily concerned with the development of the head (H) and thoracic segments T1 and T2.

The homeotic genes are transcriptional activators of other genes. Most *HOX* genes contain one or more copies of a characteristic sequence of about 180 nucleotides called a **homeobox,** which is also found in key genes concerned with the development of embryonic segmentation in organisms as diverse as segmented worms, frogs, chickens, mice, and human beings. Homeobox sequences are present in exons and code for a protein-folding domain that includes a helix-turn-helix DNA-binding motif (Chapter 11).

HOX genes function at many levels in the regulatory hierarchy.

In *Drosophila,* a *HOX* gene called *Ultrabithorax (Ubx)* is normally active in segment T3, which gives rise to the halteres, and only in segment T3. Reduced *Ubx* expression makes the haltere more wing-like, and in the absence of *Ubx* expression, the haltere disk forms a structurally normal wing (Figure 12.27). Given this information, it is tempting to say that *Ubx* "controls" the development of the haltere. But this inference is wrong, because it implies that *Ubx* expression is both necessary and sufficient for the development of the haltere. The fact is that *Ubx* homologs are expressed in T3 in probably all insects, but segment T3 carries halteres only in the dipteran insects, of which *Drosophila* is an example (Figure 12.28). The ancestral condition in insects is similar to that in today's dragonfly, in which the hind wings are virtually identical to the forewings; the dragonfly *Ubx* homolog clearly does not make a haltere. In the insect lineage leading to present-day lepidopterans, the hind wing became modified from the forewing, but not to such an extreme extent as in the dipterans; yet the lepidopteran *Ubx* homolog is expressed in the hind wing.

The resolution of the paradox of how *Ubx* can seemingly "control" the development of the haltere in dipterans without doing so in other insects is found in the gradual evolution of sets of genes that come under *Ubx* control. The situation is exemplified in Figure 12.28, in which the small circles denote genes and the arrows gene activation in

(A)

Haltere

(B)

Figure 12.27 (A) Wildtype *Drosophila* showing wings and halteres (the pair of knob-like structures protruding posterior to the wings). (B) A fly with four wings produced by mutations in the *bithorax* complex. The mutations convert the third thoracic segment into the second thoracic segment, and the halteres normally present on the third thoracic segment become converted into the posterior pair of wings. [Courtesy of E. B. Lewis.]

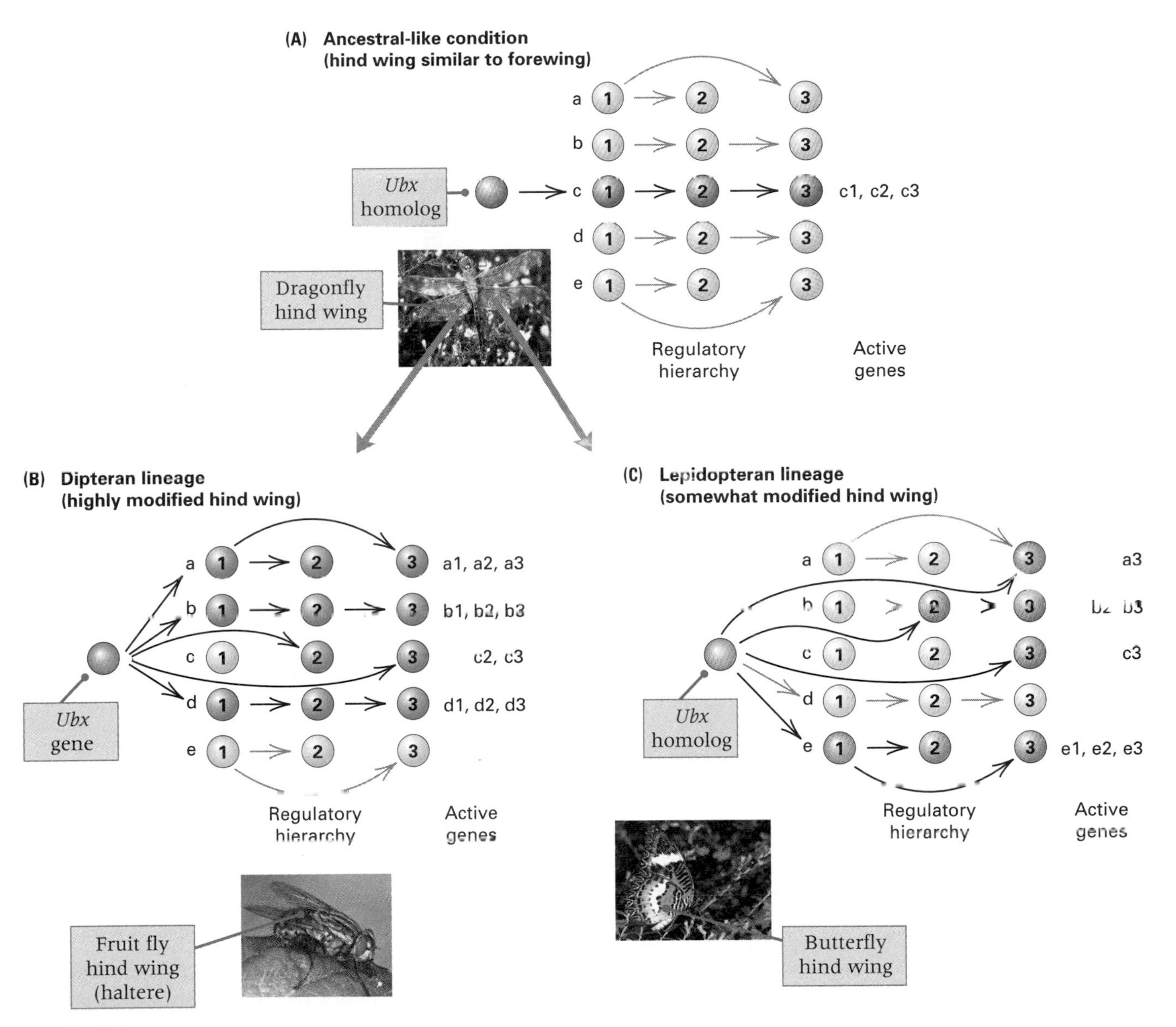

Figure 12.28 Evolutionary scheme by which the *Ubx* homolog expressed in segment T3 produces different developmental pathways for the hind wing in different insect lineages. As each lineage evolves, mutations gradually accumulate, through natural selection, that bring different sets of genes under the control of the *Ubx* homolog. [Based on ideas of Sean Carroll.]

segment T3. The condition in part A is assumed to be the ancestral state, in which the gene *c1* has evolved in such a way that it is activated by the *Ubx* homolog. In this organism, the active genes in T3 will consist of the genes activated by the *Ubx* homolog (*c1*), plus genes further along the regulatory hierarchy activated by these genes (*c2* and *c3*), plus other genes (not shown) whose expression in T3 is independent of the *Ubx* pathway. The complete set of *Ubx*-induced genes, direct and indirect, is shown at the right of the regulatory hierarchy. These are the genes that account for the slight modifications of the hind wing as compared with the forewing. In the absence of the *Ubx* homolog, the hind wing would lose these slight modifications and become even more like a forewing.

As evolution proceeds, mutations can occur that, by chance, can either release a gene from *Ubx* control or bring a gene under *Ubx* control. Such mutations would typically include the loss or gain of *Ubx*-responsive enhancers. A newly evolved *Ubx* responsiveness, or loss of *Ubx* responsiveness, would not affect the expression of the same genes in other tissues, because *Ubx* is expressed specifically in T3. If the novel pattern of T3 gene expression is favored by natural selection, then the mutation will become incorporated into the species. After a long period of evolution, numerous genes will have evolved in this way, progressively modifying the hind-wing pattern in a manner favored by natural selection. In dipterans, for example, the selection was for extreme modification of size and structure to form the haltere. The result of the evolution is shown diagrammatically in Figure 12.28B by depicting *c1* as having lost *Ubx* activation and *a1* and *a3, b1, c2* and *c3*, and *d1* as having gained it. (Note that the *Ubx* gene can control genes anywhere in the regulatory hierarchy, not only at the beginning.) Similarly, in the lepidopteran lineage (part C), there was selection for less extreme modification, which is depicted as loss of *c1* activation, but gain of *a3, b2, c3,* and *e1* and *e3* activation. The point is that, independently in each lineage, the hind wing was modified gradually while a sequence of genes was brought under *Ubx* control or released from *Ubx* control. The result is that in dipterans the *Ubx* gene "controls" haltere development, whereas in lepidopterans it "controls" a less extreme modification of the forewing. In each lineage the *Ubx* homolog is critical in developmental control, but it regulates different genes. In each lineage also, mutations in *Ubx* will make the hind wing more similar to the forewing.

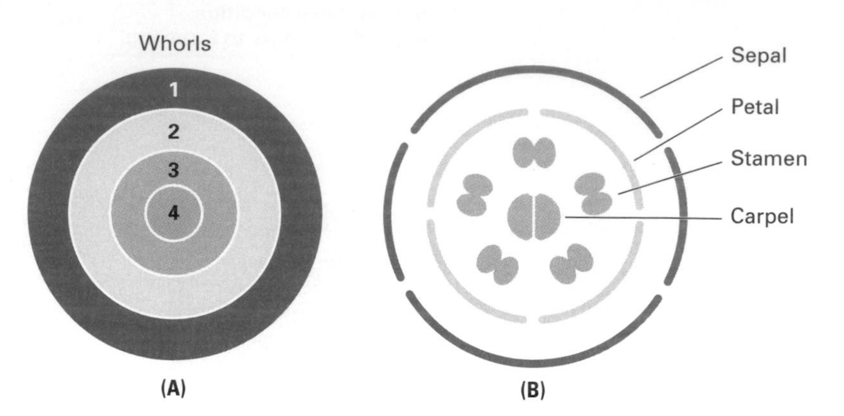

Figure 12.29 (A) The organs of a flower are arranged in four concentric rings, or whorls. (B) Whorls 1, 2, 3, and 4 give rise to sepals, petals, stamens, and carpels, respectively.

12.5 Development in higher plants uses some of the same mechanisms as in animals

As we have seen, most of the major developmental decisions in animals are made early in life, during embryogenesis. In higher plants, differentiation takes place almost continuously throughout life in regions of actively dividing cells called **meristems** in both the vegetative organs (root, stem, and leaves) and the floral organs (sepal, petal, pistil, and stamen). The shoot and root meristems are formed during embryogenesis and consist of cells that divide in distinctive geometric planes and at different rates to produce the basic morphological pattern of each organ system. The floral meristems are established by a reorganization of the shoot meristem after embryogenesis and eventually differentiate into floral structures characteristic of each particular species. One important difference between animal and plant development is that

> In higher plants, as groups of cells leave the proliferating region of the meristem and undergo further differentiation into vegetative or floral tissue, their developmental fate is determined almost entirely by their position relative to neighboring cells.

The critical role of positional information in higher plant development stands in contrast to animal development, in which cell lineage often plays a key role in determining cell fate.

The plastic or "indeterminate" growth patterns of higher plants are the result of continuous production of both vegetative and floral organ systems. These patterns are conditioned largely by day length and the quality and intensity of light. The plasticity of plant development gives plants a remarkable ability to adjust to environmental perturbations as well as to genetic aberrations.

Flower development in *Arabidopsis* is controlled by relatively few genes.

Genetic analysis in *Arabidopsis thaliana*, a member of the mustard family, has revealed some important principles in the genetic determination of floral structures. As is typical of flowering plants, the flowers of *Arabidopsis* are composed of four types or organs arranged in concentric rings or whorls. Figure 12.29 illustrates the geometry, looking down at a flower from the top. From outermost to innermost, the whorls are designated 1, 2, 3, and 4 (Figure 12.29A). In the development of the flower, each whorl gives rise to a different floral organ (Figure 12.29B). Whorl 1 yields the sepals (the green, outermost floral leaves), whorl 2 the petals (the white, inner floral leaves), whorl 3 the stamens (the male organs, which form pollen), and whorl 4 the carpels (which fuse to form the ovary).

Mutations affecting floral development fall into three major classes, each causing a characteristic phenotype (Figure 12.30). Compared with the wildtype flower (panel A), one mutant class lacks sepals and petals (panel B), another class lacks petals and stamens (panel C), and the third class lacks stamens and carpels (panel D). On the basis of crosses between homozygous mutant organisms,

(A) Wildtype

(B) *apetala-2* (*ap2*)

(C) *pistillata* (*pi*)

(D) *agamous* (*ag*)

Figure 12.30 Phenotypes of the major classes of floral mutations in *Arabidopsis.* (A) The wild-type floral pattern consists of concentric whorls of sepals, petals, stamens, and carpels. (B) The homozygous mutation *ap2 (apetala-2)* results in flowers missing sepals and petals. (C) Genotypes that are homozygous for either *ap3 (apetala-3)* or *pi (pistillala)* yield flowers that have sepals and carpels but lack petals and stamens. (D) The homozygous mutation *ag (agamous)* yields flowers with sepals and petals but lacking stamens and carpels. [Courtesy of Elliot M. Meyerowitz and John Bowman. Part B from Elliot M. Meyerowitz. 1994. The genetics of flower development. *Scientific American* 271: 56 (November 1994).]

these classes of mutants can be assigned to four complementation groups, each of which defines a different gene. The phenotype that lacks sepals and petals is caused by mutations in the gene *ap2 (apetala-2).* The phenotype that lacks petals and stamens is caused by a mutation in either of two genes, *ap3 (apetala-3)* or *pi (pistillata).* The phenotype that lacks stamens and carpels is caused by mutations in the gene *ag (agamous).* Each of these genes has been cloned and sequenced. They are all transcription factors. The transcription factors encoded by *ap3, pi,* and *ag* are members of what is called the *MADS box* family of transcription factors; each member of this family contains a sequence of 58 amino acids in which common features can be identified. MADS box transcription factors are very common in plants but are found less frequently in animals.

Development of the floral organs is under combinatorial control.

The role of the Ap2, Ap3, Pi, and Ag transcription factors in the determination of floral organs can be inferred from the phenotypes of the mutants. The logic of the inference is based on the observation that mutations in any of the genes eliminate two floral organs arising from adjacent whorls. This pattern suggests that *ap2* is necessary for sepals and petals, *ap3* and *pi* are both necessary for petals and stamens, and *ag* is necessary for stamens and carpels. Because the mutant phenotypes result from loss-of-function alleles of the genes, it may be inferred that *ap2* is expressed in whorls 1 and 2, that *ap3* and *pi* are expressed in whorls 2 and 3, and that *ag* is expressed in whorls 3 and 4.

The overlapping patterns of expression are shown at the top in Figure 12.31. This model of gene expression suggests that floral development is controlled in combinatorial fashion by the four genes. Sepals develop from tissue in which only *ap2* is active; petals are fated by a combination of *ap2*, *ap3*, and *pi*; stamens are determined by a combination of *ap3*, *pi* and *ag*; and carpels derive from tissue in which only *ag* is expressed. The developmental fates are illustrated graphically by the flower at the bottom in Figure 12.31. In order to explain the various mutant phenotypes, it is necessary to make the additional assumption that *ap2* and *ag* expression are mutually exclusive: In the presence of the Ap2 transcription factor, *ag* is repressed, and in the presence of the Ag transcription factor, *ap2* is repressed. If this were the case, then in *ap2* mutants, *ag* expression would spread into whorls 1 and 2; likewise, in *ag* mutants, *ap2* expression would spread into whorls 3 and 4. With this additional assumption, the phenotypes of all of the single and double mutants can be explained.

With the additional assumption about *ap2* and *ag* interaction, the model in Figure 12.31 fits the data, but is the model true? For these genes, the patterns of gene expression, assayed by *in situ* hybridization of RNA in floral cells with labeled probes for each of the genes, fits the patterns at the top of the figure. In particular, *ap2* is expressed in whorls 1 and 2, *ap3* and *pi* in whorls 2 and 3, and *ag* in whorls 3 and 4. (The *ap2* gene is also expressed in nonfloral tissue, but its role in these tissues is unknown.) Furthermore, the seemingly *ad hoc* assumption about *ap2* and *ag* expression being mutually exclusive turns out to be true. In *ap2* mutants, *ag* is expressed in whorls 1 and 2; reciprocally, in *ag* mutants, *ap2* is expressed in whorls 3 and 4. It is also known how *ap3* and *pi* work together. The active transcription factor corresponding to these genes is a dimeric protein composed of Ap3 and Pi polypeptides. Each component polypeptide, in the absence of the other, remains inactive in the cytoplasm. Together, they form an active dimeric transcription factor that migrates into the nucleus.

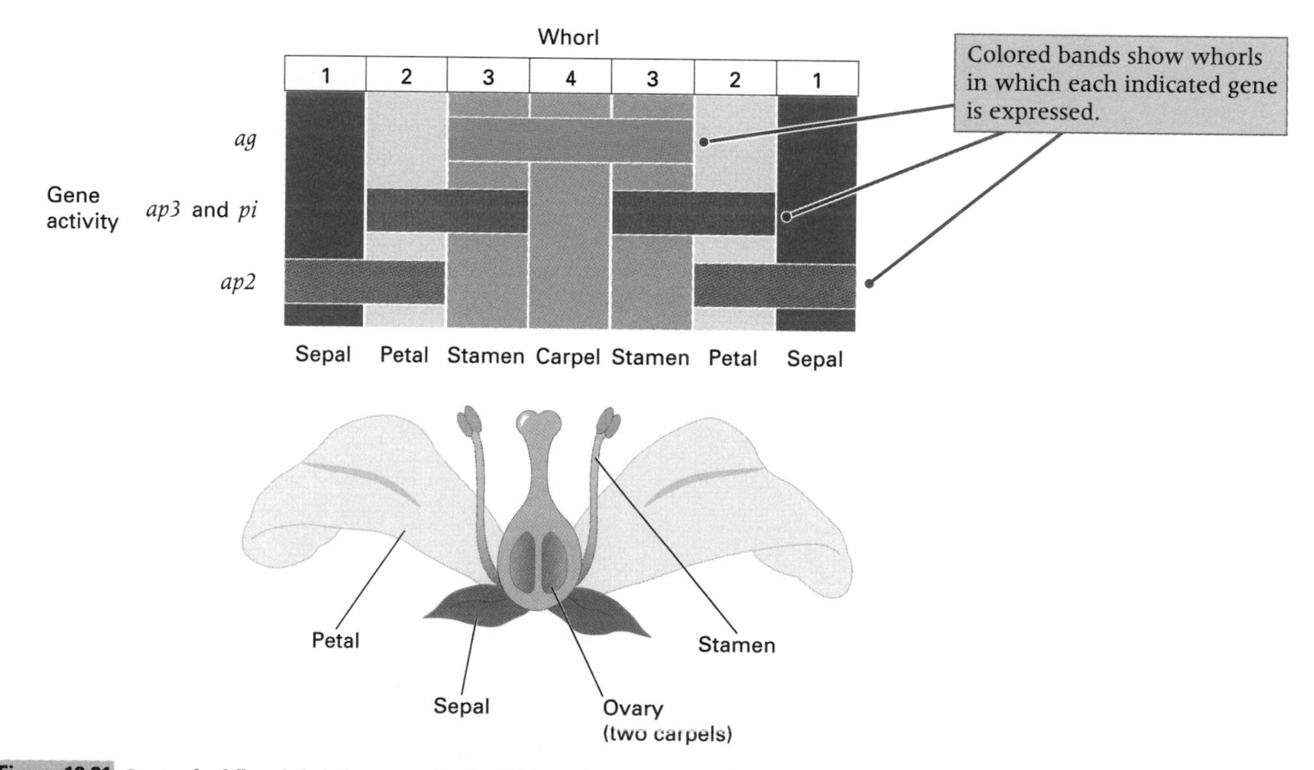

Figure 12.31 Control of floral development in *Arabidopsis* by the overlapping expression of four genes. The sepals, petals, stamens, and carpels are floral organ systems that form in concentric rings or whorls. The developmental identity of each concentric ring is determined by the genes *ap2*, *ap3* and *pi*, and *ag*, each of which is expressed in two adjacent rings. Gene *ap2* is expressed in the outermost two rings, *ap3* and *pi* in the middle two, and *ag* in the inner two. Therefore, each whorl has a unique combination of active genes.

Chapter Summary

- **The spatial organization of gene products can determine the fate of daughter cells.**

The genotype determines the developmental potential of the embryo by means of a developmental program that results in different sets of genes being expressed in different types of cells. Mutations that interrupt developmental processes identify genetic factors that control development.

- **Cell fate is progressively restricted in animal development.**

Cell fate is determined by autonomous development and/or intercellular signaling.
The zygotic genome is activated at different times in different animals.
The initial events in development are controlled by gene products in the oocyte.

In animals, the earliest events in embryonic development depend on the correct spatial organization of numerous constituents present in the oocyte. Developmental genes that are needed in the mother for proper oocyte formation and zygotic development are maternal-effect genes. Genes that are required in the zygote nucleus are zygotic genes. Fertilization of the oocyte initiates a series of mitotic cleavage divisions that form the "hollow ball" blastula, which rapidly undergoes a restructuring into the gastrula. Accompanying these early morphological events are a series of molecular events that determine the developmental fates that cells undergo. Execution of a developmental state may be autonomous (genetically programmed) or may require positional information supplied by neighboring cells or the local concentration of one or more morphogens.

The soil nematode *Caenorhabditis elegans* is used widely in studies of cell lineages, because many lineages in the organism undergo virtually autonomous development and the developmental program is identical from one organism to the next. Most lineages are affected by many genes, including genes that control the sublineages into which the lineage can differentiate.

Genes that control key points in development can often be identified by the unusual feature that recessive alleles (ideally, loss-of-function mutations) and dominant alleles (ideally, gain-of-function mutations) result in opposite effects on phenotype. For example, if loss of function results in failure to execute a developmental program in a particular anatomical position, then gain of function should result in execution of the program in an abnormal location.

- **The developmental fate of cells is often under genetic control.**

Development in the nematode follows a fixed program of lineage diversification.
Mutations can affect cell lineages in development.
Developmental mutations can be classified by their effects on lineages.
Transmembrane receptors often mediate signal transduction.

Early development in *Drosophila* includes the formation of a syncytial blastoderm by early cleavage divisions without cytoplasmic division, the setting apart of pole cells that form the germ line at the posterior of the embryo, the migration of most nuclei to the periphery of the syncytial blastoderm, cellularization to form the cellular blastoderm, and determination of the blastoderm fate map at or before the cellular blastoderm stage. Metamorphosis into the adult fly makes use of about 20 imaginal disks present in the larva that contain developmentally committed cells that divide and develop into the adult structures. Most imaginal disks include several discrete groups of cells or compartments separated by boundaries that progeny cells do not cross.

Early development in *Drosophila* to the level of segments and parasegments requires four classes of segmentation genes: (1) coordinate genes that establish the basic anterior-posterior and dorsal-ventral aspect of the embryo, (2) gap genes for longitudinal separation of the embryo into regions, (3) pair-rule genes that establish an alternating on/off striped pattern of gene expression along the embryo, and (4) segment-polarity genes that refine the patterns of gene expression within the stripes and determine the basic layout of segments and parasegments. The segmentation genes can regulate themselves, other members of the same class, and genes of other classes farther along the hierarchy. Together, the segmentation genes control the homeotic *(HOX)* genes that initiate the final stages of developmental specification.

Mutations in homeotic genes result in the transformation of one body segment into another. For example, *bithorax* causes transformation of the anterior part of the third thoracic segment T3 into the anterior part of the second thoracic segment T2. Homeotic genes act within developmental compartments to control other genes concerned with such characteristics as rates of cell division, orientation of mitotic spindles, and the capacity to differentiate bristles, legs, and other features.

- **Mutations have been isolated for many developmental genes in *Drosophila*.**

Mutations in a maternal-effect gene result in defective oocytes.
Embryonic pattern formation is under genetic control.
Coordinate genes establish the main body axes.
Gap genes regulate other genes in broad anterior-posterior regions.
Pair-rule genes are expressed in alternating segments or parasegments.
Segment-polarity genes govern differentiation within segments.
Homeotic genes function in the specification of segment identity.
HOX genes function at many levels in the regulatory hierarchy.

- **Development in higher plants uses some of the same mechanisms as in animals.**

Flower development in *Arabidopsis* is controlled by relatively few genes.
Development of the floral organs is under combinatorial control.

Developmental processes in higher plants differ significantly from those in animals in that developmental decisions continue throughout life in the meristem regions of the vegetative organs (root, stem, and leaves) and the floral organs (sepal, petal, pistil, and stamen). However, genetic control of plant development is mediated by transcription factors analogous to those in animals. For example, control of floral development in *Arabidopsis* is through the combinatorial expression of four genes in each of a series of three concentric rings, or whorls, of cells that eventually form the sepals, petals, stamens, and carpels.

Key Terms

apoptosis
autonomous determination
blastoderm
blastula
cell fate
cell lineage
cleavage division
coordinate gene
fate map
gain-of-function mutation
gap gene
gastrula
homeobox
homeotic gene (*HOX* gene)
imaginal disk
ligand
lineage diagram
loss-of-function mutation
maternal-effect genes
meristem
morphogen
pair-rule gene
parasegment
pattern formation
pole cell
positional information
segment
segmentation gene
segment-polarity gene
syncytial blastoderm
transmembrane receptor
zygotic gene

Concepts and Issues: Testing Your Knowledge

- What is meant by *positional information* in regard to development? How can positional information affect cell fate?
- How does knowledge of the complete cell lineage of nematode development demonstrate the importance of programmed cell death (apoptosis) in development?
- If a gene is both necessary and sufficient for determining a developmental pathway, why would loss-of-function mutants be expected to have a different phenotype than gain-of-function mutants?
- What is a receptor? What is a ligand? What role do these types of molecules play in signaling between cells?
- Why was the study of maternal-effect lethal genes a key to deciphering the genetic control of early embryogenesis in *Drosophila?*
- Do plants have a germ line in the same sense as animals? What does the difference in germ-cell origin imply about the potential role of "somatic" mutations in the evolution of each type of organism?
- How does the genetic determination of floral development in *Arabidopsis* illustrate the principle of combinatorial control?

Solutions: Step by Step

Problem 1

Flies of *Drosophila melanogaster* that are mutant for the gene *eyeless* are almost completely devoid of eye tissue in the head region normally occupied by the compound eyes. A deletion of *eyeless,* when homozygous, produces the eyeless phenotype. Transgenic flies that carry the wildtype *eyeless* gene under the control of a promoter that causes expression in abnormal body parts (called *ectopic expression*) develop misplaced eye tissue in these body parts. Some of the phenotypes are shown in the figure below. Part A shows the result of ectopic expression in the entire head; eyes develop in the normal place, but also eye tissue develops instead of the antennae. Other results of ectopic expression are shown in parts B through D. Part B shows a wing with eye tissue growing out from it, part C a single antenna in which most of the third segment consists of eye tissue, and part D a middle leg with an eye outgrowth at the base of the tibia. What do these results imply about the role of the *eyeless* gene in eye development?

Solution The key feature of this situation is that loss-of-function mutants and gain-of-function mutants have oppo-

Problem 1—Phenotypes

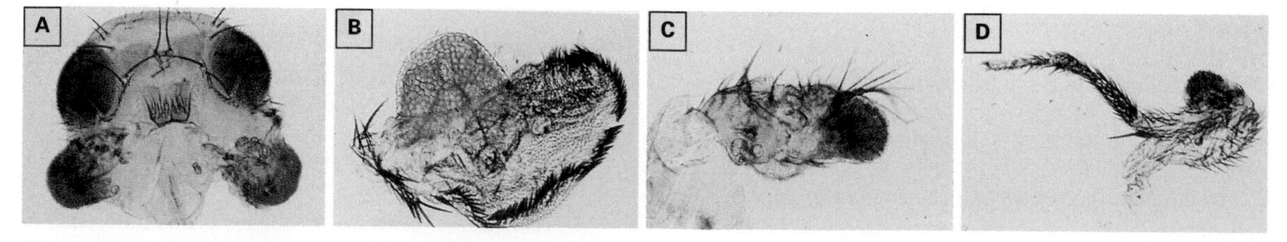

[Photos courtesy of G. Halder and W. J. Gehring. From G. Halder, P. Callaerts, and W. J. Gehring. 1995. *Science* 267: 1788.]

site phenotypes. A deletion of *eyeless* is a loss-of-function mutation, and normal eye development is completely absent. The transgenic flies with ectopic expression are the equivalent of gain-of-function mutants, and they develop eyes in the ectopic locations. The inference from these results is that *eyeless* is a key regulatory gene in eye development. The expression of *eyeless* is *necessary* for eye tissue development and also *sufficient* for eye tissue development. (Interestingly, the human genome contains a homolog of *eyeless*, which also plays a critical role in the specification of eye development.)

Problem 2

Consider a hypothetical mutant protease that affects floral development in *Arabidopsis thaliana*. The protease has an altered substrate specificity that enables it to cleave and inactivate both Ap2 and Ag proteins (the products of *ap2* and *ag*, respectively). Considering that tissue containing the Ap3/Pi dimeric protein, but neither Ap2 nor Ag, develops into floral organs intermediate between petals and stamens, what floral phenotype would be expected in the protease mutant?

Solution In whorl 1, Ap2 activity is missing, so this region will develop as a whorl of leaves. Likewise, in whorl 4, Ag activity is missing, so this region will develop as a whorl of leaves. In whorls 2 and 3, only Ap3/Pi is present, so these will develop as whorls of tissue intermediate between petals and stamens. Therefore, the flower phenotype will be leaves in whorls in 1 and 4 and will be petal/stamens intermediates in whorls 2 and 3.

Concepts in Action
Problems for Solution

12.1 Why is transcription of the zygote nucleus dispensable in early *Drosophila* development but not in early development of the mouse?

12.2 Cell death (apoptosis) is responsible in part for shaping many organs and tissues in normal development. If a group of cells in the duck leg primordium that are destined to die are transplanted from their normal leg site to another part of the embryo just before they would normally die, they still die on schedule. The same operation performed a few hours earlier rescues the cells, and they do not die. How can you explain this observation?

12.3 Classify each of the following mutant alleles as a cell-lineage mutation, a homeotic mutation, or a pair-rule mutation.

(a) A mutant allele in *C. elegans* in which a cell that normally produces two daughter cells with different fates gives rise to two daughter cells with identical fates.
(b) A mutation in *Drosophila* that causes an antenna to appear at the normal site of a leg.
(c) A lethal mutation in *Drosophila* that is responsible for abnormal gene expression in alternating segments of the embryo.

12.4 How would you prove that a newly discovered mutant allele has a maternal effect?

12.5 Distinguish between a loss-of-function mutation and a gain-of-function mutation. Can the same gene undergo both types of mutation? Can the same allele have both types of effect?

12.6 The drug actinomycin D prevents RNA transcription but has little direct effect on protein synthesis. When fertilized sea urchin eggs are immersed in a solution of the drug, development proceeds to the blastula stage, but gastrulation does not take place. How would you interpret this finding?

12.7 A mutant allele in the axolotl designated *o* is a maternal-effect lethal because embryos from *oo* females die at gastrulation, irrespective of their own genotype. However, the embryos can be rescued by injecting oocytes from *oo* females with an extract of nuclei from either $o^+ o^+$ or $o^+ o$ eggs. Injection of cytoplasm is not so effective. Suggest an explanation for these results.

12.8 The same transmembrane receptor protein encoded by the *lin-12* gene is used in the determination of different developmental fates. What is the principal difference between the two types of target cells that develop differently in response to Lin-12 stimulation?

12.9 The homeotic floral-identity genes *X*, *Y*, and *Z* in whirligigs act either separately or in pairs to control the identity of floral organs. The floral structure consists of four whorls, the outermost being whorl 1 and the innermost whorl 4. Gene product X is expressed in whorls 1 and 2, Y in 2 and 3, and Z in 3 and 4. The presence of X without Y induces sepals, X + Y together induce petals, Y + Z together induce stamens, and Z without Y induces carpels. In the absence of X, the domain of activity of Z expands to all four whorls. In the absence of Z, the domain of activity of X expands to all four whorls. Mutant alleles of *X*, *Y*, and *Z* eliminate characteristic floral organs from a specific whorl, and a different floral organ appears in its place. What would you expect for the phenotype of each of the following?

(a) A loss-of-function *X* allele
(b) A loss-of-function *Y* allele
(c) A loss-of-function *Z* allele

12.10 Using the information presented in the foregoing whirligig problem, deduce the expected floral phenotype of a double mutant homozygous for loss-of-function alleles of both *X* and *Y*.

12.11 Two classes of genes involved in segmentation of the *Drosophila* embryo are gap genes, which are expressed in one region of the developing embryo, and pair-rule genes, which are expressed in seven stripes. Homozygotes for mutations in gap genes lack a continuous block of larval segments; homozygotes for mutations in pair-rule genes lack alternating segments. You examine gene expression by mRNA *in situ* hybridization and find that (1) the embryonic expression pattern of gap genes is normal in all pair-rule mutant homozygotes and (2) the pair-rule gene expression pattern is abnormal in all gap gene mutant homozygotes. What do these observations tell you about the temporal hierarchy of gap genes and pair-rule genes in the developmental pathway of segmentation?

geNETics on the web

www.jbpub.com/genetics

These GeNETics on the Web exercises will introduce some of the most useful sites for finding genetic information on the World Wide Web. Genetic sites provide access to a rich storehouse of information on all aspects of genetics. These range from sites written in nontechnical language for the lay person to sites with sophisticated databases designed for the professional geneticist. In carrying out these exercises, you will get a taste of what the Internet can offer a student in genetics.

The keywords shown in color in the following exercises are available on the Jones and Bartlett Publishers' web site as hyperlinks to various genetic sites. To complete the exercises, visit the GeNETics on the Web home page at

http://www.jbpub.com/genetics

Select the link to Essential geNETics on the web. Then choose a chapter. You find a list of keywords that correspond to the exercises below. Select a keyword to link to a web site containing the genetic information necessary to complete the exercise. Each exercise includes a specific assignment that makes use of the information available at the site.

Exercises

- *Genetic imprinting* is the modification of gene expression in the embryo that results from egg-specific or sperm-specific changes in the gene. The mechanism of imprinting is thought to involve DNA methylation, in part because DNA methylation patterns can be inherited. Even though only a handful of loci are known to be imprinted in mammals, the existence of imprinting means that the maternal and paternal gametes do not contribute to embryonic development in completely equivalent ways. Two genes that appear to be differentially methylated depending on their parental origin are insulin-like growth factor 2 (Igf-2) and the Igf-2 receptor (Igf-2r). Use the keyword Igf-2 in the search engine at this site to obtain more information about these genes. Write a report specifying what genetic evidence suggests that the genes are differentially modified depending on their parental origin.
- The fragile-X syndrome is the single most common inherited cause of mental impairment. Its frequency may be as high as 1 per 1000 males and females of all races and ethnic groups, yet 80 to 90 percent of people with the syndrome are not correctly diagnosed. Part of the problem in diagnosis is that physical signs of fragile X are subtle. Use this keyword site to learn more about the symptoms of this disease and its genetic basis. Write a one-page description of the common features of the disorder, and explain the genetic basis of the mutation in the syndrome.

Pic Site

The Pic Site showcases some of the most visually appealing genetics sites on the World Wide Web. To visit the showcase genetics site, select the Pic Site for Chapter 12.

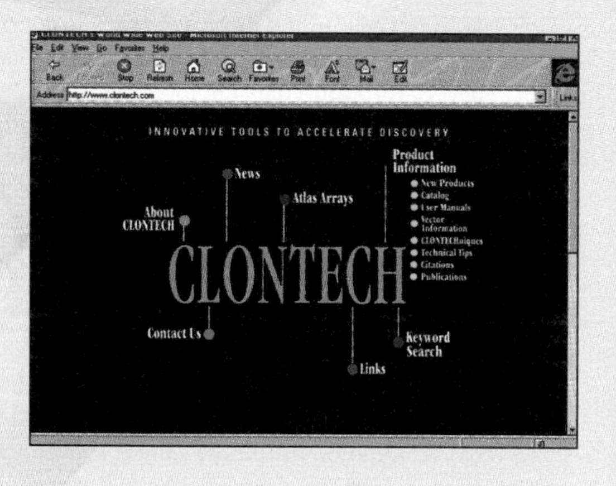

12.12 Edward B. Lewis, Christianne Nüsslein-Volhard, and Eric Wieschaus shared a 1995 Nobel Prize in physiology or medicine for their work on the developmental genetics of *Drosophila.* In their screen for developmental genes, Nüsslein-Volhard and Wieschaus initially identified 20 lines bearing maternal-effect mutations that produced embryos lacking anterior structures but having the posterior structures duplicated. When Nüsslein-Volhard mentioned this result to a colleague, the colleague was astounded to learn that mutations in 20 genes could give rise to this phenotype. Explain why his surprise was completely unfounded.

Further Readings

Avery, L., and S. Wasserman. 1992. Ordering gene function: The interpretation of epistasis in regulatory hierarchies. *Trends in Genetics* 8: 312.

Capecchi, M. R., ed. 1989. *The Molecular Genetics of Early Drosophila and Mouse Development.* Cold Spring Harbor, NY: Cold Spring Harbor Laboratory.

Chater, K., A. Downie, B. Drobak and C. Martin. 1995. Alarms and diversions: The biochemistry of development. *Trends in Genetics* 11: 79.

De Robertis, E. M., G. Oliver, and C. V. E. Wright. 1990. Homeobox genes and the vertebrate body plan. *Scientific American,* July.

Duke, R. C., D. M. Ojcius, and J. D. E. Young. 1996. Cell suicide in health and disease. *Scientific American,* December.

Gaul, U., and H. Jäckle. 1990. Role of gap genes in early *Drosophila* development. *Advances in Genetics* 27: 239.

Grunert, S., and D. St. Johnston. 1996. RNA localization and the development of asymmetry during *Drosophila* oogenesis. *Current Opinion in Genetics & Development* 6: 395.

Irish, V. 1987. Cracking the *Drosophila* egg. *Trends in Genetics* 3: 303.

Kaufman, T. C., M. A. Seeger, and G. Olsen. 1990. Molecular and genetic organization of the Antennapedia gene complex of *Drosophila melanogaster. Advances in Genetics* 27: 309.

Kennison, J. A. 1995. The Polycomb and Trithorax group proteins of *Drosophila: Trans*-regulators of homeotic gene function. *Annual Review of Genetics* 29: 289.

Kornfeld, K. 1997. Vulval development in *Caenorhabditis elegans. Trends in Genetics* 13: 55.

Lawrence, P. A. 1992. *The Making of a Fly: The Genetics of Animal Design.* Oxford, England: Blackwell.

Ma, H. 1998. To be, or not to be, a flower: Control of floral meristem identity. *Trends in Genetics* 14: 26.

McCall, K., and H. Steller. 1997. Facing death in the fly: Genetic analysis of apoptosis in *Drosophila. Trends in Genetics* 13: 222.

Meyerowitz, E. M. 1996. Plant development: Local control, global patterning. *Current Opinion in Genetics & Development* 6: 475.

Morisato, D., and K. V. Anderson. 1995. Signaling pathways that establish the dorsal-ventral pattern of the *Drosophila* embryo. *Annual Review of Genetics* 29: 371.

Nüsslein-Volhard, C. 1996. Gradients that organize embryo development. *Scientific American,* August.

Riverapomar, R., and H. Jäckle. 1996. From gradients to stripes in *Drosophila* embryogenesis: Filling in the gaps. *Trends in Genetics* 12: 478.

Sternberg, D. W. 1990. Genetic control of cell type and pattern formation in *Caenorhabditis elegans. Advances in Genetics* 27: 63.

Weigel, D. 1995. The genetics of flower development: From floral induction to ovule morphogenesis. *Annual Review of Genetics* 29: 19.

Wieschaus, E. 1996. Embryonic transcription and the control of developmental pathways. *Genetics* 142: 5.

Wolpert, L. 1996. One hundred years of positional information. *Trends in Genetics* 12: 359.

Wood, W. B., ed. 1988. *The Nematode* Caenorhabditis elegans. Cold Spring Harbor, NY: Cold Spring Harbor Laboratory.

Wright, T. R. F., ed. 1990. *Genetic Regulatory Hierarchies in Development.* New York: Academic Press.

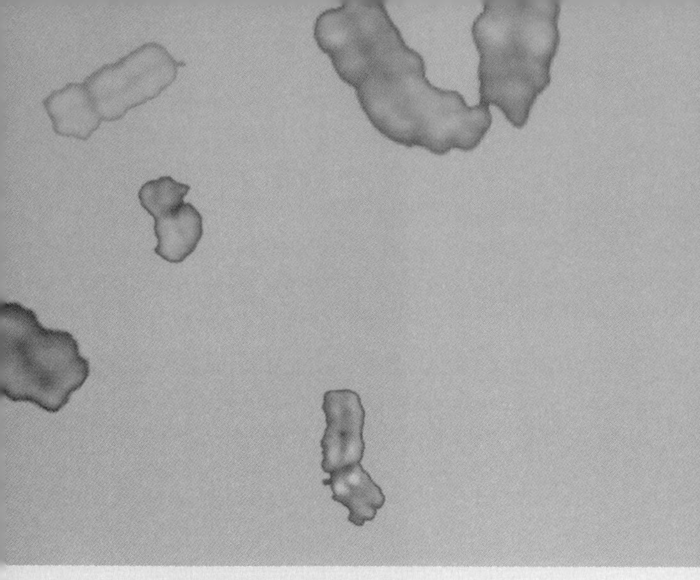

Key Concepts

- Many genes in natural populations are polymorphic; they have two or more common alleles.
- With random mating, the alleles in gametes are combined at random to form the zygotes of the next generation.
- Genetic polymorphisms can be used as genetic markers in pedigree studies and for individual identification (DNA typing).
- Relative to the frequencies of genotypes expected with random mating, inbreeding results in an excess of homozygous genotypes.
- Mutation and migration introduce new alleles into populations.
- Natural selection and random genetic drift are the usual causes of change in allele frequency; selection changes allele frequency in a systematic direction, whereas random genetic drift changes allele frequency in an unpredictable direction.
- The mitochondrial DNA in all present-day human beings probably originated from a single female who lived approximately 150,000 to 300,000 years ago, most likely in Africa.

CHEETAHS ARE GENETICALLY almost identical from one to the next, the result of severe random genetic drift due to an extreme and prolonged "bottleneck" in population size, resulting in the loss of most of the genetic variation from the population. [© M. P. Kahl/Photo Researchers, Inc.]

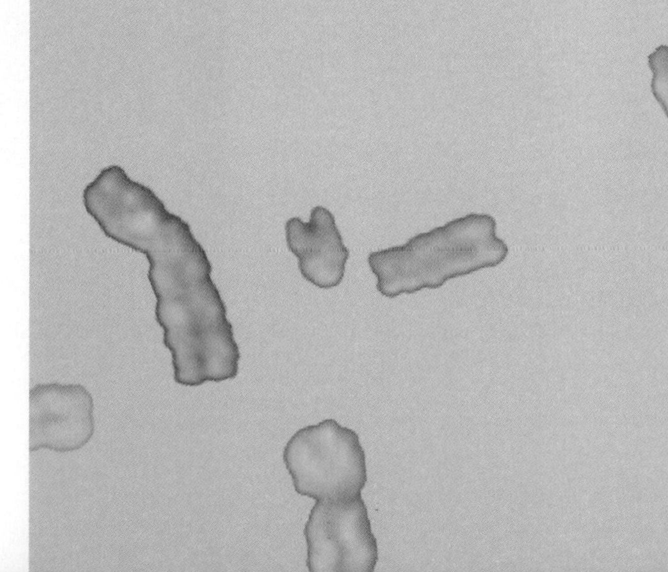

CHAPTER 13

Population Genetics and Evolution

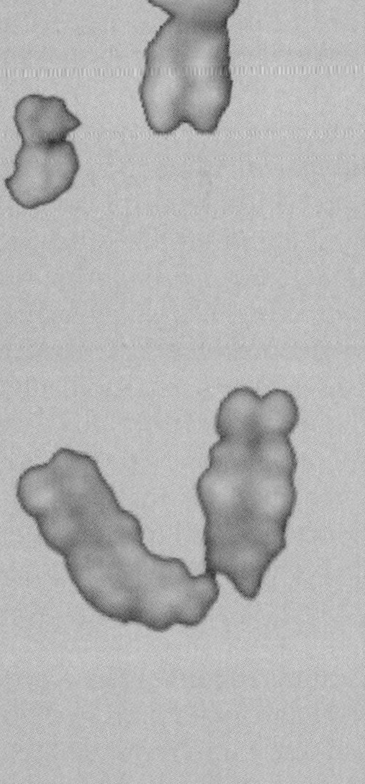

IN THE GENETIC ANALYSES examined so far, matings have been deliberately designed by geneticists. In most organisms, however, matings are not under the control of the investigator, and the familial relationships among individuals are often unknown. This situation is typical of studies of organisms in their natural habitat. Most organisms do not live in discrete family groups but exist as part of populations of individuals with unknown genealogical relationships. At first, it might seem that classical genetics with its simple Mendelian ratios could have little to say about such complex situations, but this is not the case. The principles of Mendelian genetics can be used both to interpret data collected in natural populations and to make predictions about the genetic composition of populations. Application of genetic principles to entire populations of organisms constitutes the subject of **population genetics.**

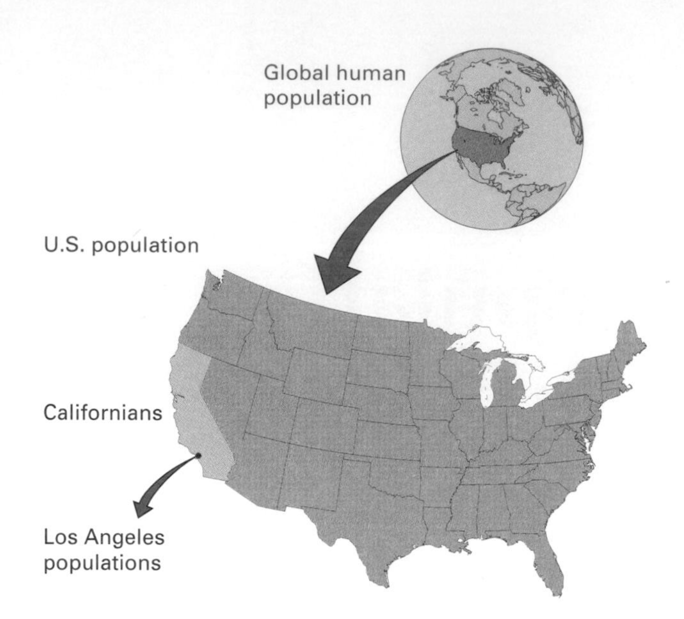

Figure 13.1 Examples of different uses of the term *population*. There is a global population of human beings, as well as a population of human beings who live in the United States or, more restricted still, who live in California. Within a larger population there may also be many subpopulations that can be distinguished, such as subpopulations composed of people of different ethnic groups who live in the greater Los Angeles area. Population geneticists are usually concerned with *local populations*—units within which individuals are likely to form mating pairs.

13.1 Genotypes may differ in frequency from one population to another.

The term **population** refers to a group of organisms of the same species living within a prescribed geographical area. Sometimes the area is large, such as that occupied by the population of sparrows in North America. The area may even include the entire earth, as when we refer to the "human population." Most widespread populations are subdivided by geographical or other features into smaller units called **subpopulations.** In the human population, for example, we may distinguish the subpopulation of people who live in the United States (Figure 13.1). More restricted subpopulations may be distinguished within this larger whole, such as the subpopulation of people who live in California. Still finer subdivisions are possible, including perhaps "Northern California" and "Southern California," or perhaps subdivision according to ethnicity in the region around Los Angeles. In many disciplines, populations are classified into subpopulations according to convenience or for purposes of comparison. The individuals in a human population may be classified into subpopulations according to annual income, for example, or according to their educational level. In genetics, the most important basis of classification is that of likelihood of finding a mating partner. The subpopulations of greatest interest are usually those within which individuals in the subpopulation are likely to find mates. A subpopulation of such a size that most individuals are likely to find their mates within it is called a **local population.** The complete set of genetic information contained within the individuals in a population constitutes the **gene pool.** This section begins with an analysis of a local population with respect to a phenotype determined by two alleles.

Allele frequencies are estimated from genotype frequencies.

The genetic composition of a population can often be described in terms of the frequencies, or relative abundances, in which alternative alleles are found. The concepts will be illustrated using the human MN blood groups because the genetic system is exceptionally simple. There are three possible phenotypes, M, MN, and N, which correspond to the combinations of M and N antigens that can be present on the surface of red blood cells. These antigens are unrelated to ABO and other red-cell antigens. The M, MN, and N phenotypes correspond to three genotypes of one gene—*MM, MN,* and *NN,* respectively. In a study of a British population, a sample of 1000 people yielded 298 M, 489 MN, and 213 N phenotypes. From the one-to-one correspondence between genotype and phenotype in this system, the genotypes can be directly inferred to be

298 *MM* 489 *MN* 213 *NN*

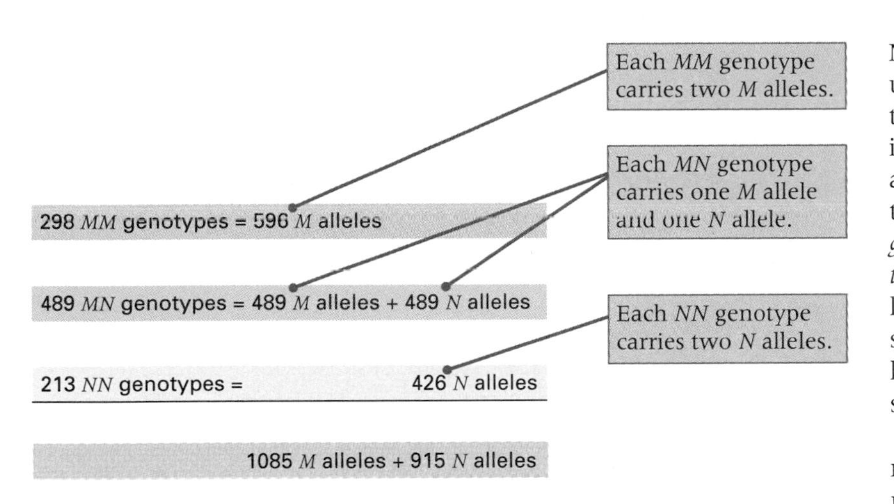

Figure 13.2 Analysis of the M and N alleles present among a sample of 1000 people typed for the MN blood groups.

These numbers contain a surprising amount of information about the population, including, for example, information on whether there may be mating between relatives. One of the goals of population genetics is to be able to interpret this information. First, note that the sample contains two types of data: (1) the number of each of the three genotypes and (2) the number of individual M and N alleles. Furthermore, the 1000 genotypes represent 2000 alleles of the gene because each genome is diploid. These alleles break down as shown in Figure 13.2. Each homozygous MM genotype represents two M alleles, each homozygous NN genotype two N alleles, and each heterozygous MN genotype one allele of each type. By the kind of allele counting in Figure 13.2, the sample of 1000 people therefore represents 1085 M alleles and 915 N alleles.

Usually, it is more convenient to analyze the data in terms of relative frequency than in terms of the observed numbers. For genotypes, the **genotype frequency** in a population is the proportion of organisms that have the particular genotype. For each allele, the **allele frequency** is the proportion of all alleles that are of the specified type. In the MN example, the genotype frequencies are obtained by dividing the observed numbers by the total sample size—in this case, 1000. Therefore, the genotype frequencies are

$$0.298\ MM \qquad 0.489\ MN \qquad 0.213\ NN$$

Similarly, the allele frequencies are obtained by dividing the observed number of each allele by the total number of alleles (in this case, 2000), so

$$\text{Allele frequency of } M = 1085/2000 = 0.5425$$

$$\text{Allele frequency of } N = 915/2000 = 0.4575$$

Note that the genotype frequencies add up to 1.0, as do the allele frequencies; this is a consequence of their definition in terms of proportions, which must add up to 1.0 when all of the possibilities are taken into account. *Allele and genotype frequencies must always be between 0 and 1.* A population with an allele frequency of 1.0 for some allele is said to be **fixed** for that allele. If an allele frequency becomes 0, the allele is said to be **lost.**

Allele frequencies can be used to make inferences about matings in a population and to predict the genetic composition of future generations. Allele frequencies are often more useful than genotype frequencies because genes, not genotypes, form the bridge between generations. Genes rarely undergo mutation in a single generation, so they are relatively stable in their transmission from one generation to the next. Genotypes are not permanent; they are disrupted in the processes of segregation and recombination that take place in each reproductive cycle. We know from simple Mendelian considerations what types of gametes must be produced from the MM, MN, and NN genotypes (Figure 13.3). The homozygous MM and NN genotypes must produce all M-bearing or all N-bearing gametes, respectively, whereas the heterozygous MN genotypes are expected to produce an equal frequency of M-bearing and N-bearing gametes. Consequently, the M-bearing gametes in the population consist of all the gametes from MM individuals and half the gametes from MN individuals; likewise, the N-bearing gametes consist of all the gametes from NN individuals and half the gametes from MN individuals.

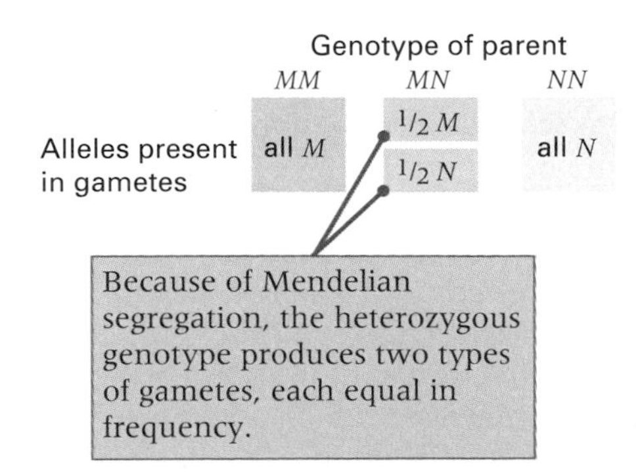

Figure 13.3 Mendelian considerations in population genetics. Each homozygous genotype produces gametes containing a single allele, but a heterozygous genotype, because of Mendelian segregation, produces two types of gametes in equal frequency.

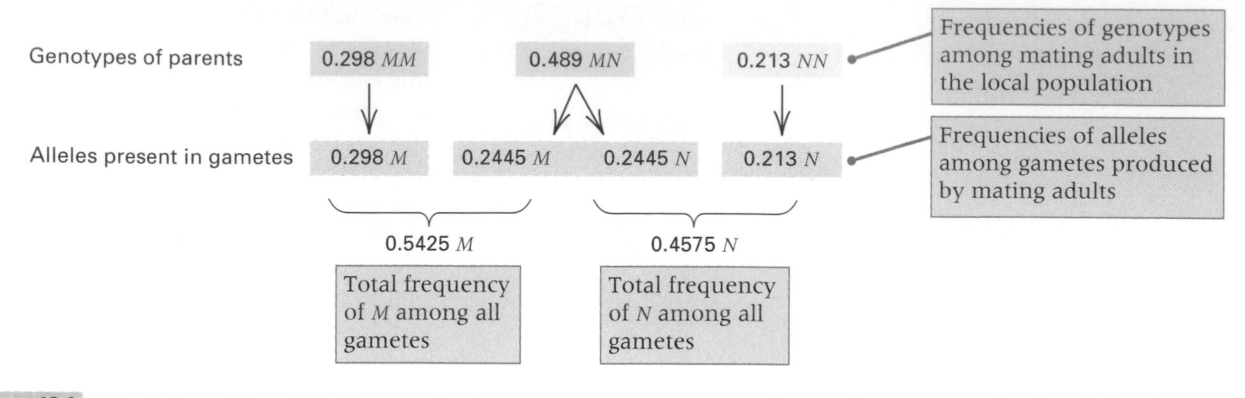

Figure 13.4 Calculation of the allele frequencies in gametes for the *MN* case. In any population, the frequency of gametes containing any particular allele equals the frequency of the genotypes that are homozygous for the allele, plus one-half the frequency of all genotypes that are heterozygous for the allele.

Taking the genotype frequencies among the parents into account, as well as Mendelian segregation in the heterozygous *MN* genotypes, the gametes produced in the local population have the composition deduced in Figure 13.4. One of the important features of this calculation is that the allele frequencies among the *gametes,* namely 0.5425 *M* and 0.4575 *N*, equal the allele frequencies among *adults* calculated earlier. This equality must be true whenever each adult in the population produces the same number of functional gametes.

Many genes have multiple polymorphic alleles in natural populations.

An important method for studying genes in natural populations is **electrophoresis,** or the separation of charged molecules in an electric field (Section 6.6). This method is commonly used to study the relative sizes of homologous DNA fragments produced by a restriction enzyme and to study genetic variation in proteins coded by alternative alleles. In the case of DNA, the allelic difference in presence or absence of a restriction site results in a difference in the size of a DNA restriction fragment produced by enzyme digestion. In the case of proteins, a genetic difference in the amino acid sequence results in a different electrophoretic mobility of the protein molecule, especially if the amino acid replacement alters the electric charge of the protein. In this chapter we will consider both types of genetic variation, beginning with proteins.

In protein electrophoresis, protein-containing tissue samples are placed near the edge of a gel, and a voltage is applied for several hours. All charged molecules in the sample move in response to the voltage, and proteins that move at different rates become separated as they move across the gel. After electrophoresis, any of a variety of techniques can be used to locate particular proteins. For example, an enzyme can be located by staining the gel with a reagent that is converted into a colored product by the enzyme; wherever the enzyme is located, a colored band appears.

Figure 13.5 is a photograph of a gel stained to reveal the enzyme phosphoglucose isomerase in tissue samples from 16 mice; it illustrates typical raw data from an electrophoretic study. The pattern of bands varies from one mouse to the next, but only three patterns are observed. Samples from mice 1, 2, 4, 8, 10, 11, and 15 have two bands: one that moves rapidly (upper band) and one that moves slowly (lower band). A second pattern is seen for the samples from mice 3, 5, 9, 12, and 16, in which only the fast band appears. The samples from mice 6, 7, 13, and 14 show a third pattern, in which only the slow band appears. Electrophoretic analysis of many enzymes in a wide variety of organisms has shown that variation in mobility almost always has a simple genetic basis: Each electrophoretic form of the enzyme contains one or more polypeptides with a genetically determined amino acid replacement that changes the electrophoretic mobility of the enzyme. Alternative forms of an enzyme coded by alleles of a single gene are known as **allozymes.** Alleles that code for allozymes are usually codominant, which means that heterozygotes express the allozyme corresponding to each allele. Therefore, in Figure 13.5, mice with only the fast allozyme are *FF* homozygotes, those with only the slow allozyme are *SS* homozygotes, and those with both fast and slow are *FS* heterozygotes. The 16 genotypes in Figure 13.5 are consequently

5 mice of genotype *FF*

7 mice of genotype *FS*

4 mice of genotype *SS*

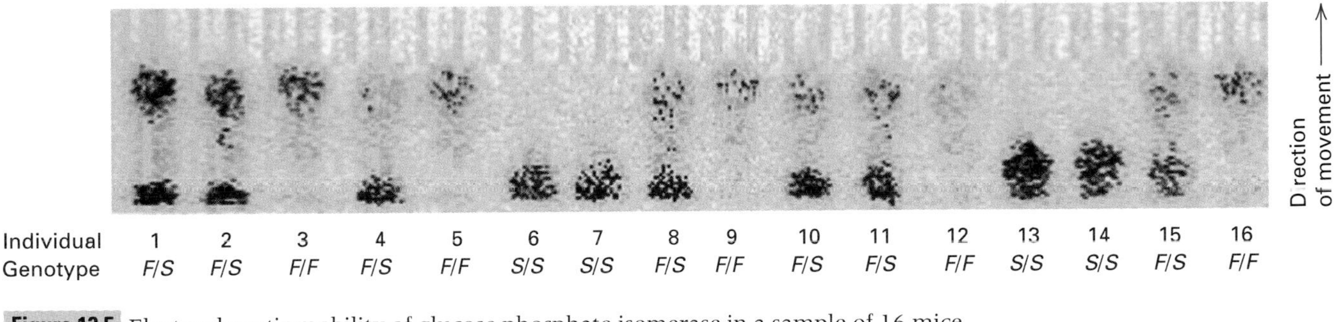

Figure 13.5 Electrophoretic mobility of glucose phosphate isomerase in a sample of 16 mice. [Courtesy of S. E. Lewis and F. M. Johnson.]

and the allele frequency of *F* in this small sample is

$$\text{Allele frequency of } F = [(2 \times 5) + 7]/(2 \times 16) = 0.53$$

Because only two alleles are represented in Figure 13.5, the allele frequency of *S* must equal $1 - 0.53 = 0.47$, which can also be calculated directly as

$$\text{Allele frequency of } S = [7 + (2 \times 4)]/(2 \times 16) = 0.47$$

Many genes in natural populations are **polymorphic,** which means that they have two or more relatively frequent alleles. Among plants and vertebrate animals, approximately 15 percent to 30 percent of enzyme genes are polymorphic, and invertebrate animals have an even greater level of polymorphism (Figure 13.6). The implication of this finding is that genetic variation among organisms is very common in natural populations of most organisms. Furthermore, among plants and vertebrate animals, the average proportion of genes that are heterozygous within an organism is between about 5 percent and 7 percent, which again emphasizes the prevalence of alternative alleles in populations. In invertebrate animals such as *Drosophila,* the proportion of heterozygous loci per individual averages approximately 11 percent.

DNA polymorphisms are important genetic markers in natural populations.

Polymorphisms can also be detected directly in DNA molecules. One method uses a *probe* DNA usually obtained from bacteria or phage cells into which DNA fragments from the organism of interest have been cloned (Section 6.6). The probe DNA is complementary in sequence to a segment of DNA in the organism of interest, but its presence in clones allows the probe to be obtained in large quantities and pure form. In the *Southern blot* procedure described in Section 6.6, genomic DNA is digested with a restriction enzyme, fragments of different size are separated by electrophoresis, and

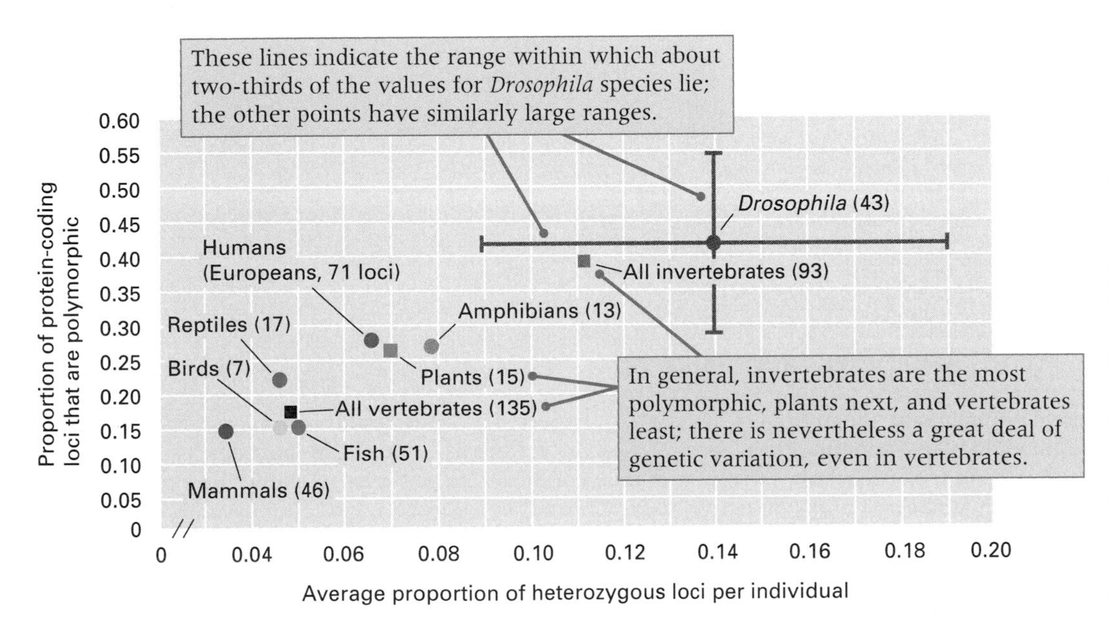

Figure 13.6 Levels of heterozygosity (*x*-axis) and proportion of polymorphic loci (*y*-axis) in a variety of organisms. The number near each point is the number of species examined. Note that there is a great deal of variation among species within any particular group of organisms.

the separated fragments are transferred by blotting onto a special type of filter paper. Radioactive probe DNA is then applied to the filter, which hybridizes only with DNA fragments containing complementary sequences and so identifies the positions of these restriction fragments.

In studies of natural populations, it is common to find restriction fragments complementary to a probe that differ in size among individuals. These differences in fragment size result from differences in the locations of restriction sites along the DNA. Each region of the genome that hybridizes with the probe generates a restriction fragment. The restriction fragments that derive from corresponding positions in homologous chromosomes are analogous to alleles in that they segregate in meiosis. Segregation means that the gametes produced by a heterozygous genotype may carry a chromosome that contains restriction sites resulting in either the larger restriction fragment or the smaller restriction fragment, but not both. Moreover, allelic restriction fragments of different sizes are codominant because both fragments are detected in heterozygotes; this is the best kind of situation for genetic studies, because genotypes can be inferred directly from phenotypes.

One type of restriction fragment length variation is illustrated in Figure 13.7A. In this example, a single nucleotide change knocks out a restriction site, so the restriction fragment complementary to the probe stretches farther downstream to the next restriction site. Therefore, hybridization with probe DNA will reveal a shorter restriction fragment produced from allele 1 than that produced from allele 2. As with any other type of allelic variation, an organism may be homozygous for allele 1, homozygous for allele 2, or heterozygous. Differing sizes of restriction fragments produced from the alleles of a

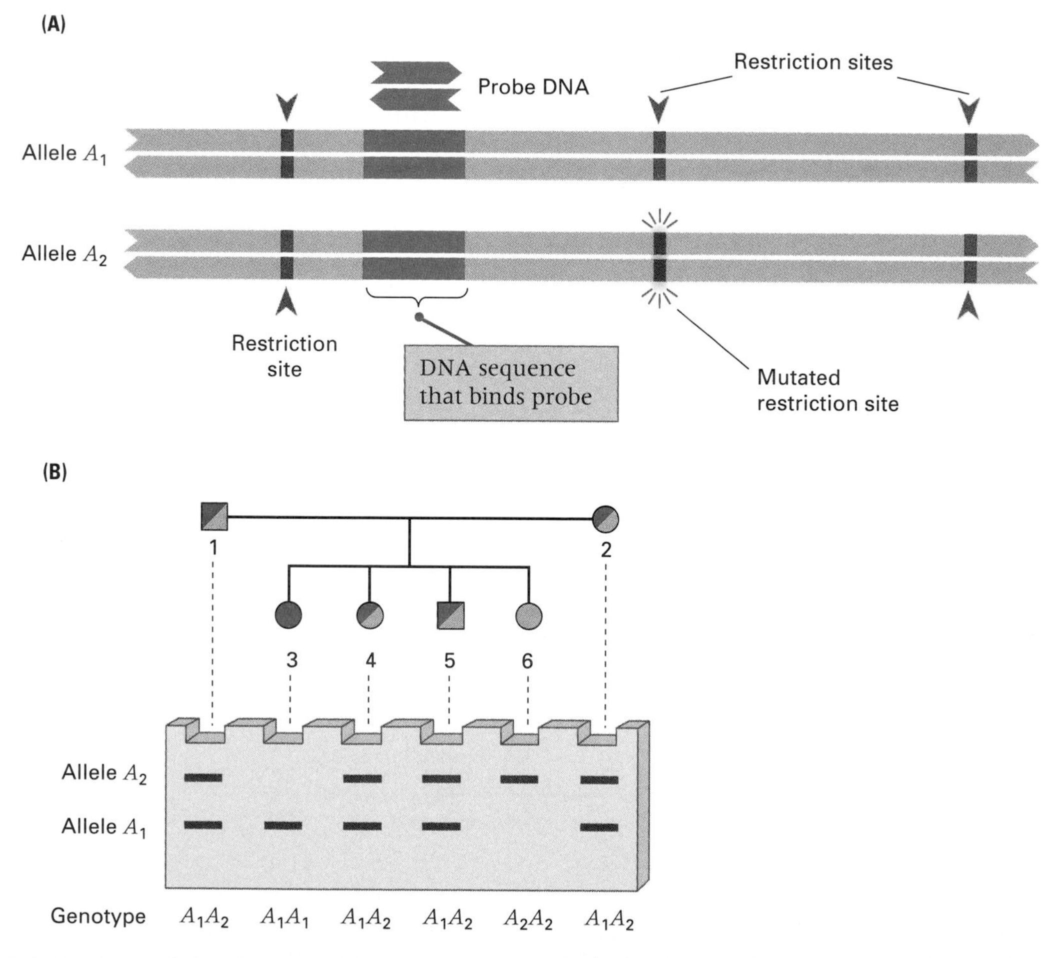

Figure 13.7 Variation in restriction fragment size in DNA. (A) Nucleotide substitution (marked with a sunburst) in a DNA molecule eliminates a restriction site. The result is that in the region detected by the probe DNA, alleles 1 and 2 differ in the size of a restriction fragment. (B) Segregation of a restriction fragment length polymorphism like that in part A. In the pedigree, both parents (individuals 1 and 2) are heterozygous A_1/A_2. Expected progeny resulting from segregation are A_1/A_1, A_1/A_2, and A_2/A_2 in the proportions 1 : 2 : 1. The diagram of the electrophoresis gel shows the expected patterns of bands in a Southern blot. Homozygous individuals numbered 3 (A_1/A_1) and 6 (A_2/A_2) yield a single restriction fragment that hybridizes with the probe DNA. The heterozygous individuals show both bands.

gene constitute a *restriction fragment length polymorphism*, or *RFLP*, of the type discussed in Section 4.4. RFLPs are extremely common in the human genome and in the genomes of most other organisms; hence they provide an extensive set of highly polymorphic, codominant genetic markers scattered throughout the genome, which can be studied with Southern blotting procedures identical except for the use of different restriction enzymes and different types of probe DNA. Figure 13.7B illustrates the segregation of alleles of a restriction fragment polymorphism within a human pedigree.

A second type of variation in DNA consists of a short sequence of nucleotides repeated in tandem a number of times, such as 5'-TGTGTGTGTGTG-3'. The number of repeats may differ from one chromosome to the next, so the number of repeats can be used to identify different chromosomes. A polymorphism of this type is called a *simple tandem repeat polymorphism* or *STRP* (Section 4.4). Although there are many places in the genome where an STRP of TG's (or any other short repeat) may occur, each place is flanked by sequences that are unique in the genome. By using oligonucleotide primers specific for the unique flanking sequences, any one of the STRPs in the genome can be amplified in the polymerase chain reaction (Section 6.7), and the number of repeats in any chromosome can be determined by examining the length of the fragment that was amplified.

13.2 Random mating means that mates pair without regard to genotype.

When a local population undergoes **random mating,** it means that organisms in the local population form mating pairs independently of genotype. Each type of mating pair is formed as often as would be expected by chance encounters. Random mating is by far the most prevalent mating system for most species of animals and plants, except for plants that regularly reproduce through self-fertilization. Self-fertilization is an extreme example of another important type of mating system, which is called *inbreeding,* or mating between relatives. Figure 13.8 represents an example of inbreeding. In this case, the female I is the offspring of a first-cousin mating (G with H). The closed loop in the pedigree is diagnostic of inbreeding, and the individuals designated A and B are called *common ancestors* of I, because they are ancestors of both of the parents of I. Because A and B are common ancestors, a particular allele present in A (or in B) could, by chance, be transmitted in inheritance down both sides of the pedigree, to meet again in the formation of I. This possibility is the most important and characteristic

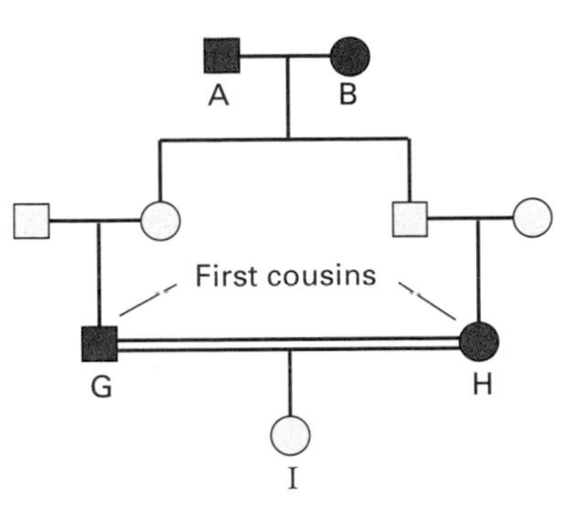

Figure 13.8 An inbreeding pedigree in which individual I is the result of a mating between first cousins.

consequence of inbreeding, and it will be discussed later in this chapter.

With random mating, the relationship between allele frequency and genotype frequency is particularly simple because of the following principle:

> Random mating of individuals is equivalent to the random union of gametes.

On the basis of this principle, we may imagine all the gametes of a population to be present in a large container. To form the zygote genotypes of the next generation, pairs of gametes are withdrawn from the container at random. To take a specific example, consider the alleles M and N in the MN blood groups, whose allele frequencies are p and q, respectively (remember that $p + q = 1$). The genotype frequencies expected with random mating can be deduced from the tree diagram in Figure 13.9. In this diagram, the gametes at the left represent the eggs and those in the middle the sperm. The genotypes that can be formed with two alleles are shown at the right. With random mating, the frequency of each zygote genotype is calculated by multiplying the allele frequencies of the corresponding gametes. Note that the genotype MN can be formed in two ways—the M allele could have come from the mother or from the father. In each case, the frequency of the MN genotype is pq, and so, considering both possibilities, the frequency of MN is $pq + pq = 2pq$. Consequently, the overall genotype frequencies expected with random mating are

$$MM: p^2 \qquad MN: 2pq \qquad NN: q^2 \qquad (13.1)$$

The frequencies p^2, $2pq$, and q^2 result from random mating for a gene with two alleles; they constitute what is called the **Hardy–Weinberg principle** after Godfrey Hardy and Wilhelm Weinberg, who derived it independently in 1908. Sometimes the Hardy–Weinberg principle is demonstrated by a

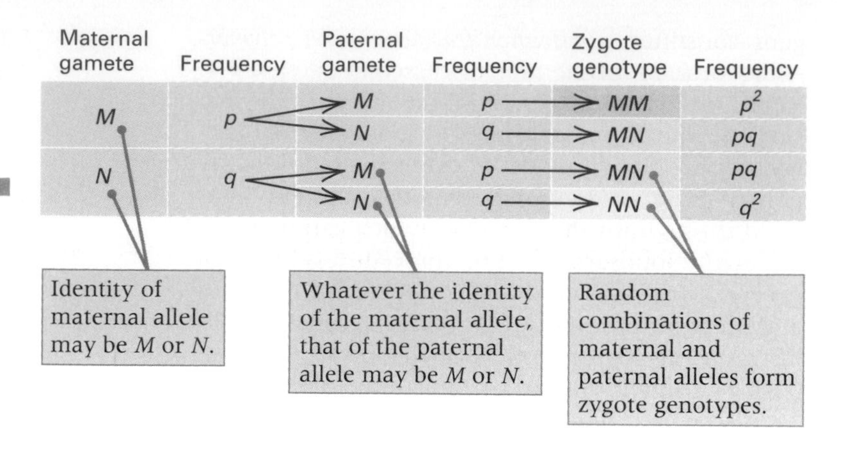

Figure 13.9 When gametes containing either of two alleles unite at random to form the next generation, the genotype frequencies among the zygotes are given by the ratio $p^2 : 2pq : q^2$. This is called the Hardy–Weinberg principle.

Punnett square, as illustrated in Figure 13.10. Such a square is completely equivalent to the tree diagram in Figure 13.9. Although the Hardy–Weinberg principle is exceedingly simple, it has a number of important implications that are not obvious. These are described in the following sections.

The Hardy–Weinberg principle has important implications for population genetics.

One important implication of the Hardy–Weinberg principle is that *the allele frequencies remain constant from generation to generation.* To understand why, consider a gene with two alleles, *A* and *a*, having frequencies *p* and *q*, respectively ($p + q = 1$). With random mating, the frequencies of genotypes *AA*, *Aa*, and *aa* among zygotes are p^2, $2pq$, and q^2, respectively. Assuming equal ability to survive among the genotypes, these frequencies equal those among adults. If, in addition, all of the adult genotypes are equally fertile, then the frequency of allele *A* among gametes that form the zygotes of the next generation can be calculated in terms of the frequency of the *A* allele in the previous generation. If we use a prime to denote the frequency of the *A* allele in the next generation, then the allele frequency is p'. In terms of the alleles present in the previous generation, p' includes all of the *A* alleles in homozygous *AA* genotypes (frequency p^2) plus half of the alleles in heterozygous *Aa* genotypes (frequency $2pq$). The *Aa* heterozygotes are weighed by half because of Mendelian segregation; only half of the gametes from *Aa* genotypes carry *A*. Putting all this together, the frequency p' of the *A* allele in the next generation is

$$p' = p^2 + 2pq/2 = p(p + q) = p$$

The final equality follows because $p + q = 1$. We have therefore shown that the frequency of allele *A* remains constant at the value of *p* through the passage of one or more complete generations. This principle depends on certain assumptions, of which the most important are the following:

- Mating is random; there are no subpopulations that differ in allele frequency.
- Allele frequencies are the same in males and females.
- All the genotypes are equal in survival and fertility; *selection* does not operate.
- Mutation does not occur.
- Migration into the population is absent.
- The population is sufficiently large that the frequencies of alleles do not change from generation to generation because of chance.

As a practical application of the Hardy–Weinberg principle, consider again the MN blood groups among British people, discussed in Section 13.1. The frequencies of alleles *M* and *N* among 1000 adults were 0.5425 and 0.4575, respectively; assuming random mating, the *expected* genotype frequencies can be calculated from Equation (13.1) as

Expected frequency of *MM*
$= (0.5425)^2 = 0.2943$

Expected frequency of *MN*
$= 2(0.5425)(0.4575) = 0.4964$

Expected frequency of *NN*
$= (0.4575)^2 = 0.2093$

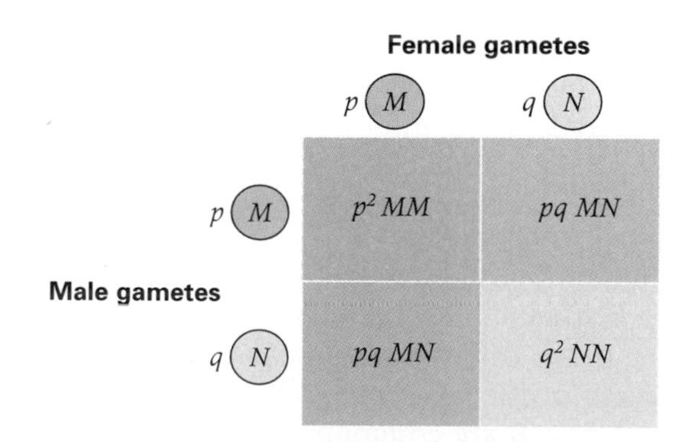

Figure 13.10 A Punnett square showing, in a cross-multiplication format, the ratio $p^2 : 2pq : q^2$ characteristic of random mating.

Because the total number of people in the sample is 1000, the expected numbers with these genotypes would be 294.3 *MM*, 496.4 *MN*, and 209.3 *NN*. The *observed* numbers are 298 *MM*, 489 *MN*, and 213 *NN*. Goodness of fit between observed numbers and expected numbers would normally be determined by means of the χ^2 test described in Chapter 3. We will not apply the test here, because the agreement between expected and observed numbers is quite good. The χ^2 test yields a probability of about 0.67, which means that the hypothesis of random mating can account for the data. On the other hand, the χ^2 test detects only deviations that are rather large, so a good fit to Hardy–Weinberg frequencies should not be overinterpreted:

> It is entirely possible for one or more assumptions of the Hardy–Weinberg principle to be violated, including the assumption of random mating, and still not produce deviations from the expected genotype frequencies that are large enough to be detected by the χ^2 test.

If an allele is rare, it is found mostly in heterozygous genotypes.

Another important implication of the Hardy–Weinberg principle is that *for a rare allele, the frequency of heterozygotes far exceeds the frequency of the rare homozygote.* For example, when the frequency of the rarer allele is $q = 0.1$, the ratio of heterozygotes to homozygotes equals $2pq/q^2 = 2(0.9)/(0.1)$, or approximately 20; when $q = 0.01$, this ratio is about 200; and when $q = 0.001$, it is about 2000. In other words,

> When an allele is rare, there are many more heterozygotes than there are homozygotes for the rare allele.

The reason for this perhaps unexpected relationship is shown in Figure 13.11, which plots the frequencies of homozygotes and heterozygotes with random mating. Note that at allele frequencies near 0 or 1, the frequency of the heterozygous genotype goes to 0 much more slowly than does the frequency of the rarer homozygous genotype.

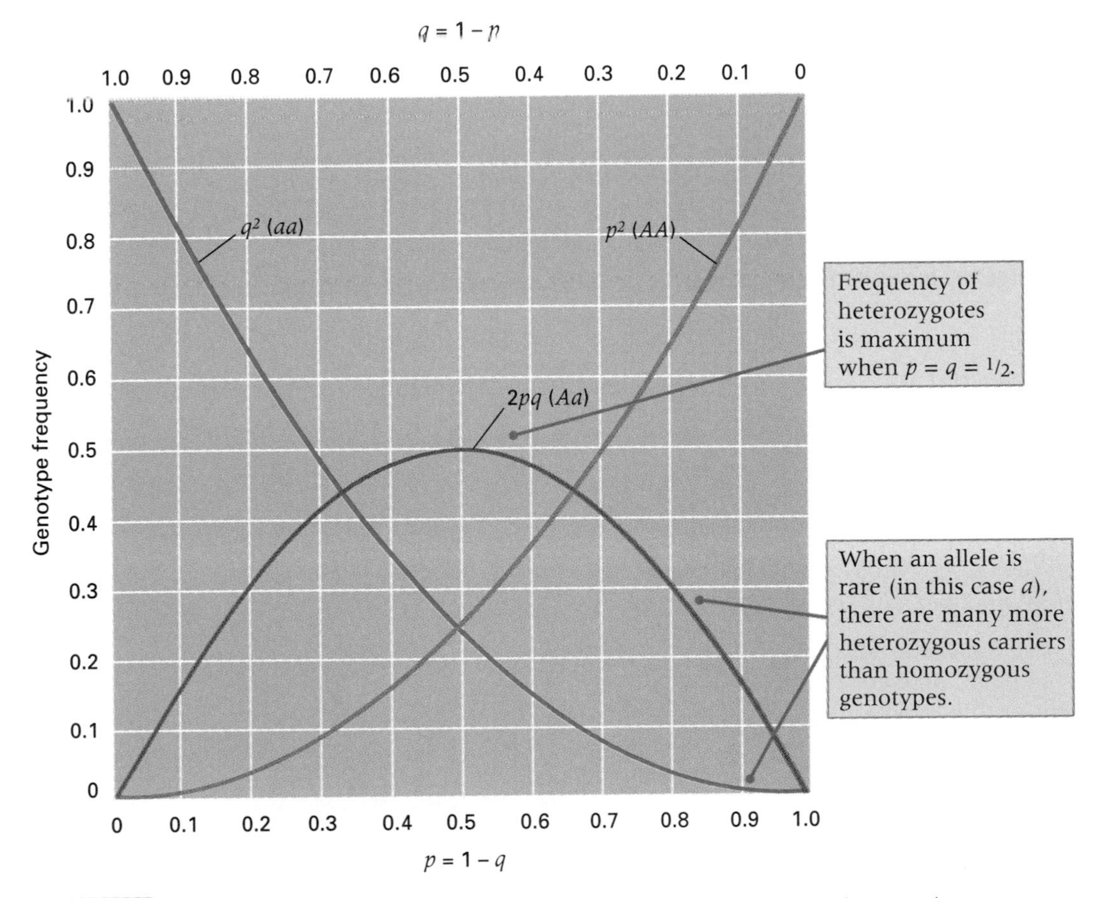

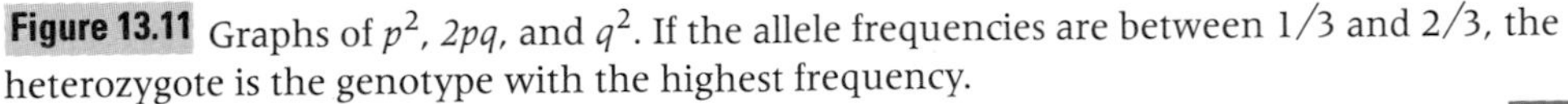

Figure 13.11 Graphs of p^2, $2pq$, and q^2. If the allele frequencies are between 1/3 and 2/3, the heterozygote is the genotype with the highest frequency.

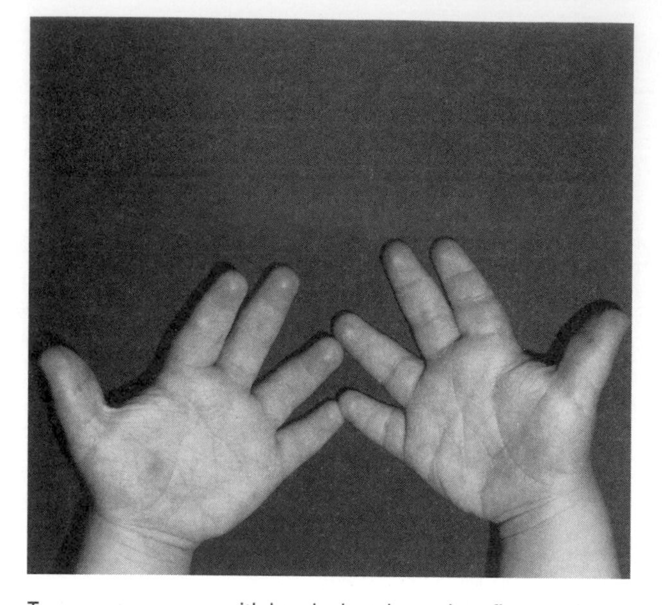

THE HANDS OF A CHILD with brachydactyly, or short fingers, due to a dominant gene. Note that the fingers are only slightly longer than the thumb.

One practical implication of this principle is seen in the example of cystic fibrosis, an inherited secretory disorder of the pancreas and lungs. Cystic fibrosis is one of the most common recessively inherited severe disorders among Caucasians; it affects about 1 in 1700 newborns. In this case, the heterozygotes cannot readily be identified by phenotype, so a method of calculating allele frequencies that is different from the gene-counting method used earlier must be used. This method is straightforward because with random mating, the frequency of recessive homozygotes must correspond to q^2. In the case of cystic fibrosis,

$$q^2 = 1/1700 = 0.00059$$

or

$$q = (0.00059)^{1/2} = 0.024$$

and consequently,

$$p = 1 - q = 1 - 0.024 = 0.976$$

The frequency of heterozygotes that carry the allele for cystic fibrosis is calculated as

$$2pq = 2(0.976)(0.024) = 0.047 = 1/21$$

This calculation implies that for cystic fibrosis, although only 1 person in 1700 is affected with the disease (homozygous), about 1 person in 21 is a carrier (heterozygous). The calculation should be regarded as approximate because it is based on the assumption of Hardy–Weinberg genotype frequencies. Nevertheless, considerations like these are important in predicting the outcome of population screening for the detection of carriers of harmful recessive alleles, which is essential in evaluating the potential benefits of such programs.

Hardy–Weinberg frequencies can be extended to multiple alleles.

Extension of the Hardy–Weinberg principle to multiple alleles of a single autosomal gene can be illustrated by a three-allele case. Figure 13.12 shows the results of random mating in which three alleles are considered. The alleles are designated A_1, A_2, and A_3, where the uppercase letter represents the gene and the subscript designates the particular allele. The allele frequencies are p_1, p_2, and p_3, respectively. With three alleles (as with any number of alleles), the allele frequencies of all alleles must sum to 1; in this case, $p_1 + p_2 + p_3 = 1.0$. As in Figure 13.10, the entry in each square is obtained by multiplying the frequencies of the alleles at the corresponding margins; any homozygote (such as A_1A_1) has a random-mating frequency equal to the square of the corresponding allele frequency (in this case, p_1^2). Any heterozygote (such as A_1A_2) has a random-mating frequency equal to twice the product of the corresponding allele frequencies (in this case, $2p_1p_2$). The extension to any number of alleles is straightforward:

Frequency of any homozygote
= square of allele frequency
Frequency of any heterozygote
= 2 × product of allele frequencies (13.2)

Multiple alleles determine the human ABO blood groups (Chapter 2). The gene has three principal alleles, designated I^A, I^B, and I^O. In one study of 3977 Swiss people, the allele frequencies were found to be 0.27 I^A, 0.06 I^B, and 0.67 I^O. Applying the rules

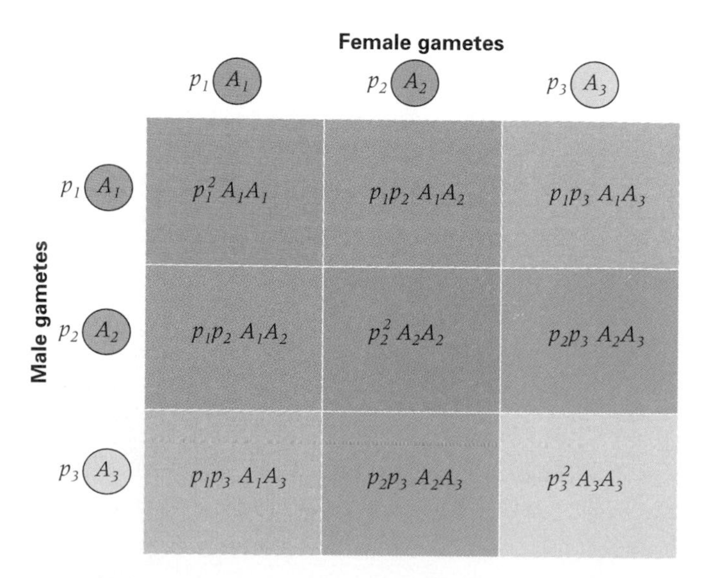

Figure 13.12 Punnett square showing the results of random mating with three alleles.

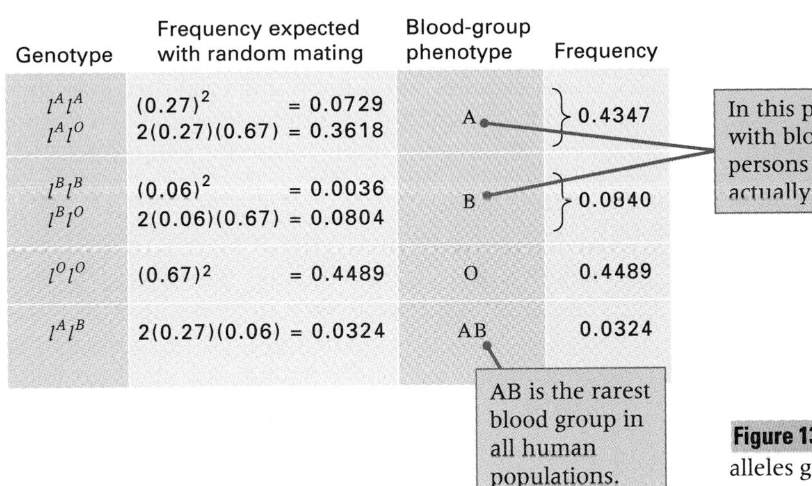

Genotype	Frequency expected with random mating	Blood-group phenotype	Frequency
I^AI^A	$(0.27)^2 = 0.0729$	A	0.4347
I^AI^O	$2(0.27)(0.67) = 0.3618$		
I^BI^B	$(0.06)^2 = 0.0036$	B	0.0840
I^BI^O	$2(0.06)(0.67) = 0.0804$		
I^OI^O	$(0.67)^2 = 0.4489$	O	0.4489
I^AI^B	$2(0.27)(0.06) = 0.0324$	AB	0.0324

Figure 13.13 Random-mating frequencies for the three alleles governing the ABO blood groups.

for multiple alleles, we can expect the genotype frequencies that result from random mating to be as shown in Figure 13.13. Because both I^A and I^B are dominant to I^O, the expected frequency of blood-group *phenotypes* is that shown at the right in the illustration. Note that the majority of A and B phenotypes are actually heterozygous for the I^O allele; this is because the I^O allele has such a high frequency in the population.

X linked genes are a special case because males have only one X-chromosome.

The implications of random mating for two X-linked alleles (*H* and *h*) are illustrated in Figure 13.14. The principles are the same as those considered earlier, but male gametes carrying the X chromosome (Figure 13.14A) must be distinguished from those carrying the Y chromosome (Figure 13.14B). When the male gamete carries an X chromosome, the Punnett square is exactly the same as that for the two-allele autosomal gene in Figure 13.10. However, because the male gamete carries an X chromosome, all the offspring in question are female. Consequently, among females, the genotype frequencies are

$$\text{Frequency of } HH \text{ females} = p^2$$
$$\text{Frequency of } Hh \text{ females} = 2pq$$
$$\text{Frequency of } hh \text{ females} = q^2$$

When the male gamete carries a Y chromosome, the outcome is quite different (Figure 13.14B). All the offspring are male, and each has only one X chromosome, which is inherited from the mother. Therefore, each male receives only one copy of each X-linked gene, and the genotype frequencies among males are the same as the allele frequencies:

$$\text{Frequency of } H \text{ males} = p$$
$$\text{Frequency of } h \text{ males} = q$$

An important implication of Figure 13.14 is that if *h* is a rare recessive allele, then there will be many more males exhibiting the trait than females because the frequency of affected females (q^2) will

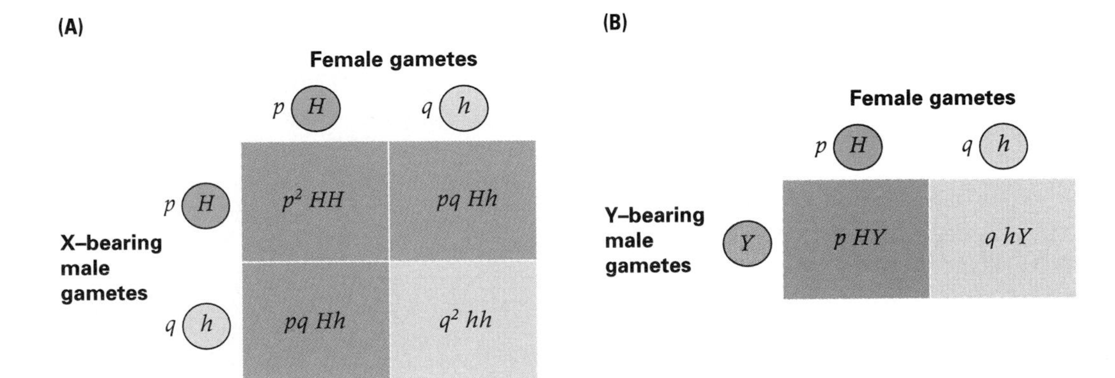

Figure 13.14 The results of random mating for an X-linked gene. (A) Genotype frequencies in females. (B) Genotype frequencies in males.

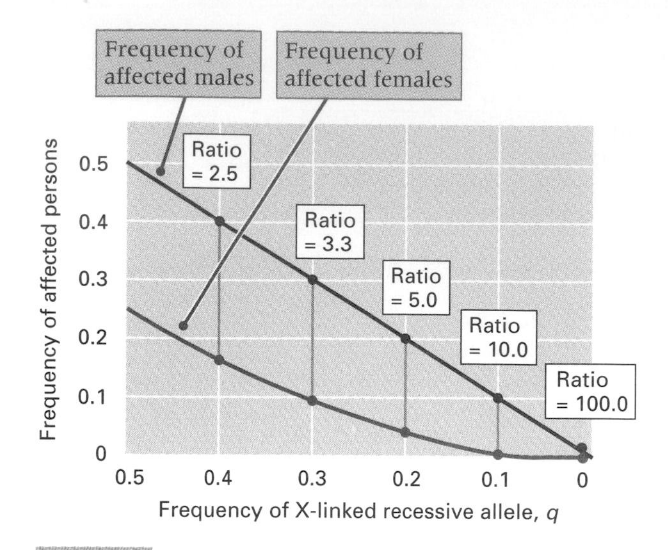

Figure 13.15 For a recessive X-linked allele, the upper curve gives the frequency of affected males (q) and the lower curve the frequency of affected females (q^2), for values of $q < 0.5$. Although both frequencies decrease as the recessive allele becomes rare, the frequency of affected females decreases faster. The result is that the ratio of affected males to affected females increases as the allele frequency decreases.

be much smaller than the frequency of affected males (q). This principle is illustrated in Figure 13.15. As the allele frequency of the recessive decreases toward 0, the frequencies of affected males and females both decrease, but the frequency of affected females decreases faster. The result is that the ratio of affected males to affected females increases. At an allele frequency of $q = 0.3$, for example, the ratio of affected males to affected females is 3.33; but for an allele frequency of $q = 0.1$, the ratio of affected males to affected females is 10.0. In general, the ratio of affected males to affected females is q/q^2, or $1/q$.

For an X-linked recessive trait, the frequency of affected males provides an estimate of the frequency of the recessive allele. A specific example is found in the common form of X-linked color blindness in human beings. This trait affects about 1 in 20 males among Caucasians, so $q = 1/20 = 0.05$. The expected frequency of color-blind females is therefore estimated as $q^2 = (0.05)^2 = 0.0025$, or about 1 in 400.

13.3 Highly polymorphic genes are used in DNA typing.

In principle, no two human beings are genetically identical. The only exceptions are identical twins, identical triplets, and so forth. Each human genotype is unique because so many genes in human populations are polymorphic. The theoretical principle of genetic uniqueness has become a practical reality through the study of DNA polymorphisms. Small samples of human material from an unknown person (for example, material left at the scene of a crime) often contain enough DNA that the genotype can be determined for a number of polymorphisms and matched against those present among a group of suspects. Typical examples of crime-scene evidence include blood, semen, hair roots, and skin cells. Even a small number of cells is sufficient, because predetermined regions of DNA can be amplified by the polymerase chain reaction (Chapter 6).

If a suspect's DNA contains sequences that are clearly not present in the crime-scene sample, then the sample must have originated from a different person. On the other hand, if a suspect's DNA *does* match that of the crime-scene sample, then the suspect could be the source. The strength of the DNA evidence depends on the number of polymorphisms that are examined and the number of alleles present in the population. The greater the number of polymorphisms that match, especially if they are highly polymorphic, the stronger the evidence linking the suspect to the sample taken from the scene of the crime. The use of polymorphisms in DNA to link suspects with samples of human material is called **DNA typing.** This method of individual identification is generally regarded as the most important innovation in criminal investigation since the development of fingerprinting more than a century ago.

Figure 13.16 illustrates one type of polymorphism used in DNA typing. The restriction fragments corresponding to each allele differ in length because they contain different numbers of units re-

Blood or other tissues found at a crime scene can often be highly incriminating if the DNA type of a suspect is found to match. [© Jerome Friar/Impact Visuals.]

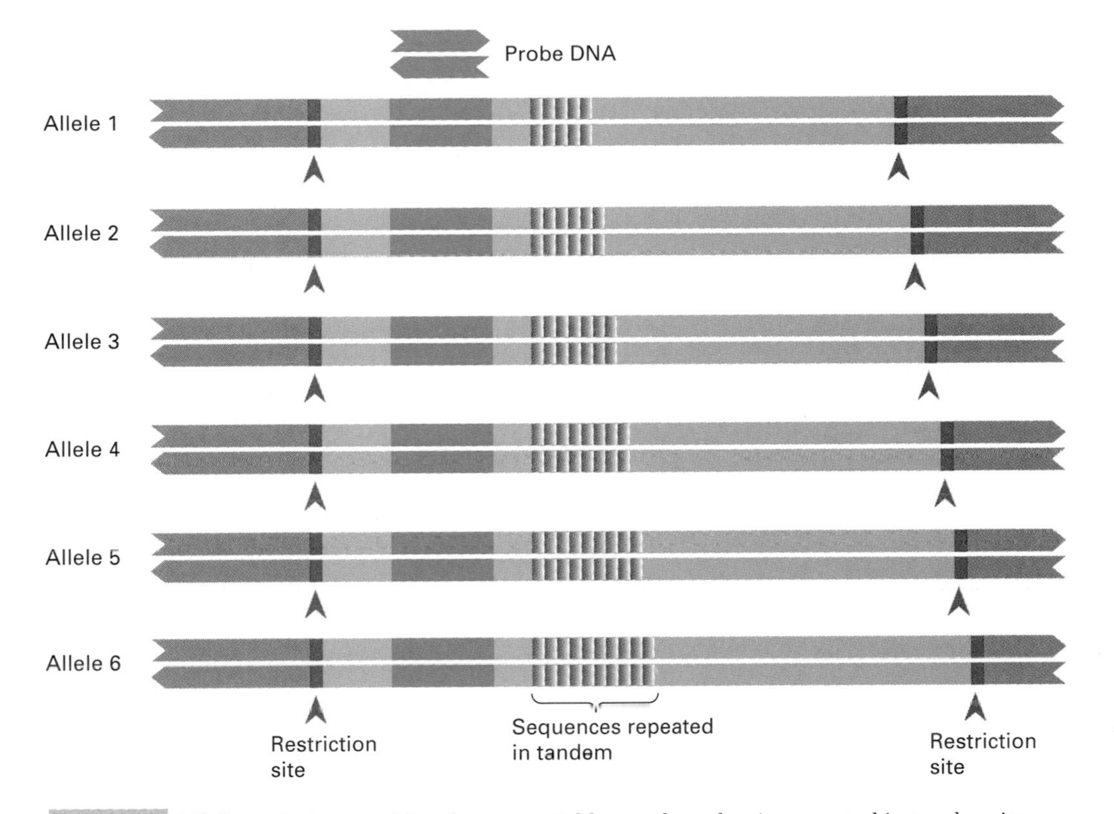

Figure 13.16 Allelic variation resulting from a variable number of units repeated in tandem in a nonessential region of a gene. The probe DNA detects a restriction fragment for each allele. The length of the fragment depends on the number of repeating units present.

peated in tandem. A polymorphism of this type is called an *STRP,* which stands for *simple tandem repeat polymorphism* (Section 4.4). STRPs are of value in DNA typing because many alleles are possible, owing to the variable number of repeating units. Although many alleles may be present in the population as a whole, each person can have no more than two alleles for each STRP polymorphism. An example of an STRP used in DNA typing is shown in Figure 13.17. The lanes in the gel labeled M contain multiple DNA fragments of known size to serve as molecular-weight markers. Each numbered lane 1 through 9 contains DNA from a single person. Two typical features of STRPs are to be noted:

1. Most people are heterozygous for STRP alleles that produce restriction fragments of different sizes. Heterozygosity is indicated by the presence of two distinct bands. In Figure 13.17, only the person numbered 1 appears to be homozygous for a particular allele.
2. The restriction fragments from different people cover a wide range of sizes. The variability in size indicates that the population as a whole contains many STRP alleles.

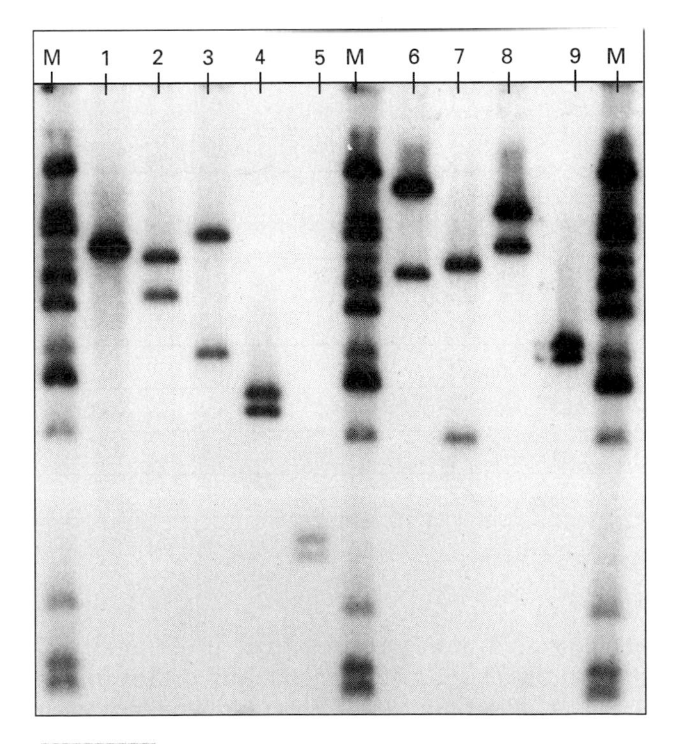

Figure 13.17 Genetic variation in an STRP used in DNA typing. Each numbered lane contains DNA from a single person; the DNA has been cleaved with a restriction enzyme, separated by electrophoresis, and hybridized with radioactive probe DNA. The lanes labeled M contain molecular-weight markers. [Courtesy of R. W. Allen.]

The reason why STRPs are useful in DNA typing is also evident in Figure 13.17: Each of the nine

Table 13.1
Allele frequencies for *D1S80* among U.S. population groups

Repeat number	Caucasian	Hispanic	African American	Asian
14	0	0	0	0
15	0	0.001	0	0
16	0.001	0.010	0.002	0.034
17	0.002	0.009	0.028	0.025
18	0.237	0.224	0.073	0.152
19	0.003	0.005	0.003	0.022
20	0.018	0.013	0.032	0.007
21	0.021	0.028	0.115	0.034
22	0.038	0.024	0.081	0.017
23	0.012	0.009	0.014	0.017
24	0.378	0.315	0.234	0.230
25	0.046	0.072	0.045	0.027
26	0.020	0.007	0.006	0
27	0.007	0.016	0.008	0.047
28	0.063	0.078	0.130	0.076
29	0.052	0.055	0.053	0.042
30	0.008	0.039	0.009	0.123
31	0.072	0.053	0.054	0.093
32	0.006	0.005	0.007	0.012
33	0.003	0.004	0.004	0.005
34	0.001	0.006	0.086	0.005
35	0.003	0	0.002	0.005
36	0.004	0.011	0.001	0.005
37	0.001	0.004	0	0.007
38	0	0	0	0
39	0.003	0.004	0.003	0.005
40	0	0	0	0
41	0	0.002	0.002	0.007
>41	0.001	0.006	0.007	0.002
Sample size	718	409	606	204

Source: Data from B. Budowle, et al. 1995. *Journal of Forensic Science* 40:38.

people tested has a different pattern of bands and thus could be identified uniquely by means of this STRP. On the other hand, the uniqueness of each person in Figure 13.17 is due in part to the high degree of polymorphism of the STRP and in part to the small sample size. If more people were examined, then pairs that matched in their STRP types by chance would certainly be found. Table 13.1 shows the frequencies of alleles for the STRP designated *D1S80* in major population groups in the United States. The allele with 24 repeats has a frequency of 0.378 in Caucasians, and so, with random mating, more that 14 percent of the population would be homozygous for this allele. The same allele is also quite frequent in the other population groups. Because of the possibility of chance matches between STRP types, applications of DNA typing are usually based on at least three highly polymorphic loci, and preferably more.

Suppose that a person is found whose DNA type for three STRP loci matches that of a sample from the scene of a crime. How is the significance of this match to be evaluated? The significance of the match depends on the likelihood of its happening by chance, because matches of rare DNA types are more significant than matches of common DNA types. If random mating proportions can be assumed, and if it can also be assumed that each STRP is independent of the others, then it is reasonable to use Equation (13.2) to calculate the expected frequency of the particular genotype for each STRP separately and then to multiply the expected frequencies for each STRP in order to calculate the expected frequency of the multilocus genotype. In practice, the calculations are carried out using estimated allele frequencies from the racial group of the suspect.

AN ARGENTINIAN GENETICIST performing DNA fingerprinting to establish kinship of the "disappeared," of children born in captivity of parents kidnapped and assassinated by the former military regime. A bank of blood samples, taken from all the potential relatives, provides the DNA for comparison with that from the child in question. Every human being produces a unique pattern of DNA bands, hence its potential as a kind of genetic "identity card."

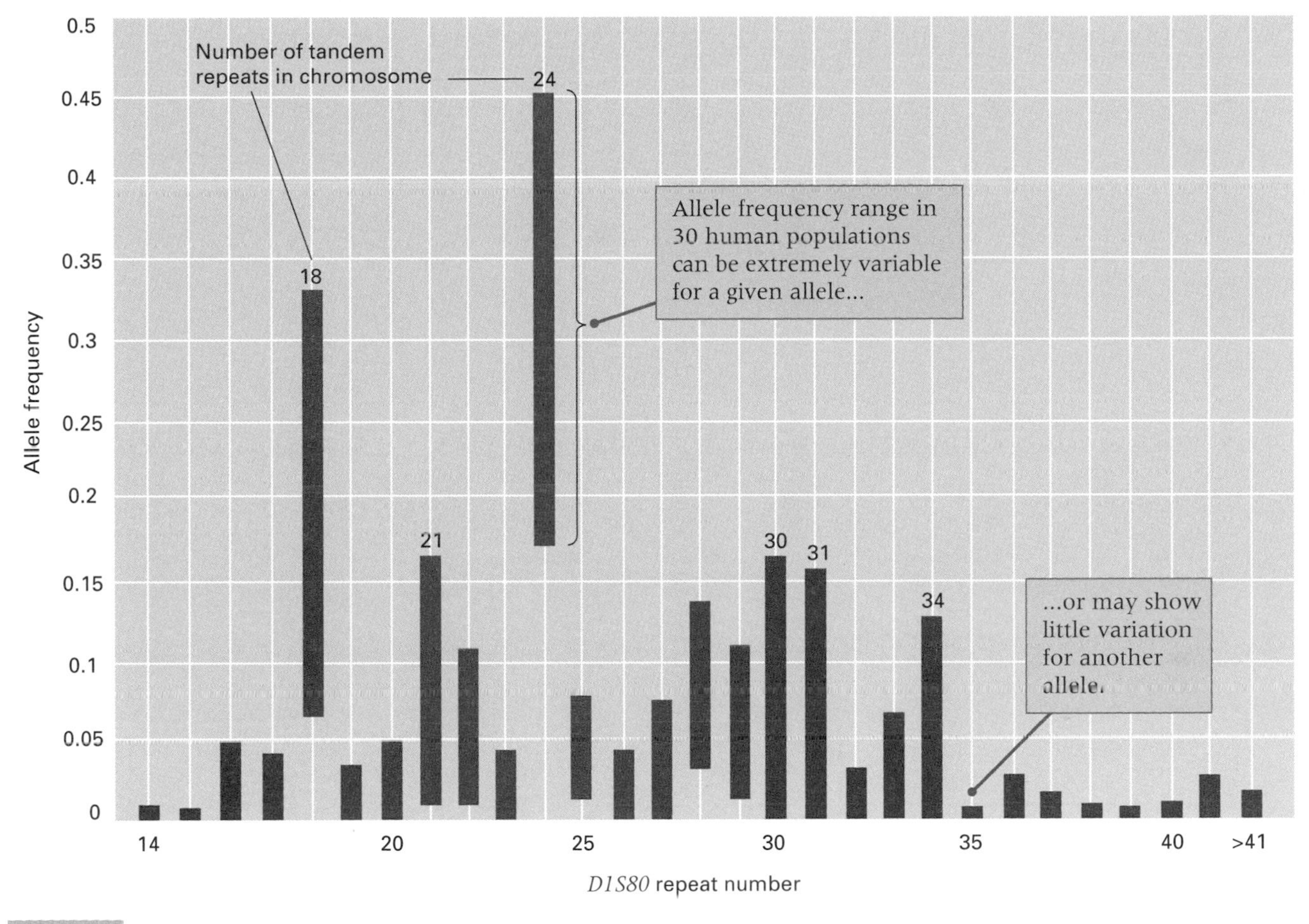

Figure 13.18 Range of allele frequencies found among human subpopulations for the STRP *D1S80*. [Data from B. Budowle, et al. 1995. *J. Forensic Science* 40:38.]

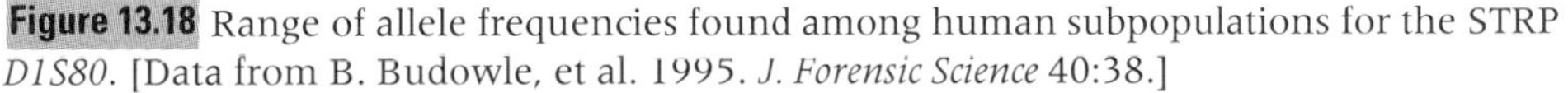

Polymorphic alleles may differ in frequency among subpopulations.

The use of Equation (13.2) in DNA typing is based on a number of assumptions about human populations: (1) that the Hardy–Weinberg principle holds for each locus; (2) that each locus is statistically independent of the others, so that the multiplication across loci is justified; and (3) that the only level of population substructure that is important is that of race. The term **population substructure** refers to the extent to which a larger population is subdivided into smaller subpopulations that may differ in allele frequencies from one subpopulation to the next.

There are, however, differences in the frequencies of STRP alleles among different human subpopulations. Figure 13.18 shows an example in which the length of each bar indicates the range in frequency (minimum to maximum) of each allele among approximately 30 human subpopulations. For some alleles, the range is very great; for example, alleles with 21, 30, 31, or 34 repeats are very rare in some subpopulations but relatively common in others. Even the common alleles, with 18 and 24 repeats, can vary substantially in frequency from one subpopulation to the next.

Because human subpopulations may differ in the allele frequencies of STRP polymorphisms, calculations of the random probability of matching DNA types based on the average allele frequencies have a considerable margin of error. For this reason, it is preferable to carry out DNA typing using as many STRP loci as feasible. The greater the number of loci, the less likely it is that a match of DNA types is due to chance. For a match of six or more highly polymorphic STRPs, it is nearly certain that the two DNA samples came from the same person. Figure 13.19 shows an example in which DNA evidence (semen samples) from seven different rape cases (U1 through U7) matches perfectly with DNA from one suspect (S2), who was also implicated on the basis of other evidence from some of the crime scenes. The suspect was convicted of 81 criminal charges and sentenced to 139 years in prison. He will be eligible for parole in the year 2087.

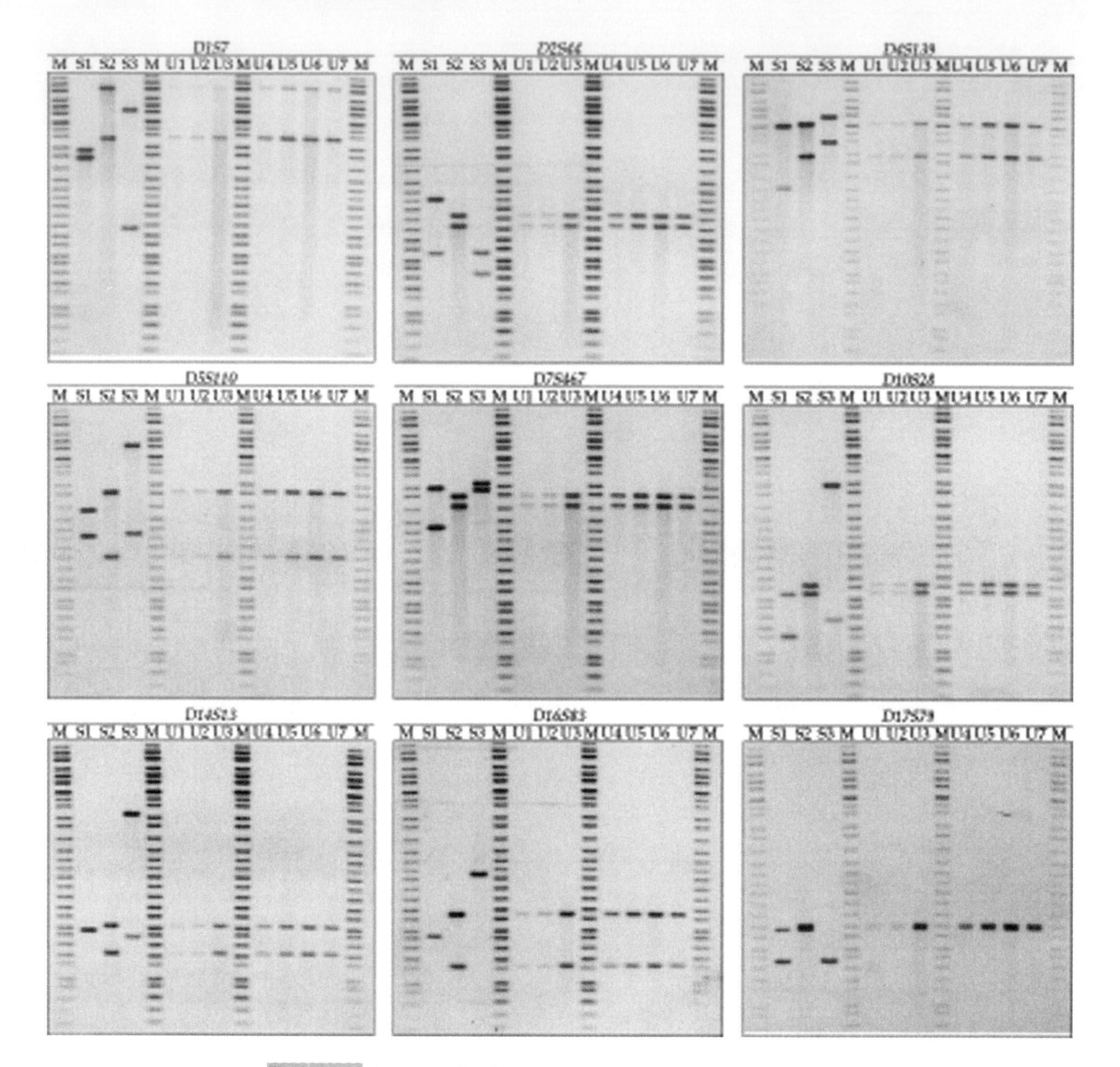

Figure 13.19 An example of DNA typing in a criminal case. Each panel is the result of DNA typing for a different STRP. The lanes marked S1, S2, and S3 contain DNA from blood samples of three male suspects, those in columns U1 through U7 contain DNA from semen samples collected from seven female victims of rape. The lanes marked M contain molecular-weight markers. In each case, the DNA from suspect S2 matches the samples obtained from the victims. [Courtesy of Steven J. Redding, Office of the Hennepin County District Attorney, Minneapolis, and Lowell C. Van Berkom and Carla J. Finis, Minnesota Bureau of Criminal Apprehension.]

DNA exclusions are definitive.

Issues of probability apply only when DNA types match. DNA types that do *not* match can exclude innocent suspects irrespective of any considerations about population substructure. For example, if the DNA type of a suspected rapist does *not* match the DNA type of semen taken from the victim, then the suspect could not be the perpetrator. Another example, which illustrates the use of DNA typing in paternity testing, is given in Figure 13.20. The numbers 1 and 2 designate different cases, in each of which a man was alleged to have fathered a child. In each case, DNA was obtained from the mother (M), the child (C), and the man (A). The pattern of bands obtained for one STRP is shown. The lanes labeled A + C contain a mixture of DNA from the man and the child. In case 1, the lower

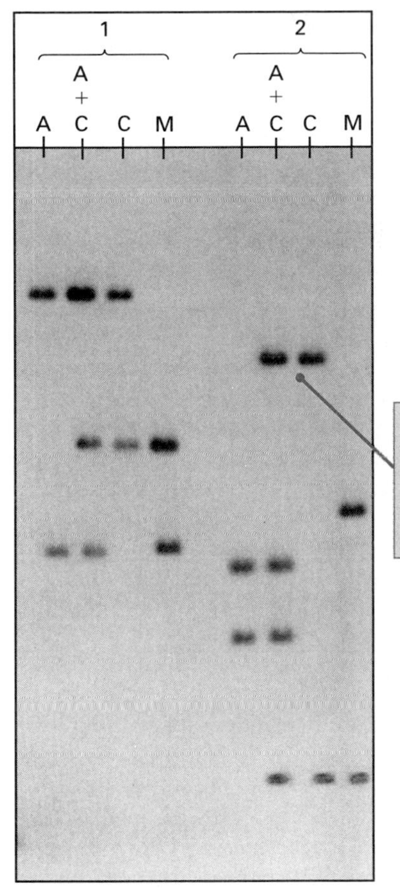

Figure 13.20 Use of DNA typing in paternity testing. The sets of lanes numbered 1 and 2 contain DNA samples from two different paternity cases. In each case, the lanes contain DNA fragments from the following sources: M, the mother; C, the child; A, the accused father. The lanes labeled A + C contain a mixture of DNA fragments from the accused father and the child. [Courtesy of R. W. Allen.]

band in the child was inherited from the mother and the upper band from the father; because the upper band is the same size as one of those in the alleged father, he could have contributed this allele to the child. This finding does not prove that this individual is the father; it says only that he cannot be excluded on the basis of this particular STRP. (However, if a large enough number of STRPs is studied, and the man cannot be excluded by any of them, it does make it more likely that he really is the father.) Case 2 in Figure 13.20 is an exclusion. The small band lowest in the gel is the band inherited by the child from the mother. The other band in the child does not match either of the bands in the accused father, so the accused man could not be the biological father. (In theory, mutation could be invoked to explain the result, but this is extremely unlikely.)

13.4 Inbreeding means mating between relatives.

Inbreeding means mating between relatives, such as first cousins. The principal consequence of inbreeding is that the frequency of heterozygous offspring is smaller than it is with random mating. This effect is seen most dramatically in repeated self-fertilization, which occurs naturally in certain plants. To understand why the frequency of heterozygous genotypes decreases with self-fertilization, consider a hypothetical initial population consisting exclusively of *Aa* heterozygotes (Figure 13.21). With self-fertilization, each plant would produce offspring in the proportions 1/4 *AA*, 1/2 *Aa*, and 1/4 *aa*. Thus one generation of self-fertilization reduces the proportion of heterozygotes from 1 to 1/2. In the second generation, only the heterozygous plants can again produce heterozygous offspring, and only half of their offspring will again be heterozygous. Heterozygosity is therefore reduced to 1/4 of what it was originally. Three generations of self-fertilization reduce the heterozygosity to $1/4 \times 1/2 = 1/8$, and so forth. The remainder of this section demonstrates how the reduction in heterozygosity because of inbreeding can be expressed quantitatively and measured.

Repeated self-fertilization is a particularly intense form of inbreeding, but weaker forms of inbreeding are qualitatively similar in that they also lead to a reduction in heterozygosity. A convenient measure of the effect of inbreeding is based on the reduction in heterozygosity. Suppose that H_I is the frequency of heterozygous genotypes in a population of inbred organisms. The most widely used measure of inbreeding is called the **inbreeding coefficient,** symbolized *F*, which is defined as the

AA	*Aa*	*aa*	Genotype
0	1	0	Initial frequency
$^1/_4$	$^1/_2$	$^1/_4$	After one generation of self-fertilization
$^3/_8$	$^1/_4$	$^3/_8$	After two generations of self-fertilization
$^7/_{16}$	$^1/_8$	$^7/_{16}$	After three generations of self-fertilization
$^{15}/_{32}$	$^1/_{16}$	$^{15}/_{32}$	After four generations of self-fertilization

With repeated self-fertilization, the frequency of heterozygous genotypes is reduced by half in each successive generation.

Figure 13.21 Effects of repeated self-fertilization on the genotype frequencies. In each generation, the proportion of heterozygous genotypes decreases to half of its value in the previous generation.

proportionate reduction in H_I compared with the value of $2pq$ that would be expected with random mating; that is,

$$F = (2pq - H_I)/2pq$$

This equation can be rearranged as

$$H_I = 2pq(1 - F)$$

which says that in a population of organisms having an inbreeding coefficient of F, the proportion of heterozygous genotypes is reduced by the fraction F relative to what it would be with random mating. As the proportion of heterozygous genotypes decreases in frequency, the proportion of homozygous genotypes increases correspondingly. The overall genotype frequencies in an inbred population are given by

$$\begin{aligned}
&\text{Frequency of } AA \text{ genotype} = p^2(1 - F) + pF \\
&\text{Frequency of } Aa \text{ genotype} = 2pq(1 - F) \qquad (13.3) \\
&\text{Frequency of } aa \text{ genotype} = q^2(1 - F) + qF
\end{aligned}$$

These frequencies are a modification of the Hardy–Weinberg principle that takes inbreeding into account. When $F = 0$ (no inbreeding), the genotype frequencies are the same as those given in the Hardy–Weinberg principle in Equation (13.1), namely p^2, $2pq$, and q^2. At the other extreme, when $F = 1$ (complete inbreeding), the inbred population consists entirely of AA and aa genotypes in the frequencies p and q, respectively. Whatever the value of F, however, the allele frequencies remain at the values of p and q because

$$\begin{aligned}
p^2(1 - F) + pF + (1/2)[2pq(1 - F)] \\
= p(p + q)(1 - F) + pF = p
\end{aligned}$$

A graphical representation of the genotype frequencies in Equation (13.3) is shown in Figure 13.22. For each organism, the population is divided conceptually into two groups. In one group (amounting to a proportion F of the population), the gene in question has been affected by the inbreeding, which means that the two alleles present in the organism are identical by descent owing to DNA replication in a common ancestor of the inbred organism. In the other group (amounting to a proportion $1 - F$ of the population), the gene in question has, by chance, escaped the effects of inbreeding, which means that the genotype frequencies are those expected with random mating. Taking both groups into account results in the genotype frequencies in Equation (13.3).

The reduction in heterozygosity due to inbreeding is shown in Figure 13.23. Each curve is given by an equation of the form $H_I = 2pq(1 - F)$. The case $F = 0$ corresponds to random mating. The curves show that as the inbreeding coefficient increases,

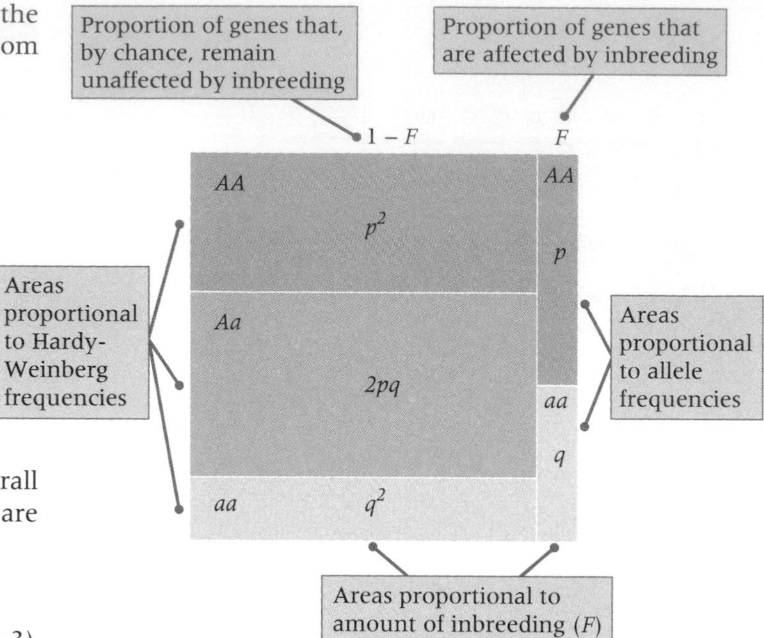

Figure 13.22 Effect of inbreeding on genotype frequencies. The large rectangles on the left pertain to alleles whose ancestries, by chance, are not affected by inbreeding and for which the genotype frequencies remain in Hardy–Weinberg proportions. The narrow rectangles on the right pertain to alleles whose ancestries are affected by the inbreeding, and in this case the genotype frequencies of AA and aa are related as $p : q$. (Note that there are no heterozygous genotypes in the latter case.)

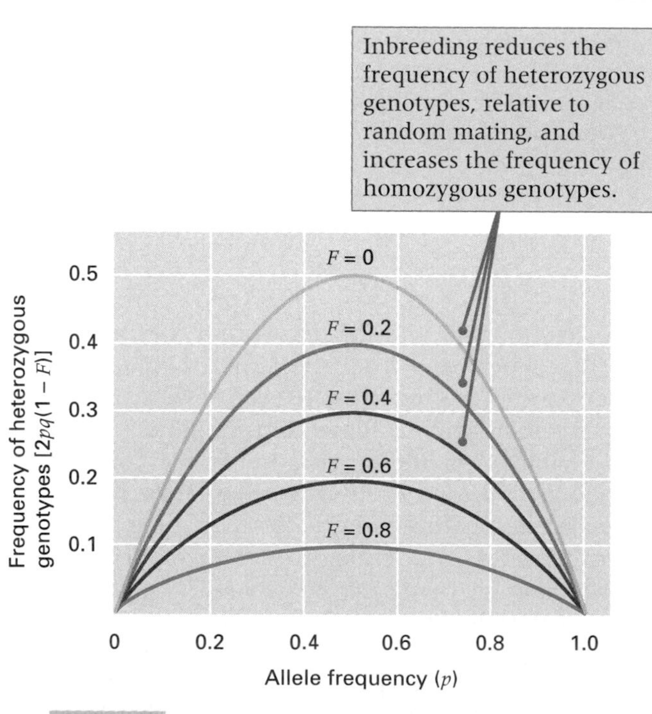

Figure 13.23 Frequency of heterozygous genotypes in an inbred population (y-axis) against allele frequency (x-axis). As the inbreeding becomes more intense (greater inbreeding coefficient F), the proportion of heterozygous genotypes decreases.

the frequency of heterozygous genotypes decreases proportionally until, when $F = 1$, there are no heterozygotes remaining in the inbred population. In other words, in a highly inbred population with $F = 1$, the genotypes consist either of AA or of aa in the relative proportions p and q, respectively.

Inbreeeding results in an excess of homozygotes compared with random mating.

The effects of inbreeding differ according to the normal mating system of an organism. At one extreme, in regularly self-fertilizing plants, inbreeding is already so intense and the organisms are so highly homozygous that additional inbreeding has virtually no effect. However,

> In most species, inbreeding is harmful, and much of the effect is due to rare recessive alleles that would not otherwise become homozygous.

Among human beings, close inbreeding is usually uncommon because of social conventions. The closest type of inbreeding usually found is mating between first cousins. In small isolated populations (such as aboriginal groups and religious communities) some inbreeding is inevitable as a result of matings between remote relatives. The effect of inbreeding is always an increase in the frequency of genotypes that are homozygous for rare, usually harmful recessives. For example, among American whites, the frequency of albinism among offspring of matings between nonrelatives is approximately 1 in 20,000, but among offspring of first-cousin matings, the frequency is approximately 1 in 2000. The reason for the increased risk may be understood by comparing the genotype frequencies of homozygous recessives in Equation (13.3) (for inbreeding) and Equation (13.1) (for random mating). In the most common form of inbreeding among human beings (mating between first cousins), $F = 0.062$ $(= 1/16)$ among the offspring. Therefore, the frequency of homozygous recessives produced by first-cousin mating is

$$q^2(1 - 0.062) + q(0.062)$$

whereas, among the offspring of nonrelatives, the frequency of homozygous recessives is q^2. For albinism, $q = 0.007$ (approximately), and the calculated frequencies are 5×10^{-4} for the offspring of first cousins and 5×10^{-5} for the offspring of nonrelatives.

The effect of first-cousin mating in causing an increased frequency of offspring homozygous for a rare recessive allele is shown in Figure 13.24. The increase occurs for any allele frequency, but the relative increase is greater for alleles that are more rare. Considering the curves in Figure 13.24, with an allele frequency $q = 0.05$, for example, the relative risk of a homozygous offspring is $0.0335938/0.0025 = 13.4$; whereas with an allele frequency $q = 0.01$, the relative risk of a homozygous offspring is $0.00634375/0.0001 = 63.4$.

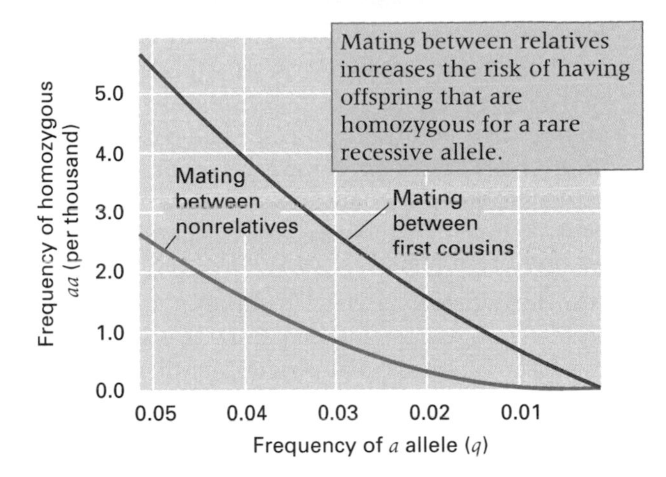

Figure 13.24 Effect of first cousin mating on the frequency of offspring genotypes that are homozygous for a rare autosomal recessive allele. Although both curves decrease as the allele becomes rare, the curve for mating between nonrelatives decreases faster. As a consequence, the more rare the recessive allele, the greater the proportion of all affected individuals that result from first-cousin matings.

13.5 Evolution is accompanied by genetic changes in species.

In its broadest sense, the term **evolution** means any change in the gene pool of a population or in the allele frequencies present in a population. Evolution is possible because genetic variation exists in populations. Four processes account for most of the evolutionary changes. They form the basis of cumulative change in the genetic characteristics of populations, leading to the descent with modification that characterizes the process of evolution. Although the point has yet to be proved, most evolutionary biologists believe that these same processes, when carried out continuously over geological time, also account for the formation of new species and higher taxonomic categories. These processes are

1. **Mutation,** the origin of new genetic capabilities in populations by means of spontaneous heritable changes in genes.

2. **Migration,** the movement of organisms among subpopulations within a larger population.
3. **Natural selection,** resulting from the different abilities of organisms to survive and reproduce in their environment. Natural selection is the primary process by which populations of organisms become progressively more adapted to their environments.
4. **Random genetic drift,** the random, undirected changes in allele frequency that happen by chance in all populations, but particularly in small ones.

Evolution is a population phenomenon, so it is conveniently discussed in terms of allele frequencies. In the following sections, we consider some of the population genetic implications of the major evolutionary processes.

13.6 Mutation and migration bring new alleles into populations.

Mutation is the ultimate source of genetic variation. It is an essential process in evolution, but it is a relatively weak force for changing allele frequencies, primarily because typical mutation rates are so low. Moreover, most newly arising mutations are harmful to the organism. Although some mutations may be **selectively neutral,** meaning that they do not affect the ability of the organism to survive and reproduce, only a very few mutations are favorable for the organism and contribute to adaptation. The low mutation rates that are observed are thought to have evolved as a compromise. The mutation rate is high enough to generate the favorable mutations that a species requires to continue evolving, but it is not so high that the species suffers excessive genetic damage from the preponderance of harmful mutations.

Migration is similar to mutation in that new alleles can be introduced into a local population, although the alleles derive from another subpopulation rather than from new mutations. In the absence of migration, the allele frequencies in each local population can change independently, so local populations may undergo considerable genetic differentiation. Genetic differentiation among subpopulations means that there are differing frequencies of common alleles among the local populations or that some local populations possess certain rare alleles not found in others. The accumulation of genetic differences among subpopulations can be minimized if some migration of organisms among the subpopulations is possible. In fact, only a relatively small amount of migration among subpopulations (on the order of just a few migrant organisms in each local population in each generation) is usually sufficient to prevent high levels of genetic differentiation. On the other hand, some genetic differentiation can accumulate in spite of migration if other evolutionary forces, such as natural selection for adaptation to the local environments, are sufficiently strong.

13.7 Natural selection favors genotypes better able to survive and reproduce.

The driving force of adaptive evolution is natural selection, which is a consequence of hereditary differences among organisms in their ability to survive and reproduce in the prevailing environment. Since first proposed by Charles Darwin in 1859, the concept of natural selection has been revised and extended, most notably by the incorporation of genetic concepts. In its modern formulation, the concept of natural selection rests on three premises:

- In all organisms, more offspring are produced than survive and reproduce.
- Organisms differ in their ability to survive and reproduce, and some of these differences are due to genotype.
- In every generation, genotypes that promote survival in the prevailing environment (favored genotypes) are present in excess at the reproductive age, and hence they contribute disproportionately to the offspring of the next generation. In this way, the alleles that enhance survival and reproduction increase in frequency from generation to generation, and the population becomes progressively better able to survive and reproduce in its environment. This progressive genetic improvement in populations constitutes the process of evolutionary **adaptation.**

Fitness is the relative ability of genotypes to survive and reproduce.

Selection over many generations can be studied in bacterial populations because of the short generation time (about 30 minutes). Figure 13.25 shows the result of competition between two bacterial genotypes, *A* and *B*. Genotype *A* is the superior competitor under the particular conditions. In the experiment, the competition was allowed to continue for 290 generations, during which time the proportion of *A* genotypes (p) increased from 0.60

to 0.9995 and that of B genotypes decreased from 0.40 to 0.0005. The data points give a satisfactory fit to an equation of the form

$$\frac{p_n}{q_n} = \left(\frac{p_0}{q_0}\right)\left(\frac{1}{w}\right)^n \tag{13.4}$$

in which p_0 and q_0 are the initial frequencies of A and B (in this case 0.6 and 0.4, respectively), p_n and q_n are the frequencies after n generations of competition, n is the number of generations of competition, and w is a measure of the competitive ability of B when competing against A under the conditions of the experiment. The theoretical derivation of Equation (13.4) is based on the definition of w as the rate of survival and/or reproduction of genotype B, relative to genotype A, under the conditions of the experiment, and assuming that w is constant through all generations. What this means is that if the relative frequencies of $A : B$ in generation 0 are $p_0 : q_0$, then in the next generation they will be in the relative frequencies $p_0 : q_0 w$; which is to say that $p_1/q_1 = (p_0/q_0)(1/w)$. Likewise, in the next generation, the relative frequencies will be given by $p_2/q_2 = (p_1/q_1)(1/w)$, and by substituting for p_1/q_1, we obtain $p_2/q_2 = (p_0/q_0)(1/w)^2$. Continuing in this manner, generation after generation, we obtain Equation (13.4). Because $q = 1 - p$ in every generation, Equation (13.4) implies that

$$p_n = \frac{p_0}{p_0 + q_0 w^n} \tag{13.5}$$

This is the smooth curve plotted in Figure 13.25 for the values $p_0 = 0.6$, $q_0 = 0.4$, and $w = 0.958$. The dots show the good fit with the experimental data.

The value of $w = 0.958$ is called the **relative fitness** of the B genotype relative to the A genotype under these particular conditions. As we have seen, the relative fitness measures the comparative contribution of each parental genotype to the pool of offspring genotypes produced in each generation. A value $w = 0.958$ means that for each offspring cell produced by an A genotype, a B genotype produces, on the average, 0.958 offspring cell.

In population genetics, relative fitnesses are usually calculated with the superior genotype (A in this case) taken as the standard with a fitness of 1.0. However, the selective disadvantage of a genotype is often of greater interest than its relative fitness. The selective disadvantage of a disfavored genotype is called the **selection coefficient** associated with the genotype, and it is calculated as the difference between the fitness of the standard (taken as 1.0) and the relative fitness of the genotype in question. In the present example, the selection coefficient against B, denoted s, is

$$s = 1.000 - 0.958 = 0.042 \tag{13.6}$$

The specific meaning of s in this example is that the selective disadvantage of strain B is 4.2 percent per generation. When the selection coefficient is known, Equation (13.5) also permits the prediction of the allele frequencies in any future generation, given the original frequencies. Alternatively, it can be used to calculate the number of generations required for selection to change the allele frequencies from any specified initial values to any later ones. For example, from the relative fitnesses of A and B just estimated, what is the number of generations required to change the frequency of A from 0.1 to 0.8? In this particular problem, $p_0/q_0 = 0.1/0.9$, $p_n/q_n = 0.8/0.2$, and $w = 0.958$. A little manipulation of Equation (13.5) gives

$$n = [\log(0.1/0.9) - \log(0.8/0.2)]/\log(0.958)$$
$$= 83.5 \text{ generations}$$

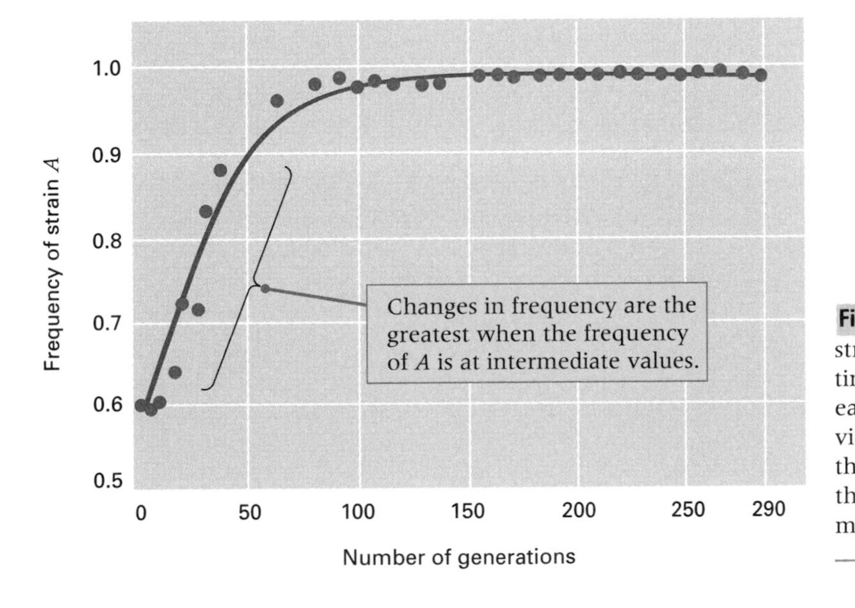

Figure 13.25 Increase in frequency of a favored strain of *E. coli* resulting from selection in a continuously growing population. The y value for each point is the number of A cells at any time divided by the total number of cells ($A + B$). Note that the changes in frequency are greatest when the frequency of the favored strain, A, is at intermediate values.

The human connection

Resistance in the Blood

Anthony C. Allison 1954
Radcliffe Infirmary,
Oxford, England
Protection Afforded by Sickle-Cell Trait Against Subtertian Malarial Infection

Malaria is the most prevalent infectious disease in tropical and subtropical regions of the world, infecting up to 250 million people each year and causing as many as 2 million deaths. The disease is characterized by recurrent episodes of fever with alternating shivering and sweating. Patients suffer anemia due to destruction of red blood cells, as well as enlargement of the spleen, inflammation of the digestive system, bronchitis, and many other severe complications. The type of malaria caused by the protozoan parasite *Plasmodium falciparum* is called "subtertian" malaria because there is less than a 3-day interval between bouts of fever. This parasite is spread through bites by the mosquito *Anopheles gambiae.* Upon transmission, the parasites multiply in the liver for about a week and then begin to infect and multiply in red blood cells (parasitemia), which are destroyed after a few days. In regions of Africa, the Middle East, the Mediterranean region, and India where *falciparum* malaria is endemic, there is also a relatively high frequency of the sickle-cell mutation affecting the amino acid sequence of the beta chain of hemoglobin. Heterozygous carriers have no severe clinical symptoms, but they have the so-called "sickle-cell trait," in which the red blood cells collapse into half-moon, or sickle, shapes after 1 to 3 days when sealed under a cover slip on a microscope slide. Homozygous affected persons have "sickle-cell anemia," in which many red blood cells sickle spontaneously while still in the bloodstream, causing severe complications and often death. Why would a genetic disease that is effectively lethal when homozygous be maintained at a frequency of 10 percent or more in a population? Allison noted the correlation between the sickle-cell mutation and malaria and speculated that the sickle-cell trait gives heterozygous carriers some protection against malaria. Key evidence supporting this hypothesis is presented here. Later work showed that the heterozygous carriers have an approximately 15 percent selective advantage as a result of their malaria resistance.

During the course of field work in Africa in 1949 I was led to question the view that the sickle-cell trait is neutral from the point of view of natural selection and to reconsider the possibility that it is associated with a selective advantage. I noted that the incidence of the sickle-cell trait was higher in regions where malaria was prevalent than elsewhere. . . . It became imperative, then, to ascertain whether sickle cells can afford some degree of protection against malarial infection. . . . Children were chosen rather than adults as subjects for the observations so as to minimize the effect of acquired immunity to malaria. The recorded incidence of parasitemia in a group of 290 Ganda children [living near Kampala, Uganda] is presented in the accompanying Table. . . .

It is apparent that the incidence of parasitemia is lower in the sickle-cell group than in the group without sickle cells. The difference is statistically significant (χ^2 = 4.8 for 1 degree of freedom). . . . The parasite density in the two groups also differed: of 12 sicklers with malaria, 66.7% had only slight parasitemia while 33.3% had a moderate parasitemia. Of the 113 non-sicklers with malaria, 34% had slight parasitemia, the parasite density in the remainder being moderate or severe. . . . [Among a group of adult males who volunteered to be bitten by heavily infected *Anopheles gambiae* mosquitoes], an infection with *Plasmodium falciparum* was established in 14 out of 15 men without the sickle-cell trait, whereas in a comparable group of 15 men without the trait only 2 developed parasites. It is concluded that the abnormal erythrocytes of individuals with the sickle-cell trait are less easily parasitized by *P. falciparum* than are normal erythrocytes. Hence those who are heterozygous for the sickle-cell gene will have a selective advantage in regions where malaria is hyperendemic. This fact may explain why the sickle-cell gene remains common in these areas in spite of the elimination of genes in patients dying of sickle-cell anemia.

IT BECAME IMPERATIVE . . . to ascertain whether sickle cells can afford some degree of protection against malarial infection.

Source: British Medical Journal 1: 290–294.

	With parasitemia	Without parasitemia	Total
Sicklers	12 (27.9%)	31 (72.1%)	43
Non-sicklers	113 (45.7%)	134 (53.3%)	247

Allele frequencies change slowly when alleles are either very rare or very common.

Selection in diploids is analogous to that in haploids, but dominance and recessiveness create additional complications. Figure 13.26 shows the change in allele frequencies for both a favored dominant and a favored recessive. The striking feature of the figure is that the frequency of the favored dominant allele changes very slowly when the allele is common, and the frequency of the favored recessive allele changes very slowly when the allele is rare. The reason is that rare alleles are found much more frequently in heterozygotes than in homozygotes. With a favored dominant at high frequency, most of the recessive alleles are present in heterozygotes, and the heterozygotes are not exposed to selection and hence do not contribute to change in allele frequency. Conversely, with a favored recessive at low frequency, most of the favored alleles are in heterozygotes, and again the heterozygotes are not exposed to selection and do not contribute to change in allele frequency. The principle is quite general:

> Selection for or against recessive alleles is very inefficient when the recessive allele is rare.

One simple example of this principle is selection against a recessive lethal. In this case, it can be shown that the number of generations required to reduce the frequency of the recessive allele from q to $q/2$ equals $1/q$ generations. For example, if $q = 0.01$, then successive halvings of the frequency of the recessive allele require 100, 200, 400, 800, 1600, . . . generations, so a recessive lethal allele is eliminated very slowly.

The inefficiency of selection against rare recessive alleles has an important practical implication. There is a widely held belief that medical treatment to save the lives of persons with rare recessive disorders will cause a deterioration of the human gene pool because of the reproduction of persons who carry the harmful genes. This belief is unfounded. With rare alleles, the proportion of homozygotes is so small that reproduction of the homozygotes contributes a negligible amount to change in allele frequency. Considering their low frequency, it matters little whether homozygotes reproduce or not. Similar reasoning applies to eugenic proposals to "cleanse" the human gene pool of harmful recessives by preventing the reproduction of affected persons. People with severe genetic disorders rarely reproduce anyway, and, even when they do, they have essentially no effect on allele frequency. In other words,

> The largest reservoir of harmful recessive alleles is in the genomes of heterozygous carriers, who are phenotypically normal.

Selection can be balanced by new mutations.

It is apparent from Figure 13.26 that selection tends to eliminate harmful alleles from a population. However, harmful alleles can never be eliminated totally because recurrent mutation of the normal allele continually creates new harmful alleles. These new mutations tend to replenish the harmful

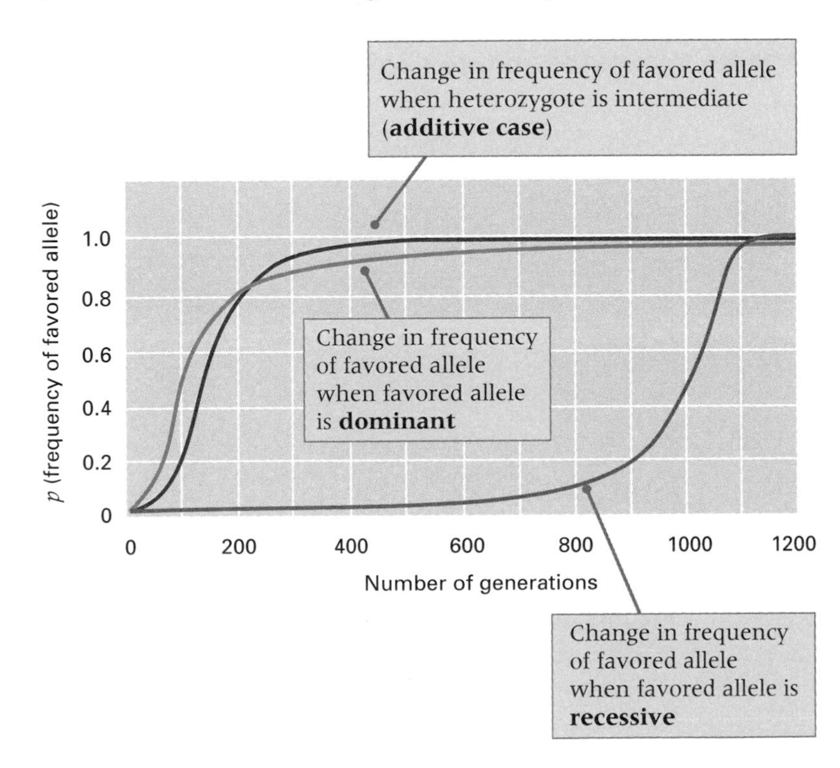

Figure 13.26 Theoretically expected change in frequency of an allele favored by selection in a diploid organism undergoing random mating when the favored allele is dominant or recessive or when the heterozygote is exactly intermediate in fitness (additive). In each case, the selection coefficient against the least fit homozygous genotype is 5 percent. The data are plotted in terms of p, the allele frequency of the beneficial allele. These curves demonstrate that selection for or against a recessive allele is very inefficient when the recessive allele is rare.

alleles eliminated by selection. Eventually the population will attain a state of equilibrium in which the new mutations exactly balance the selective eliminations. Two important cases, pertaining to a complete recessive and to partial dominance, must be considered. In both cases, equilibrium results from the balance of new mutations against those alleles eliminated by selection.

The allele frequency of a harmful allele maintained at equilibrium depends strongly on whether the allele is completely recessive. Because selection against a complete recessive is so inefficient when the allele is rare, even a small amount of selection against heterozygous carriers results in a dramatic reduction in the equilibrium allele frequency. For example, consider a homozygous-lethal allele that has an equilibrium frequency of 0.01 when completely recessive; if the heterozygous carriers had a relative fitness of 0.99, instead of 1.0, the equilibrium frequency would decrease to 0.001. In other words, a mere 1 percent decrease in the fitness of heterozygous carriers results in a tenfold decrease in the equilibrium frequency. The decrease is so dramatic because there are many more heterozygotes than homozygotes for the rare allele; therefore, a small amount of selection against heterozygotes affects such a relatively large number of organisms that the result is very great.

Occasionally the heterozygote is the superior genotype.

So far, we have considered only cases in which the heterozygote is intermediate in fitness between the homozygotes (or possibly equal in fitness to one homozygote). In these cases, the allele associated with the superior homozygote eventually becomes fixed, unless the selection is opposed by mutation. In this section, we consider the possibility of **heterozygote superiority,** in which the fitness of the heterozygote is greater than that of both homozygotes.

When there is heterozygote superiority, neither allele can be eliminated by selection. In each generation, the heterozygotes produce more offspring than the homozygotes, and the selection for heterozygotes keeps both alleles in the population. Selection eventually produces an equilibrium in which the allele frequencies no longer change. If the relative fitnesses of the genotypes *AA, Aa,* and *aa* are $1 - s : 1 : 1 - t$, then it can be shown that at equilibrium, the values of p and q are given by the ratio $p/q = t/s$. This formula makes good intuitive sense, because the greater the selection coefficient against *aa* (t), the larger the equilibrium ratio of p/q, and vice versa.

Heterozygote superiority does not appear to be a particularly common form of selection in natural populations. However, there are several well-established cases, the best known of which is the sickle-cell hemoglobin mutation (Hb^S) and its relationship to a type of malaria caused by the parasitic protozoan *Plasmodium falciparum* (Figure 13.27). The mutation in sickle-cell anemia is in the gene for β-globin, and it changes the sixth codon in the coding sequence from the normal 5'-GAG-3', which codes for glutamic acid, into the codon 5'-GUG-3', which codes for valine (Section 7.2). In the absence of effective medical care, the Hb^S allele is virtually lethal when homozygous, yet in certain parts of

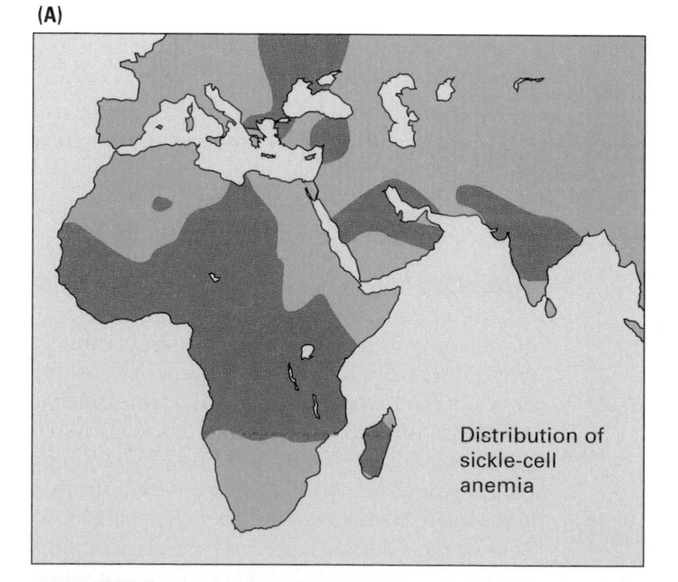

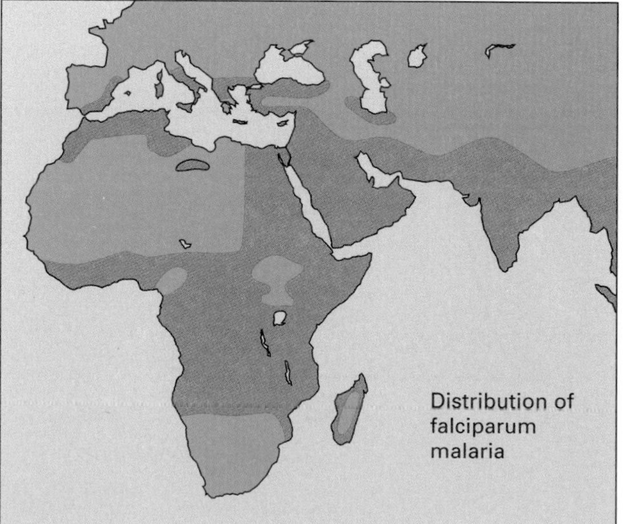

Figure 13.27 Geographic distribution of sickle-cell anemia (A) and falciparum malaria (B) in the 1920s, before extensive malaria-control programs were launched.

Africa and the Middle East, the allele frequency reaches 10 percent or even higher. The *Hb*^{S} allele is maintained because heterozygous persons are less susceptible to malaria, and when they are infected have milder infections, than homozygous-normal persons; the heterozygous genotypes therefore have the highest fitness.

13.8 Some changes in allele frequency are random.

Random genetic drift comes about because populations are not infinitely large, as we have been assuming all along, but finite (limited in size). The breeding organisms in any one generation produce a potentially infinite pool of gametes. Barring differences in fertility, the allele frequencies among gametes would equal the allele frequencies among adults. However, because of the finite size of the population, only relatively few of the gametes participate in fertilization to form the zygotes of the next generation. In other words, a process of *sampling* takes place from one generation to the next; because there is chance variation among samples, the allele frequencies among gametes and zygotes may differ.

A concrete example illustrates the essential features of random genetic drift. Table 13.2 shows the result of random genetic drift in 12 subpopulations (labeled *a* through *l*), each initially consisting of eight heterozygous genotypes. Each subpopulation therefore starts with eight copies of allele *A* and eight copies of allele *a*. A computer, programmed to simulate random mating and Mendelian segrega-

Table 13.2
Effects of random genetic drift. Each entry is the number of *A* alleles in a subpopulation of size 16.

	Population designation												Average
Generation	*a*	*b*	*c*	*d*	*e*	*f*	*g*	*h*	*i*	*j*	*k*	*l*	$\bar{p}$
0	8	8	8	8	8	8	8	8	8	8	8	8	0.500
1	11	7	9	8	8	6	8	7	8	6	11	10	0.516
2	10	9	10	8	8	8	6	7	6	7	13	11	0.536
3	7	11	6	5	12	5	8	5	8	7	14	9	0.505
4	8	11	7	4	12	5	8	8	7	4	15	6	0.495
5	8	8	5	3	13	2	8	12	9	5	15	6	0.490
6	11	5	3	1	13	3	10	13	6	7	15	3	0.469
7	11	8	4	3	11	1	8	14	3	7	15	2	0.453
8	14	7	4	3	10	1	9	14	3	9	16	1	0.474
9	15	5	3	5	7	0	12	14	2	11	16	0	0.469
10	16	6	5	9	8	0	9	13	3	10	16	0	0.495
11	16	1	5	11	6	0	10	13	3	10	16	0	0.474
12	16	0	5	12	6	0	9	13	2	9	16	0	0.458
13	16	0	3	12	7	0	9	13	1	11	16	0	0.458
14	16	0	5	15	7	0	9	11	1	12	16	0	0.479
15	16	0	3	14	7	0	8	12	3	13	16	0	0.479
16	16	0	2	14	9	0	11	14	2	14	16	0	0.510
17	16	0	1	15	6	0	12	14	2	13	16	0	0.495
18	16	0	1	15	4	0	13	15	5	13	16	0	0.510
19	16	0	1	16	2	0	14	16	6	14	16	0	0.526
20	16	0	1	16	2	0	15	16	9	16	16	0	0.557
21	16	0	2	16	3	0	15	16	10	16	16	0	0.573
	A	*a*		*A*		*a*		*A*		*A*	*A*	*a*	
	fixed	fixed		fixed		fixed		fixed		fixed	fixed	fixed	

tion, was used to calculate the number of *A* alleles in each subpopulation through time. The vertical columns show the number of *A* alleles in each population as time passes, and the dispersion of allele frequencies resulting from random genetic drift are apparent. These changes in allele frequency would be less pronounced and would require more time in larger populations than in the very small populations illustrated here, but the overall effect would be the same. That is, the dispersion of allele frequency resulting from random genetic drift depends on population size; the smaller the population, the greater the dispersion and the more rapidly it happens.

In Table 13.2, the principal effect of random genetic drift is evident in the first generation—the allele frequencies have begun to spread out. By the seventh generation, the spreading is extreme, and the number of *A* alleles ranges from 1 to 15. In general,

> Random genetic drift causes differences in allele frequency among subpopulations and therefore causes genetic differentiation among subpopulations.

Although allele frequencies among the subpopulations spread out because of random genetic drift, the *average* allele frequency among subpopulations remains approximately constant. This point is illustrated by the column at the far right in Table 13.2. The average allele frequency stays close to 0.5, its initial value. If an infinite number of subpopulations were being considered instead of the 12 in Table 13.2, the average allele frequency would be exactly 0.5 in every generation. In other words, in spite of the random drift of allele frequency among subpopulations, the average allele frequency among a large number of subpopulations remains constant and equal to the average allele frequency among the original subpopulations.

A RARE ALBINO PENGUIN (an immature emperor penguin) found on snow-covered sea ice in Antarctica. Its lack of feather pigmentation would put it at a severe selective disadvantage when swimming in the dark waters.

After a sufficient number of generations of random genetic drift, some of the subpopulations become fixed for *A* and others for *a*. Because we are ignoring the consequences of mutation, a population that becomes fixed for an allele remains fixed thereafter. After 21 generations in Table 13.2, only four of the populations—*c, e, g,* and *i*—are still segregating; eventually, these too will become fixed. Because the average allele frequency of *A* remains constant at its frequency in the initial generation (p_0), it follows that the fraction p_0 of the subpopulations will ultimately become fixed for *A*, and the fraction $1 - p_0$ will ultimately become fixed for *a*. That is, *the probability of ultimate fixation of a particular allele is equal to the frequency of that allele in the original population.* In Table 13.2, five of the fixed populations are fixed for *A* and three for *a*, which is not very different from the equal numbers expected theoretically with an infinite number of subpopulations.

An example of random genetic drift in small experimental populations of *Drosophila* that exhibit the characteristics pointed out in connection with Table 13.2 is illustrated in Figure 13.28A. The figure is based on 107 subpopulations, each initiated with eight bw^{75} / bw females (bw = brown eyes) and eight bw^{75} / bw males, and maintained at a constant size of 16 by randomly choosing eight males and eight females from among the progeny of each generation. Note how the allele frequencies among subpopulations spread out because of random genetic drift and how subpopulations soon begin to be fixed for either bw^{75} or bw. Although the data are somewhat rough because there are only 107 subpopulations, the overall pattern of genetic differentiation has a reasonable resemblance to that expected from the theory based on the binomial distribution (Figure 13.28B).

If random genetic drift were the only force at work, all alleles would become either fixed or lost and there would be no polymorphism. On the other hand, many factors can act to retard or prevent the effects of random genetic drift, of which the following are the most important: (1) large population size; (2) mutation and migration, which impede fixation because alleles lost by random genetic drift can be reintroduced by either process; and (3) natural selection, particularly those modes of selection that tend to maintain genetic diversity, such as heterozygote superiority.

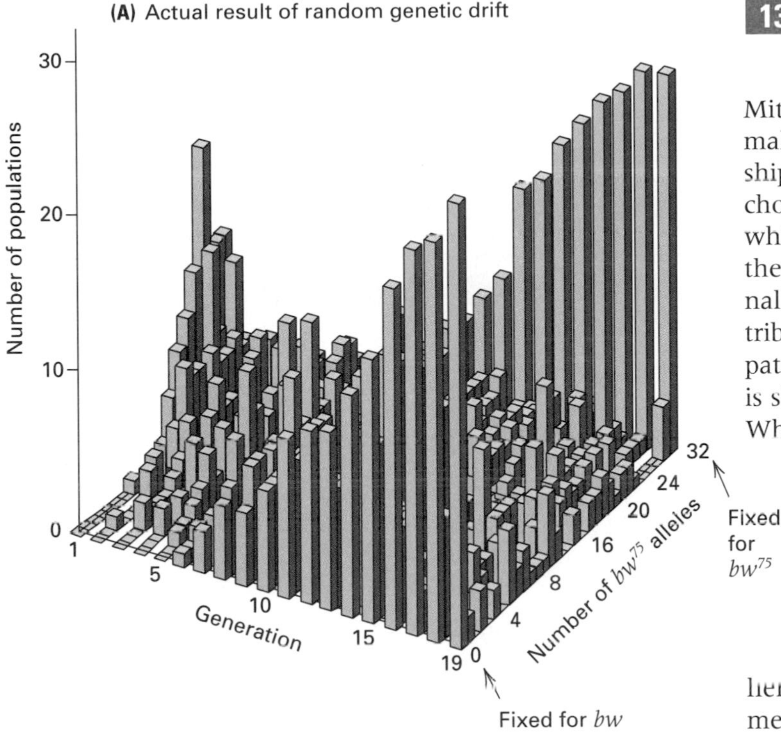

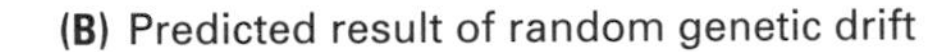

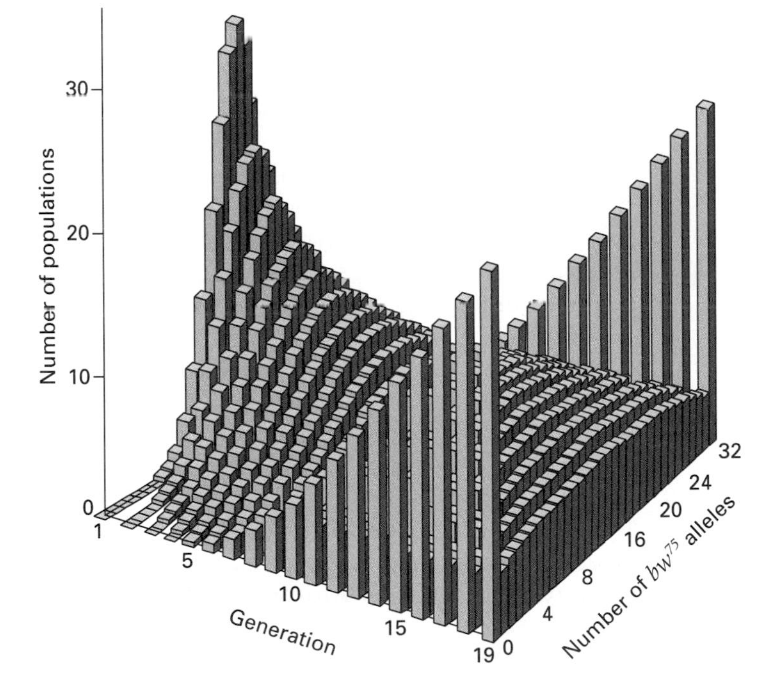

Figure 13.28 (A) Random genetic drift in 107 experimental populations of *Drosophila melanogaster,* each consisting of 8 females and 8 males. (B) Theoretical expectation of the same situation, calculated from the binomial distribution. [Data in part A from P. Buri. 1956. *Evolution* 10:367. Graphs from D. L. Hartl and A. G. Clark. 1989. *Principles of Population Genetics.* Sunderland, MA: Sinauer Associates.]

13.9 Mitochondrial DNA is maternally inherited.

Mitochondrial DNA has a number of features that make it useful for studying the genetic relationships among organisms. In higher animals, mitochondria usually show **maternal inheritance,** which means genetic transmission only through the female. The mitochondria are typically maternally inherited because the egg is the major contributor of cytoplasm to the zygote. A typical pattern of maternal inheritance of mitochondria is shown in the human pedigree in Figure 13.29. When human mitochondrial DNA is cleaved with the restriction enzyme *Hae*II, the cleavage products include either one fragment of 8.6 kb or two fragments of 4.5 kb and 4.1 kb (Figure 13.29A). The pattern with two smaller fragments is typical, and it results from the presence of a *Hae*II cleavage site within the 8.6-kb fragment. Maternal inheritance of the 8.6-kb mitochondrial DNA fragment is indicated in Figure 13.29B, because males (I-2, II-7, and II-10) do not transmit the pattern to their progeny, whereas females (I-3, I-5, and II-8) transmit the pattern to all of their progeny. Although the mutation in the *Hae*II site yielding the 8.6-kb fragment is not associated with any disease, a number of other mutations in mitochondrial DNA do cause diseases and have similar patterns of mitochondrial inheritance. Most of these conditions decrease the ATP-generating capacity of the mitochondria and affect the function of muscle and nerve cells, particularly in the central nervous system, leading to blindness, deafness, or stroke. Many of the conditions are lethal in the absence of some normal mitochondria, and there is variable expressivity because of differences in the proportions of normal and mutant mitochondria among affected persons. The condition in which two or more genetically different types of mitochondria are present in the same cell is unusual among animals. For example, a typical human cell contains from 1000 to 10,000 mitochondria, all of them genetically identical.

"Eve" was probably more lucky than special.

Another important feature of mitochondrial DNA is that it does not ordinarily undergo genetic recombination. Hence the DNA molecule in any mitochondrion derives from a single ancestral molecule. Mitochondrial DNA is also a good genetic marker for tracing human ancestry, because it evolves at a rate considerably faster than that of nuclear genes. Differences in mitochondrial DNA

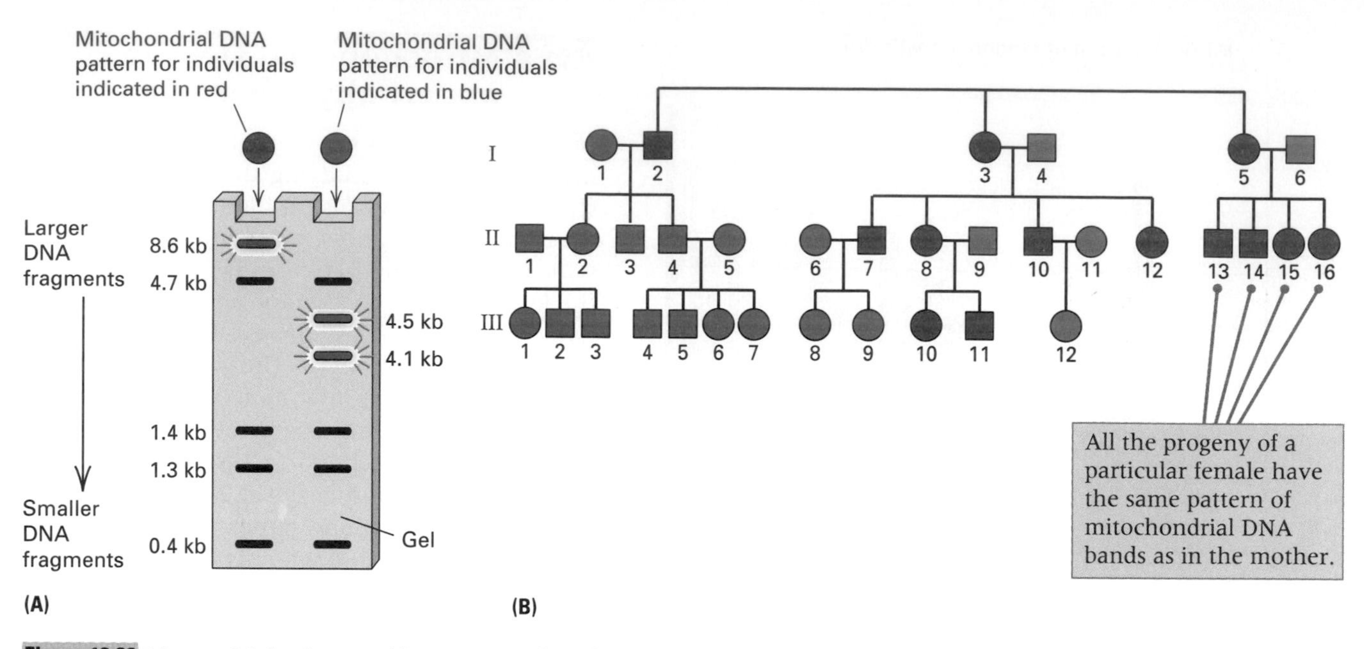

Figure 13.29 Maternal inheritance of human mitochondrial DNA. (A) Pattern of DNA fragments obtained when mitochondrial DNA is digested with the restriction enzyme *Hae*II. The DNA type at the left includes a fragment of 8.6 kb (red). The DNA type at the right contains a cleavage site for *Hae*II within the 8.6-kb fragment, which results in smaller fragments of 4.5 kb and 4.1 kb (blue). (B) Pedigree showing maternal inheritance of the DNA pattern with the 8.6-kb fragment (red symbols). The mitochondrial DNA type is transmitted only through the mother. [After D. C. Wallace. 1989. *Trends in Genetics* 5: 9.]

sequences accumulate among human lineages at a rate of approximately 1 change per mitochondrial lineage every 1500 to 3000 years. For example, the people of New Guinea have been relatively isolated genetically from the aboriginal people of Australia ever since these areas were colonized approximately 40,000 years ago and 30,000 years ago, respectively. The total time for evolution between the populations is therefore 70,000 years (40,000 years in New Guinea and 30,000 years in Australia). If one nucleotide change accumulates every 1500 to 3000 years, the total number of differences in the mitochondrial DNA of New Guineans and Australian aborigines is expected to range from 23 nucleotides to 47 nucleotides (calculated as 70,000/3000 and 70,000/1500, respectively). This example shows how the rate of mitochondrial DNA evolution can be used to predict the number of differences between populations. In practice, the calculation is done the other way around, and the observed number of differences between pairs of populations is used to estimate the rate of mitochondrial evolution.

Nucleotide differences in mitochondrial DNA have been used to reconstruct the probable historical relationships among human populations. Figure 13.30 is an example that includes mitochondrial DNA types from five native populations: African, Asian, Australian, Caucasian (European), and New Guinean. The diagram is *not* a pedigree, although it superficially resembles one. Rather, the diagram is a hypothetical **phylogenetic tree,** which depicts the lines of descent connecting the mitochondrial DNA sequences. The "true" phylogenetic tree connecting the mitochondrial DNA sequences is unknown, so the tree illustrated should be regarded as a hypothesis. It is hypothetical because it is inferred from analysis of the mitochondrial DNA, and the analysis requires a number of assumptions, including, for example, uniformity of rates of evolution in different branches. Nevertheless, the tree in Figure 13.30 has two interesting features: (1) Each population contains multiple mitochondrial types connected to the tree at widely dispersed positions, and (2) one of the primary branches issuing from the inferred common ancestor is composed entirely of Africans. These features can be explained most readily by postulating that the mitochondrial DNA type of the common ancestor was African. Furthermore, because the inferred common ancestor links mitochondrial DNA types that differ at an average of 95 nucleotides, the time since the existence of the common ancestor can be estimated as between $95 \times 1500 = 142{,}500$ years and $95 \times 3000 = 285{,}000$ years. Therefore, the common ancestor of all human mitochondrial DNA was a woman who lived in Africa between 142,500 and 285,000 years ago. This woman has been given the name "Eve" by the popular press, although she was not the only female alive at the time. Nevertheless,

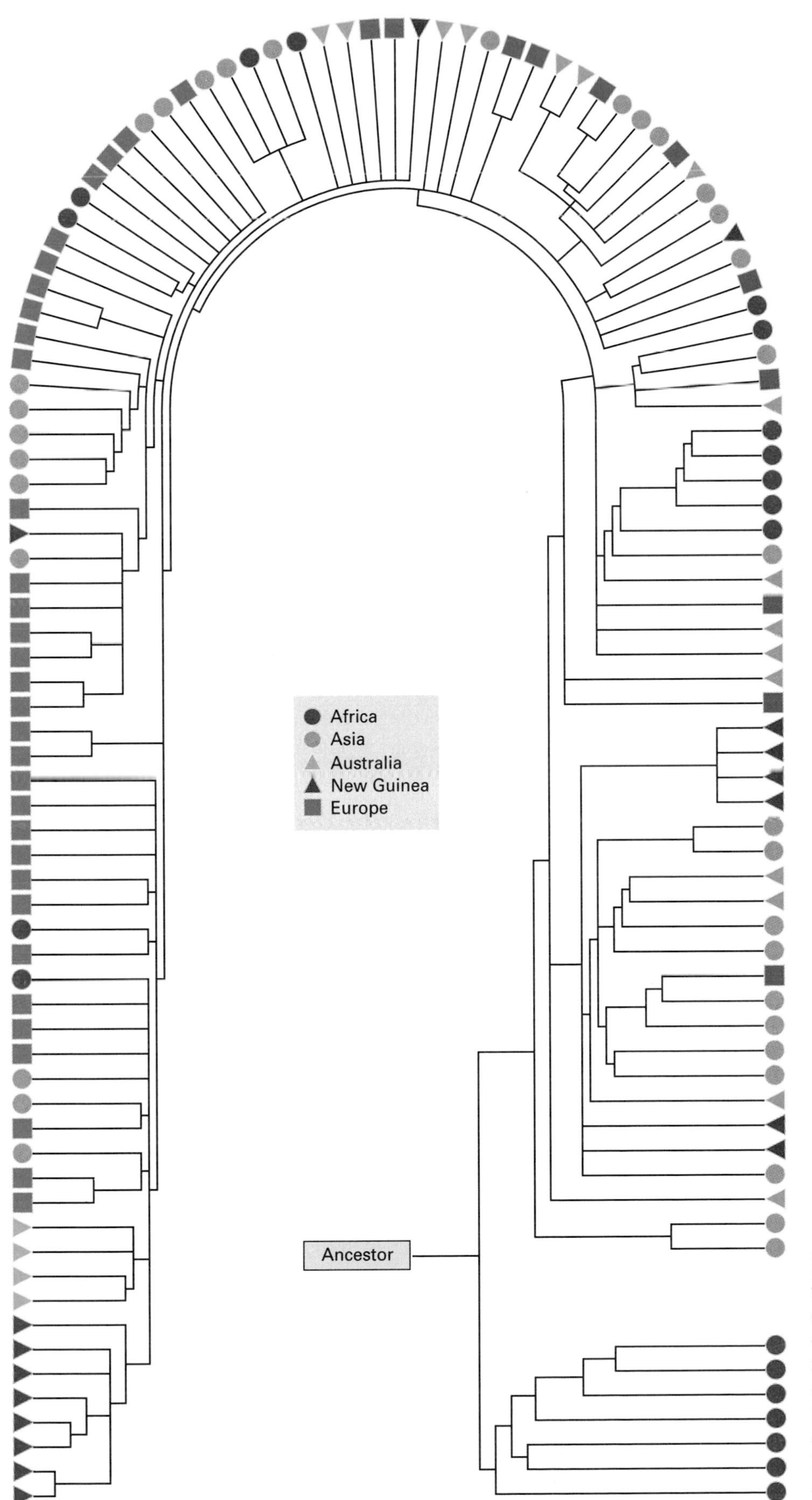

Figure 13.30 Possible phylogenetic tree of human mitochondria based on analysis of presence or absence of restriction sites for 12 different restriction enzymes in mitochondrial DNA from 134 persons. [From R. L. Cann, M. Stoneking, and A. C. Wilson. 1987. *Nature* 325: 31.]

Eve's mitochondrial DNA became the common ancestor of all human mitochondrial DNA that exists today because, during the intervening years, all other mitochondrial lineages except hers became extinct (probably by chance).

The hypothesis of an African mitochondrial Eve has been very controversial. The difficulty is not with postulating a common ancestor of human mitochondrial DNA. Considering the relatively small size of subpopulations throughout most of human history, it is quite plausible to suppose that some mitochondrial DNA lineage might have become fixed as a result of random genetic drift. But other issues are in dispute: Was the common ancestor of all mitochondria an African? When did she live? Later analysis of mitochondrial DNA data suggested that other trees are almost as good as the one in Figure 13.30. Some trees are even a little better. The main criterion for comparing hypothetical trees based on DNA sequences is the total number of mutational steps in all the branches together: Trees with fewer mutational steps are generally regarded as better. Therefore, the finding of better trees than the one in Figure 13.30 was a blow to the hypothesis of an African Eve, because none of the better trees had an earliest branch composed exclusively of Africans. On the other hand, considerable evidence of other kinds supports Africa as the cradle of human evolution and colonization.

Chapter Summary

- **Genotypes may differ in frequency from one population to another.**

  Allele frequencies are estimated from genotype frequencies.

  Many genes have multiple polymorphic alleles in natural populations.

  DNA polymorphisms are important genetic markers in natural populations.

Population genetics is the application of Mendel's laws and other principles of genetics to populations of organisms. A subpopulation, or local population, is a group of organisms of the same species living within a geographical region of such size that most matings are between members of the group. In most natural populations, many genes are polymorphic in that they have two or more common alleles. One of the goals of population genetics is to determine the nature and extent of genetic variation in natural populations.

- **Random mating means that mates pair without regard to genotype.**

  The Hardy–Weinberg principle has important implications for population genetics.

  If an allele is rare, it is found mostly in heterozygous genotypes.

  Hardy–Weinberg frequencies can be extended to multiple alleles.

  X-linked genes are a special case because males have only one X chromosome.

- **Highly polymorphic genes are used in DNA typing.**

  Polymorphic alleles may differ in frequency among subpopulations.

  DNA exclusions are definitive.

The relationship between the relative proportions of particular alleles (allele frequencies) and genotypes (genotype frequencies) is determined in part by the frequencies with which particular genotypes form mating pairs. In random mating, mating pairs are independent of genotype. When a population undergoes random mating for an autosomal gene with two alleles, the frequencies of the genotypes are given by the Hardy–Weinberg principle. If the alleles of the gene are A and a, and their allele frequencies are p and q, respectively, then the Hardy–Weinberg principle states that the genotype frequencies with random mating are AA with frequency p^2, Aa with frequency $2pq$, and aa with frequency q^2. These are often good approximations for genotype frequencies within subpopulations. An important implication of the Hardy–Weinberg principle is that rare alleles are found much more frequently in heterozygotes than in homozygotes ($2pq$ versus q^2).

- **Inbreeding means mating between relatives.**

  Inbreeding results in an excess of homozygotes compared with random mating.

Inbreeding means mating between relatives, and the extent of inbreeding is measured by the inbreeding coefficient, F. The main consequence of inbreeding is that a rare harmful allele present in a common ancestor may be transmitted to both parents of an inbred organism in a later generation and become homozygous in the inbred offspring. Among the progeny produced by inbreeding, the frequency of heterozygous genotypes is smaller, and that of homozygous genotypes greater, than it would be with random mating.

- **Evolution is accompanied by genetic changes in species.**
- **Mutation and migration bring new alleles into populations.**

Evolution is a progressive change in the gene pool of a population or in the allele frequencies present in a population. One principal mechanism of evolution is natural selection, in which genotypes superior in survival or reproductive ability in the prevailing environment contribute a disproportionate share of genes to future generations, thereby gradually increasing the frequency of the favorable alleles in the whole population. By this process, a species becomes genetically better adapted to its environment. At least three other processes also can change allele frequency: mutation (heritable change in a gene), migration (movement of organisms among subpopulations), and random genetic drift (resulting from restricted population size). Spontaneous mutation rates are generally so low that the effect of mutation on changing allele frequency is minor, except for rare alleles. Migration can have significant effects on allele frequency because migration rates may be very large. The main effect of migration is the tendency to equalize allele frequencies among the local populations that exchange migrants. Selection operates through differences in viability (the probability of survival of a genotype) and in fertility (the probability of successful reproduction).

Populations maintain harmful alleles at low frequencies as a result of a balance between selection, which tends to eliminate the alleles, and mutation, which tends to increase their frequencies. When there is selection-mutation balance, the allele frequency at equilibrium is usually greater if the allele is completely recessive than if it is partially dominant. This difference arises because selection is quite ineffective in affecting the frequency of a completely recessive allele when the allele is rare, because the allele appears almost exclusively in heterozygotes.

- **Natural selection favors genotypes better able to survive and reproduce.**

Fitness is the relative ability of genotypes to survive and reproduce.
Allele frequencies change slowly when alleles are either very rare or very common.
Selection can be balanced by new mutations.
Occasionally the heterozygote is the superior genotype.

A few examples are known in which the heterozygous genotype has a greater fitness than either of the homozygous genotypes (heterozygote superiority). Heterozygote superiority results in an equilibrium in which both alleles are maintained in the population. An example is sickle-cell anemia in regions of the world where falciparum anemia is endemic. Heterozygous persons have an increased resistance to malaria and only a mild anemia, which results in fitness greater than that of either homozygote.

- **Some changes in allele frequency are random.**

Random genetic drift is a statistical process of change in allele frequency in small populations, resulting from the inability of every organism to contribute equally to the offspring of successive generations. In a subdivided population, random genetic drift results in differences in allele frequency among the subpopulations. In an isolated population, barring mutation, an allele will ultimately become fixed or lost as a result of random genetic drift.

- **Mitochondrial DNA is maternally inherited.**

"Eve" was probably more lucky than special.

Human mitochondrial DNA types have been organized into a hypothetical phylogenetic tree in which one branch issuing from the inferred common ancestor is composed entirely of mitochondrial DNA from Africans. The woman whose mitochondrial DNA is thought to be the common ancestor of all human mitochondria has been dubbed "Eve." The all-African branch in the phylogenetic tree and the number of nucleotide differences observed in mitochondrial DNA have been interpreted to mean that Eve was a woman who lived in Africa between 142,500 and 285,000 years ago.

Key Terms

adaptation
allele frequency
allozyme
DNA typing
electrophoresis
evolution
fixed allele
gene pool
genotype frequency
Hardy–Weinberg principle
heterozygote superiority
inbreeding
inbreeding coefficient
local population
lost allele
maternal inheritance
migration
mutation
natural selection
phylogenetic tree
polymorphic gene
population
population genetics
population substructure
random genetic drift
random mating
relative fitness
selection coefficient
selectively neutral mutation
subpopulation

Concepts and Issues
Testing Your Knowledge

- What does the Hardy–Weinberg principle imply about the relative frequencies of heterozygous carriers and homozygous affected organisms for a rare, harmful recessive allele?
- Traits due to recessive alleles in the X chromosome are usually much more prevalent in males than in females. Explain why this discrepancy is expected with random mating.
- Why are the effects of inbreeding more easily observed with a rare recessive allele than with a rare dominant allele?
- What is the fitness of an organism that dies before the age of reproduction? What is the fitness of an organism that is sterile?
- Many recessive alleles are extremely harmful when homozogous, so there is selection in every generation that tends to reduce the allele frequency. Yet harmful recessive alleles are maintained at a low frequency for almost every gene. What process prevents harmful recessive alleles from being completely eliminated?
- Heterozygote superiority of the type observed with sickle-cell anemia is sometimes called *balancing selection.* Do you think this is an appropriate term? Explain why or why not.
- What is random genetic drift and why does it occur? Explain why this process implies that in the absence of other forces, the ancestry of all alleles present at a locus in a population can eventually be traced back to a single allele present in some ancestral population.

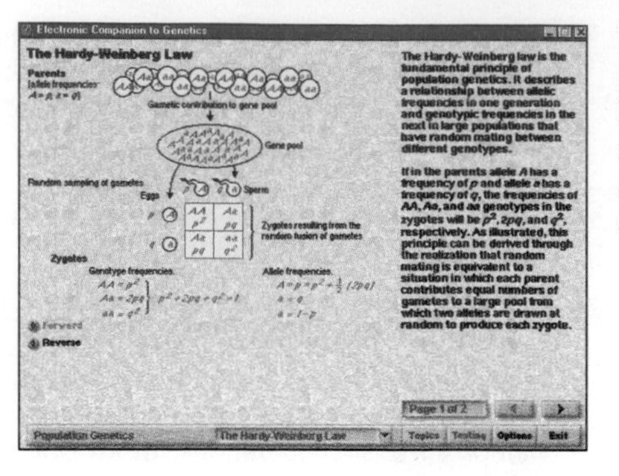

TO ENHANCE YOUR STUDY turn to *An Electronic Companion to Genetics*™, © 1998, Cogito Learning Media, Inc. This CD-ROM is a multimedia tool that provides interactive explanations of important genetic concepts through animations, diagrams, and videos. It also provides interactive test questions.

Solutions Step by Step

Problem 1

The accompanying illustration shows a human pedigree along with the pattern of restriction fragments observed in a DNA sample from each of the individuals when hybridized with a labeled probe. Examine the pedigree and the band patterns, and suggest a mode of inheritance.

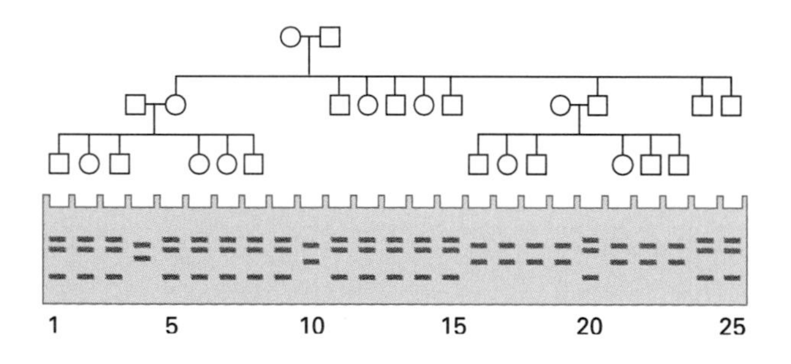

Solution The first step is to examine the pedigree carefully, looking for signs of Mendelian segregation to indicate whether the DNA fragment is inherited on an autosome, on the X chromosome, or on the Y chromosome. In this case the pattern of inheritance fits none of these possibilities. In each sibship, the phenotypes (band patterns) of all the progeny are identical to each other. Furthermore, when the maternal and paternal phenotypes differ, the progeny phenotype is exactly like that of the mother. Therefore, the pedigree implies that the DNA fragment detected by the probe shows maternal inheritance, and the most likely explanation is that the probe is a fragment of mitochondrial DNA.

Problem 2

In human populations, a locus called *secretor* determines whether the carbohydrate A and B antigens of the ABO blood groups are secreted into the saliva and other body fluids. The genotypes *Se Se* and *Se se* are secretors, and the genotype *se se* is not. Among Caucasians known to have blood group A, B, or AB because of the presence of these antigens on the red blood cells, the frequency of the nonsecretor phenotype is about 33 percent.

(a) Assuming random mating, what is the allele frequency of the *se* allele?

(b) Among Caucasians with blood group A, B, or AB, what are the expected proportions of homozygous secretors and heterozygous secretors?

Solution This is a typical problem that makes use of the Hardy–Weinberg principle for random mating. **(a)** With Hardy–Weinberg proportions, the frequency of homozygous recessive genotypes equals q^2, which in this case is given as 0.33. Hence $q = \sqrt{(0.33)} = 0.574$, so $p = 1 - q = 0.426$. **(b)** The expected proportions of *Se Se* and *Se se* genotypes are $p^2 = (0.426)^2 = 0.181$ and $2pq = 2(0.574)(0.426) = 0.489$, respectively. Note that approximately half of the genotypes in the population are heterozygous for the *secretor* locus.

Problem 3

Xeroderma pigmentosum (XP) is an often-fatal skin cancer resulting from a recessive mutant allele that affects DNA repair. In the United States, the frequency of homozygous-recessive affected people is approximately 1 in 250,000. (The mutant allele can be in any one of about eight different genes, but for the purposes of this problem it makes no difference.)

(a) What is the expected frequency of XP among the offspring of first-cousin matings?

(b) What is the ratio of XP among the offspring of first-cousin matings to that among the offspring of nonrelatives?

Solution First we must calculate the frequency of the recessive allele, q, using the information that the frequency of homozygous recessives is 1 in 250,000. Assuming random mating frequencies, $q = (1/250{,}000) = 0.002$. **(a)** Among the offspring of first-cousin matings, the expected frequency of XP equals $q^2(1 - F) + qF$, where $F = 1/16$ for the offspring of first cousins. In this case the formula yields $(0.002)^2(15/16) + (0.002)(1/16) = 0.00012875$, or about 1 in 7767. **(b)** The ratio is given in the following formula, where $q = 0.002$ and $F = 1/16$.

$$\text{Ratio} = \frac{q^2(1 - F) + qF}{q^2} = \frac{(1/7{,}767)}{(1/250{,}000)} = 32.2$$

In other words, the offspring of first cousins have more than a 32-fold greater risk of being homozygous for a recessive mutant allele causing XP.

Concepts in Action
Problems for Solution

13.1 A trait due to a harmful recessive X-linked allele in a large, randomly mating population affects 1 in 50 males. What is the frequency of carrier females? What is the expected frequency of affected females?

13.2 How many *A* and *a* alleles are present in a sample of organisms consisting of 10 *AA*, 15 *Aa*, and 4 *aa* genotypes? What are the allele frequencies in this sample?

13.3 Which of the following genotype frequencies of *AA*, *Aa*, and *aa*, respectively, satisfy the Hardy–Weinberg principle?

(a) 0.25, 0.50, 0.25
(b) 0.36, 0.55, 0.09
(c) 0.49, 0.42, 0.09
(d) 0.64, 0.27, 0.09
(e) 0.29, 0.42, 0.29

13.4 A man is known to be a carrier of the cystic fibrosis allele. He marries a phenotypically normal woman. In the general population, the incidence of cystic fibrosis at birth is approximately 1 in 1700. Assume Hardy–Weinberg proportions.

(a) What is the probability that the wife is also a carrier?
(b) What is the probability that their first child will be affected?

13.5 A man with normal parents whose brother has phenylketonuria marries a phenotypically normal woman. In the general population, the incidence of phenylketonuria at birth is approximately 1 in 10,000. Assume Hardy–Weinberg proportions.

(a) What is the probability that the man is a carrier?
(b) What is the probability that the wife is also a carrier?
(c) What is the probability that their first child will be affected?

13.6 For an X-linked gene with two alleles in a large, randomly mating population, the frequency of carrier females equals one-half of the frequency of the males carrying the recessive allele. What are the allele frequencies?

13.7 Allozyme phenotypes of alcohol dehydrogenase in the flowering plant *Phlox drummondii* are determined by codominant alleles of a single gene. In one sample of 35 plants, the following data were obtained:

Genotype	*AA*	*AB*	*BB*	*BC*	*CC*	*AC*
Number	2	5	12	10	5	1

What are the allele frequencies in this sample? With random mating, what are the expected numbers of each of the genotypes?

13.8 A population of maple trees contains a lethal allele that allows homozygous recessive seeds to germinate, but the plants produce no chlorophyll and die shortly after germination. The allele frequency of the recessive allele for this condition in the population is 0.01.

(a) If a randomly mating population produces one million seeds, how many plants will lack chlorophyll?
(b) Will natural selection eliminate the allele after one generation? Why or why not?

13.9 Hartnup disease is an autosomal recessive disorder of intestinal and renal transport of amino acids. The frequency of affected newborn infants is about 1 in 14,000. Assuming random mating, what is the frequency of heterozygotes?

13.10 Self-fertilization in the annual plant *Phlox cuspidata* results in an average inbreeding coefficient of $F = 0.66$.

(a) What frequencies of the genotypes for the enzyme phosphoglucose isomerase would be expected in a population with alleles *A* and *B* at respective frequencies 0.43 and 0.57?
(b) What frequencies of the genotypes would be expected with random mating?

13.11 Two strains of bacteria, A and B, are placed into direct competition in a chemostat. A is favored over B. If the selection coefficient per generation is constant, what is its value if, in an interval of 100 generations:

(a) The ratio of A cells to B cells increases by 10 percent?
(b) The ratio of A cells to B cells increases by 90 percent?
(c) The ratio of A cells to B cells increases by a factor of 2?

13.12 An allele *A* undergoes mutation to the allele *a* at the rate of 10^{-5} per generation. If a very large population is fixed for *A* (generation 0), what is the expected frequency of *A* in the following generation (generation 1)? What is the expected frequency of *A* in generation 2? Deduce the rule for the frequency of *A* in generation *n*.

Further Readings

Allison, A. C. 1956. Sickle cells and evolution. *Scientific American,* August.

Ayala, F. 1978. The mechanisms of evolution. *Scientific American,* September.

Bittles, A. H., W. M. Mason, J. Greene, and N. A. Rao. 1991. Reproductive behavior and health in consanguineous marriages. *Science* 252: 789.

Bodmer, W. F., and L. L. Cavalli-Sforza. 1976. *Genetics, Evolution, and Man.* New York: Freeman.

Cann, R. L., M. Stoneking, and A. C. Wilson. 1987. Mitochondrial DNA and human evolution. *Nature* 325: 31.

Cavalli-Sforza, L. L. 1974. The genetics of human populations. *Scientific American,* September.

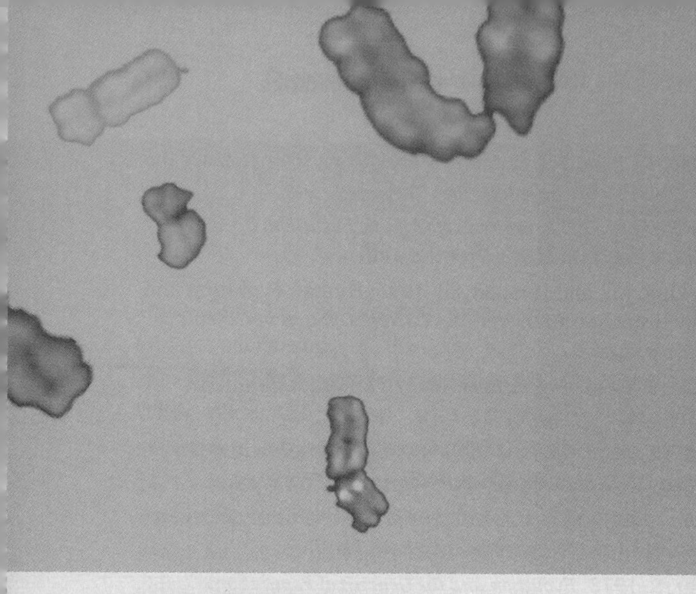

Key Concepts

- Multifactorial traits are determined by multiple genetic and environmental factors acting together.
- The relative contributions of genotype and environment to a trait are measured by the variance due to genotype (genotypic variance) and the variance due to environment (environmental variance).
- Correlations between relatives are used to estimate various components of variation, such as genotypic variance, additive variance, and dominance variance.
- Additive variance accounts for the parent-offspring correlation; dominance variance accounts for the sib-sib correlation over and above that expected from the additive variance.
- Narrow-sense heritability is the ratio of additive (transmissible) variance to the total phenotypic variance; it is widely used in plant and animal breeding.
- Genes that affect quantitative traits (QTLs) can be identified and genetically mapped using various kinds of genetic polymorphisms.
- Many complex human behaviors are affected by multiple genetic and environmental factors and the interactions between them.

AN EAR OF WILD TEOSINTE (left) looks dramatically different from an ear of modern cultivated maize (right), but the organisms are very closely related, producing fertile hybrids in both reciprocal crosses (middle). Early native Americans first domesticated teosinte and, through many generations, selected the mutant phenotypes most favorable for cultivation, leading ultimately to modern maize. [© John Doebley.]

CHAPTER 14

The Genetic Architecture of Complex Traits

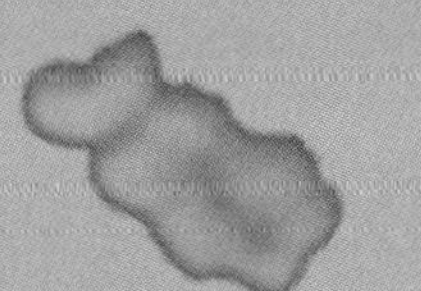

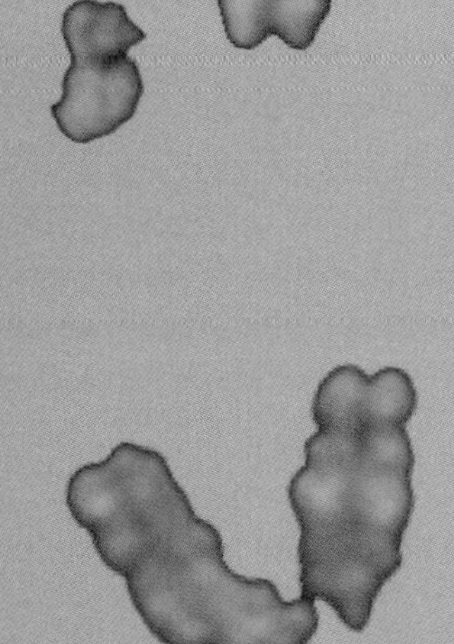

Earlier chapters have emphasized traits in which differences in phenotype result from alternative genotypes of a single gene. Examples include green versus yellow peas, red eyes versus white eyes in *Drosophila,* normal versus sickle-cell hemoglobin, and the ABO blood groups. These traits are particularly well suited for genetic analysis through the study of pedigrees, because there is a small number of genotypes and phenotypes and because there is a relatively simple correspondence between genotype and phenotype. However, many traits that are important in plant breeding, animal breeding, and medical genetics are influenced by multiple genes as well as by the effects of environment. These are known as **multifactorial traits** because of the multiple genetic and environmental factors implicated in their causation. With a multifactorial trait, a single genotype can have many possible phenotypes (depending on the environment), and a single phenotype can include many possible genotypes.

Multifactorial traits are often called **complex traits** because each factor that affects the trait contributes, at most, a modest amount to the total variation in the trait observed in the entire population. Most traits that vary in populations of humans and other organisms, including common human diseases that have a genetic component, are complex traits. For a complex trait, the **genetic architecture** consists of a description of all of the genetic and environmental factors that affect the trait, along with the magnitude of their individual effects and the magnitudes of interactions among the factors. It is, in principle, possible to define the genetic components in terms of Mendelian segregation and locations along a genetic map. Environmental factors are much less easily partitioned into separate factors whose individual effects and interactions can be sorted out. The genetic analysis of complex traits requires special concepts and methods, which are introduced in this chapter.

14.1 Multifactorial traits are determined by multiple genes and the environment.

Multifactorial traits are often called **quantitative traits** because the phenotypes in a population differ in quantity rather than in type. A trait such as height is a quantitative trait. Heights are not found in discrete categories but differ merely in quantity from one person to the next. The opposite of a quantitative trait is a "discrete trait," in which the phenotypes differ in kind—for example, brown eyes versus blue eyes.

Quantitative traits are typically influenced not only by the alleles of two or more genes but also by the effects of environment. Therefore, with a quantitative trait, the phenotype of an organism is potentially influenced by

- **Genetic factors** in the form of alternative genotypes of one or more genes.
- **Environmental factors** in the form of conditions that are favorable or unfavorable for the development of the trait. Examples include the effect of nutrition on the growth rate of animals and the effects of fertilizer, rainfall, and planting density on the yield of crop plants.

With some quantitative traits, differences in phenotype result largely from differences in genotype, and the environment plays a minor role. With other quantitative traits, differences in phenotype result largely from the effects of environment, and genetic factors play a minor role. Most quantitative traits fall between these extremes, so both genotype and environment must be taken into account in their analysis.

In a genetically heterogeneous population, many genotypes are formed by the processes of segregation and recombination. Variation in genotype can be eliminated by studying inbred lines, which are homozygous for most genes, or the F_1 progeny from a cross of inbred lines, which are uniformly heterozygous for all genes in which the parental inbreds differ (Figure 14.1). In contrast, complete elimination of environmental variation is impossible, no matter how hard the experimenter may try to render the environment identical for all members of a population. With plants, for example, small variations in soil quality or exposure to the sun will produce slightly different environments, sometimes even for adjacent plants. Similarly, highly inbred *Drosophila* still show variation in phenotype (for example, in body size) brought about by environmental differences among animals within the same culture bottle. Therefore, traits that are susceptible to small environmental effects will never be uniform, even in inbred lines.

Most traits of importance in plant and animal breeding are quantitative traits. In agricultural production, one economically important quantitative trait is yield, or size of the harvest, per unit area—whether the organism be corn, tomatoes, soybeans, or grapes. In domestic animals, important quantitative traits include meat quality, milk production per cow, egg laying per hen, fleece weight per sheep, and litter size per sow. Important quantitative traits in human genetics include infant growth rate, adult weight, blood pressure, serum cholesterol, and length of life. In evolutionary studies, fitness is the preeminent quantitative trait.

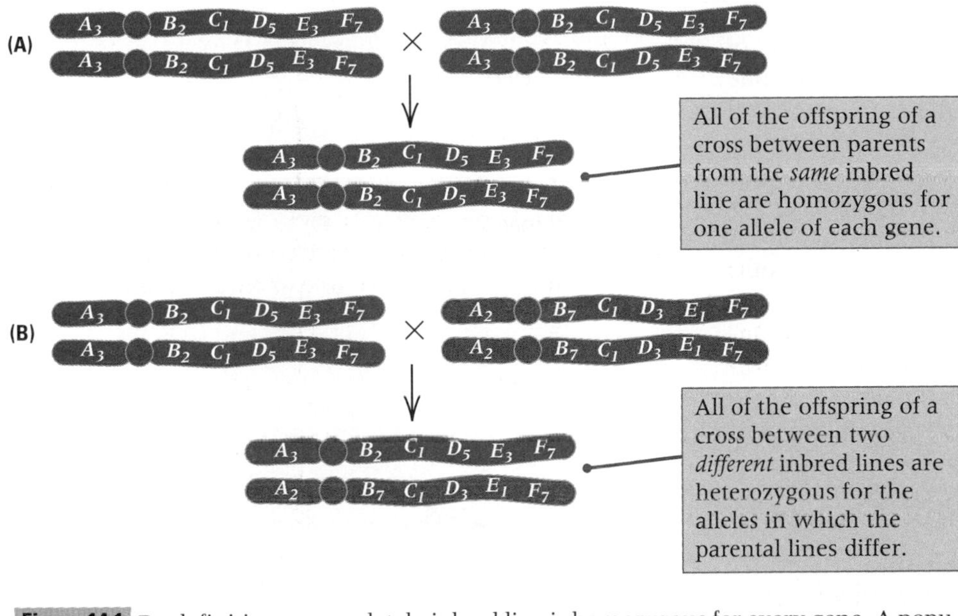

Figure 14.1 By definition, a completely inbred line is homozygous for every gene. A population of organisms, each having an identical genotype, can be created by crossing inbred parents. (A) If the parents are from the same inbred line, the progeny are all genetically identical and homozygous. (B) If the parents are from different inbred lines, the progeny are all genetically identical but heterozygous for alleles at which the parental inbred lines differ.

Most quantitative traits cannot be studied by means of the usual pedigree methods, because the effects of segregation of alleles of one gene may be concealed by effects of other genes, and because environmental effects may cause identical genotypes to have different phenotypes. Therefore, individual pedigrees of quantitative traits do not fit any simple pattern of dominance, recessiveness, or X linkage. Nevertheless, genetic effects on quantitative traits can be assessed by comparing the phenotypes of relatives who, because of their familial relationship, must have a certain proportion of their genes in common. Such studies utilize many of the concepts of population genetics discussed in Chapter 13.

Three categories of traits are frequently found to have quantitative inheritance. They are described in the following section.

Continuous, meristic, and threshold traits are usually multifactorial.

Most phenotypic variation in populations is not manifested in a few easily distinguished categories. Instead, the traits vary continuously from one phenotypic extreme to the other, with no clear-cut breaks in between. Human height is a prime example of such a trait. Other examples include milk production in cattle, growth rate in poultry, yield in corn, and blood pressure in human beings. Such traits are called **continuous traits** because there is a continuous gradation from one phenotype to the next. The range of phenotypes is continuous, from minimum to maximum, with no clear categories. Weight is an example of a continuous trait because the weight of an organism can fall anywhere along a continuous scale of weights, so the number of possible phenotypes is virtually unlimited.

Two other types of quantitative traits are not continuous:

Meristic traits are traits in which the phenotype is determined by counting. Some examples are number of skin ridges forming the fingerprints, number of kernels on an ear of corn, number of eggs laid by a hen, number of bristles on the abdomen of a fly, and number of puppies in a litter. An example of a meristic trait is the number of ears on a stalk of corn, which typically has the value 1, 2, 3, or 4 ears on a given stalk.

Threshold traits are traits that have only two, or a few, phenotypic classes, but their inheritance is determined by the effects of multiple genes together with the environment. Examples of threshold traits include twinning in cattle as well as parthenogenesis (development of unfertilized eggs) in turkeys. In a threshold trait, each organism has an underlying and not directly observable predisposition to express the trait, such as a predisposition for a cow to give birth to twins. If the underlying predisposition is high enough (above a "threshold"), the cow will actually give birth to twins; otherwise she will give birth to a single calf. In many threshold-trait disorders, the phenotypic classes are "affected" versus "not affected." Examples of threshold-trait disorders in human beings include adult diabetes, schizophrenia, and many congenital abnormalities, such as spina bifida. Threshold traits can be interpreted as continuous traits by imagining that each individual has an underlying risk or *liability* toward manifestation of the condition. A liability above a certain cutoff, or *threshold,* results in expression of the condition; a liability below the threshold results in normality. The liability of an individual to a threshold trait cannot be observed directly, but inferences about liability can be drawn from the incidence of the condition among individuals and their relatives. The manner in which this is done is discussed in Section 14.4.

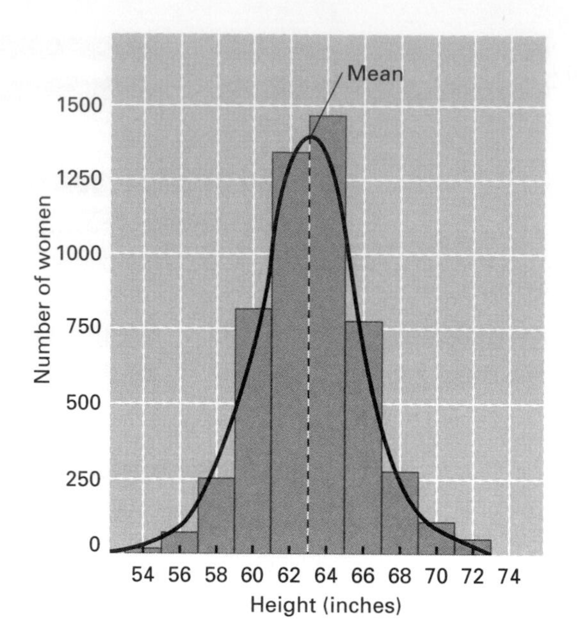

Figure 14.2 Distribution of height among 4995 British women and the smooth normal distribution that approximates the data.

The distribution of a trait in a population implies nothing about its inheritance.

The **distribution** of a trait in a population is a description of the population in terms of the proportion of individuals that have each of the possible phenotypes. Characterizing the distribution of some traits is straightforward because the number of phenotypic classes is small. For example, the distribution of progeny in a certain pea cross may consist of 3/4 green seeds and 1/4 yellow seeds, and the distribution of ABO blood groups among one sample of Greeks consisted of 42 percent O, 39 percent A, 14 percent B, and 5 percent AB. However, with continuous traits, the large number of possible phenotypes makes such summaries impractical. Often, it is convenient to reduce the number of phenotypic classes by grouping similar phenotypes together. Data for an example pertaining to the distribution of height among 4995 British women are given in Table 14.1 and in Figure 14.2. You can imagine each bar in the graph in Figure 14.2 being built step by step, as each of the women is measured, by placing a small square along the *x*-axis at the location corresponding to the height of each woman. As sampling proceeds, the squares begin to pile up in certain places, leading ultimately to the bar graph shown.

Table 14.1 Distribution of height among British women

Interval number (i)	Height interval (inches)	Midpoint (x_i)	Number of women (f_i)
1	53–55	54	5
2	55–57	56	33
3	57–59	58	254
4	59–61	60	813
5	61–63	62	1340
6	63–65	64	1454
7	65–67	66	750
8	67–69	68	275
9	69–71	70	56
10	71–73	72	11
11	73–75	74	4
		Total N =	4995

It is worth taking a moment to consider a more vivid example of what a histogram like that in Figure 14.2 really represents. The "real thing" is shown in Figure 14.3, which is a picture of 162 scholars in genetics at the University of Connecticut, Storrs, who have arranged themselves in order of their height. The women are in white, the men in blue. This is a "living histogram," in which each building block not merely represents a person, each building block *is* a person.

Displaying a distribution completely, either in tabular form, as in Table 14.1, or in graphical form, as in Figure 14.2 or even Figure 14.3, is always helpful but often unnecessary. In many cases a description of the distribution in terms of two major features is sufficient. These features are the *mean* and the *variance.* To discuss the mean and the variance in quantitative terms, we shall use the data in Table 14.1. The height intervals are numbered from 1 (53–55 inches) to 11 (73–75 inches). The symbol x_i designates the midpoint of the height interval numbered i; for example, $x_1 = 54$ inches, $x_2 = 56$ inches, and so on. The number of women in height interval i is designated f_i; for example, $f_1 = 5$ women, $f_2 = 33$ women, and so forth. The total size

Figure 14.3 Genetics scholars at the University of Connecticut in Storrs, who have helpfully arranged themselves by height to form a "living histogram." [Photo 1996 by Peter Morenus, courtesy of Linda Strausbaugh.]

of the sample, in this case 4995, is denoted N. The mean and variance serve to characterize the distribution of height among these women as well as the distribution of many other quantitative traits.

- The **mean,** or average, is the peak of the distribution. The mean of a population is estimated from a sample of individuals from the population, as follows:

$$\bar{x} = \frac{\sum f_i x_i}{N} \qquad (14.1)$$

in which $\bar{x}$ is the estimate of the mean and Σ symbolizes summation over all classes of data (in this example, summation over all 11 height intervals). In Table 14.1, the mean height in the sample of women is 63.1 inches.

- The **variance** is a measure of the spread of the distribution and is estimated in terms of the squared *deviation* (difference) of each observation from the mean. The variance is estimated from a sample of individuals as follows:

$$s^2 = \frac{\sum f_i (x_i - \bar{x})^2}{N - 1} \qquad (14.2)$$

in which s^2 is the estimated variance and x_i, f_i, and N are as in Table 14.1. Note that $(x_i - \bar{x})$ is the difference from the mean of each height category, and that the denominator is the total number of individuals minus 1. The variance describes the extent to which the phenotypes are clustered around the mean, as shown in Figure 14.4. A large value implies that the distribution is spread out, and a small value implies that it is clustered near the mean. From the data in Table 14.1, the variance of the population of British women is estimated as $s^2 = 7.24$ in^2.

A quantity closely related to the variance—the **standard deviation** of the distribution—is defined as the square root of the variance. For the data in Table 14.1, the estimated standard deviation s is obtained from Equation 14.2 as $s = (s^2)^{1/2} = (7.24 \text{ in}^2)^{1/2} = 2.69$ inches. The standard deviation has the useful feature of having the same units of dimension as the mean—in this example, inches.

When the data are symmetrical, or approximately symmetrical, the distribution of a trait can often be approximated by a smooth arching curve of the type shown in Figure 14.4. The arch-shaped curve is called the **normal distribution.** Because the normal curve is symmetrical, half of its area is determined by points with values greater than the mean and half by points with values less than the mean, and thus the proportion of phenotypes that exceed the mean is 1/2. One important characteristic of the normal distribution is that it is completely determined by the value of the mean and the variance.

The mean and standard deviation (square root of the variance) of a normal distribution provide a great deal of information about the distribution of phenotypes in a population, as is illustrated in Figure 14.5. Specifically, for a normal distribution,

1. Approximately 68 percent of the population have a phenotype within *one* standard deviation of the mean (in the symbols of Figure 14.5, between $\mu - \sigma$ and $\mu + \sigma$).
2. Approximately 95 percent lie within *two* standard deviations of the mean (between $\mu - 2\sigma$ and $\mu + 2\sigma$).
3. Approximately 99.7 percent lie within *three* standard deviations of the mean (between $\mu - 3\sigma$ and $\mu + 3\sigma$).

Applying these rules to the data in Figure 14.2, in which the mean and standard deviation are 63.1 and 2.69 inches, approximately 68 percent of the women are expected to have heights in the range from 63.1 − 2.69 inches to 63.1 + 2.69 inches (that

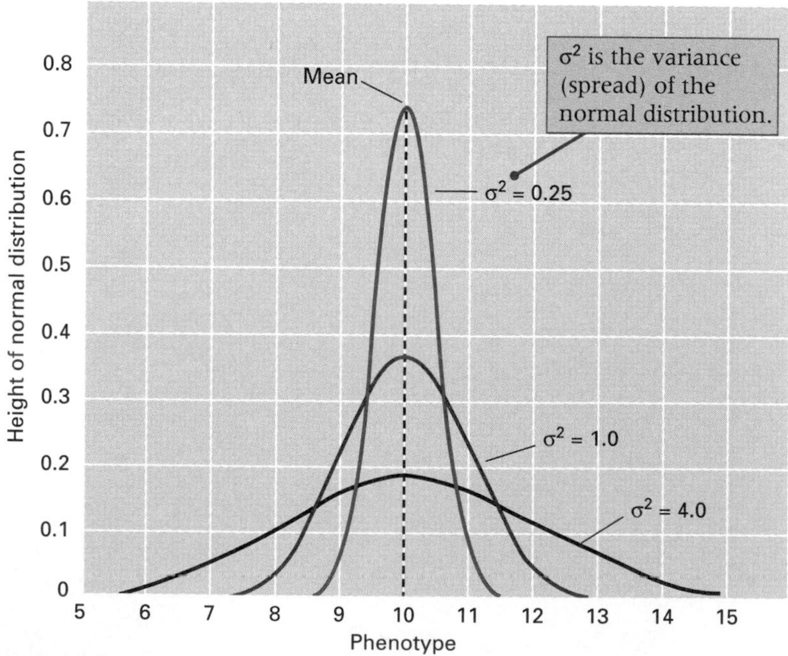

Figure 14.4 Graphs showing that the variance of a distribution measures the spread of the distribution around the mean. The area under each curve covering any range of phenotypes equals the proportion of individuals having phenotypes within the range.

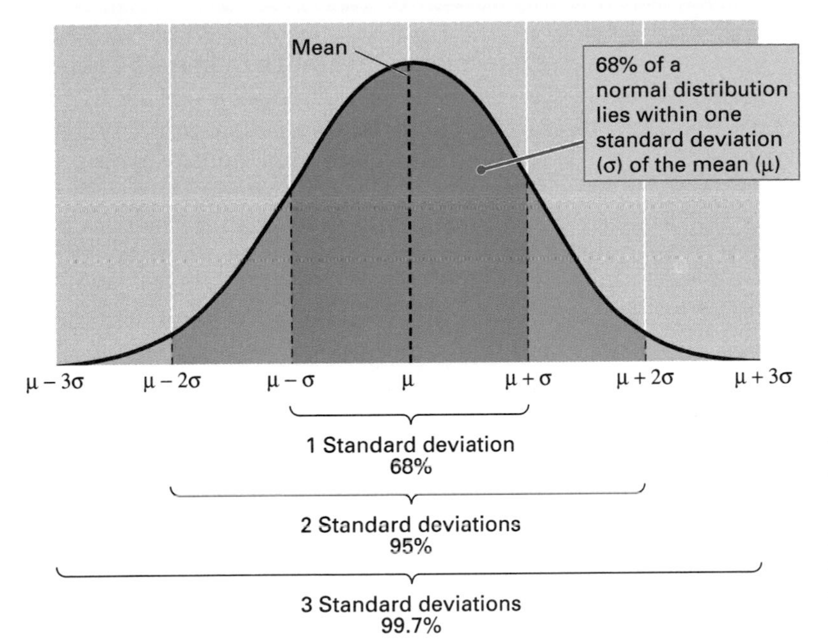

Figure 14.5 Features of a normal distribution. The proportions of individuals lying within one, two, and three standard deviations from the mean are approximately 68 percent, 95 percent, and 99.7 percent, respectively. In this normal distribution, the mean is symbolized μ and the standard deviation σ.

is, 60.4–65.8), and approximately 95 percent are expected to have heights in the range from $63.1 - 2 \times 2.69$ inches to $63.1 + 2 \times 2.69$ inches (that is, 57.7–68.5).

Real data frequently conform to the normal distribution. Normal distributions are usually the rule when the phenotype is determined by the cumulative effect of many individually small independent factors. This is the case for many multifactorial traits.

14.2 Variation in a trait can be separated into genetic and environmental components.

In considering the genetics of multifactorial traits, an important objective is to assess the relative importance of genotype versus environment. In some cases in experimental organisms, it is possible to separate genotype and environment with respect to their effects on the mean. For example, a plant breeder may study the yield of a series of inbred lines grown in a group of environments that differ in planting density or amount of fertilizer. It would then be possible

1. to compare yields of the same genotype grown in different environments and thereby rank the *environments* relative to their effects on yield, or
2. to compare yields of different genotypes grown in the same environment and thereby rank the *genotypes* relative to their effects on yield.

Such a fine discrimination between genetic and environmental effects is not usually possible, particularly in human quantitative genetics. For example, with regard to the height of the women in Figure 14.2, environment could be considered favorable or unfavorable for tall stature only in comparison with the mean height of a genetically identical population reared in a different environment. This reference population does not exist. Likewise, the genetic composition of the population could be judged as favorable or unfavorable for tall stature only in comparison with the mean of a genetically different population reared in an identical environment. This reference population does not exist, either.

Without such standards of comparison, it is impossible to determine the genetic versus environmental effects on the mean. However, it is still possible to assess genetic versus environmental contributions to the *variance,* because instead of comparing the means of two or more populations, we can compare the phenotypes of individuals within the *same* population. Some of the differences in phenotype result from differences in genotype and others from differences in environment, and it is often possible to separate these effects.

In any distribution of phenotypes, such as the one in Figure 14.2, four sources contribute to phenotypic variation:

1. Genotypic variation.
2. Environmental variation.
3. Variation due to genotype-environment interaction.
4. Variation due to genotype-environment association.

Each of these sources of variation is discussed in the following sections.

The genotypic variance results from differences in genotype.

The variation in phenotype caused by differences in genotype among individuals is termed **genotypic variance.** Figure 14.6 illustrates the genetic variation expected among the F_2 generation from a cross of two inbred lines differing in genotype for three unlinked genes. The alleles of the three genes are represented as *A* / *a*, *B* / *b*, and *C* / *c*, and the genetic variation in the F_2 generation caused by segregation and recombination is evident in the differences in color. Relative to a meristic trait (one whose phenotype is determined by counting, such as ears per stalk in corn), if we assume that each uppercase allele is favorable for the expression of the trait and adds one unit to the phenotype, whereas each lowercase allele is without effect, then the *aa bb cc* genotype has a phenotype of 0 and the *AA BB CC* genotype has a phenotype of 6. There are seven possible phenotypes (0 through 6) in the F_2 generation. The distribution of phenotypes in the F_2 generation is shown in the colored bar graph in Figure 14.7. The normal distribution approximating the data has a mean of 3 and a variance of 1.5. In this case, we are assuming that *all* of the variation in phenotype in the population results from differences in genotype among the individuals.

Figure 14.7 also includes a bar graph with diagonal lines representing the theoretical distribution when the trait is determined by 30 unlinked genes segregating in a randomly mating population, grouped into the same number of phenotypic classes as the three-gene case. We assume that 15 of the genes are nearly fixed for the favorable allele and 15 are nearly fixed for the unfavorable allele. The contribution of each favorable allele to the phenotype has been chosen to make the mean of

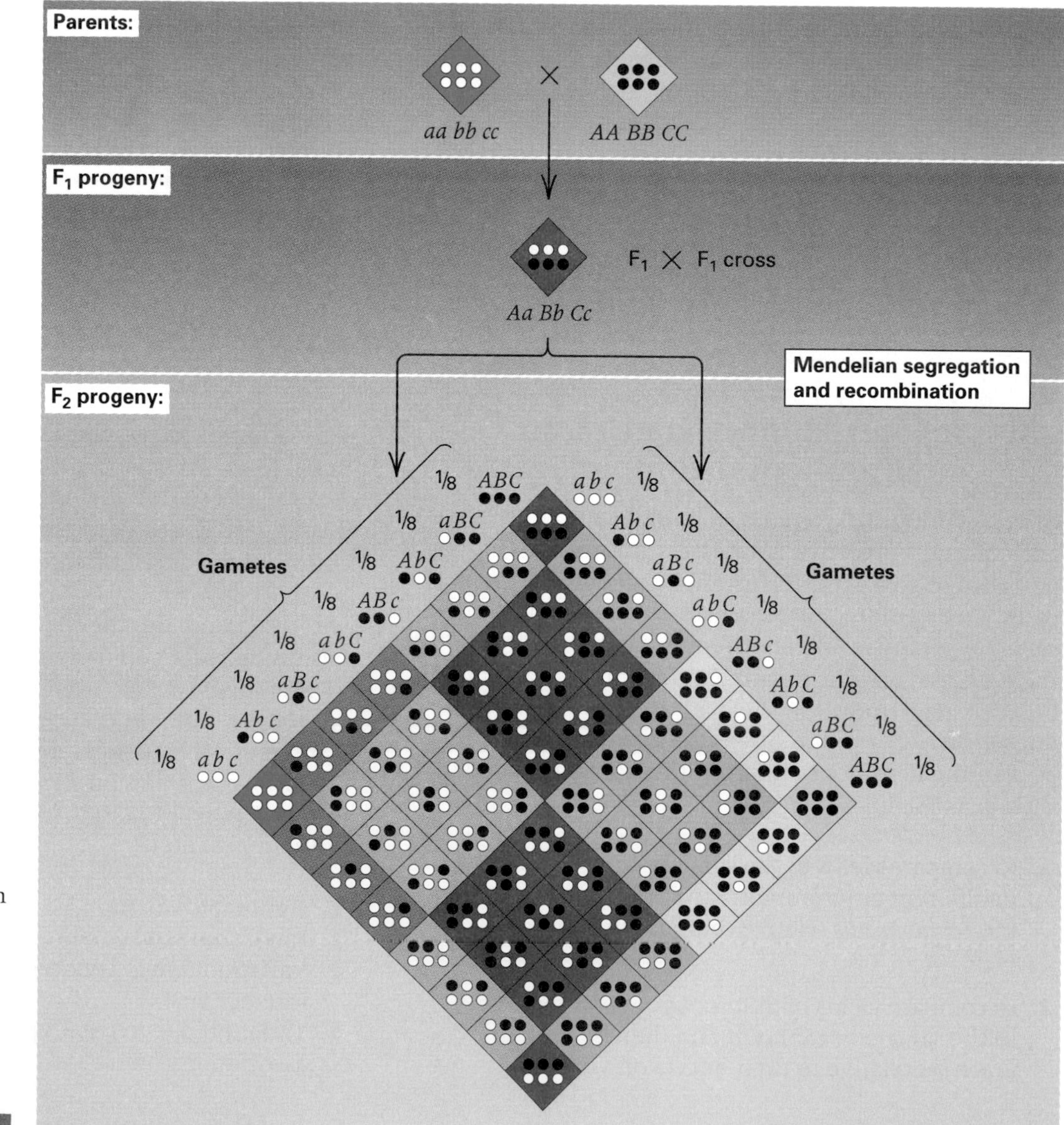

Figure 14.6 Segregation of three independent genes affecting a quantitative trait. Each uppercase allele in a genotype contributes one unit to the phenotype.

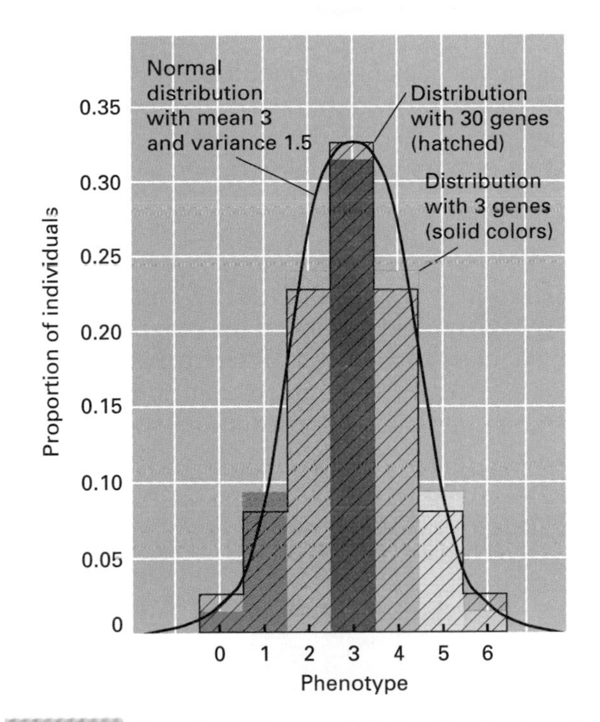

Figure 14.7 The colored bar graph is the distribution of phenotypes determined by the segregation of the three genes illustrated in Figure 14.6. The bar graph with the diagonal lines is the theoretical distribution expected from the segregation of 30 independent genes. Both distributions are approximated by the same normal distribution (black curve).

the distribution equal to 3 and the variance equal to 1.5. Note that the distribution with 30 genes is virtually identical to that with three genes and that both are approximated by the same normal curve. If such distributions were encountered in actual research, the experimenter would not be able to distinguish between them. The key point is that

> Even in the absence of environmental variation, the distribution of phenotypes, by itself, provides no information about the number of genes influencing a trait and no information about the dominance relations of the alleles.

However, the number of genes influencing a quantitative trait is important in determining the potential for long-term genetic improvement of a population by means of artificial selection. For example, in the three-gene case in Figure 14.7, the best possible genotype would have a phenotype of 6; but in the 30-gene case, the best possible genotype (homozygous for the favorable allele of all 30 genes) would have a phenotype of 30.

Later in this chapter, some methods for estimating the number of genes affecting a quantitative trait will be presented. All the methods depend on comparing the phenotypic distributions in the F_1 and F_2 generations of crosses between nearly or completely homozygous lines.

The environmental variance results from differences in environment.

The variation in phenotype caused by differences in environment among individuals is termed **environmental variance.** Figure 14.8 is an example showing the distribution of seed weight in edible beans. The mean of the distribution is 500 mg and the standard deviation is 95 mg. All of the beans in this population are genetically identical and homozygous because they are highly inbred. Therefore, in this population, *all* of the phenotypic variation in seed weight results from environmental variance. A comparison of Figures 14.7 and 14.8 demonstrates the following principle:

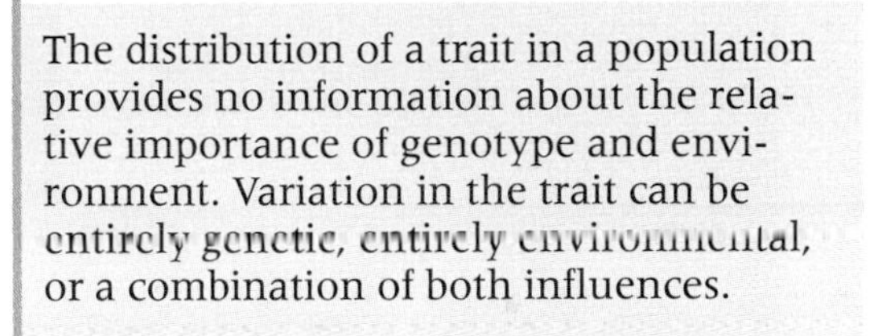
The distribution of a trait in a population provides no information about the relative importance of genotype and environment. Variation in the trait can be entirely genetic, entirely environmental, or a combination of both influences.

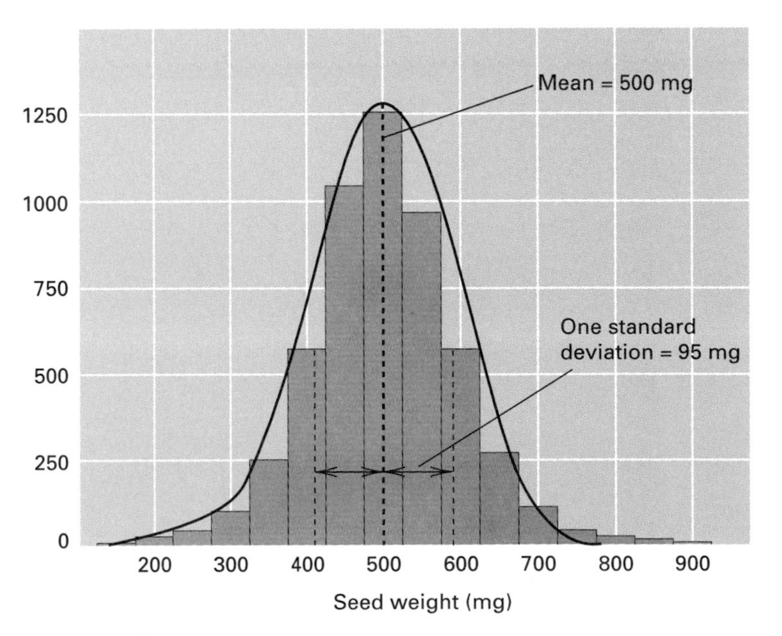

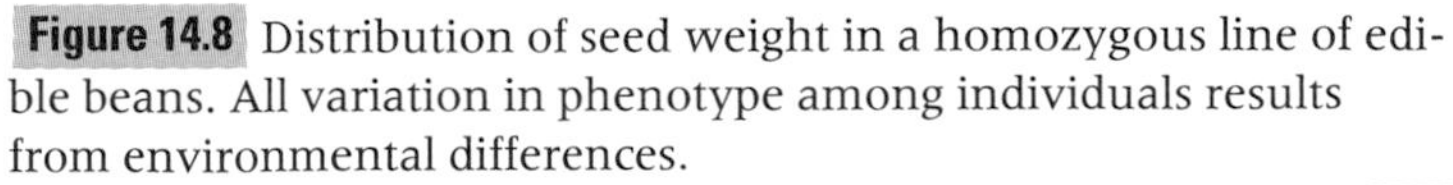
Figure 14.8 Distribution of seed weight in a homozygous line of edible beans. All variation in phenotype among individuals results from environmental differences.

Genotypic and environmental variance are seldom separated as clearly as in Figures 14.7 and 14.8, because usually they work together. Their combined effects are illustrated for a simple hypothetical case in Figure 14.9. At the upper left is the distribution of phenotypes for three genotypes assumed not to be influenced by environment. As depicted, the trait can have one of three distinct and nonoverlapping phenotypes determined by the effects of two additive alleles. The genotypes are in random-mating proportions for an allele frequency of 1/2, and the distribution of phenotypes has mean 5 and variance 2. Because it results solely from differences in genotype, this variance is *genotypic variance,* which is symbolized σ_g^2. The three panels at the upper right depict the distribution of phenotypes of each of the three genotypes in the presence of environmental variation. In each case the variance in phenotype, due to environment alone, equals 1. Because this variance results solely from differences in environment, it is *environmental variance,* which is symbolized σ_e^2. When the effects of genotype and environment are combined in the same population, then both differences in genotype and differences in environment contribute to variation in the trait, and the distribution shown in the

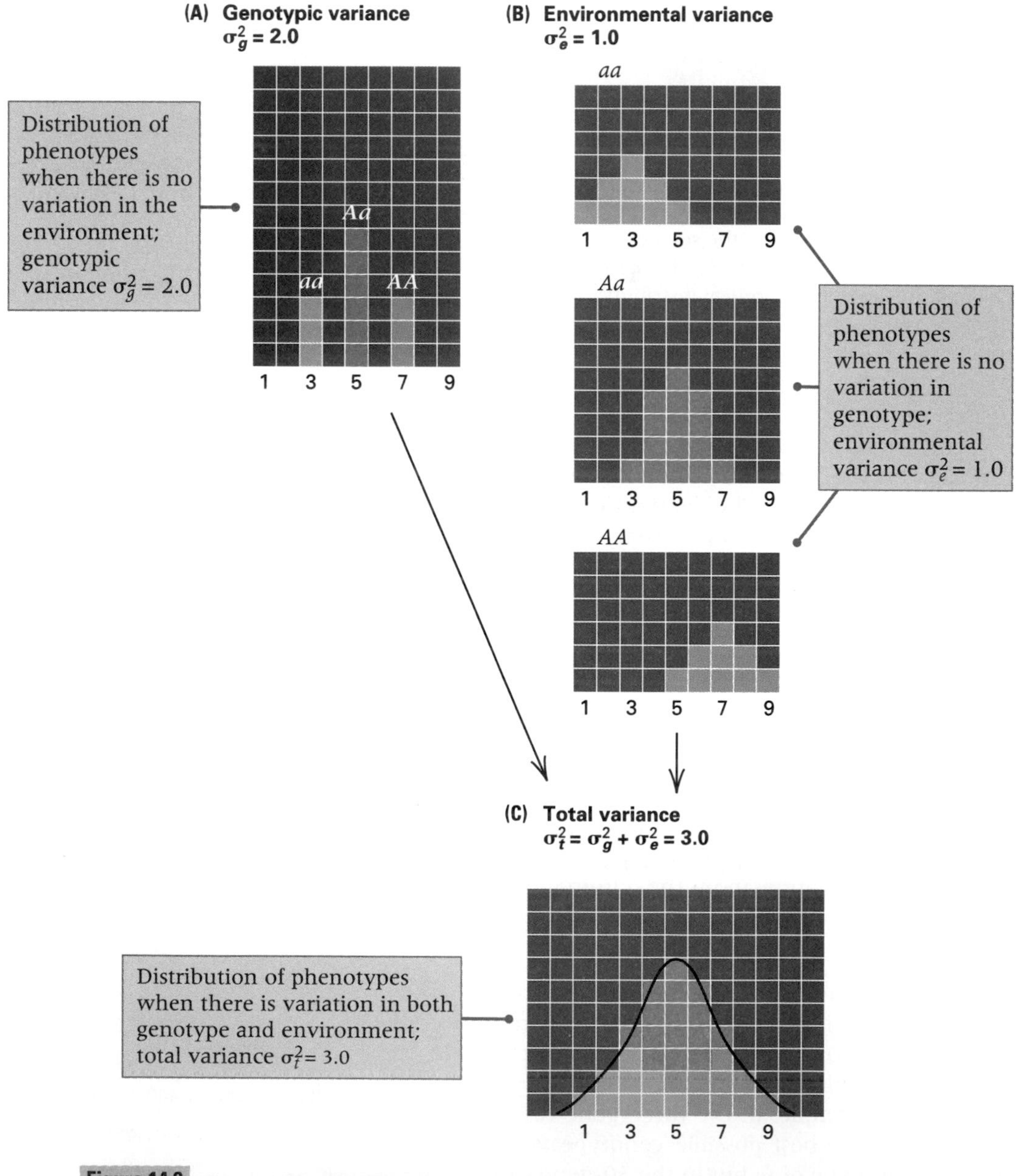

Figure 14.9 The combined effects of genotypic and environmental variance. (A) Population affected only by genotypic variance σ_g^2. (B) Populations of each genotype separately showing the effects of environmental variance σ_e^2. (C) Population affected by both genotypic and environmental variance; the total phenotypic variance σ_t^2 equals the sum of σ_g^2 and σ_e^2.

lower part of the figure results. The variance of this distribution is the **total variance** in phenotype, which is symbolized σ_t^2. Because we are assuming that genotype and environment have separate and independent effects on phenotype, we expect σ_t^2 to be greater than either σ_g^2 or σ_e^2 alone. In fact,

$$\sigma_t^2 = \sigma_g^2 + \sigma_e^2 \qquad (14.3)$$

In words, Equation (14.3) states that

> When genetic and environmental effects contribute independently to phenotype, then the total variance equals the sum of the genotypic and environmental variance.

Equation (14.3) is one of the most important relations in quantitative genetics. How it can be used to analyze data will be explained shortly. Although the equation serves as an excellent approximation in very many cases, it is valid in an exact sense only when genotype and environment are independent in their effects on phenotype. The two most important departures from independence are discussed in the next section.

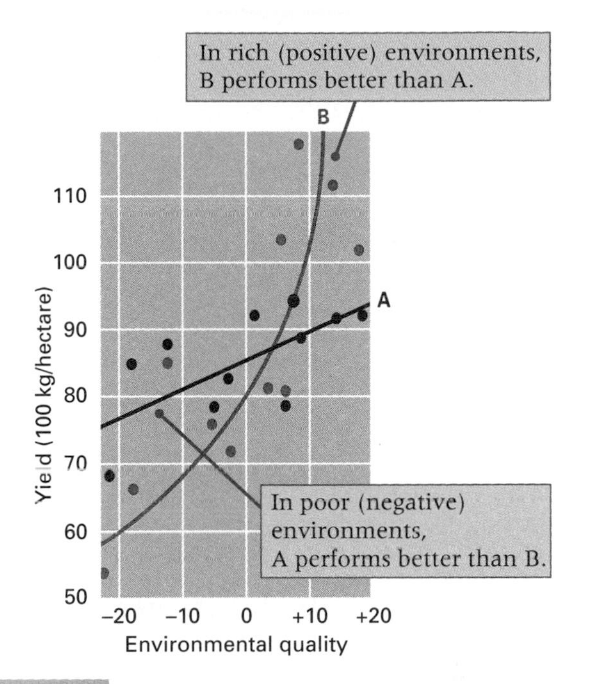

Figure 14.10 Genotype-environment interaction in maize. Strain A is superior when environmental quality is low (negative numbers), but strain B is superior when environmental quality is high. [Data from W. A. Russell. 1974. *Annual Corn & Sorghum Research Conference* 29: 81.]

Genotype and environment can interact, or they can be associated.

In the simplest cases, environmental effects on phenotype are additive, and each environment adds or detracts the same amount from the phenotype, independent of the genotype. When this is not true, then environmental effects on phenotype differ according to genotype, and a **genotype-environment interaction (G-E interaction)** is said to be present. In some cases, G-E interaction can even change the relative ranking of the genotypes, so a genotype that is superior in one environment may become inferior in another.

An example of genotype-environment interaction in maize is illustrated in Figure 14.10. The two strains of corn are hybrids formed by crossing different pairs of inbred lines, and their overall means, averaged across all of the environments, are approximately the same. However, the strain designated A clearly outperforms strain B in the negative stressful environments, whereas the performance is reversed when the environment is of high quality. (Environmental quality is judged on the basis of soil fertility, moisture, and other factors.) In some organisms, particularly plants, experiments like those illustrated in Figure 14.10 can be carried out to determine the contribution of G-E interaction to the total observed variation in phenotype. In other organisms, particularly human beings, the effect cannot be evaluated separately.

Interaction of genotype and environment is common and is very important in both plants and animals. Because of this interaction, no one plant variety outperforms all others in all types of soil and climate, and therefore plant breeders must develop special varieties that are suited to each growing area.

Another important type of interaction is **genotype-by-sex interaction,** which decribes a situation in which the same genotype results in a different phenotype according to the sex of the organism. Genotype-by-sex interactions are very common in quantitative genetics. One example is seen in the living histogram in Figure 14.3. The distribution of height among the women is clearly shifted to the left relative to the distribution of height among men. The averages differ by more than 5 inches (64.8 inches for the women, 70.1 inches for the men). Yet there is no reason for thinking that the genes that affect height are distributed differently in women and men. The genes simply have somewhat different effects depending on sex—an example of genotype-by-sex interaction.

When different genotypes in a population are not distributed at random among all the possible environments, there is **genotype-environment association (G-E association).** In these circumstances, certain genotypes are preferentially associated with certain environments, which may either increase or decrease the phenotype of these genotypes compared with what would result in the absence of G-E association. An example of deliberate genotype-environment association can be found in dairy husbandry, in which some farmers feed each of their cows in proportion to its level of milk production. Because of this practice, cows with superior genotypes with respect to milk production also have a superior environment in that they receive more feed. In plant breeding, genotype-environment association can often be eliminated or minimized by appropriate randomization of genotypes within the experimental plots. In other cases, human genetics again being a prime example, the possibility of G-E association cannot usually be controlled.

There is no genotypic variance in a genetically homogeneous population.

Equation (14.3) can be used to separate the effects of genotype and environment on the total phenotypic variance. Two types of data are required:

1. The phenotypic variance of a genetically uniform population, which provides an estimate of σ_e^2 because a genetically uniform population has a value of $\sigma_g^2 = 0$
2. The phenotypic variance of a genetically heterogeneous population, which provides an estimate of $\sigma_g^2 + \sigma_e^2$

An example of a genetically uniform population is the F_1 generation from a cross between two highly homozygous strains, such as inbred lines (Figure 14.1). An example of a genetically heterogeneous population is the F_2 generation from the same cross, as indicated in Figure 14.6. If the environments of both populations are the same, and if there is no G-E interaction, then the estimates may be combined to deduce the value of σ_g^2.

To take a specific numerical illustration, consider variation in the size of the eyes in the cave-dwelling fish *Astyanax*, reared in the same environments. The variances in eye diameter in the F_1 and F_2 generations from a cross of two highly homozygous strains were estimated as 0.057 and 0.563, respectively. Written in terms of the components of variance, these are

$$F_2\text{: } \sigma_t^2 = \sigma_g^2 + \sigma_e^2 = 0.563$$

$$F_1\text{: } \sigma_e^2 = 0.057$$

The estimate of genotypic variance σ_g^2 is obtained by subtracting the second equation from the first; that is,

$$\sigma_g^2 = 0.563 - 0.057 = 0.506$$

because

$$(\sigma_g^2 + \sigma_e^2) - \sigma_e^2 = \sigma_g^2$$

Hence the estimate of σ_g^2 is 0.506, whereas that of σ_e^2 is 0.057. In this example, the genotypic variance is much greater than the environmental variance, but this is not always the case.

The next section shows what information can be obtained from an estimate of the genotypic variance.

The number of genes affecting a quantitative trait need not be large.

When the number of genes influencing a quantitative trait is not too large, knowledge of the genotypic variance can be used to estimate the number of genes. All that is needed is the means and variances of two phenotypically divergent strains and their F_1, F_2, and backcrosses. In ideal cases, the data appear as in Figure 14.11, in which P_1 and P_2 represent the divergent parental strains (for example, inbred lines). The points lie on a triangle, with increasing variance according to the increasing genetic heterogeneity (genotypic variance) of the populations. If the F_1 and backcross means lie exactly between their parental means, then these means will lie at the midpoints along the sides of the triangle, as shown in Figure 14.11. This finding implies that the alleles affecting the trait are *addi-*

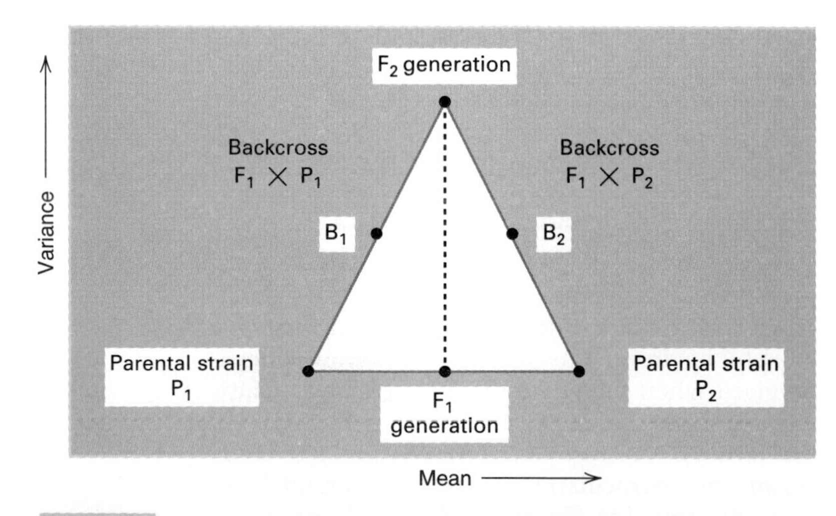

Figure 14.11 Means and variances of parents (P), backcross (B), and hybrid (F) progeny of inbred lines for an ideal quantitative trait affected by unlinked and completely additive genes. [After R. Lande. 1981. *Genetics* 99: 541.]

tive; that is, for each gene, the phenotype of the heterozygote is the average of the phenotypes of the corresponding homozygotes. In such a simple situation, it may be shown that the number, n, of genes contributing to the trait is

$$n = \frac{D^2}{8\sigma_g^2} \qquad (14.4)$$

in which D represents the difference between the means of the original parental strains, P_1 and P_2. This equation may be verified by applying it to the ideal case in Figure 14.9. In this case, the parental strains would be the homozygous genotypes *AA*, with a mean phenotype of 7, and *aa*, with a mean phenotype of 3. Consequently, $D = 7 - 3 = 4$. The genotypic variance is given in Figure 14.7 as $\sigma_g^2 = 2$. Substituting D and σ_g^2 into Equation (14.4), we obtain $n = 16/(8 \times 2) = 1$, which is correct because there is only one gene, with alleles *A* and *a*, affecting the trait.

Applied to actual data, Equation (14.4) requires several assumptions that are not necessarily valid. In addition to the assumption that all generations are reared in the same environment, the theory also makes the genetic assumptions that (1) the alleles of each gene are additive, (2) the genes contribute equally to the trait, (3) the genes are unlinked, and (4) the original parental strains are homozygous for alternative alleles of each gene. However, when the assumptions are invalid, the outcome is that the calculated n is smaller than the actual number of genes affecting the trait. The estimated number is a minimum, because almost any departure from the genetic assumptions leads to a smaller genotypic variance in the F_2 generation and so, for the same value of D, would yield a larger value of n in Equation (14.4). This is why the estimated n is the *minimum* number of genes that can account for the data. For the cave-dwelling *Astyanax* fish discussed in the preceding section, the parental strains had average phenotypes of 7.05 and 2.10, giving $D = 4.95$. The estimated value of $\sigma_g^2 = 0.506$, so the minimum number of genes affecting eye diameter is $n = (4.95)^2/(8 \times 0.506) = 6.0$. Therefore, at least six different genes affect the diameter of the eye of the fish.

The number of genes that affect a quantitative trait is important because it influences the amount by which a population can be genetically improved by selective breeding. With traits determined by a small number of genes, the potential for change in a trait is relatively small, and a population consisting of the best possible genotypes may have a mean value that is only two or three standard deviations above the mean of the original population. However, traits determined by a large number of genes have a large potential for improvement. For example, after a population of the flour beetle *Tribolium* was bred for increased pupa weight, the mean value for pupa weight was found to be 17 standard deviations above the mean of the original population. Determination of traits by a large number of genes implies that

> Selective breeding can create an improved population in which the value of *every* individual greatly exceeds that of the *best* individuals that existed in the original population.

This principle at first seems paradoxical because, in a large enough population, every possible genotype should be created at some low frequency. The explanation of the paradox is that real populations subjected to selective breeding typically consist of a few hundred organisms (at most), and thus many of the theoretically possible genotypes are never formed because their frequencies are much too rare. As selection takes place, and the allele frequencies change, these genotypes become more common and allow the selection of superior organisms in future generations.

The broad-sense heritability includes all genetic effects combined.

Estimates of the number of genes that determine quantitative traits are frequently unavailable because the necessary experiments are impractical or have not been carried out. Another attribute of quantitative traits, which requires less data to evaluate, makes use of the ratio of the genotypic variance to the total phenotypic variance. This ratio of σ_g^2 to σ_t^2 is called the **broad-sense heritability,** symbolized as H^2, and it measures the importance of genetic variation, relative to environmental variation, in causing variation in the phenotype of a trait of interest. Broad-sense heritability is a ratio of variances, specifically

$$H^2 = \frac{\sigma_g^2}{\sigma_t^2} = \frac{\sigma_g^2}{\sigma_g^2 + \sigma_e^2} \qquad (14.5)$$

Substitution of the data for eye diameter in *Astyanax*, in which $\sigma_g^2 = 0.506$ and $\sigma_g^2 + \sigma_e^2 = 0.563$, into Equation (14.5) yields $H^2 = 0.506/0.563 = 0.90$ for the estimate of broad-sense heritability. This value implies that 90 percent of the variation in eye diameter in this population results from differences in genotype among the fish.

Knowledge of heritability is useful in the context of plant and animal breeding because heritability can be used to predict the magnitude and speed of population improvement. The broad-sense heritability defined in Equation (14.5) is used in predicting the outcome of selection practiced among clones, inbred lines, or varieties. Analogous predictions for randomly bred populations utilize another type of heritability (the narrow-sense heritability), which will be discussed in the next section. Broad-sense heritability measures how much of the total variance in phenotype results from differences in genotype. For this reason, H^2 is often of interest in regard to human quantitative traits.

Twin studies are often used to assess genetic effects on variation in a trait.

In human beings, twins would seem to be ideal subjects for separating genotypic and environmental variance because **identical twins,** which arise from the splitting of a single fertilized egg, are genetically identical and are often strikingly similar in such traits as facial features and body build. **Fraternal twins,** which arise from two fertilized eggs, have the same genetic relationship as ordinary siblings, so only half of the genes in either twin are identical with those in the other. Theoretically, the variance between members of an identical-twin pair would be equivalent to σ_e^2, because the twins are genetically identical. The variance between members of a fraternal-twin pair would include not only σ_e^2, but also part of the genotypic variance (approximately $\sigma_g^2/2$, because of the identity of half of the genes in fraternal twins). Consequently, both σ_g^2 and σ_e^2 could be estimated from twin data and combined as in Equation (14.5) to estimate H^2. Table 14.2 summarizes estimates of H^2 based on twin studies of several traits.

Unfortunately, twin studies are subject to several important sources of error, most of which increase the similarity of identical twins, so the numbers in Table 14.2 should be considered very approximate and probably too high. Here are four of the potential sources of error:

1. Genotype-environment interaction, which increases the variance in fraternal twins but not in identical twins
2. Frequent sharing of embryonic membranes between identical twins, resulting in a more similar intrauterine environment
3. Greater similarity in the treatment of identical twins by parents, teachers, and peers, resulting in a decreased environmental variance in identical twins
4. Different sexes in half of the pairs of fraternal twins, in contrast with the same sex of identical twins

These pitfalls and others imply that data from human twin studies should be interpreted with caution and reservation.

14.3 Artificial selection is a form of "managed evolution."

The practice of breeders to choose a select group of organisms from a population to become the parents of the next generation is termed **artificial selection.** When artificial selection is carried out either by choosing the best organisms in a species that reproduces asexually or by choosing the best subpopulation among a series of subpopulations, each propagated by close inbreeding (such as self-fertilization), then broad-sense heritability permits an assessment of how rapidly progress can be achieved. Broad-sense heritability is important in this context, because with clones, inbred lines, or varieties, superior genotypes can be perpetuated without disruption of favorable gene combinations by Mendelian segregation. An example is the selection of superior varieties of plants that are propagated asexually by means of cuttings or grafts, or of animals that are reproduced by cloning. Because there is no sexual reproduction, each offspring has exactly the same genotype as its parent.

In sexually reproducing populations that are genetically heterogeneous, broad-sense heritability is not relevant in predicting progress resulting from artificial selection, because superior genotypes must necessarily be broken up by the processes of segregation and recombination. For example, if the best genotype is heterozygous for each of two unlinked loci, A/a; B/b, then because of segregation and independent assortment, among the progeny of a cross between parents with the best genotypes—A/a; $B/b \times A/a$; B/b—only 1/4 will have the

Table 14.2
Broad-sense heritability, in percent, based on twin studies

Trait	Heritability (H^2)	Trait	Heritability (H^2)
Longevity	29	Verbal ability	63
Height	85	Numerical ability	76
Weight	63	Memory	47
Amino acid excretion	72	Sociability index	66
Serum lipid levels	44	Masculinity index	12
Maximum blood lactate	34	Temperament index	58
Maximum heart rate	84		

same favorable A/a; B/b genotype as the parents. The rest of the progeny will be genetically inferior to the parents. For this reason, to the extent that high genetic merit may depend on particular combinations of alleles, each generation of artificial selection results in a slight setback in that the offspring of superior parents are generally not quite so good as the parents themselves. Progress under selection can still be predicted, but the prediction must make use of another type of heritability, the narrow-sense heritability, which is discussed in the next section.

The narrow-sense heritability is usually the most important in artificial selection.

Figure 14.12 illustrates a typical form of artificial selection and its result. The organism is *Nicotiana longiflora* (tobacco), and the trait is the length of the corolla tube (the corolla is a collective term for all the petals of a flower). Part A shows the distribution of phenotypes in the parental generation, and part B shows the distribution of phenotypes in the offspring generation. The parental generation is the population from which the parents were chosen for breeding. The type of selection is called **individual selection,** because each member of the population to be selected is evaluated according to its own individual phenotype. The selection is practiced by choosing some arbitrary level of phenotype—called the **truncation point**—that determines which individuals will be saved for breeding purposes. All individuals with a phenotype above the threshold are randomly mated among themselves to produce the next generation.

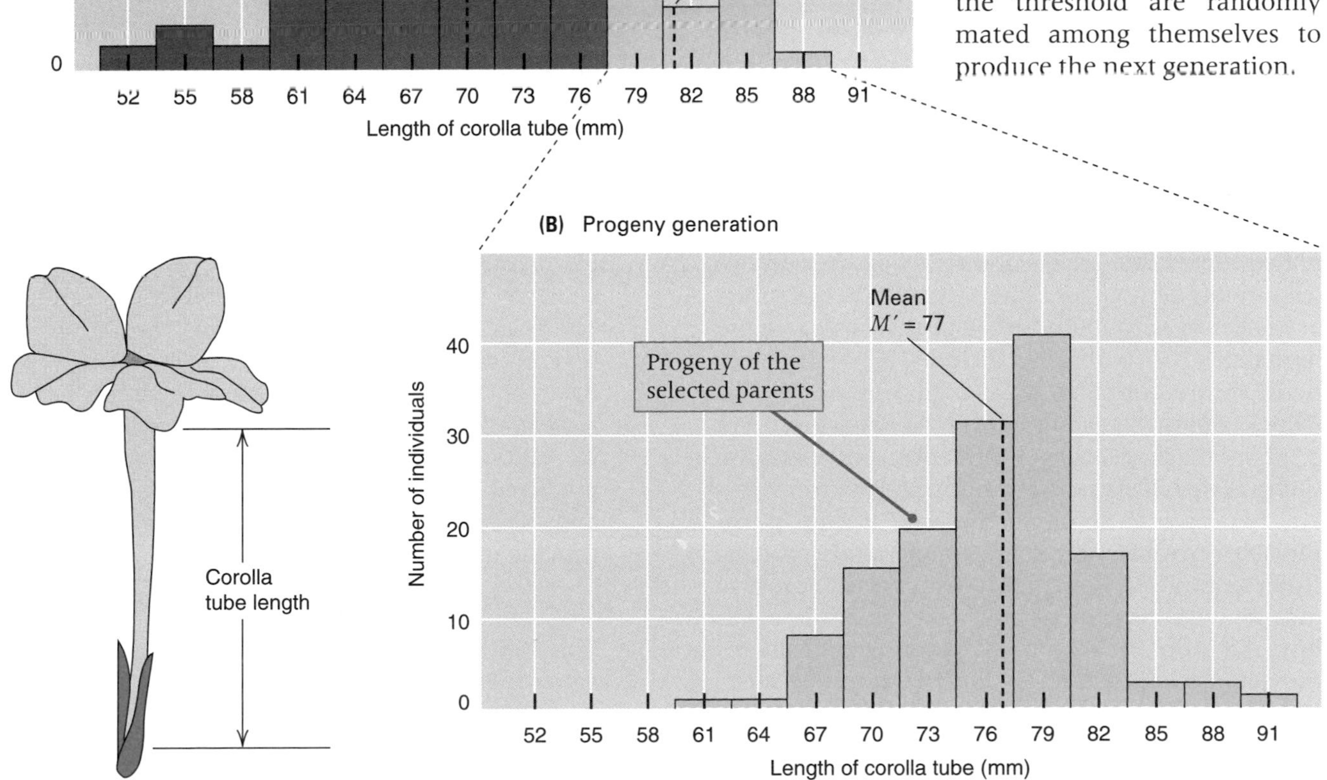

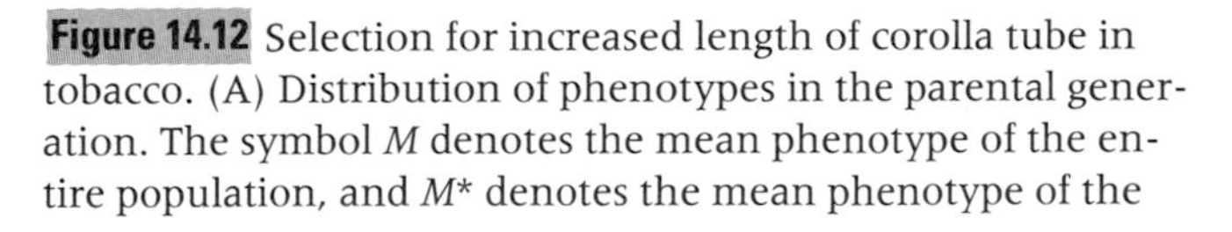

Figure 14.12 Selection for increased length of corolla tube in tobacco. (A) Distribution of phenotypes in the parental generation. The symbol M denotes the mean phenotype of the entire population, and M^* denotes the mean phenotype of the organisms chosen for breeding (organisms with a phenotype that exceeds the truncation point). (B) Distribution of phenotypes among the offspring bred from the selected parents. The symbol M' denotes the mean.

In evaluating progress through individual selection, three distinct phenotypic means are important. In Figure 14.12, these means are symbolized as M, M^*, and M', and they are defined as follows:

1. M is the mean phenotype of the entire population in the parental generation, including both the selected and the nonselected individuals.
2. M^* is the mean phenotype among those individuals selected as parents (those with a phenotype above the truncation point).
3. M' is the mean phenotype among the progeny of selected parents.

The relationship among these three means is given by

$$M' = M + h^2(M^* - M) \qquad (14.6)$$

in which the symbol h^2 is the **narrow-sense heritability** of the trait in question.

Later in this chapter, a method for estimating narrow-sense heritability from the similarity in phenotype among relatives will be explained. In Figure 14.12, h^2 is the only unknown quantity, so it can be estimated from the data themselves. Rearranging Equation (14.6) and substituting the values for the means from Figure 14.12, we get

$$h^2 = \frac{M' - M}{M^* - M} = \frac{77 - 70}{81 - 70} = 0.64$$

Analogous to the way in which total phenotypic variance can be split into the sum of the genotypic variance and the environmental variance (Equation 14.3), the genotypic variance can be split into parts resulting from the additive effects of alleles, dominance effects, and effects of interaction between alleles of different genes. The difference between the broad-sense heritability, H^2, and the narrow-sense heritability, h^2, is that H^2 includes all of these genetic contributions to variation, whereas h^2 includes only the additive effects of alleles. In ordinary language, the additive effects of alleles are the genetic effects that are transmissible from parent to offspring; dominance effects are not transmissible because of segregation, and epistatic (interaction) effects are not transmissible because of independent assortment and recombination. It follows that from the standpoint of animal or plant improvement, h^2 is the heritability of interest because

The narrow-sense heritability, h^2, is the proportion of the variance in phenotype that is transmissible from parents to offspring and that can be used to predict changes in the population mean with individual selection, according to Equation (14.6).

The distinction between the broad-sense heritability and the narrow-sense heritability can be appreciated intuitively by considering a population in which there is a rare recessive gene. In such a case, most homozygous-recessive genotypes come from matings between heterozygous carriers. Some such kindreds have more than one affected offspring. Hence affected siblings can resemble each other more than they resemble their parents. For example, if a is a recessive allele, the mating $Aa \times Aa$, in which both parents show the dominant phenotype, may yield two offspring that are aa, which both show the recessive phenotype; in this case, the offspring are more similar to each other than they are to either of the parents. It is the dominance of the wildtype allele that causes this paradox; dominance and epistasis contribute to the broad-sense heritability of the trait but not to the narrow-sense heritability. The narrow-sense heritability includes only those genetic effects that contribute to the resemblance between parents and their offspring, because narrow-sense heritability measures how similar offspring are to their parents.

In general, the narrow-sense heritability of a trait is smaller than the broad-sense heritability. For example, in the parental generation in Figure 14.12, the broad-sense heritability of corolla tube length is $H^2 = 0.82$. The two types of heritability are equal only when the alleles affecting the trait are additive in their effects; additive effects means that each heterozygous genotype shows a phenotype that is exactly intermediate between the phenotypes of the respective homozygous genotypes and that the effects are also additive across loci.

Equation (14.6) is of fundamental importance in quantitative genetics because of its predictive value. This can be seen in the following example. The selection in Figure 14.12 was carried out for several generations. After two generations, the mean of the population was 83, and parents having a mean of 90 were selected. By use of the estimate $h^2 = 0.64$, the mean in the next generation can be predicted. The information provided is that $M = 83$ and $M^* = 90$. Therefore, Equation (14.6) implies that the predicted mean is

$$M' = 83 + (0.64)(90 - 83) = 87.5$$

This value is in good agreement with the observed value of 87.9.

There are limits to the improvement that can be achieved by artificial selection.

Artificial selection is analogous to natural selection in that both types of selection cause an increase in the frequency of alleles that improve the selected trait (or traits). The principles of natural selection discussed in Chapter 13 also apply to artificial selection. For example, artificial selection is most effective in changing the frequency of alleles that are in an intermediate range of frequency ($0.2 < p < 0.8$). Selection is less effective for alleles with frequencies outside this range, and it is least effective for rare recessive alleles. For quantitative traits, including fitness, the total selection is shared among all the genes that influence the trait, and the selection coefficient affecting each allele is determined by (1) the magnitude of the effect of the allele, (2) the frequency of the allele, (3) the total number of genes affecting the trait, (4) the narrow-sense heritability of the trait, and (5) the proportion of the population that is selected for breeding.

The value of heritability is determined both by the magnitude of effects and by the frequency of alleles. If all favorable alleles were fixed ($p = 1$) or lost ($p = 0$), the heritability of the trait would be 0. Therefore, the heritability of a quantitative trait is expected to decrease over many generations of artificial selection as a result of favorable alleles becoming nearly fixed. For example, ten generations of selection for less fat in a population of Duroc pigs decreased the heritability of fatness from 73 to 30 percent because of changes in allele frequency resulting from the selection.

Population improvement by means of artificial selection cannot continue indefinitely. A population may respond to selection until its mean is many standard deviations different from the mean of the original population, but eventually the population reaches a **selection limit** in which successive generations show no further improvement. (An exception to this generalization is found in traits that are affected by a very large number of genes and in which selection is carried out in a very large population; in such a case, selective advance can be continued indefinitely because new mutations continue to add genetic variation.) In more typical cases, progress under selection may cease because all alleles that affect the trait are either fixed or lost, and so the narrow-sense heritability of the trait becomes 0. Fixation or loss of all relevant alleles is rare. The usual reason for a selection limit is that natural selection counteracts artificial selection. Many genes that respond to artificial selection as a result of their favorable effect on a selected trait also have indirect harmful effects on fitness. For example, selection for increased size of eggs in poultry results in a decrease in the number of eggs, and selection for extreme body size (large or small) in most animals results in a decrease in fertility. When one trait (for example, number of eggs) changes in the course of selection for a different trait (for example, size of eggs), the unselected trait is said to have undergone a **correlated response** to selection. Correlated response of fitness is typical in long-term artificial selection. Each increment of progress in the selected trait is partially offset by a decrease in fitness because of correlated response. Eventually, artificial selection for the trait of interest is exactly balanced by natural selection against the trait, so a selection limit is reached and no further progress is possible without changing the strategy of selection.

Inbreeding is generally harmful, and hybrids may be the best.

Inbreeding can have harmful effects on economically important traits such as yield of grain or egg production. This decline in performance is called **inbreeding depression,** and it results principally from rare harmful recessive alleles becoming homozygous because of inbreeding (Chapter 13). The degree of inbreeding is measured by the inbreeding coefficient F discussed in Chapter 13. Figure 14.13 is an example of inbreeding depression in yield of corn, in which the yield decreases linearly as the inbreeding coefficient increases.

Most highly inbred strains suffer from many genetic defects, as might be expected from the uncovering of deleterious recessive alleles. One would also expect that if two different inbred strains were crossed, the F_1 would show improved features, because a harmful recessive allele inherited from one parent would be likely to be covered up by a normal dominant allele from the other parent. In many organisms, the F_1 generation of a cross between inbred lines is superior to the parental lines, and the phenomenon is called **heterosis,** or **hybrid vigor.** This phenomenon, which is widely used in the production of corn and other agricultural products, yields genetically identical hybrid plants with traits that are sometimes more favorable than those of

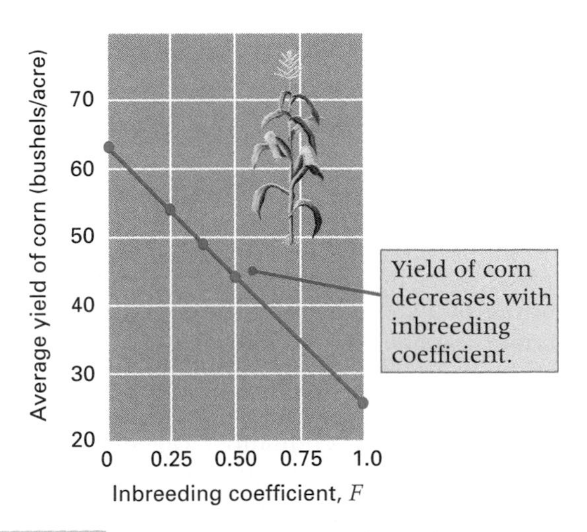

Figure 14.13 Inbreeding depression for yield in corn. [Data from N. Neal. 1935. *J. Amer. Soc. Agronomy* 27: 666.]

the ancestral plants from which the inbreds were derived. The features that most commonly distinguish hybrid plants from their inbred parents are their rapid growth, larger size, and greater yield. Furthermore, the F_1 plants have a fairly uniform phenotype; because the progeny of inbred parents all have the same genotype (Figure 14.1), the genetic variance $\sigma_g^2 = 0$, and so all of the variance in phenotype is due to variation in the environment. Genetically heterogeneous crops with high yields or certain other desirable features can also be produced by traditional plant-breeding programs, but growers often prefer hybrid plants because of this relative uniformity in phenotype. For example, uniform height and time of maturity facilitate machine harvesting, and plants that all bear fruit at the same time accommodate picking and shipping schedules.

Hybrid varieties of corn are used almost exclusively in the United States for commercial crops. A farmer cannot plant the seeds from his crop because the F_2 generation consists of a variety of genotypes, most of which do not show hybrid vigor. The production of hybrid seeds is a major industry in corn-growing sections of the United States.

14.4 Genetic variation is revealed by correlations between relatives.

Quantitative genetics relies extensively on similarity among relatives to assess the importance of genetic factors. Particularly in the study of complex behavioral traits in human beings, genetic interpretation of familial resemblance is not always straightforward because of the possibility of nongenetic, but nevertheless familial, sources of resemblance. However, in plant and animal breeding, the situation is usually less complex because genotypes and environments are under experimental control.

Covariance is the tendency for traits to vary together.

Genetic data about families are frequently pairs of numbers: pairs of parents, pairs of twins, or pairs consisting of a single parent and a single offspring. An important issue in quantitative genetics is the degree to which the numbers in each pair are associated. The usual way to measure the association is to calculate a statistical quantity called the *correlation coefficient* between the members of each pair.

The correlation coefficient among relatives is based on the covariance in phenotype among them. Much as the variance describes the tendency of a set of measurements to vary [Equation (14.2)], the covariance describes the tendency of pairs of numbers to vary together (co-vary). Calculation of the covariance is similar to that of the variance in Equation (14.2) except that the squared deviation term $(x_i - \bar{x})^2$ is replaced with the product of the deviations of the pairs of measurements from their respective means—that is, $(x_i - \bar{x})(y_i - \bar{y})$.

For example, $(x_i - \bar{x})$ could be the deviation of a particular father's height from the overall father mean, and $(y_i - \bar{y})$ could be the deviation of his son's height from the overall son mean. In symbols, let f_i be the number of pairs of relatives with phenotypic measurements x_i and y_i. Then the estimated **covariance** (***Cov***) of the trait among the relatives is

$$Cov(x,y) = \frac{\sum f_i (x_i - \bar{x})(y_i - \bar{y})}{N - 1} \qquad (14.7)$$

where N is the total number of pairs of relatives studied.

The **correlation coefficient (*r*)** of the trait between the relatives is calculated from the covariance as follows:

$$r = \frac{Cov(x,y)}{s_x s_y} \qquad (14.8)$$

in which s_x and s_y are the standard deviations of the measurements, estimated from Equation (14.2). The correlation coefficient can range from -1.0 to $+1.0$. A value of $+1.0$ means perfect association. When $r = 0$, x and y are not associated.

The additive genetic variance is transmissible; the dominance variance is not.

Covariance and correlation are important in quantitative genetics because the correlation coefficient of a trait between individuals with various degrees of genetic relationship is related fairly simply to the narrow-sense or broad-sense heritability, as shown in Table 14.3. The table gives the theoretical values of the correlation coefficient for various pairs of relatives; h^2 represents the narrow-sense heritability and H^2 the broad-sense heritability. Considering parent-offspring, half-sibling, or first-cousin pairs, narrow-sense heritability can be estimated directly by multiplication. Specifically, h^2 can be estimated as twice the parent-offspring correlation, four times the half-sibling correlation, or eight times the first-cousin correlation.

With full siblings, identical twins, and double first cousins, the correlation coefficient is related to the broad-sense heritability, H^2, because phenotypic resemblance depends not only on additive effects but also on dominance. In these relatives, dominance contributes to resemblance because the relatives can share *both* of their alleles as a result of their common ancestry, whereas parents and offspring, half siblings, and first cousins can share at most a single allele of any gene because of common ancestry. Therefore, to the extent that phenotype depends on dominance effects, full siblings can resemble each other more than they resemble their parents.

The most common disorders in human families are multifactorial.

The most common human diseases are caused by genetic and environmental factors acting together. Each of the factors adds to the risk or **liability** that a person has for manifesting the trait. For example, genetic risk factors for heart disease are revealed in the family history of the disease, and environmental risk factors include being overweight and such behaviors as cigarette smoking. The overall risk of a person for manifesting a disease is a multifactiorial trait determined by numerous genetic and environmental factors and the interactions among these factors. Such a trait is a *threshold trait;* although the underlying risk itself is not directly observable, the trait will be either present or absent according to whether the risk is above or below a critical (threshold) value.

As with other multifactorial traits, the risk of manifesting a threshold trait has a broad-sense and a narrow-sense heritability that may differ among populations according to the allele frequencies and the distribution of environmental risk factors. The heritabilities cannot be estimated directly, because the risk is not observable directly, but the heritabilities can be inferred from the incidences of the trait among individuals and their relatives. The statistical techniques for doing this are quite specialized, but some of the theoretical results are shown in Figure 14.14, along with observed data for the most common congenital abnormalities in Caucasians. The x-axis gives the incidence of the trait in the general population, and the y-axis gives the risk of the trait in brothers or sisters of affected persons. The two horizontal lines near the top yield the proportions for simple Mendelian dominance and simple Mendelian recsssive inheritance, which are 50 percent (dominance) and 25 percent (recessiveness), respectively. Note that these proportions do not depend on the incidence of the trait in the population. The other curves pertain to threshold traits with the narrow-sense heritabilities of liability as noted. Now the proportion of affected siblings *does* depend on the incidence of the trait in the general population, as well as on the heritability of liability. Consider a trait with a population incidence of 0.05 percent (1 in 2000). If the heritability of liability is 0.40, the proportion of affected siblings among affected persons is a little less than 0.5 percent (1 in 200), or about a tenfold increase over that of the general population. If the heritability is 1.00, the proportion of affected siblings among affected persons is a little less than 5 percent (1 in 20), or about a hundredfold increase over that of the general population. Note in Figure 14.14 that the most common traits tend to be threshold traits; with a few exceptions, the proportion of affected siblings tends to be moderate or low, corresponding to the heritabilities indicated.

ARTIFICIAL SELECTION over many generations has created impressive breeds of horses. These racehorses have been bred for their speed. Other breeds have been bred for stamina, strength, and agility. [© Jim Corwin/Photo Researchers, Inc.]

Table 14.3
Theoretical correlation coefficient in phenotype between relatives

Degree of relationship	Correlation coefficient*
Offspring and one parent	$h^2/2$
Offspring and average of parents	$h^2/2$
Half siblings	$h^2/4$
First cousins	$h^2/8$
Monozygotic twins	H^2
Full siblings	$\sim H^2/2$

*Contributions from interactions among alleles of different genes have been ignored. For this and other reasons, H^2 correlations are approximate.

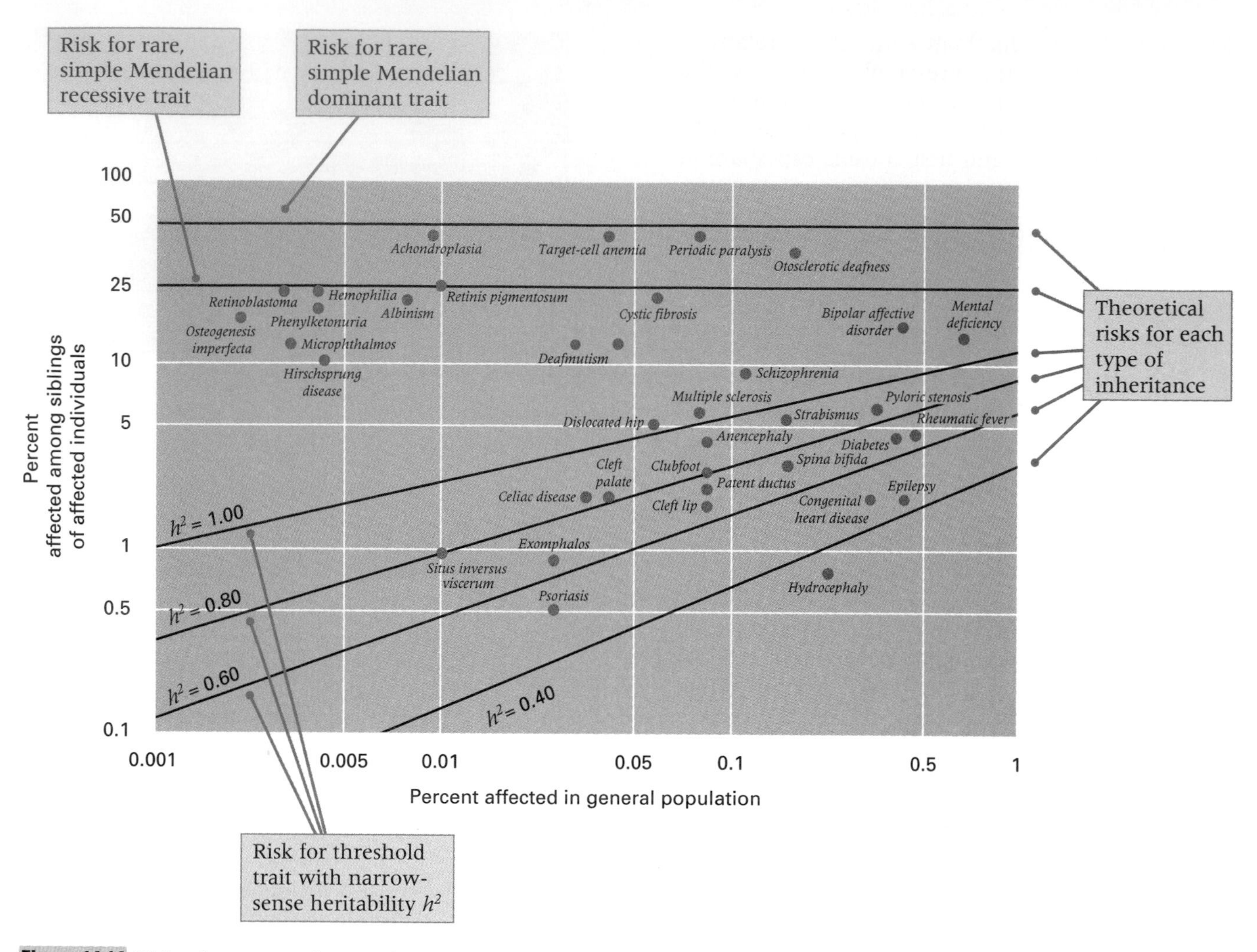

Figure 14.14 Risks of common abnormalities in the siblings of affected persons. Diagonal lines are the theoretical risks for threshold traits with the indicated values of the narrow-sense heritability of liability. Horizontal lines are the theoretical risks for simple dominant or recessive traits.

14.5 Pedigree studies of genetic polymorphisms are used to map loci for quantitative traits.

Genes affecting quantitative traits cannot usually be identified in pedigrees because their individual effects are obscured by the segregation of other genes and by environmental variation. Even so, genes affecting quantitative traits can be localized if they are genetically linked with genetic markers, such as simple tandem repeat polymorphisms (STRPs), discussed in Chapter 4, because the effects of the genotype affecting the quantitative trait are correlated with the genotype of the linked STRP. A gene that affects a quantitative trait is a **quantitative-trait locus (QTL).** Locating QTLs in the genome is important to the manipulation of genes in breeding programs and to the cloning and study of genes in order to identify their functions.

STRPs result from variation in number, n, of repeating units of a simple-sequence repeat, such as $(TG)_n$. The number of repeats can be determined by PCR amplification of the STRP using oligonucleotide primers to unique-sequence DNA flanking the repeats. The size of the amplified DNA fragment is a function of the number of repeats in the STRP. Such STRP markers are abundant, are distributed throughout the genome, and often have multiple codominant alleles—qualities that make them ideally suited for linkage studies of quantitative traits. In STRP studies, as many widely scattered STRPs as possible are monitored, along with the quantitative trait, in successive generations of a genetically heterogeneous population. Statistical studies are then carried out to identify which STRP alleles are the best predictors of phenotype of the quantitative trait because their presence in a genotype is consistently accompanied by superior performance with respect to the quantitative trait. These STRPs identify regions of the genome that contain one or more

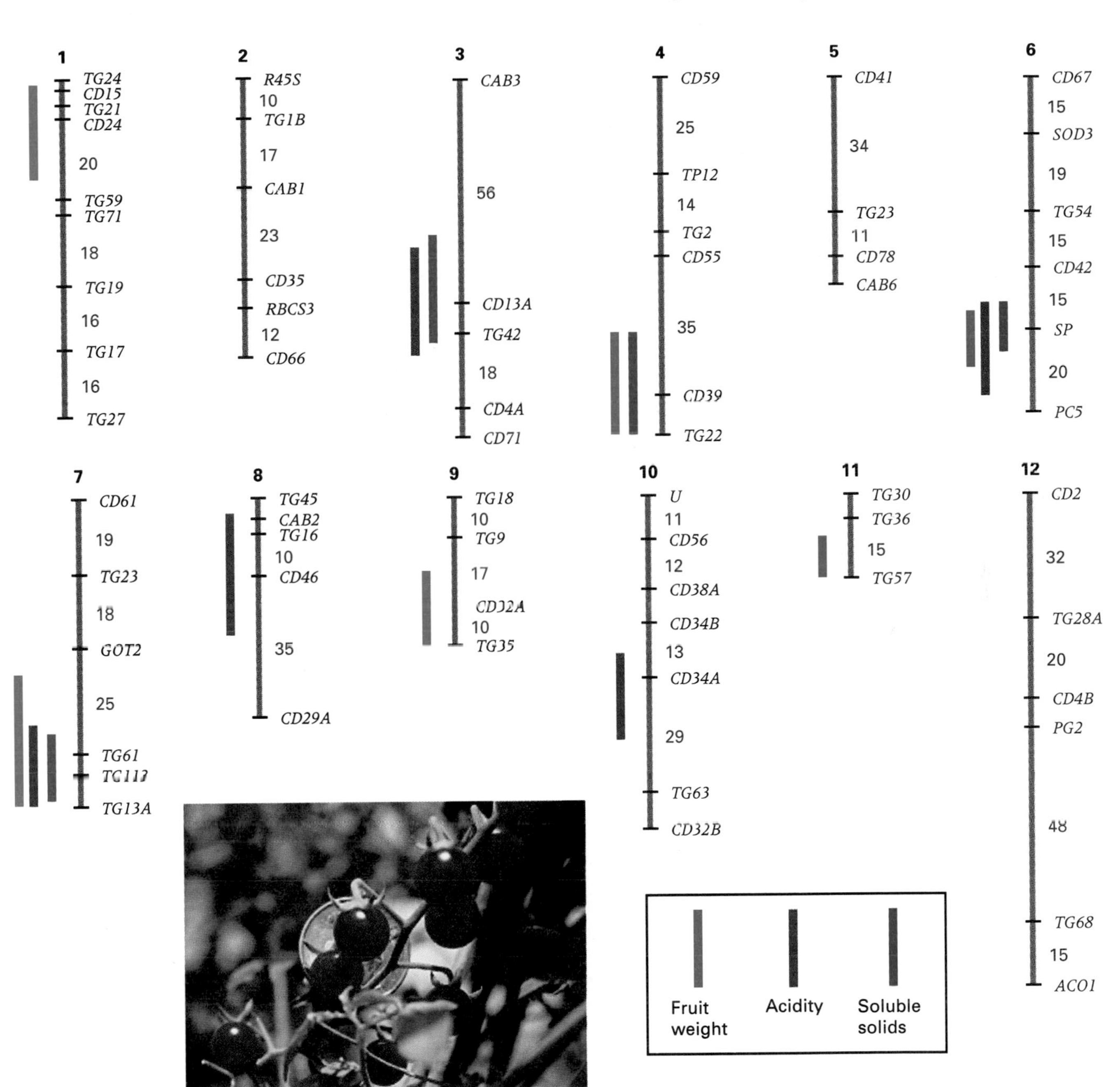

Figure 14.15 Location of QTLs for several quantitative traits in the tomato genome. The genetic markers are shown for each of the 12 chromosomes. The numbers in red are distances in map units between adjacent markers, but only map distances larger than 10 are indicated. The regions in which the QTLs are located are indicated by the bars: green bars, QTLs for fruit weight; blue bars, QTLs for content of soluble solids; and pink bars, QTLs for acidity (pH). The data are from crosses between the domestic tomato (*Lycopersicon esculentum*) and a wild South American relative with small, green fruit (*Lycopersicon chmielewskii*). The photograph is of fruits of wild tomato. Note the small size in comparison with the coin (a U.S. quarter). The F_1 generation was backcrossed with the domestic tomato, and fruits from the progeny were assayed for STRP markers and each of the quantitative traits. [Data from A. H. Paterson, E. S. Lander, J. D. Hewitt, S. Peterson, S. E. Lincoln, and S. D. Tanksley. 1988. *Nature* 335: 721. Photograph courtesy of Steven D. Tanksley.]

genes with important effects on the quantitative trait, and the STRPs can be used to trace the segregation of the important regions in breeding programs and even as starting points for cloning genes with particularly large effects.

An example of genetic mapping of quantitative trait loci for several quantitative traits in tomatoes is illustrated in Figure 14.15. More than 300 highly polymorphic genetic markers have been mapped in the tomato genome, with an average spacing between markers of 5 map units. The chromosome maps in Figure 14.15 show a subset of 67 markers that were segregating in crosses between the domestic tomato and a wild South American relative.

The human connection

Win, Place, or Show?

Hugh E. Montgomery and 18 other investigators 1998
University College, London,
and 6 other research institutions
Human Gene for Physical Performance

A professional basketball executive once said of Chicago Bulls superstar Michael Jordan, "Gene for gene, DNA for DNA, nobody can replace him." The implication that Jordan's phenomenal success may not be due exclusively to his work ethic, practice, commitment, consistency, and character is interesting. But nobody doubts that Jordan's success in basketball is associated in part with his being tall. And height is a multifactorial trait partly under genetic control. About 50 percent of the variation in height among people is due to genetic differences. But might invisible, metabolic differences also influence physical performance? Some people find this possibility unsettling, because it potentially raises all sorts of issues about the value of effort and practice, the fairness of competition, genetic testing for entry into elite physical training programs, and so forth. The extent to which genotype influences physical performance is not known, but some genetic differences may be important. In this paper, a polymorphism in the angiotensin-converting enzyme ACE is reported to have a major effect on physical endurance. It is not difficult to see how ACE could play a role. The enzyme degrades a class of vasodilator proteins and also converts angiotensin I into a vasoconstrictor. Note, however, that the authors are cautious in their conclusions and are careful not to speculate that the gene for ACE is a QTL (quantitative trait locus) affecting physical performance. It is still too early to reach firm conclusions.

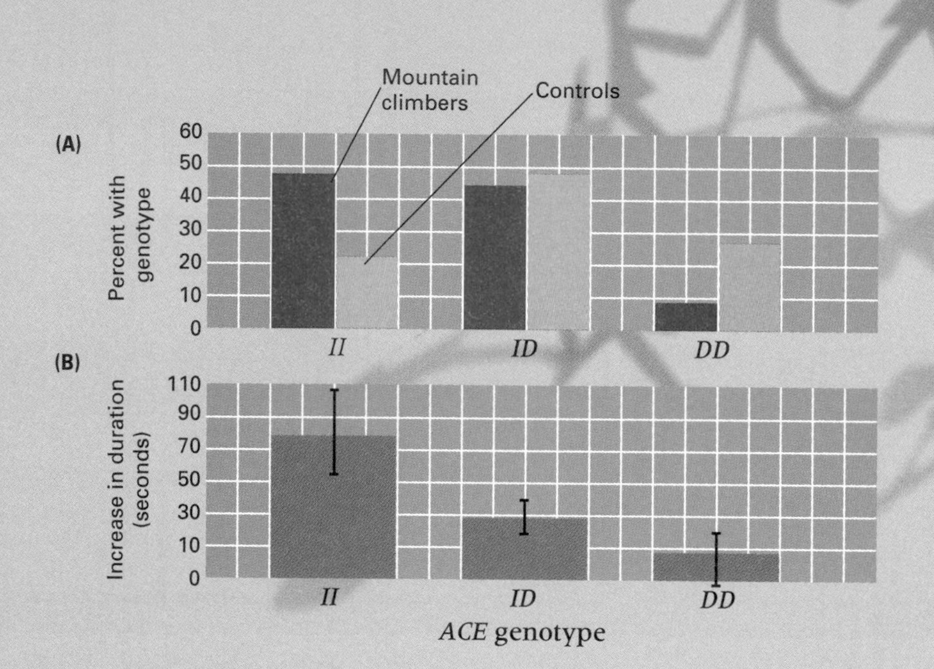

A specific genetic factor that strongly influences human physical performance has not so far been reported, but here we show that a polymorphism in the gene encoding angiotensin-converting enzyme does just that. An "insertion" allele of the gene is associated with elite endurance performance among high-altitude mountaineers. Also, after physical training, repetitive weight-lifting is improved elevenfold in individuals homozygous for the "insertion" allele compared with those homozygous for the "deletion" allele. The endocrine renin-angiotensin system is important in controlling the circulation. . . . A polymorphism in the human *ACE* gene for angiotensin-converting enzyme has been described in which the deletion (*D*) rather than insertion (*I*) allele is associated with higher activity by tissue ACE. . . . Our initial studies suggested that the *I* allele was associated with improved endurance performance. . . . High-altitude mountaineers perform extreme-endurance exercise. [The *ACE* genotypes of 25] elite unrelated male British mountaineers . . . were compared with those of 1,906 British males. . . . Both the genotype distribution and allele frequency differed significantly between climbers and controls (Figurc A).

In a second study, *ACE* genotype was determined in 123 Caucasian males recruited into the United Kingdom army consecutively. . . . The maximum duration (in seconds) for which they could perform repetitive elbow flexion while holding a 15-kg barbell was assessed both before and after a 10-week physical training period. . . . Duration of exercise improved significantly for those of *II* and *ID* genotypes, but not for those of *DD* genotype (Figure B). . . . Increased performance is likely to be due to an improvement in the endurance characteristics of the tested muscles. . . . Further work will be needed to determine whether this correlation holds beyond the limited group studied here.

WE SHOW THAT a polymorphism in the gene encoding angiotensin-converting enzyme [influences human physical performance].

Source: Nature 393: 221.

The average spacing between the markers is 20 map units. Backcross progeny of the cross $F_1 \times$ domestic tomato were tested for the genetic markers, and the fruits of the backcross progeny were assayed for three quantitative traits—fruit weight, content of soluble solids, and acidity. Statistical analysis of the data was carried out in order to detect marker alleles that were associated with phenotypic differences in any of the traits; a significant association indicates genetic linkage between the marker gene and one or more QTLs affecting the trait. A total of six QTLs affecting fruit weight were detected (green bars), as well as four QTLs affecting soluble solids (blue bars) and five QTLs affecting acidity (pink bars). Although additional QTLs of smaller effects undoubtedly remained undetected in these types of experiments, the effects of the mapped QTLs are substantial: The mapped QTLs account for 58 percent of the total phenotypic variance in fruit weight, 44 percent of the phenotypic variance in soluble solids, and 48 percent of the phenotypic variance in acidity. The genetic markers linked to the QTLs with substantial effects make it possible to trace the transmission of the QTLs in pedigrees and manipulate them in breeding programs by following the transmission of the linked marker genes. Figure 14.15 also indicates a number of chromosomal regions containing QTLs for two or more of the traits—for example, the QTL regions on chromosomes 6 and 7, which affect all three traits. Because the locations of the QTLs can be specified only to within 20–30 map units, it is unclear whether the coincidences result from pleiotropy, in which a single gene affects several traits simultaneously or from the independent effects of multiple, tightly linked genes. However, the locations of QTLs for different traits coincide frequently enough that, in most cases, pleiotropy is likely to be the explanation.

Some QTLs have been identified by examining candidate genes.

Besides genetic mapping, another approach to the identification of QTLs is through the use of candidate genes. A **candidate gene** for a trait is a gene for which there is some *a priori* basis for suspecting that the gene affects the trait. In human behavior genetics, for example, if a pharmacological agent is known to affect a personality trait, and the molecular target of the drug is known, then the gene that codes for the target molecule and any gene whose product interacts with the drug or with the target molecule, are candidate genes for affecting the personality trait.

One example of the use of candidate genes in the study of human behavior genetics is in the identification of a naturally occurring genetic polymorphism that affects neuroticism and, in particular, the anxiety component of neuroticism. The neurotransmitter substance serotonin (5-hydroxytryptamine) is known to influence a variety of psychiatric conditions, such as anxiety and depression. Among the important components in serotonin action is the serotonin transporter protein. Neurons that release serotonin to stimulate other neurons also take it up again through the serotonin transporter. This uptake terminates the stimulation and also recycles the molecule for later use.

The serotonin transporter became a strong candidate gene for anxiety-related personality traits when it was discovered that the transporter is the target of a class of antidepressants known as selective serotonin reuptake inhibitors. The widely prescribed antidepressant, Prozac, is an example of such a drug.

Motivated by the strong suggestion that the serotonin transporter might be involved in anxiety-related traits, researchers looked for evidence of genetic polymorphisms affecting the transporter gene in human populations. Such a polymorphism was found in the promoter region of the transporter. About 1 kb upstream from the transcription start site is a series of 16 tandem repeats of a nearly identical DNA sequence of about 15 base pairs. This is the *l* (*long*) allele, which has an allele frequency of 57 percent among Caucasians. There is also an *s* (*short*) allele, in which three of the repeated sequences are not present. The *s* allele has an allele frequency of 43 percent. The genotypes *l*/*l*, *l*/*s*, and *s*/*s* are found in the Hardy–Weinberg proportions of 32 percent, 49 percent, and 19 percent, respectively (Chapter 13). Further studies revealed that the polymorphism does have a physiological effect. In cells grown in culture, *l*/*l* cells had approximately 50 percent more mRNA for the transporter than *l*/*s* or *s*/*s* cells, and *l*/*l* cells had approximately 35 percent more membrane-bound transporter protein than *l*/*s* or *s*/*s* cells.

On the basis of these results, the researchers predicted that the serotonin transporter polymorphism would be found to be associated with anxiety-related personality traits. The predicted result was observed in a study of 505 people who were genotyped for the transporter polymorphism and whose personality traits were classified by a self-report questionnaire. Significant associations were found between the transporter genotype and the overall neuroticism score, and the highest correlations were with the anxiety-related traits "tension" and "harm avoidance." A comparison between the genotypes with respect to neuroticism score is shown in Figure 14.16. Because the *s* allele is dominant to *l* with respect to gene expression and

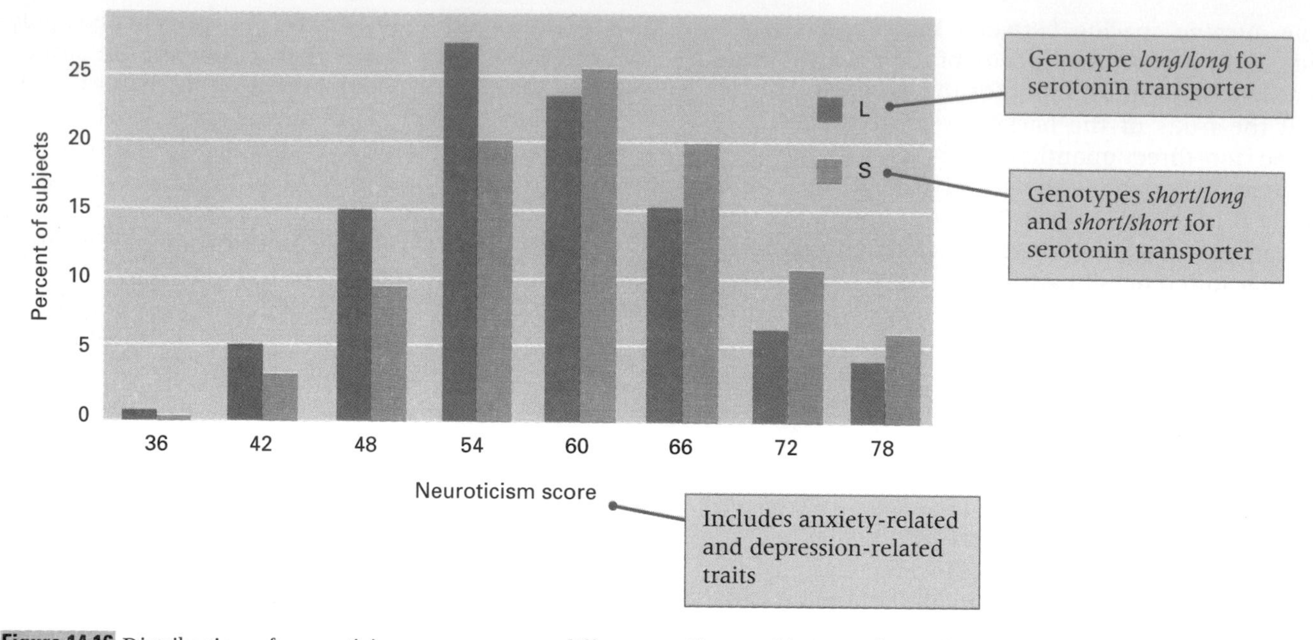

Figure 14.16 Distribution of neuroticism scores among different genotypes for the serotonin transporter polymorphism. The genotype L includes only *l/l*, whereas S includes *s/s* and *s/l*. The total is 163 subjects for the L genotype and 342 for the S genotypes. [From K.-P. Lesch, et al. 1996. *Science* 274: 1527.]

personality score, the L category includes only the *l/l* genotype, whereas the S category includes both *s/s* and *l/s*.

Note in Figure 14.16 that there is a great deal of overlap in the L and S distributions. The overlap means that the transporter polymorphism accounts for a relatively small proportion of the total variance in neuroticism score in the whole population. In quantitative terms, the transporter polymorphism accounts for 3–4 percent of the total variance in neuroticism score and for 7–9 percent of the genotypic variance in neuroticism score. If all genes affecting neuroticism had the same magnitude of effect as the serotonin transporter polymorphism, approximately 10 to 15 genes would be implicated in the 45 percent of the variance that can be attributed to genetic factors.

Chapter Summary

- **Multifactorial traits are determined by multiple genes and the environment.**

  Continuous, meristic, and threshold traits are usually multifactorial.

  The distribution of a trait in a population implies nothing about its inheritance.

Many traits that are important in agriculture and human genetics are determined by the effects of multiple genes and by the environment. Such traits are multifactorial (also called quantitative) traits, and their analysis is known as quantitative genetics. There are three types of multifactorial traits: continuous, meristic, and threshold. Continuous traits are expressed according to a continuous scale of measurement, such as height. Meristic traits are traits that are expressed in whole numbers, such as the number of grains on an ear of corn. Threshold traits have an underlying risk and are either expressed or not expressed in each individual; an example is diabetes. The genes affecting quantitative traits are no different from those affecting simple Mendelian traits, and the genes can have multiple alleles, partial dominance, and so forth. When several genes affect a trait, the pattern of genetic transmission need not fit a simple Mendelian pattern because the effects of one gene can be obscured by other genes or the environment. However, the number of genes can be estimated, and many of them can be mapped using linkage to molecular polymorphisms.

- **Variation in a trait can be separated into genetic and environmental components.**

  The genotypic variance results from differences in genotype.

  The environmental variance results from differences in environment.

  Genotype and environment can interact, or they can be associated.

  There is no genotypic variance in a genetically homogeneous population.

  The number of genes affecting a quantitative trait need not be large.

  The broad-sense heritability includes all genetic effects combined.

  Twin studies are often used to assess genetic effects on variation in a trait.

Many quantitative and meristic traits have a distribution that approximates the bell-shaped curve of a normal distribution. A normal distribution can be completely described by two quantities: the mean and the variance. The standard deviation of a distribution is the square root of the variance. In a normal distribution, approximately 68 percent of the individuals have a phenotype within one standard deviation from the mean, and approximately 95 percent of the individuals have a phenotype within two standard deviations from the mean.

- **Artificial selection is a form of "managed evolution."**

  The narrow-sense heritability is usually the most important in artificial selection.

  There are limits to the improvement that can be achieved by artificial selection.

  Inbreeding is generally harmful, and hybrids may be the best.

- **Genetic variation is revealed by correlations between relatives.**

  Covariance is the tendency for traits to vary together.

  The additive genetic variance is transmissible; the dominance variance is not.

  The most common disorders in human families are multifactorial.

Variation in phenotype of multifactorial traits among individuals in a population derives from four principal sources: (1) variation in genotype, which is measured by the genotypic variance; (2) variation in environment, which is measured by the environmental variance; (3) variation resulting from the interaction between genotype and environment (G-E interaction); and (4) variation resulting from nonrandom association of genotypes and environments (G-E association). The ratio of genotypic variance to the total phenotypic variance of a trait is called the broad-sense heritability; this quantity is useful in predicting the outcome of artificial selection among clones, inbred lines, or varieties. When artificial selection is carried out in a randomly mating population, the narrow-sense heritability is used for prediction. The value of the narrow-sense heritability can be determined from the correlation in phenotype among groups of relatives.

One common type of artificial selection is called truncation selection, in which only those individuals that have a phenotype above a certain value (the truncation point) are saved for breeding the next generation. Artificial selection usually results in improvement of the selected population. The general principle is that the deviation of the progeny mean from the mean of the previous generation equals the narrow-sense heritability times the deviation of the parental mean from the mean of their generation: $M' = M + h^2(M^* - M)$. When selection is carried out for many generations, progress often slows or ceases because (1) some of the favorable genes become nearly fixed in the population, which decreases the narrow-sense heritability, and (2) natural selection may counteract the artificial selection.

- **Pedigree studies of genetic polymorphisms are used to map loci for quantitative traits.**

  Some QTLs have been identified by examining candidate genes.

A gene affecting a quantitative trait is a quantitative-trait locus, or QTL. The locations of QTLs in the genome can be determined by genetic mapping with respect to simple-sequence length polymorphisms or other types of genetic markers. The mapping of QTLs aids in the manipulation of genes in breeding programs and in the identification of the genes. An alternative approach to identifying QTLs is the study of candidate genes known to function in the biochemical pathway that affects the trait.

Key Terms

artificial selection
broad-sense heritability
candidate gene
complex trait
continuous trait
correlated response
correlation coefficient
covariance
distribution
environmental variance
fraternal twins
genetic architecture
genotype-by-sex interaction
genotype-environment association
genotype-environment interaction
genotypic variance
heterosis
hybrid vigor
identical twins
inbreeding depression
individual selection
liability
mean
meristic trait
multifactorial trait
narrow-sense heritability
normal distribution
quantitative trait
quantitative-trait locus (QTL)
selection limit
standard deviation
threshold trait
total variance
truncation point
variance

Concepts and Issues
Testing Your Knowledge

- Give an example of a trait in human beings that is affected both by environmental factors and by genetic factors. Specify some of the environmental factors that affect the trait.
- Does the distribution of phenotypes of a trait in a population tell you anything about the relative importance of genes and environment in causing differences in phenotype among individuals? Does it tell you anything about the number of genes that may affect the trait? Explain.
- In regard to the genotypic variance, what is special about an inbred line or about the F_1 progeny of a cross between two inbred lines?
- The distribution of bristle number on one of the abdominal segments in a population of *Drosophila* ranges from 12 to 26. The narrow-sense heritability of the trait is approximately 50 percent. Do you think that it would be possible, by practicing artificial selection over a number of generations, to obtain a population in which the *mean*

bristle number was greater than 26? A population in which the *minimum* bristle number was greater than 26? Explain why or why not.

- Why are correlations between relatives of interest to the quantitative geneticist?
- In the context of multifactorial inheritance, what is a QTL? Does a QTL differ from any other kind of gene? How can QTLs be detected?
- In the context of multifactorial inheritance, what is a candidate gene? Would you regard the human *PAH* gene for phenylalanine hydroxylase as a natural candidate gene for mental retardation? Why or why not?

Solutions

Step by Step

Problem 1

The weight in pounds at the time of weaning in a representative sample of lambs taken from a large flock is shown in the accompanying table. Estimate the mean, the variance, and the standard deviation of weaning weight in this population.

37	37	38	46	39
30	31	35	30	42
43	39	48	27	41
43	41	37	29	26

Solution The mean and variance of weaning weight are estimated from the sum of the individual values and the sum of the deviations from the mean, as expressed in the formulas. Note that the variance has units of pounds squared. The original units of pounds are restored by taking the square root, which is the standard deviation $s = \sqrt{(40.16 \text{ lb})^2} = 6.24$ lb.

$$\bar{x} = \frac{\Sigma x_i}{N}$$

$$= \frac{739}{20} = 36.95 \text{ pounds}$$

$$s^2 = \frac{\Sigma(x_i - \bar{x})^2}{N - 1}$$

$$= \frac{762.95}{19} = 40.16 \text{ pounds}$$

An equivalent formula for the variance, which is much easier to use in calculations, is shown below. The symbol Σx_i^2 means the sum of the squares of the individual values, and $(\Sigma x_i)^2$ means the square of the sum of the individual values.

$$s^2 = \frac{\Sigma x_i^2}{N-1} - \frac{(\Sigma x_i)^2}{N(N-1)}$$

$$= 1477.32 - 1437.16$$

$$= 40.16 \text{ pounds}$$

Problem 2

Artificial selection for fruit weight is carried out in a tomato population. From each plant, all the fruits are weighed, and the average weight per tomato is regarded as the phenotype of the plant. In the population as a whole, the average fruit weight per plant is 75 grams. Plants whose fruit weight averages 100 grams per tomato are selected and mated at random to produce the next generation. The narrow-sense heritability of average fruit weight in this population is estimated to be 20 percent.

(a) What is the expected average fruit weight among plants after one generation of selection?

(b) What is the expected average fruit weight among plants after five consecutive generations of selection when, in each generation, the chosen parents have an average fruit weight 25 grams above the population average in that generation? Assume that the narrow-sense heritability remains constant during this time.

Solution In the first generation of selection, the mean of the population is $M = 75$ grams, that of the selected parents is $M^* = 100$ grams, and the narrow-sense heritability h^2 is given as 0.20. **(a)** The formula below gives the prediction equation for this case, and the predicted average fruit weight is 80 grams per tomato. **(b)** It is clear from the formula that if $M^* - M = 25$ in each generation, and h^2 remains at 0.20, then the expected gain per generation is 5 grams. After five generations the expected gain is 25 grams, yielding an expected average fruit weight after five generations of 100 grams per tomato. In the supermarket, this represents a decrease from about 6 tomatoes per pound to about 4 to 5 tomatoes per pound.

$$M' = M + h^2(M^* - M)$$

$$= 75 + (0.20)(100 - 75) = 75 + 5$$

$$= 80 \text{ grams}$$

Problem 3

A restriction fragment polymorphism (RFLP) in *Drosophila* is linked to a QTL (quantitative trait locus) affecting the number of abdominal bristles. A probe for the RFLP hybridizes with a 6.5-kb restriction fragment, with a 4.0-kb restriction fragment, or with both (in heterozygotes). The QTL genotypes may be designated *QQ*, *Qq*, and *qq*. The average abdominal bristle numbers in these genotypes are 20, 18, and 16, respectively. The typical standard deviation in bristle number is about 2, so the effect of the QTL is equal to the phenotypic standard deviation. This would usually be regarded as a rather large effect. A mating is carried out as shown, and each of the progeny is examined for bristle number and then subjected to electrophoresis and DNA hybridization to classify the RFLP genotype.

$$\frac{6.5\ Q}{4.0\ q} ♀♀ \times \frac{6.5\ Q}{4.0\ q} ♂♂$$

As expected, the progeny phenotypes with respect to the RFLP fall into three classes: those with the 4.0 band only (homozygous 4.0), those with both bands (4.0/6.5 heterozygotes), and those with the 6.5 band only (homozygous 6.5). In the boxes under the lanes in the gel, write the average bristle number for each of the three RFLP phenotypes, assuming that the frequency of recombination between the

RFLP locus and the QTL is 10 percent. Remember that there is no recombination in *Drosophila* males.

Solution Solving this kind of problem absolutely requires a Punnett square so that all the progeny genotypes and phenotypes can be displayed with their respective frequencies. The appropriate format for this problem is shown in the accompanying illustration, where the frequencies of the female gametes result from linkage with a recombination frequency of 0.1, and those in males result from random segregation with no recombination. The average bristle number in each class of progeny is shown in bold type. From this tabulation it is clear that among flies with the 4.0 band only, the ratio of *Qq* : *qq* genotypes is 1 : 9, so the average bristle number in this class is $(1/10) \times 18 + (9/10) \times 16 = 16.2$. Among flies with the 4.0 and the 6.5 band, the ratio of *QQ* : *Qq* : *qq* genotypes is 1 : 18 : 1, so the average bristle number in this class is $(1/20) \times 20 + (18/20) \times 18 + (1/20) \times 16 = 18.0$. Finally, among flies with the 6.5

Problem 3

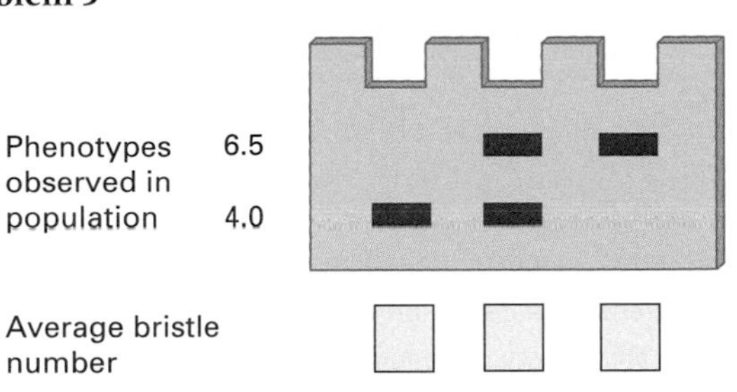

band only, the ratio of *QQ* : *Qq* genotypes is 9 : 1, so the average bristle number in this class is $(9/10) \times 20 + (1/10) \times 18 = 19.8$. Note that the difference in mean phenotype between the RFLP genotypes is less than 1 phenotypic standard deviation, which means that even a difference this large would be very difficult to detect and that doing so would require large sample sizes.

Problem 3 Solution

Female Gamete	Frequency	0.5 (6.5 *Q*)			0.5 (4.0 *q*)		
							Male Gamete Frequency
6.5 *Q*	(1 – 0.1)/2 = 0.45	0.225	6.5 *Q* / 6.5 *Q*	**20**	0.225	6.5 *Q* / 4.0 *q*	**18**
6.5 *q*	0.1/2 = 0.05	0.025	6.5 *q* / 6.5 *Q*	**18**	0.025	6.5 *q* / 4.0 *q*	**16**
4.0 *Q*	0.1/2 = 0.05	0.025	4.0 *Q* / 6.5 *Q*	**20**	0.025	4.0 *Q* / 4.0 *q*	**18**
4.0 *q*	(1 – 0.1)/2 = 0.45	0.225	4.0 *q* / 6.5 *Q*	**18**	0.225	4.0 *q* / 4.0 *q*	**16**

Concepts in Action
Problems for Solution

14.1 Two varieties of corn, A and B, are field-tested in Indiana and North Carolina. Strain A is more productive in Indiana, but strain B is more productive in North Carolina. What phenomenon in quantitative genetics does this example illustrate?

14.2 Distinguish between the broad-sense heritability of a quantitative trait and the narrow-sense heritability. If a population is fixed for all genes that affect a particular quantitative trait, what are the values of the narrow-sense and broad-sense heritabilities?

14.3 Some estimates of broad-sense heritabilities of human traits are 0.85 for stature, 0.62 for body weight, 0.57 for systolic blood pressure, 0.44 for diastolic blood pressure, 0.50 for twinning, and 0.1 to 0.2 for overall fertility. Which of these characters is most likely to "run in families"? If one of your parents and one of your grandparents has high blood pressure, should you be concerned about the likelihood of your having the same problem?

14.4 Tabulated below are the numbers of eggs laid by 50 hens over a 2-month period. The hens were selected at random from a much larger population. Estimate the mean, variance, and standard deviation of the distribution of egg number in the entire population from which the sample was drawn.

48	50	51	47	54	45	50	38	40	52
58	47	55	53	54	41	59	48	53	49
51	37	31	47	55	46	49	48	43	68
59	51	52	66	54	37	46	55	59	45
44	44	57	51	50	57	50	40	63	33

14.5 Two highly inbred strains of mice are crossed, and the F_1 generation has a mean tail length of 5 cm and a standard deviation of 1.5 cm. The F_2 generation has a mean tail length of 5 cm and a standard deviation of 4 cm. What are the environmental variance, the genetic variance and the broad-sense heritability of tail length in this population?

14.6 For the difference between the domestic tomato, *Lycopersicon esculentum*, and its wild South American relative, *Lycopersicon chmielewskii*, the environmental variance (σ_e^2) accounts for 13 percent of the total phenotypic variance (σ_t^2) of fruit weight, 9 percent of σ_t^2 of soluble-solid content, and 11 percent of σ_t^2 of acidity. What are the broad-sense heritabilities of these traits?

14.7 Two homozygous genotypes of *Drosophila* differ in the number of abdominal bristles. In genotype *AA*, the mean bristle number is 20 with a standard deviation of 2. In genotype *aa*, the mean bristle number is 23 with a standard deviation of 3. Both distributions conform to the normal curve, in which the proportions of the population with a pheno-

geNETics on the web

www.jbpub.com/genetics

These GeNETics on the Web exercises will introduce some of the most useful sites for finding genetic information on the World Wide Web. Genetic sites provide access to a rich storehouse of information on all aspects of genetics. These range from sites written in nontechnical language for the lay person to sites with sophisticated databases designed for the professional geneticist. In carrying out these exercises, you will get a taste of what the Internet can offer a student in genetics.

The keywords shown in color in the following exercises are available on the Jones and Bartlett Publishers' web site as hyperlinks to various genetic sites. To complete the exercises, visit the GeNETics on the Web home page at

http://www.jbpub.com/genetics

Select the link to Essential geNETics on the web. Then choose a chapter. You find a list of keywords that correspond to the exercises below. Select a keyword to link to a web site containing the genetic information necessary to complete the exercise. Each exercise includes a specific assignment that makes use of the information available at the site.

Exercises

- Many human disorders are complex, multifactorial traits with significant genetic components as well as multiple environmental factors in their causation. Identifying the specific genetic and environmental causes of such conditions is extremely difficult. A case in point is schizophrenia, a relatively common mental disease with an estimated incidence of 1 to 1.5 percent in the U. S. population. Though serious, schizophrenia is usually treatable with drug therapy. Schizophrenia appears to have a significant genetic component, and there are probably multiple forms of the disease. At this keyword site you can learn more about this condition, and you can also find a useful glossary of terms related to mental illness. Write a one-page report on the main symptoms of the disease, the usual age of onset, and the evidence that there is a genetic component in the causation. Explain the meaning of the following statement: "People do not cause schizophrenia; they merely blame each other for doing so."
- Human fingerprints are remarkable because each pattern is unique, yet one component is highly heritable. Fingerprint ridges are the raised skin ridges that carry sweat glands and that vary so much in pattern from one person to the next that each person has unique fingerprints suitable for identification. Each finger may have a ridge pattern forming an arch, a loop, or a whorl, and certain conventions are used to identify the center and edge of the pattern. The number of ridges between the center and edge constitutes the ridge count for any finger, and the sum of the ridge counts for all ten fingers constitutes the total fingerprint ridge count. Total fingerprint ridge count is a convenient human quantitative trait because it is expressed at all ages, it is stable throughout life, and the phenotype can be determined by noninvasive procedures. Although the ridge patterns differ even between identical twins, the total fingerprint ridge count is highly heritable. The narrow-sense heritability is about 86 percent. At the keyword site fingerprint you can learn more about fingerprint patterns and their classification. Follow the instructions to take your own fingerprints, and then classify the pattern you find on each finger.

Pic Site

The Pic Site showcases some of the most visually appealing genetics sites on the World Wide Web. To visit the showcase genetics site, select the Pic Site for Chapter 14.

type within an interval defined by the mean ±1, ±1.5, ±2, and ±3 standard deviations are 68, 87, 95, and 99.7 percent, respectively.

(a) In genotype *AA*, what is the proportion of flies with a bristle number between 20 and 23?
(b) In genotype *aa*, what is the proportion of flies with a bristle number between 20 and 23?
(c) What proportion of *AA* flies have a bristle number greater than the mean of *aa* flies?
(d) What proportion of *aa* flies have a bristle number greater than the mean of *AA* flies?

14.8 The narrow-sense heritability of withers height in a population of quarterhorses is 30 percent. (Withers height is the height at the highest point of the back, between the shoulder blades.) The average withers height in the population is 17 hands. (A "hand" is a traditional measure equal to the breadth of the human hand, now taken to equal 4 inches.) From this population, studs and mares with an average withers height of 16 hands are selected and mated at random. What is the expected withers height of the progeny? How does the value of the narrow-sense heritability change if withers height is measured in meters rather than hands?

14.9 Consider a complex trait in which the phenotypic values in a large population are distributed approximately according to a normal distribution with mean 100 and standard deviation 15. What proportion of the population has a phenotypic value above 130? Below 85? Above 85?

14.10 In an experimental population of the flour beetle *Tribolium castaneum*, the pupal weight is distributed normally with a mean of 2.0 mg and a standard deviation of 0.2 mg. What proportion of the population is expected to have a pupal weight between 1.8 and 2.2 mg? Between 1.6 and 2.4 mg? Would you expect to find an occasional pupa weighing 3.0 mg or more? Explain your answer.

14.11 Estimate the minimum number of genes affecting fruit weight in a population of the domestic tomato produced by crossing two inbred strains. Measured as the logarithm of fruit weight in grams, the inbred lines have average fruit weights of −0.137 and 1.689. The F_1 generation has a variance of 0.0144, and the F_2 generation has a variance of 0.0570.

14.12 In human beings, the correlation coefficient between first cousins in the total fingerprint ridge count is 10 percent. On the basis of this value, what is the estimated narrow-sense heritability of this trait?

Further Readings

Black, D. M., and E. Solomon. 1993. The search for the familial breast/ovarian cancer gene. *Trends in Genetics* 9: 22.

Bouchard, T. J., Jr., D. T. Lykken, M. McGue, N. L. Segal, and A. Tellegen. 1990. Sources of human psychological differences: The Minnesota study of twins reared apart. *Science* 250: 223.

Crow, J. F. 1993. Francis Galton: Count and measure, measure and count. *Genetics* 135: 1.

Devor, E. J., and C. R. Cloninger. 1989. The genetics of alcoholism. *Annual Review of Genetics* 23: 19.

East, E. M. 1910. Mendelian interpretation of inheritance that is apparently continuous. *American Naturalist* 44: 65.

Falconer, D. S., and T. F. C. Mackay. 1996. *Introduction to Quantitative Genetics*, 4th ed. Essex, England: Longman.

Feldman, M. W., and R. C. Lewontin. 1975. The heritability hang-up. *Science* 190: 1163.

Gottesman, I. I., 1997. Twins: En route to QTLs for cognition. *Science* 276: 1522.

Gottesman, I. I., and J. Shields. 1982. *Schizophrenia: The Epigenetic Puzzle.* Cambridge, England: Cambridge University Press.

Greenspan, R. J. 1995. Understanding the genetic construction of behavior. *Scientific American*, April.

Haley, C. S. 1996. Livestock QTLs: Bringing home the bacon. *Trends in Genetics* 11: 488.

Hartl, D. L., and A. G. Clark. 1997. *Principles of Population Genetics.* 3d ed. Sunderland, MA: Sinauer.

Lander, E. S., and N. J. Schork. 1994. Genetic dissection of complex traits. *Science* 265: 2037.

Lesch, K.-P., D. Bengel, A. Heils, S. Z. Sabol, B. D. Greenberg, S. Petri, J. Benjamin, C. R. Muller, D. H. Hamer, and D. L. Murphy. 1996. Association of anxiety-related traits with a polymorphism in the serotonin transporter gene regulatory region. *Science* 274: 1527.

Paterson, A. H., E. S. Lander, J. D. Hewitt, S. Peterson, S. E. Lincoln, and S. D. Tanksley. 1988. Resolution of quantitative traits into Mendelian factors by using a complete linkage map of restriction fragment length polymorphisms. *Nature* 335: 721.

Pericakvance, M. A., and J. L. Haines. 1995. Genetic susceptibility to Alzheimer disease. *Trends in Genetics* 11: 504.

Pirchner, F. 1983. *Population Genetics in Animal Breeding.* 2d ed. New York: Plenum.

Plomin, R., and J. C. Defries. 1998. The genetics of cognitive abilities and disabilities. *Scientific American*, May.

Risch, N., and H. Zhang. 1995. Extreme discordant sib pairs for mapping quantitative trait loci in humans. *Science* 268: 1584.

Stigler, S. M. 1995. Galton and identification by fingerprints. *Genetics* 140: 857.

Stuber, C. W. 1996. Mapping and manipulating quantitative traits in maize. *Trends in Genetics* 11: 477.

Tanksley, S. D. and S. R. McCouch. 1997. Seed banks and molecular maps: Unlocking genetic potential from the wild. *Science* 277:1063.

Answers to Even-Numbered Problems

Chapter 1

1.2 The importance of the nucleus in inheritance was implied by its prominence in fertilization. The discovery of chromosomes inside the nucleus, their behavior during cell division, and the observation that each species has a characteristic chromosome number made it likely that chromosomes were the carriers of the genes. Microscopic studies showed that DNA and proteins are both present in chromosomes, but whereas nearly all cells of a given species contain a constant amount of DNA, the amount and kinds of proteins differ greatly in different cell types.

1.4 Because the mature T2 phage contains only DNA and protein, the labeled RNA was left behind in material released by the burst cells.

1.6 RNA differs from DNA in that the sugar–phosphate backbone contains ribose rather than deoxyribose, RNA contains the base uracil (U) instead of thymine (T), and RNA usually exists as a single strand (although any particular molecule of RNA may contain short regions of complementary base pairs that can come together to form duplexes).

1.8 Because A ≠ T and G ≠ C in this DNA, it seems likely that the DNA molecule present in this particular virus is single-stranded.

1.10 The repeating Asn results from translation in the reading frame

5'-AAUAAUAAUAAU · · · -3'

and the repeating Ile results from translation in the reading frame

5'-AUAAUAAUAAUA · · · -3'

There is no product corresponding to the third reading frame (5'-UAAUAAUAAUAA · · · -3') because 5'-UAA-3' is a stop codon.

1.12 (a) Met–Ser–Thr–Ala–Val–Leu–Glu–Asn–Pro–Gly.
(b) The mutation changes the initiation codon into a noninitiation codon, so translation will not start with the first AUG; translation will start either with the next AUG farther along the mRNA or, if this is too distant, not at all.
(c) Met–Ser–Thr–Ala–Val–Leu–Glu–Asn–Pro–Gly; there is no change, because both 5'-UCC-3' and 5'-UCG-3' code for serine.
(d) Met–Ser–Thr–Ala–Ala–Leu–Glu–Asn–Pro–Gly; there is a Val → Ala amino acid replacement because 5'-GUC-3' codes for Val, whereas 5'-GCC-3' codes for Ala.
(e) Met–Ser–Thr–Ala–Val–Leu; translation is terminated at the UAA because 5'-UAA-3' is a "stop" (termination) codon.

Chapter 2

2.2 The existing data enable us to group the mutants into three complementation groups as follows: {a, c, d}, {b, f}, and {e}. The missing entries are shown in the accompanying table.

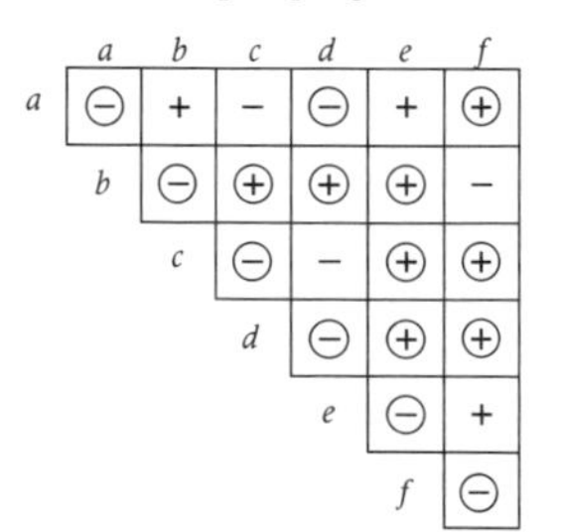

	a	*b*	*c*	*d*	*e*	*f*
a	⊖	+	–	⊖	+	⊕
b		⊖	⊕	⊕	⊕	–
c			⊖	–	⊕	⊕
d				⊖	⊕	⊕
e					⊖	+
f						⊖

2.4 We expect a 3 : 1 ratio of the dominant to the recessive phenotype when two heterozygotes are crossed, but in this case we observe 2 : 1. A cross of two *Cy*/+ heterozygotes is expected to yield a progeny genotypic ratio of 1 *Cy*/*Cy* : 2 *Cy*/+ : 1 +/+. Because we see only two curly-winged flies for every wildtype fly, one possible explanation is that the *Cy*/*Cy* homozygotes die. In other words, *Cy* is dominant with respect to wing phenotype but recessive with respect to lethality.

2.6 (a) Two phenotypic classes are expected for each of the *A*, *a* and *B*, *b* pairs of alleles, and three are expected for the *R*, *r* pair, yielding a total number of 12. **(b)** 1/64. **(c)** 1/8.

2.8 (a) Because the trait is rare, it is reasonable to assume that the affected father is heterozygous *HD* / *hd*, where *hd* represents the normal allele. Half of the father's gametes contain the mutant *HD* allele, so the probability is 1/2 that the son received the allele and will later develop the disorder. **(b)** We do not know whether the son is heterozygous *HD* / *hd*, but the probability is 1/2 that he is; and if he is heterozygous, half of his gametes will contain the *HD* allele. Therefore, the overall probability that the grandchild has the *HD* allele is $(1/2) \times (1/2) = 1/4$.

2.10 In the functional female gametes, the ratio of *A* : *a* is 1/2 : 1/2; and in the functional male gametes, the ratio of *A* : *a* is 1/4 : 3/4 (because only half of the *A*-bearing products of meiosis are functional). The F_2 ratio of *AA* : *Aa* : *aa* is therefore 1/8 : 1/2 : 3/8.

2.12 $1/2 \times 1/2 \times 1/4 = 1/16$.

Chapter 3

3.2 Panel B is anaphase of mitosis, because the homologous chromosomes are not paired. Panel A is anaphase I of meiosis, because the homologous chromosomes are paired. Panel C is anaphase II of meiosis, because the chromosome number has been reduced by half.

3.4 (a) Ratio X/A = 0.5, phenotype male. **(b)** Ratio X/A = 1.0, phenotype female. **(c)** Ratio X/A > 1.0, phenotype female.
(d) Ratio X/A = 0.66, phenotype intersex.

3.6 (a) The boy's mother must be $cb\ cb^+$ for color blindness, because her father was color blind. If her father was also *BB*, then her genotype would be *Bb*, because we are assuming that the maternal grandmother was *bb*. Because the segregation of the $cb\ cb^+$ and *Bb* allele pairs is independent, the probability that the son will be color blind and bald is $1/2 \times 1/2 = 1/4$. **(b)** In this case the mother has the genotype *Bb* or the genotype *bb*, each with probability 1/2, so the probability that the son will be color blind and bald is $1/2 \times 1/2 \times 1/2 + 1/2 \times 1/2 \times 0 = 1/8$.

3.8 The chance that the second gene is on a different chromosome than the first is 6/7; the chance that the third gene is on a different chromosome than either of the first two is 5/7. Continuing in this manner, we find that the overall chance is $6/7 \times 5/7 \times 4/7 \times 3/7 \times 2/7 \times 1/7 = 0.00612$, or less than 1 percent. The likelihood of having every gene on a different chromosome is quite small, and for this reason Mendel has been accused of deliberately choosing unlinked genes. In fact, it is now known that not all of Mendel's genes are on different chromosomes. Three of the genes (*fa*, determining axial versus terminal flowers; *le*, determining tall versus short plants; and *v*, determining smooth versus constricted pods) are all located on chromosome 4. The *le* and *v* genes undergo recombination at the rate of about 12 percent, but Mendel apparently did not study this particular combination for independent assortment. The *fa* gene is very distant from the other two and shows independent assortment with them.

3.10 The chi-square value equals

$$(188 - 186)^2/186 + (203 - 186)^2/186 + (175 - 186)^2/186 + (178 - 186)^2/186 = 2.56$$

It has 3 degrees of freedom (four classes of data), and the *P* value is about 0.5. We therefore conclude that there is no reason, on the basis of these data, to reject the hypothesis of 1 : 1 : 1 : 1 segregation.

3.12 The chi-square value is 7.33 with 3 degrees of freedom, which yields a *P* value of about 0.08. The difference from expectation is not statistically significant, so the genetic hypothesis cannot be rejected even though the discrepancies from the expected numbers may seem to be large.

Chapter 4

4.2 Only **(b)** is true; all the others are false.

4.4 This is a testcross, so the phenotypes of the progeny reveal the ratio of meiotic products from the doubly heterozygous parent. The expected progeny are 86 percent parental (43 percent wildtype and 43 percent scarlet, spineless) and 14 percent recombinant (7 percent scarlet and 7 percent spineless).

4.6 (a) The parental types are evidently $v^+\ pr\ bm$ and $v\ pr^+\ bm^+$ and the double-recombinant types $v^+\ pr^+\ bm^+$ and $v\ pr\ bm$. This puts *v* in the middle. The *pr*–*v* recombination frequency is $(69 + 76 + 36 + 41)/1000 = 22.2$ percent, and the *v*–*bm* recombination frequency is

$(175 + 181 + 36 + 41)/1000 = 43.3$ percent. The expected number of double crossovers equals $0.222 \times 0.433 \times 1000 = 96.13$, so the coincidence is $(36 + 41)/96.13 = 0.80$. The interference is therefore $1 - 0.80 = 0.20$. **(b)** The true map distances are certainly larger than 22.2 and 43.3 centimorgans. The frequencies of recombination between these genes are so large that there are undoubtedly many undetected double crossovers in each region.

4.8 (a) The three-allele hypothesis predicts that the matings will be $FS \times FS$ and should yield $FF : FS : SS$ offspring in a ratio of 1 : 2 : 1. **(b)** These ratios are not observed, and furthermore, some of the progeny show the "null" pattern. **(c)** One possibility is that the *F* and *S* bands are from unlinked loci and that there is a "null" allele of each—say, *f* and *s*. Because *f* and *s* are common and *F* and *S* are rare, most persons with two bands would have the genotype *Ff Ss*. The cross should therefore yield 9/16 *F– S–* (two bands), 3/16 *F– ss* (fast band only), 3/16 *ff S–* (slow band only), and 1/16 *ff ss* (no bands). **(d)** The data are consistent with this hypothesis (chi-square = 2.67 with 3 degress of freedom, *P* value approximately 0.50).

4.10 In this case one must start by calculating the number of double-crossover gametes that would be observed, given the interference. The number of observed doubles equals the number of expected doubles times the coincidence, or $0.15 \times 0.20 \times 0.20 = 0.006$, or among 1000 gametes, 3 each of *o* + + and + *ci p*. The single recombinants in the *o–ci* interval would therefore number $0.15 \times 1000 - 6 = 144$, or 72 each of *o* + *p* and + *ci* +. The single recombinants in the *ci–p* interval would number $0.20 \times 1000 - 6 = 194$, or 97 each of *o ci p* and + + +. The remaining 656 gametes are nonrecombinant, 328 each of *o ci* + and + + *p*.

4.12 The gene-centromere map distance equals 1/2 the frequency of second-division segregation, which also equals the frequency of crossing-over in the region. In this problem it is easiest to answer the questions by taking the cases out of order, considering the second-division segregations at the beginning. **(c)** The frequency of second-division segregation of *cys-1* must be 14 percent, because the map distance is 7 cM. Because of the complete interference, a crossover on one side of the centromere precludes a crossover on the other side, so these asci must have first-division segregation for *pan-2*. **(b)** Similarly, the frequency of second-division segregation of *pan-2* must be 6 percent, because the map distance is 3 cM; these asci must have first-division segregation for *cys-1*. **(d)** Because of the complete interference, second-division segregation of both markers is not possible. **(a)** The only remaining possibility is first-division segregation of both markers, which must have a frequency of $1 - 0.14 - 0.06 = 80$ percent. **(e)** First-division segregation of both markers yields a PD tetrad, and second-division segregation for one of the markers yields a TT tetrad. Because there are no double crossovers, there can be no NPD tetrads. Hence the frequencies are PD = 80 percent and TT = 20 percent.

Chapter 5

5.2 The inversion has the sequence *A B E D C F G*, the deletion *A B F G*. The possible translocated chromosomes are (1) *A B C D E T U V* and *M N O P Q R S F G* or (2) *A B C D E S R Q P O N M* and *V U T F G*. One of these possibilities includes two monocentric chromosomes, and the other includes a dicentric and an acentric. Only the translocation with two monocentrics is genetically stable.

5.4 One approach would be to treat the diploid with colchicine to yield a tetraploid, which undergoes meiosis, producing diploid gametes ($2n$). Cross the tetraploid with the diploid to obtain the triploid ($3n$), and again treat with colchicine to yield the hexaploid ($6n$), which produces triploid gametes ($3n$). Now cross the hexaploid with the tetraploid to obtain the pentaploid ($3n + 2n$).

5.6 Species A ($n = 6$) hybridizes with species B ($n = 6$). The F_1 progeny will have 12 chromosomes and be sterile. The sterility can be overcome by endoreduplication in an F_1 organism, which creates a fertile new species with a chromosome number of 24. This new species ($n = 12$) hybridizes with a third species, C ($n = 6$), yielding another sterile F_1 progeny with 18 chromosomes. Endoreduplication in one of these sterile F_1 organisms gives rise to a fully fertile new species with 36 chromosomes. Note that this scenario is very similar to that which produced hexaploid wheat.

5.8 The chromosomes underwent endoreduplication, resulting in an autotetraploid.

5.10 (a) In homozygotes, there is no impediment to crossing-over. Hence, for a map distance of 12 map units, one should expect to observe 12 percent recombination, because over this length of genetic interval, multiple crossing-over can be neglected. **(b)** In heterozygotes, the products of recombination would not be recovered, and if the whole region were involved in the inversion, the frequency of recombination would be 0. Because only 1/3 of the interval is inverted in this case, the recombination frequency is expected to be $(1/3) \times 12 = 8$ percent.

5.12 The original strain carried an inversion with a breakpoint between *A* and the centromere. Single crossovers within the inversion loop created chromosomally abnormal ascospores, which were the ascospores that failed to survive. The ascospore that was recovered from second-division segregation carried the inversion as well as the *a* allele. When crossed with the original wildtype strain, the inversion became homozygous, restoring the normal 20 percent second-division segregation, which corresponds to 10 map units from the centromere.

Chapter 6

6.2 Because of semiconservative replication, equal amounts of $^{14}N^{14}N$ and $^{14}N^{15}N$ would be expected.

6.4 Ten percent of the haploid genome is 10,000 kb, or 10^7 base pairs. Each base pair in the double helix extends 3.4 angstrom units, so 10^7 base pairs equals 3.4×10^7 angstrom units. Because there are 10^{-4} micrometers per angstrom unit, the required length is 3.4×10^3 micrometers or, dropping the scientific notation, 3400 micrometers. This length is 3.4 mm, which, if the molecule were not so thin, would be long enough to see with the naked eye.

6.6 (a) The distance between nucleotide pairs is 3.4 Å. The length of the molecule is 68×10^4 Å. Therefore, the number of nucleotide pairs equals $(68 \times 10^4)/3.4 = 2 \times 10^5$. **(b)** There are 10 nucleotide pairs per turn of the helix, so the total number of turns equals $2 \times 10^5/10 = 2 \times 10^4$.

6.8 Replication is bidirectional from a single origin of replication.

6.10 In a double-stranded DNA molecule, A = T and G = C because of complementary base pairing. Therefore, if A/C = 1/3, then C = G = 3 × A. Because A + T + G + C = 1, everything can be put in terms of A as follows: A + A + 3A + 3A = 1. Hence A = T = 0.125. This makes C = G = (1 − 0.25)/2 = 0.375. In other words, the DNA is 12.5 percent A, 12.5 percent T, 37.5 percent G, and 37.5 percent C.

6.12 The difference in sizes of the restriction fragment could be due either to the insertion of DNA between the two restriction sites or to the elimination of one of the two flanking *Xho*I restriction sites. The pattern of hybridization of the 10-kb fragment to multiple locations along the polytene chromosomes, which differ between strains, suggests not only that the mutation is caused by insertion of DNA but also that the inserted DNA is a transposable element. Transposable elements are often present in multiple copies in a genome, and because of the lack of specificity for insertion sites that many transposable elements display, one would not expect to find elements inserted at the same genomic locations in two independent wildtype isolates.

Chapter 7

7.2 There are nine different mutant codons: (1) UGC → CGC (Arg), (2) UGC → AGC (Ser), (3) GGC → GGC (Gly), (4) UGC → UUC (Phe), (5) UGC → UCC (Ser), (6) UGC → UAC (Tyr), (7) UGC → UGU (Cys), (8) UGC → UGA (Stop), and (9) UGC → UGG (Trp). Number 7 is silent, number 8 is nonsense (termination codon), and the rest are missense.

7.4 (a) 2; **(b)** 2; **(c)** 3; **(d)** 1.

7.6 If the +1 frameshift mutation shifts the reading frame to produce a nonsense codon, no −1 frameshift beyond the nonsense codon can prevent the termination of translation.

7.8 (a) There are 1×10^6 replications and a rate of mutation of 1×10^{-6} new mutants per replication, so the expected number of new mutants is $(1 \times 10^6) \times (1 \times 10^{-6}) = 1$. **(b)** The total number of cells after division is 2×10^6. The probability of any one cell being a mutant is therefore $2/(2 \times 10^6) = 5 \times 10^{-7}$. The probability that any cell is a nonmutant is thus $1 - (5 \times 10^{-7})$, and the probability that each of the cells in the population is nonmutant is this quantity raised to the power of 2×10^6. The answer is P = 0.368. [The log P = $(2 \times 10^6) \times \log[1 - (5 \times 10^{-7})] = (2 \times 10^6) \times (-2.17 \times 10^{-7}) = -0.43429$, so P = 0.368. Students familar with the Poisson

Answer 8.12—Map

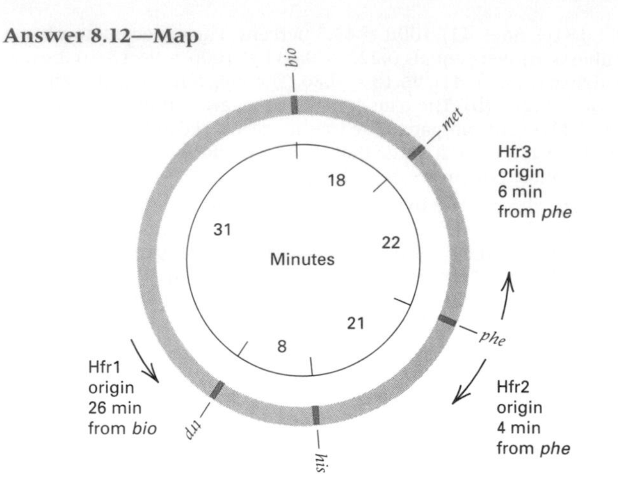

distribution may recognize this as $e^{-1} = 0.368$, the "Poisson zero term."]

7.10 (a) The wildtype anticodon is 5'-CAG-3'; the mutant anticodon is 5'-CAC-3'. **(b)** The mutational event was a CG to GC transversion. **(c)** Val. **(d)** Leu.

7.12 (a) Acsi of types (4) and (5) show typical 3 : 1 patterns of segregation expected with gene conversion, which strongly suggests that mismatch repair does take place. The ratios of m : + in these asci are 6 : 2 and 2 : 6, respectively. Asci (2) and (3) also show evidence of mismatch repair because the segregation ratios are, respectively, 5 : 3 and 3 : 5. **(b)** In asci of types (1), (2), and (3), adjacent pairs of ascospores, which result from post-meiotic mitosis, do not have the same genotype; this result implies that the immediate product of meiosis was a heteroduplex, resulting in genetically different ascospores after the mitotic division. The persistence of the heteroduplexes implies that mismatch repair is not 100 percent efficient.
(c) Half the asci with a heteroduplex across the $m/+$ region will have recombination of the outside markers. Among these, 4 : 4 segregation of m : + will be found either if neither heteroduplex was corrected or if both heteroduplexes were corrected but in opposite directions. The probability of neither heteroduplex being corrected is $(0.30)^2 = 0.09$, and that of both being corrected is $(0.70)^2 = 0.49$. Among those in which both heteroduplexes are corrected, the proportion that are corrected in opposite directions is $2 \times 0.40 \times 0.30 = 0.24$. Putting all this together yields the required answer: $0.50 \times (0.09 + 0.24) = 0.165$, or 16.5 percent.

Chapter 8

8.2 Yes, as long as it has a replication origin.

8.4 Apparently *h* and *tet* are closely linked, so recombinants that contain the h^+ allele of the Hfr tend also to contain the *tet-s* allele of the Hfr, and these recombinants are eliminated by the counterselection for *tet-r.*

8.6 The attachment site, *att*, must be between genes *f* and *g*.

8.8 The three possible orders are (1) *pur–pro–his*, (2) *pu–his–pro*, and (3) *pro–pur–his*. The predictions of the three orders are as follows: (1) Virtually all pur^+ his^- transductants should be pro^-, but this is not true. (2) Virtually all pur^+ pro^- transductants should be his^-, but this is not true. (3) Some pur^+ pro^- transductants will be his^-, and some pur^+ his^- transductants will be pro^- (depending on the locations of the exchanges). Therefore, order (3) is the only one that is not contradicted by the data.

8.10 Any phage transduces one small, contiguous piece of DNA. Cotransduction therefore indicates very close linkage. Hence *G* and *H* are close, and *G* and *I* are close, but *H* and *I* are not close (because they do not cotransduce), so the order of these three genes must be *I G H* (or the reverse). The location of the other three genes can be deduced similarly: *A* is close to *H*, *B* is close to *I*, and *T* must be close to *A* but not close to *H*. Hence the gene order is *B I G H A T* or, equivalently, *T A H G I B*.

8.12 The map implied by the time-of-entry data is shown at the top of the page. The entries that correspond to the question marks should be, for *his* in Hfr1, 87 min; for *phe* in Hfr2, 4 min; at 35 min in Hfr3, *trp*; and for *bio* in Hfr3, 66 min.

Chapter 9

9.2 The 5' end of the RNA transcript is transcribed first, so **(a)** through **(c)** are all wrong. The 5' end of the RNA transcript is transcribed from the 3' end of the DNA template, so **(e)** is wrong. The additional statement in **(e)** that the amino end of the polypeptide is synthesized first serves as confirmation of the correctness of statement **(d)**.

9.4 (a) Methionine has 1 codon, histidine 2, and threonine 4, so the total number is $1 \times 2 \times 4 = 8$. The sequence AUGCAYACN encompasses them all. **(b)** Because arginine has six codons, the possible number of sequences coding for Met–Arg–Thr is $1 \times 6 \times 4 = 24$. The sequences are AUGCGNACN (16 possibilities) and AUGAGRACN (8 possibilities).

9.6 With a G at the 5' end, the first codon is GUU, which codes for valine; and with a G at the 3' end, the last codon is UUG, which codes for leucine.

9.8 (a) With four kinds of nucleotides and only five kinds of amino acids, the smallest codon size is 2, which gives 16 possible different combinations of four bases taken two at a time ($4^2 = 16$). **(b)** If the codon size is 2, it would take $2 \times 10 = 20$ bases to code for one protein, or $20 \times 100 = 2000$ to code for all the proteins the organism makes. Therefore, the minimum size of the genome of the organism is 2000 base pairs, assuming that the genes are nonoverlapping.

9.10 (a) The deletion must have fused the amino-coding terminus of the *B* gene with the carboxyl-coding terminus of the *A* gene. The nontranscribed strand must therefore be oriented 5'-*B*–*A*-3'. **(b)** The number of bases deleted must be a multiple of three, otherwise the carboxyl terminus would not have the correct reading frame.

9.12 The wildtype and double-mutant mRNA codons are:

Wildtype:

5'-AAR AAR UAY CAY CAR UGG ACN UGY AAY-3'

Double Mutant:

5'-AAR CAR AUH CCN CCN GUN GAY AUG AAY-3'

Comparing these sequences, it can be deduced that the first frameshift addition is the C at position 4 in the double-mutant gene. Realigning the sequences to take this into account yields

Wildtype:

5'-AAR AAR UAY CAY CAR UGG ACN UGY AAY-3'

Double Mutant:

5'-AARC ARA UHC CNC CNG UNG AYA UGA AY-3'

Now it is also clear that the single-nucleotide deletion in the wildtype gene must be the Y in the fourth nucleotide position from the 3' end. Realigning again to take this into account, we have

Wildtype:

5'-AAR AAR UAY CAY CAR UGG ACN UGY AAY-3'

Double Mutant:

5'-AARC ARA UHC CNC CNG UNG AYA UG AAY-3'

Hence the complete sequences, as thoroughly as they can be resolved from the data, are

Wildtype:

5'-AAR AAA UAC CAC CAG UGG ACA UGY AAY-3'

Double Mutant:

5'-AAR CAA AUA CCA CCA GUG GAC AUG AAY-3'

Chapter 10

10.2 It occurs 23 times there also, because the sequence is a "palindrome."

10.4 You must use cDNA under a suitable promoter that functions in a prokaryotic system. Eukaryotic genomic DNA includes regulatory elements that will not work in bacteria and introns that cannot be spliced out of RNA in prokaryotic cells.

10.6 The primer sequences are included in the PCR product at both ends of the gene fragment. The total length of the amplified piece of DNA is therefore $19 + 303 + 19 = 341$ base pairs.

10.8 Six-base cutters, such as *Bam*HI, find their recognition sequences every 4096 (4^6) bases, on the average. In a genome of 4,600,000 bases, about 1123 fragments are expected, assuming that A, G, C, and T are in equal proportions in the genome and are randomly distributed.

10.10 The hint says that if the genome were represented *x* times in the library, the probability that a particular sequence would be missing is e^{-x}, which we want to equal 0.01. Hence $x = 4.6$. Because one haploid representation of the genome equals $(6 \times 10^9)/2 = 3 \times 10^9$ base pairs, and the average insert size is 2×10^4 base pairs, one requires $[(3 \times 10^9)/(2 \times 10^4)] \times 4.6 = 6.9 \times 10^5$ clones.

10.12 The only restriction map consistent with the data is a circular one:

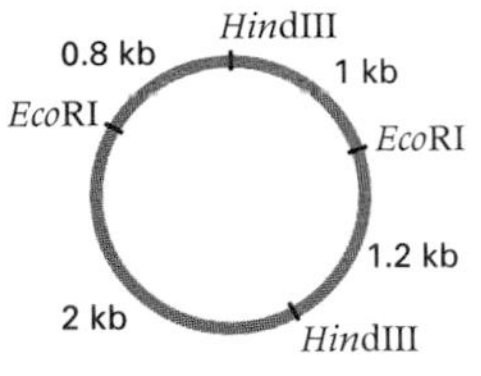

Chapter 11

11.2 The repressor gene need not be near because the repressor is diffusible. The *trp* repressor is one example in which the repressor gene is located quite far from the structural genes.

11.4 In the presence of glucose, bacterial cells have very low levels of cAMP. Without cAMP, the binding of the CRP protein cannot take place, transcription cannot be activated, and no *lac* proteins are made. Lack of induction is true even if lactose is present. The system ensures that primary carbon sources are exhausted before the cell expends energy for the synthesis of new enzymes to utilize secondary carbon sources.

11.6 In the absence of lactose or glucose, two proteins are bound: the *lac* repressor and CRP-cAMP. In the presence of glucose, only the repressor is bound.

11.8 In the first cross, active repressor is already present in the recipient cells, so no activation of *lac* genes is possible. In the second cross, when the *lac* genes enter the recipient, no repressor is present, so *lac* mRNA is produced and β-galactosidase is synthesized. The effect is transient, because after a short period of time, repressor is also made using the donor $lacI^+$ gene, and transcription of *lac* genes is repressed.

11.10 The predicted levels are shown in the table.

Genotype	Uninduced level Z	Y	Induced level Z	Y
(a)	100	100	100	100
(b)	0.1	0.1	100	100
(c)	25	25	100	100
(d)	25	25	100	100

11.12 (a) Yes; the repressor is functional, and the presence of lactose activates transcription of the *lac* genes. **(b)** and **(d)** Yes; at 42°C, the repressor cannot bind the operator, which means that the *lac* operon is transcribed whether or not the inducer is present. **(c)** No; at this temperature, the repressor functions normally, and because lactose is absent, the *lac* operon is repressed.

Chapter 12

12.2 The first experiment suggests that the genetic program of cell death is already activated and cannot be turned off. The second experiment shows that intercellular signaling must be responsible for inducing the cell-death fate, because when uninduced cells are transplanted to a new site, they do not die.

12.4 Perform reciprocal crosses. A maternal-effect mutation results in phenotypically mutant progeny when the mother is homozygous.

12.6 The proteins required for cleavage and blastula formation are translated from mRNAs present in the mature oocyte, but transcription of zygotic genes is required for gastrulation.

12.8 Distinct sets of genes are activated in the different target cells.

12.10 An *X Y* double mutant has only *Z* function, and gene product Z expands into all four whorls. The mutant flower would therefore have carpels induced in all four whorls.

12.12 The 20 mutations need not be mutations of 20 different genes; indeed, many of them proved to be alleles of one or another of a small set of genes.

Chapter 13

13.2 The 10 *AA* genotypes contribute 20 *A* alleles to the sample, the 15 *Aa* genotypes contribute 15 *A* and 15 *a* alleles to the sample, and the 4 *aa* genotypes contribute 8 *a* alleles to the sample. Altogether, there are 35 *A* and 23 *a* alleles in the sample, for a total of 58. In this sample the allele frequency of *A* is 35/58 = 0.603 and that of *a* is 23/58 = 0.397.

13.4 (a) The probability that the wife is a carrier is the probability that a phenotypically normal person is a carrier. This probability is $2pq/(p^2 + 2pq)$, where $q = \sqrt{(1/1700)} = 0.024$ and $p = 0.976$. The numerator $2pq = 0.047$ and the denominator $p^2 + 2pq = 0.999$ (close enough to 1 that it can be ignored). Hence the probability that the wife is a carrier equals 0.047. **(b)** If she is a carrier, the probability that the first child will be affected is 1/4, so the overall probability that the first child will be affected equals $0.047 \times 1/4 = 0.012$, or about 1 in 84.

13.6 Set $q = 2 \times 2pq$; hence $p = 1/4$ and $q = 3/4$.

13.8 (a) $(0.01)^2 \times 1{,}000{,}000 = 100$ seedlings. **(b)** The lethal recessive allele will persist for many generations at a low frequency, because it is present primarily in heterozygotes and is shielded from selection by the dominant allele.

13.10 (a) Here $F = 0.66$, $p = 0.43$, and $q = 0.57$. The expected genotype frequencies are as follows: *AA*, $(0.43)^2(1 - 0.66) + (0.43)(0.66) = 0.347$; *AB*, $2(0.43)(0.57)(1 - 0.66) = 0.167$; and *BB*, $(0.57)^2(1 - 0.66) + (0.57)(0.66) = 0.487$. **(b)** With random mating, the expected genotype frequencies are as follows: *AA*, $(0.43)^2 = 0.185$; *AB*, $2(0.43)(0.57) = 0.490$; and *BB*, $(0.57)^2 = 0.325$.

13.12 Let p_n be the allele frequency of *A* in generation *n*, and for convenience, set $\mu = 10^{-5}$. The probability of a particular *A* allele not mutating in any one generation is $1 - \mu$, hence $p_1 = (1 - \mu)p_0$, but because $p_0 = 1$, $p_1 = 1 - \mu = 0.99999$. The relationship between p_2 and p_1 is the same as that between p_1 and p_0, and so $p_2 = (1 - \mu)p_1 = (1 - \mu)(1 - \mu)p_0 = (1 - \mu)^2 p_0 = (1 - \mu)^2 = 0.99998$. In general, the rule is $p_n = (1 - \mu)^n p_0$, which, in this example, is $(0.99999)^n$.

Chapter 14

14.2 Broad-sense heritability is the proportion of the phenotypic variance attributable to all differences in genotype, which includes dominance effects and interactions between alleles. Narrow-sense heritability is the proportion of the phenotypic variance due only to additive ("transmissible") gene effects, which is used to predict the resemblance between parents and offspring in artificial selection. If all allele frequencies equal 1 or 0, both heritabilities equal zero.

14.4 $\Sigma x_i = 2480$, so the estimated mean equals 2480/50 = 49.60 eggs. To estimate the variance, subtract the mean from each value, square, sum these together, and divide by 49; in symbols, $s^2 = \Sigma(x_i - 49.60)^2/49 = 60.3265$. The standard deviation is the square root of the variance, hence $s = \sqrt{(60.3265)} = 7.76702$. (*Note:* An alternative method of calculating the estimate of the variance is to use the sum of the values and the sum of the squares as follows: $s^2 = N/(N - 1) \times [\Sigma x^2/N - (\Sigma x/N)^2]$. In the present example, this expression evaluates to $(49/50) \times [125{,}964/50 - (49.60)^2] = 60.3265$, the same as calculated earlier.)

14.6 For fruit weight, $\sigma_e^2/\sigma_t^2 = 0.13$ is given. Because $\sigma_t^2 = \sigma_g^2 + \sigma_e^2$, then $H^2 = \sigma_g^2/\sigma_t^2 = 1 - \sigma_e^2/\sigma_t^2 = 1 - 0.13 = 0.87$. Therefore, 87 percent is the broad-sense heritability of fruit weight. The values of H^2 for soluble-solid content and acidity are 91 percent and 89 percent, respectively.

14.8 The expected average height of the progeny equals $17 + 0.30 \times (16 - 17) = 16.7$ hands. The heritability is a ratio of variances and is independent of the unit of measurement. Thus whether the withers height is measured in meters or in hands makes no difference in the heritability.

14.10 The range 1.8 to 2.2 mg is the mean ± 1 standard deviation, and 68 percent of the population is expected to fall within this range. Similarly, the range 1.6 to 2.4 mg is the mean ± 2 standard deviations, and 95 percent of the population is expected to fall within this range. A beetle with a pupal weight of 3.0 mg or more deviates by more than +5 standard deviations from the mean. Because only 0.15 percent of the population deviates by more than +3 standard deviations, animals with pupae weighing more than 3.0 mg are expected to be exceedingly rare and, in practice, are not found. (The expected proportion weighing ≥ 3.0 mg actually equals 3×10^{-7}, or roughly 1 animal in 3 million.)

14.12 The narrow-sense heritability equals 8 times the first-cousin correlation coefficient, so the narrow-sense heritability of total fingerprint ridge count is 80 percent.

Glossary

acentric chromosome A chromosome with no centromere.
acridine A chemical mutagen that intercalates between the bases of a DNA molecule, causing single-base insertions or deletions.
acrocentric chromosome A chromosome with the centromere near one end.
adaptation Any characteristic of an organism that improves its chance of survival and reproduction in its environment; the evolutionary process by which a species undergoes progressive modification favoring its survival and reproduction in a given environment.
addition rule The principle that the probability that any one of a set of mutually exclusive events is realized equals the sum of the probabilities of the separate events.
adenine (A) A nitrogenous purine base found in DNA and RNA.
adjacent segregation Type of segregation from a heterozygous reciprocal translocation in which a structurally normal chromosome segregates with a translocated chromosome. In adjacent-1 segregation, homologous centromeres go to opposite poles of the first-division spindle; in adjacent-2 segregation, homologous centromeres go to the same pole of the first-division spindle.
albinism Absence of melanin pigment in the iris, skin, and hair of an animal; absence of chlorophyll in plants.
alkaptonuria A recessively inherited metabolic disorder in which a defect in the breakdown of tyrosine leads to excretion of homogentisic acid (alkapton) in the urine.
alkylating agent An organic compound capable of transferring an alkyl group to other molecules.
allele Any of the alternative forms of a given gene.
allele frequency The relative proportion of all alleles of a gene that are of a designated type.
allopolyploid A polyploid formed by hybridization between two different species.
allozyme Any of the alternative electrophoretic forms of a protein coded by different alleles of a single gene.
alpha satellite Highly repetitive DNA sequences associated with mammalian centromeres.
alternate segregation Segregation from a heterozygous reciprocal translocation in which both parts of the reciprocal translocation separate from both nontranslocated chromosomes in the first meiotic division.
Ames test A bacterial test for mutagenicity; also used to screen for potential carcinogens.
amino acid Any one of a class of organic molecules that have an amino group and a carboxyl group; 20 different amino acids are the usual components of proteins.
aminoacylated tRNA A tRNA covalently attached to its amino acid; charged tRNA.
amino terminus The end of a polypeptide chain at which the amino acid bears a free amino group ($-NH_2$).
amniocentesis A procedure for obtaining fetal cells from the amniotic fluid for the diagnosis of genetic abnormalities.
anaphase The stage of mitosis or meiosis in which chromosomes move to opposite ends of the spindle. In anaphase I of meiosis, homologous centromeres separate; in anaphase II, sister centromeres separate.
aneuploid A cell or organism in which the chromosome number is not an exact multiple of the haploid number; more generally, aneuploidy refers to a condition in which particular genes or chromosomal regions are present in extra or fewer copies compared with wildtype.
antibiotic-resistant mutant A cell or organism that carries a mutation conferring resistance to an antibiotic.
antibody A blood protein produced in response to a specific antigen and capable of binding with the antigen.
anticodon The three bases in a tRNA molecule that are complementary to the three-base codon in mRNA.
antigen A substance able to stimulate the production of antibodies.
antiparallel The chemical orientation of the two strands of a double-stranded nucleic acid molecule; the 5'-to-3' orientations of the two strands are opposite one another.
AP endonuclease An endonuclease that cleaves a DNA strand at any site at which the deoxyribose lacks a base.
apoptosis Genetically programmed cell death, especially in embryonic development.
aporepressor A protein converted into a repressor by binding with a particular molecule.
Archaea One of the three major classes of organisms; also called archaebacteria, they are unicellular microorganisms, usually found in extreme environments, that differ as much from bacteria as either group differs from eukaryotes. *See also* **Bacteria.**
artificial selection Selection, imposed by a breeder, in which organisms of only certain phenotypes are allowed to breed.
ascospore *See* **ascus.**
ascus A sac containing the spores (ascospores) produced by meiosis in certain groups of fungi, including *Neurospora* and yeast.
attached-X chromosome A chromosome in which two X chromosomes are joined to a common centromere; also called a compound-X chromosome.
attenuation *See* **attenuator.**
attenuator A regulatory base sequence near the beginning of an mRNA molecule at which transcription can be terminated; when an attenuator is present, it precedes the coding sequences.
autonomous determination Cellular differentiation determined intrinsically and not dependent on external signals or interactions with other cells.
autopolyploid organism An organism whose cells contain more than two sets of homologous chromosomes.
autoregulation Regulation of gene expression by the product of the gene itself.
autosomes All chromosomes other than the sex chromosomes.
auxotroph A mutant microorganism that is unable to synthesize a compound required for its growth but is able to grow if the compound is provided.

BAC *See* **bacterial artificial chromosome.**
backcross The cross of an F_1 heterozygote with a partner that has the same genotype as one of its parents.
Bacteria One of the major kingdoms of living things; includes most bacteria. *See also* **Archaea.**
bacterial artificial chromosome A plasmid vector with regions derived from the F plasmid that contains a large fragment of cloned DNA.
bacteriophage A virus that infects bacterial cells; commonly called a phage.
Barr body A darkly staining body found in the interphase nucleus of certain cells of female mammals; consists of the condensed, inactivated X chromosome.
base Single-ring (pyrimidine) or double-ring (purine) component of a nucleic acid.
base analog A chemical so similar to one of the normal bases that it can be incorporated into DNA.
base pair A pair of nitrogenous bases, most commonly one purine and one pyrimidine, held together by hydrogen bonds in a double-stranded region of a nucleic acid molecule; commonly abbreviated bp, the term is often used interchangeably with the term *nucleotide pair.* The normal base pairs in DNA are A—T and G—C.
base-substitution mutation Incorporation of an incorrect base into a DNA duplex.
β-galactosidase An enzyme that cleaves lactose into its glucose and galactose constituents; produced by a gene in the *lac* operon.
biochemical pathway A diagram showing the order in which intermediate molecules are produced in the synthesis or degradation of a metabolite in a cell.
bivalent A pair of homologous chromosomes, each consisting of two chromatids, associated in meiosis I.
blastoderm Structure formed in the early development of an insect larva; the syncytial blastoderm is formed from repeated cleavage of the zygote nucleus without cytoplasmic division; the cellular blasto-

derm is formed by migration of the nuclei to the surface and their inclusion in separate cell membranes.
blastula A hollow sphere of cells formed early in development.
block in a biochemical pathway Stoppage in a reaction sequence due to a defective or missing enzyme.
blunt ends Ends of a DNA molecule in which all terminal bases are paired; the term usually refers to termini formed by a restriction enzyme that does not produce single-stranded ends.
broad-sense heritability The ratio of genotypic variance to total phenotypic variance.

cAMP-CRP complex A regulatory complex consisting of cyclic AMP (cAMP) and the CAP protein; the complex is needed for transcription of certain operons.
candidate gene A gene proposed to be involved in the genetic determination of a trait because of the role of the gene product in the cell or organism.
cap A complex structure at the 5' termini of most eukaryotic mRNA molecules, having a 5'-5' linkage instead of the usual 3'-5' linkage.
carbon-source mutant A cell or organism that carries a mutation preventing the use of a particular molecule or class of molecules as a source of carbon.
carboxyl terminus The end of a polypeptide chain at which the amino acid has a free carboxyl group (—COOH).
carrier A heterozygote for a recessive allele.
cassette In yeast, either of two sets of inactive mating-type genes that can become active by relocating to the *MAT* locus.
cDNA *See* **complementary DNA**.
cell cycle The growth cycle of a cell; in eukaryotes, it is subdivided into G_1 (gap 1), S (DNA synthesis), G_2 (gap 2), and M (mitosis).
cell fate The pathway of differentiation that a cell normally undergoes.
cell lineage The ancestor-descendant relationships of a group of cells in development.
cellular oncogene A gene coding for a cellular growth factor whose abnormal expression predisposes to malignancy. *See also* **oncogene**.
centimorgan A unit of distance in the genetic map equal to 1 percent recombination; also called a map unit.
central dogma The concept that genetic information is transferred from the nucleotide sequence in DNA to the nucleotide sequence in an RNA transcript to the amino acid sequence of a polypeptide chain.
centromere The region of the chromosome that is associated with spindle fibers and that participates in normal chromosome movement during mitosis and meiosis.
chain elongation The process of addition of successive amino acids to the growing end of a polypeptide chain.
chain initiation The process by which polypeptide synthesis is begun.
chain termination The process of ending polypeptide synthesis and releasing the polypeptide from the ribosome; a chain-termination mutation creates a new stop codon, resulting in premature termination of synthesis of the polypeptide chain.
charged tRNA A tRNA molecule to which an amino acid is linked; acylated tRNA.
chiasma The cytological manifestation of crossing-over; the cross-shaped exchange configuration between nonsister chromatids of homologous chromosomes that is visible in prophase I of meiosis. The plural can be either *chiasmata* or *chiasmas*.
chimeric gene A gene produced by recombination, or by genetic engineering, that is a mosaic of DNA sequences from two or more different genes.
chi-square (χ^2) A statistical quantity calculated to assess the goodness of fit between a set of observed numbers and the theoretically expected numbers.
chromatid Either of the longitudinal subunits produced by chromosome replication.
chromatid interference In meiosis, the effect that crossing-over between one pair of nonsister chromatids may have on the probability that a second crossing-over in the same chromosome will involve the same or different chromatids; chromatid interference does not generally occur.
chromatin The aggregate of DNA and histone proteins that makes up a eukaryotic chromosome.
chromomere A tightly coiled, bead-like region of a chromosome most readily seen during the pachytene substage of meiosis; the beads are in register in a polytene chromosome, resulting in the banded appearance of the chromosome.
chromosome In eukaryotes, a DNA molecule that contains genes in linear order to which numerous proteins are bound and that has a telomere at each end and a centromere; in prokaryotes, the DNA is associated with fewer proteins, lacks telomeres and a centromere, and is often circular; in viruses, the chromosome is DNA or RNA, single-stranded or double-stranded, linear or circular, and often free of bound proteins.
chromosome complement The set of chromosomes in a cell or organism.
chromosome map A diagram showing the locations and relative spacing of genes along a chromosome.
chromosome painting Use of differentially labeled, chromosome-specific DNA strands for hybridization with chromosomes to label each chromosome with a different color.
chromosome theory of heredity The theory that chromosomes are the cellular objects that contain the genes.
circular permutation A permutation of a group of elements in which the elements are in the same order but the beginning of the sequence differs.
***cis* configuration** The arrangement of linked genes in a double heterozygote in which both mutations are present in the same chromosome—for example, $a_1\ a_2/+\ +$; also called coupling.
***cis*-dominant** Of or pertaining to a mutation that affects the expression of only those genes on the same DNA molecule.
***cis* heterozygote** *See* ***cis* configuration**.
cistron A DNA sequence specifying a single genetic function as defined by a complementation test; a nucleotide sequence coding for a single polypeptide.
***ClB* method** A genetic procedure used to detect X-linked recessive lethal mutations in *Drosophila melanogaster;* so named because one X chromosome in the female parent is marked with an inversion (*C*), a recessive lethal allele (*l*), and the dominant allele for Bar eyes (*B*).
cleavage division Mitosis in the early embryo.
clone A collection of organisms derived from a single parent and, except for new mutations, genetically identical to that parent; in genetic engineering, the linking of a specific gene or DNA fragment to a replicable DNA molecule, such as a plasmid or phage DNA.
cloned DNA sequence A DNA fragment inserted into a vector and transformed into a host organism.
cloned gene A DNA sequence incorporated into a vector molecule capable of replication in the same or a different organism.
cloning The process of producing cloned genes.
coding region The part of a DNA sequence that codes for the amino acids in a protein.
coding sequence A region of a DNA strand with the same sequence as is found in the coding region of a messenger RNA, except that T is present in DNA instead of U.
codominance The expression of both alleles in a heterozygote.
codon A sequence of three adjacent nucleotides in an mRNA molecule, specifying either an amino acid or a stop signal in protein synthesis.
coefficient of coincidence An experimental value obtained by dividing the observed number of double recombinants by the expected number calculated under the assumption that the two events take place independently.
cohesive end A single-stranded region at the end of a double-stranded DNA molecule that can adhere to a complementary single-stranded sequence at the other end or in another molecule.
colchicine A chemical that prevents formation of the spindle in nuclear division.
colinearity The linear correspondence between the order of amino acids in a polypeptide chain and the corresponding sequence of nucleotides in the DNA molecule.
colony A visible cluster of cells formed on a solid growth medium by repeated division of a single parental cell and its daughter cells.
colony hybridization assay A technique for identifying colonies that contain a particular cloned gene; many colonies are transferred to a filter, lysed, and exposed to radioactive DNA or RNA complementary to the DNA sequence of interest, after which colonies that contain a sequence complementary to the probe are located by autoradiography.

color blindness In human beings, the usual form of color blindness is X-linked red-green color blindness. Unequal crossing-over between the adjacent red and green opsin pigment genes results in chimeric opsin genes that cause mild or severe green-vision defects (deuteranomaly or deuteranopia, respectively) and mild or severe red-vision defects (protanomaly or protanopia, respectively).
combinatorial control Strategy of gene regulation in which a relatively small number of time- and tissue-specific positive and negative regulatory elements are used in various combinations to control the expression of a much larger number of genes.
complementary DNA (cDNA) A DNA molecule made by copying RNA with reverse transcriptase.
complementation The phenomenon in which two recessive mutations with similar phenotypes result in a wildtype phenotype when both are heterozygous in the same genotype; complementation means that the mutations are in different genes.
complementation group A group of mutations that fail to complement one another.
complementation test A genetic test to determine whether two mutations are alleles (are present in the same functional gene).
complex trait A multifactorial trait influenced by multiple genetic and environmental factors, each of relatively small effect, and their interactions.
conditional mutation A mutation that results in a mutant phenotype under certain (restrictive) environmental conditions but results in a wildtype phenotype under other (permissive) conditions.
conjugation A process of DNA transfer in sexual reproduction in certain bacteria; in *E. coli*, the transfer is unidirectional, from donor cell to recipient cell. Also, a mating between cells of *Paramecium*.
consensus sequence A generalized base sequence derived from closely related sequences found in many locations in a genome or in many organisms; each position in the consensus sequence consists of the base found in the majority of sequences at that position.
conserved sequence A base or amino acid sequence that changes very slowly in the course of evolution.
constant antibody region The part of the heavy and light chains of an antibody molecule that has the same amino acid sequence among all antibodies derived from the same heavy-chain and light-chain genes.
constitutive mutant A mutant in which synthesis of a particular mRNA molecule (and the protein that it encodes) takes place at a constant rate independent of the presence or absence of any inducer or repressor molecule.
contig A set of cloned DNA fragments overlapping in such a way as to provide unbroken coverage of a contiguous region of the genome; a contig contains no gaps.
continuous trait A trait in which the possible phenotypes have a continuous range from one extreme to the other rather than falling into discrete classes.
coordinate gene Any of a group of genes that establish the basic anterior-posterior and dorsal-ventral axes of the early embryo.
coordinate regulation Control of synthesis of several proteins by a single regulatory element; in prokaryotes, the proteins are usually translated from a single mRNA molecule.
core particle The aggregate of histones and DNA in a nucleosome, without the linking DNA.
co-repressor A small molecule that binds with an aporepressor to create a functional repressor molecule.
correlated response Change of the mean in one trait in a population accompanying selection for another trait.
correlation coefficient A measure of association between pairs of numbers, equaling the covariance divided by the product of the standard deviations.
cotransduction Transduction of two or more linked genetic markers by one transducing particle.
cotransformation Transformation in bacteria of two genetic markers carried on a single DNA fragment.
counterselected marker A mutation used to prevent growth of a donor cell in an Hfr $\times$ F^- bacterial mating.
coupled transcription-translation In prokaryotes, the translation of an mRNA molecule before its transcription is completed.
coupling *See* ***cis*** **configuration.**
covariance A measure of association between pairs of numbers that is defined as the average product of the deviations from the respective means.
crossing-over A process of exchange between nonsister chromatids of a pair of homologous chromosomes that results in the recombination of linked genes.
cryptic splice site A potential splice site not normally used in RNA processing unless a normal splice site is blocked or mutated.
cyclin One of a group of proteins that participates in controlling the cell cycle. Different types of cyclins interact with the p34 kinase subunit and regulate the G_1/S and G_2/M transitions. The proteins are called cyclins because their abundance rises and falls rhythmically in the cell cycle.
cytological map Diagrammatic representation of a chromosome.
cytosine (C) A nitrogenous pyrimidine base found in DNA and RNA.

daughter strand A newly synthesized DNA or chromosome strand.
deficiency *See* **deletion.**
degeneracy *See* **redundancy.**
degrees of freedom An integer that determines the significance level of a particular statistical test. In the goodness-of-fit type of chi-square test in which the expected numbers are not based on any quantities estimated from the data themselves, the number of degrees of freedom is one less than the number of classes of data.
deletion Loss of a segment of the genetic material from a chromosome; also called deficiency.
deoxyribonucleic acid *See* **DNA.**
deoxyribose The five-carbon sugar present in DNA.
depurination Removal of purine bases from DNA.
diakinesis The substage of meiotic prophase I, immediately preceding metaphase I, in which the bivalents attain maximum shortening and condensation.
dicentric chromosome A chromosome with two centromeres.
dideoxyribose A deoxyribose sugar that lacks the 3' hydroxyl group; when incorporated into a polynucleotide chain, it blocks further chain elongation.
dideoxy sequencing method Procedure for DNA sequencing in which a template strand is replicated from a particular primer sequence and terminated by the incorporation of a nucleotide that contains dideoxyribose instead of deoxyribose; the resulting fragments are separated by size via electrophoresis.
dihybrid Heterozygous at each of two loci; progeny of a cross between true-breeding, or homozygous, strains that differ genetically at two loci.
diploid A cell or organism with two complete sets of homologous chromosomes.
diplotene The substage of meiotic prophase I, immediately following pachytene and preceding diakinesis, in which pairs of sister chromatids that make up a bivalent (tetrad) begin to separate from each other and chiasmata become visible.
direct repeat Copies of an identical or very similar DNA or RNA base sequence in the same molecule and in the same orientation.
distribution In quantitative genetics, the mathematical relation that gives the proportion of members in a population that have each possible phenotype.
DNA Deoxyribonucleic acid, the macromolecule, usually composed of two polynucleotide chains in a double helix, that is the carrier of the genetic information in all cells and many viruses.
DNA chip A small plate of silicon, glass, or other material containing an array of oligonucleotides to which DNA samples are hybridized.
DNA cloning *See* **cloned gene.**
DNA ligase An enzyme that catalyzes formation of a covalent bond between adjacent 5'-P and 3'-OH termini in a broken polynucleotide strand of double-stranded DNA.
DNA looping A mechanism by which enhancers that are distant from the immediate proximity of a promoter can still regulate transcription; the enhancer and promoter, both bound with suitable protein factors, come into indirect physical contact by means of the looping out of the DNA between them. The physical interaction stimulates transcription.
DNA methylase An enzyme that adds methyl groups ($-CH_3$) to certain bases, particularly cytosine.
DNA polymerase Any enzyme that catalyzes the synthesis of DNA from deoxynucleoside 5'-triphosphates, using a template strand.
DNA repair Any of several different processes for restoration of the correct base sequence of a DNA molecule into which incorrect bases have been incorporated or whose bases have been chemically modified.

DNA replication The semiconservative copying of a DNA molecule.
DNA typing Electrophoretic identification of individual persons by the use of DNA probes for highly polymorphic regions of the genome, such that the genome of virtually every person exhibits a unique pattern of bands; sometimes called DNA fingerprinting.
DNA uracyl glycosylase An enzyme that removes uracil bases when they occur in double-stranded DNA.
dominant gene An allele whose presence in a heterozygous genotype results in a phenotype characteristic of the allele.
dosage compensation A mechanism regulating X-linked genes such that their activities are equal in males and females; in mammals, random inactivation of one X chromosome in females results in equal amounts of the products of X-linked genes in males and females.
double-stranded DNA A DNA molecule consisting of two antiparallel strands that are complementary in nucleotide sequence.
double-Y syndrome The clinical features of the karyotype 47,XYY.
Down syndrome The clinical features of the karyotype 47,+21 (trisomy 21).
duplex DNA A double-stranded DNA molecule.
duplication A chromosome aberration in which a chromosome segment is present more than once in the haploid genome; if the two segments are adjacent, the duplication is a tandem duplication.

editing function The activity of DNA polymerases that removes incorrectly incorporated nucleotides; also called the proofreading function.
electrophoresis A technique used to separate molecules on the basis of their different rates of movement in response to an applied electric field, typically through a gel.
embryoid A small mass of dividing cells formed from haploid cells in anthers that can give rise to a mature haploid plant.
embryonic stem cells Cells in the blastocyst that give rise to the body of the embryo.
endonuclease An enzyme that breaks internal phosphodiester bonds in a single- or double-stranded nucleic acid molecule; usually specific for either DNA or RNA.
endoreduplication Doubling of the chromosome complement because of chromosome replication and centromere division without nuclear or cytoplasmic division.
enhancer A base sequence in eukaryotes and eukaryotic viruses that increases the rate of transcription of nearby genes; the defining characteristics are that it need not be adjacent to the transcribed gene and that the enhancing activity is independent of orientation with respect to the gene.
environmental variance The part of the phenotypic variance that is attributable to differences in environment.
enzyme A protein that catalyzes a specific biochemical reaction and is not itself altered in the process.
epistasis A term referring to an interaction between nonallelic genes in their effects on a trait. Generally, *epistasis* means any type of interaction in which the genotype at one locus affects the phenotypic expression of the genotype at another locus. In a more restricted sense, it refers to a situation in which the genotype at one locus determines the phenotype in such a way as to mask the genotype present at a second locus.
equational division Term applied to the second meiotic division because the haploid chromosome complement is retained throughout.
EST *See* **expressed sequence tag.**
euchromatin A region of a chromosome that has normal staining properties and undergoes the normal cycle of condensation; relatively uncoiled in the interphase nucleus (compared with condensed chromosomes), it apparently contains most of the genes.
Eukarya One of the major kingdoms of living organisms, in which the cells have a true nucleus and divide by mitosis or meiosis.
eukaryote A cell with a true nucleus (DNA enclosed in a membranous envelope) in which cell division takes place by mitosis or meiosis; an organism composed of eukaryotic cells.
euploid A cell or an organism having a chromosome number that is an exact multiple of the haploid number.
evolution Cumulative change in the genetic characteristics of a species through time.
excisionase An enzyme that is needed for prophage excision; works together with an integrase.
excision repair Type of DNA repair in which segments of a DNA strand that are chemically damaged are removed enzymatically and then resynthesized, using the other strand as a template.
exon The sequences in a gene that are retained in the messenger RNA after the introns are removed from the primary transcript.
exon shuffle The theory that new genes can evolve by the assembly of separate exons from preexisting genes, each coding for a discrete functional domain in the new protein.
exonuclease An enzyme that removes a terminal nucleotide in a polynucleotide chain by cleavage of the terminal phosphodiester bond; nucleotides are removed successively, one by one; usually specific for either DNA or RNA and for either single-stranded or double-stranded nucleic acids. A 5'-to-3' exonuclease cleaves successive nucleotides from the 5' end of the molecule; a 3'-to-5' exonuclease cleaves successive nucleotides from the 3' end.
expressed sequence tag A partial or complete cDNA sequence.

factorial Of a number, the product of all integers from 1 through the number itself; 0! is defined to equal 1.
fate map A diagram of the insect blastoderm identifying the regions from which particular adult structures derive.
F_1 generation The first generation of descent from a given mating.
F_2 generation The second generation of descent from a given mating, produced by intercrossing or self-fertilizing F_1 organisms.
first-division segregation Separation of a pair of alleles into different nuclei in the first meiotic division; happens when there is no crossing-over between the gene and the centromere in a particular cell.
first meiotic division The meiotic division that reduces the chromosome number; sometimes called the reduction division.
fitness A measure of the average ability of organisms with a given genotype to survive and reproduce.
5' end The end of a DNA or RNA strand that terminates in a free phosphate group not connected to a sugar farther along.
fixed allele An allele whose allele frequency equals 1.0.
F^- strain A strain, typically of *Escherichia coli*, lacking the F plasmid.
folding domain A short region of a polypeptide chain within which interactions between amino acids result in a three-dimensional conformation that is attained relatively independently of the folding of the rest of the molecule.
forward mutation A change from a wildtype allele to a mutant allele.
F plasmid A bacterial plasmid—often called the F factor, fertility factor, or sex plasmid—that is capable of transferring itself from a host (F^+) cell to a cell not carrying an F factor (F^- cell); when an F factor is integrated into the bacterial chromosome (in an Hfr cell), the chromosome becomes transferrable to an F^- cell during conjugation.
F^+ strain A strain, typically of *Escherichia coli*, containing an unintegrated F plasmid.
F' plasmid An F plasmid that contains genes obtained from the bacterial chromosome in addition to plasmid genes; formed by aberrant excision of an integrated F factor, taking along adjacent bacterial DNA.
fragile-X chromosome A type of X chromosome containing a site toward the end of the long arm that tends to break in cultured cells that are starved for DNA precursors; causes fragile-X syndrome.
frameshift mutation A mutational event caused by the insertion or deletion of one or more nucleotide pairs in a gene, resulting in a shift in the reading frame of all codons following the mutational site.
fraternal twins Twins that result from the fertilization of separate ova and are genetically related as siblings; also called dizygotic twins.
free radical A highly reactive molecule produced when ionizing radiation interacts with water; free radicals are potent oxidizing agents.
frequency of cotransduction The proportion of transductants carrying a selected genetic marker that also carry a nonselected genetic marker.
frequency of recombination The proportion of gametes carrying combinations of alleles that are not present in either parental chromosome.

gain-of-function mutation Mutation in which a gene is overexpressed or inappropriately expressed.
gamete A mature reproductive cell, such as a sperm or egg in animals.
gametophyte In plants, the haploid part of the life cycle that produces the gametes by mitosis.

gap gene Any of a group of genes that control the development of contiguous segments or parasegments in *Drosophila* such that mutations result in gaps in the pattern of segmentation.
gastrula Stage in early animal development marked by extensive cell migration.
gel electrophoresis *See* **electrophoresis.**
gene The hereditary unit defined experimentally by the complementation test. At the molecular level, a region of DNA containing genetic information, usually transcribed into an RNA molecule that is processed and either functions directly or is translated into a polypeptide chain; a gene can mutate to various forms called alleles.
gene amplification A process in which certain genes undergo differential replication either within the chromosome or extrachromosomally, increasing the number of copies of the gene.
gene cloning *See* **cloned gene.**
gene conversion The phenomenon in which the products of a meiotic division in an *Aa* heterozygous genotype are in some ratio other than the expected 1*A* : 1*a*—for example, 3*A* : 1*a*, 1*A* : 3*a*, 5*A* : 3*a*, or 3*A* : 5*a*.
gene dosage Number of gene copies.
gene expression The multistep process by which a gene is regulated and its product synthesized.
gene pool The totality of genetic information in a population of organisms.
gene product A term used for the polypeptide chain translated from an mRNA molecule transcribed from a gene; if the RNA is not translated (for example, ribosomal RNA), the RNA molecule is the gene product.
generalized transducing phage *See* **transducing phage.**
general transcription factor A protein molecule needed to bind with a promoter before transcription can proceed; transcription factors are necessary, but not sufficient, for transcription, and they are shared among many different promoters.
gene regulation Processes by which gene expression is controlled in response to external or internal signals.
gene targeting Disruption or mutation of a designated gene by homologous recombination.
gene therapy Deliberate alteration of the human genome for alleviation of disease.
genetic architecture Of a complex trait, specification of the genetic and environmental factors that contribute to the trait, and their interactions.
genetic code The set of 64 triplets of bases (codons) that correspond to the 20 amino acids in proteins and the signals for initiation and termination of polypeptide synthesis.
genetic engineering The linking of two DNA molecules by *in vitro* manipulations for the purpose of generating a novel organism with desired characteristics.
genetic map *See* **linkage map.**
genetic marker Any pair of alleles whose inheritance can be traced through a mating or through a pedigree.
genetics The study of biological heredity.
genome The total complement of genes contained in a cell or virus; commonly used to refer to all genes present in one complete haploid set of chromosomes in eukaryotes.
genotype The genetic constitution of an organism or virus, typically with respect to one or a few genes of interest, as distinguished from its appearance, or phenotype.
genotype-by-sex interaction Genetic determination that differs between the sexes to result in different phenotypes for the same genotype depending on the sex of the individual.
genotype-environment association The condition in which genotypes and environments are not in random combinations.
genotype-environment interaction The condition in which genetic and environmental effects on a trait are not additive.
genotype frequency The proportion of members of a population that are of a prescribed genotype.
genotypic variance The part of the phenotypic variance that is attributable to differences in genotype.
germ cell A cell that gives rise to reproductive cells.
germ line Cell lineage consisting of germ cells.
germ-line mutation *See* **germinal mutation.**
goodness of fit The extent to which observed numbers agree with the numbers expected on the basis of some specified genetic hypothesis.
G_1 period *See* **cell cycle.**
G_2 period *See* **cell cycle.**
guanine (G) A nitrogenous purine base found in DNA and RNA.
guide RNA The RNA template present in telomerase.

haploid A cell or organism of a species containing the set of chromosomes normally found in gametes.
Hardy–Weinberg principle The genotype frequencies expected with random mating.
hemophilia A One of two X-linked forms of hemophilia; patients are deficient in blood-clotting factor VIII.
heritability A measure of the degree to which a phenotypic trait can be modified by selection. *See also* **broad-sense heritability** and **narrow-sense heritability.**
heterochromatin Chromatin that remains condensed and heavily stained during interphase; commonly present adjacent to the centromere and in the telomeres of chromosomes. Some chromosomes are composed primarily of heterochromatin.
heteroduplex All or part of a double-stranded nucleic acid molecule in which the two strands have different hereditary origins; produced either as an intermediate in recombination or by the *in vitro* annealing of single-stranded complementary molecules.
heterosis The superiority of hybrids over either inbred parent with respect to one or more traits; also called hybrid vigor.
heterozygote superiority The condition in which a heterozygous genotype has greater fitness than either of the homozygotes.
heterozygous Carrying dissimilar alleles of one or more genes; not homozygous.
hexaploid A cell or organism with six complete sets of chromosomes.
H4 histone *See* **histone.**
Hfr strain An *E. coli* strain in which an F plasmid is integrated into the chromosome, making possible the transfer of part or all of the chromosome to an F^- cell.
histone Any of the small basic proteins bound to DNA in chromatin; the five major histones are designated H1, H2A, H2B, H3, and H4. Each nucleosome core particle contains two molecules each of H2A, H2B, H3, and H4. The H1 histone forms connecting links between nucleosome core particles.
Holliday junction resolving enzyme An enzyme that catalyzes the breakage and rejoining of two DNA strands in a Holliday junction to generate two independent duplex molecules.
Holliday model A molecular model of genetic recombination in which the participating duplexes contain heteroduplex regions of the same length.
Holliday structure A cross-shaped configuration of two DNA duplexes formed as an intermediate in recombination.
homeobox A DNA sequence motif found in the coding region of many regulatory genes; the amino acid sequence corresponding to the homeobox has a helix-loop-helix structure.
homeotic (HOX) gene Any of a group of genes in which a mutation results in the replacement of one body structure by another body structure.
homogentisic acid Substance excreted in the urine of alkaptonurics that turns black upon oxidation.
homologous In reference to DNA, having the same or nearly the same nucleotide sequence as a result of common ancestry.
homologous chromosomes Chromosomes that pair in meiosis and have the same genetic loci and structure; also called homologs.
homothallism The capacity of cells in certain fungi to undergo a conversion in mating type to make possible mating between cells produced by the same parental organism.
homozygous Having the same allele of a gene in homologous chromosomes.
hormone A small molecule in higher eukaryotes, synthesized in specialized tissue, that regulates the activity of other specialized cells. In animals, hormones are transported from their source to a target tissue in the blood.
hot spot A site in a DNA molecule at which the mutation rate is much higher than the rate for most other sites.
housekeeping gene A gene that is expressed at the same level in virtually all cells and whose product participates in basic metabolic processes.
human genome project A worldwide project to map genetically and sequence the human genome.
Huntington disease Dominantly inherited degeneration of the neuromuscular system, with onset in middle age.

hybrid An organism produced by the mating of genetically unlike parents; also, a duplex nucleic acid molecule produced of strands derived from different sources.
hybrid vigor *See* **heterosis.**
hydrogen bond A weak noncovalent linkage between two negatively charged atoms in which a hydrogen atom is shared.

identical twins Twins developed from a single fertilized egg that splits into two embryos at an early division; also called monozygotic twins.
imaginal disk Structures present in the body of insect larvae from which the adult structures develop during pupation.
immunity A general term for resistance of an organism to specific substances, particularly agents of disease.
imprinting A process of DNA modification in gametogenesis that affects gene expression in the zygote; a probable mechanism is the methylation of certain bases in the DNA.
inborn error of metabolism A genetically determined biochemical disorder, usually in the form of an enzyme defect that produces a metabolic block.
inbreeding Mating between relatives.
inbreeding coefficient A measure of the genetic effects of inbreeding in terms of the proportionate reduction in heterozygosity in an inbred organism compared with the heterozygosity expected with random mating.
incomplete penetrance Condition in which a mutant phenotype is not expressed in all organisms with the mutant genotype.
independent assortment Random distribution of unlinked genes into gametes, as with genes in different (nonhomologous) chromosomes or genes that are so far apart on a single chromosome that the recombination frequency between them is 1/2.
individual selection Selection based on each organism's own phenotype.
induced mutation A mutation formed under the influence of a chemical mutagen or radiation.
inducer A small molecule that inactivates a repressor, usually by binding to it and thereby altering the ability of the repressor to bind to an operator.
inducible transcription Transcription of a gene, or a group of genes, only in the presence of an inducer molecule.
induction Activation of an inducible gene; prophage induction is the derepression of a prophage that initiates a lytic cycle of phage development.
initiation The beginning of protein synthesis.
inosine (I) One of a number of unusual bases found in transfer RNA.
insertion sequence A DNA sequence capable of transposition in a prokaryotic genome; such sequences usually code for their own transposase.
integrase An enzyme that catalyzes the site-specific exchange when a prophage is inserted into or excised from a bacterial chromosome; in the excision process, an accessory protein, excisionase, also is needed.
integrated phage A state in which the phage DNA molecule is inserted intact into the bacterial chromosome; the integrated phage is called a prophage.
intercalation Insertion of a planar molecule between the stacked bases in duplex DNA.
interference The tendency for crossing-over to inhibit the formation of another crossover nearby.
interphase The interval between nuclear divisions in the cell cycle, extending from the end of telophase of one division to the beginning of prophase of the next division.
interrupted-mating technique In an Hfr $\times$ F^- cross, a technique by which donor and recipient cells are broken apart at specific times, allowing only a particular length of DNA to be transferred.
intervening sequence *See* **intron.**
intron A noncoding DNA sequence in a gene that is transcribed but is then excised from the primary transcript in forming a mature mRNA molecule; found primarily in eukaryotic cells. *See also* **exon.**
inversion A structural aberration in a chromosome in which the order of several genes is reversed from the normal order. A pericentric inversion includes the centromere within the inverted region, and a paracentric inversion does not include the centromere.
inversion loop Loop structure formed by synapsis of homologous genes in a pair of chromosomes, one of which contains an inversion.
inverted repeat Either of a pair of base sequences present in the same molecule that are identical or nearly identical but are oriented in opposite directions; often found at the ends of transposable elements.
ionizing radiation Electromagnetic or particulate radiation that produces ion pairs when dissipating its energy in matter.
IS element *See* **insertion sequence.**

karyotype The chromosome complement of a cell or organism; often represented by an arrangement of metaphase chromosomes according to their lengths and the positions of their centromeres.
kilobase pair (kb) Unit of length of a duplex DNA molecule; equal to 1000 base pairs.
kinetochore The cellular structure, formed in association with the centromere, to which the spindle fibers become attached in cell division.
Klinefelter syndrome The clinical features of human males with the karyotype 47,XXY.

lactose permease An enzyme responsible for transport of lactose from the environment into bacteria.
lagging strand The DNA strand whose complement is synthesized in short fragments that are ultimately joined together.
lariat structure Structure of an intron immediately after excision in which the 5' end loops back and forms a 5'-2' linkage with another nucleotide.
leader *See* **leader polypeptide.**
leader polypeptide A short polypeptide encoded in the leader sequence of some operons coding for enzymes in amino acid biosynthesis; translation of the leader polypeptide participates in regulation of the operon through attenuation.
leading strand The DNA strand whose complement is synthesized as a continuous unit.
leptotene The initial substage of meiotic prophase I during which the chromosomes become visible in the light microscope as unpaired thread-like structures.
liability Risk, particularly toward a threshold type of quantitative trait.
library *See* **gene library.**
ligand The molecule that binds to a specific receptor.
lineage diagram A diagram of cell lineages and their developmental fates.
linkage The tendency of genes located in the same chromosome to be associated in inheritance more frequently than expected from their independent assortment in meiosis.
linkage group The set of genes present together in a chromosome.
linkage map A diagram of the order of genes in a chromosome in which the distance between adjacent genes is proportional to the rate of recombination between them; also called a genetic map.
linker DNA In genetic engineering, synthetic DNA fragments that contain restriction-enzyme cleavage sites used to join two DNA molecules. *See also* **nucleosome.**
local population A group of organisms of the same species occupying an area within which most individual members find their mates; synonymous terms are *deme* and *Mendelian population.*
locus The site or position of a particular gene on a chromosome.
loss-of-function mutation A mutation that eliminates gene function; also called a null mutation.
lost allele An allele no longer present in a population; its frequency is 0.
lysis Breakage of a cell caused by rupture of its cell membrane and cell wall.
lysogen Clone of bacterial cells that have acquired a prophage.
lysogenic cycle In temperate bacteriophage, the phenomenon in which the DNA of an infecting phage becomes part of the genetic material of the cell.
lytic cycle The life cycle of a phage, in which progeny phage are produced and the host bacterial cell is lysed.

major groove In B-form DNA, the larger of two continuous indentations running along the outside of the double helix.
map-based cloning A strategy of gene cloning based on the position of a gene in the genetic map; also called positional cloning.
mapping function The mathematical relation between the genetic map distance across an interval and the observed percentage of recombination in the interval.

map unit A unit of distance in a linkage map that corresponds to a recombination frequency of 1 percent. Technically, the map distance across an interval in map units equals one-half the average number of crossovers in the interval, expressed as a percentage. Map units are sometimes called centimorgans (cM).
masked mRNA Messenger RNA that cannot be translated until specific regulatory substances are available; present in eukaryotic cells, particularly eggs; storage mRNA.
maternal effect gene A gene that influences early development through its expression in the mother and the presence of the gene product in the oocyte.
maternal inheritance Extranuclear inheritance of a trait through cytoplasmic factors or organelles contributed by the female gamete.
maternal PKU A condition that resembles phenylketonuria and results from embryonic development in the uterus of a woman deficient in phenylalanine hydroxylase.
mating-type interconversion Phenomenon in homothallic yeast in which cells switch mating type as a result of the transposition of genetic information from an unexpressed cassette into the active mating-type locus.
MCS Multiple cloning site. *See* **polylinker.**
mean The arithmetic average.
megabase pair Unit of length of a duplex nucleic acid molecule; equal to 1 million base pairs.
meiocyte A germ cell that undergoes meiosis to yield gametes in animals or spores in plants.
meiosis The process of nuclear division in gametogenesis or sporogenesis in which one replication of the chromosomes is followed by two successive divisions of the nucleus to produce four haploid nuclei.
Mendelian genetics The mechanism of inheritance in which the statistical relations between the distribution of traits in successive generations result from (1) particulate hereditary determinants (genes), (2) random union of gametes, and (3) segregation of unchanged hereditary determinants in the reproductive cells.
meristem The mitotically active growing point of plant tissue.
meristic trait A trait in which the phenotype is determined by counting, such as number of ears on a stalk of corn and number of eggs laid by a hen.
messenger RNA (mRNA) An RNA molecule transcribed from a DNA sequence and translated into the amino acid sequence of a polypeptide. In eukaryotes, the primary transcript undergoes elaborate processing to become the mRNA.
metabolic pathway A set of chemical reactions that take place in a definite order to convert a particular starting molecule into one or more specific products.
metabolite Any small molecule that serves as a substrate, an intermediate, or a product of a metabolic pathway.
metacentric chromosome A chromosome with its centromere about in the middle so that the arms are equal or almost equal in length.
metaphase In mitosis, meiosis I, or meiosis II, the stage of nuclear division in which the centromeres of the condensed chromosomes are arranged in a plane between the two poles of the spindle.
metaphase plate Imaginary plane, equidistant from the spindle poles in a metaphase cell, on which the centromeres of the chromosomes are aligned by the spindle fibers.
migration Movement of organisms among subpopulations; also, the movement of molecules in electrophoresis.
minimal medium A growth medium consisting of simple inorganic salts, a carbohydrate, vitamins, organic bases, essential amino acids, and other essential compounds; its composition is precisely known. Minimal medium contrasts with complex medium or broth, which is an extract of biological material (vegetables, milk, meat) that contains a large number of compounds and the precise composition of which is unknown.
minor groove In B-form DNA, the smaller of two continuous indentations running along the outside of the double helix.
mismatch repair Removal of one nucleotide from a pair that cannot properly hydrogen-bond, followed by replacement with a nucleotide that can hydrogen-bond.
missense mutation An alteration in a coding sequence of DNA that results in an amino acid replacement in the polypeptide.
mitosis The process of nuclear division in which the replicated chromosomes divide and the daughter nuclei have the same chromosome number and genetic composition as the parent nucleus.
monohybrid A genotype that is heterozygous for one pair of alleles; the offspring of a cross between genotypes that are homozygous for different alleles of a gene.
monoploid The basic chromosome set that is reduplicated to form the genomes of the species in a polyploid series; the smallest haploid chromosome number in a polyploid series.
monosomic Condition of an otherwise diploid organism in which one member of a pair of chromosomes is missing.
morphogen Substance that induces differentiation.
mosaic An organism composed of two or more genetically different types of cells.
M period *See* **cell cycle.**
mRNA *See* **messenger RNA.**
multifactorial trait A trait determined by the combined action of many factors, typically some genetic and some environmental.
multiple alleles The presence in a population of more than two alleles of a gene.
multiple cloning site *See* **polylinker.**
multiplication rule The principle that the probability that all of a set of independent events are realized simultaneously equals the product of the probabilities of the separate events.
mutagen An agent that is capable of increasing the rate of mutation.
mutant Any heritable biological entity that differs from wildtype, such as a mutant DNA molecule, mutant allele, mutant gene, mutant chromosome, mutant cell, mutant organism, or mutant heritable phenotype; also, a cell or organism in which a mutant allele is expressed.
mutant screen A type of genetic experiment in which the geneticist seeks to isolate multiple new mutations that affect a particular trait.
mutation A heritable alteration in a gene or chromosome; also, the process by which such an alteration happens. Used incorrectly, but with increasing frequency, as a synonym for *mutant,* even in some excellent textbooks.
mutation rate The probability of a new mutation in a particular gene, either per gamete or per generation.

narrow-sense heritability The fraction of the phenotypic variance revealed as resemblance between parents and offspring; technically, the ratio of the additive genetic variance to the total phenotypic variance.
natural selection The process of evolutionary adaptation in which the genotypes genetically best suited to survive and reproduce in a particular environment give rise to a disproportionate share of the offspring and so gradually increase the overall ability of the population to survive and reproduce in that environment.
negative regulation Regulation of gene expression in which mRNA is not synthesized until a repressor is removed from the DNA of the gene.
nick A single-strand break in a DNA molecule.
nitrous acid HNO_2, a chemical mutagen.
nondisjunction Failure of chromosomes to separate (disjoin) and move to opposite poles of the division spindle; the result is loss or gain of a chromosome.
nonparental ditype An ascus containing two pairs of recombinant spores.
nonselective medium A growth medium that allows growth of wildtype and of one or more mutant genotypes.
nonsense mutation A mutation that changes a codon specifying an amino acid into a stop codon, resulting in premature polypeptide chain termination; also called a chain-termination mutation.
normal distribution A symmetrical bell-shaped distribution characterized by the mean and the variance; in a normal distribution, approximately 68 percent of the observations are within 1 standard deviation from the mean, and approximately 95 percent are within 2 standard deviations.
nuclease An enzyme that breaks phosphodiester bonds in nucleic acid molecules.
nucleic acid A polymer composed of repeating units of phosphate-linked five-carbon sugars to which nitrogenous bases are attached. *See also* **DNA** and **RNA.**

nucleic acid hybridization The formation of duplex nucleic acid from complementary single strands.
nucleolus (*pl.* nucleoli) Nuclear organelle in which ribosomal RNA is made and ribosomes are partially synthesized; usually associated with the nucleolar organizer region. A nucleus may contain several nucleoli.
nucleoside A purine or pyrimidine base covalently linked to a sugar.
nucleosome The basic repeating subunit of chromatin, consisting of a core particle composed of two molecules each of four different histones around which a length of DNA containing about 145 nucleotide pairs is wound, joined to an adjacent core particle by about 55 nucleotide pairs of linker DNA associated with a fifth type of histone.
nucleotide A nucleoside phosphate.
nucleotide analog A molecule that is structurally similar to a normal nucleotide and that is incorporated into DNA.
nutritional mutation A mutation in a metabolic pathway that creates a need for a substance to be present in the growth medium or that eliminates the ability to utilize a substance present in the growth medium.

Okazaki fragment Any of the short strands of DNA produced during discontinuous replication of the lagging strand; also called a precursor fragment.
oligonucleotide primer A short, single-stranded nucleic acid synthesized for use in DNA sequencing or as a primer in the polymerase chain reaction.
oncogene A gene that can initiate tumor formation.
open reading frame (ORF) In the coding strand of DNA or in mRNA, a region containing a series of codons uninterrupted by stop codons and therefore capable of coding for a polypeptide chain; abbreviated ORF.
operator A regulatory region in DNA that interacts with a specific repressor protein in controlling the transcription of adjacent structural genes.
operon A collection of adjacent structural genes regulated by an operator and a repressor.
ORF *See* **open reading frame.**

PAC *See* **P1 artificial chromosome.**
pachytene The middle substage of meiotic prophase I in which the homologous chromosomes are closely synapsed.
pair-rule gene Any of a group of genes active early in *Drosophila* development that specifies the fates of alternating segments or parasegments. Mutations in pair-rule genes result in loss of even-numbered or odd-numbered segments or parasegments.
paracentric inversion An inversion that does not include the centromere.
parasegment Developmental unit in *Drosophila* consisting of the posterior part of one segment and the anterior part of the next segment in line.
parental combination Alleles present in an offspring chromosome in the same combination as that found in one of the parental chromosomes.
parental ditype An ascus containing two pairs of nonrecombinant spores.
parent strand In DNA replication, the strand that served as the template in a newly formed duplex.
partial diploid A cell in which a segment of the genome is duplicated, usually in a plasmid.
Pascal's triangle Triangular configuration of integers in which the nth row gives the binomial coefficients in the expansion of $(x + y)^{n-1}$. The first and last numbers in each row equal 1, and the others equal the sum of the adjacent numbers in the row immediately above.
pattern formation The creation of a spatially ordered and differentiated embryo from a seemingly homogeneous egg cell.
PCR *See* **polymerase chain reaction.**
pedigree A diagram representing the familial relationships among relatives.
penetrance The proportion of organisms having a particular genotype that actually express the corresponding phenotype. If the phenotype is always expressed, penetrance is complete; otherwise, it is incomplete.
peptide bond A covalent bond between the amino group ($-NH_2$) of one amino acid and the carboxyl group ($-COOH$) of another.
pericentric inversion An inversion that includes the centromere.
permissive condition An environmental condition in which the phenotype of a conditional mutation is not expressed; contrasts with the nonpermissive or restrictive condition.
P_1 generation The parents used in a cross, or the original parents in a series of generations; also called the P generation if there is no chance of confusion with the grandparents or more remote ancestors.
phage *See* **bacteriophage.**
phage repressor Regulatory protein that prevents transcription of genes in a prophage.
phenotype The observable properties of a cell or an organism, which result from the interaction of the genotype and the environment.
phenylalanine hydroxylase The enzyme, deficient in phenylketonuria, that converts phenylalanine into tyrosine.
phenylketonuria (PKU) A hereditary human condition resulting from the inability to convert phenylalanine into tyrosine; causes severe mental retardation unless treated in infancy and childhood via a low-phenylalanine diet; abbreviated PKU.
phosphodiester bond In nucleic acids, the covalent bond formed between the 5'-phosphate group (5'-P) of one nucleotide and the 3'-hydroxyl group (3'-OH) of the next nucleotide in line; these bonds form the backbone of a nucleic acid molecule.
photoreactivation The enzymatic splitting of pyrimidine dimers produced in DNA by ultraviolet light; requires visible light and the photoreactivation enzyme.
phylogenetic tree A diagram showing the genealogical relationships among a set of genes or species.
physical map A diagram showing the relative positions of physical landmarks in a DNA molecule; common landmarks include the positions of restriction sites and particular DNA sequences.
plaque A clear area in an otherwise turbid layer of bacteria growing on a solid medium, caused by the infection and killing of the cells by a phage; because each plaque results from the growth of one phage, plaque counting is a way of counting viable phage particles. The term is also used for animal viruses that cause clear areas in layers of animal cells grown in culture.
plasmid An extrachromosomal genetic element that replicates independently of the host chromosome; it may exist in one or many copies per cell and may segregate in cell division to daughter cells in either a controlled or a random fashion. Some plasmids, such as the F factor, may become integrated into the host chromosome.
pleiotropic effect Any phenotypic effect that is a secondary manifestation of a mutant gene.
pleiotropy The condition in which a single mutant gene affects two or more distinct and seemingly unrelated traits.
polarity The 5'-to-3' orientation of a strand of nucleic acid.
pole cell Any of a group of cells, set off at the posterior end of the *Drosophila* embryo, from which the germ cells are derived.
poly-A tail The sequence of adenines added to the 3' end of many eukaryotic mRNA molecules in processing.
polycistronic mRNA An mRNA molecule from which two or more polypeptides are translated; found primarily in prokaryotes.
polylinker A short DNA sequence that is present in a vector and that contains a number of unique restriction sites suitable for gene cloning.
polymerase chain reaction (PCR) Repeated cycles of DNA denaturation, renaturation with primer oligonucleotide sequences, and replication, resulting in exponential growth in the number of copies of the DNA sequence located between the primers.
polymorphic gene A gene for which there is more than one relatively common allele in a population.
polymorphism The presence in a population of two or more relatively common forms of a gene, chromosome, or genetically determined trait.
polynucleotide chain A polymer of covalently linked nucleotides.
polypeptide *See* **polypeptide chain.**
polypeptide chain A polymer of amino acids linked together by peptide bonds.
polyploidy The condition of a cell or organism with more than two complete sets of chromosomes.
polysome A complex of two or more ribosomes associated with an mRNA molecule and actively engaged in polypeptide synthesis; a polyribosome.
polysomy The condition of a diploid cell or organism that has three or more copies of a particular chromosome.

polytene chromosome A giant chromosome consisting of many identical strands laterally apposed and in register, exhibiting a characteristic pattern of transverse banding.
P1 artificial chromosome A plasmid vector containing regions of the bacteriophage P1 and a large inserted DNA fragment.
population A group of organisms of the same species.
population genetics Application of Mendel's laws and other principles of genetics to entire populations of organisms.
population substructure Organization of a population into smaller breeding groups between which migration is restricted. Also called population subdivision.
positional cloning A strategy of gene cloning based on the position of a gene in the genetic map; also called map-based cloning.
positional information Developmental signals transmitted to a cell by virtue of its position in the embryo.
positive regulation Mechanism of gene regulation in which an element must be bound to DNA in an active form to allow transcription. Positive regulation contrasts with negative regulation, in which a regulatory element must be removed from DNA.
postreplication repair DNA repair that takes place in nonreplicating DNA or after the replication fork is some distance beyond a damaged region.
precursor fragment *See* **Okazaki fragment.**
primary transcript An RNA copy of a gene; in eukaryotes, the transcript must be processed to form a translatable mRNA molecule.
primer In nucleic acids, a short RNA or single-stranded DNA segment that functions as a growing point in polymerization.
primosome The enzyme complex that forms the RNA primer for DNA replication in eukaryotic cells.
probe A radioactive DNA and RNA molecule used in DNA-RNA and DNA-DNA hybridization assays.
product molecule The end result of a biochemical reaction or a metabolic pathway.
programmed cell death Cell death that happens as part of the normal developmental process. *See also* **apoptosis.**
prokaryote An organism that lacks a nucleus; prokaryotic cells divide by fission.
promoter A DNA sequence at which RNA polymerase binds and initiates transcription.
proofreading function *See* **editing function.**
prophage The form of phage DNA in a lysogenic bacterium; the phage DNA is repressed and is usually integrated into the bacterial chromosome, but some prophages are in plasmid form.
prophage induction Activation of a prophage to undergo the lytic cycle.
prophase The initial stage of mitosis or meiosis, beginning after interphase and terminating with the alignment of the chromosomes at metaphase; often absent or abbreviated between meiosis I and meiosis II.
prototroph Microbial strain capable of growth in a defined minimal medium that ideally contains only a carbon source and inorganic compounds. The wildtype genotype is usually regarded as a prototroph.
pseudoautosomal region In mammals, a small region of the X and Y chromosome containing homologous genes.
***P* transposable element** A *Drosophila* transposable element used for the induction of mutations, germ-line transformation, and other types of genetic engineering.
Punnett square A cross-multiplication square used for determining the expected genetic outcome of a mating.
purine An organic base found in nucleic acids; the predominant purines are adenine and guanine.
pyrimidine An organic base found in nucleic acids; the predominant pyrimidines are cytosine, uracil (in RNA only), and thymine (in DNA only).
pyrimidine dimer Two adjacent pyrimidine bases, typically a pair of thymines, in the same polynucleotide strand, between which chemical bonds have formed; the most common lesion formed in DNA by exposure to ultraviolet light.

quantitative trait A trait—typically measured on a continuous scale, such as height or weight—that results from the combined action of several or many genes in conjunction with environmental factors.
quantitative trait locus (QTL) A locus segregating for alleles that have different, measurable effects on the expression of a quantitative trait.

random genetic drift Fluctuation in allele frequency from generation to generation resulting from restricted population size.
random mating System of mating in which mating pairs are formed independently of genotype and phenotype.
random spore analysis In fungi, the genetic analysis of spores collected at random rather than from individual tetrads.
reading frame The phase in which successive triplets of nucleotides in mRNA form codons; depending on the reading frame, a particular nucleotide in an mRNA could be in the first, second, or third position of a codon. The reading frame actually used is defined by the AUG codon that is selected for chain initiation.
recessive Refers to an allele, or the corresponding phenotypic trait, expressed only in homozygotes.
reciprocal cross A cross in which the sexes of the parents are the reverse of those in another cross.
reciprocal translocation Interchange of parts between nonhomologous chromosomes.
recombinant A chromosome that results from crossing-over and that carries a combination of alleles differing from that of either chromosome participating in the crossover; the cell or organism that contains a recombinant chromosome.
recombinant DNA A DNA molecule composed of one or more segments from other DNA molecules.
recombination Exchange of parts between DNA molecules or chromosomes; recombination in eukaryotes usually entails a reciprocal exchange of parts, but in prokaryotes it is often nonreciprocal.
recruitment The process in which a transcriptional activator protein interacts with one or more components of the transcription complex and attracts it to the promoter.
red-green color blindness *See* **color blindness.**
reductional division Term applied to the first meiotic division because the chromosome number (counted as the number of centromeres) is reduced from diploid to haploid.
redundancy The feature of the genetic code in which an amino acid corresponds to more than one codon; also called degeneracy.
relative fitness The fitness of a genotype expressed as a proportion of the fitness of another genotype.
replica plating Procedure in which a particular spatial pattern of colonies on an agar surface is reproduced on a series of agar surfaces by stamping them with a template that contains an image of the pattern; the template is often produced by pressing a piece of sterile velvet upon the original surface, which transfers cells from each colony to the cloth.
replication *See* **DNA replication; θ replication.**
replication fork In a replicating DNA molecule, the region in which nucleotides are added to growing strands.
replication origin The base sequence at which DNA synthesis begins.
replication slippage The process in which the number of copies of a small tandem repeat can increase or decrease during replication.
replicon A DNA molecule that has a replication origin.
repressible transcription A regulatory process in which a gene is temporarily rendered unable to be transcribed.
repressor A protein that binds specifically to a regulatory sequence adjacent to a gene and blocks transcription of the gene.
repulsion *See* ***trans*** **configuration.**
restriction endonuclease A nuclease that recognizes a short nucleotide sequence (restriction site) in a DNA molecule and cleaves the molecule at that site; also called a restriction enzyme.
restriction enzyme *See* **restriction endonuclease.**
restriction fragment A segment of duplex DNA produced by cleavage of a larger molecule by a restriction enzyme.
restriction fragment length polymorphism (RFLP) Genetic variation in a population associated with the size of restriction fragments that contain sequences homologous to a particular probe DNA; the polymorphism results from the positions of restriction sites flanking the probe, and each variant is essentially a different allele.
restriction map A diagram of a DNA molecule showing the positions of cleavage by one or more restriction endonucleases.
restriction site The base sequence at which a particular restriction endonuclease makes a cut.

restrictive condition A growth condition in which the phenotype of a conditional mutation is expressed.
retinoblastoma An inherited cancer caused by a mutation in the tumor-suppressor gene located in chromosome band *13q14*. Inheritance of one copy of the mutation results in multiple malignancies in retinal cells of the eyes in which the mutation becomes homozygous—for example, through a new mutation or mitotic recombination.
retrovirus One of a class of RNA animal viruses that cause the synthesis of DNA complementary to their RNA genomes on infection.
reverse genetics Procedure in which mutations are deliberately produced in cloned genes and introduced back into cells or the germ line of an organism.
reverse mutation A mutation that undoes the effect of a preceding mutation.
reverse transcriptase An enzyme that makes complementary DNA from a single-stranded RNA template.
reverse transcriptase PCR (RT-PCT) Amplification, using an RNA template, of a duplex DNA molecule originally produced by reverse trascriptase.
reversion Restoration of a mutant phenotype to the wildtype phenotype by a second mutation.
RFLP *See* **restriction fragment length polymorphism.**
R group *See* **side chain.**
ribonucleic acid *See* **RNA.**
ribose The five-carbon sugar in RNA.
ribosomal RNA (rRNA) RNA molecules that are components of the ribosomal subunits; in eukaryotes, there are four rRNA molecules—5S, 5.8S, 18S, and 28S; in prokaryotes, there are three 5S, 16S, and 23S.
ribosome The cellular organelle on which the codons of mRNA are translated into amino acids in protein synthesis. Ribosomes consist of two subunits, each composed of RNA and proteins. In prokaryotes, the subunits are 30S and 50S particles; in eukaryotes, they are 40S and 60S particles.
ribosome-binding site The base sequence in a prokaryotic mRNA molecule to which a ribosome can bind to initiate protein synthesis; also called the Shine–Dalgarno sequence.
ribosome tRNA-binding sites The tRNA-binding sites on the ribosome to which tRNA molecules are bound. The aminoacyl site receives the incoming charged tRNA, the peptidyl site holds the tRNA with the nascent polypeptide chain, and the exit site holds the outgoing uncharged tRNA.
ribosomal translocation Movement of the ribosome along a molecule of messenger RNA in translation.
ribozyme An RNA molecule able to catalyze one or more biochemical reactions.
ring chromosome A chromosome whose ends are joined; one that lacks telomeres.
RNA Ribonucleic acid, a nucleic acid in which the sugar constituent is ribose; typically, RNA is single-stranded and contains the four bases adenine, cytosine, guanine, and uracil.
RNA polymerase An enzyme that makes RNA by copying the base sequence of a DNA strand.
RNA processing The conversion of a primary transcript into an mRNA, rRNA, or tRNA molecule; includes splicing, cleavage, modification of termini, and (in tRNA) modification of internal bases.
RNA splicing Excision of introns and joining of exons.
Robertsonian translocation A chromosomal aberration in which the long arms of two acrocentric chromosomes become joined to a common centromere.
rolling-circle replication A mode of replication in which a circular parent molecule produces a linear branch of newly formed DNA.
R plasmid A bacterial plasmid that carries drug-resistance genes; commonly used in genetic engineering.
rRNA *See* **ribosomal RNA.**
RT-PCR *See* **reverse transcriptase PCR.**

satellite DNA Eukaryotic DNA that forms a minor band at a different density from that of most of the cellular DNA in equilibrium density gradient centrifugation; consists of short sequences repeated many times in the genome (highly repetitive DNA) or of mitochondrial or chloroplast DNA.
scaffold A protein-containing material in chromosomes, believed to be responsible in part for the compaction of chromatin.
second-division segregation Segregation of a pair of alleles into different nuclei in the second meiotic division, the result of crossing-over between the gene and the centromere of the pair of homologous chromosomes.
second meiotic division The meiotic division in which the centromeres split and the chromosome number is not reduced; also called the equational division.
segment Any of a series of repeating morphological units in a body plan.
segmentation gene Any of a group of genes that determines the spatial pattern of segments and parasegments in *Drosophila* development.
segment-polarity gene Any of a group of genes that determines the spatial pattern of development within the segments of *Drosophila* larvae.
segregation Separation of the members of a pair of alleles into different gametes in meiosis.
selected marker A genetic mutation that allows growth in selective medium.
selection In evolution, intrinsic differences in the ability of genotypes to survive and reproduce; in plant and animal breeding, the choosing of organisms with certain phenotypes to be parents of the next generation; in mutation studies, a procedure designed in such a way that only a desired type of cell can survive, as in selection for resistance to an antibiotic.
selection coefficient The amount by which relative fitness is reduced or increased.
selection limit The condition in which a population no longer responds to artificial selection for a trait.
selectively neutral mutation A mutation that has no (or negligible) effects on fitness.
selective medium A medium that allows growth only of cells with particular genotypes.
selfish DNA DNA sequences that do not contribute to the fitness of an organism but are maintained in the genome through their ability to replicate and transpose.
semiconservative replication The usual mode of DNA replication, in which each strand of a duplex molecule serves as a template for the synthesis of a new complementary strand, and the daughter molecules are composed of one old (parental) and one newly synthesized strand.
semisterility A condition in which a significant proportion of the gametophytes produced by a plant or of the zygotes produced by an animal are inviable, as in the case of a translocation heterozygote.
sequence-tagged site (STS) A DNA sequence, present once per haploid genome, that can be amplified by the use of suitable oligonucleotide primers in the polymerase chain reaction in order to identify clones that contain the sequence.
sex chromosome A chromosome, such as the human X or Y, that has a role in the determination of sex.
sib *See* **sibling.**
sibling A brother or sister, each having the same parents.
sibship A group of brothers and sisters.
sickle-cell anemia A severe anemia in human beings inherited as an autosomal recessive and caused by an amino acid replacement in the β-globin chain; heterozygotes tend to be more resistant to *falciparum* malaria than are normal homozygotes.
significant *See* **statistically significant.**
silent mutation A mutation that has no phenotypic effect.
simple tandem repeat polymorphism (STRP) A DNA polymorphism in a population in which the alleles differ in the number of copies of a short, tandemly repeated nucleotide sequence.
single-active-X principle In mammals, the genetic inactivation of all X chromosomes but one in each cell lineage, except in the very early embryo.
single-nucleotide polymorphism A site in the DNA occupied by a different nucleotide pair among a significant fraction of the individuals in a population.
single-stranded DNA A DNA molecule that consists of a single polynucleotide chain.
sister chromatids Chromatids produced by replication of a single chromosome.
small ribonucleoprotein particles Small nuclear particles that contain short RNA molecules and several proteins. They are involved in intron excision and splicing and in other aspects of RNA processing.

snRNP Any of several classes of small ribonucleoprotein particles involved in RNA splicing.
somatic cell Any cell of a multicellular organism other than the gametes and the germ cells from which gametes develop.
somatic mutation A mutation arising in a somatic cell.
Southern blot A nucleic acid hybridization method in which, after electrophoretic separation, denatured DNA is transferred from a gel to a membrane filter and then exposed to radioactive DNA or RNA under conditions of renaturation; the radioactive regions locate the homologous DNA fragments on the filter.
specialized transducing phage *See* **transducing phage.**
S period *See* **cell cycle.**
spindle A structure composed of fibrous proteins on which chromosomes align during metaphase and move during anaphase.
splice acceptor The 5' end of an exon.
splice donor The 3' end of an exon.
spliceosome An RNA-protein particle in the nucleus in which introns are removed from RNA transcripts.
spontaneous mutation A mutation that happens in the absence of any known mutagenic agent.
spore A unicellular reproductive entity that becomes detached from the parent and can develop into a new organism upon germination; in plants, spores are the haploid products of meiosis.
sporophyte The diploid, spore-forming generation in plants, which alternates with the haploid, gamete-producing generation (the gametophyte).
standard deviation The square root of the variance.
start codon An mRNA codon, usually AUG, at which polypeptide synthesis begins.
statistically significant Said of the result of an experiment or study that has only a small probability of happening by chance on the assumption that some hypothesis is true. Conventionally, if results as bad or worse would be expected less than 5 percent of the time, the result is said to be statistically significant; if less than 1 percent of the time, the result is called statistically highly significant; both outcomes cast the hypothesis into serious doubt.
sticky end A single-stranded end of a DNA fragment produced by certain restriction enzymes capable of reannealing with a complementary sequence in another such strand.
stop codon One of three mRNA codons—UAG, UAA, and UGA—at which polypeptide synthesis stops.
STRP *See* **simple tandem repeat polymorphism.**
STS *See* **sequence-tagged site.**
submetacentric chromosome A chromosome whose centromere divides it into arms of unequal length.
subpopulation Any of the breeding groups within a larger population between which migration is restricted.
substrate molecule A substance acted on by an enzyme.
synapsis The pairing of homologous chromosomes or chromosome regions in the zygotene substage of the first meiotic prophase.
syncytial blastoderm Stage in early *Drosophila* development formed by successive nuclear divisions without division of the cytoplasm.

tandem duplication A pair of identical or closely related DNA sequences that are adjacent and in the same orientation.
TATA binding protein (TBP) A protein that binds to the TATA motif in the promoter region of a gene.
TATA box The base sequence 5'-TATA-3' in the DNA of a promoter.
***T* DNA** Transposable element found in *Agrobacterium tumefaciens,* which produces crown gall tumors in a wide variety of dicotyledonous plants.
telomerase An enzyme that adds specific nucleotides to the tips of the chromosomes to form the telomeres.
telomere The tip of a chromosome, containing a DNA sequence required for stability of the chromosome end.
telophase The final stage of mitotic or meiotic nuclear division.
temperature-sensitive mutation A conditional mutation that causes a phenotypic change only at certain temperatures.
template A strand of nucleic acid whose base sequence is copied in a polymerization reaction to produce either a complementary DNA or an RNA strand.
terminal redundancy In bacteriophage T4, a short duplication found at opposite ends of the linear molecule; the duplicated region differs from one phage to the next.
testcross A cross between a heterozygote and a recessive homozygote, resulting in progeny in which each phenotypic class represents a different genotype.
testis-determining factor (TDF) Genetic element on the mammalian Y chromosome that determines maleness.
tetrad The four chromatids that make up a pair of homologous chromosomes in meiotic prophase I and metaphase I; also, the four haploid products of a single meiosis.
tetraploid A cell or organism with four complete sets of chromosomes; in an autotetraploid, the chromosome sets are homologous; in an allotetraploid, the chromosome sets consist of a complete diploid complement from each of two distinct ancestral species.
tetratype An ascus containing spores of four different genotypes—one each of the four genotypes possible with two alleles of each of two genes.
θ replication Bidirectional replication of a circular DNA molecule, starting from a single origin of replication.
30-nm fiber The level of compaction of eukaryotic chromatin resulting from coiling of the extended, nucleosome-bound DNA fiber.
three-point cross Cross in which three genes are segregating; used to obtain unambiguous evidence of gene order.
3'end The end of a DNA or RNA strand that terminates in a sugar and so has a free hydroxyl group on the number-3' carbon.
threshold trait A trait with a continuously distributed liability or risk; organisms with a liability greater than a critical value (the threshold) exhibit the phenotype of interest, such as a disorder.
thymine (T) A nitrogenous pyrimidine base found in DNA.
thymine dimer *See* **pyrimidine dimer.**
time of entry In an Hfr $\times$ F^- bacterial mating, the earliest time that a particular gene in the Hfr parent is transferred to the F^- recipient.
***Ti* plasmid** A plasmid that is present in *Agrobacterium tumefaciens* and is used in genetic engineering in plants.
topoisomerase An enzyme that introduces or eliminates either underwinding or overwinding of double-stranded DNA. It acts by introducing a single-strand break, changing the relative positions of the strands, and sealing the break.
total variance Summation of all sources of genetic and environmental variation.
trait Any aspect of the appearance, behavior, development, biochemistry, or other feature of an organism.
***trans* configuration** The arrangement in linked inheritance in which a genotype heterozygous for two mutant sites has received one of the mutant sites from each parent—that is, $a_1+/+a_2$.
transcript An RNA strand that is produced from, and is complementary in base sequence to, a DNA template strand.
transcription The process by which the information contained in a template strand of DNA is copied into a single-stranded RNA molecule of complementary base sequence.
transcriptional activator protein Positive control element that stimulates transcription by binding with particular sites in DNA.
transcription complex An aggregate of RNA polymerase (consisting of its own subunits) along with other polypeptide subunits that makes transcription possible.
transducing phage A phage type capable of producing particles that contain bacterial DNA (transducing particles). A specialized transducing phage produces particles that carry only specific regions of chromosomal DNA; a generalized transducing phage produces particles that may carry any region of the genome.
transduction The carrying of genetic information from one bacterium to another by a phage.
transfer RNA (tRNA) A small RNA molecule that translates a codon into an amino acid in protein synthesis; it has a three-base sequence, called the anticodon, complementary to a specific codon in mRNA, and a site to which a specific amino acid is bound.
transformation Change in the genotype of a cell or organism resulting from exposure of the cell or organism to DNA isolated from a different genotype; also, the conversion of an animal cell, whose growth is limited in culture, into a tumor-like cell whose pattern of growth is different from that of a normal cell.
transgenic organism An animal or plant in which novel DNA has been incorporated into the germ line.
***trans* heterozygote** *See* ***trans* configuration.**
transition mutation A mutation resulting from the substitution of one purine for another purine or that of one pyrimidine for another pyrimidine.

translation The process by which the amino acid sequence of a polypeptide is synthesized on a ribosome according to the nucleotide sequence of an mRNA molecule.
translocation Interchange of parts between nonhomologous chromosomes; also, the movement of mRNA with respect to a ribosome during protein synthesis. *See also* **reciprocal translocation.**
transmembrane receptor A receptor protein containing amino acid sequences that span the cell membrane.
transposable element A DNA sequence capable of moving (transposing) from one location to another in a genome.
transposase Protein necessary for transposition.
transposition The movement of a transposable element.
transposon A transposable element that contains bacterial genes—for example, for antibiotic resistance; also used loosely as a synonym for *transposable element.*
transposon tagging Insertion of a transposable element that contains a genetic marker into a gene of interest.
transversion mutation A mutation resulting from the substitution of a purine for a pyrimidine or that of a pyrimidine for a purine.
trinucleotide repeat A tandemly repeated sequence of three nucleotides; genetic instability in trinucleotide repeats is the cause of a number of human hereditary diseases.
triplet code A code in which each codon consists of three bases.
triploid A cell or organism with three complete sets of chromosomes.
trisomic A diploid organism with an extra copy of one of the chromosomes.
trisomy-X syndrome The clinical features of the karyotype 47,XXX.
trivalent Structure formed by three homologous chromosomes in meiosis I in a triploid or trisomic chromosome when each homolog is paired along part of its length with first one and then the other of the homologs.
tRNA *See* **transfer RNA.**
true-breeding Refers to a strain, breed, or variety of organism that yields progeny like itself; homozygous.
truncation point In artificial selection, the value of the phenotype that determines which organisms will be retained for breeding and which will be culled.
tumor-suppressor gene A gene whose absence predisposes to malignancy; also called an anti-oncogene.
Turner syndrome The clinical features of human females with the karyotype 45,X.

unequal crossing-over Crossing-over between nonallelic copies of duplicated or other repetitive sequences—for example, in a tandem duplication, between the upstream copy in one chromosome and the downstream copy in the homologous chromosome.
univalent Structure formed in meiosis I in a monoploid or a monosomic when a chromosome has no pairing partner.
uracil (U) A nitrogenous pyrimidine base found in RNA.

variable expressivity Differences in the severity of expression of a particular genotype.
variable antibody region The portion of an immunoglobulin molecule that varies greatly in amino acid sequence among antibodies in the same subclass.
variance A measure of the spread of a statistical distribution; the mean of the squares of the deviations from the mean.
vector A DNA molecule, capable of replication, into which a gene or DNA segment is inserted by recombinant DNA techniques; a cloning vehicle.
viral oncogene A class of genes found in certain viruses that predispose to cancer. Viral oncogenes are the viral counterparts of cellular oncogenes. *See also* **cellular oncogene, oncogene.**

Watson–Crick base pairing Base pairing in DNA or RNA in which A pairs with T (or U in RNA) and G pairs with C.
wildtype The most common phenotype or genotype in a natural population; also, a phenotype or genotype arbitrarily designated as a standard for comparison.
wobble The acceptable pairing of several possible bases in an anticodon with the base present in the third position of a codon.

X chromosome A chromosome that plays a role in sex determination and that is present in two copies in the homogametic sex and in one copy in the heterogametic sex.
xeroderma pigmentosum An inherited defect in the repair of ultraviolet-light damage to DNA, associated with extreme sensitivity to sunlight and multiple skin cancers.
X-linked gene A gene located in the X chromosome. X-linked inheritance is usually evident from the production of nonidentical classes of progeny from reciprocal crosses.

YAC *See* **yeast artificial chromosome.**
Y chromosome The sex chromosome present only in the heterogametic sex; in mammals, the male-determining sex chromosome.
yeast artificial chromosome (YAC) In yeast, a cloning vector that can accept very large fragments of DNA; a chromosome introduced into yeast derived from such a vector and containing DNA from another organism.

zygote The product of the fusion of a female gamete and a male gamete in sexual reproduction; a fertilized egg.
zygotene The substage of meiotic phrophase I in which homologous chromosomes synapse.
zygotic gene Any of a group of genes that control early development through their expression in the zygote.

Index